D0302982

FORENSIC GENETIC APPROACHES FOR IDENTIFICATION OF HUMAN SKELETAL REMAINS

FORENSIC GENETIC APPROACHES FOR IDENTIFICATION OF HUMAN SKELETAL REMAINS

Challenges, Best Practices, and Emerging Technologies

Edited by

Angie Ambers PhD

Henry C. Lee Institute of Forensic Science, University of New Haven, West Haven, CT, United States

Institute for Human Identification, LMU College of Dental Medicine, Knoxville, TN, United States

Academic Press is an imprint of Elsevier
125 London Wall, London EC2Y 5AS, United Kingdom
525 B Street, Suite 1650, San Diego, CA 92101, United States
50 Hampshire Street, 5th Floor, Cambridge, MA 02139, United States
The Boulevard, Langford Lane, Kidlington, Oxford OX5 1GB, United Kingdom

ISBN 978-0-12-815766-4

For information on all Academic Press publications
visit our website at https://www.elsevier.com/books-and-journals

Cover Credit: Dr. Ambers

Publisher: Stacy Masucci
Acquisitions Editor: Elizabeth Brown
Editorial Project Manager: Fernanda Oliveira
Production Project Manager: Maria Bernard
Cover Designer: Miles Hitchen

Typeset by STRAIVE, India

Dedication

To Mom
(1947–2009)

For your unconditional love and always believing in me. And for exemplifying a life of integrity, kindness, and compassion for everyone you encountered, including the underprivileged, oppressed, abused, and disadvantaged.

Your unwavering altruism was an inspiration to me, and I now realize that I got the best parts of who I am from you. Thank you for instilling in me a fervent commitment to justice and truth. I love you more than life's breath. I miss you every day.

To Dr. Harrell Gill-King

For inspiring me to accept the challenge of this career path and mission. And for believing in me more than I believed in myself. Words cannot express how grateful I am for your mentorship, guidance, generosity, kindness, and friendship throughout the years. You truly are one of my favorite human beings on this planet, and I am honored to have someone of your caliber and character in my life. Thank you for everything. You are in my heart always.

To Dr. Art Eisenberg
(1956–2018)

Who will always be remembered for his kindness as much as for pioneering the field of forensic DNA in the United States. His contributions to resolving missing persons and unidentified human remains (UHR) cases, both domestically and internationally, are immeasurable. I was fortunate to have been mentored early in my career by such an amazing person and scientist. Gone from this Earth, but not forgotten.

In Memory of: Paula C. Brumit, DDS, D-ABFO (1956–2021)

This book honors the memory and legacy of Dr. Paula Brumit—an exceptional scientist, humanitarian, mentor, and friend. She left an indelible mark on not just the forensic human identification community but also on all who were fortunate enough to have crossed her path personally or professionally. Her generous spirit, kindness, altruism, selflessness, and willingness to share knowledge will never be forgotten. She was the best friend one could ever hope for, and her sense of humor was unmatched—she could make you laugh in even the most despondent of times. I was fortunate to have the gift of her friendship, and I will miss her every day for the rest of my life.

Dr. Brumit was remarkably accomplished, immensely respected by her peers, and devoted her life in service to others. She earned a Bachelor of Science degree in dental hygiene and a Doctor of Dental Surgery (DDS) degree from Texas A&M Health Science Center's Baylor College of Dentistry. During her career, she made innumerable contributions to forensic odontology and human identification casework in various professional roles, including as consulting forensic odontologist for the Dallas County Medical Examiner's Office (Dallas, Texas); consultant for the UNT Center for Human Identification's Laboratory of Forensic Anthropology (Fort Worth, Texas); member of the National Disaster Mortuary Operational Response Team (DMORT) Region 6 (in which she assisted in the identification of victims from Hurricane Katrina and the 9/11 World Trade Center terrorist attacks); forensic odontologist for the Bexar County Medical Examiner's Office (Bexar County, Texas); and consultant for the FBI, the Texas Rangers, Immigration and Customs Enforcement (ICE), Child Protective Services (CPS), Department of Public Safety (DPS), and various law enforcement agencies.

Dr. Brumit was a member of the American Society of Forensic Odontology (ASFO) and served on the ASFO's Board of Governors. She was a fellow of the American Academy of Forensic Sciences (AAFS) and served as Secretary, Program Chair, and Odontology Section Chair as well as on the AAFS Board of Directors. Dr. Brumit was only the second woman to hold the office of Odontology Section Chair in the 71-year history of the American Academy of Forensic Sciences. In addition, she was a Diplomate of the American Board of Forensic Odontology (ABFO), served two consecutive terms on the ABFO Board of Directors, and one term as ABFO President. During her career, she was appointed to the Texas Forensic Science Commission's Bitemark Case Review Team, the National Institute of Standards and Technology (NIST) Scientific Area Committees (SACs) Forensic Odontology Subcommittee, and the Criminal Justice Information Services (CJIS) National Dental

Image Repository Review Panel for the Federal Bureau of Investigation (FBI). In 2020, she was elected to the Forensic Specialties Accreditation Board for a 4-year term and was hired by the National Center for Missing & Exploited Children (NCMEC) to assist in the fight against the abuse and exploitation of children.

Academically, Dr. Brumit held an appointment as Assistant Professor in the Department of General Dentistry (Division of Forensic Odontology) at the University of Tennessee Graduate School of Medicine (UTGSM) in Knoxville, Tennessee. She taught in both UTGSM's *Forensic Odontology Fellowship* program and in the *Forensic Dentistry* master's program. Dr. Brumit had recently been promoted to Associate Professor in the College of Dental Medicine at Lincoln Memorial University (LMU), and was hired as a forensic odontologist for LMU's Institute for Human Identification. In addition, she was Co-Director of the Southwest Symposium on Forensic Dentistry, and for nearly two decades was an adjunct assistant professor in the *Forensic Fellowship Program* within the Center for Education and Research in Forensics (CERF) at the University of Texas Health Science Center (San Antonio, Texas).

Contents

5. Skeletal microstructure, bone diagenesis, optimal sample selection, and pre-processing preparation techniques for DNA testing

Angie Ambers PhD

6. DNA extraction methods for human skeletal remains

Angie Ambers PhD

7. Quantitative and qualitative assessment of DNA recovered from human skeletal remains

Jodie Ward PhD and Jeremy Watherston PhD

II

Types of DNA markers and applications for identification

8. Autosomal short tandem repeat (STR) profiling of human skeletal remains

Jeremy Watherston PhD and Jodie Ward PhD

9. Y-chromosome analysis for unidentified human remains (UHR) investigations

Angie Ambers PhD

10. Mitochondrial DNA and its use in the forensic analysis of skeletal material

Brandon Letts PhD

16. Emerging technologies for DNA analysis of challenged samples

Nicole Novroski PhD

IV

Analysis of genetic data recovered from skeletonized human remains

17. Best practices in the development and effective use of a forensic DNA database for identification of missing persons and unidentified human remains

Michael Hennessey BGS, MBA

18. Software and database functionality for direct identification and kinship analysis: The Mass Fatality Identification System (M-FISys)

Howard D. Cash CEO

19. Bioinformatic tools for interrogating DNA recovered from human skeletal remains

Frank R. Wendt PhD

20. The emerging discipline of forensic genetic genealogy

Colleen Fitzpatrick PhD

V

Complementary and multidisciplinary approaches to assist in identification of unidentified human skeletal remains

21. Forensic anthropology in a DNA world: How anthropological methods complement DNA-based identification of human remains

Eric J. Bartelink PhD, D-ABFA

22. Generation of a personal chemical profile from skeletonized human remains

Suni M. Edson PhD

23. Forensic odontology: Historical perspectives and current applications for identification of human remains

Thomas J. David DDS, D-ABFO and
James M. Lewis DMD, D-ABFO

24. Forensic facial reconstruction of skeletonized and highly decomposed human remains

Stephen Missal CFA

Contributors

Angie Ambers PhD Henry C. Lee Institute of Forensic Science; Forensic Science Department, Henry C. Lee College of Criminal Justice and Forensic Sciences, Center for Forensic Investigation of Trafficking in Persons, University of New Haven, West Haven, CT; Institute for Human Identification, LMU College of Dental Medicine, Knoxville, TN, United States

Eric J. Bartelink PhD, D-ABFA Department of Anthropology, Human Identification Laboratory, California State University, Chico, CA, United States

Howard D. Cash CEO Gene Codes Forensics, Inc., Ann Arbor, MI, United States

Thomas J. David DDS, D-ABFO Georgia Bureau of Investigation, Division of Forensic Sciences, Decatur, GA; Department of General Dentistry, Division of Forensics, University of Tennessee Graduate School of Medicine; Institute for Human Identification, LMU College of Dental Medicine, Knoxville, TN, United States

Suni M. Edson PhD Assistant Technical Leader, Past Accounting Section, Armed Forces DNA Identification Laboratory (AFDIL), Armed Forces Medical Examiner System, Dover, DE, United States

Colleen Fitzpatrick PhD Identifinders International LLC, Fountain Valley, CA, United States

Ranjana Grover PhD ANDE Corporation, Waltham, MA, United States

Michael Hennessey BGS, MBA Human Identification Projects, Gene Codes Forensics, Inc., Ann Arbor, MI, United States

Irene Kahline MSFS Past Accounting Section, Armed Forces DNA Identification Laboratory (AFDIL), Armed Forces Medical Examiner System, Dover, DE, United States

Bobby L. LaRue, PhD Department of Forensic Science, Sam Houston State University, Huntsville, TX, United States; Verogen Inc., San Diego, CA, United States

Brandon Letts PhD Federal Bureau of Investigation (FBI) Laboratory, Quantico, VA, United States

James M. Lewis DMD, D-ABFO Department of General Dentistry, Division of Forensics, University of Tennessee Graduate School of Medicine; Institute for Human Identification, LMU College of Dental Medicine, Knoxville, TN; Alabama Department of Forensic Sciences, Madison, AL, United States

Odile Loreille PhD Federal Bureau of Investigation (FBI) Laboratory, Quantico, VA, United States

Murray K. Marks, PhD, D-ABFA Knox County Regional Forensic Center, Department of General Dentistry, Division of Forensic Odontology, University of Tennessee Graduate School of Medicine; National Forensic Academy Outdoor Decomposition Training Facility, University of Tennessee; Department of Pathology; Institute for Human Identification, LMU College of Dental Medicine, Knoxville, TN, United States

Timothy P. McMahon PhD Department of Defense (DoD) DNA Operations, Armed Forces Medical Examiner System, Defense Health Agency, Dover, DE, United States

Darinka Mileusnic-Polchan MD, PhD, D-ABP Knox County Regional Forensic Center, Department of Pathology, University of Tennessee Graduate School of Medicine; Institute for Human Identification, LMU College of Dental Medicine, Knoxville, TN, United States

Stephen Missal CFA Maricopa County Medical Examiner Office, Forensic Science Center; Art Institute of Phoenix, Phoenix, AZ, United States

Nicole Novroski PhD Department of Anthropology, Forensic Science Program, University of Toronto Mississauga, Toronto, ON, Canada

Vivek Sahajpal, PhD State Forensic Science Laboratory, Directorate of Forensics Services, Shimla, Himachal Pradesh, India

Richard F. Selden MD, PhD ANDE Corporation, Waltham, MA, United States

Eugene Tan PhD ANDE Corporation, Waltham, MA, United States

Rosemary Turingan Witkowski PhD ANDE Corporation, Waltham, MA, United States

Jodie Ward PhD Centre for Forensic Science, University of Technology Sydney, Sydney, NSW, Australia

Jeremy Watherston PhD Forensic and Analytical Science Service, NSW Health Pathology, Lidcombe, NSW, Australia

Frank R. Wendt PhD Department of Psychiatry, Yale School of Medicine, U.S. Department of Veteran Affairs Connecticut Healthcare System, New Haven, CT, United States

Author's biographies

Angie Ambers, PhD

Dr. Angie Ambers is an internationally recognized DNA expert and Director of Forensic Genetics at the *Institute for Human Identification* in the College of Dental Medicine at Lincoln Memorial University (LMU). She also is a Professor of Forensic Genetics in the "Forensic Human Identification" graduate program. She has a PhD in molecular biology (with an emphasis in forensic genetics and human identification) as well as master's degrees both in forensic genetics and criminology. Dr. Ambers conducts research at the National Forensic Academy (NFA) Cumberland Forest Outdoor Decomposition Facility, a research and training center devoted to studying the effects of environmental exposure on postmortem human remains. In collaboration with a team of board-certified forensic anthropologists and odontologists, her research mission is to develop and optimize methodologies to improve identification efforts in missing person cases, mass disasters, and unidentified human remains (UHR) investigations.

Prior to LMU, Dr. Ambers previously served as Assistant Director of the *Henry C. Lee Institute of Forensic Science* in Connecticut, a world-class institute that specializes in interdisciplinary research, training, testing, casework consulting, and education in forensic science. She also held an Associate Professor (Forensic DNA) appointment in the *Henry C. Lee College of Criminal Justice and Forensic Sciences* at the University of New Haven (UNH), teaching forensic biology and DNA analysis methods, in addition to specialty courses on advanced DNA topics. During this time, she also was a team member with UNH's *Center for Forensic Investigation of Trafficking in Persons.* She worked as a forensic geneticist for the University of North Texas (UNT) Center for Human Identification for 8 years and, prior to pursuing her doctorate, was lead DNA analyst and lab manager of UNT's DNA Sequencing Core Facility. Her doctoral research involved an investigation of methods (e.g., whole-genome amplification, DNA repair) for improving autosomal and Y-STR typing of degraded and low copy number (LCN) DNA from human skeletal remains and environmentally damaged biological materials. Her master's thesis research included the development and optimization of a DNA-based multiplex screening tool for genetics-based separation of fragmented and commingled skeletal remains in mass graves.

Dr. Ambers specializes in the characterization and identification of contemporary, historical, and archaeological human skeletal remains. Her casework has involved DNA testing of an American Civil War guerrilla scout; several Finnish World War II soldiers; unidentified late-19-century skeletal remains discovered by a construction crew in Deadwood, South Dakota; unidentified skeletal remains of Special Operations soldiers killed during the 1974 Turkish invasion of Cyprus; skeletal remains exhumed from Prague Castle in the Czech Republic; skeletal remains of soldiers from the seven years' war (1756–63); bone samples purported to belong to a member of Jesse James gang (killed during a bank robbery in 1876); and the exhumed remains of the wife of a Yale medical school professor. She co-presented a workshop on "Advanced Methods for DNA-based Analysis of Skeletal Remains" at the 26th International Symposium on Human Identification (ISHI), and has been an invited speaker at an international bone workshop/conference in Prague, Czech Republic. In 2017, she traveled twice to India to train scientists from various Indian states and the Maldives Police Service on the processing of bone samples in forensic DNA casework. During her visit to India, she performed autosomal STR analysis on human skeletal remains discovered along a hiking route in the Himalayas (in the northern state of Himachal Pradesh) to assist local officials in the investigation of a missing person case. Additionally, she performed DNA analysis on a female homicide victim recovered from a clandestine grave in New Delhi.

Dr. Ambers' casework and research have been published in various peer-reviewed journals, including *Forensic Science International: Genetics (FSI: Genetics), Forensic Science International (FSI), International Journal of Legal Medicine (IJLM), Legal Medicine, BMC Genomics, the Croatian Medical Journal, The Journal of Heredity*, and *Journal of Biological and Clinical Anthropology (Anthropologischer Anzeiger)*. Her work has received press in numerous local and national newspapers (including *The Washington Times, NBC News, Criminal Legal News, The Root)* and has been featured on several podcasts (*Truth and Justice, Crime Waves*).

Her most recent casework (a collaboration with the Texas Historical Commission) includes DNA analyses of human skeletal remains associated with the French explorer La Salle's last expedition. This case involved massively parallel sequencing (MPS) of DNA from two adult male skeletons recovered from the La Belle shipwreck, skeletal remains recovered from a mass grave at Fort St. Louis, and the putative remains of Sieur de Marle. Additionally, Dr. Ambers performed DNA extractions and mitochondrial DNA (mtDNA) sequencing on numerous sets of skeletal remains associated with Spanish royalty and the House of Aragon (recovered from tombs within the Royal Pantheon of San Juan de la Peña archaeological site in Spain). She is currently working on the putative remains of William Townsend Washington, the nephew of former U.S. President George Washington.

In addition to skeletal remains cases and research, Dr. Ambers collaborated with the National Institute of Justice (NIJ) and the Forensic Technology Center of Excellence (FTCoE) to develop and disseminate a formal report on the use of familial DNA searching (FDS) in casework. She is an active cold case consultant, an advocate of postconviction DNA testing, and an educator/advisor on DNA testing or retesting of old, degraded, or challenging evidentiary samples. Dr. Ambers also was the 2017–18 Project Lead on a U.S. State Department grant to combat human trafficking in Central America through the application of forensics. As part of the program objectives, she traveled to three Northern Triangle countries (Guatemala, El Salvador, Honduras) to perform gap assessments of government laboratories and train personnel in forensic DNA analysis, with the goal of promoting quality casework methods based on ISO 17025 standards. As Project Lead, she helped in training forensic DNA scientists at the *Instituto Nacional de Ciencias Forenses de Guatemala* (INACIF, the National Institute of Forensic Sciences of Guatemala) in Guatemala City, Guatemala; the *Instituto de Medicina Legal* (Institute of Legal Medicine) in San Salvador, El Salvador; and the *Instituto de Medicina Forenses: Centro de Medicina Legal y Ciencias Forenses* (the Institute of Forensic Medicine: Center of Legal Medicine and Forensic Sciences) in Tegucigalpa, Honduras.

In addition to providing both lecture and tactile training on DNA analysis methods to Northern Triangle laboratories, Dr. Ambers was part of a consortium to help these countries develop and maintain forensic DNA databases to assist in the identification of missing persons related to human trafficking.

Eric Bartelink, PhD, D-ABFA

Dr. Eric J. Bartelink is a Professor in the Department of Anthropology and Co-Director of the Human Identification Laboratory at California State University, Chico. He received his BS in Anthropology at Central Michigan University (1995), his MA in Anthropology at California State University, Chico (2001), and his PhD in Anthropology at Texas A&M University (2006). He is a board-certified Diplomate of the American Board of Forensic Anthropology (ABFA).

Dr. Bartelink's interests are in forensic anthropology and bioarchaeology, and he has conducted research focused on skeletal trauma, taphonomy, paleopathology, and stable isotope analysis. He maintains an extensive research program focused on central California bioarchaeology and also has conducted work in American Samoa. Previously, he assisted with the excavation of mass graves in Bosnia-Herzegovina through the United Nations International Criminal Tribunal for the Former Yugoslavia, and also assisted in the identification of victims from the World Trade Center 9/11 disaster.

He has published articles in *Journal of Forensic Sciences, Forensic Science International, Forensic Sciences Research, American Journal of Physical Anthropology, Journal of Archaeological Science, International Journal of Osteoarchaeology, International Journal of Paleopathology, Journal of Archaeological Method and Theory, Archaeometry,* and *California Archaeology*. He is also a co-author on the books *Forensic Anthropology: Current Methods and Practice* (Elsevier, 1st and 2nd editions), *Essentials of Physical Anthropology 10e* (Cengage), *Introduction to Physical Anthropology 15e* (Cengage), and a co-editor on *New Perspectives in Forensic Human Skeletal Identification* (Elsevier). Dr. Bartelink teaches courses in introductory physical anthropology, as well as forensic science, human osteology, forensic anthropology, bioarchaeology, and statistics. He is a Fellow of the American Academy of Forensic Sciences (AAFS), and a member of the American Association of Physical Anthropologists (AAPA), Society of American Archaeology (SAA), Paleopathology Association (PPA), and the Society for California Archaeology (SCA). He is a past Board Member and President of the American Board of Forensic Anthropology (ABFA), and former Board Member of the American Academy of Forensic Sciences (AAFS), the Scientific Working Group for Forensic Anthropology (SWGANTH), and the Organization of Scientific Area Committees (OSAC) Anthropology Subcommittee.

Howard Cash, CEO

Howard Cash is the CEO of Gene Codes Corporation. Before venturing into the field of computational biology, he studied music at the University of Pennsylvania and also served as Assistant Conductor with the Pennsylvania Opera Theater, Psychoacoustics at Stanford.

At the forefront of commercial bioinformatics since 1984, Cash was Senior Engineer and headed the Expert Systems Group at IntelliGenetics, Inc., where many seminal biotechnology software tools were developed. One of the programs he worked on during this period was key to the original sequencing of the human immunodeficiency (HIV/AIDS) virus. In 1988, he founded Gene Codes Corporation, one of the most successful bioinformatics companies in the world. He designed and developed the Sequencher program used in thousands of academic and commercial DNA sequencing labs in more than 90 countries. Specialized versions of Sequencher have been developed for applications including human identification, mitochondrial DNA (mtDNA) typing, therapy review based on HIV strain dominance in AIDS patients, and rapid characterization of H1N1 flu variants from the 2009 worldwide outbreak.

After the 9/11 terrorist attacks, New York City asked Cash to develop DNA analysis software for identifying the remains of those killed at the World Trade Center. A daunting task from a technical standpoint—with a database and set of analysis tools integrating primary sequence, SNP, and STR data—it also raised ethical and legal issues. One result of this effort was the Mass-Fatality Identification System, or M-FISys (pronounced like *emphasis*). Cash has applied bioinformatics to numerous other humanitarian genetic identification projects, including (1) the Boxing Day Tsunami; (2) the sexual assault and murder of hundreds of women in Juarez, Mexico; (3) postconflict identifications from Guatemala's 35-year civil war; and (4) an international collaboration to detect and interdict child trafficking.

Thomas David, DDS, D-ABFO—Forensic Odontologist

Dr. Thomas J. David earned his dental degree from Emory University School of Dentistry in 1977. He has maintained a dental practice since that time in the Atlanta metropolitan area. He is a member of the American Society of Forensic Odontology (ASFO), a Fellow of the American Academy of Forensic Sciences (AAFS), and a Diplomate of the American Board of Forensic Odontology (ABFO). Dr. David has served as Chair of the Odontology Section of AAFS and President of the ABFO, as well as on the Editorial Board of the *Journal of Forensic Sciences* (JFS). He has authored articles in various periodicals (including the JFS) and a number of textbook chapters (including multiple editions of the *Manual of Forensic Odontology* and the 2nd Edition of *Forensic Dentistry*). He also co-edited the book *Forensic Odontology: Principles and Practice*, which was released in 2018.

Dr. David is a consultant in Forensic Odontology to the Georgia Bureau of Investigation (GBI), Division of Forensic Sciences and the State of Georgia Medical Examiner's Office. He also holds a faculty appointment as a Clinical Assistant Professor in the Department of General Dentistry of the Graduate School of Medicine at the University of Tennessee Health Science Center. He provides instruction for the Forensic Odontology Fellowship program and the University of Tennessee biennial Symposium "All That Remains."

Suni M. Edson, PhD

Dr. Suni M. Edson started her career in science working in a research laboratory while an undergraduate at Texas Tech University. She worked with graduate students and professors on projects ranging from the insertion of traceable genetic material into cotton seeds to the analysis of transposable elements in *Peromyscine* rodents. Upon graduation with a BS in Biology, Dr. Edson migrated to Syracuse, New York to attend the State University of New York (College of Environmental Science and Forestry). She spent several of the snowiest-winters-on-record traipsing through Adirondack Park to study the introgressive hybridization of Mallards and Black Ducks. The project involved the collection of blood samples from numerous pairs of ducks and their broods, and subsequent RFLP analysis to determine if these dabbling ducks are truly "monogamous" in their pair bonds.

After earning her MS degree, Dr. Edson moved to the Washington, D.C. area to work briefly at the Lombardi Cancer Center at Georgetown University. There she spent time as a research assistant examining human leukocyte antigen (HLA) genes, garnering five publications as a coauthor. During the first few months of work at Georgetown, she answered an ad in the paper for a research assistant at the Armed Forces DNA Identification Laboratory (AFDIL) in Rockville, Maryland. While she was not hired for that position, she did become a DNA technician in the Past Accounting Section, and thus began her almost 22-year tenure with the laboratory.

Dr. Edson moved from DNA technician to DNA analyst to supervisory DNA analyst in the short span of 3.5 years. She remained a productive supervisor for 9 years before transitioning in 2010 to her current role of assistant technical leader. During the time as a supervisor, she returned to her researcher roots and started to examine the success of DNA testing of different skeletal elements. While research is not her primary role, Dr. Edson has published over 20 peer-reviewed publications and book chapters regarding DNA testing of human remains and the intricacies surrounding human identification. She has presented at numerous professional conferences in the United States and internationally. In 2015, she was accepted into a PhD program at Flinders University in South Australia. Her dissertation on improving the extraction of DNA from skeletal materials generated a novel method using GC/MS to analyze skeletal materials for inhibitors and materials deposited on the remains peri- and postmortem, as well as nine publications to date. Dr. Edson graduated in 2019 and remains in her position at AFDIL, where she continues to examine trends in DNA testing of skeletonized human remains.

Colleen Fitzpatrick, PhD

Dr. Colleen Fitzpatrick is widely recognized as the founder of modern Forensic Genetic Genealogy (FGG). She pioneered the use of genetic genealogy (Y-STR and autosomal SNP analysis) for generating forensic intelligence on cold cases that are sometimes decades old. Dr. Fitzpatrick's expertise includes the application of whole-genome sequencing to low level and highly degraded DNA, enabling identifications that otherwise are believed to have gone beyond the reach of modern technology.

Dr. Fitzpatrick has twice been awarded fifth place in the prestigious international Gordon Thomas Honeywell *Cold Case Hit of the Year* competition—first in 2018 for solving the 1992–93 Phoenix Canal Murders (out of 61 entries from 14 countries). This was the first case solved using genetic genealogy (2015). Her second *Cold Case Hit* award came in 2020 for her work on the 1991 Sarah Yarborough Homicide (out of 50 entries from 20 countries). This was the first case where genetic genealogy was used to generate investigative leads (2011).

In addition to her work in forensic identification, Dr. Fitzpatrick had been a key member of the teams that have exposed two international Holocaust literary frauds. She has also served as the forensic genealogist on several historical projects including the identification of the unknown child on the Titanic and the Abraham Lincoln DNA Project. Her collaborations include the Armed Forces DNA Identification Laboratory (AFDIL), the U.S. Army Casualty and Mortuary Affairs Office, the Simon Wiesenthal Center in Jerusalem, the University of Arizona, the University of Adelaide, the Henry C. Lee Institute of Forensic Science, and other noted professional organizations. She is a member of the Vidocq Society and an associate member of the American Academy of Forensic Science (AAFS).

Dr. Fitzpatrick is the founder of *Identifinders International* and co-founder of the *DNA Doe Project*. She lectures widely in the United States, Canada, Europe, Australia, and New Zealand. She has appeared in hundreds of newspapers and magazines (both domestically and internationally), and on international radio and television programs. She is the author of three books: *Forensic Genealogy, DNA & Genealogy*, and *The Dead Horse Investigation: Forensic Photo Analysis for Everyone*.

Ranjana Grover, PhD

Dr. Ranjana Grover is Vice President of Molecular Biology at ANDE Corporation. She received a BS in Zoology (summa cum laude), an MS in Genetics, and a PhD in Molecular Human Genetics from Delhi University. She is an expert in the field of developing rapid, highly multiplexed PCR assays for diverse applications ranging from human identification to in vitro diagnostic assays for human diseases. Dr. Grover has worked in a high-throughput human genetics research laboratory exploring the genetics of complex traits and has a strong background in statistical methods related to molecular genetics. She received postdoctoral training at Johns Hopkins University and then joined the Henry M. Jackson Foundation, where she worked on identifying genetic and epigenetic markers related to Down syndrome, traumatic brain injury, and the toxic effects of metals employed in bullets. While working on these projects, she received several Department of Defense (DoD) research grants and developed expertise in highly advanced next-generation sequencing (NGS) technologies and in second-generation sequencing analysis.

Michael Hennessey, MBA

Mr. Hennessey was the onsite Project Manager for Gene Codes Forensics (GCF) during the identification of victims of the 9/11 terrorist attacks in 2001. For 3 years after the attacks, he reported daily to the Office of Chief Medical Examiner (OCME) in New York City and provided operational consulting expertise on the World Trade Center identification project. In 2005, Mr. Hennessey also served on the DNA Subcommittee of the Thailand Tsunami Victim Identification Center in Phuket, Thailand. He has helped lead Mass Fatality Identification System (M-FISys) customer training for users in Guatemala, Kenya, El Salvador, Costa Rica, Libya, Israel, Mexico, and Saudi Arabia. Additionally, Mr. Hennessey has conducted human identification workshops at national and international forensics meetings, published peer-reviewed journal articles, and serves on the Scientific Advisory Board of the Forensic Anthropology Foundation of Guatemala (FAFG). He has a Master's of Business Administration (MBA) from the University of Michigan with a concentration in operations management.

Irene Kahline, MS

Irene Liunoras Kahline received her Bachelor of Science (BS) in Biochemistry at Florida State University. Following her undergraduate studies, she worked in a research lab studying the evolution and phylogeny of shark and ray DNA. During her time in the research lab, she decided to pursue a career in forensic DNA. She moved to the Washington, D.C. area to obtain her Master of Forensic Science in Forensic Molecular Biology at George Washington University. While pursuing her master's degree, she interned with the research department (now Emerging Technology section) at the Armed Forces DNA Identification Laboratory (AFDIL).

Irene has steadily moved through different positions at AFDIL from research intern to DNA technician to DNA analyst, and now as a supervisory DNA analyst for the Past Accounting Section. She is proficient in mitochondrial DNA (mtDNA) Sanger sequencing, next-generation sequencing (NGS) of whole-genome mitochondrial DNA, autosomal STR typing, and Y-STR technologies working on dried skeletal remains from past military conflicts. Her dedication to the mission of AFDIL was so immense that she agreed to move to Dover, Delaware from the Rockville, Maryland location.

Bobby LaRue, PhD

Dr. Bobby LaRue earned a PhD in Molecular Genetics from the University of North Texas (UNT) and completed an Intelligence Community postdoctoral fellowship in which he studied the feasibility of small-amplicon, di-allelic markers for forensic applications. Dr. LaRue recently left his position as an Associate Professor of Forensic Science at Sam Houston State University (SHSU) to accept a position in private industry (with Verogen, Inc.).

Brandon Letts, PhD

Dr. Brandon Letts is a Research Biologist at the FBI Laboratory in Quantico, Virginia. He possesses a BS in Forensic Science from Baylor University and a PhD in Genetics from Pennsylvania State University. His current duties include research and validation of new techniques and technologies for implementation at the FBI and troubleshooting laboratory issues arising in casework. Prior to becoming a research biologist, he was a forensic examiner for the U.S. Army Criminal Investigation Laboratory (USACIL) and both a forensic examiner and technical specialist for the FBI.

Dr. Letts has a background in ancient DNA, having performed his dissertation work under the supervision of Dr. Beth Shapiro, a leader in the study of ancient DNA and molecular evolution. In her lab, he analyzed hundreds of samples up to 100,000 years old and used the information he obtained to investigate the effects of environmental changes on animal populations. His research interests continue to revolve around the retrieval and analysis of DNA from highly degraded samples.

James M. Lewis, DMD, D-ABFO

Dr. James M. Lewis earned his dental degree from the University of Alabama School of Dentistry (Birmingham) in 1985. He maintained a general dentistry practice exclusively in Madison, Alabama from 1986 to 2018. Dr. Lewis completed a fellowship in Forensic Odontology from the Center for Education and Research in Forensics at the University of Texas Health Science Center at San Antonio in 2001. As a board-certified forensic odontologist, he assisted in victim identification following the World Trade Center attack in New York City and, since 2003, has served as a consultant to the Alabama Department of Forensic Sciences (ADFS) and a volunteer to the Alabama Office of Emergency Preparedness in relation to its mass disaster response group.

Dr. Lewis is a Fellow of the Odontology Section of the American Academy of Forensic Sciences (AAFS), became board-certified by the American Board of Forensic Odontology (ABFO) in 2008, has served on the Board of Governors of the American Society of Forensic Odontology (ASFO), and became President of the organization in 2012. For the American Board of Forensic Odontology (ABFO), he served as member and Chair of the Dental Age Assessment Committee (2008–20); as a member of the Certification and Examination Committee (2011–15); Bitemark Evidence and Patterned Injury Committee (2008–15; 2017–19); and currently holds the office of ABFO President. He is currently appointed to the Odontology Subcommittee of the Organization of Scientific Area Committees (OSAC) for Forensic Science with the National Institute of Standards and Technology (NIST).

Dr. Lewis previously held the position of Director of Forensic Dentistry (Division of Forensics) in the Department of General Dentistry at the University of Tennessee Graduate School of Medicine. He is currently Director of Forensic Odontology at the *Institute for Human Identification* in the College of Dental Medicine at Lincoln Memorial University (LMU). He is also a professor of Forensic Dentistry at LMU and an instructor for the National Forensic Academy (NFA) in Oak Ridge, Tennessee. He has edited numerous textbooks, authored textbook chapters, and published a variety of articles in peer-reviewed journals on forensic odontology.

Odile Loreille, PhD

Dr. Odile Loreille holds a PhD in Paleogenetics from the University of Claude Bernard in Lyon, France. During her PhD research and two postdoctoral positions, she specialized in extracting and analyzing DNA from extinct species (cave bears, cave lions), mummies, and decades-old parasites. When she joined the Armed Forces DNA Identification Laboratory (AFDIL) in 2004, she used her expertise in ancient DNA (aDNA) analysis to tackle AFDIL's most challenging identification effort—the so-called Punchbowl samples, a large set of Korean War remains buried as unknown in the National Memorial Cemetery of the Pacific. After optimizing the DNA extraction protocol with a technique that is now used in many forensic laboratories around the world, she introduced next-generation sequencing (NGS) to the laboratory. Using Illumina technology and hybridization capture, she successfully retrieved mitochondrial DNA (mtDNA) from these human remains.

Dr. Loreille is now a Research Biologist in the DNA Support Unit at the FBI Laboratory and continues to work on improving techniques for the FBI's DNA Casework Unit. The focus of her current research is the retrieval of nuclear DNA from decades-old human rootless hair shafts and characterization of single nucleotide polymorphisms (SNPs) to identify the hair's owner. Another research interest of hers is DNA analysis of remains attributed to historical figures. These projects involved the tentative identification of W.A Mozart, Friedrich Schiller, the unknown child of the Titanic, Everett Ruess, The Romanov family (including the two children found in 2007), Sir George Yeardley, and Henry Rathbone.

Dr. Loreille has been a member of the International Society for Forensic Genetics (ISFG) for the past 13 years and has published 30 peer-reviewed research articles or book chapters. Her 2007 publication on DNA extraction using total demineralization has been cited more than 250 times, and her 2018 research article on the DNA analysis of a 4000-year-old Egyptian mummy is one of the most-read articles of the journal.

Murray K. Marks, PhD, D-ABFA

Dr. Murray Marks is the board-certified forensic anthropologist for the Regional Forensic Center (RFC) in Knoxville, Tennessee, and for the Hamilton County Medical Examiner's Office in Chattanooga. In these respective appointments, he performs recovery, biological assessment, skeletal trauma, and identification casework on human remains. Dr. Marks also serves as Director of Forensic Anthropology at the *Institute for Human Identification* in the College of Dental Medicine at Lincoln Memorial University (LMU). Additionally, he is a Professor of Forensic Anthropology and Human Identification at LMU and manages the *National Forensic Academy (NFA) Outdoor Decomposition Training Facility*, where he teaches archaeological clandestine grave recovery methods and forensic anthropology. Dr. Marks previously held an Associate Research Professor appointment in the Graduate School of Medicine's Department of General Dentistry (Division of Forensics) at the University of Tennessee Medical Center Knoxville (UTMCK), instructing students in the M.S. program on forensic dentistry, head and neck anatomy, mineralized tissue histology, and research methods. He currently holds adjunct visiting professorships in the Department of Anatomy and Neurobiology at Boston University and in the Department of Anatomy at the University of Western Australia.

Timothy McMahon, PhD

Dr. Timothy McMahon received his PhD in Biomedical Sciences from the School of Public Health at the University of Albany, New York. Dr. McMahon's graduate studies and postdoctoral research were performed in the division of Infectious Disease and Immunology at the New York State Department of Health. His research included the identification and interaction of the Human Cytomegalovirus Helicase-Primase replication proteins and a pro-death late transcript that caused cell death in infected cells. From 2002 to 2007, Dr. McMahon established the U.S. Department of Defense (DoD) DNA Operation's Quality Control and Validation section; in this role, he was responsible for developing and forensically validating all new human remains DNA identification instrumentation, reagents, and testing procedures.

In March 2007, Dr. McMahon joined Applied Biosystems, where he established the first commercial Global DNA Forensic Professional Service organization whose primary function was to design and establish new DNA laboratories (as well as aid local, state, and federal crime laboratories with implementing new forensic DNA technologies and instrumentation). During his time at Applied Biosystems, Dr. McMahon was responsible for developing a state of the art, fully automated Criminal Casework Line that allowed for the processing of 640 crime scene samples per day, developed an automated differential extraction line, and performed over 200 validations for both North American and international crime laboratories. In February 2012, Dr. McMahon returned to the Armed Forces Medical Examiner System (AFMES) as the Chief of Forensic Services, where his role as the scientific subject matter expert was to advise, guide, and help maintain the AFMES as a leader in the forensic community. He oversaw all aspects of casework performed at DoD DNA operations while directing and carrying out projects and programs to develop innovative advances in DNA-based forensic testing methods. In this role, Dr. McMahon also established AFMES-AFDIL as the first laboratory in the United States to develop and implement a forensically validated next-generation sequencing (NGS) method for human remains identification.

In 2017, Dr. McMahon became the first civilian Director of the DoD DNA operations section of the AFMES, where he is in charge of the Armed Forces DNA Identification Laboratory (AFDIL) and the Armed Forces Repository of Specimen Samples for the Identification of Remains. Dr. McMahon oversees over 150 scientists and is in charge of all human remains DNA scientific initiatives in support of the AFMES' primary mission to identify fallen U.S. Service members from present-day operations and from past conflicts.

Darinka Mileusnic-Polchan, MD, PhD, D-ABP

Dr. Darinka Mileusnic-Polchan is a graduate of the University of Rijeka Medical School in Croatia. After earning her medical degree, she graduated from Loyola University's Neuroscience Graduate Program in Chicago, Illinois. She completed her training in anatomic pathology in the Department of Pathology, Loyola University Stritch School of Medicine. Dr. Mileusnic completed forensic pathology training at the Cook County Office of the Medical Examiner in Chicago, Illinois, subsequently receiving board certification in *Anatomic and Forensic Pathology*. Upon completion, she accepted the position of Assistant Medical Examiner for the Cook County Office of the Medical Examiner for several years.

In 2002, Dr. Mileusnic-Polchan relocated to Knoxville, Tennessee, where she continued the same line of forensic casework in the Department of Pathology at The University of Tennessee Graduate School of Medicine (UTMCK). Currently, she is a Clinical Associate Professor of Pathology at UTMCK. In 2008, she accepted the position of Chief Medical Examiner for Anderson County and Knox County in the state of Tennessee. Dr. Mileusnic is the Medical Director of the Knox County Regional Forensic Center (RFC), which has been accredited by the National Association of Medical Examiners since 2010. She chaired the State Medical Examiner Advisory Council and has been an invited speaker at numerous local and national conferences.

Stephen Missal, MFA, CFA

Stephen Missal graduated with a BFA and MFA in painting from Wichita State University and taught art at the college level for over four decades. He later received his training in forensic art by world-renowned forensic artists Karen T. Taylor and Betty Pat Gatliff. Stephen has specialized in forensic facial reconstruction, and for the past 12 years, he has been the sole forensic facial reconstruction artist working with the Maricopa County Medical Examiner in Arizona (home to Phoenix, Mesa, Scottsdale, and a number of other large cities). In addition, he does forensic reconstructions for several other Arizona counties, has worked with law enforcement in Georgia, and is certified by the International Association for Identification (IAI). Stephen actively pursues a reconstruction caseload and contributes to educating the public about forensic art through lectures at various colleges and to high school audiences. As an educator, Stephen achieved full Professor at the Art Institute of Phoenix, where he taught with an elite faculty that included two Emmy winners and several nationally recognized illustrators and designers. In 2016, he worked with the NRK (the national Norwegian broadcasting network) as the forensic artist responsible for a new reconstruction of the *Isdal Woman*, which was included in the documentary about the case which aired that same year. Stephen's facial reconstruction for this case has since appeared in *Die Zeit* and many other European news outlets over the past 4 years. Aside from his forensic casework, Stephen's paintings and drawings can be found in hundreds of private collections nationally. He also has co-authored and illustrated two college textbooks on drawing for animation, and has appeared on local and national cable television demonstrating his forensic art techniques.

Scores of Stephen's forensic reconstructions can be found in the National Missing and Unidentified Persons System (NamUs), as well as on the Maricopa County Medical Examiner's site where unknown cases are available for public view.

Nicole Novroski, PhD

Dr. Nicole Novroski is a forensic geneticist and tenure-stream faculty member at the University of Toronto in the Department of Anthropology, with a primary teaching appointment in the Forensic Science Program at the University of Toronto Mississauga (UTM). Her research program focuses on novel and innovative approaches to improve upon current forensic genetic methodologies using traditional (PCR-CE) and MPS-based instrumentation.

Dr. Novroski's academic accolades include a Bachelor of Science (Hons) in Forensic Science and Biology from the University of Toronto, a Master of Science in Forensic Biology from the University at Albany (SUNY), and a Doctor of Philosophy in Molecular Genetics from the University of North Texas Health Science Center (UNTHSC) at Fort Worth, Texas. Her educational training is largely focused on the exploration of previously uncharacterized genetic markers for improved DNA mixture de-convolution of complex forensic profiles. Dr. Novroski is the recipient of numerous awards including the Forensic Sciences Foundation (AAFS) Douglas M. Lucas Grant, the Connaught New Researcher Award, and the University of Toronto Mississauga Dean's Award for Excellence (Faculty). She is an associate member of the American Academy of Forensic Sciences (AAFS), an active member of the International Society of Forensic Genetics (ISFG), and an active member of the Scientific Research Honor Society Sigma Xi. Her publication record includes over 20 peer-reviewed manuscripts, 35 scientific communications, public abstracts, and oral presentations as well as two book chapters. She is an ad hoc reviewer for multiple international journals in the areas of human and forensic genetics and serves as a committee member for LabRoots Forensic Science, the Swiss National Science Foundation, and the American Society of Human Genetics "DNA Day" Essay Contest.

Dr. Novroski is passionate about mentoring, volunteerism, and continuing education. She actively holds mentorship positions with the University of Toronto Mississauga Alumni Mentor Program as well as the Women in Science Mentorship Program (as part of an ongoing NSERC CREATE PROMOTE Grant). Within her laboratory, she has a growing team of undergraduate and graduate trainees. Further, Dr. Novroski assists legal professionals and the public as a forensic DNA expert through her firm 4NGen Consulting. In her limited spare time, she volunteers with the Junior League of Toronto and Toronto Counter Human Trafficking Network and is actively pursuing a degree in Forensic Accounting to further diversify her portfolio.

Vivek Sahajpal, PhD

Dr. Vivek Sahajpal is Assistant Director of the Directorate of Forensics Services in Himachal Pradesh, India. In addition, he is the Nodal Officer for the Disaster Victim Identification (DVI) Cell at the Directorate of Forensic Services. Dr. Sahajpal has performed forensic DNA analyses in more than 2000 criminal cases, as well as for numerous cases involving the identification of human skeletal remains and disaster victim identification. Dr. Sahajpal attained his master's degree in Forensic Science from Punjabi University Patiala, and his concern for wildlife conservation and his interest in the field of Wildlife Forensics brought him to the Wildlife Institute of India, Dehradun. At the Wildlife Institute of India, Dr. Sahajpal carried out some of the pioneering research work in the field of Wildlife Forensics and genetics. He earned his PhD in Forensic Science from Punjabi University, Patiala. Identification of human skeletal remains has been an important area of work for Dr. Sahajpal because Himachal Pradesh is a popular tourist destination for hikers, trekkers, and pilgrims—who often go missing at high altitudes in the Himalayas (and along mountain passes and rivers) after getting lost or due to abrupt weather changes. Skeletonized remains are quite often recovered in these cases.

Dr. Sahajpal is currently working on the genetic diversity of different forensically significant markers across populations in India. He has experience in the field of quality assurance (QA) and has worked as the quality manager (QM) for the forensics laboratory in Himachal Pradesh. Dr. Sahajpal is a member of the Indian Hair Research Society and the International Society for Forensic Genetics (ISFG). He serves on the editorial panel of *Forensic Science International: Reports*, and works for the "DNA Fights Rape—Save the Evidence" program run by Gordon Thomas Honeywell (GTH) Government Affairs. Dr. Sahajpal has authored two books, contributed chapters to four books, and has published 25 research papers in the field of forensic genetics.

Richard Selden, MD, PhD

Dr. Richard Selden is the Founder and Chief Scientific Officer (CSO) of ANDE Corporation. He founded the company in 2004 with a vision to move DNA analysis from sophisticated laboratories to the field, where it could have a daily impact in the military, disaster victim identification (DVI), law enforcement, homeland security, and clinical diagnostic applications. He received his BA degree from Harvard University, MA and PhD degrees in Genetics from the Harvard Graduate School of Arts and Sciences, an MD from Harvard Medical School, and trained as a pediatrician at Massachusetts General Hospital.

Dr. Selden is an author of 45 scientific publications and an inventor on 47 U.S. patents. He was named an honorary member of the Butte County Sheriff's Department for his humanitarian work on the 2018 Camp Fire, the deadliest wildfire in California history. He has received numerous awards, including the *Ernst & Young New England Entrepreneur of the Year* award, the *R&D 100 Award*, the *Not Impossible Award*, and the 2020 *CES Innovation Award*. Additionally, Dr. Selden has been the principal investigator (PI) on a broad range of government grants and contracts for clinical and forensic programs.

Eugene Tan, PhD

Dr. Eugene Tan is Senior Vice President of Product Development at ANDE Corporation. He received a BS in Engineering (summa cum laude), an MS in Engineering, and a PhD in Engineering Physics from McMaster University. He has extensive experience in realizing commercial products from early-stage research programs. Among these commercialized products include the ANDE Rapid DNA system, the Genebench FX ruggedized microfluidic electrophoresis system, the DeNOVA 5000HT (Shimadzu) high-throughput DNA sequencer, the XTM-72 (Xanoptix) highly parallel optical transceiver for optical interconnection, and the ML-20 Tunable Laser (CoreTek/Nortel) for telecommunications. At ANDE, Dr. Tan leads the development of Rapid DNA Systems for clinical diagnostics, human forensic identification, and biothreat detection. He is responsible for the design, construction, testing, and optimization of the optical, mechanical, and electrical systems; microfluidic design and testing; and development and optimization of chip fabrication processes. Additionally, he is responsible for the development of data processing and expert system algorithms. Dr. Tan was the Program Manager and Chief System Architect for the Joint DoD/DHS/FBI ANDE program. He was also the principal investigator on three National Institute of Justice (NIJ) programs on Forensic DNA Research and Development. Dr. Tan is an inventor on 21 ANDE-issued U.S. patents and an author on numerous peer-reviewed publications.

Rosemary Turingan Witkowski, PhD

Dr. Rosemary Turingan Witkowski is Vice President of Forensic Biology at ANDE Corporation. She earned a BS degree in Chemistry (with Honors) at the University of the Philippines (Diliman, Quezon City). After moving to the United States, she earned an MS and PhD in Chemistry from the University of Massachusetts, Amherst. Her graduate research focused on understanding transcription kinetics and DNA-T7 RNA polymerase structural dynamics through fluorescence resonance energy transfer for which she was recognized at the Biophysical Society Annual Meeting for Student Research Achievement. She joined ANDE in 2006 and has become an integral player in the development, validation, and commercialization of the ANDE Rapid DNA Identification System. She is responsible for developing protocols for analysis of a wide range of sample types for Rapid DNA Identification, including Disaster Victim Identification (DVI) and sexual assault evidence applications. She was directly involved in the identification of the remains of the 2018 California Wildfires, and her expertise in the processing of human remains has led to the resolution of investigative cases at the request of various law enforcement agencies across the country. At ANDE, Dr. Turingan Witkowski is also responsible for the development of biological threat assays and related clinical diagnostics based on highly multiplexed PCR amplification and sequencing methods. She is an author of several peer-reviewed scientific publications and an inventor on a number of U.S. patents.

Jodie Ward, PhD

Associate Professor Jodie Ward is Program Lead for the Australian Federal Police's National DNA Program for Unidentified and Missing Persons. This multidisciplinary forensic science program serves to aid the investigation of Australia's current unidentified and missing person cases. In addition, Dr. Ward is Director of the Australian Facility for Taphonomic Experimental Research (AFTER) and an Associate Professor in the Centre for Forensic Science at the University of Technology Sydney. This unique joint appointment facilitates her ability to lead the research, development, and application of forensic human identification techniques for missing person casework in Australia. Her specific research interests include investigating novel DNA technologies (e.g., massively parallel sequencing, Rapid DNA testing) and their integration with complementary forensic techniques to optimize the identification of unknown deceased persons in humanitarian forensic operations. Previously, she has held forensic DNA specialist roles with the New South Wales (NSW) Forensic and Analytical Science Service and NSW Police Force, and forensic biology academic roles with the National Center for Forensic Studies. She is considered an expert in mitochondrial DNA testing and provides expert evidence for both criminal and coronial cases in Australia.

Dr. Ward holds a PhD in forensic molecular biology from The Australian National University and has postgraduate qualifications in management and higher education. Her career highlight includes being awarded a prestigious 2015 Churchill Fellowship to investigate world-leading DNA identification techniques for missing persons and disaster victims. Following her fellowship, she devised, published, and promoted a number of international best-practice recommendations for the DNA-led identification of human remains, which provide the foundation for the National DNA Program for Unidentified and Missing Persons. She was also recognized as one of Science and Technology Australia's *2017 Superstars of STEM* and Australian Financial Review's 2018 *100 Women of Influence* for pioneering a specialist nuclear- and mitochondrial DNA identification laboratory in NSW, which was used nationally by police, forensic, and military agencies to identify challenging human skeletal samples from contemporary and historical contexts. She is a member of the Australian Academy of Forensic Sciences (AAFS), Australian and New Zealand Forensic Science Society

(ANZFSS), and International Society of Forensic Genetics (ISFG), and has been a Board Member of the Missing Persons Advocacy Network since 2018.

Jeremy Watherston PhD

Jeremy Watherston holds a Bachelor of Science in Biomedical Science (Forensic Biology), a Bachelor of Social Science (Psychology), a Master of Forensic Studies in Forensic Science (MForStForSc), and a PhD from the University of Technology Sydney (Australia). An experienced Senior Forensic Biologist currently working in the Case Management Unit at the NSW Health Pathology, Forensic, and Analytical Science Service (Sydney, Australia), he reports on nuclear DNA, Y-chromosome DNA, mitochondrial DNA (mtDNA), and mixed DNA profiles (including probabilistic genotyping analyses), as well as genetic identification using paternity and kinship analysis. Jeremy's research focuses on the recovery of DNA from compromised human remains and DNA-based disaster victim identification (DVI). His research covers optimal sample selection, collection, and preservation, as well as novel profiling approaches and the application to Rapid DNA platforms. Dr. Watherston's research is being carried out in collaboration with the Australian Facility for Taphonomic Experimental Research (AFTER).

Frank Wendt, PhD

Dr. Frank Wendt holds a Bachelor's Degree in Forensic Science (emphasis: biology) from Pennsylvania State University (PSU) and a PhD in Biomedical Science (emphasis: molecular genetics) from the University of North Texas Health Science Center (UNTHSC). His educational training is largely focused on the population genetics of forensically relevant marker types, including insertion–deletion, single nucleotide, and tandem repetitive polymorphisms in the human genome, including mitochondrial DNA elements. As an undergraduate, Dr. Wendt was involved in the initial testing of Rapid DNA instrumentation now readily employed in law enforcement agencies globally. At UNTHSC, he developed pharmacokinetic-driven machine learning models of response to the synthetic opioid agonist tramadol. This exposure to complex trait genetics brought him to the Yale School of Medicine (Department of Psychiatry) to study the cause and consequence of shared genetic architecture across complex traits related to psychiatry, cognition, and human behavior. Dr. Wendt is currently a National Research Service Award Fellow through the National Institute of Mental Health. His work focuses on understanding sex differences in psychiatric disorders and related comorbidities using sophisticated computational multiomics methods. To understand these differences, Dr. Wendt works primarily with large-scale genetic data as an analyst for the Psychiatric Genomics Consortium Posttraumatic Stress Disorder and Substance Use Disorder Working Groups (PGC-PTSD) and a genetic epidemiologist within the U.S. Department of Veterans Affairs *Million Veteran* Program.

Dr. Wendt has received numerous prestigious awards from regional, national, and international organizations including the *Million Veteran Program* and the *World Congress of Psychiatric Genetics Early Career Investigator* Awards. Dr. Wendt is a criminalistics trainee affiliate of the American Academy of Forensic Sciences (AAFS) and a member of the American Society of Human Genetics. He serves as an ad hoc reviewer for 10 journals for the international forensic science and human genetics communities. He has published a total of 50 peer-reviewed research articles, preprints, review articles, and book chapters; has contributed to 41 conference abstracts; and has given

15 invited lectures across the United States and internationally.

Dr. Wendt is passionate about mentorship, STEM outreach, and bioinformatics education. He holds an associate certification of College Teaching Preparation from the multiinstitutional Center for the Integration of Research, Teaching, and Learning. He has served on three master's thesis committees for the Penn State Forensic Science Program and is an active participant in the Penn State Alumni Mentor Network where he mentors undergraduate forensic science students.

Editor biography

Dr. Angie Ambers is an internationally recognized DNA expert and the Director of Forensic Genetics at the *Institute for Human Identification* in the College of Dental Medicine at Lincoln Memorial University (LMU). She is also a Professor of Forensic Genetics and Human Identification at LMU. She has a PhD in Molecular Biology (with an emphasis in Forensic Genetics and human identification) as well as master's degrees in both Forensic Genetics and Criminology. Dr. Ambers conducts research at the National Forensic Academy (NFA) Cumberland Forest Outdoor Decomposition Facility, a research and training center devoted to the scientific study of the effects of environmental exposure on postmortem human remains. In collaboration with a team of board-certified forensic anthropologists and odontologists, her research mission is to develop and optimize methodologies to improve identification efforts in missing person cases, mass disasters, and unidentified human remains (UHR) investigations.

Dr. Ambers previously served as Assistant Director of the *Henry C. Lee Institute of Forensic Science* in Connecticut, a world-class institute that specializes in interdisciplinary research, training, testing, casework consulting, and education in forensic science. She has also held an Associate Professor (Forensic DNA) appointment in the *Henry C. Lee College of Criminal Justice and Forensic Sciences* at the University of New Haven (UNH), teaching forensic biology and DNA analysis methods, in addition to specialty courses on advanced DNA topics. During this time, she also was a team member of UNH's *Center for Forensic Investigation of Trafficking in Persons.* She worked as a forensic geneticist for the University of North Texas (UNT) Center for Human Identification for 8 years and, prior to pursuing her doctorate, was lead DNA analyst and lab manager of UNT's DNA Sequencing Core Facility. Her doctoral research involved an investigation of methods (e.g., whole genome amplification, DNA repair) for improving autosomal and Y-STR typing of degraded and low copy number (LCN) DNA from human skeletal remains and environmentally damaged biological materials. Her master's thesis research included the development and optimization of a DNA-based multiplex screening tool for genetics-based separation of fragmented and commingled skeletal remains in mass graves.

Dr. Ambers specializes in the characterization and identification of contemporary, historical, and archaeological human skeletal remains. Her casework involved DNA testing of an American Civil War guerrilla scout; several Finnish World War II soldiers; unidentified late-19th century skeletal remains discovered by a construction crew in Deadwood, South Dakota; unidentified skeletal remains of Special Operations soldiers killed during the 1974 Turkish invasion of Cyprus; skeletal remains exhumed from Prague Castle in the Czech Republic; skeletal remains of soldiers from the 7 Years' War (1756–63); bone samples purported to belong to a member of Jesse James gang (killed during a bank robbery in 1876); and the exhumed remains of the wife of a Yale medical school professor. She copresented a workshop on "Advanced Methods for DNA-based Analysis of Skeletal Remains" at the 26th International Symposium on Human Identification (ISHI), and has been an invited speaker at an international bone workshop/conference in Prague, Czech Republic. In 2017, she traveled twice to India to train scientists from various Indian states and the Maldives Police Service on the processing of bone samples in forensic DNA casework. During her visit to India, she performed autosomal STR analysis on human skeletal remains discovered along a hiking route in the Himalayas (in the northern state of Himachal Pradesh) to assist local officials in the investigation of a missing person case. Additionally, she performed DNA analysis on a female homicide victim recovered from a clandestine grave in New Delhi.

Dr. Ambers' casework and research have been published in various peer-reviewed journals, including *Forensic Science International: Genetics* (*FSI:Genetics*), *Forensic Science International, International Journal of Legal Medicine, Legal Medicine, BMC Genomics, the Croatian Medical Journal, The Journal of Heredity, and Journal of Biological and Clinical Anthropology* (*Anthropologischer Anzeiger*). Her work has received press in numerous local and national newspapers (including *The Washington Times, NBC News, Criminal Legal News, The Root*) and has been featured on several podcasts (*Truth and Justice, Crime Waves*).

Among her most recent casework (a collaboration with the Texas Historical Commission) includes DNA analyses of human skeletal remains associated with the French explorer La Salle's last expedition. This case involved massively parallel sequencing (MPS) of DNA from two adult male skeletons recovered from the *La Belle* shipwreck, skeletal remains recovered from a mass grave at Fort St. Louis, and the putative remains of Sieur de Marle. Additionally, Dr. Ambers performed DNA extractions and mitochondrial DNA (mtDNA) sequencing on numerous sets of skeletal remains associated with Spanish royalty and the House of Aragon (recovered from tombs within the Royal Pantheon of San Juan de la Peña archaeological site in Spain). She is currently working on the putative remains of William Townsend Washington, the nephew of former U.S. President George Washington.

In addition to skeletal remains cases and research, Dr. Ambers collaborated with the National Institute of Justice (NIJ) and the Forensic Technology Center of Excellence (FTCoE) to develop and disseminate a formal report on the use of familial DNA searching (FDS) in casework. She is an active cold case consultant, an advocate of postconviction DNA testing, and an educator/advisor on DNA testing or retesting of old, degraded, or challenging evidentiary samples. Dr. Ambers also was the 2017–18 project lead on a U.S. State Department grant to combat human trafficking in Central America through the application of forensics. As part of the program objectives, she traveled to three Northern Triangle countries (Guatemala, El Salvador, Honduras) to perform gap assessments of government laboratories and train personnel in forensic DNA analysis, with the goal of promoting quality casework methods based on ISO 17025 standards. As Project Lead, she helped train forensic DNA scientists at the *Instituto Nacional de Ciencias Forenses de Guatemala* (INACIF, the National Institute of Forensic Sciences of Guatemala) in Guatemala City, Guatemala; the *Instituto de Medicina Legal* (Institute of Legal Medicine) in San Salvador, El Salvador; and the *Instituto de Medicina Forenses: Centro de Medicina Legal y Ciencias Forenses* (the Institute of Forensic Medicine: Center of Legal Medicine and Forensic Sciences) in Tegucigalpa, Honduras. In addition to providing both lecture and tactile training on DNA analysis methods to Northern Triangle laboratories, Dr. Ambers was part of a consortium to help these countries develop and maintain forensic DNA databases to assist in the identification of missing persons related to human trafficking.

Dr. Ambers has mentored hundreds of students during her career and for multiple years served as the faculty advisor for *Scientista*, the largest network of college and graduate women innovating science, technology, engineering, and math (STEM) in the United States. The Scientista Foundation is a national organization that empowers women in science, technology, engineering, and math (STEM) through content, communities, and conferences. During her free time, Dr. Ambers enjoys traveling, hiking, kayaking, reading, spending time with her dogs, and volunteering to help victims of abuse as well as minority and oppressed populations.

In 2017–18, Dr. Ambers was Project Lead on a U.S. State Department grant to combat human trafficking in Central America's Northern Triangle, where she helped train forensic DNA scientists at the *Instituto Nacional de Ciencias Forenses de Guatemala* (INACIF, the National Institute of Forensic Sciences of Guatemala) in Guatemala City, Guatemala; the *Instituto de Medicina Legal* (Institute of Legal Medicine) in San Salvador, El Salvador; and the *Instituto de Medicina Forenses: Centro de Medicina Legal y Ciencias Forenses* (the Institute of Forensic Medicine: Center of Legal Medicine and Forensic Sciences) in Tegucigalpa, Honduras

Dr. Ambers at the 2016 funeral ceremony of unidentified human skeletal remains unearthed by a construction crew in Deadwood, South Dakota. The remains were discovered in a residential neighborhood that was built atop the original site of Ingleside Cemetery (Deadwood's first burial ground, 1876–78). The remains were reinterred in Mt. Moriah Cemetery, the final resting place of "Wild Bill" Hickok, Calamity Jane, and the City of Deadwood's first Sheriff Seth Bullock.

Preface

"I hope my achievements in life shall be these—that I will have fought for what was right and fair, that I will have risked for that which mattered, that I will have given help to those who were in need, and that I will have left this Earth a better place for what I've done and who I've been."

—Carl Hoppe

The impetus for this book arose during an international trip that I took in 2017, on behalf of Thermo Fisher Scientific's Human Identification (HID) University, to train scientists from various states in India and the Maldives Police Service on the processing and extraction of DNA from human skeletal remains. During my time in India, I was approached by a forensic DNA specialist from Himachal Pradesh, the northernmost state in India which borders the Himalayan Mountain Range and is the residence of the Dalai Lama in exile. This scientist's team had been tasked with identifying a set of skeletal remains discovered in the Himalayas along a popular hiking/trekking route—and he had brought the remains with him to the bone workshop. Each day, after the workshop training ended, we spent the evenings in the Thermo Fisher Scientific laboratory facility in New Delhi attempting to recover sufficient genetic material from the remains in hopes of making a positive identification and returning the remains to the family. During some of the long incubation periods throughout the DNA extraction process, discussions ensued about the possible identity of the decedent. I learned that Himachal Pradesh is a popular travel destination for American and European tourists who are seeking adventure and spiritual enlightenment amongst the majesty of the Himalayas.

One evening, in my room at the Trident Gurgaon Hotel, I initiated an internet search to learn more about this magical region of northern India. Despite its allure to adventure seekers and spiritualists, and its beautiful landscape, many travelers underestimate the sheer vastness of the mountain range and the unpredictable weather conditions in the area. As such, there is a popular hiking region in the Himalayas within Himachal Pradesh (called Parvati Valley) that has been unofficially nicknamed the "Valley of Death" because of the large number of American and European tourists who have gone missing while hiking in the area over the past few decades. In fact, the number of missing American and European tourists is in the dozens. During my search to learn more about the missing tourists, I ran across a website titled "The Adventures of Justin," which chronicled the adventures of a 32-year-old man named Justin Alexander Shetler, who quit a lucrative job in corporate America, sold most of his belongings, and left behind the security of a salary and a life of consumerism in search of a life with deeper meaning. As I read Justin's posts, I felt a kinship to his expression that experiences are more valuable than material things. On his website, I was particularly drawn to a photograph he had posted of the Catacombs in Paris, along with the caption *"A reminder that we are impermanent. One day we will be nothing but bones and stories."* There was a strange irony to this statement, and I remember getting goosebumps thinking about his insight and the ostensible premonition of his own impending mortality. Although Justin was a seasoned traveler and an experienced outdoorsman, he had disappeared in the Parvati Valley of the Himalayas in 2016, nearly a year before I arrived in India to teach the workshop. As I sat on the bed in my hotel room perusing his story on my laptop, I wondered if the remains recovered in the Indian Himalayas were those of Justin Shetler.

After numerous evenings in the laboratory, and after processing a number of different skeletal elements for DNA, a complete short tandem repeat (STR) profile was obtained using the FBI's expanded core CODIS loci and the GlobalFiler™ PCR Amplification Kit (Thermo Fisher Scientific). For this case, we had the luxury of using state-of-the-art facilities and equipment at the Thermo Fisher Scientific laboratory in New Delhi to conduct DNA testing on these samples, which should have facilitated the process of making a positive identification of the remains. Whenever I talk about this case to colleagues and students—and once I disclose that we obtained a complete DNA profile—the immediate assumption is that we were able to positively identify this individual. However, at the time of this case, India did not have a national DNA database, nor had they collected reference samples from the families of the missing foreign nationals, so there was no genetic data available for comparison to aid in the identification or for reassociating the remains with any particular family. This is a common issue encountered in forensic genetic investigations of unidentified human remains (UHR). It exemplifies the fact that, even when using the most advanced technology and laboratory techniques, the process of identification is multifaceted and requires that certain infrastructure is in place in order to be successful. Ultimately, the DNA profile obtained from the skeletal remains is only one piece of the puzzle.

At this stage in the investigation (and armed with a complete 24-locus STR profile), one potential option was to contact the respective Embassies of each foreign national who had disappeared in the "Valley of Death," to notify the families of the missing and request reference samples for comparison to the DNA profile obtained from the remains. However, this option was far from ideal, as only *one* of those families (or possibly *none* of them) would receive the news that their missing loved one had finally been found. There was a distinct possibility that the skeletal remains discovered did not belong to any of the missing American or European tourists. Furthermore, this type of investigative approach lacks sensitivity to families who have already endured years (or even decades) of grief regarding their missing loved one. Rather than notifying the families and unnecessarily raising their hopes that their loved one's remains had finally been found, I suggested that we first investigate the racial background of the decedent to see if it aligned with the Anglo-European descriptions of those missing in the "Valley of Death." Although a forensic anthropologist can often predict the race of an unknown individual by examining various characteristics of the skull, that possibility was precluded in this case by the partial nature of the skeletal remains recovered. The skull was disarticulated, incomplete, and heavily fragmented. Fortunately, recent advances in DNA technology provide a mechanism through which to make an assessment of race in partial skeletal remains cases. With additional testing of a panel of ancestry-informative single nucleotide polymorphisms (aiSNPs) using massively parallel sequencing (MPS), and then comparing this aiSNP data to the Forensic Research/Reference on Genetics knowledge base (FROG-kb) and the ALele FREquency Database (ALFRED), it was determined that the remains recovered from the hiking trail in the Indian Himalayas did not belong to an individual of Anglo-European descent.

Rather, the decedent was Asian (most likely Sri Lankan or Indian) based on likelihood ratio calculations generated during the FROG-kb genetic ancestry assessment. This finding changed the trajectory of the investigation and prevented unnecessary anguish among the families of the missing American and European tourists.

At the time of the publishing of this book, the remains of Justin Shetler have still not been found. His story—and his disappearance—still haunts me. Even more notable is that the disappearance of Justin Shetler is only one of literally millions of unsolved missing person cases across the globe. Although the hope will always be to find missing persons alive, forensic DNA testing can assist in reunifying loved ones with their families after remains are located and recovered. This book is intended to serve as a repository of established forensic DNA techniques for unidentified human remains (UHR) investigations, and to provide a detailed overview of recent advancements in the field that can assist in identifying the unknown.

Furthermore, the Himalayas case emphasizes the necessity for DNA databasing to assist in missing persons and unidentified human remains (UHR) investigations. Although many countries are performing forensic DNA analyses and have validated procedures for casework, genetic data obtained from bones and teeth have no context without reference samples for comparison. The best testing methods and approaches in the world will not suffice in making positive identifications without concurrent accessibility to the appropriate exemplars for comparison. In the absence of a DNA database or reference samples, recent advances in DNA technology and expansion of the types of genetic markers tested can provide additional investigational leads about the identity of a decedent and can assist in the investigation. These expanded DNA markers systems, technological advances, and their associated applications are discussed throughout this book.

I have been involved in skeletal remains work since 2005. As a student, I became fascinated with bones and their ability to survive for such long periods of time (even in harsh environmental conditions), and I have long been a history buff. There is often substantial publicity tied to current, modern cases (understandably due to the emotional trauma and impact on victims and their families). However, it is important to remember that a primary goal of forensics is to answer questions and solve mysteries. Sometimes historical cases get overlooked because they fall into the category of being "nonprobative," i.e., any persons involved in the individual's death may have long been deceased and are therefore not available to stand trial in criminal court proceedings. However, as practitioners and fellow human beings we should certainly ask the question—does the mere passage of time negate the value of a person's life or lessen the impact of solving a mystery? Historical and archaeological investigations are the coldest of cold cases and still deserve our attention.

Ultimately, there is a broad range of scenarios in which skeletonized, damaged, or highly decomposed human remains are presented for identification. Regardless of whether these cases are contemporary or historical, the remains of these individuals deserve to be dignified with proper recovery and identification. Although DNA often garners the spotlight in skeletal remains cases, the identification process is typically a cooperative and collective effort between experts in various forensic disciplines and subdisciplines (e.g., anthropology, odontology, forensic art/sculpting, forensic genetic genealogy).

The famous adage "dead men do tell tales" is accurate and relevant in the era of modern forensic technology. What lies preserved deep within the skeleton is our molecular signature and genetic "story." As forensic geneticists, we must engage in a "forensic interview" of sorts with the remains that are presented to us for analysis. We now have the tools and technological advancements to provide answers to a number of questions about the missing and unidentified: Who are you? Where are you from? What did you look like?

Acknowledgments

A work of this magnitude would not have been possible without the encouragement and support of friends, family, and colleagues, and without the collaboration of numerous forensic experts who so willingly donated their time and expertise to this endeavor. To Eric Bartelink D-ABFA (Human Identification Laboratory, California State University, Chico); Howard Cash (Gene Codes Corporation); Thomas David D-ABFO (University of Tennessee, Knoxville, Georgia Bureau of Investigation); Suni M. Edson (AFDIL); Colleen Fitzpatrick (Identifinders International, Vidocq Society); Ranjana Grover (ANDE Corporation); Michael Hennessey (Gene Codes Corporation); Irene Kahline (AFDIL); Bobby LaRue (Verogen); Brandon Letts (FBI); James M. Lewis D-ABFO (University of Tennessee, Knoxville); Odile Loreille (FBI); Murray K. Marks (University of Tennessee, Knoxville, Knox County Regional Forensic Center, National Forensic Academy); Timothy McMahon (AFDIL); Darinka Mileusnic-Polchan MD (University of Tennessee, Knoxville, Knox County Regional Forensic Center); Stephen Missal CFA (Maricopa Forensic Laboratory and Medical Examiner's Office); Nicole Novroski (University of Toronto, Mississauga); Vivek Sahajpal (Directorate of Forensic Services, Himachal Pradesh); Richard Selden (ANDE Corporation); Eugene Tan (ANDE Corporation); Rosemary Turingan Witkowski (ANDE Corporation); Jodie Ward (Australia National DNA Program for Unidentified and Missing Persons, Australian Facility for Taphonomic Experimental Research); Jeremy Watherston (New South Wales Pathology, Forensic, and Analytical Science Service); and Frank Wendt (Yale University)—you all have my sincerest gratitude, respect, and admiration. This book would not have been possible without you. Your contributions and expertise are invaluable. Although we are spread across the globe and distributed throughout a number of jurisdictions and organizations, we do operate as part of the same sentient team whose collective goal is identifying the dead, the nameless, and the forgotten. I am beyond honored to have you all as colleagues, collaborators, and friends.

To Gary and Michele Cox—our 30 years of friendship is incomparable and is one of the most treasured things in my life. To Allison Silveus—for 15 years of unwavering true friendship and support. Our journeys and adventures have been many, and you are my sister for life. To my Texas family—Allison, Jorge, Angie, Jeffrey, and Gigi—your love, kindness, support, companionship, and generosity over the years mean so very much to me. Family is not always genetic, and I am forever grateful to be an adopted part of yours. And to my wonderful godsons Hadrian and Emory: Live your life with integrity, kindness, and compassion—for it is through these three virtues—not perceived power, prestige, or success—that you will positively impact the lives of others.

To Paula Brumit D-ABFO—I could never express in words how much your friendship and support has meant to me. I look forward to a lifetime of research collaboration and sisterhood. To Murray Marks, Jim Lewis, and Tom David for embracing me as the DNA nerd in your anthropology and odontology group. Together we are the "dream team" and can change the world of forensic human identification.

Finally, to Dr. Henry C. Lee for your kindness and generosity, and for welcoming me into your Institute family. Your contributions to the field of forensic science are immeasurable. The dedication, altruism, and excellence that you embody set the highest of standards for professionals in our field. Thank you for your faith in me, and for your friendship. I am forever grateful to you and for the experience I gained at the *Henry C. Lee Institute of Forensic Science*. I will never forget our long conversations over dinner or in your office, and your shared wisdom about life. Working for you has been one of the great honors of my career.

CHAPTER

1

Missing persons and unidentified human remains: The world's silent mass disaster

Angie Ambers PhD[a,b,c]

[a]Henry C. Lee Institute of Forensic Science, University of New Haven, West Haven, CT, United States [b]Forensic Science Department, Henry C. Lee College of Criminal Justice and Forensic Sciences, Center for Forensic Investigation of Trafficking in Persons, University of New Haven, West Haven, CT, United States [c]Institute for Human Identification, LMU College of Dental Medicine, Knoxville, TN, United States

There are 206 bones and 32 teeth in the human body, and each has a story to tell. – ***Clyde Snow (1928–2014)***

Introduction

Missing persons and unidentified human remains (UHR) cases pose one of the greatest challenges that law enforcement agencies, investigators, and forensic scientists will encounter in their careers. In the United States alone, over 600,000 individuals are reported missing each year, according to the FBI's National Crime Information Center (NCIC). Between 2007 and 2020, an average of 664,776 missing persons were reported annually in the United States (https://www.fbi.gov/services/cjis/ncic). Although many missing children and adults are eventually found alive, tens of thousands of individuals remain unaccounted for after being missing for more than a year.

In addition to missing persons cases, it is estimated that 4400 unidentified human remains are recovered during a typical year in the United States, with approximately 1000 of them becoming "cold cases" and remaining unidentified even after a full year of investigation. Furthermore, a 2007 Special Report by the Bureau of Justice Statistics (U.S. Department of Justice) noted that the actual number of UHRs may be substantially higher because only approximately half of medical examiners' and coroners' offices across the country had policies

Forensic Genetic Approaches for Identification of Human Skeletal Remains
https://doi.org/10.1016/B978-0-12-815766-4.00001-7

in place for retaining records on unidentified decedents and/or for mandated reporting of the cases to the NCIC (Hickman and Strom, 2007). That same year, attempts to remedy this issue were made with the introduction of the National Missing and Unidentified Persons System (NamUs) by the National Institute of Justice (NIJ). Since 2007, NamUs has served as a national information repository and resource center for missing, unidentified, and unclaimed persons cases across the United States (https://namus.nij.ojp.gov/). However, although NamUs was developed to improve resolution of missing and unidentified persons cases, law enforcement agencies do not use it consistently, and mandatory reporting to NamUs is currently only in effect for 10 U.S. states (Arkansas, Illinois, Michigan, New Mexico, New York, North Carolina, Oklahoma, Tennessee, Washington, and West Virginia). Two additional states—Connecticut and Texas—either have formal protocols in place or are considering enacting legislation for mandatory reporting to NamUs (Spencer, 2021). Many experts agree that mandatory reporting to both NamUs and NCIC could greatly improve the chances of solving missing and unidentified persons cases.

The number of missing persons and unidentified human remains cases are likely even higher in many other countries around the world, especially in countries that have been ravaged by years of civil war or in regions with a history of "forced disappearances" under oppressive, totalitarian regimes.

Forensic genetic investigations of human skeletal remains: Challenges and considerations

Regardless of origin or postmortem interval (PMI), the bones and teeth of the skeleton house and protect the molecular signature of a decedent. DNA within skeletal elements survives long after soft tissue decomposes and far beyond the timeframe when an individual may be recognizable by external features and physical appearance. Although accessing and analyzing the DNA contained within a skeleton can provide a wealth of information about an unidentified individual—including their family lineage, biogeographic ancestry, and even phenotypic characteristics such as hair color, eye color, and skin tone—there are a number of challenges and barriers to recovering sufficient genetic data to make a positive identification. Despite advancements and improvements in DNA technology over the past few decades, skeletonized human remains pose exceptional challenges for identification efforts due to the condition and/or limited nature of the samples available for genetic testing. Remains can be significantly damaged, fragmented, commingled with other remains, and/or may be recovered as partial skeletons with key skeletal elements missing that could have aided in identification. Additionally, skeletonized remains typically have been exposed to the environment for extended periods of time, which compromises not only the integrity of bone microstructure, but also the quality and quantity of endogenous DNA.

DNA in skeletal remains often is highly degraded, and the molecular chemistry of DNA damage can be quite complex. A variety of different types of molecular lesions exist—single-strand breaks (SSBs), double-strand breaks (DSBs), abasic (AP) sites, interstrand crosslinks, intrastrand crosslinks, deaminated bases, oxidized bases, alkylated bases—and each type of lesion presents its own unique challenge in genetic testing efforts. Although the presence of even one type of lesion can pose problems for DNA typing, skeletonized human remains often contain a combination of all (or most) of these types of molecular lesions, further complicating our ability to obtain DNA of sufficient quality and quantity for identification.

In addition to DNA degradation, skeletal samples often arrive in the laboratory infiltrated with a variety of substances that can inhibit downstream DNA typing efforts. Skeletonized or highly decomposed remains often contain both endogenous and exogenous (environmental) inhibitors that must be purified away from the genetic material in order for DNA typing to be successful. Ironically, some of the chemicals that are necessary for isolating DNA from bones or teeth will themselves interfere with downstream DNA testing efforts if these substances are co-purified during extraction procedures. Hence, in addition to the type of DNA extraction method used, the precision and accuracy of the analyst in carrying out the steps of the procedure can have a dramatic impact on success.

Ultimately, a variety of factors—degree of DNA degradation, quantity of recovered DNA, presence of co-extracted inhibitors, and the skills and training of the analyst—will contribute collectively to the success or failure of forensic DNA analyses. Moreover, every skeletal remains case is different and will pose its own unique set of challenges for identification.

Why so many unidentified human remains?

Although skeletonized human remains are not the most common types of samples encountered in forensic DNA casework, there are a number of scenarios—across a broad range of contemporary, historical, and archaeological contexts—in which these samples may appear at the forefront of investigation and identification efforts. Among these scenarios include homicides, fires, explosions, aviation accidents, terrorist attacks, humanitarian crises, natural disasters, genocide, and mass graves. In addition, court-ordered cemetery exhumations and funeral business malpractice cases also present instances in which skeletal samples may be submitted for genetic testing.

Natural disasters

A number of contemporary and historical natural disasters have resulted in large numbers of missing persons and unidentified human remains. Among the deadliest types of natural disasters include earthquakes, tsunamis, hurricanes, floods, tornados, avalanches (landslides), and wildfires—all of which may result in loss of life for a large number of individuals and pose enormous challenges for disaster victim identification (DVI) efforts. Two notable natural disasters in recent history that resulted in mass casualties were: (1) the 2004 Indian Ocean earthquake and tsunami (also referred to as the "Boxing Day" Tsunami and the Sumatra-Andaman Earthquake) and (2) the 2010 Haiti earthquake. Death tolls for these two events alone were approximately 275,000 and 316,000 persons, respectively (Sadiq and McEntire, 2012; Bilham, 2010; Margesson and Taft-Morales, 2010; Dolan et al., 2009). The 2011 Tōhoku earthquake and tsunami in Japan claimed the lives of approximately 20,000 people.

In addition to earthquakes and tsunamis, hurricanes also can often result in large numbers of casualties. Despite modern improvements in infrastructure and the existence of advanced warning systems, a few recent hurricanes have killed thousands of people, including: (1) Hurricane Jeanne, which pummeled the Caribbean and Eastern U.S. in 2004; (2) Hurricane Katrina, which destroyed parts of the U.S. Gulf Coast and Bahamas in 2005; (3) Hurricane Stan, which affected Mexico and various regions of Central America in 2005; and (4) Hurricane

Maria, which devastated Puerto Rico in 2017. These hurricanes collectively caused the deaths of approximately 10,000 people (3037, 1836, 1668, and 3059 deaths, respectively).

When natural disasters occur, electricity often is not available to power facilities that could be used for refrigeration (preservation) and storage of human remains. Decomposition of soft tissue progresses rapidly, and identification efforts must typically center around dental record comparisons and forensic DNA analysis.

In other scenarios, such as the 2018 "Camp Fire" wildfire in Northern California (the deadliest in California history), the human remains recovered were so damaged and fragmented that DNA testing was the only viable option to make positive identifications in a large number of the cases (Glynn and Ambers, 2021; Gin et al., 2020). Of the 85 victims who perished in this wildfire, 2 were identified via serial numbers on artificial joints, 15 from dental records, 5 from fingerprints, and more than 50 required DNA.

Epidemics and pandemics—Mass graves and communal burials

Even more deadly than some of the worst natural disasters are epidemics and global pandemics caused by infectious pathogens. Although the bubonic plague ("Black Death") during the 14th century was the deadliest pandemic in human history (claiming the lives of an estimated 200 million people worldwide), it was caused by a bacterium, and the invention of antibiotics in the 1920s has largely mitigated the possibility of another similar catastrophe. However, more recent pandemics have been caused by viruses and are much more difficult to control. The Spanish Flu (1918–1920), also referred to as the Great Influenza Epidemic, claimed the lives of approximately 100 million individuals, and the coronavirus pandemic (which began in 2019) has already caused the death of more than 5 million people globally.

When large numbers of individuals in a population die within a short period of time from an infectious pathogen, it is not uncommon for cities and authorities to cremate or bury victims quickly in an effort to prevent additional spread of disease. This approach may even involve communal burials, especially in the case of homeless or unclaimed decedents. Although mass graves are more typically associated with war conflicts, genocide, or murder under oppressive regimes, communal burials of large numbers of individuals have been carried out throughout history for the purposes of infection control and to assuage sanitation concerns.

Hart Island in the Bronx borough of New York City is an example of a mass grave whose origin began out of necessity when high death tolls occurred during pandemics in densely populated areas of the city. Measuring approximately 1 mile (1.6 km) long by 0.33 miles (0.53 km) wide, Hart Island is America's largest mass grave, containing the burials of more than 1 million individuals. In addition to being the final resting place of epidemic and pandemic victims, the island also has become a "Potter's Field," serving as a burial site for New York's indigent, unknown, unidentified, or unclaimed. Regardless of cause of death, the dead are buried in trenches, in pine boxes stacked 3–5 coffins deep, each labeled with an identification number as well as the decedent's name and age (if known). Thousands of victims from the 1980s AIDS epidemic, as well as from the COVID-19 pandemic, are buried on the island.

It has become a relatively frequent occurrence in recent years to exhume the bodies of adults buried on Hart Island as families make attempts to locate their relatives through DNA, photographs, and fingerprints kept on file at the Office of the Chief Medical Examiner (OCME) in New York City. There were an average of 72 disinterments per year from 2007 to 2009. In 2011, the Hart Island Project, a private charitable foundation, was formed to organize and catalog burial records (for interments between 1980 and present), provide historical information about the island to the public, and assist with family reunification efforts (https://www.hartisland.net/).

Oppressive regimes, forced disappearances, genocide, and war conflicts

Although a plethora of ancient and medieval wars have been fought on this planet—and not all human remains from those wars have been discovered—countless civil wars, international conflicts, genocides, and forced disappearances have occurred in modern times that have resulted in the loss of almost unfathomable numbers of human lives. Table 1 provides an overview of more contemporary conflicts and the estimated death tolls per event. Many of the individuals who perished during these events have not yet been recovered, identified, and repatriated.

TABLE 1 Overview of the number of deaths resulting from civil wars, international conflicts, genocides, and forced disappearances since the early 1900s.

Name of conflict	Location	Date(s)	Estimated # of deaths
Russo-Japanese War	Northeast Asia	1904–1905	123,000–157,000
Mexican Revolution	Central America (Mexico)	1910–1920	500,000–2 million
1911 Revolution	Asia (China)	1911	220,000
Balkan Wars	Europe (Balkan Peninsula)	1912–1913	140,000+
World War I	Worldwide	1914–1918	16–40 million
Russian Civil War	Europe (Russia)	1917–1922	5–9 million
Kurdish separatism in Iran	Middle East (Iran)	1918–present	15,000–58,000
Iraqi-Kurdish Conflict	Middle East (Iraq)	1918–2003	138,000–320,000
Rif War	Africa (Morocco)	1921–1926	30,500+
Kurdish rebellions in Turkey	Middle East	1921–present	100,000+
Second Italo-Senussi War	Africa (Libya)	1923–1932	40,000+
Chinese Civil War	Asia (China)	1927–1949	8–11.6 million
Chaco War	South America	1932–1935	85,000–130,000
Second Italo-Ethiopian War	Africa (Ethiopia)	1935–1936	278,000+
Spanish Civil War	Europe (Spain)	1936–1939	500,000–1 million
Second Sino-Japanese War	Asia (China)	1937–1945	20–25 million
World War II	Worldwide	1939–1945	85 million
Winter War	Europe (Finland)	1939–1940	154,000–195,000
Greco-Italian War	Southeast Europe	1940–1941	27,000+

Continued

TABLE 1 Overview of the number of deaths resulting from civil wars, international conflicts, genocides, and forced disappearances since the early 1900s—cont'd

Name of conflict	Location	Date(s)	Estimated # of deaths
Continuation War	Northern Europe	1941–1944	387,000+
Soviet-Japanese War	Asia (China)	1945	33,000–96,000
First Indo-China War	Southeast Asia	1946–1954	400,000+
Greek Civil War	Europe (Greece)	1946–1949	158,000+
Malagasy Uprising	Africa (Madagascar)	1947–1948	11,000–89,000
Kashmir Conflict	Asia (India/Pakistan)	1947–present	80,000–110,000
La Violencia	South America (Colombia)	1948–1958	192,000–195,000
Internal Conflict-Myanmar	Asia (Myanmar)	1948–present	130,000–250,000
Arab-Israeli Conflict	Middle East	1948–present	116,000+
Korean War	Asia (Korea)	1950–1953	1.5 million–4.5 million
Algerian War	Africa (Algeria)	1954–1962	400,000–1.5 million
Ethnic conflict in Nagaland	Asia (India)	1954–present	34,000+
Vietnam War	Asia (Vietnam)	1955–1975	1.3 million–4.3 million
First Sudanese Civil War	Africa (Sudan)	1955–1972	500,000+
Congo Crisis	Central Africa (Congo)	1960–1965	100,000+
Angola War of Independence	Africa (Angola)	1961–1974	83,000–103,000
North Yemen Civil War	Asia (Yemen)	1962–1970	100,000–200,000
Mozambican War of Independence	Africa (Mozambique)	1964–1974	63,000–89,000
Insurgency in Northeast India	Asia (India)	1964–present	25,000+
Colombian Conflict	South America (Colombia)	1964–present	220,000+
Nigerian Civil War	Africa (Nigeria)	1967–1970	1 million–3 million
Moro Conflict	Asia (Philippines)	1969–2019	120,000+
Communist Rebellion (Philippines)	Asia (Philippines)	1969–present	30,000–43,000
Bangladesh Liberation War	Asia (Bangladesh)	1971	300,000–3 million +
Ethiopian Civil War	Africa (Ethiopia)	1974–1991	500,000–1.5 million
Angolan Civil War	Africa (Angola)	1975–2002	504,000+
Lebanese Civil War	Asia (Lebanon)	1975–1990	120,000–150,000
Insurgency in Laos	Asia (Laos)	1975–2007	100,000+
Afghanistan Conflict	Middle East (Afghanistan)	1978–present	1.4 million–2 million
Kurdish-Turkish Conflict	Middle East	1978–present	45,000+
Soviet-Afghan War	Middle East (Afghanistan)	1979–1989	600,000–2 million

TABLE 1 Overview of the number of deaths resulting from civil wars, international conflicts, genocides, and forced disappearances since the early 1900s—cont'd

Name of conflict	Location	Date(s)	Estimated # of deaths
Salvadoran Civil War	Central America (El Salvador)	1979–1992	70,000–80,000
Iran-Iraq War	Middle East	1980–1988	500,000
Internal Conflict in Peru	South America (Peru)	1980–present	70,000+
Ugandan Bush War	Africa (Uganda)	1981–1986	100,000–500,000
Second Sudanese Civil War	Africa (Sudan)	1983–2005	1 million–2 million
Sri Lankan Civil War	Asia (Sri Lanka)	1983–2009	80,000–100,000
Somali Civil War	Africa (Somalia)	1986–present	300,000–500,000
Lord's Resistance Army Insurgency	Central Africa	1987–present	100,000–500,000
Iraq War (First Gulf War)	Middle East (Iraq)	1990–1991	25,000–41,000
Algerian Civil War	Africa (Algeria)	1991–2002	44,000–200,000
Bosnian War	Europe (Bosnia)	1991–1995	97,000–105,000
Iraqi Uprisings	Middle East (Iraq)	1991	85,000–235,000
Sierra Leone Civil War	Africa (Sierra Leone)	1991–2002	50,000–300,000
Burundian Civil War	Africa (Burundi)	1993–2005	300,000+
Rwandan Genocide	Africa (Rwanda)	1994	800,000
First Congo War	Central Africa (Congo)	1996–1997	250,000–800,000
Second Congo War	Central Africa (Congo)	1998–2003	2.5 million–5.4 million
Ituri Conflict	Central Africa (Congo)	1999–2003	60,000+
Iraq War (Second Gulf War)	Middle East (Iraq)	2003–2011	405,000–655,000
Tigray War	Africa (Ethiopia)	2020–present	100,000+
Global War on Terrorism	Worldwide	2001–present	272,000–1.3 million
War in Darfur	Africa (Sudan)	2003–present	300,000+
Kivu Conflict	Central Africa (Congo)	2004–present	100,000+
War in Pakistan	Asia (Pakistan)	2004–2017	46,000–79,000
War in Afghanistan	Middle East (Afghanistan)	2001–2021	213,000
Mexican Drug War	Central America	2006–present	150,000–250,000
Boko Haram Insurgency	Africa (Nigeria)	2009–present	350,000+
Syrian Civil War	Asia (Syria)	2011–present	606,000+
Rojava-Islamist Conflict	Asia (Syria)	2013–present	50,000+
Iraqi Civil War	Middle East (Iraq)	2014–2017	200,000+
Yemeni Civil War	Asia (Yemen)	2014–present	377,000+

Cemetery mismanagement and funeral malpractice

Although many people are familiar with stories involving recovery of skeletal remains in homicide cases, aviation accidents, or mass disasters, lesser-known scenarios include mismanagement of cemeteries and funeral malpractice. These types of cases may involve: (1) overburials in cemeteries (in which the same gravesite is sold repeatedly over time, and multiple individuals become interred in a burial plot intended for a single person); (2) unauthorized services (e.g., removal of body parts without consent from the decedent or family members, to be sold for profit for use in transplants or other medical procedures); or (3) improper disposal of human remains (in which a decedent's remains are not handled in accordance with a pre-defined contractual agreement or which is not in alignment with legal, health, and ethical guidelines for handling of the deceased). In these types of scenarios, DNA testing may be requested to reassociate and identify the remains.

Overburial—Eastern Cemetery (Louisville, Kentucky)

One of the most egregious cases of cemetery mismanagement and overburial involves Eastern Cemetery in Louisville, Kentucky (U.S.A.). Eastern Cemetery spans 28 acres of land and technically contains approximately 16,000 burial plots; however, the cemetery made national news in 1989 when a whistleblower alerted authorities about the mistreatment of human remains and reselling of occupied gravesites (New York Times, 1989; AP News, 1989). An investigation revealed that over 138,000 bodies had been buried in the cemetery since its inception in the 1840s. Subsequent review of cemetery records exposed intentional efforts to reconstruct maps and rename sections of the cemetery, so that headstones could be removed and previously occupied burial plots could be re-sold for additional new burials (Bailey, 2015). Investigators discovered that some graves in the cemetery contained up to six individuals, and human bones were found uninterred in scattered locations across the cemetery grounds, including in a storage shed, in the glove compartment of a truck, in a toolbox, in excavated dirt piles, and even inside discarded fast-food packaging. Cemetery management officials were indicted on over 60 charges, including unauthorized reuse of graves, breach of contract, and abuse of corpses. In 2014, a nonprofit volunteer group called "Friends of Eastern Cemetery" was formed to restore and maintain the grounds for the dignity and respect of those buried there (https://friendsofeasterncemetery.com/). In 2020, the documentary "Facing East: The True Story of the Most Overburied Cemetery in America" (by Ronin Noir Films) was released and exposes appalling details of the mismanagement and desecration committed by the directors in charge of Eastern Cemetery (Austin, 2020).

Unauthorized services (selling of body parts)—The New York Bone Snatchers

In addition to the mismanagement of cemetery grounds and re-selling of occupied burial plots, other types of cases have involved illicit funeral practices, including the unauthorized removal and sale of body parts from decedents during preparation for burial or cremation. One of the most infamous cases regarding these types of crimes involved an elaborate tissue and bone harvesting operation in Brooklyn, New York that spanned the years of 2001–2005 and included the illegal sale of body parts from more than 1000 individuals (Hamilton, 2006). In 2005, the national news story broke about the case and identified Dr. Michael Mastromarino as the mastermind behind the highly organized scheme. Dr. Mastromarino had once been a prominent maxillofacial surgeon whose medical license was suspended after numerous lawsuits for malpractice were filed, and amidst allegations that he performed surgical procedures while under the influence of controlled substances. After losing his license to practice, Mastromarino formed a company named Biomedical Tissue Services (BTS) and established cooperative business relationships with funeral homes under the pretense of being a legal human graft tissue distributor (although BTS was not an accredited member of the American Association of Tissue Banks). The racketeering operation involved conspiring with funeral home directors to gain access to bodies of recently deceased persons, removing body parts without permission from the families and without proper medical screening, and then selling the parts to hospitals across the United States for transplants and other medical procedures. An audit performed during the investigation revealed that BTS had sold illegally obtained body parts to approximately 10,000 unsuspecting patients across the United States. BTS employees had removed human allograft tissue, bones, ligaments, and other cadaver material by forging donor forms without actual authorized consent, and often against the written directives of families. The probe into the case led to a number of exhumations and included the discovery that many of the victims had had bones removed and replaced with PVC piping. In 2010, the documentary "Body Snatchers of New York" (FilmRise Studios) detailed the crimes committed in this case (Morales and Blau, 2013). Ironically, Mastromarino died in 2013 of bone cancer.

Improper disposal of human remains—The Tri-State Crematory Scandal (Georgia)

Yet another egregious case involving funeral malpractice was uncovered in 2002 at the Tri-State Crematory in Noble, Georgia (U.S.A.). Early that year, an anonymous tip to the Environmental Protection Agency (EPA) initiated an investigation into the facility's operations and a search of the grounds surrounding the crematorium. Nearly 350 bodies of deceased persons whose remains had been consigned to the facility for cremation were discovered to have never been cremated. An organized search by a federal disaster response team located piles of decomposing human bodies in a storage shed, in vaults, and scattered across the property grounds. By the end of the search, 339 bodies had been recovered. DNA testing assisted in the identification of 226 of these individuals (Steadman et al., 2008; Horton, 2003; Engel, 2002).

Organizations and laboratories with specialization in missing persons and unidentified human remains investigations

International Commission on Missing Persons (ICMP)

The International Commission on Missing Persons (ICMP) was established in 1996 as an intergovernmental agency whose original mandate was to address and resolve over 40,000 missing persons cases related to various conflicts that occurred between 1991 and 1995 in Bosnia and Herzegovina, the Republic of Croatia, and the former Federal Republic of Yugoslavia. The ICMP has since expanded on this mission and now works to secure the cooperation of governments and other authorities in locating and identifying missing persons associated with human rights abuses, armed conflicts, natural disasters, and organized crime around the world. The organization assists in the exhumation of mass graves, and in both anthropological and DNA-based identification of human remains, and provides technical training and assistance in capacity building. Although originally based in Sarajevo (the capital of Bosnia and Herzegovina), the ICMP headquarters relocated to The Hague, Netherlands in 2016 (https://www.icmp.int/).

Armed Forces DNA Identification Laboratory (AFDIL)

The Armed Forces DNA Identification Laboratory (AFDIL), part of the Armed Forces Medical Examiner System (AFMES), provides forensic DNA testing services for the U.S. Department of Defense (DoD), with the primary mission of identifying and repatriating the remains of soldiers and military personnel from current and past war conflicts. AFDIL has become a global leader in the field of human remains identification and, in addition to processing the remains of U.S. military service members, provides worldwide scientific consultation, research, and education services in forensic DNA analysis. Established in 1991, AFDIL is located in Dover, Delaware.

Forensic Anthropology Foundation of Guatemala (FAFG)

Formally established in 1997, the Forensic Anthropology Foundation of Guatemala (FAFG, Fundacion de Antropologia Forense de Guatemala) is a non-governmental, nonprofit scientific and technical organization that applies various forensic science disciplines to locate, exhume, and identify the victims of Guatemala's Internal Armed Conflict, a civil war that spanned a period of 36 years (1960–1996) and involved "forced disappearances" of more than 40,000 people. The FAFG's Forensic Genetics Laboratory opened its doors in 2008 and gained accreditation in 2010 (https://fafg.org/home/). To date, over 3500 victims have been identified, and the FAFG continues its mission to help re-unify families with their missing loved ones.

Argentine Forensic Anthropology Team—Equipo Argentino de Antropologia Forense (EAAF)

The Argentine Forensic Anthropology Team (Equipo Argentino de Antropología Forense, EAAF) is a scientific, non-governmental (and nonprofit) organization created in 1986 to assist in the identification of victims of human rights violations associated with the military dictatorship during Argentina's "Dirty War" (1976–1983). Since then, the EAAF has conducted field work in over 30 countries and assisted in the investigation of a broad range of human rights violations. The headquarters of the EAAF is located in Buenos Aires, Argentina, and they have implemented a forensic genetics laboratory to assist with identifications.

Gene Codes Forensics and the Mass Fatality Identification System (M-FISys)

Gene Codes Forensics, Inc. is a subsidiary of Gene Codes Corporation (Ann Arbor, Michigan U.S.A.), a privately owned bioinformatics company that has been developing cutting-edge software programs for forensic DNA sequence analysis since 1997 (https://www.genecodesforensics.com). Gene Codes Corporation's flagship product, Sequencher®, is the most widely used sequence assembly software program in academic, biotechnology, and pharmaceutical life sciences laboratories around the world. Additionally, Sequencher® has been widely adopted for mitochondrial DNA (mtDNA) sequence analysis by a number of forensic and government laboratories, including at the Armed Forces DNA Identification Laboratory (AFDIL) and the FBI (https://www.genecodesforensics.com/sequencher/).

Gene Codes Forensics, Inc. was established in response to the terrorist attacks on the World Trade Center in New York on September 11, 2001. At the request of the New York City Office of the Chief Medical Examiner (OCME) and as part of a collaborative initiative to help identify victims of the attacks, Gene Codes Forensics, Inc. developed an innovative software program capable of databasing, organizing, and comparing DNA profiles from a large number of individuals for human identification purposes. This software program is called the Mass Fatality Identification System (M-FISys) (Gene Codes Forensics, Inc., 2019; Cash and Hennessey, 2004; Cash et al., 2003). Since its inception, Gene Codes Forensics, Inc. has used M-FISys software and provided consulting services to help identify human remains associated with a broad range of events, including commercial airline disasters, the 2004 Thailand Tsunami, Hurricane Katrina, missing persons cases in Mexico, human trafficking investigations in Chile (Hennessey et al., 2018), and human rights violations (forced disappearances) in Guatemala (Garcia et al., 2009).

University of North Texas Center for Human Identification (UNTCHI)

In 2004, the University of North Texas Center for Human Identification (UNTCHI) was co-founded by forensic genetics pioneer Dr. Arthur J. Eisenberg (1956–2018) and forensic anthropologist Dr. Harrell Gill-King, D-ABFA (Eisenberg and Planz, 2008). UNTCHI is located on the University of North Texas Health Science Center (UNTHSC) campus in Fort Worth, Texas, and offers both DNA and forensic anthropology services to law enforcement, investigative, and medicolegal agencies across the United States (https://www.untchi.org). Under the leadership of Dr. Eisenberg and Dr. Gill-King, the UNTCHI has assisted in the identification of victims from some of the most infamous cases in U.S. history, including the 9/11 terrorist attacks on the World Trade Center in New York City and serial killers John Wayne Gacy ("The Clown Killer") and Gary Ridgway ("The Green River Killer"), as well as individuals who perished in natural disasters such as Hurricane Katrina.

National Missing and Unidentified Persons System (NamUs)

In 2003, the National Institute of Justice (NIJ) began funding major efforts to maximize the use of forensic DNA technology in the U.S. criminal justice system, including in the investigation of missing and unidentified person cases. In 2005, NIJ hosted a national summit in which criminal justice practitioners, forensic scientists, policymakers, and victim advocates convened to define the major challenges encountered in solving missing persons and unidentified human remains cases. One result of the symposium was the realization that there was a need to improve access to information that could help solve missing and unidentified person cases. The National Missing and Unidentified Persons System (NamUs) was created to meet that need. In 2007, the Unidentified Persons (UP) database was launched, followed by the Missing Persons (MP) database in 2008. In 2009, the UP and MP databases were combined to form what it now known as NamUs. Since then, NamUs has served as a national information repository, clearinghouse, and resource center for missing, unidentified, and unclaimed persons cases throughout the United States.

NamUs continues to be funded and administered by NIJ (a subsidiary of the U.S. Department of Justice), and the database is currently managed through a contract with RTI International (headquartered in Research Triangle Park, North Carolina, U.S.A.). NamUs resources are provided at no cost to law enforcement, medical examiners, coroners, forensic science professionals, and family members of missing persons. Data regarding missing or unidentified persons can be entered into NamUs by law enforcement personnel, missing persons advocates, family members, and the general public.

The Doe Network—International Center for Unidentified and Missing Persons

The Doe Network—International Center for Unidentified and Missing Persons (https://www.doenetwork.org/) is a volunteer organization devoted to assisting investigators and law enforcement agencies in the resolution of national and international cold cases involving missing and unidentified persons. Through its website, The Doe Network's mission is to catalog information about missing and unidentified persons cases, generate media exposure, and provide credible potential matches to investigative agencies and authorities who can mobilize the information to assist in making positive identifications. One of the co-founders of The Doe Network, Todd Matthews, is the former Director of Communications and Outreach for NamUs and now also serves as the Director of Project EDAN, a team of experienced, certified forensic artists who volunteer to assist law enforcement, medical examiners, and coroners with unidentified human remains investigations. Forensic artists with Project EDAN create composite sketches and clay reconstructions from human skulls and postmortem photographs to assist in identifications.

Identifinders International LLC

Identifinders International LLC (Fountain Valley, California, U.S.A.) is a privately-owned genetic genealogy company that specializes in solving cold cases and identifying human remains associated with criminal, contemporary, historical, and archaeological investigations. Identifinders was founded in 2011 by Dr. Colleen Fitzpatrick, who is widely recognized as the pioneer of the use of Forensic Genetic Genealogy (FGG) for solving violent crimes and cold cases. Identifinders regularly collaborates with law enforcement agencies across the United States and internationally, as well as with academic institutions and government organizations, on "cold case" unidentified human remains (UHR) investigations. To date, Identifinders has provided genetic genealogy services to 223 jurisdictions across the United States and to 53 countries worldwide (https://identifinders.com/).

In 2021, the Identifinders team identified the remains of a young male hitchhiker who perished in an automobile accident in 1961, representing the oldest case of a National Center for Missing and Exploited Children (NCMEC) subject ever solved by genetic genealogy (Identifinders International LLC, 2021; Patel, 2021). In addition to serving as President and Lead Forensic Genetic Genealogist at Identifinders, Fitzpatrick also is a member of the Vidocq Society (Philadelphia, Pennsylvania, U.S.A.), a group of forensic experts and investigators who work collaboratively with law enforcement to promote the resolution of unsolved homicides and cold cases (https://www.vidocq.org).

Institute for Human Identification—Lincoln Memorial University (College of Dental Medicine)

Founded in 2022, the Institute for Human Identification in the College of Dental Medicine at Lincoln Memorial University is a state-of-the-art facility that conducts research in forensic anthropology, forensic odontology, and forensic genetics with a fundamental mission of developing and optimizing methods for human remains identification. The Institute for Human Identification is partnered with the Regional Forensic Center (RFC) (which houses the medical examiner's office) and Identifinders International, and provides expert services in forensic anthropology, odontology, and DNA analysis for cold cases and unidentified human remains investigations throughout the United States and internationally. In addition to casework, the Institute team provides consulting services and training through the National Forensic Academy (NFA) and conducts research at its outdoor decomposition research facility ("body farm").

Summary

Ultimately, although skeletal remains are not the most common type of sample submitted for forensic DNA testing, there are a number of scenarios in which bones or teeth may be presented for analysis. Some of these instances may involve deaths by natural causes, while others are a result of violent crimes, natural disasters, humanitarian crises, or human rights abuses. Regardless of the circumstances or cause of death, investigators should be cognizant that DNA recovery from skeletal remains is far more challenging than working with traditional sample types such as blood, semen, or saliva. Processing bones and teeth for DNA requires specialized equipment and facilities, as well as specialized training and knowledge. The same DNA extraction methods used for body fluids and soft tissues often will not suffice in recovering adequate amounts of DNA from skeletal remains. Modifications to existing methods, or the use of alternative (usually more time-consuming and labor-intensive) approaches, are often necessary. Knowledge of postmortem changes to bone microstructure—and how this correlates to DNA preservation—can assist scientists in selecting and developing the best strategies and approaches for maximum genetic data recovery. The chapters in this book provide details on the complexities of recovering DNA from skeletal remains and highlight some of the technological advances available to forensic geneticists for unidentified human remains investigations, including: (1) new and expanded DNA marker systems; (2) massively parallel sequencing (MPS); (3) forensic DNA phenotyping; (4) field-deployable Rapid DNA testing platforms; and (5) forensic genetic genealogy (FGG). Also discussed are complementary disciplines (e.g., anthropology, odontology, forensic art) that are often used in concert with forensic DNA analysis to assist in the identification process.

References

AP News, November 27, 1989. Up to 48,000 People Buried in Already Occupied Graves, Investigators Say. Associated Press. https://apnews.com/article/4ac61b74d07872ac52652f790fdef556.

Austin, E., 2020. New documentary explores malpractice, overcrowding at Louisville's Eastern Cemetery. Louisville Cour. J. https://www.courier-journal.com/story/entertainment/movies/2020/03/03/facing-east-documentary-explores-nations-most-overcrowded-cemetery/4927406002/.

Bailey, C., 2015. Fields of the forgotten: Abandoned cemetery law in Kentucky. J. Anim. Environ. Law 7, 91.

Bilham, R., 2010. Lessons from the Haiti earthquake. Nature 463, 878–879.

Cash, H.D., Hennessey, M.J., 2004. Human identification software for missing persons, scalable for a mass fatality incident: Building on lessons learned over the course of a major disaster victim identification project. In: Promega 15th Symposium for Human Identification. https://www.genecodesforensics.com/news/Promega%202004%20M-FISys%20for%20Missing%20Persons%20paper%20revised.pdf.

Cash, H.D., Hoyle, J.W., Sutton, A.J., 2003. Development under extreme conditions: Forensic bioinformatics in the wake of the World Trade Center disaster. In: Pacific Symposium on Biocomputing, pp. 638–653. 12603064.

Dolan, S., Saraiya, D., Donkervoort, S., Rogel, K., Lieber, C., Sozer, A., 2009. The emerging role of genetics professionals in forensic kinship DNA identification after a mass fatality: Lessons learned from Hurricane Katrina volunteers. Genet. Med. 11, 414–417. https://doi.org/10.1097/GIM.0b013e3181a16ccc.

Eisenberg, A.J., Planz, J.V., 2008. University of North Texas Center for Human Identification Project: The Anthropological, mtDNA, and STR Analysis of Unidentified Human Remains and Family Reference Samples for Entry into CODIS and the Field Testing and Implementation of New Technologies to Facilitate Additional Identifications. National Institute of Justice (NIJ). NCJ Number 223976.

Engel, M., February 17, 2002. More Than 100 Bodies Found as U.S. Crematorium Gives Up Grisly Secret. The Guardian. https://www.theguardian.com/world/2002/feb/18/matthewengel.

Garcia, M., Martinez, L., Stephenson, M., Crews, J., Peccerelli, F., 2009. Analysis of complex kinship cases for human identification of civil war victims in Guatemala using M-FISys software. Forensic Sci. Int. Genet. Suppl. Ser. 2, 250–252.

Gene Codes Forensics Inc, 2019. The Mass Fatality Identification System (M-FISys): The Complete Solution for DVI and Missing Persons Casework. https://www.genecodesforensics.com/software/.

Gin, K., Tovar, J., Bartelink, E.J., Kendell, A., Milligan, C., Willey, P., Wood, J., Tan, E., Turingan, R.S., Selden, R.F., 2020. The 2018 California wildfires: Integration of rapid DNA to dramatically accelerate victim identification. J. Forensic Sci. 65 (3). https://doi.org/10.1111/1556-4029.14284.

Glynn, C.L., Ambers, A., 2021. Rapid DNA analysis—Need, technology, and applications. In: Crocombe, R., Leary, P., Kammrath, B. (Eds.), Portable Spectroscopy and Spectrometry. John Wiley & Sons.

Hamilton, B., February 26, 2006. Inside Bowels of Gory Cadaver Scheme. The New York Post. https://nypost.com/2006/02/26/inside-bowels-of-gory-cadaver-scheme/.

Hennessey, M.J., Kroll, J., Lorente, J.A., Cash, H.D., 2018. Addressing technical, legal, and diplomatic obstacles to combatting human trafficking using DNA databases across national boundaries. In: ISMB-LA 2018, Vina del Mar, Chile. https://www.genecodesforensics.com/news/ISCB-LA-2018-Poster-Proof.pdf.

Hickman, M.J., Strom, K.J., 2007. Medical Examiners and Coroners' Offices, 2004. Bureau of Justice Statistics Special Report. U.S. Department of Justice, Office of Justice Programs.

Horton, K.E., 2003. Who's watching the crypt keeper: The need for regulation and oversight in the crematory industry. Elder Law J. 11 (2), 425–458.

Identifinders International LLC, 2021. 1961 Alabama Car Accident Victim Identified After 60 Years: Oldest Identification on Record for National Center for Missing and Exploited Children. PRNewswire. https://www.prnewswire.com/news-releases/1961-alabama-car-accident-victim-identified-after-60-years-301413949.html.

Margesson, R., Taft-Morales, M., 2010. Haiti Earthquake: Crisis and Response. Congressional Research Service. https://apps.dtic.mil/sti/pdfs/ADA516429.pdf.

Morales, M., Blau, R., July 7, 2013. Bone Snatcher Michael Mastromarino Dies of Bone Cancer as Victims See Sad Irony. New York Daily News. https://www.nydailynews.com/new-york/body-snatcher-michael-mastromarino-dead-article-1.1392503.

New York Times, November 28, 1989. Thousands Buried in Old Graves, Investigators in Kentucky Report. The New York Times. https://www.nytimes.com/1989/11/28/us/thousands-buried-in-old-graves-investigators-in-kentucky-report.html.

Patel, V., November 3, 2021. Killed in a 1961 Crash, 'Unknown Boy' is Finally Identified. The New York Times. https://www.nytimes.com/2021/11/03/us/unknown-boy-danny-armantrout-identified.html?auth=login-google.

Sadiq, A., McEntire, D., 2012. Challenges in mass fatality management: A case study of the 2010 Haiti earthquake. J. Emerg. Manag. 10 (6), 459–471.

Spencer, C., April 29, 2021. Is NamUs Reporting Required in Your State? Here's a Look at State Legislation on the Missing & Unidentified Persons Clearinghouse. Biometrica. https://www.biometrica.com/is-namus-reporting-required-in-your-state.

Steadman, D.W., Sperry, K., Snow, F., Fulginit, L., Craig, E., 2008. Anthropological investigations of the Tri-State Crematorium incident. In: Adams, B.J., Byrd, J.E. (Eds.), Recovery, Analysis, and Identification of Commingled Human Remains. Humana Press, pp. 81–96, https://doi.org/10.1007/978-1-59745-316-5_5.

CHAPTER

2

Challenges in forensic genetic investigations of decomposed or skeletonized human remains: Environmental exposure, DNA degradation, inhibitors, and low copy number (LCN)

Angie Ambers PhD[a,b,c]

[a]Henry C. Lee Institute of Forensic Science, University of New Haven, West Haven, CT, United States [b]Forensic Science Department, Henry C. Lee College of Criminal Justice and Forensic Sciences, Center for Forensic Investigation of Trafficking in Persons, University of New Haven, West Haven, CT, United States [c]Institute for Human Identification, LMU College of Dental Medicine, Knoxville, TN, United States

"The river of my title is a river of DNA—a river of information, not a river of bones and tissues".
– Richard Dawkins

Introduction

DNA is perhaps the most essential molecule in humans. It contains the *in vivo* blueprint for physical features and, after death, is the remaining "molecular signature" of a person's hereditary and evolutionary history. Postmortem survival of DNA within bones and teeth occurs due to the compact microstructure of the skeleton and its ability to provide a strong, protective physical barrier to environmental and intentional insults. Nonetheless, over time

https://doi.org/10.1016/B978-0-12-815766-4.00002-9

endogenous DNA becomes damaged, limiting our ability to detect it and affecting its utility in making a positive identification.

Forensic genetic investigations are limited by the quality, quantity, and purity of DNA recovered from biological samples. Significant damage or alteration to the molecular structure of DNA is problematic because polymerases stall at damaged/altered sites, preventing PCR amplification (and subsequent analysis) of target loci. Further (often concurrent) complications arise from endogenous and/or environmental inhibitors that co-extract with DNA and impede or completely block downstream polymerase-based reactions. In order to assess potential strategies for improving genetic typing of degraded samples, it is necessary to understand the nature and variety of DNA damage as well as the conditions that cause it. Although the mechanisms of DNA damage can be divided into four major categories (depurination/depyrimidination, strand breakage, crosslinking, base alteration), the molecular chemistry of resultant nucleic acid modifications can be quite complex and the variety of possible lesions in any given sample is almost limitless. Moreover, the degree and spectrum of DNA damage (as well as its rate of incidence) depends largely on the sample source, the environment to which it was exposed, and the length of exposure time (Dabney et al., 2013; Geacintov and Broyde, 2010; Shafirovich and Geacintov, 2010; Jun, 2010; De Grujil and Rebel, 2008; Cooke et al., 2003; Gates, 2009; Lindahl and Nyberg, 1972). Human remains that have naturally decomposed to a skeletonized state (due to environmental exposure) or remains that have been intentionally damaged (e.g., via fire, physical forces, or chemical treatments) pose a particularly unique challenge in forensic casework due to the prevalence of DNA damage and degradation.

DNA structure and susceptibility to damage

In order to understand the challenges associated with DNA identification of highly decomposed or skeletonized human remains, it is necessary to understand: (1) the molecular structure of DNA; (2) the various types of chemical and structural modifications that compromise both the quality and the quantity of DNA available for genetic profiling; and (3) the endogenous and exogenous factors that contribute to degradation.

In human cells, nuclear DNA is tightly coiled inside structures called chromosomes. This compact, condensed packaging is facilitated by electrostatic interactions between the negatively charged DNA backbone and positively charged proteins called histones. DNA is tightly wrapped around histone proteins, forming DNA-protein complexes referred to as chromatin. The basic repeating structural unit of chromatin is the nucleosome, which contains 8 histone proteins (an octamer) and approximately 146 base pairs (bp) of DNA (Kawane et al., 2014; Annunziato, 2008). Humans have 46 chromosomes (23 pairs), and each chromosome contains hundreds-to-thousands of nucleosomes ("beads on a string") connected by what is referred to as "linker DNA." The length of "linker DNA" between adjacent nucleosomes varies among different tissues and cell types, but these regions are particularly vulnerable to postmortem attack by endogenous nucleases (Dabney et al., 2013; Annunziato, 2008).

On a micromolecular level, DNA is a polymer consisting of basic monomeric subunits called nucleotides (also referred to as deoxyribonucleotide triphosphates, or dNTPs). A DNA molecule consists of billions of these nucleotides (dNTPs) and hence is often referred to as a polynucleotide. One nucleotide monomer is comprised of a phosphate group, a 5-carbon sugar [deoxyribose (dR)], and a nitrogenous base. The backbone of DNA is a chain of nucleotides, with alternating phosphate groups and deoxyribose sugars held together by phosphodiester bonds. Attached to each sugar moiety is a nitrogenous base composed of rings of carbon and nitrogen atoms. DNA contains four different nitrogenous bases—adenine (A), guanine (G), cytosine (C), and thymine (T)—and a dNTP is designated by the specific base it contains (i.e., as dATP, dGTP, dCTP, or dTTP). Nitrogenous bases are subdivided into purines and pyrimidines based on differences in chemical structure. Purines (adenine, guanine) are composed of two aromatic rings, whereas pyrimidines (cytosine, thymine) are single-ringed structures. Purines on one strand of DNA form hydrogen bonds with corresponding pyrimidines on the opposing strand, according to traditional Watson-Crick base pairing rules. This hydrogen bonding between bases holds the two DNA strands together in its native double-stranded conformation. The sequence of DNA bases constitutes an organism's molecular signature, and modification to or loss of this sequence information impedes (and sometimes completely prevents) forensic DNA typing.

Inherent instability of DNA and endogenous contributing factors to damage

A major consideration in understanding DNA's susceptibility to damage is to acknowledge the inherent instability of the DNA molecule itself, which is largely due to the fact that an aqueous environment favors hydrolysis of polynucleotides. An aqueous environment exists naturally within cells of the human body, and also can be derived from moisture in the external environment. Aside from its propensity to be hydrolyzed in the presence of water, the DNA molecule is subject to postmortem enzymatic and chemical damage by endonucleases and free radicals that are naturally produced within the cell (Geacintov and Broyde, 2010; Onori et al., 2006). These free radicals, known as reactive oxygen species (ROS) and reactive nitrogen species (RNS), are chemical intermediates generated during the course of a cell's normal metabolic activity (i.e., a consequence of aerobic metabolism, in which inhaled oxygen is converted to highly reactive intermediates). *In vivo*, the harmful effects of these highly reactive intermediates are mitigated by enzymatic pathways (e.g., superoxidase dismutase, catalase) and by nonenzymatic mechanisms involving antioxidants. However, when a cell dies, these free radicals immediately attack endogenous biomolecules such as DNA and can induce significant damage (Kawane et al., 2014; Dabney et al., 2013; Shafirovich and Geacintov, 2010; Gates, 2009; Valko et al., 2007). Ultimately, endogenous DNA damage is an unavoidable consequence of the presence of oxygen, water, and alkylating agents in human cells.

Environmental (exogenous) influences on DNA degradation

Environmental (exogenous) factors tend to cause more diverse lesions and more dramatic alteration to the molecular structure of DNA than those which are endogenously induced (Geacintov and Broyde, 2010). Human remains in an advanced state of decomposition or that are completely skeletonized typically have been exposed to harsh environmental conditions (e.g., excessive heat, humidity, acidic soil) for extended periods of time, facilitating the generation of a variety of different types of DNA damage. Temperature has a profound effect on biological/chemical processes and is particularly relevant in discussions regarding DNA damage. As a general rule, the rate of chemical reactions (including those responsible for hydrolysis of bonds within DNA) doubles for every 10°C increase in temperature (Prangnell and McGowan, 2009).

Burial in soil is a common method of disposal of human remains, either legitimately (in cemetery plots or in mausoleums) or clandestinely (to conceal criminal activity). Human remains often have been submerged or immersed in soil for prolonged time periods before being exhumed or discovered and excavated; and soil conditions (e.g., pH, moisture level, microbial content) play a role in the destruction (or preservation) of DNA. Research has shown that climate heavily influences soil chemistry, particularly soil pH (acidity or alkalinity). Acidity not only accelerates decomposition of soft tissues and bones/teeth, but also causes significant damage to endogenous DNA. In general, soil tends to be alkaline in dry environments and acidic in wet climates (Slessarev et al., 2016; Cohen, 2016). A global soil pH map was constructed in 2016 by a team at the University of California Santa Barbara and, in addition to onsite soil testing during recovery of remains, can serve as an informative resource for forensic teams investigating buried human remains (Cohen, 2016).

In addition to soil acidity, microbes (e.g., bacteria, fungi), the moisture content of soil, environmental humidity, ambient (or artificial) heat, and sunlight also contribute to DNA damage. High temperatures are encountered in fires, bombings, terrorist attacks, and in the environment in general (particularly in deserts and tropical climates). Above-ground deposition of human remains and "surface scatters" (the latter of which may occur in high-impact events such as aviation accidents or over time due to scavenger activity) can subject the remains to extended periods of sunlight, which also contributes to increased heat exposure. Human remains also may be recovered from freshwater or saltwater environments (e.g., due to drowning, criminal disposal, shipwrecks, aviation disasters), and water promotes hydrolysis of bonds within the DNA molecule. All of these environmental factors—soil acidity, microbes, heat, ambient humidity, sunlight, water submersion—and their respective degradative effects should be considered in DNA recovery efforts.

Molecular biology of DNA damage: Types of lesions and causes

Structural and chemical damage to a DNA molecule can manifest in a variety of ways, including: (1) loss of bases or primary sequence information (via hydrolysis of glycosidic bonds within nucleotide monomers); (2) strand breaks (due to hydrolysis of phosphodiester bonds in the backbone of DNA); (3) crosslinking of nitrogenous bases on the same or opposing DNA strand (intrastrand or interstrand crosslinks, respectively), and (4) modification of nitrogenous bases (via deamination, oxidation, or alkylation processes).

Abasic (AP) sites: Depurination and depyrimidination

Within the nucleotide monomeric subunit of DNA, a nitrogenous base (A, G, C, T) is attached to the 5-carbon sugar (deoxyribose) via a glycosidic bond. An apurinic/apyrimidinic (AP) site, also referred to as an abasic ("without a base") site, is formed when the glycosidic bond between nitrogenous base and deoxyribose sugar is hydrolyzed or broken (Fig. 2.1). AP sites can be subdivided into: (1) apurinic sites, in which a purine base (adenine or guanine) is missing, and (2) apyrimidinic sites, which result from the separation of a pyrimidine base

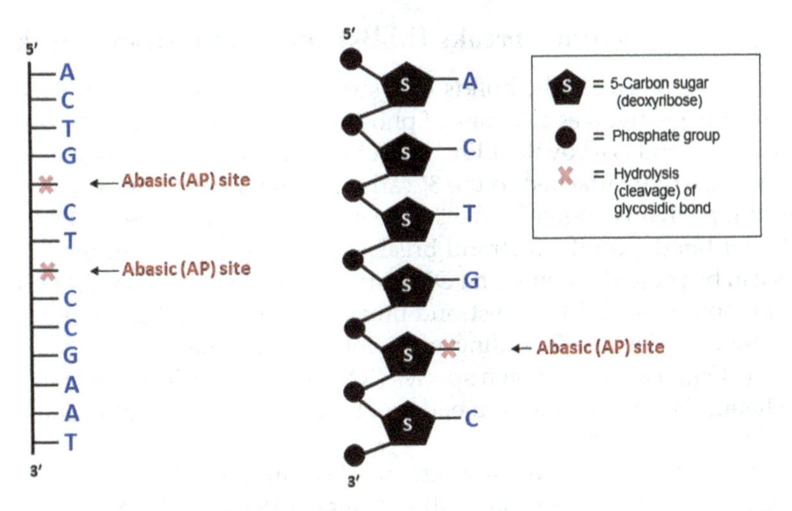

FIG. 2.1 Schematic representation of an abasic (AP) site in a DNA strand, resulting in the loss of primary sequence information. This type of DNA damage is caused by hydrolysis (cleavage) of the glycosidic bond between the 5-carbon sugar (deoxyribose) and the nitrogenous base of a nucleotide within the DNA molecule. *Graphic design credit: Angie Ambers, MA, MS, PhD (Henry C. Lee Institute of Forensic Science; Henry C. Lee College of Criminal Justice and Forensic Sciences; Henry C. Institute for Human Identification, LMU College of Dental Medicine).*

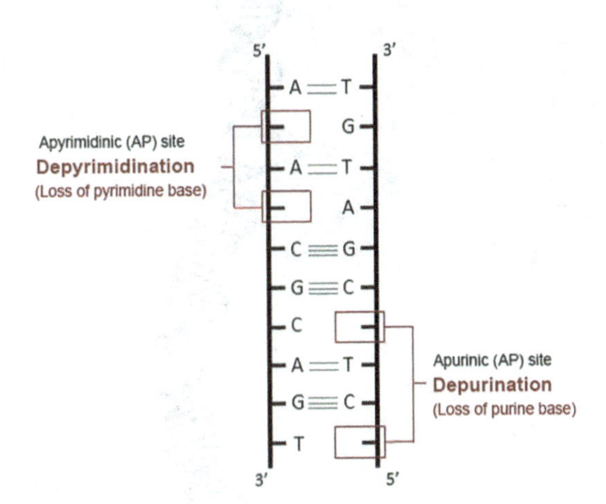

FIG. 2.2 Depurination generates an apurinic (AP) site on a DNA strand and involves loss of a purine base (adenine or guanine). Conversely, depyrimidination results in an apyrimidinic (AP) site that is missing a pyrimidine base (cytosine or thymine). *Graphic design credit: Morgan Barrett, MS (Henry C. Lee College of Criminal Justice and Forensic Sciences).*

(cytosine or thymine) from the DNA strand (Fig. 2.2). DNA is prone to depurination (and to a lesser extent depyrimidination) when exposed to high temperatures and acidic pH levels. The presence of AP sites results in the loss of primary sequence information, and polymerases stall at these regions during PCR (thereby inhibiting amplification of that region of DNA). In addition, accumulation of AP sites destabilizes the DNA backbone, leading to strand breaks (Dabney et al., 2013; Geacintov and Broyde, 2010; Gates, 2009; Lindahl, 1993; Lindahl and Nyberg, 1972).

Strand breakage: Single-strand breaks (SSBs) and double-strand breaks (DSBs)

Besides hydrolysis of glycosidic bonds and subsequent generation of abasic (AP) sites, another type of damage involves cleavage of phosphodiester bonds in the backbone of DNA. The phosphodiester bond is a covalent linkage between the phosphate of one nucleotide and the hydroxyl (–OH) group attached to the 3′ carbon of the deoxyribose sugar in an adjacent nucleotide, forming what is known as the "sugar-phosphate backbone" of DNA. Hydrolysis of phosphodiester bonds results in strand breaks and fragmentation of the DNA molecule. Strand breaks can be present on only one DNA strand [single-strand breaks (SSBs)] (Fig. 2.3) or adjacently on both strands [double-strand breaks (DSBs)] (Fig. 2.4). Strand breaks can be caused by a variety of factors, including ultraviolet (UV) radiation, free radicals [reactive oxygen species (ROS), reactive nitrogen species (RNS)], excessive heat, alkylating agents, environmental chemicals, and postmortem endonuclease activity (Geacintov and Broyde, 2010; Gates, 2009; Friedberg et al., 2006).

Cellular death (apoptosis) occurs at a rate of a few million cells per second in the postmortem human body (Kawane et al., 2014; Nagata, 1997, 2005; Wyllie, 1980). During

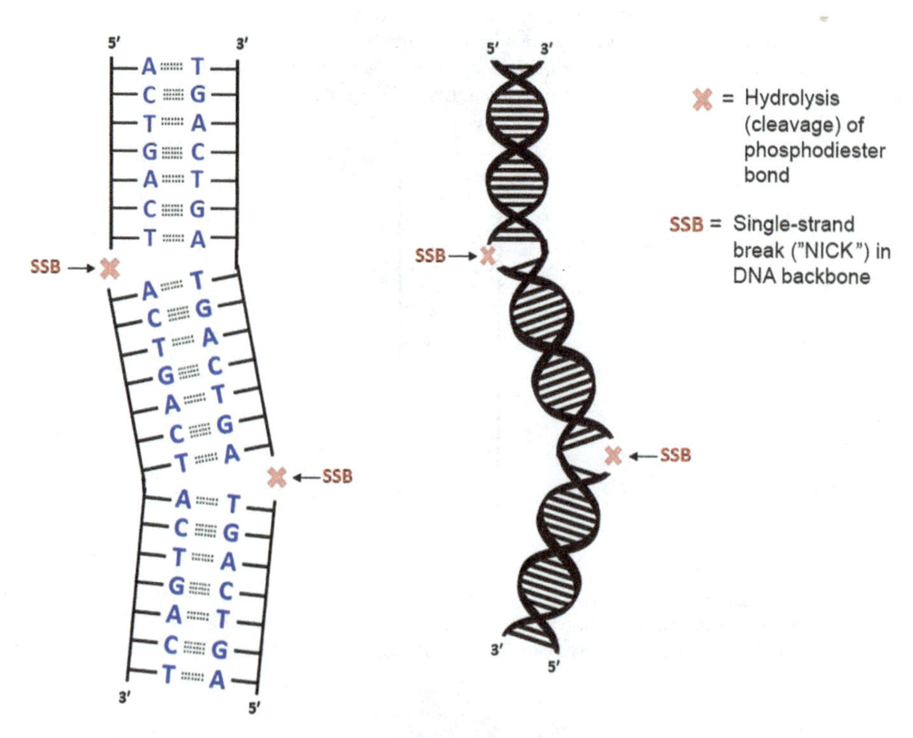

FIG. 2.3 Fragmentation results from hydrolysis of phosphodiester bonds in the sugar-phosphate backbone of DNA. A single-strand break (SSB) occurs when the phosphodiester bond on one DNA strand is broken, as opposed to a double-strand break (DSB) which involves both DNA strands (Fig. 2.4). SSBs are sometimes referred to as "nicks" in the DNA backbone. *Graphic design credit: Angie Ambers, MA, MS, PhD (Henry C. Lee Institute of Forensic Science; Henry C. Lee College of Criminal Justice and Forensic Sciences; Henry C. Institute for Human Identification, LMU College of Dental Medicine).*

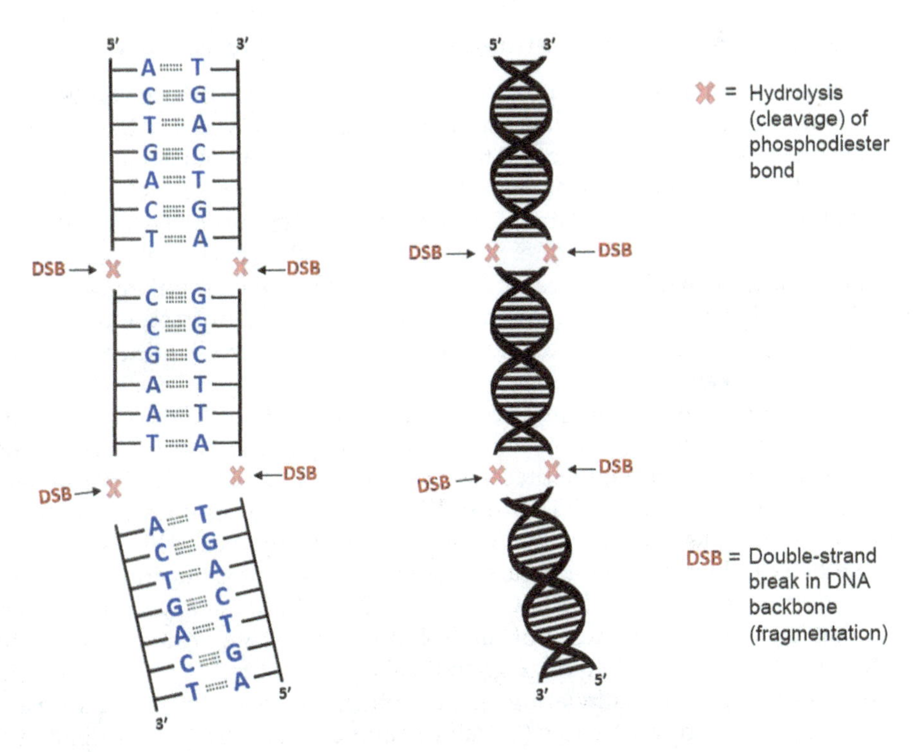

FIG. 2.4 Schematic representation of a double-strand break (DSB), in which adjacent phosphodiester bonds are broken in the sugar-phosphate backbone, resulting in fragmentation of the DNA molecule into smaller and smaller pieces. Fragmentation restricts the size (length) of target loci that can be examined in forensic analyses. *Graphic design credit: Angie Ambers, MA, MS, PhD (Henry C. Lee Institute of Forensic Science; Henry C. Lee College of Criminal Justice and Forensic Sciences; Henry C. Institute for Human Identification, LMU College of Dental Medicine).*

apoptosis, DNA is one of many biomolecules degraded by endonuclease activity. The primary endogenous nuclease (DNase) responsible for apoptotic DNA degradation is caspase-activated DNase (CAD), also referred to as DNA fragmentation factor 40 (DFF-40). In caspase-activated apoptosis pathways, chromosomal DNA is degraded in two steps: first into larger (50–100 kb) fragments and then into smaller nucleosomal units. Typically, enzymatic cleavage initiates in "linker DNA" regions between adjacent nucleosomes, and then later progresses with hydrolysis of the DNA directly interacting with histones (Kawane et al., 2014; Nagata, 2005).

DNA in forensic, historical, and ancient biological samples (particularly skeletonized remains) is often highly fragmented due to the passage of time and long-term exposure to both endogenous and environmental destructive factors. Fragmentation significantly hinders the success of PCR amplification and restricts the size (length) of target loci that can be examined. For successful amplification to occur, both the target region and its associated primer-binding sites must be intact (Dabney et al., 2013; Geacintov and Broyde, 2010; Nelson, 2009; Golenberg et al., 1996).

Crosslinking of DNA bases: Interstrand and intrastrand crosslinks

Exposure to solar ultraviolet (UV) radiation can generate several different types of damage in the DNA molecule. Although ultraviolet radiation consists of UV-A, UV-B, and UV-C rays, the latter is absorbed by the atmosphere and therefore is not likely to cause substantial damage to DNA (Tuchinda et al., 2006; Cadet et al., 2005; Hall and Ballantyne, 2004). UV-A and UV-B rays cause indirect and direct DNA damage, respectively. UV-A rays (320–400nm) create free radicals which subsequently cause indirect damage to the DNA molecule (e.g., bond hydrolysis, base modifications), while UV-B rays (290–320nm) result in crosslinking. Crosslinks are covalent linkages between nucleobases on the same DNA strand (intrastrand crosslinks) or between bases on opposite strands (interstrand crosslinks) (Fig. 2.5). Additionally, crosslinks can form between DNA and proteins.

The most common types of intrastrand crosslinks induced by UV irradiation are cyclopyrimidine dimers (CPDs) (i.e., thymine dimers) (Fig. 2.6A) and 6-4 photoproducts. Regardless of origin or causation, the presence of crosslinks can cause a physical deformation or "kink" in the double helix (Fig. 2.6B). Polymerases stall at intrastrand crosslinks, and interstrand crosslinks are problematic because their presence inhibits denaturation of the double helix (which is the necessary first step in PCR amplification of target loci) (Geacintov and Broyde, 2010; Jun, 2010; Onori et al., 2006; Cadet et al., 2005). It is important to note that there are other causes of nucleobase crosslinking besides UV irradiation. Exposure to formalin or formaldehyde (e.g., in the case of medical or museum specimens) and exposure to environmental alkylating agents (which are ubiquitous in nature) also can cause crosslinking between DNA bases (Geacintov and Broyde, 2010; De Grujil and Rebel, 2008; Friedberg et al., 2006).

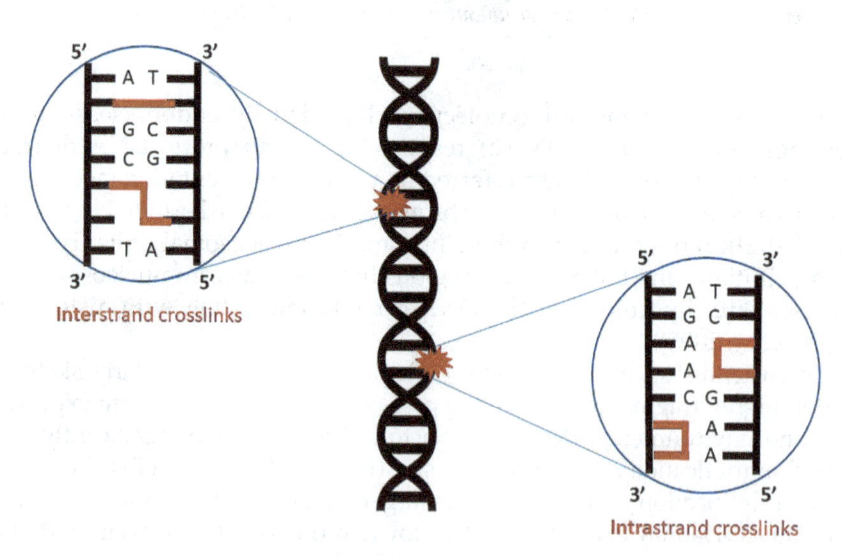

FIG. 2.5 Basic schematic representation of interstrand and intrastrand crosslinks in DNA, two types of direct damage that can be induced by exposure to ultraviolet (UV) irradiation, treatment with formalin or formaldehyde, and environmental alkylating agents. *Graphic design credit: Lily Kate Josephs, MS (Henry C. Lee College of Criminal Justice and Forensic Sciences).*

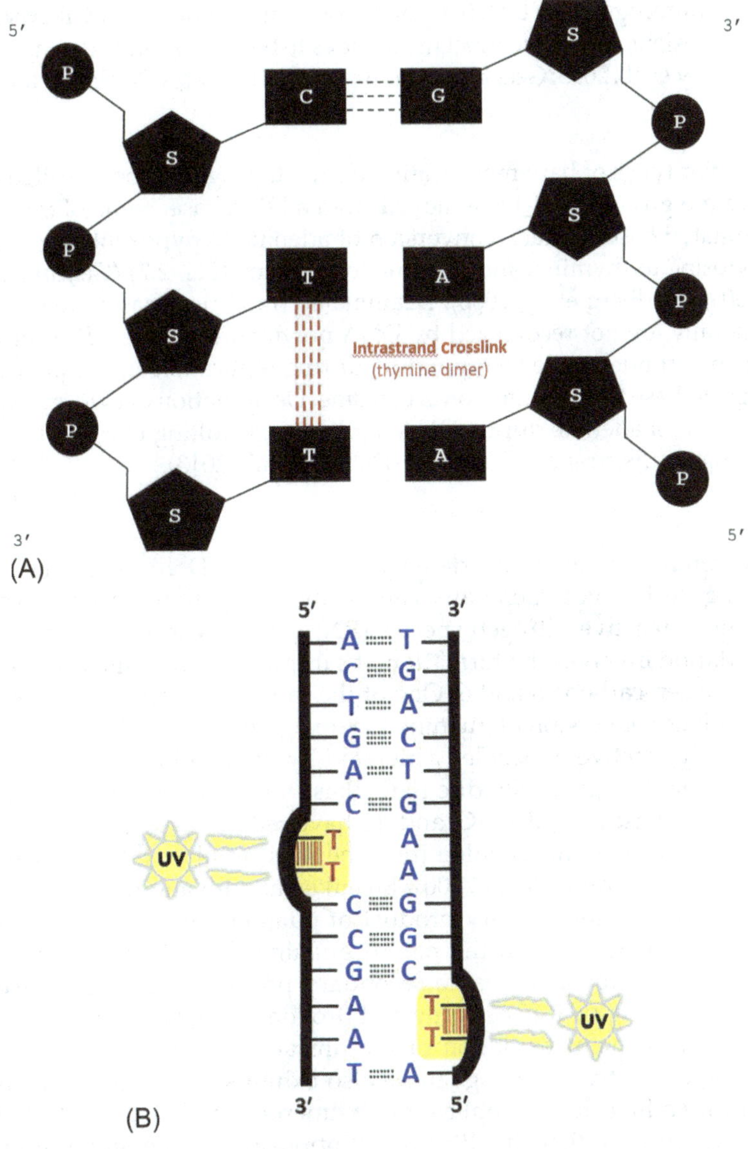

FIG. 2.6 (A) Diagram of a pyrimidine (thymine) dimer, one of the most common forms of intrastrand crosslinks in DNA, induced by exposure to UV irradiation. (B) Intrastrand crosslinks cause a physical deformation or "kink" (bulge) in the double helix, and DNA polymerases stall at these sites during PCR amplification. *Graphic design credits: (A) Ryan Durkee (Henry C. Lee College of Criminal Justice and Forensic Sciences); (B) Angie Ambers, MA, MS, PhD (Henry C. Lee Institute of Forensic Science; Henry C. Lee College of Criminal Justice and Forensic Sciences; Henry C. Institute for Human Identification, LMU College of Dental Medicine).*

DNA base modifications: Deamination, oxidation, and alkylation

In addition to abasic (AP) sites, strand breaks, and crosslinks, there are various mechanisms that can alter or modify DNA nucleobases, including deamination, oxidation, and alkylation. These chemical processes convert standard Watson-Crick nucleobases into modified versions that are generally unrecognizable by DNA polymerases, thus inhibiting PCR. Also, the presence of some modified lesions causes incorrect nucleotides to be incorporated during replication and sequencing (Dabney et al., 2013; Geacintov and Broyde, 2010; Gates, 2009; Friedberg et al., 2006).

Deamination

One of the major types of base modification occurs through a process called deamination, in which the amino group ($-NH_2$) is removed from a DNA base. Some of the most common forms of deaminated bases include conversion of adenine to hypoxanthine, cytosine to uracil, 5-methylcytosine to thymine, and guanine to xanthine (Fig. 2.7) (Geacintov and Broyde, 2010; Gates, 2009; Friedberg et al., 2006). Deaminated (modified) bases are noncoding derivatives that generally are not recognized by DNA polymerases during PCR or, alternatively, can result in incorrect nucleotide incorporation during replication and sequencing. The most frequently targeted base for deamination is cytosine. Deamination of cytosine to uracil results in the incorporation of adenine during DNA replication, resulting in a $C \rightarrow T$ substitution (or $G \rightarrow A$, depending on the strand sequenced) (Dabney et al., 2013).

Oxidation

Similar to deamination, oxidative damage can occur to DNA bases, resulting in additional noncoding derivatives. Generally caused by endogenous reactive oxygen species (ROS), endogenous reactive nitrogen species (RNS), chemicals, or free radicals in the environment, oxidation involves the formation of saturated pyrimidine rings and loss of the double bond between carbons 5 and 6. One of the most common types of oxidative damage in DNA involves conversion of guanine to 8-oxoguanine (Fig. 2.8) (Gates, 2009; Cooke et al., 2003). Highly reactive molecules generated by ROS and RNS exhibit selectivity for the oxidation of guanine, primarily due to the base's simple reduction–oxidation (redox) chemistry. The DNA bases dG, dA, dC, and dT have reduction potentials of 1.29, 1.42, 1.6, and 1.7 V, respectively, making guanine (dG) the most easily oxidized base by ROS and RNS intermediates (Shafirovich et al., 2001; Steenken and Jovanovic, 1997; Buettner, 1993). More importantly, the major primary product of guanine oxidation (8-oxoguanine) is at least 1000-fold more reactive than the parent guanine toward further oxidation, which gives rise to a spectrum of more stable secondary products that are unrecognizable by DNA polymerases during PCR (Geacintov and Broyde, 2010; Margolin et al., 2006).

In addition to subsequent oxidation of the unstable 8-oxoguanine molecule to form stable, noncoding derivatives, 8-oxoguanine also exhibits dual coding potential and has a tendency for miscoding during replication. Numerous studies have demonstrated that DNA polymerases insert either dCTP or dATP opposite 8-oxoguanine (Eoff et al., 2007; Zang et al., 2006; Brieba et al., 2004; Haracska et al., 2003; Einolf and Guengerich, 2001; Haracska et al., 2000; Furge and Guengerich, 1997, 1998; Shibutani et al., 1991). The mechanism of miscoding by 8-oxoG has been extensively investigated, and it is thought that in order to alleviate a steric clash between the C8 oxygen and the O4′ atom of the deoxyribose sugar, 8-oxoG adopts a *syn* conformation (as opposed to its normal *anti* conformation in duplex DNA), thereby allowing dATP to form two stable hydrogen bonds with the Hoogsteen face of 8-oxoG (Geacintov and Broyde, 2010; McAuley-Hecht et al., 1994). More importantly, the dATP:8-oxoG base pair is geometrically similar to the dTTP:dATP pair and, as such, often evades the proofreading activity of some DNA polymerases. In order

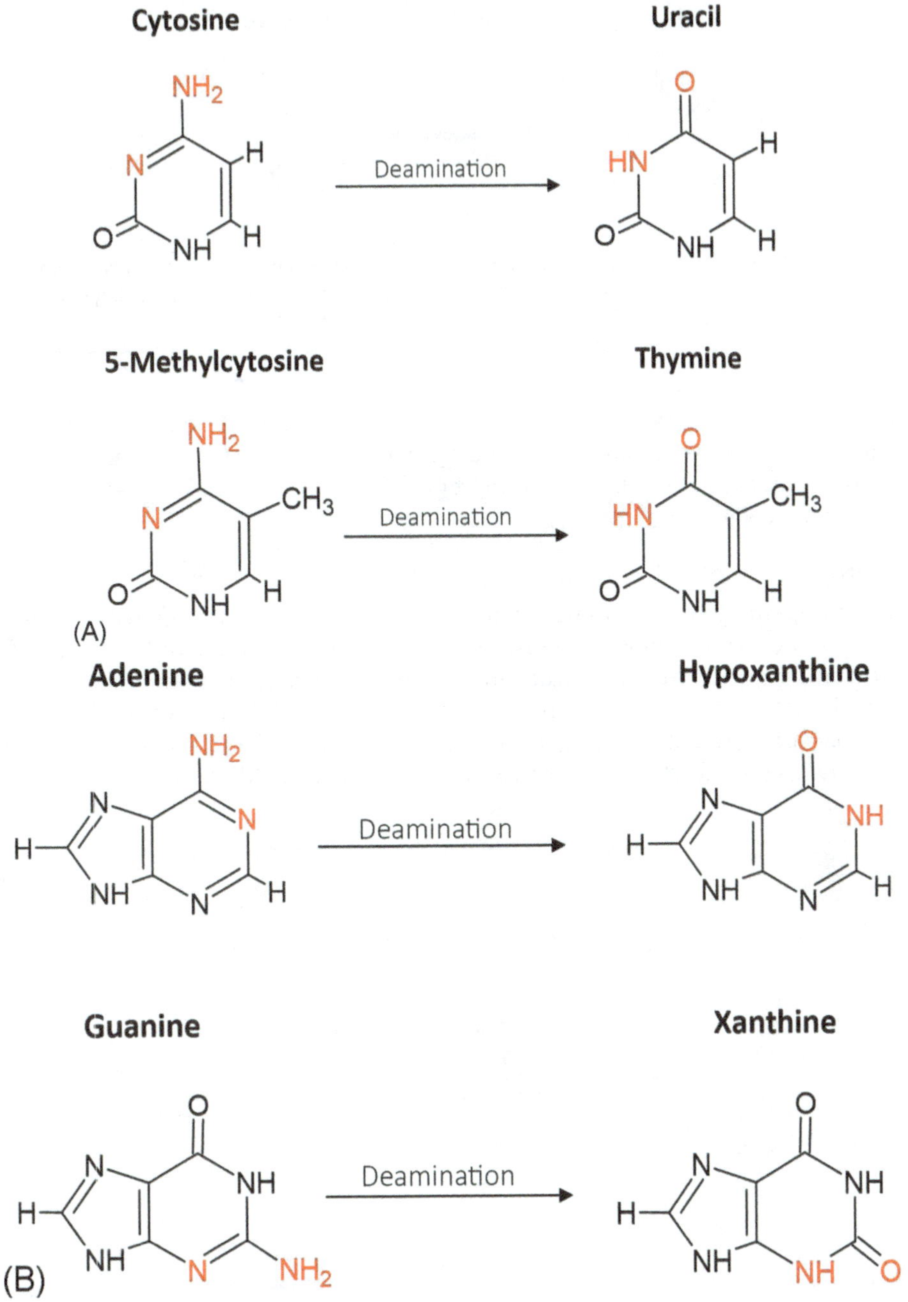

Cytosine → Deamination → Uracil

5-Methylcytosine → Deamination → Thymine

(A)

Adenine → Deamination → Hypoxanthine

Guanine → Deamination → Xanthine

(B)

FIG. 2.7 Examples of common DNA base modifications resulting from deamination :(A) Deamination of pyrimidine bases (i.e., conversion of cytosine to uracil; conversion of 5-methylcytosine to thymine), and (B) deamination of purine bases (i.e., conversion of adenine to hypoxanthine, conversion of guanine to xanthine). *Graphic design credit: Erin E. Dimino, MS (Henry C. Lee College of Criminal Justice and Forensic Sciences).*

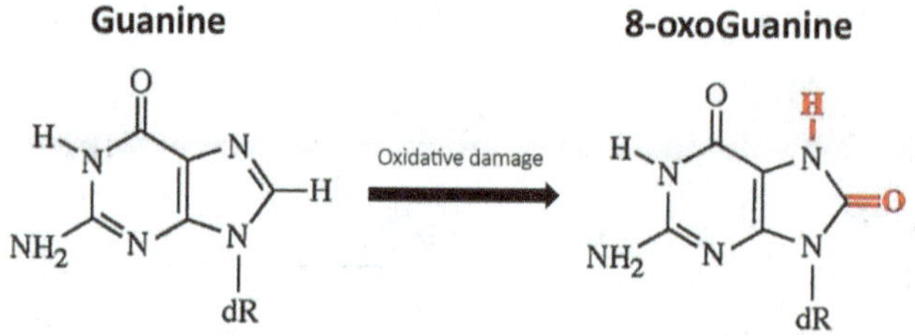

Guanine

Oxidative damage

8-oxoGuanine

FIG. 2.8 One of the most common modifications to a DNA base via oxidative damage involves conversion of guanine to 8-oxoguanine, which readily undergoes subsequent oxidation to form stable noncoding derivatives. *Graphic design credit: Angie Ambers, MA, MS, PhD (Henry C. Lee Institute of Forensic Science; Henry C. Lee College of Criminal Justice and Forensic Sciences; Henry C. Institute for Human Identification, LMU College of Dental Medicine).*

to incorporate dCTP opposite 8-oxoguanine, a polymerase must be capable of overcoming the thermodynamic barrier that favors the *syn* configuration of the lesion (Geacintov and Broyde, 2010; Rechkoblit et al., 2006; Zang et al., 2006; Brieba et al., 2004; Hsu et al., 2004; Freisinger et al., 2004).

Alkylation

Alkylating agents provide another means of base modification, primarily resulting in the attachment of methyl- or other alkyl groups to the N- and O- atoms of DNA bases (Fig. 2.9). These alkylating agents are produced endogenously during cellular metabolism and are ubiquitous in nature (i.e., are found in air, water, and food, although generally in small concentrations). Variation exists in alkylation patterns because the exact pattern exhibited depends on the precise alkylating agent (or agents) involved. Alkylated bases are especially

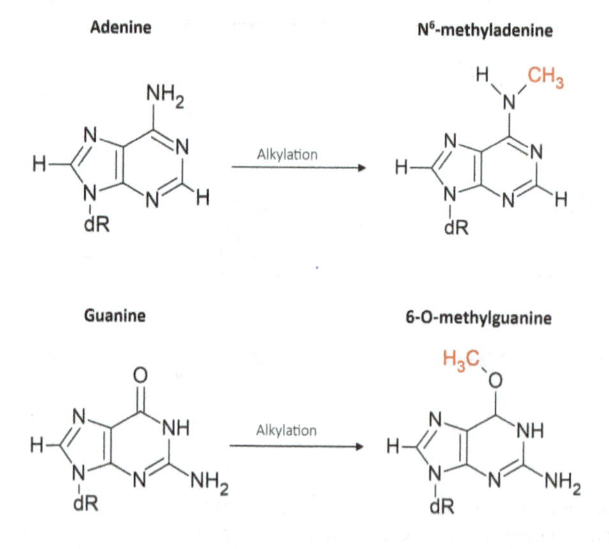

FIG. 2.9 Alkylation is a type of modification that involves attachment of methyl or other alkyl groups to the nitrogen (N-) and oxygen (O-) atoms of DNA bases, forming noncoding derivatives. Alkylated bases are also prone to spontaneous depurination (resulting in abasic sites) and secondary damage (including strand breaks and crosslinks). *Graphic design credit: Erin E. Dimino, MS (Henry C. Lee College of Criminal Justice and Forensic Sciences).*

TABLE 2.1 Summary of principal sources of DNA damage (left) and major types of DNA lesions (right) in forensic, historical, and archaeological biological samples.

Sources of DNA damage	Types of DNA lesions
Inherent instability (aqueous cellular environment)	Abasic (AP) sites
Postmortem nuclease activity (endogenous DNases)	Single-strand breaks (SSBs)
Excessive heat (high temperatures)	Double-strand breaks (DSBs)
Excessive ambient humidity	Interstrand DNA crosslinks
Acidic pH levels (pH < 7.0)	Inrastrand DNA crosslinks
Exposure to UV radiation (UV-A, UV-B)	DNA-protein crosslinks
Geochemical properties of soil (humic/fulvic acids)	Deaminated nucleobases
Microorganism digestion (bacteria, fungi)	Oxidized nucleobases
Environmental toxins and chemicals	Alkylated nucleobases

problematic because they are prone to spontaneous depurination and hydrolysis, and secondary damage (e.g., strand breaks, crosslinks) often accompanies the presence of alkylation adducts (Geacintov and Broyde, 2010; Sedgwick, 2004).

Ultimately, the extensive spectrum of DNA damage and the nearly limitless combinations of lesions that can be present in any particular sample pose a unique challenge for forensic analyses. Table 2.1 provides a summary of the principal causes/sources of DNA damage and a synopsis of the major types of lesions that occur in forensic, historical, and archaeological samples.

Inhibitors

In addition to the challenges associated with molecular DNA damage, compounds that inhibit PCR amplification can co-extract with DNA and present further complications for analysis. Since the majority of modern DNA typing technologies involve real-time PCR quantification and PCR-based amplification of target loci, substances which can interfere with these processes must be considered and effectively mitigated (or completely eliminated). There are a variety of different PCR inhibitors that will be encountered during forensic genetic investigations of decomposed and/or skeletonized human remains. The consequences of their presence include partial or total PCR inhibition (specifically decreased sensitivity or false-negative results, respectively).

PCR inhibitors can affect quantification and amplification of DNA directly or indirectly via: (1) competitive binding to DNA polymerases (preventing the polymerase from binding to template DNA); (2) competitive binding of polymerase cofactors (rendering them unavailable to the polymerase); (3) molecular interactions with the DNA molecule and/or oligonucleotide primers; (4) degradation of DNA polymerases during PCR; or (5) quenching of fluorescence (Sidstedt et al., 2018; Schrader et al., 2012; Opel et al., 2010; Eilert and Foran, 2009). All of these mechanisms slow or completely prevent polymerase activity and

TABLE 2.2 Summary of PCR inhibitors commonly encountered in forensic genetic investigations of decomposed or skeletonized human remains. Inhibitors can be endogenous to the tissue or bone sample, exogenous (environmentally-derived), or introduced during routine laboratory processing for DNA.

PCR inhibitor	Origin	Source
Heme (hemoglobin)	Endogenous	Blood, bone marrow
Melanin	Endogenous	Hair, epithelial tissue (skin)
Calcium (Ca^{2+})	Endogenous	Mineral component of bone
Collagen	Endogenous	Protein component of bone
Humic acids	Exogenous	Soil, groundwater
Fulvic acids	Exogenous	Soil, groundwater
Calcium (Ca^{2+})	Exogenous	Soil, groundwater
Tannic acids (tannins)	Exogenous	Plants, soil, groundwater, wood coffins
Vivianite [$Fe_3(PO_4)_2 \cdot 8\,H_2O$]	Exogenous	Specific burial conditions
Metal ions (Fe, Cu, Al, Pb)	Exogenous	Caskets, burial vaults, soil
Phenol chloroform	Exogenous	Laboratory (DNA extraction)
Ethanol (EtOH)	Exogenous	Laboratory (DNA extraction)
Ionic detergenets (SDS, sarkosyl)	Exogenous	Laboratory (DNA extraction)
Protease enzymes (proteinase K)	Exogenous	Laboratory (DNA extraction)
Ethylenediaminetetraacetic acid (EDTA)	Exogenous	laboratory (DNA extraction)
Chaotropic salts (guanidinium)	Exogenous	Laboratory (DNA extraction)
Buffer salts (NaCl, KCl)	Exogenous	Laboratory (DNA extraction)

processivity (depending on concentrations present), and can result in failure to detect DNA and/or to obtain genetic data even in the presence of a sufficient quantity of DNA. Relevant inhibitors fall into one of three categories: (1) endogenous inhibitors derived from the actual tissue or skeletal element being interrogated; (2) exogenous inhibitors which originate in the physical environment from which the remains were recovered; or (3) exogenous inhibitors introduced during sample processing and/or DNA extraction (Table 2.2). With skeletal remains, it is probable that the analyst must deal with the presence of a combination of different types of inhibitors. As mentioned previously in this chapter, these substances must be effectively removed and purified away from DNA in order to maximize genotyping and sequencing success. The DNA isolation method chosen, as well as the scientist's attention to detail and micropipetting technique, will be instrumental in obtaining high purity DNA for downstream analyses.

Endogenous inhibitors

PCR inhibitors are contained within a variety of biological specimens, including human organs, tissues, body fluids, bones, and teeth. For decomposed human remains with intact soft tissue, the primary endogenous inhibitory substances that must be purified away from

DNA include heme [from the hemoglobin component of erythrocytes (red blood cells)] and melanin [present in epithelial tissues (skin) and hair] (Sidstedt et al., 2018; Schrader et al., 2012; Opel et al., 2010; Eckhart et al., 2000; Abu Al-Soud and Radstrom, 1998; Wilson, 1997; Akane, 1996; Akane et al., 1994).

With skeletonized remains, the primary endogenous inhibitors encountered are calcium (a mineral) and collagen (a protein). Bone is a connective tissue with an extracellular matrix composed of organic collagenous proteins (predominantly Type I collagen) and an inorganic mineralized matrix [calcium hydroxyapatite, $Ca_{10}(PO_4)_6(OH)_2$] (Florencio-Silva et al., 2015; Tzaphlidou, 2008; Collins et al., 2002). The negatively charged backbone of DNA interacts with positively charged calcium residues within bone and, in order to effectively isolate DNA, this interaction must be broken. Most successful DNA extraction methods for skeletal remains involve a demineralization (i.e., decalcification) step to separate DNA from calcium residues. However, if calcium is carried over into the final elution, its presence can inhibit downstream PCR-based testing (Sidstedt et al., 2018, 2019; Opel et al., 2010; Loreille et al., 2007).

Exogenous (environmental) inhibitors

In addition to endogenous inhibitors in bone (i.e., collagen, calcium), exogenous components of soil further complicate forensic genetic testing. Humic substances (e.g., humic and fulvic acids) are naturally occurring organic acids present in soils throughout the world (at varying concentrations depending on geographical location), and both are known PCR inhibitors (Bleam, 2017; Monroe et al., 2013; Matheson et al., 2010). Humic substances are relatively resistant to both chemical degradation and biological degradation and, on average, comprise 5–8 milligrams (mg) per gram (g) of soil (depending on soil type). Humic acids (HAs) are a major portion of humic substances in soil, and are characterized as dark-colored, alkali-soluble, acid-insoluble, and high-molecular-weight organic matter. Fulvic acids are light brown, water-soluble compounds (Matheson et al., 2010; Piccolo, 2002; Watson and Blackwell, 2000; Schnitzer, 1991). In addition to humic acids and fulvic acids, tannins are another common PCR inhibitor associated with buried remains. Tannins (tannic acids) are complex chemical substances contained within many plants, including coniferous trees and many flowering plant species. Tannins leach out of plants into the surrounding soil and groundwater (and can even drain into surrounding lakes and rivers) (Sidstedt et al., 2019; Monroe et al., 2013; Matheson et al., 2010; Eilert and Foran, 2009). Tannins also can be introduced when human remains are buried in wooden (pine) coffins, which are used frequently in Potter's fields (paupers' graves), historical burials, and in accordance with some contemporary religious and cultural practices. The structure of a pinewood coffin will decompose over time due to moisture, acidity, and microbial activity, leaching tannins into the soil surrounding the remains. Ultimately, humic acids, fulvic acids, and tannins are present in all climates and all regions of the world, and since most recovered or exhumed skeletal remains have been buried in soil (or at least have superficially interacted with soil over time, as may be the case with surface deposition or "surface scatters"), these acids are pervasive in bone samples presented for DNA testing.

Other exogenous inhibitors that may be encountered in skeletal remains casework include metal ions, such as iron (Fe), copper (Cu), aluminum (Al), and lead (Pb). Metal ions may already be present in the soil in which the remains are buried, or can be derived from caskets, burial vaults, or other metal artifacts contained within the grave (Pokines and Baker, 2014).

The gradual breakdown or rusting of a casket or burial vault can expose remains and allow infiltration of sediment/soil. As diagenesis progresses and the bone matrix becomes more porous, metal ions from the soil and/or from decomposition of the casket or vault can diffuse into skeletal elements. The presence of metal ions on the surface of bones/teeth and/or within porous bone microstructure must be dealt with during initial processing and DNA extraction procedures in order to avoid inhibition during downstream analyses (Combs et al., 2015; Schrader et al., 2012).

Aside from soil acids and metal ions, another inhibitor derived from the burial environment is a substance called vivianite. Vivianite [$Fe_3(PO_4)_2 \cdot 8H_2O$] is a common and complex product of the interaction of phosphates, iron, and water in the natural environment. It is found worldwide and forms in sedimentary environments, particularly in aquatic ecosystems (in freshwater and marine sediments) and in waterlogged terrestrial soil or bogs. Vivianite formation requires very specific conditions: (1) presence of a phosphate source (e.g., human bones, decomposition products); (2) a low-oxygen (anaerobic), acidic environment (e.g., acidic soil or groundwater, deep water, waterlogged soil); and (3) an iron (Fe) source (e.g., iron-rich soil, manufactured objects such as casket hardware or weapons). All three of these conditions must exist in concert in order for vivianite crystals to manifest (Rothe et al., 2016; Ern et al., 2010; McGowan and Prangnell, 2006; Cox and Bell, 1999; Kvaal and During, 1999; Mann et al., 1998).

Vivianite forms on the surfaces of buried human skeletal remains. In the absence of oxygen (i.e., in its unoxidized form), vivianite mineral crystals are colorless, white, or light gray in color. However, as vivianite oxidizes upon exposure to air (e.g., during an excavation or exhumation), the crystals change to a vibrant blue color (Cronyn, 1990; Guthrie, 1990). Although most of the scientific literature about vivianite is in reference to archaeological contexts, this blue staining has been observed on skeletal remains of missing American soldiers from the Vietnam War recovered from mass graves (Mann et al., 1998) and on other human remains buried for only 15–20 years (Dupras and Schultz, 2014). This indicates that vivianite formation (and its associated inhibitory effects) should be considered a possibility in modern forensic cases involving remains with much shorter postmortem intervals (PMIs) than would be observed in an archaeological setting.

Exogenous (laboratory-introduced) inhibitors

Besides the endogenous inhibitors inherent in human body tissues and bone, as well as environmental inhibitors derived from soil or burial conditions, additional PCR inhibitors are introduced in the laboratory during the DNA extraction process itself. Among these inhibitors include ionic detergents [e.g., sodium dodecyl sulfate (SDS), sodium lauroyl sarcosinate (sarkosyl)], phenol chloroform isoamyl alcohol (PCIA), proteases (e.g., proteinase K), buffer salts [e.g., sodium chloride (NaCl), potassium chloride (KCl)], chaotropic salts [e.g., guanidinium isothiocyanate (GITC)], ethylenediaminetetraacetic acid (EDTA), and ethanol (EtOH). Although these substances are required components for effective isolation of genetic material (e.g., for the purposes of membrane lysing, protein digestion, stabilization, or precipitation), they must be effectively removed and separated from DNA during the final elution step to promote and maximize successful downstream testing (including DNA quantification and PCR amplification of target loci) (Schrader et al., 2012; Demeke and Jenkins, 2010; Opel et al., 2010).

Low copy number (LCN) DNA

The term low copy number (LCN) typically describes samples that contain less than 100 picograms (pg) of DNA. Decomposed and skeletonized human remains often present LCN templates for genetic analyses. A number of challenges exist when trying to obtain DNA profiles from such samples. Traditional strategies used to address such challenges [e.g., increased number of PCR cycles, increased capillary electrophoresis (CE) injection time, reduced volume PCR, post-PCR sample desalting] each bring their own set of advantages and limitations. For example, increasing the number of PCR cycles enables short tandem repeat (STR) typing to routinely obtain results from LCN samples, but this approach also increases the risk of allele drop-in due to amplification of contaminant or exogenous DNA (Balding and Buckelton, 2009; Caddy et al., 2008; Gill, 2001; Gill et al., 2000).

When testing LCN samples, three types of artifacts typically arise: (1) allele drop-in (due to sporadic contamination from exogenous sources); (2) allele dropout (which commonly occurs due simply to stochastic sampling during PCR); and (3) increased stutter product formation (often higher than the typical 5%–10% of the nominal allele) (Balding and Buckelton, 2009; Caddy et al., 2008; Gill, 2001; Gill et al., 2000). Heterozygote peak imbalance and allele dropout also are often exacerbated due to stochastic variation and/or preferential amplification of one of the alleles during the early rounds of PCR. The presence of these artifacts can make interpretation of STR profiles from LCN samples difficult and unreliable (Budowle et al., 2009; Gill, 2001) and, aside from the inherent limitations of LCN templates, forensic and ancient DNA samples often suffer from the dual problem of low quantity and low quality (which further complicates analyses).

Summary

The robustness and reliability of forensic genetic analyses are directly correlated to the quality, quantity, and purity of DNA available for testing. Samples containing degraded, inhibited, and/or low copy number (LCN) templates can be particularly problematic for forensic, historical, and archaeological investigations.

Decomposed or skeletonized human remains often have been exposed to destructive taphonomic conditions and extensive environmental (or intentional) insults prior to being recovered and submitted for genetic analyses. Consequently, the DNA encompassed within these forensic samples frequently contains multiple complex lesions and may be highly fragmented. Ultimately, DNA recovered from decomposed or skeletonized remains is typically characterized by: (1) the presence of lesions which block replication (and therefore analysis) of target loci; (2) modified lesions which cause incorrect nucleotides to be incorporated during replication; and/or (3) a reduction in fragment sizes available for interrogation. In addition to molecular degradation, a number of both endogenous and exogenous PCR inhibitors can affect replication if not sufficiently purified from the DNA during processing, and recovered DNA quantity may be low. An understanding and awareness of these complex issues can serve as valuable information for strategic planning of initial and downstream testing approaches.

Glossary

Abasic (AP) site site on a DNA strand "without a base"; caused by hydrolysis (cleavage) of the glycosidic bond between a sugar moiety and its corresponding nitrogenous base
Alkylation attachment of methyl- or other alkyl groups to the N- and O-atoms of DNA bases
Apurinic (AP) site section of DNA missing a purine (adenine, guanine)
Apyrimidinic (AP) site section of DNA missing a pyrimidine (cytosine, thymine)
Calcium hydroxyapatite mineral (inorganic) component of bone; endogenous PCR inhibitor
Chromatin coiled complexes of DNA and protein in the nucleus of human cells; purpose is to facilitate tight packaging of long DNA molecules into compact, dense conformation
Collagen protein (organic) component of bone; endogenous PCR inhibitor
Deaminated base modified base, via removal of an amino group (–NH_2)
Deoxyribose (dR) the 5-carbon sugar moiety of the DNA molecule
Depurination loss of a purine base (adenine or guanine)
Depyrimidination loss of a pyrimidine base (cytosine, thymine)
Deoxyribonucleic acid (DNA) polymer of nucleotides; genetic material in cells
DNase nuclease that degrades/digests DNA molecules
Glycosidic bond bond between the sugar and base in a nucleotide
Humic acid exogenous (environmental) inhibitor in soil
Hydrolysis breakage of chemical bonds via reaction with water
Interstrand crosslink crosslink between bases on the opposing DNA strands
Intrastrand crosslink crosslink between bases on the same DNA strand
Linker DNA short segments of DNA that connect nucleosomes in chromatin
Low copy number (LCN) low quantity DNA, defined as $\leq$100 picograms (pg)
Monomer basic individual subunit of a polymer
Nuclease enzyme that degrades nucleic acids (DNA, RNA)
Nucleosome basic repeating structural unit of chromatin; contains an octamer of histone proteins and approximately 146 base pairs (bp) of DNA
Nucleotide basic monomeric subunit of the DNA polymer; consists of a phosphate group, a 5-carbon sugar (deoxyribose), and one of four nitrogenous bases (adenine, guanine, cytosine, or thymine)
Oxidation a type of DNA base modification (damage) involving formation of saturated pyrimidine rings and loss of the double bond between carbons 5 and 6
Phosphodiester bond bond that connects nucleotides in DNA backbone; a covalent linkage between the phosphate of one nucleotide and the hydroxyl (–OH) group attached to the 3′ carbon of the deoxyribose sugar in an adjacent nucleotide
Polymer large molecule consisting of similar monomeric subunits, e.g., DNA is a polymer of nucleotide (dNTP) monomers
Purines single-ringed bases in DNA (guanine, adenine)
Pyrimidines double-ringed bases in DNA (cytosine, thymine)
Single-strand break (SSB) a lesion caused by hydrolysis of the phosphodiester bond in one DNA strand; often referred to as a "nick"
Tannin (tannic acid) acid in plants that can leech into soil, groundwater, and surrounding lakes or rivers; exogenous PCR inhibitor
Double-strand break (DSB) lesion caused by hydrolysis phosphodiester bonds on adjacent strands of DNA, resulting in fragmentation
Vivianite exogenous PCR inhibitor derived from the burial environment; forms on bones/teeth under very specific conditions and manifests as vivid blue crystals when exposed to air during excavation or exhumation

Acknowledgments and graphic design credits

I acknowledge several undergraduate and graduate students in my "*Advanced Forensic DNA Applications for Missing Persons, Unidentified Human Remains, and Mass Death Investigations*" course for their valuable contributions to the graphics used in this chapter. These students are the next generation of forensic DNA scientists, and I am grateful to have been part of their journey. Design credits for each figure illustrating the various types of DNA lesions are detailed below:

Fig. 2.2	Depurination vs. depyrimidination	Morgan Barrett, M.S.
Fig. 2.5	Interstrand vs. intrastrand crosslinks	Lily Kate Josephs, M.S.
Fig. 2.6A	Pyrimidine (thymine) dimer	Ryan Durkee
Fig. 2.7A	Deamination (pyrimidines)	Erin E. Dimino, M.S.
Fig. 2.7B	Deamination (purines)	Erin E. Dimino, M.S.
Fig. 2.9	Alkylation (alkylated bases)	Erin E. Dimino, M.S.

References

Abu Al-Soud, W., Radstrom, P., 1998. Capacity of nine thermostable DNA polymerases to mediate DNA amplification in the presence of PCR-inhibiting samples. Appl. Environ. Microbiol. 64, 3748–3753.

Akane, A., 1996. Hydrogen peroxide decomposes the heme compound in forensic specimens and improves the efficiency of PCR. Biotechniques 21, 392–394.

Akane, A., Matsubara, K., Nakamura, H., Takahashi, S., Kimura, K., 1994. Identification of the heme compound copurified with deoxyribonucleic acid (DNA) from bloodstains, a major inhibitor of polymerase chain reaction (PCR) amplification. J. Forensic Sci. 39 (2), 362–372.

Annunziato, A., 2008. DNA packaging: nucleosomes and chromatin. Nature Educ. 1 (1), 26.

Balding, D.J., Buckelton, J., 2009. Interpreting low template DNA profiles. Forensic Sci. Int. Genet. 4, 1–10.

Bleam, W., 2017. Soil and Environmental Chemistry, second ed. Academic Press.

Brieba, L.G., Eichman, B.F., Kokoska, R.J., Doublie, S., Kunkel, T.A., Ellenberger, T., 2004. Structural basis for the dual coding potential of 8-oxoguanosine by a high fidelity DNA polymerase. EMBO J. 23, 3452–3461.

Budowle, B., Eisenberg, A.J., van Daal, A., 2009. Validity of low copy number typing and applications to forensic science. Croat. Med. J. 50, 207–217.

Buettner, G.R., 1993. The pecking order of free radicals and antioxidants: lipid peroxidation, α-tocopherol, and ascorbate. Arch. Biochem. Biophys. 300, 535–543.

Caddy, B., Taylor, G.R., Linacre, A.M.T., 2008. A Review of the Science of Low Template DNA Analysis. UK Home Office Report.

Cadet, J., Sage, E., Douki, T., 2005. Ultraviolet radiation-mediated damage to cellular DNA. Mutat. Res. 571, 3–17.

Cohen, J., 2016. Soil fertility: Global map of soil pH: How soil pH changes between wet and dry climates. Science Daily, 8 December 2016. <www.sciencedaily.com/releases/2016/12/161208143456.htm>.

Collins, M.J., Nielsen-Marsh, C.M., Hiller, J., Smith, C.I., Roberts, J.P., 2002. The survival of organic matter in bone: a review. Archaeometry 44 (3), 383–394.

Combs, L.G., Warren, J.E., Huynh, V., Castenada, J., Golden, T.D., Rhoby, R.K., 2015. The effects of metal ion PCR inhibitors on results obtained with the Quantifiler® Human DNA Quantification Kit. Forensic Sci. Int. Genet. 19, 180–189.

Cooke, M.S., Evans, M.D., Dizdaroglu, M., Lunec, J., 2003. Oxidative DNA damage: Mechanisms, mutation, and disease. FASEB J. 17, 1195–1214.

Cox, M., Bell, L., 1999. Recovery of human skeletal elements from a recent UK murder inquiry: Preservational signatures. J. Forensic Sci. 44 (5), 945–950.

Cronyn, J.M., 1990. The Elements of Archeological Conservation. Routledge, London, UK.

Dabney, J., Meyer, M., Paabo, S., 2013. Ancient DNA damage. Cold Spring Harb. Perspect. Biol. 5, a012567.

De Grujil, F., Rebel, H., 2008. Early events in UV carcinogenesis: DNA damage, target cells, and mutant p53 foci. J. Photochem. Photobiol. 84, 382–387.

Demeke, T., Jenkins, G.R., 2010. Influence of DNA extraction methods, PCR inhibitors, and quantification methods on real-time PCR assay of biotechnology-derived traits. Anal. Bioanal. Chem. 396, 1977–1990.

Dupras, T.L., Schultz, J.J., 2014. Taphonomic bone staining and color changes in forensic contexts. In: Pokines, J.T., Symes, S.A. (Eds.), Manual of Forensic Taphonomy. CRC Press, Taylor and Francis, Boca Raton, FL.

Eckhart, L., Bach, J., Ban, J., Tschachler, E., 2000. Melanin binds reversibly to thermostable DNA polymerase and inhibits its activity. Biochem. Biophys. Res. Commun. 271, 726–730.

Eilert, K.D., Foran, D.R., 2009. Polymerase resistance to polymerase chain reaction inhibitors in bone. J. Forensic Sci. 54 (5), 1001–1007.

Einolf, H.J., Guengerich, F.P., 2001. Fidelity of nucleotide insertion at 8-oxo-7,8-dihydroguanine by mammalian DNA polymerase δ: Steady-state and pre-steady-state kinetic analysis. J. Biol. Chem. 276, 3764–3771.

Eoff, R.L., Irimia, A., Angel, K.C., Egli, M., Guengerich, F.P., 2007. Hydrogen bonding of 7,8-dihydro-8-oxoguanosine with a charged residue in the little finger domain determines miscoding events in Sulfolobus solfataricus DNA polymerase Dpo4. J. Biol. Chem. 282, 19831–19843.

Ern, S.I.E., Trombino, L., Cattaneo, C., 2010. Micromorphological aspects of forensic geopedology: Can vivianite be a marker of human remains permanence in soil? Geophysical Res. Abs. 12. EGU2010-4423.

Florencio-Silva, R., da Silve Sasso, G.R., Sasso-Cerri, E., Simoes, M.J., Cerri, P.S., 2015. Biology of bone tissue: Structure, function, and factors that influence bone cells. Biomed. Res. Int. Article ID 421746.

Freisinger, E., Grollman, A.P., Miller, H., Kisker, C., 2004. Lesion intolerance reveals insights into DNA replication fidelity. EMBO J. 23, 1494–1505.

Friedberg, E.C., Walker, G.C., Siede, W., Wood, R.D., Schultz, R.A., Ellenberger, T., 2006. DNA Repair and Mutagenesis. ASM Press, Washington, DC.

Furge, L.L., Guengerich, F.P., 1997. Analysis of nucleotide insertion and extension at 8-oxo-7,8-dihydroguanine by replicative T7 polymerase *exo-* and human immunodeficiency virus-1 reverse transcriptase using steady-state and pre-steady-state kinetics. Biochemistry 36, 6475–6487.

Furge, L.L., Guengerich, F.P., 1998. Pre-steady-state kinetics of nucleotide insertion following 8-oxo-7,8-dihydroguanine base pair mismatches by bacteriophage T7 DNA polymerase exo. Biochemistry 37, 3567–3574.

Gates, K.S., 2009. An overview of chemical processes that damage cellular DNA: Spontaneous hydrolysis, alkylation, and reactions with radicals. Chem. Res. Toxicol. 22, 1747–1760.

Geacintov, N.E., Broyde, S., 2010. The Chemical Biology of DNA Damage. Wiley-VCH Publishing, Weinheim, Germany.

Gill, P., 2001. Application of low copy number DNA profiling. Croat. Med. J. 42, 229–232.

Gill, P., Whitaker, J., Brown, N., Buckleton, J., 2000. An investigation of the rigor of interpretation rules from STRs derived from less than 100pg of DNA. Forensic Sci. Int. 112, 17–40.

Golenberg, E.M., Bickel, A., Weihs, P., 1996. Effect of highly fragmented DNA on PCR. Nucleic Acids Res. 24, 5026–5033.

Guthrie, R.D., 1990. Frozen Fauna of the Mammoth Steppe: The Story of Blue Babe. University of Chicago Press, Chicago, IL.

Hall, A., Ballantyne, J., 2004. Characterization of UVC-induced DNA damage in bloodstains: Forensic implications. Anal. Bioanal. Chem. 380, 72–83.

Haracska, L., Yu, S.L., Johnson, R.E., Prakash, L., Prakash, S., 2000. Efficient and accurate replication in the presence of 7,8-dihydro-8-oxoguanine by DNA polymerase η. Nat. Genet. 25, 458–461.

Haracska, L., Prakash, S., Prakash, L., 2003. Yeast DNA polymerase ζ is an efficient extender of primer ends opposite from 7,8-dihydro-8-oxoguanine and O6-methylguanine. Mol. Cell. Biol. 23, 1453–1459.

Hsu, G.W., Ober, M., Carell, T., Beese, L.S., 2004. Error-prone replication of oxidatively damaged DNA by a high fidelity DNA polymerase. Nature 431, 217–221.

Jun, S., 2010. DNA chemical damage and its detection. Int. J. Chem. 2, 261–265.

Kawane, K., Motani, K., Nagata, S., 2014. DNA degradation and its defects. Cold Spring Harb. Perspect. Biol. 6 (6), a016394.

Kvaal, S.I., During, E.M., 1999. A dental study comparing age estimations of the human remains from the Swedish warship *Vasa*. Int. J. Osteoarchaeol. 9, 170–181.

Lindahl, T., 1993. Instability and decay of the primary structure of DNA. Nature 362, 709–715.

Lindahl, T., Nyberg, B., 1972. Rate of depurination of native deoxyribonucleic acid. Biochem. 11, 3610–3618.

Loreille, O.M., Diegoli, T.M., Irwin, J.A., Coble, M.D., Parsons, T.J., 2007. High efficiency DNA extraction from bone by total demineralization. Forensic Sci. Int. Genet. 1, 191–195.

Mann, R.W., Feather, M.E., Tumosa, C.S., Holland, T.D., Schneider, K.N., 1998. A blue encrustation found on skeletal remains of Americans missing in action in Vietnam. Forensic Sci. Int. 97, 79–86.

Margolin, Y., Cloutier, J.F., Shafirovich, V., Geacintov, N.E., Dedon, P.C., 2006. Paradoxical hotspots for guanine oxidation by a chemical mediator of inflammation. Nat. Chem. Biol. 3, 365–366.

Matheson, C.D., Gurney, C., Esau, N., Lehto, R., 2010. Assessing PCR inhibition from humic substances. Open Enzyme Inhib. J. 3, 38–45.

McAuley-Hecht, K.E., Leonard, G.A., Gibson, N.J., Thomson, J.B., Watson, W.P., Hunter, W.N., Brown, T., 1994. Crystal structure of a DNA duplex containing 8-hydroxydeoxyguanine-adenine base pairs. Biochemistry 33, 10266–10270.

McGowan, G., Prangnell, J., 2006. The significance of vivianite in archeological settings. Geoarchaeol.: Int. J. 21 (1), 93–111.

Monroe, C., Grier, C., Kemp, B.M., 2013. Evaluating the efficacy of various thermo-stable polymerases against co-extracted PCR inhibitors in ancient samples. Forensic Sci. Int. 228, 142–153.

Nagata, S., 1997. Apoptosis by death factor. Cell 88, 355–365.

Nagata, S., 2005. DNA degradation in development and programmed cell death. Annu. Rev. Immunol. 23, 853–875.

Nelson, J., 2009. Repair of Damaged DNA for Forensic Analysis. National Institute of Justice.

Onori, N., Onofri, V., Alessandrini, F., Buscemi, L., Pesaresi, M., Turchi, C., Tagliabracci, A., 2006. Post-mortem DNA damage: a comparative study of STRs and SNPs typing efficiency in simulated forensic samples. Int. Congr. Ser. 1288, 510–512.

Opel, K.L., Chung, D., McCord, B.R., 2010. A study of PCR inhibition mechanisms using real-time PCR. J. Forensic Sci. 55 (1), 25–33.

Piccolo, A., 2002. The supramolecular structure of humic substances: a novel understanding of humus chemistry and implications in soil science. In: Advances in Agronomy. Academic Press, pp. 57–134.

Pokines, J.T., Baker, J.E., 2014. Effects of burial environment on osseous remains. In: Pokines, J.T., Symes, S.A. (Eds.), Manual of Forensic Taphonomy. CRC Press, Taylor and Francis, Boca Raton, FL.

Prangnell, J., McGowan, G., 2009. Soil temperature calculation for burial site analysis. Forensic Sci. Int. 191, 104–109.

Rechkoblit, O., Malinina, L., Cheng, Y., Kuryavyi, V., Broyde, S., Geacintov, N.E., Patel, D.J., 2006. Stepwise translocation of Dpo4 polymerase during error-free bypass of an oxoG lesion. PLoS Biol. 4, e11.

Rothe, M., Kleeberg, A., Hupfer, M., 2016. The occurrence, identification, and relevance of vivianite in waterlogged soils and aquatic sediments. Earth Sci. Rev. 158, 51–64.

Sedgwick, B., 2004. Repairing DNA-methylation damage. Nat. Rev. Mol. Cell Biol. 5, 148–157.

Schnitzer, M., 1991. Soil organic matter—the next 75years. Soil Sci. 151, 41–58.

Schrader, C., Schielke, A., Ellerbroek, L., Johne, R., 2012. PCR inhibitors—occurrence, properties, and removal. J. Appl. Microbiol. 113, 1014–1026.

Shafirovich, V., Geacintov, N.E., 2010. Role of free radical reactions in the formation of DNA damage. In: The Chemical Biology of DNA Damage. Wiley-VCH Publishing, Weinheim, Germany, pp. 81–104.

Shafirovich, V., Dourandin, A., Huang, W., Geacintov, N.E., 2001. The carbonate radical is a site-selective oxidizing agent of guanine in double-stranded oligonucleotides. J. Biol. Chem. 276, 24621–24626.

Shibutani, S., Takeshita, M., Grollman, A.P., 1991. Insertion of specific bases during DNA synthesis past the oxidation-damaged base 8-oxodG. Nature 349, 431–434.

Sidstedt, M., Hedman, J., Romsos, E.L., Waitara, L., Wadso, L., Steffen, C.R., Vallone, P.M., Radstrom, P., 2018. Inhibition mechanisms of hemoglobin, immunoglobulin G, and whole blood in digital and real-time PCR. Anal. Bioanal. Chem. 410, 2569–2583.

Sidstedt, M., Steffen, C.R., Kiesler, K.M., Vallone, P.M., Radstrom, P., Hedman, J., 2019. The impact of common PCR inhibitors on forensic MPS analysis. Forensic Sci. Int. Genet. 40, 182–191.

Slessarev, E.W., Lin, Y., Bingham, N.L., Johnson, J.E., Dai, Y., Schimel, J.P., Chadwick, O.A., 2016. Water balance creates a threshold in soil pH at the global scale. Nature. https://doi.org/10.1038/nature20139.

Steenken, S., Jovanovic, S.V., 1997. How easily oxidizable is DNA? One-electron reduction potentials of adenosine and guanosine radicals in aqueous solution. J. Am. Chem. Soc. 119, 617–618.

Tuchinda, C., Srivannaboon, S., Lim, H.W., 2006. Photoprotection by window glass, automobile glass, and sunglasses. J. Am. Acad. Dermatol. 55, 845–854.

Tzaphlidou, M., 2008. Bone architecture: Collagen structure and calcium/phosphorus maps. J. Biol. Phys. 34 (1–2), 39–49.

Valko, M., Leibfritz, L.D., Moncol, J., Cronin, M.T., Mazur, M., Tesler, J., 2007. Free radicals and antioxidants in normal physiological functions and human disease. Int. J. Biochem. Cell Biol. 39, 44–84.

Watson, R.J., Blackwell, B., 2000. Purification and characterization of a common soil component which inhibits the polymerase chain reaction. Can. J. Microbiol. 46, 633–642.

Wilson, I.G., 1997. Inhibition and facilitation of nucleic acid amplification. Appl. Environ. Microbiol. 63, 3741–3751.

Wyllie, A.H., 1980. Glucocorticoid-induced thymocyte apoptosis is associated with endogenous endonuclease activity. Nature 284, 555–556.

Zang, H., Irimia, A., Choi, J.Y., Angel, K.C., Loukachevitch, L.V., Egli, M., Guengerich, F.P., 2006. Efficient and high fidelity incorporation of dCTP opposite 7,8-dihydro-8-oxodeoxyguanosine by Sulfalobus solfataricus DNA polymerase Dpo4. J. Biol. Chem. 281, 2358–2372.

SECTION I

Guidelines and best practices for handling and processing human skeletal remains for genetic studies

CHAPTER

3

Facilities design and workflow considerations for processing unidentified human skeletal remains

Odile Loreille PhD

Federal Bureau of Investigation (FBI) Laboratory, Quantico, VA, United States

Introduction

DNA-based identification of human skeletal remains began in the early 1990s with the identification of murder victims (Ginther et al., 1992; Hagelberg et al., 1991; Stoneking et al., 1991), victims of wars (Holland et al., 1993; King, 1991), and historical figures (Gill et al., 1994; Jeffreys et al., 1992). In 1991, the first high-throughput laboratory devoted to the identification of human skeletal remains opened in the United States. The Armed Forces DNA Identification Laboratory (AFDIL), a division of the Armed Forces Medical Examiner System, was established as the only Department of Defense (DoD) forensic DNA testing laboratory for the identification of human remains. Scientists working with human skeletal remains quickly realized that minuscule levels of contaminants could lead to erroneous results and that strict measures of quality control were needed. In July 1995, a DoD Science Board report on the "Use of DNA Technology for Identification of Ancient Remains" recognized the special complexities associated with the identification of war casualties from 50-plus years ago (DTIC ADA301521, 1995). The DoD quality assurance program for the mitochondrial DNA (mtDNA) identification of ancient remains listed seven recommendations for laboratory facilities, all of which are still valid today. The report recommended that the lab has:

1. A space dedicated to skeletal remains processing
2. Controlled and limited access to the laboratory
3. Separate pre-PCR and post-PCR areas
4. Separate areas for extraction and PCR set-up in the pre-PCR areas
5. Dedicated equipment for the amplification room and dedicated areas for PCR product

Forensic Genetic Approaches for Identification of Human Skeletal Remains
https://doi.org/10.1016/B978-0-12-815766-4.00003-0

6. Air handling systems appropriate for the prevention of sample contamination
7. A minimum of 4 linear feet of bench space per analyst during processing in both pre-PCR and post-PCR areas, or sufficient space to prevent sample contamination when more than one analyst is working.

During subsequent years, other large missing persons (MP) DNA laboratories opened. The International Commission on Missing Persons (ICMP) was established in 1996 to identify the ~40,000 victims of conflict in the former Yugoslavia in the mid-1990s (Parsons et al., 2019). Recently relocated to the Netherlands, the ICMP has been active in some 40 countries that have confronted large numbers of MP as a result of natural and man-made disasters, wars, widespread human rights abuses, organized crime, and other causes. Currently, the laboratory almost exclusively employs short tandem repeat (STR) typing for MP identification.

The first center for the identification of American civilians was established in 2002 at the University of North Texas (UNT). Texas was the first state in the United States with an MP DNA database capable of analyzing both mitochondrial and STR systems and was the first state to participate in the U.S. federal database for MP (the National Missing Persons DNA Database, powered by the Combined DNA Index System, or CODIS; Alvarez-Cubero et al., 2012). CODIS provides a powerful tool for investigators seeking to locate MP or identify remains by allowing federal, state, and local crime laboratories to electronically exchange and compare DNA profiles. The UNT Center for Human Identification (UNTCHI) Missing Persons Unit is part of an accredited laboratory specializing in DNA analysis of skeletal remains and other types of evidence for missing and unidentified persons cases.

Several other forensic DNA laboratories around the world have developed capabilities to analyze mitochondrial and/or nuclear DNA from human skeletal remains. In many instances, however, the MP lab is part of a crime lab and only one or two rooms are devoted to analyses of skeletal remains. Generally, a minority of the scientists are qualified to work on hard tissues in these laboratories. At the FBI Laboratory, for example, only 25% of the DNA Casework Unit biologists have been trained and are qualified to analyze human skeletal remains.

Today, no official guidance exists on how to set up a DNA laboratory for MP identification, and each active MP center operates somewhat differently. This chapter aims to cover aspects of facility design and workflows that should be considered when establishing a new MP laboratory, where limiting contamination should be the foremost priority. While funding constraints may prevent the creation of the ideal laboratory in many cases, the following recommendations should help scientists make decisions that are based on best practices.

Laboratory infrastructure

Laboratory layout

General considerations

Starting a molecular laboratory requires extensive planning, particularly regarding space and design. The infrastructure of the workspace should be divided into two separate zones: one containing office space and the other used for laboratory activities. Each zone should have its own ventilation system. On the laboratory side, each room should have adequate overall space and enough bench space for multiple scientists and to accommodate all the necessary instruments and equipment (e.g., refrigerators, freezers, hoods, robotics). Other necessary infrastructure requirements include:

- A fully integrated laboratory control system to control the temperature, ventilation rate, and room pressurization
- Fume hoods, sinks with deionized water, several electrical outlets, backup generators, and telephone lines
- Sealed and durable benches and floors. These should be easy to clean, resistant to chemicals such as strong acids and bleach, and should be able to withstand frequent cleaning
- Enough storage space to reduce clutter and make it easy to disinfect all surfaces
- Nonoperating windows
- An adequate number and placement of safety showers, eyewash units, and fire extinguishers for operations

For more detailed guidelines that include site design, general building design, security design, plumbing system recommendations, and other associated information, refer to the National Institute of Justice (NIJ) report "Forensic Laboratories: Handbook for Facility Planning, Design, Construction, and Moving" (NIJ, 1998) or the National Institute of Standards and Technology (NIST) report titled "Forensic Science Laboratories: Handbook for Facility Planning, Design, Construction, and Relocation"(NIST, 2013).

Additionally, it is recommended to:

- Favor round corners between floor and wall, and between wall and ceiling, to minimize particle retention (Worrilow, 2017)
- Favor laboratories that have windows between the lab and the hallway. Not only do windows allow guests to view work being undertaken in the laboratory (without entering and possibly contaminating the lab), they are also a safety feature for staff working alone in the laboratory
- Avoid wood and nonsealed light fixtures, as well as porous flooring

Separation of pre-PCR and post-PCR areas

Some of the greatest features of PCR are the method's high specificity, sensitivity, and ability to generate 10^9 to 10^{13} identical copies of a target sequence from a small amount of DNA. The drawback is that even very minor template contamination can create problems. In a standard PCR reaction producing 10^{12} copies of the starting template, a single aerosol droplet (10^{-6} μL) may contain up to 10^5 potential targets (Persing, 1991). The risk of contaminating gloves and/or pipettes is especially high when opening tubes containing PCR product. Contaminants present on gloves can easily be transferred to anything a scientist might touch in the laboratory. Although the biggest release of amplicon aerosols happens when PCR tubes are opened after the amplification step, aerosolization can also occur when tubes containing PCR product remain open for some time. As aerosolized PCR products are airborne, they may also contaminate areas that have not been touched, as well as a scientist's hair and clothes. Without strict precautions in place, PCR products present on an employee can easily be inadvertently carried into a clean room, potentially causing false positive results down the road.

The design of a MP laboratory should, thus, revolve around limiting PCR contamination. The best way to limit contamination is to divide the complete laboratory space into two major areas: (1) the pre-PCR area (or clean room) which should be devoted to the isolation of nucleic acids and the assembly of the reaction to amplify the samples, and (2) the post-PCR area (a.k.a. dirty lab), where DNA is amplified and PCR product is handled (Dieffenbach and Dveksler, 1993). The pre-PCR and post-PCR areas must be physically separated. Ideally, they should either be far from each other on the same building floor, or on different building floors. It is critical that the workflow between the two areas be **unidirectional from pre-PCR areas to post-PCR areas**, and never in reverse. No one should be permitted to enter a pre-PCR area if they have walked through a post-PCR area, unless they have showered, shampooed, and changed clothes. Materials used in a post-PCR lab (e.g., reagents, consumables, pens, notebooks, calculators, etc.) should never be brought back to a clean lab. Rigid adherence to this rule is essential.

Fig. 1 provides an example of laboratory layout with unidirectional workflow. It is acknowledged that this example layout requires large facilities that are not always available or financially feasible. A less expensive way to construct a MP laboratory would be to purchase modular clean rooms. Modular construction offers several advantages: it is fast, relatively cheap, flexible, and the materials selected for each system are designed with clean room process operation in mind (Traver and Wiker, 2017). Modular clean rooms are commercially available at any size and can be built in single or multi-room configurations. For example, the BioSafe tempered glass cleanroom from Terra Universal (Fullerton, CA) is an ISO 5 to 8 clean room with tempered glass, stainless steel frames, HEPA or ULPA filter units, and bi-swing doors with self-closing hinges to maintain room pressure. Many other American companies sell modular clean rooms, including Abtech (Santa Ana, CA), American Cleanroom Systems (Rancho Santa Margarita, CA), Atmos-tech Industries (Ocean, NJ), Belcher equipment, LLC (Mequon, WI), Careforde (Chicago, IL), Clean Air (Minneapolis, MN), Clean Air Technology, Inc. (Canton, MI), CleanRooms International (Grand Rapids, MI), Modular Cleanrooms (Denver, CO), Portafab (Chesterfield, MO), Starrco (Maryland Heights, MO), and Technical Air Products (Belmont, MI), to cite just a few. Another option is to design and construct your own miniature laboratory. For example, Matsvay et al. (2018) performed work on ancient skeletal remains in a homemade module that consisted of four glove boxes connected by antechambers and special ventilation.

Ventilation

Pre-PCR and post-PCR areas should have independent environmental control and should not use common ductwork for air conditioning. All ventilation systems should have a device that readily permits the user to monitor whether the total system and its essential components are functioning properly. Every air entry point must be fitted with a High-Efficiency Particulate Air (HEPA) filter or with an Ultra Low Particulate Air (ULPA) filter and, if possible, a prefilter that should be changed regularly. A HEPA filter is a filter designed to remove 99.97% of airborne particles measuring 0.3 μm or greater in diameter passing through it, while a ULPA filter removes at least 99.999% of dust, pollen, mold, bacteria, and any airborne particles with a size 0.1 μm or larger from the air.

Laboratory heating, ventilating, and air conditioning (HVAC) systems can be classified into two categories of air pressurization: positive and negative. Positive air pressure means that the room is "pumped up" with more filtered air than the surrounding outside space.

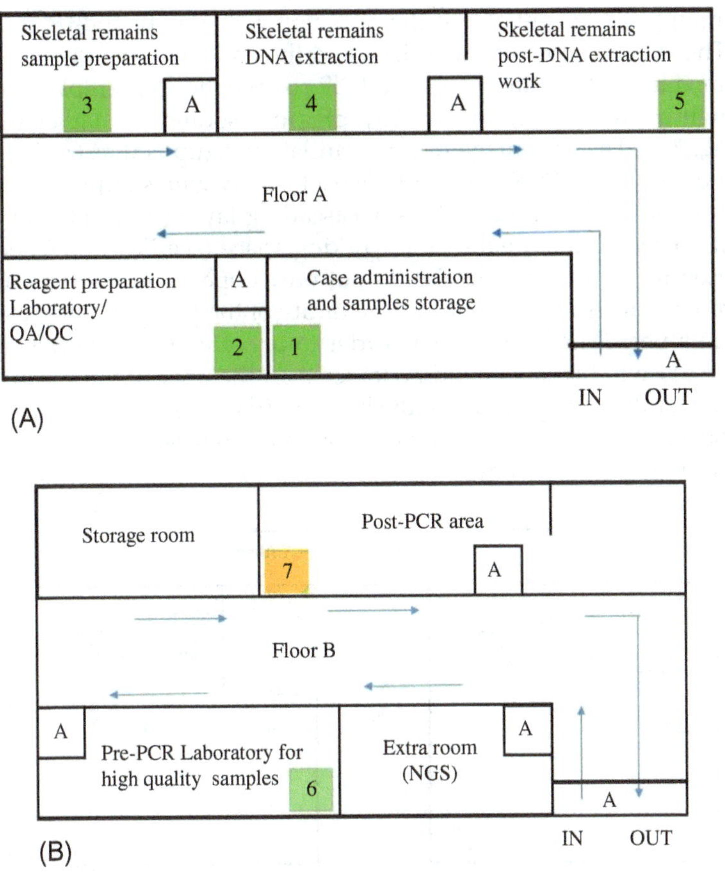

FIG. 1 An example of MP laboratory layout (A stands for Airlock). Clean rooms (pre-PCR areas) are numbered in *green squares*, while the post-PCR area (a.k.a. dirty lab) is numbered in an *orange square*. The flow must be unidirectional from room 1 to 7. Installing a pass-through window between rooms 3 and 4 is recommended to limit the transport of powdered samples through the hallway and the airlock. For laboratories using next-generation sequencing (NGS) technologies, an extra room to set up indexing PCRs would be useful.

Negative pressure, on the other hand, causes air to be removed at a higher rate than supplied so that the room is under negative pressure relative to surrounding rooms and corridors. Negative pressure prevents air from migrating from the laboratory into adjacent areas.

In a MP laboratory, pre-PCR rooms should be positively pressurized so that air does not flow in from the adjoining rooms, vestibule, or hallways when the door is opened; post-PCR rooms should be negatively pressurized to contain contaminants inside the room.

Airlocks and pressure

An airlock (or anteroom) is an intermediate chamber generally located between a laboratory and the hallway. It has controlled pressure and two airtight doors fitted with interlocks so that both doors cannot be opened at the same time.

In pre-PCR rooms, it is essential that the pressure in the lab be higher than the pressure in the airlock. The airflow between the airlock and the hallway differs according to the type of work performed in the laboratory. In a pre-PCR room handling samples that contain high quantities of high quality DNA, the hallway pressure should be higher than the airlock pressure (see Fig. 2A). The advantage of this ventilation setup is that the hallways are kept DNA-free at all times. In pre-PCR labs working exclusively with samples that contain highly degraded and/or very little (human) DNA, a cascading layout from the lab to the hallway will ensure that no contaminant enters the working space (see Fig. 2B). Choosing the right type of ventilation for your clean lab should not only depend on the type of samples processed in that room, but also on the relative proximity of the pre-PCR and post-PCR areas.

Conversely, in a post-PCR laboratory, in order to prevent any contaminant from leaving the room, the pressure in the lab should be lower than the pressure in the airlock, which in turn should be lower than the pressure in the hallway (Fig. 2C).

A pre-PCR lab airlock should be treated as another clean room and every effort should be made to limit the accumulation of dust:

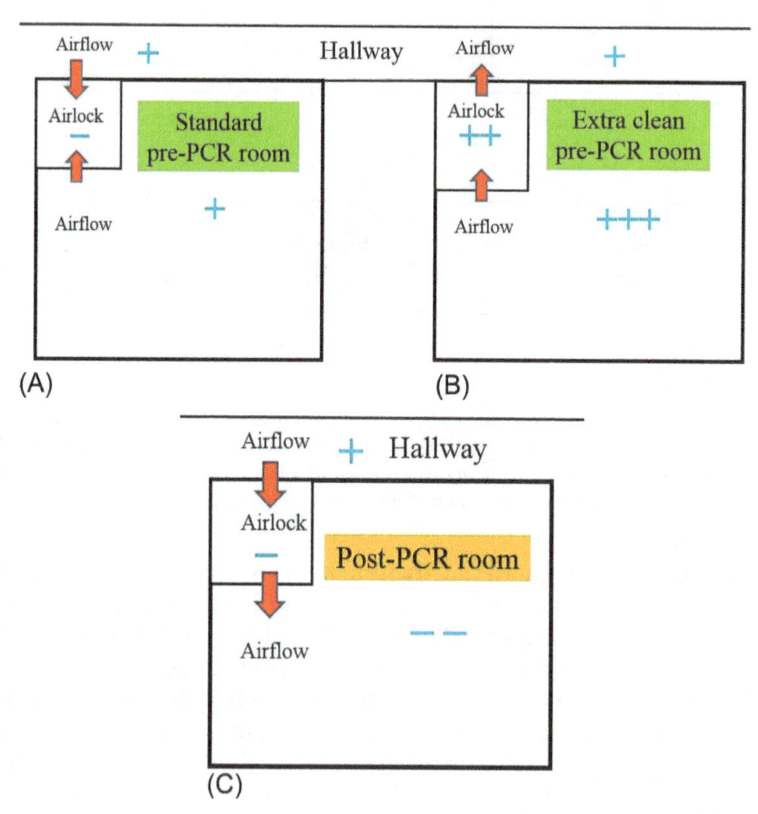

FIG. 2 Schematic of airlocks and pressure in an MP laboratory ("+++" and "– –" represent the highest and lowest pressure, respectively; direction of airflow is symbolized with *red arrows*): (A) standard layout for a pre-PCR laboratory, (B) the pressurization recommended in a very clean lab, such as a reagent preparation room or a laboratory dedicated to the analysis of degraded samples, and (C) the pressurization for a post-PCR laboratory and in any laboratory that handles hazardous material.

- Sticky floor mats (a.k.a. tacky mats) should be placed in front of each door to prevent dust or unwanted particles that may be stuck on shoes from entering the laboratory. The top adhesive plastic film layer should be periodically peeled off and discarded
- Avoid bringing cardboard boxes and packaging accessories such as foam peanuts into the airlock
- Immediately remove any dust ball found, as these generally contain (among other things) hairs, dandruff, and dead skin cells
- Since reducing clutter will lower dust accumulation, lockers or cabinets may be installed to store personal protective equipment (PPE)

Pre-PCR areas

A room where PCR product was previously handled should not be converted into a new clean room. A minimum of three (and ideally six) clean rooms are ideal for a MP laboratory. Some of these are rooms where no DNA/RNA in solution should ever be introduced.

DNA-free rooms

In an ideal layout, three areas should be considered DNA-free rooms: (1) the case administration room, (2) the reagent preparation room, and (3) the bone sampling room.

Case administration and sample storage room

This room is where samples are transported upon arrival at the laboratory, labeled, weighed, potentially photographed, and stored. Required equipment includes (but may not be limited to) refrigerators, freezers, at least one computer fitted with a Laboratory Information Management System (LIMS), a camera, a scale, and measuring devices. A LIMS is particularly important as it will allow the lab to collect all information pertaining to the samples, whether they are unidentified person samples or reference samples.

The first step includes the assignment of a unique case number for each sample by an evidence custodian. From this point on, all information about a sample will be tracked electronically within the system. The stored data shall include things like the provenance of the sample and the sample type (e.g., bone, buccal swab, soft tissue, reference material). Once entered into the LIMS, samples must be stored in a cool place (refrigerator or freezer). High quality samples should be stored separately from low quality samples.

Reagent preparation room (QA/QC area)

This is the cleanest possible laboratory space and should be used exclusively for reagent preparation. No sample material or work product should ever be brought into the room. Following accreditation guidelines, new vendor purchased kits or reagents will first be brought to this room and will be quality-controlled before being used for casework. This lab is also the place where homemade or custom reagents will be prepared and aliquoted.

All reagents should be aliquoted into volumes suitable for a small number of samples. When a stock solution/bottle is first opened, it should be labeled with the date, the scientist's name, and the expiration date. The reagent preparation scientist should make an aliquot for testing, rather than working directly from the stock. It may also be useful to assign one member of the staff the task of monitoring stock levels, ordering, and ensuring that reagents and consumables are stored appropriately (Viana and Wallis, 2011).

Sampling room

Hard tissues present a greater technical challenge than soft tissues, as they require mechanical disruption prior to DNA extraction. The sample preparation room should be devoted to cleaning/decontamination of hard tissues and pulverization of the samples for DNA extraction.

Molecular biology pre-PCR laboratories

Pre-PCR areas should be dedicated for molecular biology work. Ideally, one lab should be dedicated to handling high quality, high quantity DNA (HQHQ) samples and the other(s) to low quality/low quantity (LQLQ) samples.

Identifications of skeletonized remains are made by comparing a specimen's DNA profile to reference profiles. Direct reference DNA samples can be obtained either from personal items from the deceased (e.g., toothbrush, hairbrush, razors, baby teeth) or from banked biological samples (e.g., newborn screening cards, sperm, or biopsy tissues; Ward, 2015). If no direct reference is available, DNA of at least two of the victim's close relatives should be collected (Ward, 2015).

High quality/high quantity DNA (HQHQ) pre-PCR room

The HQHQ room should be dedicated to processing samples that contain large quantities of pristine DNA.

Buccal swabs may be collected from relatives of a MP, but could also originate from people who cannot self-identify, such as people with certain neurological disorders, babies, and young children. Since DNA typing of buccal swabs may be automated, this room should be large enough to accommodate robotics if high-throughput testing is expected.

Reference samples are not the only type of samples that should be processed in a HQHQ lab. For example, if an unidentified deceased person was found in an early stage of decomposition, blood and/or soft tissues (e.g., red muscle, brain, or heart tissues) may be available for DNA typing. These types of high DNA quantity samples should also be processed in a HQHQ lab. It is, however, essential not to process MP samples and their associated reference samples at the same time

Low quality/low quantity DNA (LQLQ) pre-PCR rooms

The LQLQ rooms are basically used to analyze samples that contain very small amounts of genetic material. The state of the DNA in these samples can vary from relatively high quality (e.g., reference samples like hairs, touch DNA on glasses) to highly degraded. The purity of the DNA can also vary from being a single-source sample to a sample contaminated with exogenous DNA. Skeletal remains are among the most challenging samples because they generally contain very small amounts of degraded human DNA mixed with very high amounts of nonhuman DNA originating mostly from the burial environment. If nuclear DNA is too degraded or too scarce to yield forensically informative STR profiles, mitochondrial DNA (mtDNA) testing is performed. The high copy number of the mitochondrial genome (mtGenome) per cell significantly improves the chances of recovering a profile. On the other hand, this high copy number of mtGenomes/cell makes it easy for people to contaminate their environment with mtDNA. Consequently, a LQLQ laboratory working on highly degraded human skeletal remains should treat their laboratory like a clean room or an ancient DNA laboratory. They should be positively pressurized, preferably have an airlock, and be restricted to a small number of employees wearing adequate PPE and who have been fully trained to work in such space.

Ideally, the LQLQ area should be divided into two distinct spaces and should have separate, dedicated supplies and equipment. The first area should be used exclusively for DNA extraction. Once extracted, the DNA should be moved to the second area where it will be used to set up amplifications or to prepare libraries.

Equipment and reagents

Dedicated personal protective equipment (PPE)

In the United States, the Occupational Safety and Health Administration (OSHA) requires that employers protect their employees from workplace hazards that can cause injury. Personal protective equipment, commonly referred to as "PPE," is equipment worn to minimize exposure to a variety of hazards. In a clean MP lab, PPE is also essential to protect the samples from the lab practitioner.

Lab coats/coveralls and sleeves

Lab coats are mandatory in every molecular biology laboratory. They should produce little or no particulate emission, i.e., the fabric or material should be stable and possess a high ability to resist breakdown.

In a post-PCR lab, the main function of lab coats is to protect employees from hazardous products, though they also protect street clothing from being heavily contaminated with amplicon aerosols. Lab coats used in these spaces are generally made out of cotton, or polyester mixed with cotton, and should at least cover a person from neck to knees. Post-PCR lab coats should be washed on a regular basis in a specialized cleaning facility.

Lab coats serve dual functions in a MP pre-PCR lab. They protect the employee from hazardous products and protect the samples from the employee's DNA. In a HQHQ pre-PCR lab, lab coats/frocks with elastic wrists and no pocket are convenient. Cotton lab coats can also be used, but should be washed on a regular basis in a specialized facility and should never be laundered with lab coats used in post-PCR areas.

In LQLQ rooms, such as a reagent preparation room or areas where highly degraded samples are handled, contamination risks increase as soon as a person enters the room. Each day, an individual loses 500,000 skin cells and 50–100 hairs, so extreme care must be taken to avoid spreading these sources of genetic material into the laboratory. The best protection is offered by a disposable hooded coverall that will be discarded at the end of the day. Suits made of Tyvek, such as Tyvek IsoClean single-use garments, offer efficient protection. Tyvek is a unique nonwoven material from DuPont (Wilmington, Delaware) made entirely of flash-spun continuous polyethylene fibers with no binders or fillers. Tyvek suits are generally recommended for ISO class 5–8 clean rooms. They are very lightweight and have a special antistatic coating applied during manufacturing.

The use of an unhooded coverall with a separate protective hood is probably as effective as wearing a hooded coverall as both combinations will prevent the spread of hairs and shed skin cells into the lab space. Some good rules for lab coat use have been described by the University of Alabama (University of Alabama in Huntsville, 2013). These are applicable to any lab, including a MP one.

- DO wear lab coats that cover the knees and have full-length sleeves, preferably tight sleeves rather than loose sleeves
- DO keep lab coats completely "buttoned" up. Snap closures are preferred over buttons or zippers to keep the body covered and to allow quick removal in an emergency
- DO NOT wear disposable lab coats more than 8 hours or if it has become contaminated
- DO NOT wear lab coats unbuttoned
- DO NOT roll up the sleeves on lab coats for comfort or ventilation
- DO NOT wear someone else's lab coat

Whatever the garment choices, the addition of disposable IsoClean sleeves that will protect the frock's arms from aerosol dispersion is strongly recommended. If aerosol contamination (or another type of contamination) is suspected, changing a sleeve is a more practical and economical solution than changing a coverall. The easiest way to guarantee that arms and wrists will never be exposed is to slide the thumbs through a small cut made close to the cuff of the sleeves (whether they are from the frock or from disposable sleeves; Fig. 3).

Gloves

Most gloves are slightly permeable, so wearing two pairs of gloves (a.k.a. "double gloving") is strongly recommended, especially when working in a pre-PCR area. The inner glove should stay on at all times, while the outside glove should be changed often and/or when it is compromised (i.e., contaminated or punctured). Since the outer gloves do not directly touch the skin, they will not accumulate sweat the way inner-layer gloves do. Consequently, double gloving is an excellent way to reduce the potential dispersion of sweat aerosols when changing gloves.

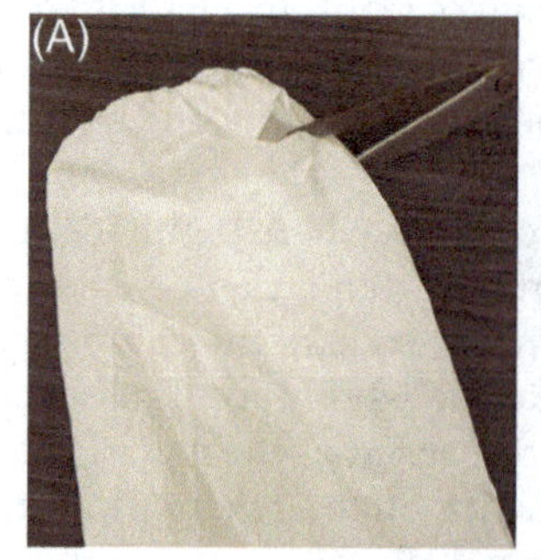

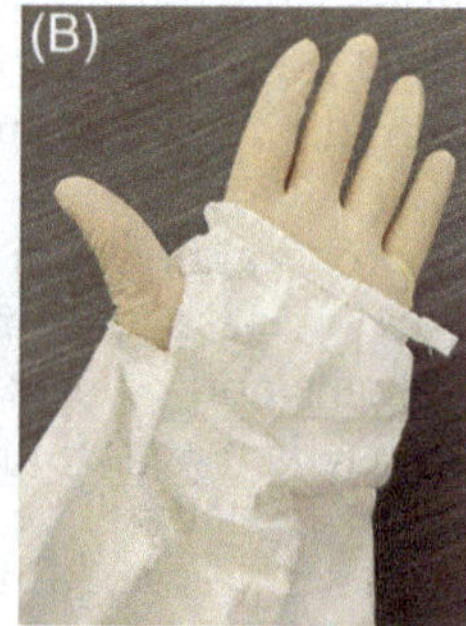

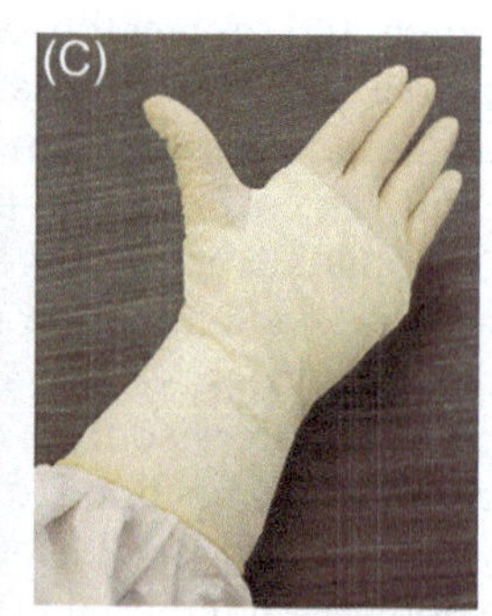

FIG. 3 How to "double glove" while wearing disposable sleeves. Using this method will guarantee that no skin will become exposed when the outer pair of gloves is changed. Put the inner (first) pair of gloves on. Using decontaminated scissors, make 2–3 in. cut close to each sleeve cuff (A); (B) Put the sleeves on and slide your thumb through the cut; and (C) Put on the outer (second) pair of gloves.

Double gloving can be done in two ways: by wearing two pairs of gloves of the same size, or wearing an outside glove slightly larger than the inner one. The latter option is only recommended if the larger/outer pair is still a close fit. Trying different combinations will help you find the right balance between feel and protection.

Several tricks will guarantee that your gloves always overlap the cuff area of your sleeves. The first one, described briefly in "Lab coats/coveralls and sleeves" section, is shown in Fig. 3. Other options to maintain the positioning of the gloves involve taping the gloves to the sleeves' cuffs or wearing longer outer gloves (12–16 in. surgical gloves are commercially available).

Masks

Disposable surgical face masks with either ear-loops or ties should be worn in pre-PCR areas and should cover the nose, mouth, and chin. Best practices when donning a surgical mask on include:

- Identifying the front and back of the mask. The colored side of the mask is usually the front and should face outward, while the white side should be positioned against your face
- Determining which side of the mask is the top. The side of the mask that has a stiff but bendable edge is the top and is designed to mold to the shape of your nose
- Molding or pinching the stiff edge to the shape of your nose and adjusting the mask to cover the nose, mouth, and chin

Disposable face masks can be used once and then discarded at the end of the day. Antifog surgical face masks are a more expensive option than regular masks, but will provide more comfort than goggles for eye protection.

Gowning protocol

After entering the laboratory airlock, individuals with long hair should secure their hair away from the face and shoulders (such as in a bun or ponytail). Lab-specific protective clothing must be worn by *all* individuals who will enter the lab, including staff, visitors, and service engineers.

The order in which PPE items should be donned will depend on the specific PPE selected for the lab (e.g., lab coat or coverall, mask, hairnet, or hood). Use common sense with respect to the potential for contamination. For example, applying a mask and hairnet cap after putting on gloves can lead to the latter being contaminated (Gall and Syndercombe Court, 2016). If a hooded coverall was selected, the following order can be used: inner pair of gloves, mask, safety glasses (goggles or face shield); then wipe the inner gloves, put the coverall on, the sleeves, and lastly, the outer pair of gloves that will be generously sprayed with diluted bleach after entering the clean lab.

Instruments and equipment

To minimize contamination, each room must have dedicated supplies and equipment. This includes refrigerators, freezers, flammable cabinets, centrifuges, microcentrifuges, vortexers, positive-displacement pipettes, tips, trash cans, racks, carousel stands for pipettes, scissors, etc.

Equipment necessary in a post-PCR lab

A MP post-PCR laboratory is similar to any molecular biology post-PCR lab. Adequate space is needed for benches and to accommodate standard instruments such as thermocyclers, sequencers, and instruments to analyze amplified DNA. In addition, if electrophoresis on agarose gels will be performed, the following instruments are necessary: a microwave, a gel box with power supply, a UV transilluminator, and a specialized camera. It is also possible to use an apparatus like a FlashGel Device Pack (Lonza, Basel, Switzerland).

Other options to analyze (and sometimes quantify) PCR product and/or next-generation sequencing libraries include (but are not limited to) the 2100 Bioanalyzer, the 5200 Fragment Analyzer, and the 4200 TapeStation (all three from Agilent technologies, Santa Clara, CA); the QIAxcel Advanced system (Qiagen, Germantown, MD); and Pippin instruments (Sage science, Beverly, MA). A Speed-Vac concentrator (i.e., centrifugal evaporator) may be necessary to dry or concentrate DNA.

Equipment necessary in all pre-PCR rooms

Laminar flow hood

Every pre-PCR room should have at least one (but preferably more than one) laminar flow hood and/or Class II biosafety cabinets. Both contain a HEPA filter that removes airborne contaminants from the air flowing into the hood, while preventing unfiltered air from the room from permeating the workspace.

- The flow hood or biosafety cabinet should be fitted with germicidal ultraviolet light. To avoid burns and eye damage, the hood should have an interlocking mechanism so that the UV lamp is only activated when the cabinet is not in use
- Germicidal lights should be on a timer to reduce the amount of ozone they produce. Leaving the ultraviolet light on for very long periods of time (such as overnight) will rapidly decrease the lamp's life and efficiency with no added benefit
  Avoid clutter in the hood to avoid having spaces not reached by the UV lights. If three-dimensional objects need to be decontaminated, they should be irradiated on every side and/or wiped with a cleaning product
- Adjust the working height of the stool or stand so that the worker's face is above the front opening
- Work at a moderate pace to prevent airflow disruption that occurs with rapid movements.
- Turn the cabinet on at least 20–30 minutes prior to use and wipe the bottom and sides of the cabinet surfaces with disinfectant when work is completed.

Crosslinkers

As will be described further in "Irradiation of consumables and buffers with UVC" section, ultraviolet irradiation requires that the object and the light source be very close to one another. As benchtop crosslinkers are generally small, they are ideal for decontaminating consumables and reagents within the lab. Their germicidal light sources are generally low-pressure mercury lamps that emit between 30% and 35% of their input energy at ~254 nm (Brais, 2017).

Fume hood

Functionally, a fume hood is a fire-resistant and chemical-resistant enclosure with one opening (face) in the front with a movable window (sash) to allow user access to the interior. When turned on, unhealthy fumes are vented out of the lab through a chimney and inhalations by the user are minimized.

Fume hoods are mandatory when handling hazardous reagents, and some forensic DNA extraction protocols do indeed include toxic solutions. For example, organic extraction protocols using a 25:24:1 mixture of phenol:chloroform:isoamyl alcohol (PCIA) are necessary when working with samples that still contain high amounts of proteins (such as collagen). Other toxic products often used in forensic laboratories include phenol, chloroform, *n*-butanol, formaldehyde, dimethyl sulfoxide (DMSO), strong acids, concentrated sodium hydroxide, and xylene. Fume hoods can also be used to handle foul smelling reagents such as DTT. When selecting a fume hood location, operator convenience, workflow, and exhaust duct locations should all be considered. Never place your face or head inside the hood and keep hands out as much as possible.

Area for skeletal remains sample preparation

This room is where skeletal samples will be decontaminated, cut, sanded, and powdered. It should include the following equipment:

- A bone sanding hood where bones will be cut and sanded, and where teeth will be drilled. A sanding hood should be a vented enclosure with a vertical flow that carries the bone powder away from the scientist to a receptacle. These often require customization. For example, the sanding hood used by the DNA Support Unit at the FBI Laboratory was custom-made by Flow Sciences, Inc. (Leland, NC)
- As a more economical option, small dental grinding boxes can be used or small glove boxes resistant to bleach and ultraviolet C (UVC) light. Whatever enclosure is used, disposable sleeves and gloves should be changed between samples
- A variable speed rotary tool. The Dremel rotary tool (Bosch, Germany) and accessories such as sanding wheels, sanding drums, and various Dremel diamond or tungsten cutters are convenient. The rotary tool can be used for both bones and teeth
- A ceramic mortar, chisels, and a dental mallet to break down pieces of bone too large to fit in a freezer mill
- An ultrasonic sonicator water bath

Any one of the following pieces of equipment can be used to powder/pulverize bones (note that access to liquid nitrogen is necessary when using freezer mills or cryomills):

- Freezer mills/cryogenic grinders (e.g., 6775 Freezer/Mill from SPEX SamplePrep, Metuchen, NJ)
- Waring blender 700S/700G with appropriately sized mini containers (MC1, MC2, or MC3) (Waring, Torrington, CT, United States)
- Mixer Mill MM 400 or a CryoMill (Retsch, Haan, Germany) or Bead Beater Tehtnica-Domel MillMix 20 (Gerhardt Analytical Systems, Northamptonshire, United Kingdom)

Optional equipment:

- Scalpel, tweezers, spatulas
- A low speed dental handpiece and burs

- Dental excavator
- Saw

Consumables and reagents:

- Bleach and/or alcohol wipes
- Weigh boats, weigh paper; antistatic weigh boats
- Bleach or sodium hypochlorite (NaOCl or NaClO)
- Benchcote or versi-dry sheets

Area for degraded DNA extraction and amplification set-up

Most protocols for DNA extraction from hard tissues require a demineralization step that can take up to 48 hours. Tubes are generally placed on a nutator (a.k.a., lab mixer, rocker, or nutating mixer), in an oven at 37–56°C. Hybridization ovens and hybridization tubes can be used as well. Equipment in this area includes:

- A precision scale or analytical balance (also necessary in the reagent preparation room)
- At least one centrifuge for large conical tubes (15 and 50 mL)
- At least one microcentrifuge for tubes (max RCF 16,000 × *g*)
- At least one laminar flow hood for DNA extraction and another one for handling extracted DNA (i.e., to set up PCR, qPCR, library preparation reactions, etc.)
- Vortexers
- Benchkote or Versi-Dry surface protectors

Optional equipment may include:

- An ice maker and ice bucket, or small portable coolers (e.g., DyNA Chill portable cooler, LabSource, Northlake, IL)
- Qubit Fluorometer (ThermoFisher Scientific)
- Dispenser seripetter for solutions such as phenol choloroform isoamyl alcohol
- Multitube mini-centrifuges (e.g., LabMini(TM) 6M multitube mini-centrifuge, Southwest Science, Hamilton Township, NJ)
- Thermomixer
- Magnet (e.g., DynaMag-2 magnet and DynaMag-96 side magnet, ThermoFisher Scientific)

Consumables and reagents:

- Kimwipes
- Bleach or sodium hypochlorite (NaOCl or NaClO)
- Parafilm M Barrier Film
- 15 and/or 50 mL polypropylene conical tubes (Sarstedt, Newton, NC, United States)
- Low binding micro-centrifuge tubes (e.g., Corning Costar, Sigma-Aldrich, St. Louis, MO)

Commercial reagents and consumables

Although contamination with human DNA does sometimes occur during the production of laboratory consumables and reagents, literature on the topic is scarce (Gill et al., 2010; Neuhuber et al., 2009; Schmidt et al., 1995).

As a starting point, however, it is important to note that sterile products are not equivalent to DNA-free products. As Gall and Syndercombe Court (2016) explain, "sterilization will kill or destroy microorganisms and damage the DNA. The agents used for sterilization will not, however, under the conditions required for sterilization, denature DNA to the extent necessary to prevent detection of contaminating alleles following the PCR process." Despite the fact that some manufacturers do sell (human) DNA-free products, the potential of low-level DNA contamination in consumables remains an issue.

To address these concerns, the European Network of Forensic Science Institutes (ENFSI), the Scientific Working Group on DNA Analysis Methods (SWGDAM), and the Biology Specialist Advisory Group (BSAG) published a statement requesting that manufacturers of disposable plastic ware and other reagents used in a forensic laboratory take additional precautions to prevent human contamination during their manufacturing processes (Gill et al., 2010). The ISO 18385 standard specifies requirements for the production of products used in the collection, storage, and analysis of biological material for forensic DNA purposes, but not for those used in postamplification analyses. Several major companies such as Life Technologies, Promega, and Qiagen are now ISO 18385 compliant. Unfortunately, the required testing specifies (negative) amplification of nuclear DNA and does not address mitochondrial DNA. Thus, while the ISO 18385 standard is a step in the right direction, it may be insufficient for laboratories where very small amounts of human mitochondrial DNA are routinely analyzed.

Anticontamination measures and best practices

This section describes ways to efficiently decontaminate the laboratory space as well as some consumables/reagents.

Sodium hypochlorite

Sodium hypochlorite (NaOCl or NaClO) is a liquid chlorine formed by the reaction of chlorine gas with sodium hydroxide. It is very reactive and unstable and contains between 4% and 15% of free, available chlorine. Temperature, light, oxygen, and certain heavy metals (e.g., copper, nickel, cobalt) are all catalysts for decomposition of sodium hypochlorite. When left exposed to the atmosphere, chlorine gas evaporates from the solution at a considerable rate; if heated, sodium hypochlorite breaks down into salt and oxygen, leaving white salt crystals.

Bleach is the generic name for any chemical product which is used industrially and domestically to whiten clothes, lighten hair color, and remove stains. Several types of bleach exist: chlorine bleach, oxygen bleach, and powdered bleach. Chlorine bleach, often referred to as "liquid bleach," is the only type that contains sodium hypochlorite. Formulations for household use generally contain ~5.25%–6.15% sodium hypochlorite by weight.

Despite the fact that bleach is widely available and relatively inexpensive compared to NaOCl, NaOCl is the preferred option for laboratory decontamination. The purity of laboratory grade NaOCl is higher than that of common chlorine bleach. Be vigilant, as additives are sometimes included in household bleach and may interfere with the laboratory instrumentation and/or chemical reactions. Additionally, the label showing the composition of the bleach product is sometimes incomplete and the expiration date may be hard to find. This is of importance because sodium hypochlorite has a very short shelf life of 3–6 months.

Because several units can be used to describe the strength of sodium hypochlorite, authors should always specify what unit they are using. The two units most commonly used in forensic publications are the "weight percent of available chlorine (wt% AC)" and the "weight percent of sodium hypochlorite (wt% NaOCl)." While close, these units are not equivalent and the following formula can be used to convert one into the other: wt% NaOCl = 1.05 × wt% AC. For example, 5.25 wt% sodium hypochlorite is equivalent to 5.0 wt% available chlorine.

A final percentage of 0.05–0.5 wt% NaOCl is considered an intermediate level disinfectant (i.e., it kills mycobacteria, most viruses, and all types of bacteria). However, this concentration is not sufficient for DNA decontamination purposes, and in most forensic laboratories, the recommended concentration is closer to 1–3 wt% NaOCl (Ballantyne et al., 2015; Fregeau et al., 2008; Gall and Syndercombe Court, 2016; Kampmann et al., 2017; Kemp and Smith, 2005). As the potency of diluted bleach decreases very quickly, dilutions should be made fresh daily with cold and soft distilled (DI) water. Avoid tap water. These measures will help minimize the amount of impurities and the cool temperature will reduce the decomposition rate.

Sodium hypochlorite is known to be corrosive, and precautions must be taken when handling it. The pH of a 5%–6% sodium hypochlorite solution is approximately 11, making it mildly irritating to the skin. However, the pH of a 10%–15% sodium hypochlorite is highly alkaline (pH ~13) and is so corrosive that it can burn skin on contact. Other health hazards of bleach include irritation of the eyes, mouth, lungs, and skin. Several studies have found that bleach and other cleaning products are associated with respiratory irritation and respiratory problems, such as bronchitis. Additionally, mixing bleach with certain chemicals common in laboratories can produce toxic fumes. For example, special care should be taken when handling guanidine salts, which are commonly used in many DNA extraction and PCR purification kits. Mixing bleach with guanidine salts will result in the formation of cyanogen chloride, a highly volatile and toxic chemical asphyxiant that interferes with the body's ability to use oxygen. Bleach mixed with ammonia will release chloramines that can cause severe respiratory irritation, and bleach mixed with acids will release chlorine gas. The signs of acute chlorine gas poisoning are primarily respiratory and include difficulty breathing and coughing, as well as possible nausea, vomiting, or headache. Finally, mixing bleach with ethanol or isopropanol will create chloroform and hydrochloric acid, as well as chloroacetone or dichloroacetone. Residual bleach thus should always be removed with water instead of ethanol (Ballantyne et al., 2015).

Good laboratory housekeeping practices are strongly recommended, especially for pre-PCR rooms where periodic laboratory bleach cleanups will help keep contamination issues at a minimum level. The frequency of regular cleaning will depend on several factors, but special cleanup days should be enforced when highly degraded bones need to be processed following the processing of bones that produced high concentration DNA extracts. This may occur with well-preserved samples (e.g., remains that were preserved in cold climates; remains that are recently skeletonized).

In the author's experience, the following protocol has proven effective for decontaminating benches, walls, floors, and other similar surfaces. Soak the surface with unexpired, freshly diluted bleach or NaOCl containing 1%–3% of available chlorine and let it soak for 5–15 minutes. Remove the bleach with clean Kimwipes dampened with deionized water (to avoid salt crystals). Then, wipe the surface with Kimwipes dampened with 70%–96% ethanol (which will help dry the surface quickly). Store the bleach in the fridge (do not freeze) or in a cool, dark, and dry place.

To decontaminate pipettes, start by wiping with a Kimwipe soaked with diluted bleach or use wet wipes such as Hype-Wipe disinfecting towels with bleach (Current Technologies, Crawfordsville, IN); then rinse them using a wipe dampened with water, followed by a wipe dampened with ethanol.

Tube racks can be submerged in fresh diluted bleach after use for at least 15 minutes, rinsed with water, and then air-dried.

For delicate metal instruments and equipment, consider avoiding bleach altogether and replace it with a less corrosive product such as DNA-OFF (Takara Bio, Mountain view, CA), DNA AWAY (Molecular Bio-products Inc., San Diego, CA), DNA *ZAP* (ThermoFisher Scientific, Waltham, MA), Decon90 (Perkin Elmer, Waltham, MA), Steriplex SD (sBioMed LLC, West Valley City, UT), DNA-Exitus-Plus (PanReac AppliChem, Darmstad, Germany), or similar. Note though that none of these products are as efficient as chlorine to get rid of contaminating DNA, and thus, several cleanups may be required (Fischer et al., 2016).

Irradiation of consumables and buffers with UVC

Ultraviolet (UV) light describes a band of the electromagnetic spectrum with wavelengths between 10 and 400 nm, shorter than that of visible light but longer than X-rays. This spectrum can be subdivided into a number of ranges: vacuum UV (10–100 nm), UVC (100–280 nm); UVB (280–315 nm), and UVA (315–400 nm). The germicidal properties of UVC were discovered at the end of the 19th century and have since been used for disinfecting water, sterilizing surfaces, and destroying harmful microorganisms in food and in the air.

The UVC spectrum, especially the range of 250–270 nm, is strongly absorbed by nucleic acids of microorganisms and is, therefore, the most lethal range of wavelengths. In brief, UVC kills germs by destroying the molecular bonds that hold their DNA together. More specifically, it forms cyclobutane rings between neighboring pyrimidine bases, thymidine or cytidine (Cone and Fairfax, 1993). On top of creating pyrimidine dimers, it leads to the oxidization of bases and the introduction of single-strand breaks and double-strand breaks (Champlot et al., 2010). Once cross-linked, pyrimidine dimers cannot be excised and either the DNA polymerase is sterically blocked or the DNA cannot completely denature, and the synthesis reaction is effectively halted (Mifflin, 2007).

While using UVC irradiation as a *disinfestation* method is very efficient, fast, and easy, using it as a *decontamination* method is not always effective, for the following reasons:

- The hydration status of DNA appears to have a great influence on its susceptibility to UV irradiation in that dry DNA seems much more resistant to the damaging effect of UV (Sarkar and Sommer, 1991). According to Hall and Ballantyne (2004), the UVC radiation dose needed for DNA profile loss was 90 times higher in the case of dried DNA sample as compared to DNA in solution
- The efficiency of UVC irradiation decreases with the square of the distance between the light source and the irradiated agent (Aslanzadeh, 2004; Champlot et al., 2010; Preuße-Prange et al., 2009). The decontamination effect of UVC lamps installed on the laboratory

ceiling is, thus, probably very low on surfaces such as lab benches and floors. For this reason, and because germicidal UVC lamps produce ozone which can be toxic in high concentrations, the installation of such UVC lamps is not recommended

- Very short DNA molecules (which can represent the majority of the molecules found in degraded human remains) may not contain adequate numbers of neighboring pyrimidines to make them susceptible targets to irradiation (Cone and Fairfax, 1993; Sarkar and Sommer, 1991)
- For efficient decontamination, surfaces must be perpendicular to the light source to achieve maximal light intensity. According to Cone and Fairfax (1993), "skewed surfaces dilute the intensity, and three-dimensional objects, such as pipettes, cannot be effectively decontaminated by UV light because only a fraction of the surface actually faces the light source"
- UVC light is blocked by materials such as glass and PVC. In high doses, UVC can damage/alter consumables (Burgess and Hall, 1999)

Since the efficiency of UVC irradiation depends on many factors (e.g., distance and orientation of the object, intensity, wavelength, exposure-time, age of the bulb, UV absorption characteristics of consumables, size and internal sequence of the contaminating DNA fragment as well as its hydration state), the amount of irradiation that reaches an item/surface is difficult to quantify accurately.

One solution to decrease the number of factors involved is to use benchtop crosslinkers since the distance between light and object will remain the same. UVC irradiation in crosslinkers has effectively been used to "decontaminate" consumables and reagents for decades. To degrade 107bp DNA fragments, Champlot et al. (2010) recommended irradiating the object at 10cm distance for 1hour, which corresponds to a measured energy of $1.45\,J/cm^2$ (Joules are defined as J = Watts × time in *seconds*). At the AFDIL, the recommended amount of energy to crosslink thin-walled tubes, such as 0.2mL PCR tubes, is $2\,J/cm^2$. For all other types of plastic ware, $6\,J/cm^2$ are used (Edson and McMahon, 2016). However, since a variety of crosslinkers are commercially available, each laboratory should conduct internal experiments to determine the appropriate distance, time, and energy necessary for decontamination of their consumables and reagents. Additionally, as the light emitted by the UV bulbs will decrease over time, it is important to calibrate crosslinkers several times per year. Pipettes should not be irradiated in the crosslinker.

The position of the tubes in the crosslinker matters. Tamariz et al. (2006) compared the decontamination efficiency on 1.5 and 0.2mL tubes irradiated upright and open inside a rack with closed tubes lying on their side. Results showed that the amount of contaminating DNA recovered in the closed, lying tubes was significantly lower than that in the tubes that were positioned upright and open inside the rack. They also showed that raising 96-well plates to 1in. of the UV light and lining the crosslinker with aluminum foil (to increase the amount of reflection) improved the decontamination process. Hall and Ballantyne (2004) also conducted experiments on closed tubes which contained DNA placed on their sides in the crosslinker. The tubes were low binding tubes made of polypropylene and were certified UVC transparent. Although the amount of energy that crossed the plastic was much lower than that which reached the outside surface of the tubes, experiments demonstrated that the amount of irradiation that passed through the tube's wall was sufficient to decontaminate the tubes.

Other decontamination options

A variety of other decontamination methods exist including autoclaving, radiation (e.g., gamma, X-ray, beta), and chemical treatments (e.g., enzymatic treatments with uracil-DNA glycosylase (UDG) or DNAses; treatments with ethylene oxide, ethylene glycol, hydrogen peroxide, or hydrochloric acid (HCl)). There is, however, some disagreement in the literature as to the effectiveness of these methods (Gefrides et al., 2010; Shaw et al., 2008). For the reasons previously mentioned, many also consider UVC irradiation to be an unreliable decontamination method (Gefrides et al., 2010; Grskovic et al., 2013; Lin, 2002; Shaw et al., 2008). Gamma and beta irradiation, as well as ethylene oxide gas treatments, are generally considered the most efficient methods to neutralize/destroy DNA contaminants. Unfortunately, these treatments must be performed in specialized facilities and are, thus, impractical for use in forensic laboratories (Gall and Syndercombe Court, 2016).

Staff training and laboratory management

All employees must be trained, and competency assessments should be conducted on a regular basis for all casework staff members. An up-to-date record should be maintained of the training that each staff member has received. For a MP lab, this training should extend beyond specific technical processes and standard safety training; it should include training aimed at minimizing contamination. For instance, it is essential that the "unidirectional workflow" rule is understood and enforced.

Staff training should emphasize anticontamination measures, to include the unidirectional workflow and general good laboratory hygiene. Bringing as few shed epithelial skin cells as possible into the clean labs is important. In 2002, Lowe et al. demonstrated that the proportion of donor DNA deposited on a clean DNA-free tube during a hold of 10 seconds increased with increasing time since handwashing (Lowe et al., 2002). Employees should shower and wear clean clothes on days when pre-PCR work will be performed and should take additional measures to mitigate potential contamination (e.g., comb hair and wash hands to minimize shedding).

Special care should be taken to ensure that laboratory workers learn how to minimize/prevent contact between their fingers and their face (e.g., nose, eyes, hairs) once they are gowned for work in the laboratory.

Make sure that an employee with a cold or other medical condition that risks compromising forensic casework (such as persistent coughing or sneezing) avoids working in clean areas. In 2003, Rutty et al. showed that a static and talking person deposited DNA in front of them within a 15-minute period. It is thus best to speak as little as possible while working in a clean lab (Rutty et al., 2003).

Additional best practices include:

- Personal items (e.g., keys, watches, rings) that may have been worn while working in a dirty lab should be removed before going to the lab or should be stored in an airlock cabinet
- Regular wipe tests should be performed in all clean rooms
- Anticipate your needs and gather all necessary materials, tools, and supplies you will need to perform your experiment

- Treat all consumables and reagents that are brought into the reagent preparation lab as if they were contaminated. Discard as much packaging as possible outside the airlock and wipe the outside of plastic packaging
- When pipetting samples, even the most seasoned technician can generate aerosols without the appropriate pipetting technique and pipette tips. Filtered tips must be used to eliminate sample-to-sample cross-contamination from aerosols. These tips have a barrier, which performs as a seal when exposed to potential liquid contaminants, trapping them inside the filter
- If the number of samples that need to be amplified isn't too high, strip-capped PCR tubes should be chosen over 96 wells plates. When using plates, favor strip caps to cover the wells instead of sticky seals and put the caps on as soon as you are done with the sample. Never leave any tube/well open longer than necessary
- Proper pipetting technique ensures that the accurate volume is aspirated and dispensed and avoids splashing when dispensing liquid. For recommendations on proper pipetting techniques, refer to "Good Pipetting Techniques Training" videos by Anachem (https://www.anachem.co.uk/Pipetting-Technique-Videos)
- The risk of contamination can also be reduced by careful handling of waste disposal in order to prevent aerosol formation (Borst et al., 2004). It is recommended to put contaminated plastic ware (e.g., tubes, tips) in a disposable bag while working. When the experiment is completed, remove the small trash bag, seal it, and discard it. Place a new bag in the trash can and turn the UVC lamp on for 30 minutes
- During long interruptions in the work cycle and at the end of the work day, make sure that all products, supplies, and materials are covered and stored properly
- Do not forget to decontaminate equipment such as centrifuges. In order to know what product should (or more importantly, should not) be used, read the instruction manual
- In pre-PCR labs, keep reagents and DNA extracts in separate refrigerators/freezers
- While contamination is not generally an issue in a post-PCR laboratory, bleach cleaning the lab (e.g., lab benches, equipment, walls, floors) at least once or twice a year will help limit dispersion of amplified PCR products to other sections of the laboratory.

An MP lab is essentially a forensic laboratory that should be designed like an ancient DNA laboratory. While most ancient DNA laboratories are academic, and thus, rarely follow validated protocols and standard operating procedures (SOPs), forensic laboratories are bound to these requirements. MP laboratories should at least be ISO/IEC 17025 accredited and should follow the SWGDAM Guidelines for Missing Persons Casework (2014). All SOPs and other documents in a MP laboratory should be reviewed and approved by the laboratory manager or technical leader on a regular basis to certify that all procedures used in the laboratory are up-to-date and accurate. In addition:

- Assays should be validated before introduction into casework
- Methods for quality control should be fully documented
- Instruments must be checked/calibrated every year

- Access to pre-PCR laboratories should be restricted. Limit the number of visitors/technicians/cleaning staff to a select few that provided a reference DNA sample.

Their profiles (both mitochondrial DNA and any nuclear DNA markers typed by the lab) should be added and stored in an elimination database
- Use nontemplate controls (e.g., negative controls, reagent blanks) during DNA extractions and amplifications
- If possible, when handling highly degraded samples, replicate DNA extractions independently, using a different skeletal element. For example, a tooth can be used for the first DNA extraction and a piece of petrous bone for the second one. If the remains come from a mass grave where bodies were commingled, or if it is uncertain whether the samples collected belong to a single individual, replicate the DNA extraction with the same skeletal element

In conclusion, setting and running a laboratory dedicated to the identification of degraded human remains is not a small task. Missing persons laboratories have very specific requirements that should be carefully considered when evaluating what is needed to start and operate this kind of facility successfully.

Acknowledgments

The author would like to thank Dr. Rebecca Just, Dr. Brandon Letts, Dr. Jodi Irwin, and Dr. Thomas Callaghan for critical review of the manuscript and helpful discussions.

Disclaimer

This is publication number 19-18 of the Laboratory Division of the Federal Bureau of Investigation (FBI). Names of commercial manufacturers are provided for identification only and inclusion does not imply endorsement of the manufacturer, or its products or services, by the FBI or the U.S. Government. The views expressed are those of the author and do not necessarily reflect the official policy or position of the FBI or the U.S. Government.

References

Alvarez-Cubero, M.J., Saiz, M., Martinez-Gonzalez, L.J., Alvarez, J.C., Eisenberg, A.J., Budowle, B., Lorente, J.A., 2012. Genetic identification of missing persons: DNA analysis of human remains and compromised samples. Pathobiology 79 (5), 228–238.

Aslanzadeh, J., 2004. Preventing PCR amplification carryover contamination in a clinical laboratory. Ann. Clin. Lab. Sci. 34 (4), 389–396.

Ballantyne, K.N., Salemi, S., Guarino, F., Pearson, J.R., Garlepp, D., Fowler, S., van Oorschot, R.A.H., 2015. DNA contamination minimisation—finding an effective cleaning method. Aust. J. Forensic Sci. 47 (4), 428–439.

Borst, A., Box, A.T., Fluit, A.C., 2004. False-positive results and contamination in nucleic acid amplification assays: suggestions for a prevent and destroy strategy. Eur. J. Clin. Microbiol. Infect. Dis. 23 (4), 289–299.

Brais, N., 2017. Air disinfection for ART clinics using ultraviolet germicidal irradiation. In: Esteves, S.C., Varghese, A.C., Worrilow, K.C. (Eds.), Clean Room Technology in ART Clinics. CRC Press, pp. 119–132 (Chapter 10).

Burgess, L.C., Hall, J.O., 1999. UV light irradiation of plastic reaction tubes inhibits PCR. Biotechniques 27 (2), 252–256.

Champlot, S., Berthelot, C., Pruvost, M., Bennett, E.A., Grange, T., Geigl, E.M., 2010. An efficient multistrategy DNA decontamination procedure of PCR reagents for hypersensitive PCR applications. PLoS One. 5 (9), e13042.

Cone, R.W., Fairfax, M., 1993. Protocol for ultraviolet irradiation of surfaces to reduce PCR contamination. PCR Methods Appl. 3 (3), S15–S17.

Dieffenbach, C.W., Dveksler, G.S., 1993. Setting up a PCR laboratory. PCR Methods Appl. 3 (2), S2–S7.

DTIC ADA301521, 1995. Use of DNA Technology for Identification of Ancient Remains. https://archive.org/details/DTIC_ADA301521.

Edson, S.M., McMahon, T.P., 2016. Extraction of DNA from skeletal remains (Chapter 6). Methods Mol. Biol. 1420, 69–87.

Fischer, M., Renevey, N., Thur, B., Hoffmann, D., Beer, M., Hoffmann, B., 2016. Efficacy assessment of nucleic acid decontamination reagents used in molecular diagnostic laboratories. PLoS One 11 (7), e0159274.

Fregeau, C.J., Lett, C.M., Elliott, J., Yensen, C., Fourney, R.M., 2008. Automated processing of forensic casework samples using robotic workstations equipped with nondisposable tips: contamination prevention. J. Forensic Sci. 53 (3), 632–651.

Gall, J.A.M., Syndercombe Court, D., 2016. DNA contamination—a pragmatic clinical view. In: Current Practice in Forensic Medicine. vol. 2. John Wiley & Sons, pp. 3–36 (Chapter 1).

Gefrides, L.A., Powell, M.C., Donley, M.A., Kahn, R., 2010. UV irradiation and autoclave treatment for elimination of contaminating DNA from laboratory consumables. Forensic Sci. Int. Genet. 4 (2), 89–94.

Gill, P., Ivanov, P.L., Kimpton, C., Piercy, R., Benson, N., Tully, G., et al., 1994. Identification of the remains of the Romanov family by DNA analysis. Nat. Genet. 6 (2), 130–135.

Gill, P., Rowlands, D., Tully, G., Bastisch, I., Staples, T., Scott, P., 2010. Manufacturer contamination of disposable plastic-ware and other reagents—an agreed position statement by ENFSI, SWGDAM and BSAG. Forensic Sci. Int. Genet. 4 (4), 269–270.

Ginther, C., Issel-Tarver, L., King, M.C., 1992. Identifying individuals by sequencing mitochondrial DNA from teeth. Nat. Genet. 2 (2), 135–138.

Grskovic, B., Zrnec, D., Popovic, M., Petek, M.J., Primorac, D., Mrsic, G., 2013. Effect of ultraviolet C radiation on biological samples. Croat. Med. J. 54 (3), 263–271.

Hagelberg, E., Gray, I.C., Jeffreys, A.J., 1991. Identification of the skeletal remains of a murder victim by DNA analysis. Nature 352 (6334), 427–429.

Hall, A., Ballantyne, J., 2004. Characterization of UVC-induced DNA damage in bloodstains: forensic implications. Anal. Bioanal. Chem. 380 (1), 72–83.

Holland, M.M., Fisher, D.L., Mitchell, L.G., Rodriquez, W.C., Canik, J.J., Merril, C.R., Weedn, V.W., 1993. Mitochondrial DNA sequence analysis of human skeletal remains: identification of remains from the Vietnam War. J. Forensic Sci. 38 (3), 542–553.

Jeffreys, A.J., Allen, M.J., Hagelberg, E., Sonnberg, A., 1992. Identification of the skeletal remains of Josef Mengele by DNA analysis. Forensic Sci. Int. 56 (1), 65–76.

Kampmann, M.-L., Borsting, C., Morling, N., 2017. Decrease DNA contamination in the laboratories. FSI Genet. Supp 6, e577–e578.

Kemp, B.M., Smith, D.G., 2005. Use of bleach to eliminate contaminating DNA from the surface of bones and teeth. Forensic Sci. Int. 154 (1), 53–61.

King, M.C., 1991. An application of DNA sequencing to a human rights problem. Mol. Genet. Med. 1, 117–131.

Lin, X.Q., 2002. UV Lamps in Laminar Flow and Biological Safety Cabinets. ESCO. www.escoglobal.com/resources/pdf/white-papers/UV_lamps.pdf.

Lowe, A., Murray, C., Whitaker, J., Tully, G., Gill, P., 2002. The propensity of individuals to deposit DNA and secondary transfer of low level DNA from individuals to inert surfaces. Forensic Sci. Int. 129 (1), 25–34.

Matsvay, A.D., Alborova, I.E., Pimkina, E.V., Markelov, M.L., Khafizov, K., Mustafin, K.K., 2018. Experimental approaches for ancient DNA extraction and sample preparation for next generation sequencing in ultra-clean conditions. Conserv. Genet. Resour., 1–9.

Mifflin, T.E., 2007. Setting up a PCR laboratory. CSH Protoc. https://doi.org/10.1101/pdb.top14.

National Institute of Justice (NIJ), 1998. Forensic Laboratories: Handbook for Facility Planning, Design, Construction, and Moving. https://www.ncjrs.gov/pdffiles/168106.pdf.

National Institute of Standards and Technology (NIST), 2013. Forensic Science Laboratories: Handbook for Facility Planning, Design, Construction, and Relocation. https://ws680.nist.gov/publication/get_pdf.cfm?pub_id=913987.

Neuhuber, F., Dunkelmann, B., Hockner, G., Kiesslich, J., Klausriegler, E., Radacher, M., 2009. Female criminals—it's not always the offender! Forensic Sci. Int. Genet. Suppl. Ser. 2, 145–146.

Parsons, T.J., Huel, R.M.L., Bajunovic, Z., Rizvic, A., 2019. Large scale DNA identification: the ICMP experience. Forensic Sci. Int. Genet. 38, 236–244.

Persing, D.H., 1991. Polymerase chain reaction: trenches to benches. J. Clin. Microbiol. 29 (7), 1281–1285.

Preuße-Prange, A., Renneberg, R., Schwark, T., Poetsch, M., Simeoni, E., von Wurmb-Schwark, N., 2009. The problem of DNA contamination in forensic case work—how to get rid of unwanted DNA? Forensic Sci. Int. Genet. Suppl. Ser. 2, 185–186.

Rutty, G.N., Hopwood, A., Tucker, V., 2003. The effectiveness of protective clothing in the reduction of potential DNA contamination of the scene of crime. Int. J. Legal Med. 117 (3), 170–174.

Sarkar, G., Sommer, S.S., 1991. Parameters affecting susceptibility of PCR contamination to UV inactivation. BioTechniques 10 (5), 590–594.

Schmidt, T., Hummel, S., Herrmann, B., 1995. Evidence of contamination in PCR laboratory disposables. Naturwissenschaften 82 (9), 423–431.

Shaw, K., Sesardic, I., Bristol, N., Ames, C., Dagnall, K., Ellis, C., et al., 2008. Comparison of the effects of sterilisation techniques on subsequent DNA profiling. Int. J. Legal Med. 122 (1), 29–33.

Stoneking, M., Hedgecock, D., Higuchi, R.G., Vigilant, L., Erlich, H.A., 1991. Population variation of human mtDNA control region sequences detected by enzymatic amplification and sequence-specific oligonucleotide probes. Am. J. Hum. Genet. 48 (2), 370–382.

SWGDAM, 2014. Guidelines for Missing Persons Casework. https://www.swgdam.org/publications.

Tamariz, J., Voynarovska, K., Prinz, M., Caragine, T., 2006. The application of ultraviolet irradiation to exogenous sources of DNA in plasticware and water for the amplification of low copy number DNA. J. Forensic Sci. 51 (4), 790–794.

Traver, P.J., Wiker, G.H., 2017. Modular clean rooms. In: Esteves, S.C., Varghese, A.C., Worrilow, K.C. (Eds.), Clean Room Technology in ART Clinics. CRC Press, pp. 91–98 (Chapter 8).

University of Alabama in Huntsville, 2013. Lab Coat Guidelines. https://www.uah.edu/images/administrative/policies/07.07.03-VPR_OEHS_Lab_Coat_Guidelines_Policy.pdf.

Viana, V.V., Wallis, C.L., 2011. Good clinical laboratory practice (GCLP) for molecular based tests used in diagnostic laboratories. In: Akyar, I. (Ed.), Wide Spectra of Quality Control. InTech. Available from: http://www.intechopen.com/books/wide-spectra-of-quality-control/goodclinical-laboratory-practice-gclp-for-molecular-based-tests-used-in-diagnostic-laboratories.

Ward, J., 2015. To Investigate Specialist DNA Techniques for the Identification of Compromised Human Remains. The Winston Churchill Memorial Trust of Australia.

Worrilow, K.C., 2017. The critical role of air quality: novel air purification technologies to achieve and maintain the optimal in vitro culture environment for IVF. In: Esteves, S.C., Varghese, A.C., Worrilow, K.C. (Eds.), Clean Room Technology in Art Clinics. CRC Press, pp. 33–46 (Chapter 3).

Further reading

Anon., Good pipetting techniques training videos. https://www.anachem.co.uk/Pipetting-Technique-Videos.

OxyChem, 2014. Sodium Hypochlorite Handbook. https://www.oxy.com/OurBusinesses/Chemicals/Products/Documents/sodiumhypochlorite/bleach.pdf.

University of California Lab Safety Design Manual, 2017. General Laboratory Ventilation Design Issues. https://lsdm.ucop.edu/sections/general-laboratory-ventilation-design-issues.

CHAPTER

4

Location, recovery, and excavation of human remains for forensic testing

Murray K. Marks PhD, D-ABFA[a,b,c,d] *and Darinka Mileusnic-Polchan MD, PhD, D-ABP*[c,d]

[a]Knox County Regional Forensic Center, Department of General Dentistry, Division of Forensic Odontology, University of Tennessee Graduate School of Medicine, Knoxville, TN, United States, [b]National Forensic Academy Outdoor Decomposition Training Facility, University of Tennessee, Knoxville, TN, United States, [c]Knox County Regional Forensic Center, Department of Pathology, University of Tennessee Graduate School of Medicine, Knoxville, TN, United States, [d]Institute for Human Identification, LMU College of Dental Medicine, Knoxville, TN, United States

Introduction

In the United States, forensic archaeology is a loosely defined protocol borrowed from traditional excavation and survey techniques taught to university archaeology students. It is not a stand-alone applied academic discipline, and with scant exception, few professional archaeologists perform forensic surface recoveries or clandestine grave excavations. This is unfortunate given the archaeologists' in-depth understanding of regional geomorphology, soil stratigraphy, and human disturbances in those contexts in which they are trained to decipher. Rather, forensic anthropologists have modified and co-opted those traditional methods into their skillsets because they have had "contextual" field school, contract, and/or prehistoric or historic mortuary site experience. The forensic anthropologist is a skeletal biologist with training in bioarchaeology and mortuary site excavation, anatomy, odontology, histology, entomology, zoology, and soft tissue decomposition. Besides possessing adequate technical archeological training, forensic anthropologists understand the underlying taphonomic and entomological principles guiding surface and buried decomposition for estimating time since death. In addition, forensic anthropologists appreciate the rules regarding evidence and the precise legal responsibility of handling forensic human remains.

Though a rare occurrence, clandestine graves are a labor-intensive scene to process and demand a temporal understanding. The forensic anthropologist is the most qualified expert

https://doi.org/10.1016/B978-0-12-815766-4.00004-2

to recognize, expose, remove, and process/package the evidence for transport to other specialists for their analyses. This holistic training requires their broad and focused crime scene attention as detective, taphonomist, entomologist, geologist, archaeologist, dentist, medical examiner, coroner, and pathologist until these specialists and other laboratory experts can receive evidence. The work of all subsequent specialists is contingent upon the specimen quality prepared at the scene. The crime scene is the ultimate responsibility of the medical examiner (ME) who must receive the remains without alteration or contamination. The forensic anthropologist serves as the medical examiner's scene investigator.

Unfortunately, the time necessary to process and excavate a burial properly can tax the patience and resources of law enforcement personnel managing the scene, which may compromise the quality of the excavation and recovery. Often these scene-processing duties have become "farmed out" to law enforcement taught by anthropologists in a "short course" because there are not enough anthropologists to perform the task, nor resources to employ them in medical examiner offices. Frequently, a "time investment vs. excavation strategy compromise" deal is struck between what is acceptable and defensible in court to the anthropologist and what is practical to law enforcement scene management. Paradoxically, many law enforcement scene personnel are among the most eager and helpful to the anthropologist in all aspects of scene development, excavation, and processing. It is at this time when the anthropologist should always ask, "What would an astute defense lawyer inquire about my involvement during this phase of the investigation?" At each stage of scene processing, that question serves to justify most activities.

In the United States, outside the military's state-of-the-art field and forensic laboratory operations, standard operating procedures (SOPs) governing recovery of decomposed or skeletonized human remains do not exist (see Moran and Gold, 2019, for a review of current methods). Most forensic anthropologists create and follow their own procedures beginning with undergraduate/graduate school training, postgraduate conference courses, collaboration, and courtroom testimony experience. In other parts of the world, standardized procedures exist where a merger of archaeology and forensic science guides the investigation and recovery of complex, premeditated mass-burial sites containing multiple bodies. This work is performed under much more stringent guidelines (see Adams and Byrd, 2014; Brickley and Ferlilini, 2007; Cox et al., 2008; Haglund and Sorg, 1997). However, whether a single interment or genocide mass-burial site excavated by an international team, each clandestine grave scene (like all crime scenes) will dictate on-site modifications to the excavation strategy that preserves and documents evidence in the best manner.

This chapter does not explain technical excavation techniques, how they evolved, or who performs them. Nor is it about the discovery and excavation of mass or complex graves with multiple bodies or soft tissue decomposition [see Cox et al., 2008; Schotsmans et al., 2017; and Pokines and Symes, 2014 (respectively) for current accounts]. Rather, it is an historic introduction, a simple primer on an excavation strategy, and the basic tools utilized by the forensic anthropologist to maximize evidence recovery and preservation.

The scene

In the same manner that analyses of human and nonhuman mortuary site skeletal remains may answer many archaeologically derived hypotheses, forensic skeletons (single or multiple) provide an "osteobiography" (Saul, 1972) that is crucial for identification of ante-, peri-, and postmortem events. Because of the postmortem longevity in preserving perimortem trauma, bones, many times, are more valuable than the decomposing moist or desiccated soft tissues that contain them.

Ubelaker's (1980) volume introduced a generation of bioarchaeologists and mortuary site archaeologists to the importance of excavating skeletal remains within a context, which helped to legitimize the transition of those skills from prehistoric and historic burials to a forensic burial. Shortly after, Morse et al. (1983) formally defined the topic of the archaeological treatment of the forensic skeleton, and Dirkmaat and Adovasio (1997) illustrated the importance of the site in the first Haglund and Sorg (1997) volume. Komar and Buikstra (2008) warn that while the methods used are not entirely synonymous, they underscore integrating the principled perspective of the bioarchaeologist to forensic archaeology (see also Martin et al., 2013). Top-tier introductory forensic anthropology texts (e.g., Byers, 2017; Christensen et al., 2019) and texts on advanced treatment of forensic burials (e.g., Dirkmaat, 2012; Houck, 2017; Iscan and Steyn, 2013) include chapters and sections devoted entirely to the discovery and recovery of clandestine burials.

The methodical skillset of the forensic anthropologist is required for the correct recovery of remains. Just one season of leaf litter covering a ground 'surface scatter' of bones, desiccated remains, or suspicious disturbed soil constitutes an archeological event requiring temporal appreciation (Fig. 1). A forensic gravesite is like any crime scene and similar to an autopsy. Once the grave is discovered, deterioration of evidence begins. The anthropologist has one opportunity to document the process of excavation or dissection, and any weaknesses in those procedures are unwelcome burdens to later forensic laboratory analyses or in courtrooms.

Every forensic practitioner recognizes that all activity surrounding the search, discovery, and recovery of evidence at the scene becomes a chain-of-custody responsibility that is admissible and defensible in court. The scene investigator must be acquainted with the rules regarding evidence, the completeness of their scene activities, processing/packaging, and delivery of the remains. How the remains are delivered can directly influence the quality of their colleagues' ensuing special analyses. All soft tissue, teeth, bone, or clothing may contain trace evidence or be subjected to specialized analyses that mandate the scene investigator to always remember: "How do I get this into the hands of the expert with minimal evidentiary?" The answer to this question requires the scene anthropologist to be first a forensic scientist who realizes that the pathologist, entomologist, dentist, or DNA expert expects zero-contaminated specimens to perform their analyses satisfactorily.

Many times, clandestine graves are a hastily prepared after-thought resulting in a shallow deposition depth. Such carelessly created burials can be subjected initially to discovery and modification/destruction by carnivores attracted to decomposition (Fig. 2). Given the correlation between victim contact and perpetrator contamination, the criminal has to get rid of the body expeditiously. In contrast, a premeditated burial prepared for several bodies may be meters deep. However, many forensic grave preparations are impulsive and dug for single interments ranging from shallow to several meters (refer to Connor, 2007).

Two important scene tasks, if not followed, will haunt the forensic anthropologist: (1) insufficient note-taking during the excavation that will jeopardize report writing and undermines

FIG. 1 Disorganized and differentially exposed bones adjacent and partially encased within extensively scavenged soft tissue by canids. UV exposure caused "bleaching" and drying of some bones. Note the relationship of the remains to leaf litter.

courtroom testimony, and (2) packaging errors that hamper, and sometimes prevent, analyses by other specialists. The opposing legal team and its experts can review all notes, images, and evidence handling records. The process of collecting all evidence and chain-of-command is subject to an explanation in court. The expert witness dealing with remains from initial exposure to handling, analyses, and results can become a target for dispute with a legal defense team.

In the United States, depending on the investigating agency, scene size, and location (whether rural or urban), uniformed officers may guard access points to prevent public or media interference. Some counties set up mobile operation posts that coordinate activities between scene managers and investigation collaborators and act as the staging area for the public and media. The closer to the scene, the more intense security becomes with roped or taped off areas-of-interest that contain the remains or burial. Only authorized personnel, detectives, death investigators, and the excavating anthropologist may enter this region. At times, the medical examiner or dentist may assist, and a district attorney

FIG. 2 Canids exposed the grave of this victim who had been missing for 11 months. Soft tissue scavenging of the arms, torso, and lower back also removed the left hand and right forearm. Drying of the scalp, neck, and trace periosteal bone covering the ribs and arm bones occurred between the time of the last canid visit to the grave and discovery. During scavenging, the viscera were removed between the ribs (important given multiple sharp-force defects present), and the clothing (left foreground) was also pulled off of the victim.

and other officials from adjoining jurisdictions of law enforcement may participate. The forensic anthropologist collaborates directly with the scene detectives to inform them how the excavation will proceed. As mentioned previously, law enforcement often is eager to assist in any aspect of the excavation, and they allow the anthropologist to direct the work. Keeping foot traffic to a minimum and away from the grave ensures preservation until the excavation is complete.

Discovery

Stratigraphy is the soil history of deposition and erosion. A shallow test pit dug in an adjacent undisturbed area will reveal the natural arrangement or horizon profile at the scene. When soil is vertically displaced or disturbed to any depth and mixed, the layers never can be reconfigured into the pre-disturbed condition. Disturbances which result in a mixing or mottling of horizon color, texture, and density are long lasting and are easily discerned throughout the grave fill. The mosaic result of this mixing on the ground surface provides clear visualization of a pit outline. Troweling and probing of the fill help provide verification of differences in the soil texture and density. With time, grave fill will collapse, begin to compact, and then will gradually regain its original density after years of water and soil percolation. However, a permanent scar of that disturbance will remain always.

Surface troweling and probing can delineate the outline and depth of the grave fill, and the excavator can see the lower horizon mottled on the surface. In addition, vegetation and mounding or depression of the grave may indicate a disturbance. The excavator must be careful when probing to prevent the remains from receiving multiple and unnecessary perforations (Fig. 3). Furthermore, decomposition odors will transfer readily onto the metal probe or within a soil core sample and is easily noticed when withdrawn. A cadaver dog will immediately respond to an opened probe hole in the grave fill overlaying a decomposing body.

Anthropologists have successfully used ground penetrating radar (GPR) to discern and map subsoil disturbances. While the GPR does not have the ability to define a precise burial outline or body detail, it does provide informative scans indicating pit outlines and exactly where to dig. GPR also allows a more accurate three-dimensional understanding of the fill (especially to discern depth and orientation) and should be used prior to gridding the excavation site. In addition, GPR can indicate if more than one body exists in the grave. GPR devices, typically on wheels, are a luxury piece of expensive equipment requiring a trained technician to operate and interpret the results. Use of GPR devices can alleviate probing the grave and doing damage to remains.

FIG. 3 Vertical grave probing allows easy sensing of a density and pressure change when compared to undisturbed soil. This signals a subsurface event of disturbed grave fill. A probe cannot penetrate dense undisturbed soil to the degree a subsurface disturbance will allow.

(Continued)

FIG. 3, CONT'D

Excavation procedures

Scene documentation begins when the anthropologist establishes the location of the grave. This may result from carnivore disturbance revealing remains, by troweling to discern soil discrepancies, probing to reveal density, or from an informant's account. Besides photography of the "undisturbed disturbance," the site becomes an archaeological endeavor, including gridding the surface to facilitate and guide vertical excavation. First, the anthropologist needs to establish a datum point to any permanent structure for reference to the excavation. Since there will be no nearby U.S. Geological Survey benchmark to measure for location, the datum point should tie that grave outline permanently to nearby structure, e.g., house, fence, largest tree, boulder, etc. State-of-the-art mapping technology using a Geographic Information Systems (GIS) total station may be used to establish a "data" location and map the entire excavation. However, on the ground, the traditional grid is established to visualize and control all horizontal and vertical activity to guide the excavation (Fig. 4).

There are numerous ways to set up a grid to monitor the excavation. Quadrants or a multiple-square design to cover the grave outline and just beyond along a north-south, east-west intersect provide the most simple and expedient method for precise vertical removal and body/bone exposure. Each bone, body part, and artifact must be recorded in the three-dimensional space of the grave fill or given an x, y, z notation (Fig. 5). If fill removal proceeds in level manner (with careful evidence plotting throughout), the excavator does not need extra grid lines, squares, and bulk walls since the site is a temporary exercise just to remove the grave fill dirt. A much greater time-investment method to expose grave contents is vertical sectioning of the fill and grave wall-undisturbed margins to expose how the remains are preserved in profile. This technique is rarely performed unless the burial is partially exposed in a cliff dimension or in a washed-out area.

The excavation method used on most archaeological features is horizontally troweling the soil in the grave fill above and adjacent to the body using a pointed or rectangular-nosed metal trowel. The trowel is not a garden shovel or hoe for blindly sinking or randomly probing into grave fill to pop out dirt. Cortical bone, decomposing soft tissues, and clothing are fragile when moist and are easily damaged from metal slicing or probe contact. The invasive probing technique is the only exercise where damage to buried remains is justified to discern soil density and buried items. The fill is removed evenly in layers throughout the grave, preserving the borders. Multiple excavators working independently on their hands and knees with assigned quadrants will adjust their speed to ensure the fill is removed in even layers.

Grave fill is not the tightly packed virgin or undisturbed soil that was originally removed and replaced when creating the grave. Rather, it is a more loose, semi-packed, discontinuous matrix, re-situated in random clumps (not layers) which contain air pockets that surface water collects in and percolates through. After the grave outline is discerned and the grid is in place, troweling begins from the center of the surface disturbance by pulling soil outwards to the edge or grave

FIG. 4 Annual clandestine grave recovery exercise at the University of Tennessee's Forensic Anthropology Center outdoor training facility. Note the initial six-unit grid setup, final mapping, note-taking, and excavation tools.

FIG. 5 Horizontal measurement from the east wall of an excavation using a plumb bob for exact landmark position. Note the extensive root growth that permeated into the grave fill after burial. Severed rootlets are botanical evidence that may help time the disturbance. Additionally, note the dark air space in the thoracic cavity where the viscera have decomposed.

outline. Trowels are held with the blade at a 45-degree angle to gently peel or skim off horizontally several millimeters of soil into a dustpan, which is then emptied into a bucket labeled by quadrant. All soil is sifted through ¼″ hardware screen to recover any tiny items missed during troweling. Troweling of a two- to three-millimeter (mm) thickness will dislodge or reveal all the contaminating inclusions in the fill that may be valuable evidence. When items are discovered during the troweling process, they are not removed until their x, y, and z coordinates are measured, photographed, mapped, and notated. The excavation halts to accommodate documentation. All photographs require a north arrow and site identification marker. After documentation, the items may be removed for examination and logged as evidence unless it is near the remains.

Mapping and note-taking are performed at every 10- or 15-cm (cm) depth while photography should be continuous. If items or bones have not been discovered in the fill at the prescribed mapping level, mention of soil changes are relevant. Note-taking is crucial to any crime scene in order to reconstruct any and all components of the excavation for use in subsequent laboratory analyses, report writing, and courtroom testimony. Like all aspects of the documentation process, being overzealous is more important than trying to rely on memory. A general overview of the scene before excavation, as well as drawings and a narrative of each measured depth, will heighten the accuracy and reproduction of the excavation. In addition, even though numerous photographs will document the progress of the excavation, note-taking with a hand-drawn or computer-generated map will enhance an understanding of depositional events.

Troweling is the method for removing the majority of the grave fill above and adjacent to the remains. A bamboo pick with a prepared (sanded) curved end, a thin tongue depressor, wooden dowel rod, or clay modeling tools are ideal to dislodge and separate the soil touching bone. Like the trowel, metal dental tools/picks and dissection probes can easily perforate bone, skin, or clothing. The excavator must take care not to undermine the soil beneath any bone so that the entire body is supported by a soil pedestal. Likely, the body is positioned or resting at the bottom of the grave and troweling of undisturbed soil beneath is tight and dense. The soil beneath the remains may contain moist decomposition fluids. Three things are performed after complete removal of the body or skeleton: (1) a metal detector is passed across the bottom surface; (2) a soil sample is taken; and (3) the temperature is recorded (Fig. 6). During the stages of the excavation, protection is afforded to preserve soft tissues, bone surfaces, and clothing from further deterioration/drying. In low-humidity or sunlight-exposed sites, dehydration can desiccate these surfaces, which can result in cracking

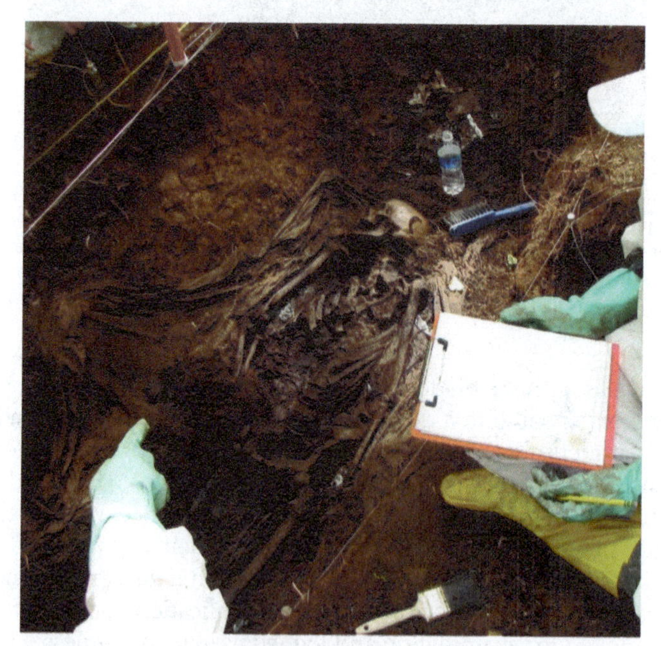

FIG. 6 Note-taking and final documentation at the end of an excavation. All dirt has been removed, with full exposure of the skeleton on the undisturbed floor of the grave.

and separation from the surrounding tissue. The excavation team can cover these exposed bones with damp cloth or a sunshade to prevent UV damage.

Often in the past, the prosecution and defense stipulated clandestine scene dynamics. Recently, both sides have begun requesting an entire walk-through with the jury to explain the scene. This is best accomplished with images of the entire excavation, including maps and drawings. The forensic anthropologist knows to be overzealous with documenting, not just for courtroom appearances, but also for the preceding analysis and report writing. From discovery to packaging, the jury and judge deserve to understand the total involvement of the anthropologist. This process demands exhaustive documentation (from mapping to imaging), and potentially having to explain a careless cause for damaged bone when the victim's family may be in the gallery is unforgivable. Furthermore, a forfeit of any of this documentation can be embarrassing and damaging to legal proceedings and to expert witness credibility.

Recovery and transport

All scene participants should see the dismantling and removal procedures of the buried remains. An attending death investigator may discover non-body/skeletal items of valuable interest to the pathologist. As mentioned, retention of body, limb, and skull to torso is paramount for the pathologist receiving the remains, so careful packaging of them together is crucial. If decomposition has skeletonized and segmented limbs, torso, and/or skull and mandible, then package accordingly. Since many specialists mentioned who perform laboratory analyses do not frequent the scene, the forensic anthropologist should know ahead of time any special packaging requirements and instructions given the condition of the body. There are technical volumes addressing these tasks (e.g., Byrd and Castner, 2010; Cox et al., 2008; Dirkmaat, 2012), although it is best to establish contact and receive directions from these specialists prior to or during scene involvement.

Rules regarding evidence mandate "chain of custody" documentation so that the location and safekeeping of the remains are archived. This involves providing a record of when and how the remains leave the site, to whether the laboratory door was unlocked for an unwitnessed period. Always, be sure the proper documents are provided "in the field" for transport of remains to the laboratory. Death investigators and law enforcement are astute regarding such detail.

Finally, there are exceptions to many excavation rules during the procedure. Like each crime, each grave is as unique as the remains it contains, and flexibility is therefore the rule. While the excavation procedure belongs to the anthropologist and those enlisted to help or volunteer, the rules regarding evidence (once exhumed) may vary depending on who oversees the scene and remains. A detective, district attorney, Disaster and Mortuary Operational Response Team (DMORT) Commander, coroner, or medical examiner (ME) may grant or deny permissions to scene investigators to proceed in a different manner. There is a hierarchy of command to follow, and jurisdiction differs from place to place.

Packaging

Only after the remains and associated grave items are fully documented *in situ*, and after all vested scene personnel have voiced consent for removal, should the remains be moved, examined, or taken from the grave and packaged for transport. Images *in situ* will have to suffice for the medical examiner to understand the relationship of the body/skeleton in the grave. Every bag, envelope, vial, or container used for packaging evidence must be labeled by site, date, time, contents, and signature(s) following the rules regarding chain-of-custody. Paperwork accompanying the remains in transport and remaining with the submitting agency needs completion at the scene prior to transport.

The anatomical position of the body or limbs may be preserved so that the medical examiner can examine or dissect them without excavation or post-excavation (transport) distortion. The team packaging the remains in the field must be aware of who will be examining the remains later and if special sampling will be part of the analysis. The field requirements needed by downstream experts to secure sensitive samples from damage or contamination are the responsibility of the anthropologist at the scene. The burial environment, the destination of the remains, and the types of future planned analyses will dictate the type of packaging the remains require. The anthropologist needs to keep in mind who is examining the remains (e.g., the pathologist, entomologist, or dentist) and if special sampling analyses (e.g., DNA) will or might be performed. Decomposing soft tissues may contain surface and deep evidence of trauma that requires preservation for autopsy inspection. As mentioned previously, the pathologist deserves to examine the remains in the most unaltered state of recovery, and only he/she can dissect or dismantle the body.

The moist or desiccated content of decomposing soft tissue or the wet, damp, or dry cortical bone quality and degree of completeness will dictate the packaging criteria for transport to the medical examiner's facility or laboratory. Wet or damp adhering soft tissue or moist bone may be wrapped in plastic for exceedingly short-term transport. Dry or desiccated limbs connected at the joints by soft tissue or bones with any degree of moist periosteal covering should be secured first in a paper wrapping to prevent crumbling. Loose and fragmented (traumatized) moist bone and teeth should be isolated and placed in paper bags or wrapped in paper and allowed to dry slowly. These bags should be secured or immobilized in a cardboard box within a body bag. Even after arrival in a secure laboratory, slow drying to prevent longitudinal splitting of the long bone diaphyses may be necessary. Loose teeth should never be removed from their sockets except by the forensic dentist performing an analysis; however, if teeth are loose, they can be securely wrapped and packaged like small skeletal remains. Paper coin envelopes and a variety of white bond paper envelopes may be used for fragments, small pieces, and loose teeth, but loose teeth should never be mixed with other loose bones. Finally, the entire body should be placed in a body bag and sealed by the death investigator.

In the case of burned remains (rare in grave settings though more common in a surface scatter), a spray adhesive should be applied to stabilize fragile regions (especially cranio-oro-facial structures) for transport. Burned remains should be completely wrapped so that fragmentation during transport is minimized. Paint cans can isolate fragile remains in danger of crumbling. If the remains are fragmented, a block of supporting soil may be lifted and wrapped together for laboratory processing. Plastic bags should never be used to package and transport dry or damp skeletal materials, as bacteria-producing mold can grow and propagate very quickly.

Finally, clothing as part of the victim is as important as any skin surface to the pathologist's inspection and interpretation of perimortem events (Fig. 7). The intersection between the victim and such events may hold trace evidence and affect the decomposition status by enhancing or slowing the tempo. Only the pathologist may remove garments to interpret the clothing/skin relationship. In addition, clothing may contain valuable entomological evidence (both wet and dry) that may be significant in buried or surface bodies and can contribute to time-since-death estimates (refer to Byrd and Castner, 2010). Law enforcement personnel may take custody of the grave fill items-of-interest above or adjacent to the victim, once they are measured, photographed, documented, removed, and packaged separately.

FIG. 7 Surface remains uncovered and scavenged by canids. Clothing and other non-clothing items (e.g., sleeping bags, tarps, plastic sheeting, carpet) may enhance or slow decomposition. Note the anterior tooth loss, which indicates the cranium and mandible have been moved after deterioration of the periodontal ligament. Also note the root-infestation of both garments that covered the body.

Surface remains recovery

Decomposing complete or partial bodies and skeletal surface remains require an approach nearly identical to a burial. As mentioned, one season of leaf litter creates an archaeological site, and the botanical evidence (including a top horizon soil sample) deserves special attention and preservation before moving or mapping of any evidence. Before human discovery, canids and rodents have facilitated dispersal by reducing soft tissue and modifying bone. The attention and disturbance by scavenging mammals to human remains (initially to decomposing soft tissues and later to bones) may last during the entire decomposition process. While canids (e.g., a wolf, dog, coyote, fox) show a major interest in scavenging decomposing tissues, bear, raccoons, and rats (and later, squirrels and mice) also reduce and modify bone. Canid soft tissue ingestion (including viscera) may begin within days, depending on environmental temperature and the predation activity of local animal populations. If decomposition is advanced, the damage can include bone. As bone becomes exposed, rats particularly are attracted to trabecular regions that may still contain grease, while other rodents seemingly prefer dry remains. Canids may revisit the body and bone cluster and cause further or an initial dispersal of bones. Most rodent damage is primarily after the skeletal remains have dried from soft tissue loss and exposure. Widespread movement of separated bones by foraging rodents is typically less than during canid scavenging activity. However, the damage from long-term exposure may be extensive with reduction and loss of diagnostic landmarks well into the trabecular and deeper layers (Fig. 8).

The human discoverers of skeletal surface remains can complicate this disturbance. Their curiosity invites compulsions to move bones (especially the skull) before contacting authorities (Fig. 9). By the time law enforcement, the anthropologist, and investigators arrive at the scene, bones have been moved, stepped on, and/or damaged. Unlike a preserved burial scene, surface remains are subject to continual changing postmortem dynamics. Best practice for dealing with surface remains is to conduct a comprehensive scene survey walkover with pin flagging of bones to decipher dispersal. This will facilitate later measurement to the torso location. While law enforcement may not recognize human bones, they do tend to recognize all bone, clothing, and other nonhuman items that the anthropologist may not deem important.

The discovery of vertebrae and ribs indicates a remarkably close approximation to the exact place of deposition because these bones do not readily move after decomposition and initial canid disturbance. Pelvic and pectoral limbs may be scattered from the canid activity before, during, or after bone appears. Skulls move easily after cervical detachment and mandibles separate from crania anywhere between the cervical vertebrae and cranium discovery. Single-rooted teeth (i.e., incisors, canines, and some premolars) may drop anywhere the skull has traveled from the original location. Dark(er) and/or differential soil staining, moisture, mold growth, and odor will indicate the original location. Additionally, careful shallow troweling may discover puparium, which indicates a generalized decomposition region, and depending on the timeframe, pupa casings may contain larvae (refer to entomology references for this chapter).

The same documentation standards exist for the two-dimensional decomposing body with loose elements and a 'surface scatter' of bones. Photography, mapping, and note-taking are mandatory during the survey, initial discovery, exposing of remains, and recovery stages. Distances measured and mapped between the main decomposition region and all bones will reflect canid activity and ground slope. Finding a datum and making a quadrant grid will facilitate the troweling procedure in the same manner for a burial. However, only a shallow depth is required for removal to recover teeth or small hand and foot bones. The underlying undisturbed or virgin soil is dense and easily discerned. Screening of the soil will provide a means of recovering small bones and fragments.

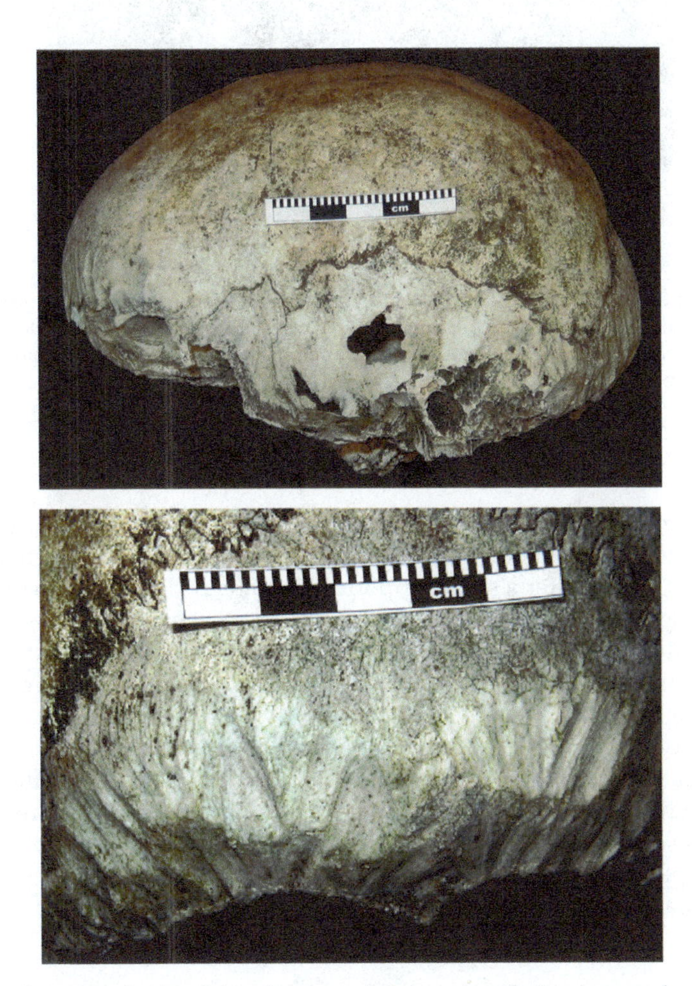

FIG. 8 Right lateral cranium (top) and dorsal closeup of that cranium (bottom) exposed on the ground surface of a dense wooded acreage for three years. Postmortem rodent scavenging is extensive with loss of the facial skeleton (lower left corner in upper image) and cranial base. The modified surfaces, especially in cortical bone contain sharp, well-defined margins from tooth abrasion. This damage, when recent, is similar to other postmortem disturbances where damaged bone appears brighter and not the homogenous color and weathering of perimortem bone breakage.

FIG. 9 Forest setting of surface-scattered remains discovered after 19 months. Note the entire loss of soft tissue with subsequent UV "bleaching" of exposed surfaces. Additionally, there is minimal bone scatter, and soil staining is present on the left side of the cranium. After discovery and examination of the skull by the landowner, it was incorrectly positioned prior to notification of law enforcement.

Conclusions

The recovery of complete or incomplete decomposed bodies and scattered skeletal remains is a challenging time-consuming and labor-intensive endeavor. Both challenges are archaeological crime scenes, which offer a single opportunity to get it right with accurate recovery of remains. The simple protocol for discovery, excavation, recovery, and preservation outlined here is an easily adopted, processing example. Coupled with detailed and exhaustive documentation, an accurate rendition of all activities will allow a confident presentation in any medico-legal setting.

The availability of more advanced tools and techniques using GPR and the GIS total station to discern subsurface burial disturbances and 'surface scatters' provides the most accurate and expedient method to digitally archive all scene activities. Both tools require advanced training, and while drastically reducing documentation time, are not a replacement for performing the principles of discovery, excavation, recovery, and packaging of remains for scientific laboratory collaborators.

References

Adams, B.A., Byrd, J.E. (Eds.), 2014. Commingled Human Remains: Methods in Recovery, Analysis and Identification. Academic Press, New York, NY.

Brickley, M.B., Ferlilini, R. (Eds.), 2007. Forensic Anthropology: Case Studies From Europe. Charles C. Thomas, Springfield, IL.

Byers, S.N., 2017. Introduction to Forensic Anthropology, fifth ed. Routledge, New York, NY.

Byrd, J.H., Castner, J.L. (Eds.), 2010. Forensic Entomology, The Utility of Arthropods in Legal Investigations. CRC Press, Boca Raton, FL.
Christensen, A.M., Passalacqua, N.V., Bartelink, E.J., 2019. Forensic Anthropology: Current Methods and Practice, second ed. Academic Press, New York, NY.
Connor, M.A., 2007. Forensic Methods, Excavation for the Archaeologist and Investigator. AltaMira Press, Lanham, MD.
Cox, M., Flavel, A., Hanson, I., Laver, J., Wessling, R., 2008. The Scientific Investigation of Mass Graves. Towards Protocols and Standard Operating Procedures. Cambridge University Press, Cambridge UK.
Dirkmaat, D.C., 2012. A Companion to Forensic Anthropology. Wiley Blackwell, Hoboken, NJ.
Dirkmaat, D.C., Adovasio, J.M., 1997. The role of archaeology in the recovery and interpretation of human remains from an outdoor forensic setting. In: Haglund, W.D., Sorg, M.H. (Eds.), Forensic Taphonomy, The Postmortem Fate of Human Remains. CRC Press, Boca Raton, FL, pp. 39–64.
Haglund, W.D., Sorg, M.H., 1997. Forensic Taphonomy: The Postmortem Fate of Human Remains. CRC Press, Boca Raton, FL.
Houck, M.M., 2017. Forensic Anthropology. Academic Press, New York, NY.
Iscan, M.Y., Steyn, M., 2013. The Human Skeleton in Forensic Medicine, third ed. Charles C. Thomas, Springfield, IL.
Komar, D.A., Buikstra, J.E., 2008. Forensic Anthropology, Contemporary Theory and Practice. Oxford University Press, New York, NY.
Martin, D.L., Harrod, R.P., Perez, V.R., 2013. Bioarcheology: an integrated approach to working with human remains. In: Manuals in Archaeological Method, Theory and Technique. Springer Nature, Switzerland, AG.
Moran, K.S., Gold, C.L., 2019. Forensic Archaeology, Multidisciplinary Perspectives. Springer Nature, Switzerland, AG.
Morse, D., Duncan, J., Stoutamire, J., 1983. Handbook of Forensic Archeology. Rose Printing Company, Tallahassee, FL.
Pokines, J.T., Symes, S.A. (Eds.), 2014. Manual of Forensic Taphonomy. CRC Press, Boca Raton, FL.
Saul, F.P., 1972. The Human Remains of Altar de Sacrificios: An Osteobiographic Analysis. Cambridge University Press, Peabody Museum, New York, NY.
Schotsmans, E.M.J., Marquez-Grant, N., Forbes, S.L., 2017. Taphonomy of Human Remains: Forensic Analysis of the Dead and the Depositional Environment. John Wiley & Sons, Hoboken, NJ.
Ubelaker, D.H., 1980. Human Skeletal Remains, Excavation, Analysis, Interpretation. Manuals on Archaeology. Taraxacum, Washington, DC.

CHAPTER

5

Skeletal microstructure, bone diagenesis, optimal sample selection, and pre-processing preparation techniques for DNA testing

Angie Ambers PhD[a,b,c]

[a]Henry C. Lee Institute of Forensic Science, University of New Haven, West Haven, CT, United States [b]Forensic Science Department, Henry C. Lee College of Criminal Justice and Forensic Sciences, Center for Forensic Investigation of Trafficking in Persons, University of New Haven, West Haven, CT, United States [c]Institute for Human Identification, LMU College of Dental Medicine, Knoxville, TN, United States

Introduction

When highly decomposed or skeletonized human remains are discovered, the ensuing forensic investigation seeks to answer a variety of fundamental questions about the decedent(s). Examination of the remains by a forensic anthropologist and odontologist can assess the individual's (1) chronological age at the time of death; (2) biological sex; (3) stature; (4) biogeographic ancestry; (5) antemortem and/or perimortem trauma; (6) dental structure (including restorations and/or defects); and (7) osseous pathology (if any). However, when partial, fragmented, significantly damaged, or intentionally altered skeletal remains are recovered, the ability to determine these features is dramatically reduced or may even be rendered impossible. In addition to anthropological and odontological analyses (or sometimes in lieu of it, depending on the completeness and condition of human remains recovered), genetic testing often is employed in an effort to gain sufficient information to make a positive identification.

Long after soft tissue decomposes, skeletal remains continue to house and protect the molecular signature of the decedent(s). Successful DNA typing of bones and teeth is largely dependent on the specific environmental conditions the remains were exposed to, the duration of exposure, the testing approaches employed, and the experience and expertise of the

https://doi.org/10.1016/B978-0-12-815766-4.00005-4

laboratory scientist processing the remains. Endogenous DNA within skeletal remains can be substantially degraded due to prolonged exposure to acidic soil, environmental heat, humidity, microbes (bacteria, fungi), ultraviolet (UV) light, chemicals, fire, saltwater, and/or freshwater. Although investigators have no control over the environmental conditions that human remains were exposed to prior to discovery, there are a number of steps that can be taken in the laboratory to maximize DNA recovery and promote success of downstream genetic testing efforts. It is imperative that laboratory personnel have knowledge of bone microstructure and the non-uniform (heterogeneous) nature in which it changes over time, as well as an understanding of the molecular interactions that occur between bone microstructure and DNA (and the role this plays in DNA preservation). Additional important considerations include: (1) selection of optimal skeletal elements for testing; (2) proper surface decontamination to remove exogenous DNA and environmental contaminants; and (3) conversion of the rigid bone matrix to a physical state that is conducive to extraction of DNA.

Bone composition, microstructure, and diagenesis

Knowledge of the major constituents of bone—and understanding changes that occur to bone microstructure over time—can assist scientists in making informed decisions regarding sampling procedures for genetic testing. Bone is an intricate composite material (a mineralized connective tissue) containing both organic and inorganic components. Bone contains four different types of cells (osteoblasts, osteoclasts, bone lining cells, osteocytes) and an extracellular matrix composed of two major structural components. The organic extracellular matrix of bone is comprised primarily of collagen, while the inorganic extracellular component is a mineralized matrix that consists predominantly of phosphate ions (PO_4^{3-}) and calcium ions (Ca^{2+}). These phosphate and calcium ions nucleate to form hydroxyapatite crystals called calcium hydroxyapatite [$Ca_{10}(PO_4)_6(OH)_2$]. Collagen and non-collagen proteins form a scaffold for hydroxyapatite deposition, and this association between structural matrix components is responsible for the rigidity and resistance of bone tissue to physical damage (Florencio-Silva et al., 2015).

Taphonomy and diagenesis: Microscopic destruction of bone

Decomposition of human remains is a complex series of biological and physiochemical processes that are directly linked to the surrounding environment. The study of changes to biological remains from the time of death until recovery and analysis is referred to as *taphonomy*. Taphonomy was originally defined by a paleontologist in 1940, generically as "the science of burial" and more specifically as "the scientific study of environmental phenomena that affect organic remains throughout their entire postmortem history" (Pokines and Symes, 2014; Efremov, 1940). Although many traditionally consider taphonomy as a process which involves decomposition of the body's *soft* tissues, taphonomic processes also produce macroscopic and microscopic changes to the skeleton (Pokines and Symes, 2014; Jans, 2008; Nielsen-Marsh et al., 2007; Smith et al., 2007; Trueman and Martill, 2002; Nielsen-Marsh and Hedges, 2000; Lyman, 1994; Micozzi, 1991). As soon as soft tissue decomposes, the skeleton begins to interact directly with the surrounding environment, as it is no longer protected by internal viscera and skin. Although the rigidity of bones and teeth initially provides a strong physical barrier to protect endogenous DNA from environmental exposure and/or intentional damage, the level of protection afforded begins to wane as the microstructural components of bone begin to change and decompose over time. Taphonomic alterations to bone microstructure are an important consideration in genetic testing efforts. Just as endogenous DNA contained within bone degrades over time, so too does the structural matrix of the bone itself. The complex process of destruction of bone microstructure and the associated chemical modifications due to exchange with the depositional environment, called **diagenesis**, has important implications for DNA preservation and recovery.

Histological studies of bone have revealed a multitude of changes that occur to the microscopic morphology of bone as a result of environmental exposure and increasing postmortem interval (PMI). **Bone diagenesis** is formally defined as "postmortem alterations in the physical, chemical, and microstructural composition of bone following its deposition in the environment" (Pokines and Symes, 2014). Among the various types of diagenetic changes that may occur to bone include: (1) microscopic cracking (development of micro-fissures); (2) collagen hydrolysis; (3) bioerosion of bone microstructural components by bacteria, fungi, and other microorganisms in terrestrial and aquatic environments; and (4) infiltration of foreign material into the bone matrix (e.g., humic acids, metallic ions) (Pokines and Symes, 2014; Jans, 2008; Nielsen-Marsh et al., 2007; Smith et al., 2007; Collins et al., 2002).

Bone diagenesis is affected by both intrinsic and extrinsic (environmental) factors. Intrinsic factors that contribute to the rate of diagenesis include the size of the bone, porosity and condition of the bone at the time of death, and bone chemistry (Pokines and Symes, 2014). Among the environmental factors that affect bone preservation include temperature, moisture, altitude, latitude, seasonality, exposure to direct sunlight, burial depth, soil type, vegetation, topography, air circulation, microorganisms, and immersion in water. Two of the most important factors influencing the preservation of bone microstructure are temperature and soil pH (Pokines and Symes, 2014; Hopkins, 2008; Hopkins et al., 2000; Nielsen-Marsh et al., 2007). Temperature is a primary driver of all biological activity and biochemical reactions. Warm ambient and/or soil temperatures accelerate microbial growth and facilitate bioerosion of bone microstructure by microorganisms (Jans et al., 2004). Soil acidity (low soil pH) dissolves the mineral components of bone and promotes leaching (Pokines and Baker, 2014). Acidic, aerated, well-drained soils are generally corrosive to bone microstructure. Under these soil conditions, leaching of bone mineral into the surrounding environment occurs, which further exposes the organic components of bone to additional alteration and degradation. Conversely, alternating wet/dry cycles and strongly alkaline soils can cause extensive cracking of bone microstructure (Nielsen-Marsh et al., 2007; Smith et al., 2007). All of these environmental factors and associated taphonomic changes in bone can compromise the quality and quantity of DNA contained within the skeleton.

Non-uniform, heterogeneous diagenesis of bone microstructure

Alteration of bone macrostructure and microstructure in the archaeological burial environment has been extensively documented. These changes to bone were once considered a phenomenon only observed in archaeological and historical samples. However, recent findings indicate that diagenetic processes and decomposition of bone microstructure can occur early on in the PMI, making it a relevant factor for consideration in contemporary forensic contexts (Pokines and Symes, 2014; Pokines and Baker, 2014; Casallas and Moore, 2012).

It is important to note that bones do not always pass through a singular taphonomic trajectory. Practically speaking, they almost never do. Significant variation in taphonomic alteration may exist between skeletonized remains, within a single skeleton, and even between regions of the same skeletal element. This is in large part due to the non-uniform (heterogeneous) manner in which diagenesis progresses through the same skeleton, as well as even within the same bone (Fig. 1).

A number of studies have focused on assessing variation in DNA quality and DNA quantity recovered from different bones within the same skeleton and even from different regions of the same skeletal element (Antinick and Foran, 2018; Andronowski et al., 2017; Hollund et al., 2017; Mundorff and Davoren, 2014; Mundorff et al., 2013). Additional case studies have inadvertently revealed heterogeneity in DNA recovery from within the same bone, further supporting that diagenetic processes progress in a non-uniform manner (Ambers et al., 2014, 2016, 2019, 2020, 2021).

Correlation between bone diagenesis and DNA preservation

Molecular interactions between the inorganic hydroxyapatite matrix of bone and DNA have been strongly correlated to DNA preservation. Prior to diagenetic alteration, the negatively charged backbone of DNA molecularly interacts with positively charged calcium (Ca^{2+}) residues in hydroxyapatite. This association is purported to protect DNA from physical and chemical damage. However, as the duration of environmental exposure and PMI increases, diagenesis of bone microstructure affects this affinity of DNA for the mineral matrix. As diagenesis progresses, ionic substitution and infilling displaces calcium ions and causes DNA to dissociate from the hydroxyapatite matrix (Fig. 2). When this occurs, DNA becomes more susceptible to damage by environmental factors (e.g., heat, moisture, microbes, soil acidity).

Understanding that diagenesis progresses in a non-uniform manner both within the same skeleton and often within the same skeletal element is important in casework, as it supports the need to consider multi-site sampling and testing of multiple cuttings as a strategic approach for maximizing chances of recovering sufficient genetic data for a positive identification.

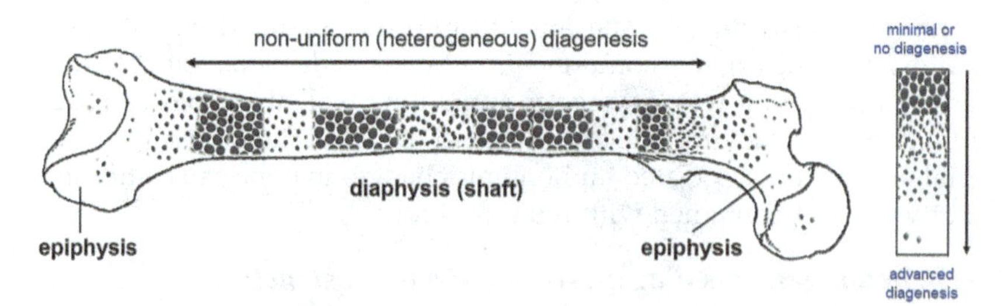

FIG. 1 Simplified schematic representation of the non-uniform manner in which bone microstructure decomposes as result of environmental exposure and with increasing postmortem interval (PMI). Due to this heterogeneous pattern of diagenesis, the diaphysis of a long bone may contain regions of intact (unaltered) bone microstructure adjacent to areas with advanced diagenetic alteration. Non-uniform, heterogeneous diagenesis is a significant contributing factor to the variation in quantity and quality of DNA often observed from different bones within the same skeleton and even from different regions of the same skeletal element. *Graphic by: Angie Ambers, PhD (Henry C. Lee Institute of Forensic Science, Henry C. Lee College of Criminal Justice and Forensic Sciences; Institute for Human Identification, LMU College of Dental Medicine).*

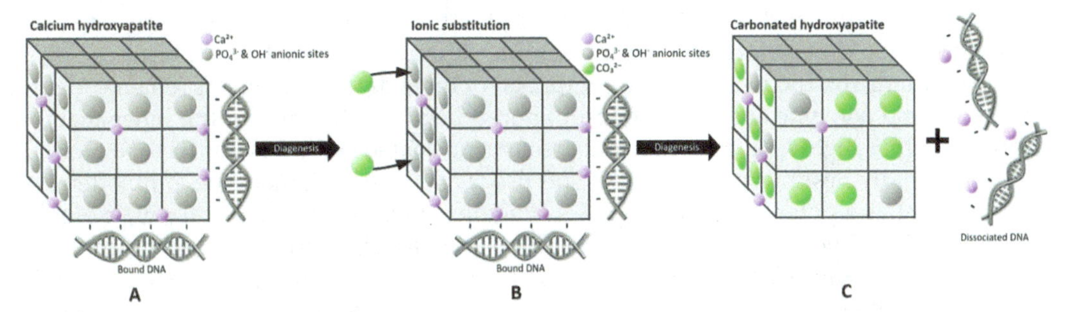

FIG. 2 Basic schematic of the mineralized hydroxyapatite matrix of bone, the interactions between DNA molecules and this microstructural component of bone, and the ionic substitutions that occur during diagenesis: (A) prior to diagenetic alteration, the negatively charge DNA backbone molecularly interacts with the positively charged calcium (Ca^{2+}) residues in the inorganic mineralized bone matrix, and this association has been shown to provide protection to DNA; (B) as diagenesis progresses, ionic substitution and infilling with carbonate (CO_3^{2-}) and other ions occurs within the hydroxyapatite matrix; and (C) as hydroxyapatite becomes more carbonated, calcium ions (Ca^{2+}) are displaced and leach out of the mineralized matrix, which causes DNA to dissociate and become more susceptible to damage by environmental factors (e.g., heat, moisture, microbes, soil acidity). *Graphic by: Paul M. Yount, MSFS (Henry C. Lee College of Criminal Justice and Forensic Sciences).*

Microanatomy of a human tooth

Human teeth are a preferred source of DNA in missing persons and unidentified human remains (UHR) investigations. The physical rigidity of teeth, as well as the fact that a portion of their structure is embedded in alveolar bone of the maxilla and mandible, provides exceptional protection from environmental damage and supports DNA preservation. The visible portion of a tooth is called the *crown*, while the *roots* are typically not visible in highly decomposed or skeletonized remains because they are encased by the alveolar bone of the maxilla and mandible (Fig. 3). However, although only the crown is visible, the roots

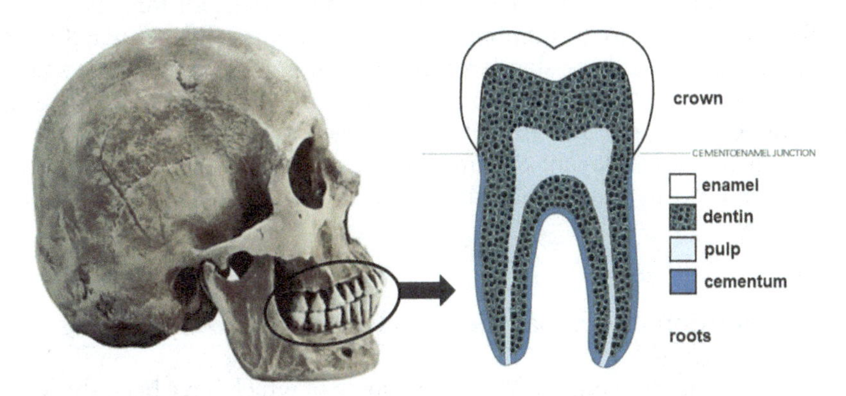

FIG. 3 Schematic of the basic microstructure of a human tooth. Teeth are comprised of three hard tissue layers (enamel, dentin, cementum) and one soft tissue (vascularized pulp). Enamel, the hardest substance in the human body, is anucleated and contains no DNA. Dentin, cementum, and vascularized pulp tissue are sources of DNA in unidentified human remains (UHR) investigations, although the latter decomposes rapidly postmortem. *Graphic by: Angie Ambers, PhD (Henry C. Lee Institute of Forensic Science; Henry C. Lee College of Criminal Justice and Forensic Sciences; Institute for Human Identification, LMU College of Dental Medicine).*

contain a substantial portion of the DNA housed within a tooth; therefore, recovery of the entire tooth should be prioritized during sample sourcing for genetic testing.

In terms of microstructure, teeth are comprised of three hard tissue layers (enamel, dentin, cementum) and one type of soft tissue (pulp) housed within a hollow, central inner chamber. Although pulp is highly vascularized and contains nucleated cells, this portion of a tooth decomposes rapidly postmortem (like other soft tissues of the body); therefore, pulp often is no longer present as a potential source of DNA (Higgins et al., 2013; Adler et al., 2011; Malaver and Yunis, 2003). The remaining three hard tissues—enamel, dentin, and cementum—must be considered for sampling. Enamel, composed primarily of calcium phosphate, covers the crown of a tooth and is the hardest substance in the human body; however, enamel is anucleated and therefore contains no cellular DNA. The remaining two hard tissues of a tooth, dentin and cementum, contain DNA (Higgins et al., 2011, 2013; Adler et al., 2011; Corte-Real et al., 2006; Malaver and Yunis, 2003). Although certain components of teeth are known PCR inhibitors (e.g., calcium, collagen), most DNA extraction methods have been developed and optimized to effectively purify endogenous DNA away from such substances. Hence, many laboratories choose to use the entire tooth for DNA extraction (Ambers et al., 2014, 2016, 2020; Ambers, 2016). This approach maximizes recovery of the total amount of DNA present within all tissues of the tooth, as well as reduces the amount of manual manipulation that is necessary to process the tooth (the latter of which increases the probability of contamination).

Adult humans typically have 32 teeth (16 in the maxilla, 16 in the mandible). Fig. 4 outlines the various types of teeth in the human mouth (canines, incisors, premolars, molars) and the Universal Numbering System approved by the American Dental Association (ADA). The Universal Numbering System is the most widely used dental notation system in the United States (hence the synonymous label as the "American System") (David and Lewis, 2018; Harris, 2018; Manjunatha, 2013). However, the Fédération Dentaire Internationale (FDI) World Dental Federation (often referred to simply as FDI) is the most commonly used tooth numbering system globally. The FDI system is designated by the International Organization for Standardization (ISO) as ISO Standard 3950:2016 (Dentistry—Designation System for Teeth and Areas of the Oral Cavity) (ISO, 2016a). It's important for forensic geneticists to be familiar with standardized tooth numbering systems because disarticulated teeth intended for DNA testing often have these standardized denotations included in the sample submission paperwork. Knowing the type of tooth provides *in situ* data about the sample, which can be informative and useful in making testing decisions. Considered in concert with other samples available and any relevant circumstances or details associated with recovery of the remains, it can be used to strategically develop a testing approach that will maximize the chances of success.

Selection of optimal skeletal elements for DNA testing

An adult human skeleton contains 206 bones, some of which have been shown to be better at preserving DNA than others. Studies suggest that prudent sample selection can significantly impact the success of genetic testing, and also can save time and resources by eliminating or reducing the need for resampling and retesting. If a complete skeleton is recovered, investigators should choose the optimal skeletal element for DNA testing based on empirical

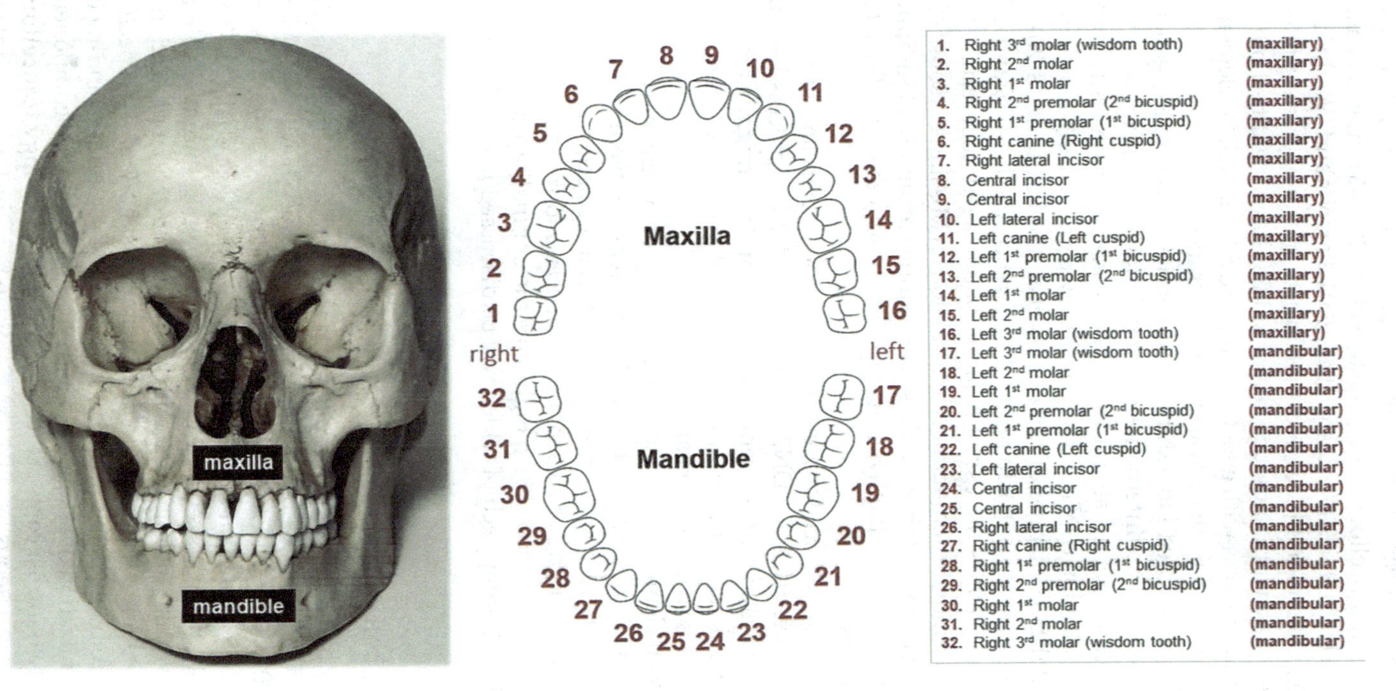

FIG. 4 Teeth are a preferred type of sample for DNA testing in missing persons and unidentified human remains (UHR) cases. In general, adult humans have 32 permanent teeth, including cuspids (canines), incisors, premolars (bicuspids), and molars. Although various classification systems exist, this figure displays the Universal Tooth Numbering System approved by the American Dental Association (ADA) and is the most commonly used notation system in the United States. *Graphic by: Angie Ambers, PhD (Henry C. Lee Institute of Forensic Science; Henry C. Lee College of Criminal Justice and Forensic Sciences; Institute for Human Identification, LMU College of Dental Medicine).*

research data and established casework recommendations. In general, dense compact bone preserves DNA better than cancellous bone (Kulstein et al., 2017; Mundorff et al., 2009; Milos et al., 2007; Rohland and Hofreiter, 2007; Edson et al., 2004; Parsons and Weedn, 1997). Weight-bearing long bones (e.g., femora, tibiae) and molar teeth are the current preferred sample types in forensic genetic casework, although recent studies have identified other sites in the skeleton that may provide success with genetic testing; therefore, skeletal elements other than weight-bearing long bones and teeth do warrant consideration.

Three internationally renowned laboratories that specialize in forensic genetic testing of skeletonized human remains—the International Commission on Missing Persons (ICMP), the Armed Forces DNA Identification Laboratory (AFDIL), and the University of North Texas Center for Human Identification (UNTCHI)—prioritize weight-bearing long bones and teeth for DNA sampling (ICMP SOP Manual, 2015; Ambers et al., 2014, 2016, 2018, 2019, 2020, 2021; Ambers, 2016, 2021; Edson et al., 2004). In order of preference, the ICMP Standard Operating Procedure (SOP) Manual dictates the use of healthy (undamaged, non-restored) teeth, femora, and tibiae for DNA testing (International Commission on Missing Persons (ICMP), 2015). For weight-bearing long bones, all three of these laboratories recommend sampling from the compact (cortical) bone along the diaphysis as opposed to from the cancellous (spongy, trabecular) bone of the epiphyses.

Despite the traditional preference for weight-bearing long bones and teeth, these skeletal elements are not always available for DNA testing. The selection process becomes increasingly difficult when recovered human remains are incomplete, fragmented, or otherwise damaged. Often, skeletal elements that may have been ideal for genetic testing have been lost due to dispersal by water, scavengers, or carnivores.

Ultimately, the skeletal element that is optimal may vary on a case-by-case basis, depending upon the precise environmental conditions that the remains were subjected to and the duration of exposure prior to recovery. Unfortunately, this information is often not known to investigators, so an understanding of current best practices and recommendations is imperative to maximize chances of success. Fig. 5 highlights the optimal types of samples from the human skeleton for DNA testing, each of which will be discussed in detail in the subsequent sections of this chapter.

Weight-bearing long bones

Weight-bearing long bones (e.g., femora, tibiae) are among the optimal sample types for genetic testing (Obal et al., 2019; Siriboonpiputtanaa et al., 2018; ICMP SOP Manual, 2015; Ambers et al., 2014, 2016, 2018, 2019, 2020, 2021; Ambers, 2016, 2021; Zupanic et al., 2012; Mundorff et al. 2009, 2013, Mundorff and Davoren, 2014; Andelovic et al., 2007; Milos et al., 2007; Edson et al., 2004). If these skeletal elements are recovered and available for testing, there are specific areas within the long bone which provide substantial protection from environmental insult and therefore ideally should be targeted for sampling. The specific region or regions of a long bone from which to sample for DNA requires an understanding of the basic microstructure and organization of osseous tissue (i.e., the hard mineral component of bone). *Osseous tissue* exists in two forms: **compact (cortical) bone** and **spongy (cancellous/trabecular) bone.** The two forms of osseous tissue differ mainly in how the bone mineral is organized and the amount of empty space present in the solidified extracellular matrix (Fig. 6).

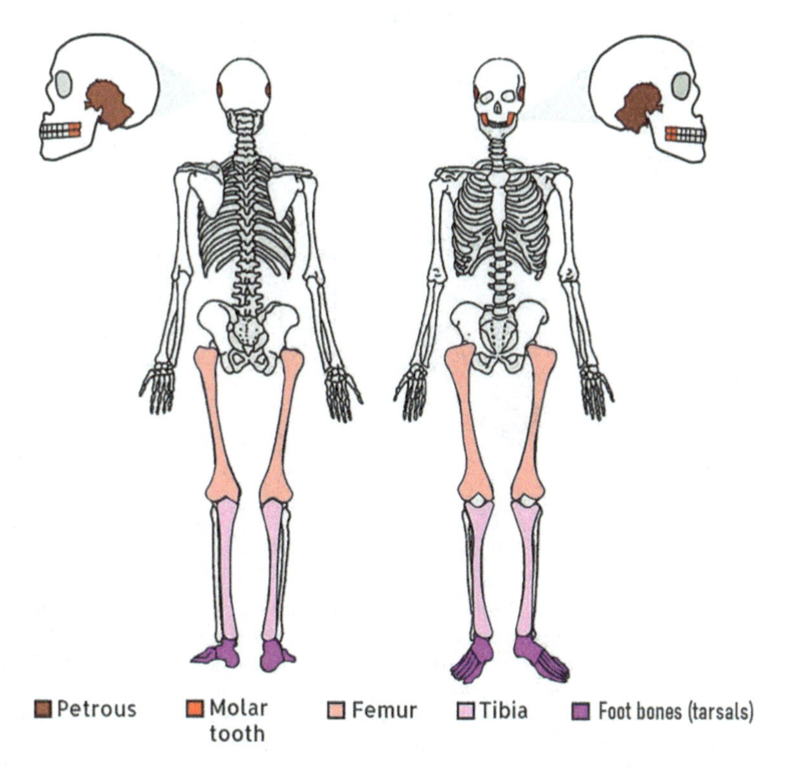

FIG. 5 Overview of optimal skeletal elements for DNA testing. Weight-bearing long bones (e.g., femora, tibiae) and molar teeth are the current preferred sample types in forensic genetic casework, although studies have identified other sites in the skeleton that often yield sufficient quantity and quality DNA (e.g., the petrous region of the temporal bone in the cranium; tarsal bones of the foot). Therefore, skeletal elements other than weight-bearing long bones and teeth do warrant consideration in identification efforts. *Graphic by: Paul M. Yount, MSFS (Henry C. Lee College of Criminal Justice and Forensic Sciences).*

The diaphysis (shaft) of a long bone consists primarily of dense, compact (cortical) bone, which contains the highest concentration and greatest number of nucleated bone cells (osteocytes) from which to source DNA. Osteocytes comprise 90%–95% of bone cells, and are located within lacunae surrounded by the mineralized bone matrix (Florencio-Silva et al., 2015; Schaffler et al., 2014; Dallas et al., 2013; Pinhasi et al., 2010). In addition to containing a high concentration of nucleated bone cells, the density and compactness of cortical bone along the diaphysis provides a strong physical barrier to the external environment, which promotes preservation of endogenous DNA.

The epiphyses (ends) of long bones are composed of spongy (cancellous, trabecular) bone, which is far less dense than the cortical bone along the shaft. Because of its microstructure, spongy bone is more prone to environmental contaminant (e.g., soil acids, microbes, chemicals) absorption, tends to be more susceptible to destruction by carnivores or scavengers, and is more vulnerable to diagenesis (particularly acidic soil corrosion and microbial bioerosion). In general, the more compact (dense) the bone or area of bone is, the stronger the barrier to environmental insult and the greater number of nucleated bone cells (osteocytes) will be present from which to obtain DNA.

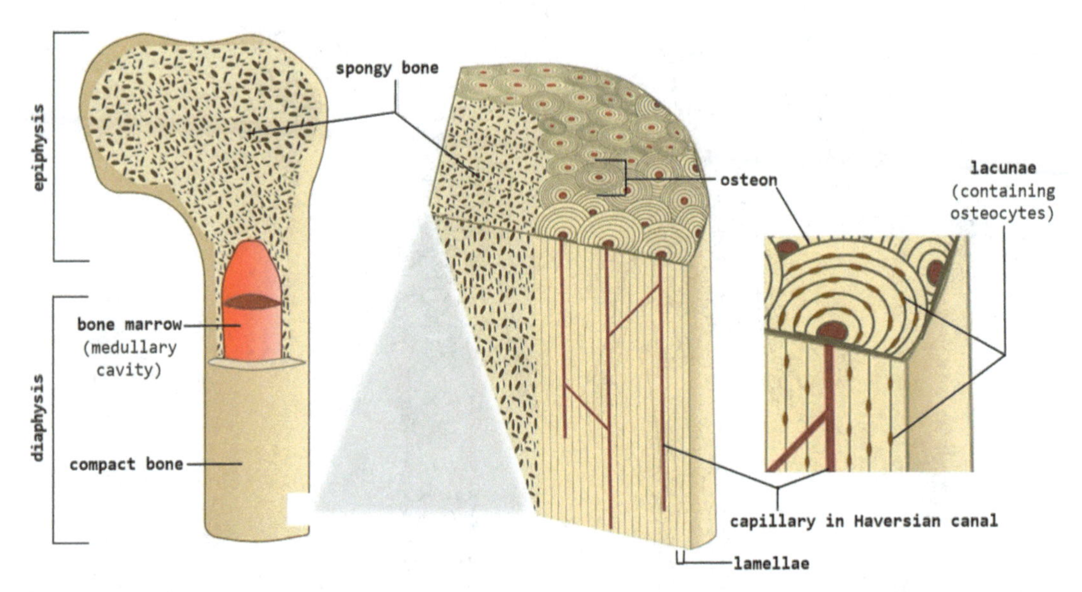

FIG. 6 Overview of the microstructure of a human long bone. The hard mineral component of bone (referred to as osseous tissue) differs primarily in how it is organized and the amount of space present in the solidified extracellular matrix. *Osseous tissue* is divided into two types: **compact (cortical) bone** and **spongy (cancellous/trabecular) bone**. The diaphysis (shaft) is comprised primarily of dense cortical bone and is preferred for genetic testing because (1) it provides a strong physical barrier against environmental insults, which promotes preservation of endogenous DNA, and (2) it contains the highest concentration of nucleated bone cells (osteocytes) from which to source DNA. Spongy (cancellous/trabecular) bone in the epiphyses and in the medullary cavity is far less dense, and is infilled with marrow. Spongy bone is targeted (and destroyed) by carnivores and animal scavengers, and by nature its microstructure is more vulnerable to the processes of diagenesis (e.g., acidic soil corrosion, microbial bioerosion). *Graphic by: Paul M. Yount, MSFS (Henry C. Lee College of Criminal Justice and Forensic Sciences).*

The epiphyses (ends) of long bones become fragile over time and are easily damaged and destroyed. Figs. 7–9 illustrate the often-observed absence or deterioration of epiphyses in skeletonized human remains recovered from a burial environment. Damage to or complete loss of epiphyseal regions of long bones typically results from acidic soil corrosion, microbial bioerosion, or carnivore activity (or a combination of these factors) (Casallas and Moore, 2012; Crow, 2008). Bone deterioration and loss in the epiphyses of long bones progresses more quickly than along the diaphysis due to the thin layer of cortical bone and high concentration of spongy (trabecular) bone in these areas. Many laboratories that are highly experienced in genetic testing of degraded human remains, such as the Armed Forces DNA Identification Laboratory (AFDIL), will not process bones or regions of bones that consist largely of spongy or trabecular bone due to their low success rates (Edson et al., 2004). If available, compact (cortical) bone is preferred.

In addition to the epiphyses, other less dense bones of the skeleton are particularly prone to acidic soil corrosion, including the vertebrae, sternum, and innominates (i.e., the fused bones of the pelvis on either side of the sacrum) (Pokines and Baker, 2014; Casallas and Moore, 2012). Hence, in addition to long bone epiphyses, vertebrae, sterna, and pelvic bones are typically not optimal sources of DNA from human remains that have been buried or otherwise exposed to soil for extended periods of time.

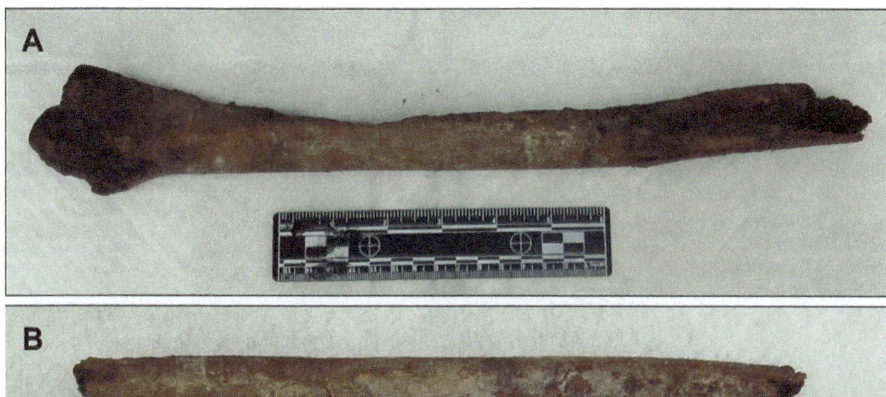

FIG. 7 Destruction of long bone epiphyses in buried human remains often caused by acidic soil corrosion and microbial bioerosion: (A) right femur (proximal epiphysis absent; partial distal epiphysis present, but heavily damaged), and (B) tibiae (proximal and distal epiphyses missing for both the left and right tibia). These skeletal remains were exhumed in 2011 after a 100+ year burial period. The anthropologist who performed the exhumation at a cemetery in West Virginia (United States) collected soil samples during the excavation for pH testing (EPA Method 9045D). The soil was found to be rather acidic; results ranged from pH 4.67 to 5.16 across three samplings: control, topsoil, and "at remains" level (Ambers et al., 2014). *Photo credit: Angie Ambers, PhD (Henry C. Lee Institute of Forensic Science; Henry C. Lee College of Criminal Justice and Forensic Sciences; Institute for Human Identification, LMU College of Dental Medicine).*

Acidic soil in particular has been described as the most pervasive long-term destructive force acting upon bone. Macroscopic and microscopic destruction of bone by acidic soil is an important factor to consider for genetic testing of human remains because skeletal elements in a burial environment have been directly exposed to soil for an extended period of time and, even in cases involving surface deposition, bones may have been in prolonged contact with topsoil after soft tissue decomposition. In both of these scenarios, the skeleton is likely to possess diagenetic alterations caused by soil acidity, which compromises the quality and quantity of DNA contained within its microstructure.

Acidic soil corrosion is concerning for skeletal remains cases because extensive global studies on soil pH levels have revealed that the majority of soils around the world are in the acidic range. Data obtained from these research studies are catalogued in the International Geosphere-Biosphere Program—Data and Information System (IGBP-DIS). In collaboration with The Nelson Institute for Environmental Studies' Center for Sustainability and the Global Environment (SAGE), the dataset has been used to construct a global soil pH map (IGBP-DIS 1998; Nelson Institute for Environmental Studies, n.d., https://nelson.wisc.edu/sage/data-and-models/atlas/maps/soilph/atl_soilph.jpg). Although a large percentage of soils are acidic, soil pH is heavily influenced by climate. In general, soil tends to be alkaline in

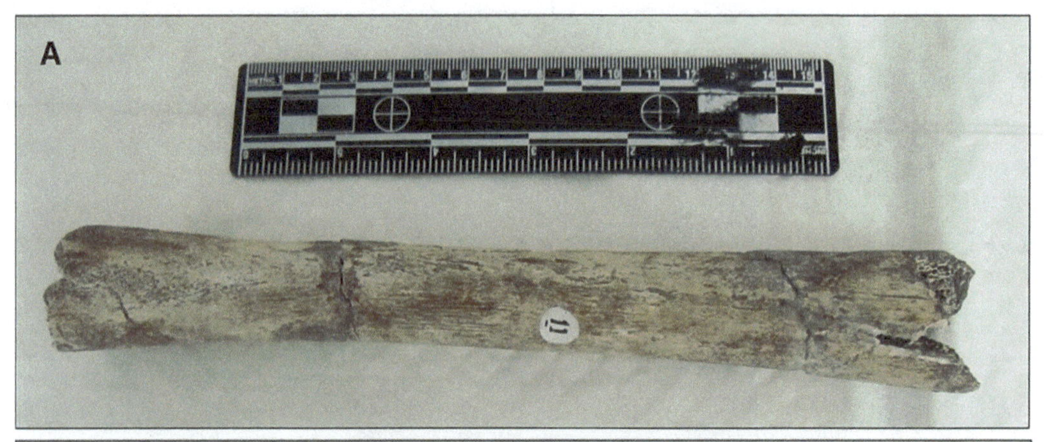

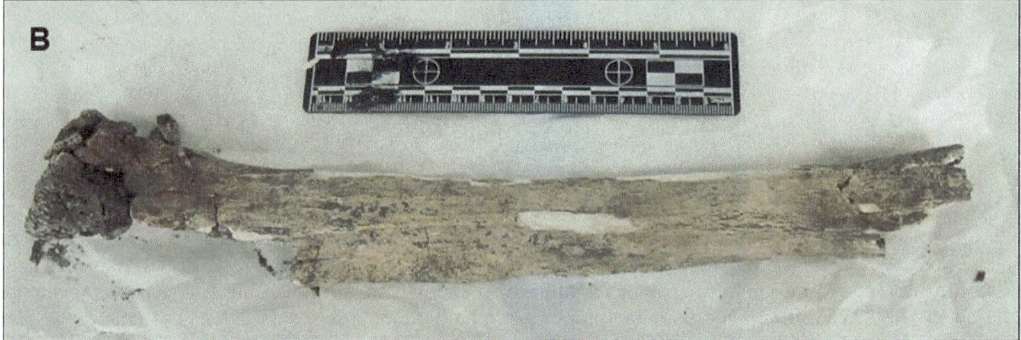

FIG. 8 Example of deterioration and loss of epiphyses in long bones after an extended period of time in the burial environment: (A) left femur of an adult female (both epiphyses absent); and (B) left tibia of an adult male (one epiphysis missing, one epiphysis in advanced stage of deterioration). These skeletal remains were exhumed from a commingled mass grave and are associated with the case described in Ambers et al. (2020, 2021) and Ambers (2021). *Photo credit: Angie Ambers, PhD (Henry C. Lee Institute of Forensic Science; Henry C. Lee College of Criminal Justice and Forensic Sciences; Institute for Human Identification, LMU College of Dental Medicine).*

dry climates, whereas in wet climates (e.g., in forested regions), soil is more acidic. However, scientists now understand that it only takes a small change in climate (e.g., during seasonal transitions) for soil pH to shift between alkalinity and acidity (Slessarev et al., 2016). This has important implications for skeletonized remains recovered from soil, as well as for preservation of DNA within bone microstructure.

Moreover, the spongy bone located in the epiphyses of long bones can be quite porous and is prone to absorbing destructive environmental contaminants (e.g., soil acids, microbes, chemicals), which can damage endogenous DNA. The spongy nature of epiphyseal bone precludes the use of some laboratory methods which are used to destroy exogenous DNA on the bone's surface. For example, it is common practice to immerse and wash bone cuttings in sodium hypochlorite (NaOCl, or bleach) to remove exogenous DNA that may be present from handling; however, if this decontamination step is performed on the epiphyses (particularly epiphyses in an advanced state of diagenesis), it may also result in damage to endogenous DNA because the bleach can easily diffuse into these porous regions of the bone. A femur

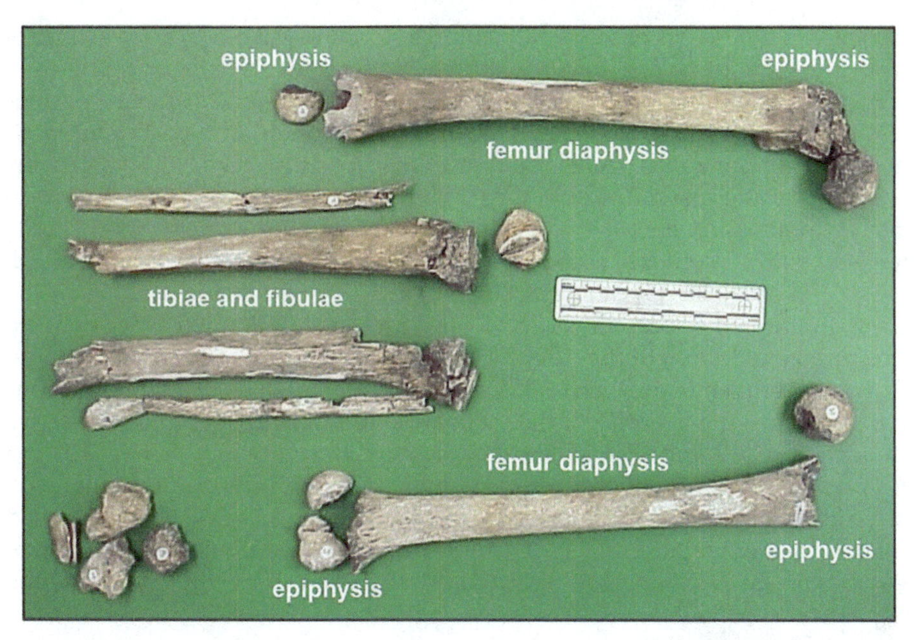

FIG. 9 The epiphyses of long bones (e.g., femora, tibiae, fibulae) are comprised predominantly of spongy (cancellous/trabecular) bone, which is more vulnerable to acidic soil corrosion, microbial bioerosion, and animal activity. Missing or deteriorated epiphyses are often encountered in human skeletal remains that have been exposed to a burial environment for an extended period of time. *Photo credit: Harrell Gill-King, PhD (Cofounder, UNT Center for Human Identification; Director, Laboratory of Forensic Anthropology).*

diaphysis is more likely to survive long-term exposure to burial or a depositional environment than the epiphyses, primarily due to the greater density of the cortical bone comprising the femur shaft (Pokines and Symes, 2014).

Teeth

As previously discussed, teeth are covered by enamel (the hardest substance in the human body) as well as alveolar bone of the maxilla and mandible (Figs. 3 and 4), providing a strong physical barrier from environmental insults, which is ideal for the preservation of endogenous DNA. Although teeth are often the most durable part of the skeleton, they are not immune to damage and loss. Single-rooted teeth (e.g., canines, incisors) often fall out of the maxilla and mandible as the body's soft tissues (e.g., gingivae, periodontal ligaments) decompose; hence, disarticulated single-rooted teeth are often not recovered in the field (Pokines and Symes, 2014). Moreover, even if disarticulated teeth are recovered, it should be considered that these teeth were more exposed in the burial or depositional environment and therefore have been more vulnerable to acidic soil erosion and microbial bioerosion over time. Teeth that have remained embedded in the alveolar bone of the maxilla and mandible are better choices for DNA testing because of the additional physical barrier provided to protect from environmental insult. Also of consideration in sample selection is the fact that anterior teeth are particularly prone to thermal destruction as a body burns (Pokines and Symes, 2014).

Among the different types of teeth in the human mouth—cuspids (canines), incisors, premolars (bicuspids), and molars—molar teeth are optimal for DNA testing due to fact that they are bulkier in size and provide the investigator with a greater amount of dentin and cementum from which to source genetic material. In addition, the location of molar teeth in the mouth (posterior to the front of the face) may position them to be better protected against physical or thermal damage in fires, vehicular accidents, or aviation disasters. The International Commission for Missing Persons (ICMP) recommends the following types of teeth for DNA testing, in order of preference: (1) first molar; (2) second molar; (3) third molar; (4) first or second premolar; (5) canine; or (6) incisor (ICMP SOP Manual, 2015).

If present, unerupted or partially erupted third molars (also known as wisdom teeth) are optimal because, in addition to the external enamel covering and other rigid hard tissues of the tooth (i.e., dentin, cementum), the DNA housed within these teeth has an additional layer of protection because the majority of the tooth structure (if unerupted or partially erupted) is still embedded within the alveolar bone of the maxilla and mandible (Fig. 10). However, one caveat to the selection of unerupted or partially erupted third molars is that root apices should be completely or almost completely formed, which can easily be confirmed using X-rays. Root formation in third molars is completed post-onset of puberty, typically between 18 and 25 years of age (Draft and Kasper, 2019). Alternatively, age estimation is routinely performed by a forensic odontologist or anthropologist when unidentified human remains are

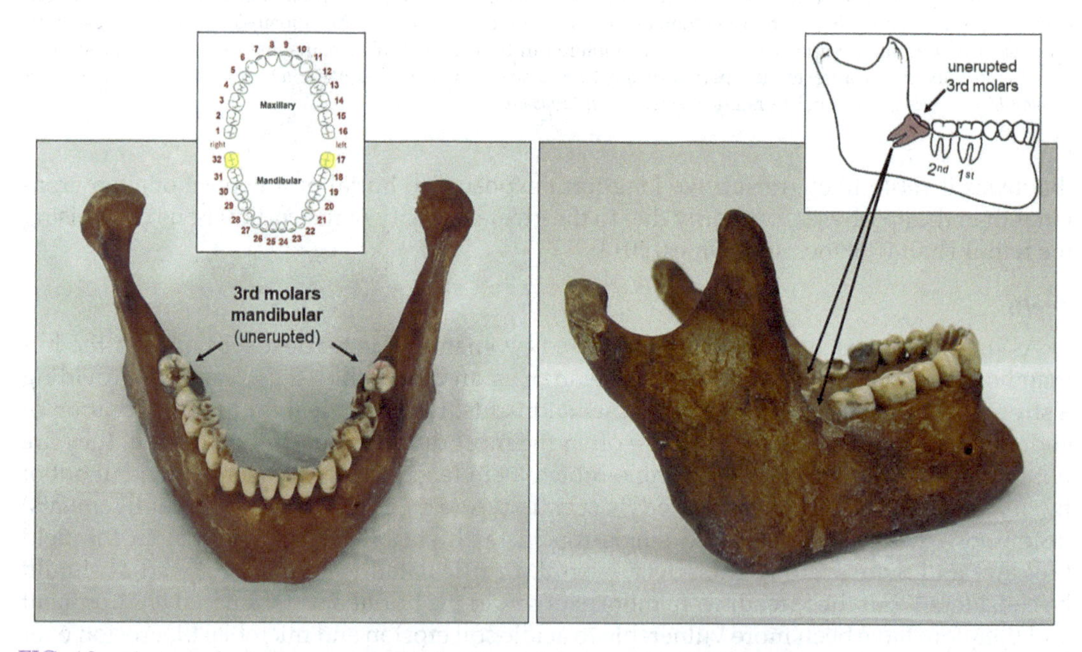

FIG. 10 Photograph of a human mandible (adult male) discovered in 2012 by a construction crew in Deadwood, South Dakota, United States (Ambers et al., 2016; Ambers, 2016). The unerupted third molars (with fully formed root apices) were selected for DNA testing. In addition to the physical barrier provided to endogenous DNA by the tooth's hard tissues (enamel, dentin, cementum), additional protection from environmental insult is afforded by the surrounding alveolar bone of the mandible. *Photo by: Angie Ambers, PhD (Henry C. Lee Institute of Forensic Science; Henry C. Lee College of Criminal Justice and Forensic Sciences).*

discovered. Hence, formal written reports from forensic odontologists and anthropologists regarding the estimated chronological age of a decedent can serve as a resource to inform on the degree of root formation present in third molars, which can be used to assess the value of this type of sample for genetic testing. If third molar root formation is absent or substantially incomplete, first or second molars are preferable.

Removal of molars from the mandible or maxilla should be performed very carefully, using an electric rotary tool (e.g., Dremel®) and a small, pre-sterilized drill bit (Fig. 11). Moreover, this process should be performed in a hood to prevent aerosolization of fine tooth powder and contamination of workbenches. Using this approach, the alveolar bone encasing the tooth within the mandible or maxilla can slowly be drilled away to ensure that the entire tooth is removed for DNA testing and to preserve the structural integrity of the remains. Pliers or other manual tools should not be used in an attempt to pull the tooth out of the mandible or maxilla. Although the use of pliers or similar manual, handheld tools is a quicker method for removing a tooth, this approach often results in fracturing the tooth as well as the mandibular or maxillary bone in which the tooth is embedded. Also, it can cause the roots of the tooth to break off and remain retained within the alveolar bone, excluding this valuable portion of a tooth from being sourced for DNA.

Regardless of the type of tooth selected for DNA testing, it is necessary to choose tooth samples that are free of caries (cavities) and that have not been subjected to antemortem restorative work. In addition, teeth with intact macrostructure (i.e., with no visible cracks or deformations) are preferred over those with compromised structural integrity. In general, the presence of caries, fractures, and/or dental restorations in teeth may reduce DNA yield and compromise the quality of endogenous DNA. Another consideration for avoiding teeth with dental restorations is that these particular teeth may be especially informative in downstream analyses should antemortem radiographs become available for comparison to make a positive identification (ICMP SOP Manual, 2015).

Petrous region of the temporal bone (skull)

The human skull consists of 22 bones (8 cranial bones and 14 facial bones). The upper and back portion of the skull, which forms a protective case around the brain, is referred to as the neurocranium. The neurocranium is comprised of eight cranial bones: one ethmoid bone, one frontal bone, one occipital bone, two parietal bones, one sphenoid bone, and two temporal bones. Two of the eight cranial bones—the temporal bones—are of particular interest for DNA testing. Sections of the temporal bones are very robust and are often more resistant to destruction and diagenesis than other parts of the cranial vault. More specifically, the petrous portion of the temporal bone is one of the densest areas in the human skeleton, and it provides an ideal environment for protection of endogenous DNA (Fig. 12). The name "petrous" originates from the Latin word *petrosus*, which means "stone-like, hard." The petrous (also referred to as *pars petrosa*) is pyramid- shaped and houses on its internal surface the bones of the inner ear (White and Folkens, 2005).

Numerous recent studies have demonstrated that the petrous region of the temporal bone often outperforms weight-bearing long bones and teeth in regards to DNA recovery (Zupanič Pajnič et al., 2021; Parker et al., 2020; Obal et al., 2019; Hansen et al., 2017; Kulstein et al., 2017; Pinhasi et al., 2015; Rollefson et al., 2015). However, the petrous is often not prioritized for sampling in casework, unless DNA testing of other skeletal elements (e.g., femora, tibiae, teeth) is unsuccessful. The location of the petrous (on the inner surface of the temporal bone of

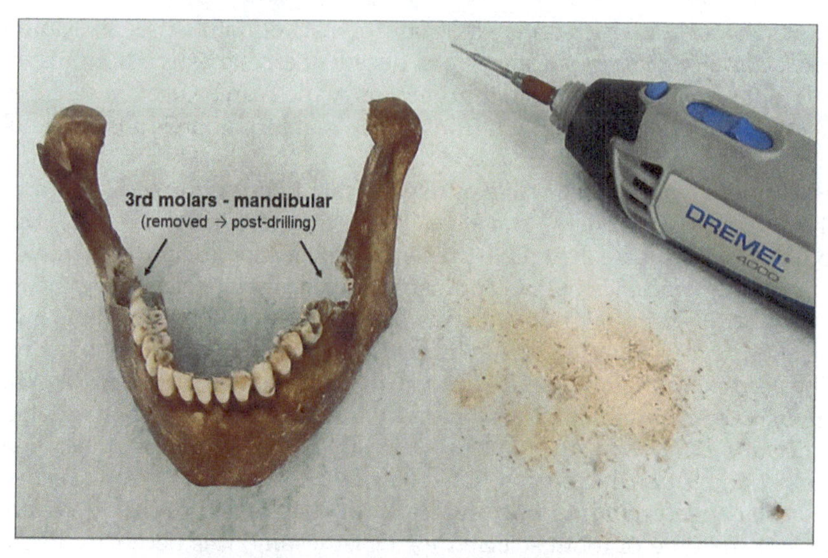

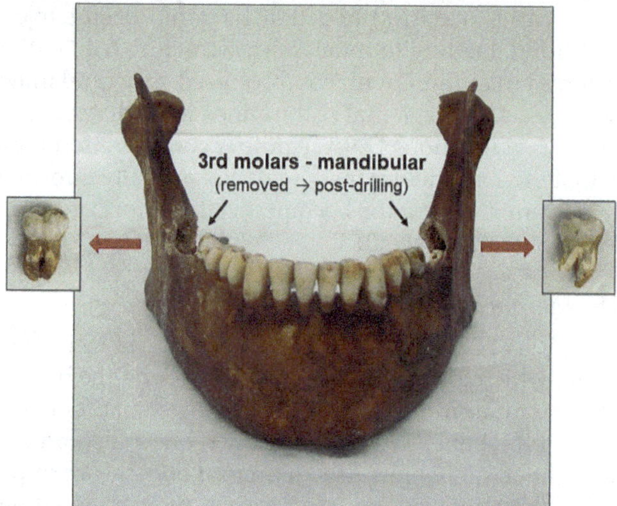

FIG. 11 Teeth can be removed from the mandible using an electric rotary tool (e.g., Dremel®) and a small, pre-sterilized drill bit. This should be performed in a hood to prevent aerosolization of fine tooth powder and laboratory contamination. Using this approach, the alveolar bone encasing the tooth within the mandible (or maxilla) can slowly and carefully be drilled away to ensure that the entire tooth is removed for DNA testing and to preserve the structural integrity of the remains. Pliers or other manual tools should not be used in an attempt to pull the tooth out, as this often results in fracturing the tooth as well as the mandible. More importantly, if the tooth breaks during removal and the roots are retained within the alveolar bone, this valuable portion of the tooth cannot be sourced for DNA. *Photos by: Angie Ambers, PhD (Henry C. Lee Institute of Forensic Science; Henry C. Lee College of Criminal Justice and Forensic Sciences; Institute for Human Identification, LMU College of Dental Medicine).*

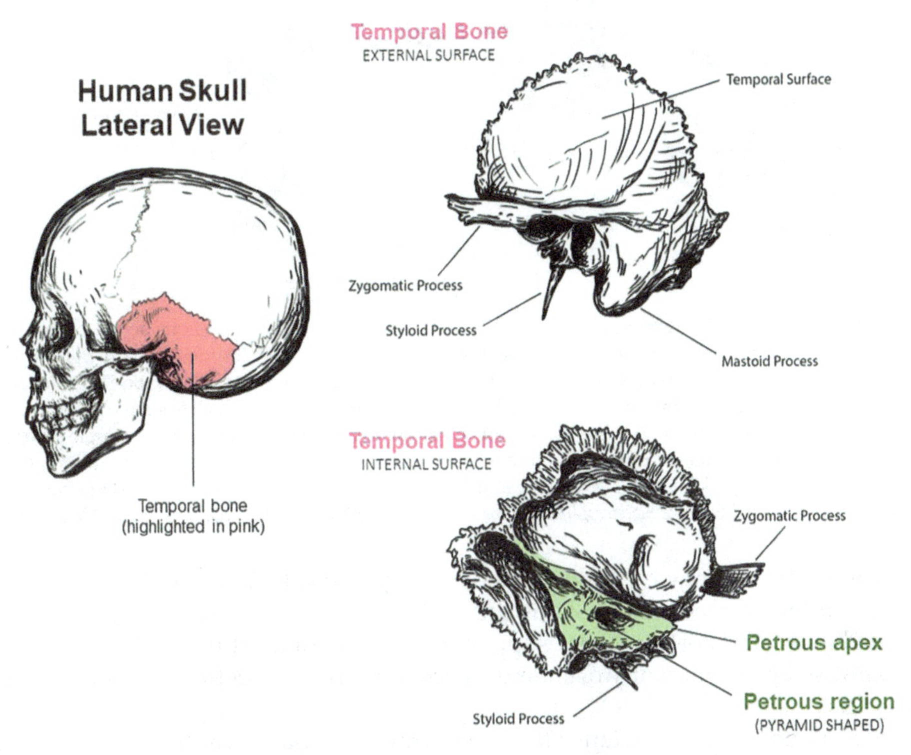

FIG. 12 Lateral view of a human skull, highlighting the external and internal surfaces of the temporal bone. The petrous (*pars petrosa*) is located on the endocranial surface of the temporal bone and is one of the densest areas within the human skeleton, thus providing an ideal environment for endogenous DNA preservation. Numerous studies have demonstrated that the petrous outperforms weight-bearing long bones and teeth in terms of DNA quantities recovered as well as the state (quality) of the DNA. However, skull bones in general are one of the least optimal skeletal elements for genetic testing; hence, conferring with a human skeletal anatomist or forensic anthropologist is necessary in order to properly identify and isolate petrous bone for DNA analysis. *Sketches by: Amelia Stockman (Henry C. Lee College of Criminal Justice and Forensic Sciences).*

the skull) and its relative inaccessibility compared to other skeletal elements is perhaps part of the reason that it is often not selected during initial rounds of genetic testing. Moreover, most geneticists are not trained in skeletal anatomy or anthropology, and therefore may not be able to accurately locate or identify the petrous. An additional consideration involves situations in which the cranial bones are recovered in a disarticulated, fragmented, or crushed state. When the cranial plates are no longer connected by sutures to form an intact cranium, locating and identifying both the temporal bone and the petrous region is even more challenging (especially for the molecular biologist or geneticist who is untrained in human skeletal anatomy and/or physical anthropology). If a skeletal anatomist or anthropologist is not available to work in conjunction with geneticists to properly identify and take samplings directly from the petrous, the skull should be avoided because studies indicate that, in general, skull bones are one of the least suitable skeletal elements for DNA testing (Milos et al., 2007; Edson et al., 2004). This is primarily due to the fact that, aside from the petrous pyramid, cranial bone is comprised predominantly of spongy cancellous bone surrounded by thin layers of compact

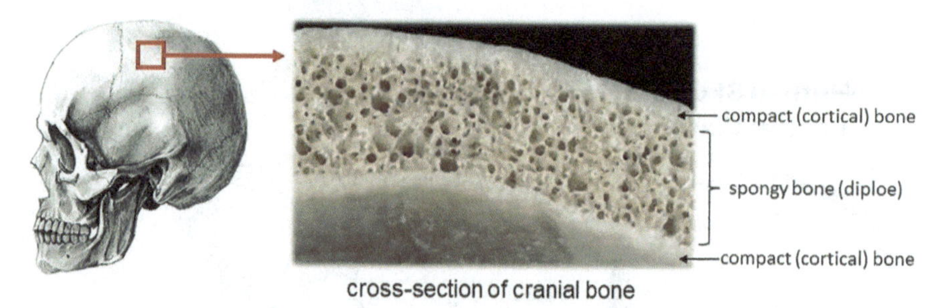

FIG. 13 Cross-section of cranial bone, revealing a large proportion of spongy cancellous bone (diploe) sandwiched between thin layers of cortical bone. Compact (cortical) bone is preferred for DNA testing due to its rigidity and density, which provides a strong structural barrier against environmental insults and supports preservation of endogenous DNA, in addition to possessing a greater number of nucleated bone cells (osteocytes) from which to source DNA. Due to the preponderance of spongy cancellous bone in the skull, cranial fragments generally are not optimal sample types for genetic testing (with the exception of the petrous region of the temporal bone). *Sketch by: Amelia Stockman (Henry C. Lee College of Criminal Justice and Forensic Sciences). Photo by: Angie Ambers, PhD (Henry C. Lee Institute of Forensic Science; Henry C. Lee College of Criminal Justice and Forensic Sciences; Institute for Human Identification, LMU College of Dental Medicine).*

(cortical) bone. The spongy cancellous bone that separates the inner and outer compact layers of bone in the skull is referred to as diploe (White and Folkens, 2005) (Fig. 13). Current recommendations and best practices support the use of compact (cortical) bone rather than bones comprised predominantly of spongy cancellous bone, for better DNA preservation and higher yields.

Another reason for the reluctance to sample from the cranial vault is that the skull is the most recognizable and relatable part of the human skeleton. Minimizing destructive sampling and preservation of structural integrity in unidentified human remains cases is an important component of forensic and archaeological casework. This is especially important with regards to sensitivity to the family members of the decedent, for the ultimate goal of genetic testing is a positive identification, in which the remains will be returned to the family for proper burial. Accessing the petrous for DNA testing often involves the removal of a portion of the temporal bone from the skull using an oscillating autopsy saw, leaving behind a large hole or missing section, the sight of which may cause emotional distress to the decedent's family. In addition, structural preservation of the skull in historical or archaeological remains is often highly desired (e.g., as is the case for museum specimens or precious/rare archaeological collections), so this may preclude destructive sampling of the petrous region of the temporal bone for genetic testing purposes.

Foot bones

There are a total of 26 bones in the human foot. The skeletal elements of the foot are divided into three types—tarsals, metatarsals, and phalanges. Specifically, there are seven tarsal bones (talus, calcaneus, cuboid, navicular, medial cuneiform, intermediate cuneiform, lateral cuneiform); five metatarsals; and fourteen phalanges (five proximal phalanges, four intermediate phalanges, five distal/terminal phalanges) (Fig. 14). The two largest bones of the foot are the calcaneus (heel bone) and the talus. The superior surface of the talus articulates with the distal surface of the tibia and fibula at the ankle joint. The calcaneus is the largest bone in the foot and supports the talus (White and Folkens, 2005) (Fig. 15). The calcaneus and talus are

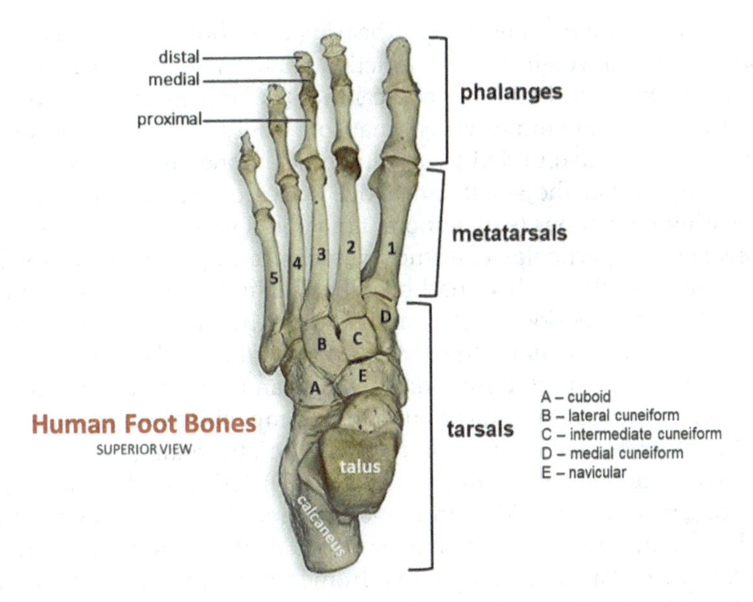

FIG. 14 Superior view of the 26 bones of the human foot: 7 tarsals, 5 metatarsals, and 14 phalanges. The two largest foot bones—the calcaneus and the talus—are the most weight-bearing of all bones within the foot and are structurally robust, which facilitates their survival in the burial or depositional environment. *Photo and graphic design by: Angie Ambers, PhD (Henry C. Lee Institute of Forensic Science; Henry C. Lee College of Criminal Justice and Forensic Sciences; Institute for Human Identification, LMU College of Dental Medicine).*

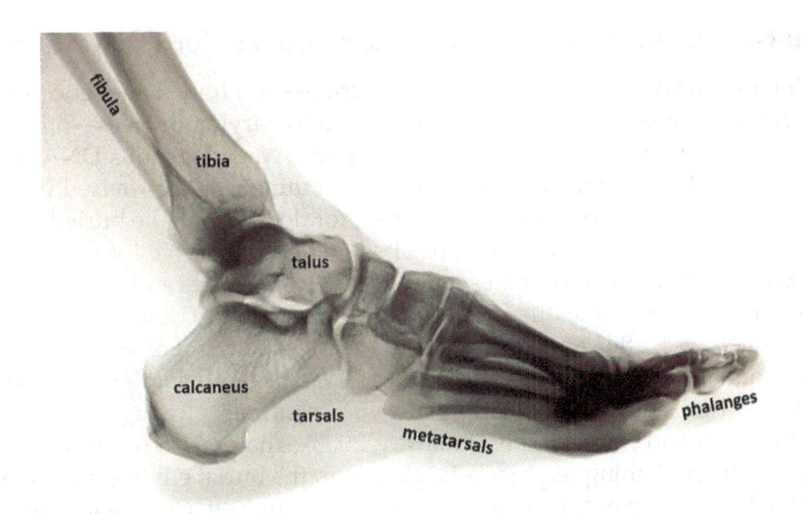

FIG. 15 Lateral view of the bones in the human foot. The two largest bones—the calcaneus (heel bone) and the talus—are the most weight-bearing of all bones within the foot and are structurally robust. The superior surface of the talus articulates with the distal surface of the tibia and fibula at the ankle joint; the calcaneus forms the heel of the foot and supports the talus. Recent studies (albeit limited) have demonstrated that tarsal bones outperform weight-bearing long bones and the petrous in terms of DNA yield.

structurally robust and are the most weight-bearing of all bones within the foot. Due to their structural robustness and weight-bearing function, these two bones are more likely to survive in the burial environment and may be good candidates for genetic testing (if available).

Although cross-sectional studies reveal that foot bones are comprised predominantly of spongy cancellous bone surrounded by a thin cortical bone layer, some recent research (albeit limited) purports that the small cancellous bones of the foot outperform the petrous and weight-bearing long bones (e.g., femora, tibiae) in terms of total DNA yield and genetic profile completeness. In particular, one study demonstrating such results involved only three sets of human remains, all of which had been buried in the same geographic location, and for a relatively short time period (i.e., 4 years) before exhumation and genetic testing. In this study, the tarsal bones (particularly the cuneiforms) performed best. The authors assert that, because foot bones are exposed to mechanical stress and possess less tissue biomass during life, they are more prone to mummification than decomposition, which perhaps confers some level of additional protection to bone microstructure and endogenous DNA compared to other components of the skeleton (Emmons et al., 2020). Other studies also support the value of foot bones for genetic testing (Mundorff and Davoren, 2014; Hines et al., 2014).

It should be noted, however, that smaller bones of the skeleton (e.g., foot bones, hand bones) are easily washed away and separated from larger components of the skeleton during heavy rains or flooding and, additionally, these bones may not be recovered due to predatory animal or scavenger activity in the depositional environment. The smaller skeletal elements of the foot also are more difficult to accurately reassociate with remains in commingled mass burial scenarios, so their use in genetic testing should be limited in these situations.

Additional considerations in selecting skeletal samples for DNA testing

In addition to bone composition and microstructure—and how diagenetic processes afflict bones within the same skeleton at different rates and to varying degrees—other circumstantial factors associated with burial can potentially preserve endogenous DNA and should be considered in sample selection. For example, it is common in improvised burials (such as those associated with homicide, genocide, or war) for decedents to be buried in clothing and footwear. Leather footwear, in particular, has been noted to survive for extended periods of time in the burial environment (Pokines and Baker, 2014; Pokines and Symes, 2014; Janaway, 2002, 2008). The resistance of leather to decay (in comparison to other types of clothing) has been documented even after approximately 100 years of burial. For example, excavation of mass graves containing the remains of the World War I (WWI) soldiers in Europe revealed complete dissolution and decomposition of fabric clothing, although leather boots were still relatively intact and recognizable (Janaway, 2008). The persistence of leather boots and other types of leather clothing (e.g., jackets, gloves) in the burial environment may provide a protective barrier (albeit temporary) against the destructive effects of acidic soil. It has been noted that leather boots preserve not only the foot bones but also the distal portions of the tibia and fibula (Pokines, 2009).

Correlation between skeletal weathering and DNA quality and quantity

Some studies have attempted to assess if there is a correlation between the quality and/or quantity of DNA present within a skeletal element and the visual appearance or physical condition of a bone or skeleton (Misner et al., 2009; Foran, 2006). These studies examined 36 human skeletons exhumed from a historical cemetery in Pennsylvania, with burials dating back to the early 1800s. Each skeleton and every bone within each skeleton were assigned a "weathering score" based on a previously devised system used by the Smithsonian Institution (Ubelaker et al., 2003; Behrensmeyer, 1978). The weathering score (ranging from 1 to 5) classified each bone's physical appearance and denoted specific features or defects that may affect DNA preservation and yield (e.g., degree of cracking, flaking, splintering). The overarching goal of this research was to determine if the degree of weathering and external visual appearance of skeletal remains could serve as a reliable predictor of downstream DNA testing success. Although a correlation between the degree of skeletal weathering and endogenous DNA quantity or quality was not demonstrated, this study inadvertently provided support that the type of skeletal element selected for testing does impact DNA yield. Moreover, the type of osseous tissue assayed had a notable and statistically significant influence on DNA typing success, with compact cortical bone yielding statistically significant higher quantities of DNA than spongy cancellous bones.

Pre-processing of skeletal remains for DNA extraction: Surface-sanding, sectioning, decontamination, and pulverization

Skeletonized remains from depositional or burial environments can have a variety of contaminants on their surfaces that must be removed prior to pulverization and DNA extraction. Otherwise, contaminants and exogenous DNA become incorporated into the bone powder fractions subjected to DNA extraction procedures and can affect the purity, reliability, and relevance of genetic data obtained. Aside from soil acids and environmental chemicals or pollutants, exogenous DNA from anyone who has come into contact with the remains can be present on bone or tooth surfaces. This may include (but is not limited to) forensic anthropologists, archaeologists, medical examiners, forensic odontologists, and laboratory personnel.

Contamination during recovery or exhumation: DNA transfer from direct contact and handling of human remains

During excavation, recovery, packaging, and transport of skeletonized remains for DNA testing, a variety of persons may come in contact with or handle the samples, including forensic anthropologists, odontologists, archaeologists, Disaster Mortuary Operational Response Team (DMORT) personnel, Disaster Victim Identification (DVI) team members, detectives, crime scene technicians, and research scientists. Forensic anthropologists and odontologists typically examine skeletal remains first, prior to submitting samples for DNA analysis. During examination and physical handling of the remains, exogenous DNA can inadvertently be transferred onto the external surfaces of bones or teeth. This is commonly referred to as "touch DNA," and this DNA transfer mechanism via shedding of epithelial cells is well documented in the literature (Szkuta et al., 2015, 2018; Buckingham and van Oorschot, 2016; Goray et al., 2016; Fonnelop et al., 2015; Oldoni et al., 2015; Oleiwi et al., 2015; Templeton et al., 2015; van Oorschot et al., 2014, 2019; Lehmann et al., 2013; Meakin and Jamieson, 2013; Phipps and Petricevic, 2007; Lowe et al., 2002). To further complicate the probability of epithelial cell shedding during direct handling or indirect contact, advancements in DNA detection technology (particularly with regards to sensitivity) have increased laboratories' capability to detect small amounts of contamination derived from a transfer event. Hence, it is imperative to carry out extensive decontamination procedures to remove any contaminant or exogenous DNA from the surface(s) of skeletal elements that are intended for genetic testing (Basset and Castella, 2019). This helps ensure that any DNA profile obtained is endogenous to the decedent and not derived from an external source.

Contamination prevention in the laboratory

If genetic testing of human skeletal remains is desirable or necessary for identification, extensive contamination control procedures should be used, regardless of whether the remains are contemporary, historical, or archaeological in nature. In general, "best practices" for processing skeletal remains are aligned with the standardized contamination prevention measures utilized for processing ancient DNA specimens, including: (1) use of protective suits, gloves, and facemasks; (2) bleach decontamination and UV irradiation of work benches and associated equipment; (3) physical removal and chemical destruction of contaminant or exogenous DNA on external surfaces of bones or teeth; (4) sterilization and UV irradiation of saw blades, drill bits, grinding stones, and Freezer/Mill accessories (e.g., end caps, impactors, grinding vials); (5) preparation and processing of bone samples in a designated low template (LT) laboratory; and (6) preparation and processing of all reference samples in a designated "high copy" laboratory that is physically separate from the space where human remains are handled (Gilbert et al., 2006; Kemp and Smith, 2005; Yang and Watt, 2005; Poinar, 2003; Cooper and Poinar, 2000). To prevent cross-contamination, bones and teeth should never be processed in the same location as the reference samples that have been submitted for comparison. In addition, every laboratory should maintain an internal employee elimination database containing DNA profiles of all laboratory staff and personnel for comparison to DNA results obtained from unidentified human remains (UHR). It also is prudent to collect reference samples from all individuals involved in recovery or exhumation for comparison and for exclusionary purposes in the event of suspected contamination.

Quality Assurance (QA) standards and Forensic DNA Grade (ISO 18385) products for identification of human skeletal remains

In addition to following the previously described contamination prevention measures during handling of remains, laboratories involved in processing contemporary, historical, or archaeological skeletal remains should meet or exceed guidelines established by the Scientific Working Group on DNA Analysis Methods (SWGDAM), the DNA Advisory Board (DAB), the European Network of Forensic Science Institutes (ENFSI), and the Biology Specialist Advisory Group (BSAG) of Australia and New Zealand. These organizations work collectively and collaboratively in both national and international capacities to promote quality assurance (QA) and accuracy in forensic genetic testing (Butler, 2012; SWGDAM, 2017; ENFSI, 2017; Gill et al., 2010). In combination with these guidelines and recommendations, laboratories involved in human identification cases should always use *Forensic DNA Grade* products to further support the integrity of genetic testing results, in accordance with Standard 18385 established by the International Organization for Standardization (ISO). The ISO 18385 Standard specifies requirements for the manufacturing of products used in collection, storage, and analysis of biological material for forensic DNA purposes. Consumables covered by this international standard include those used for evidence collection (e.g., containers, packaging materials, swabs), as well as products used for processing (e.g., disposable laboratory coats, gloves) and storage of DNA samples (e.g., tubes and other plasticware) (International Organization for Standardization (ISO), 2016b). ISO 18385 requires manufacturers to minimize the risk of occurrence of detectable human DNA contamination in products used by the global forensic community. Additional guidance on the control and avoidance of contamination both at the scene and in the laboratory have been published by the UK government's Forensic Science Regulator (FSR, 2020a,b).

Sterilization and decontamination of saw blades, tools, and accessories

In addition to the mechanical and chemical methods used to clean the surfaces of bones and teeth prior to DNA extraction, all tools and accessories should be effectively and systematically sterilized prior to coming into contact with any bones or teeth intended for DNA testing. This includes saw blades, Freezer/Mill accessories (e.g., stainless steel end caps and impactors), drill bits, and grinding vials. Prior to pulverization of bone cuttings or teeth into powder, all metal tools and accessories—such as saw blades, end caps, impactors, and drill bits—should be subjected to extensive sterilization procedures. This helps ensure that the DNA recovered is endogenous to the decedent and not from an external source as a result of contamination during handling or the use of non-sterile equipment/tools. It should be noted that "sterile" in the forensic genetics context implies the removal of exogenous or contaminant DNA (as compared to the word "sterile" as used in reference to medical or surgical tools, which focuses solely on destruction of pathogenic bacteria and viruses contained within body fluids or tissues).

Saw blades, Freezer/Mill accessories (e.g., stainless steel end caps and impactors), and drill bits can be effectively sterilized using a sonicator equipped with a water bath and heating component. Sonication should be performed in molecular grade (sterile) water and Tergazyme™ for 1 h at 60°C. Post-sonication, each tool/accessory should be sequentially rinsed with molecular grade water followed by 95%–100% ethanol and then placed in a sterile (UV-crosslinked) dead-air hood overnight to dry naturally. Clean tools and accessories that will not be used immediately can be stored in UV-crosslinked conical tubes or other pre-sterilized containers and labeled accordingly.

Furthermore, the polycarbonate grinding vials (center cylinders) for use in Freezer/Mill instruments do not meet quality assurance (QA) standards for skeletal remains casework. These vials are not manufactured as Forensic DNA Grade (ISO 18385) products for human identification and therefore must be sterilized prior to use in DNA testing of bones and teeth. The single-unit packaging of the grinding vials seemingly implies sterility, although they have been shown to contain exogenous human DNA likely as a result of handling and packing during the manufacturing process (data unpublished). Polycarbonate grinding vials (center cylinders) can be effectively sterilized by liberally spraying the internal surface sequentially with a 10% commercial bleach solution (3% sodium hypochlorite), followed by molecular grade water, and then 95%–100% ethanol. Clean polycarbonate grinding vials (center cylinders) can then be placed in sterile 50-mL polypropylene conical tubes (with caps off) to dry in a dead-air hood. As an additional QA/QC measure, the sterilized/washed grinding vials can be UV-crosslinked immediately prior to addition of a bone cutting or tooth for pulverization in a Freezer/Mill instrument.

Mechanical and physical pre-processing: Surface-sanding and sectioning of long bones

Because skeletal remains recovered from a burial or depositional environment have accumulated a variety of contaminants on their surfaces, steps must be taken to remove these substances prior to pulverization of the bone into powder for DNA extraction. Among these contaminants include soil acids (e.g., humic acids, fulvic acids), environmental chemicals or pollutants, accelerants or fuel, and exogenous DNA deposited during handling. These contaminants, if not effectively removed during pre-processing, will become incorporated into the bone powder fractions subjected to DNA extraction procedures and can affect the purity, reliability, and relevance of genetic data obtained. Moreover, soil acids (particularly humic acids) are known PCR inhibitors and may co-extract with DNA if not effectively removed or reduced in quantity, thereby inhibiting successful downstream amplification of target loci.

Often substances on the external surfaces of bones are dried and so rigidly adhered to the bone that simply rinsing or washing will not suffice to remove them. One of the best methods for cleaning the surfaces of the diaphyses of long bones in preparation for sectioning involves the use of an electric rotary tool (e.g., Dremel®) and an aluminum oxide grinding stone (Fig. 16). Most Dremel® tools have various speed settings, which provide the user with the

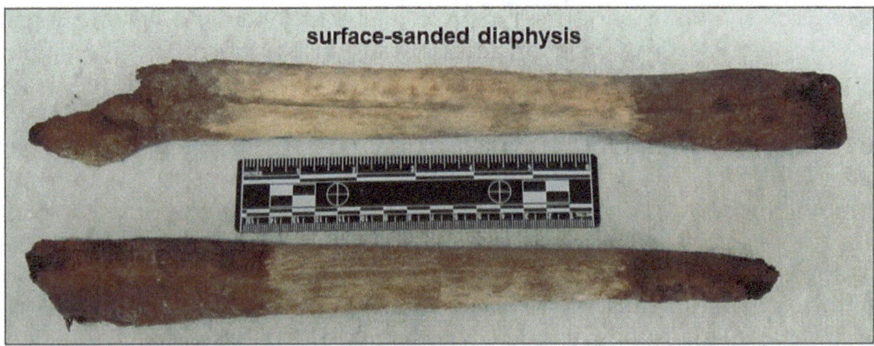

FIG. 16 Example of an electric Dremel® rotary tool equipped with an aluminum oxide grinding stone accessory for surface-sanding of the diaphysis of a long bone to remove soil acids, environmental contaminants, and exogenous DNA. Most electric rotary tools have variable speed settings, permitting the user to concurrently select the necessary speed to remove the surface layer without generating too much heat during the sanding process (which can damage endogenous DNA). Sanding procedures should always be performed in a laminar flow hood to contain aerosolized fine bone powder and prevent laboratory contamination. *Photo by: Angie Ambers PhD (Henry C. Lee Institute of Forensic Science; Henry C. Lee College of Criminal Justice and Forensic Sciences; Institute for Human Identification, LMU College of Dental Medicine).*

ability to select the setting which most effectively sands and cleans the surface of the diaphysis without simultaneously generating too much heat (which can damage endogenous DNA). In addition, aluminum oxide grinding stones come in a variety of sizes and shapes, which accommodate the broad and diverse range of bones in the human skeleton (Fig. 17). Laboratory personnel can select a grinding stone of the appropriate size and shape for the specific skeletal element submitted for DNA testing, as well as on a case-by-case basis. The entire sanding process should be performed inside a laminar flow hood, in order to contain aerosolization of fine bone powder and contamination of laboratory workbenches and equipment. A cheaper alternative to surface-sanding is sandpaper, although it often falls apart during the process and offers substantially lower capacity to remove hardened or encrusted layers of soil or other contaminants from the bone surface.

After surface-sanding, an autopsy saw (e.g., MOPEC Model BD040) and oscillating saw blade can be used to remove a "window" of compact (cortical) bone from the diaphysis of a long bone (Fig. 18). This is standard procedure in forensic DNA casework. Cutting a window in the diaphysis retains the length and physical structure of the long bone, which is important for taking measurements that anthropologists use to estimate the stature (height) of an unknown decedent. After the window is removed, it can further be sectioned into the appropriately sized (smaller) fragments necessary to fit into the grinding vials (center cylinders) used in Freezer/Mill instrumentation (Fig. 19). If preservation of the longitudinal structure of a long bone is not necessary or desired, a cross-section of the diaphysis can be removed even more quickly and then subsectioned into smaller pieces accordingly (Fig. 20).

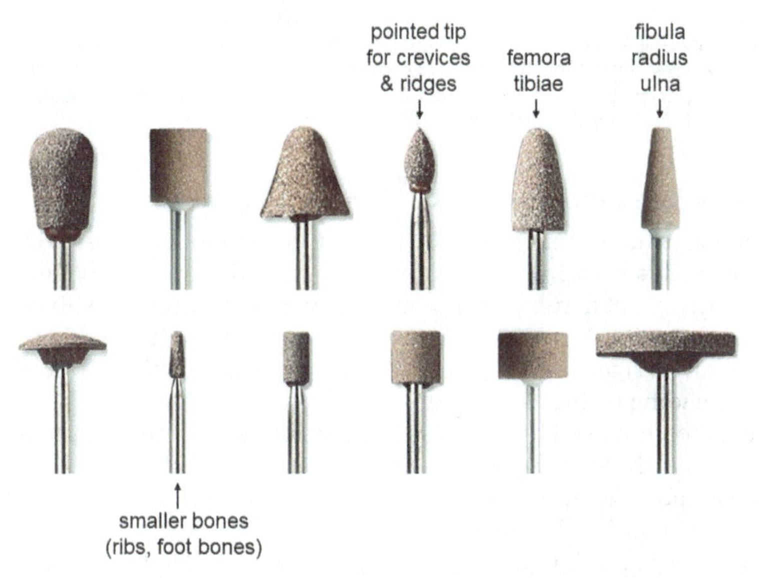

FIG. 17 Aluminum oxide grinding stones—single-use accessories for an electric rotary tool (e.g., Dremel®)—are available in a variety of different sizes and shapes, which accommodate the broad range of diverse bones in the human skeleton. Laboratory personnel should choose the size and shape of grinding stone based on the specific type of bone submitted for DNA testing. For long bones, selection of grinding stone(s) should take into consideration that both the external surface of the diaphysis and the inner medullary cavity should be surface-sanded.

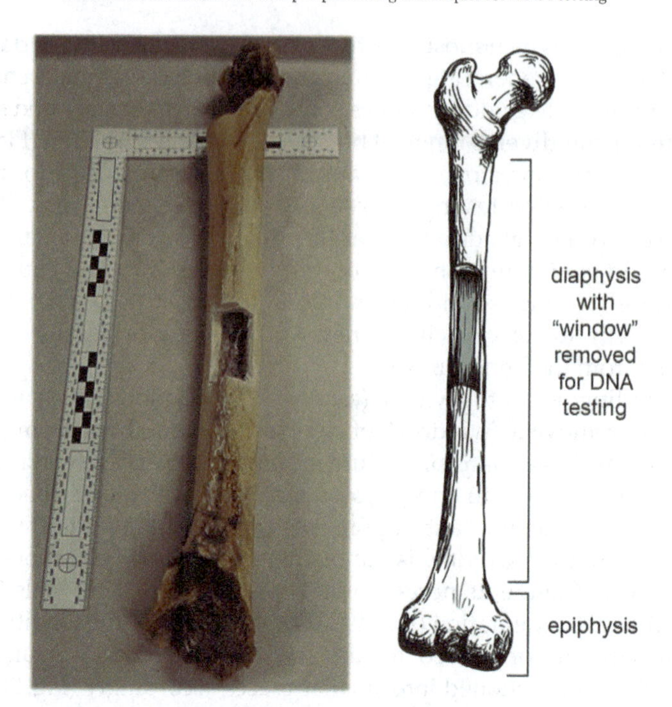

FIG. 18 After sanding the surface of a long bone to remove adhered soil acids, environmental contaminants, and exogenous DNA, standard forensic DNA casework practice involves cutting a "window" of compact (cortical) bone from the diaphysis. Removing a portion of the diaphysis in this manner preserves the length and overall physical structure of the long bone, which is important for the measurements used by forensic anthropologists in estimating the stature (height) of a decedent. *Photo by: Angie Ambers PhD (Henry C. Lee Institute of Forensic Science; Henry C. Lee College of Criminal Justice and Forensic Sciences); Sketch by: Amelia Stockman (Henry C. Lee College of Criminal Justice and Forensic Sciences; Institute for Human Identification, LMU College of Dental Medicine).*

Chemical washing of external surfaces of bone cuttings or teeth

After sectioning, bone cuttings or whole (intact) teeth should be subjected to further cleaning prior to placing in a grinding vial (center cylinder) and pulverization into powder using a Freezer/Mill instrument and liquid nitrogen. Each bone cutting or tooth can be placed individually in a sterile 50-mL polypropylene conical tube for a 3-step washing procedure, sequentially as follows: (1) immersion and mild agitation in a 50% bleach dilution (3% NaOCl) for 5–15 min (depending on the age, porosity, and condition of the bone); (2) prompt draining of the bleach solution followed by 4–5 rinses with molecular grade water to remove residual bleach; and (3) a final, brief immersion in 95%–100% ethanol. After pouring off the ethanol, the 50-mL conical tubes (each containing a clean bone cutting or tooth) should be placed in a pre-sterilized, dead-air hood with caps off to allow the bone cuttings and/or teeth to dry (usually for 24–48 h) (Fig. 21). Once the bone cuttings and/or teeth are completely dry, they can either be directly added to pre-sterilized grinding vials (center cylinders) and pulverized in a Freezer/Mill, or the conical tubes can be capped to store the samples for pulverization at a later date. Note that most commercial bleach brands contain 6% sodium hypochlorite (NaOCl), so the 50% bleach dilution described herein should yield a solution with 3% NaOCl concentration. If the laboratory purchases a concentrated version of commercial bleach, the

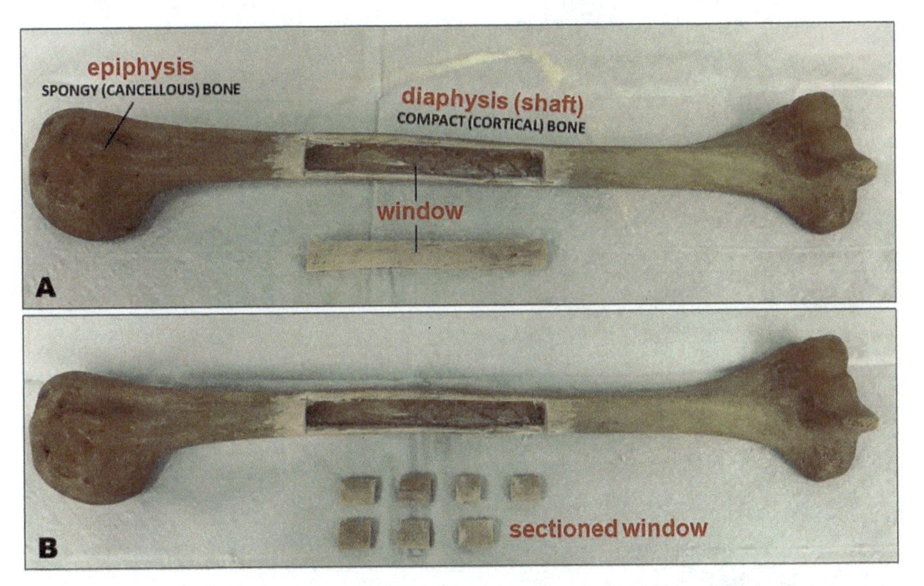

FIG. 19 (A) Example of a "window" of compact (cortical) bone removed from the diaphysis of a weight-bearing long bone. Note that the diaphysis was sanded with an electric rotary tool (e.g., Dremel®) and aluminum oxide grinding stone prior to cutting the window (as evidenced by the lighter coloration of the diaphysis surrounding the window). (B) After removing the window from the diaphysis, it should be further sectioned into smaller fragments in preparation for pulverization in a Freezer/Mill and DNA extraction. *Photos by: Angie Ambers PhD (Henry C. Lee Institute of Forensic Science; Henry C. Lee College of Criminal Justice and Forensic Sciences; Institute for Human Identification, LMU College of Dental Medicine).*

starting concentration of NaOCl should be considered in the dilution factor to ultimately yield a 3% NaOCl solution. In addition, the NaOCl concentration as well as the immersion time should be decreased for older bones or for samples with advanced diagenetic alteration. Although not an exact science, the goal of the three-step wash is to effectively remove exogenous (contaminant) DNA from the surface of the bone or tooth without damaging or destroying endogenous DNA.

Pulverization of sectioned bones and teeth into powder

Clean, dry bone cuttings or teeth are typically pulverized into a fine powder using a Freezer/Mill (e.g., SPEX Freezer/Mill 6750, SPEX Freezer/Mill 6770), liquid nitrogen, and associated grinding accessories. First, a dry bone cutting or tooth is placed into a pre-sterilized grinding vial (center cylinder) along with a stainless steel impactor, and the tube is sealed on either side with stainless steel end caps (Fig. 22). In order for the grinding process to proceed properly, the bone cutting or tooth and the impactor must be able to freely move around on the inside of the grinding vial (center cylinder). If the bone cutting or tooth is too large and the impactor cannot move from side-to-side, the sample will not be able to be pulverized during the Freezer/Mill run.

After thawing completely (typically overnight), one end cap can be removed and bone powder can be weighed and distributed into sterile, 15-mL conical tubes in preparation for DNA extraction. Typical bone or tooth powder quantities used in forensic DNA casework range from 0.2 to 1.0 g of powder per extraction (although less powder may be sufficient for

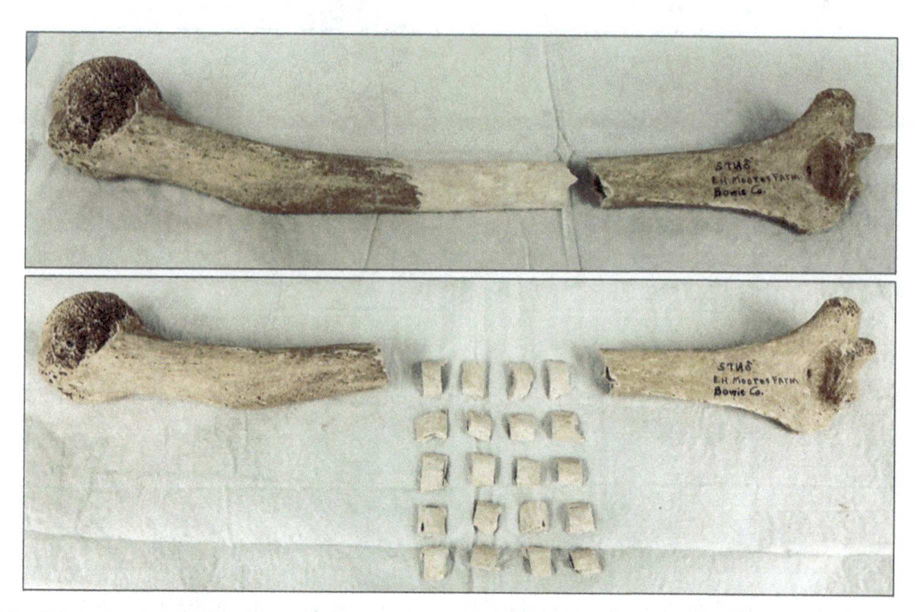

FIG. 20 Example of removal of a cross-section of the diaphysis of a long bone (humerus) recovered from an archaeological site in Texas (Ambers et al., 2021). The top photo clearly depicts the portion of the diaphysis that was sanded with an electric rotary tool (e.g., Dremel®) and an aluminum oxide grinding stone prior to removal of the cross-section (as evidenced by the lighter coloration of the diaphysis in that area). The bottom photo depicts further sectioning of the cortical bone from the diaphysis into smaller fragments in preparation for pulverization in a Freezer/Mill and subsequent DNA extraction procedures. If retention of the longitudinal physical structure (i.e., length) of the long bone is not required or necessary, this method of sampling from the diaphysis is much quicker and easier to achieve than the standard "window" cutting approach. *Photos by: Angie Ambers PhD (Henry C. Lee Institute of Forensic Science; Henry C. Lee College of Criminal Justice and Forensic Sciences; Institute for Human Identification, LMU College of Dental Medicine).*

fresh bones). Weighing of bone powder should always be performed inside a dead-air hood in the designated low-copy area of the laboratory. Weigh boats and weigh paper should be avoided if possible, as these products are not completely sterile and may contain exogenous DNA from direct handling. A "best practice" methodology for aliquoting and weighing bone powder involves first weighing a 15-mL conical tube prior to addition of bone powder and then re-weighing the tube after tapping bone powder into the tube from the grinding vial, until the desired quantity is obtained. This reduces the number of transfer steps and the number of surfaces the bone or tooth powder comes in contact with, minimizing the chances of contamination.

Summary and conclusions

In forensic casework and in historical investigations involving human remains, soft tissue and viscera may be heavily damaged, in an advanced state of decomposition, or completely absent. In these cases, hard skeletal tissues (bones, teeth) are the most viable samples for genetic testing. However, DNA recovery from human skeletal remains is challenging, time-consuming, and labor-intensive, and requires specialized equipment and training. It is

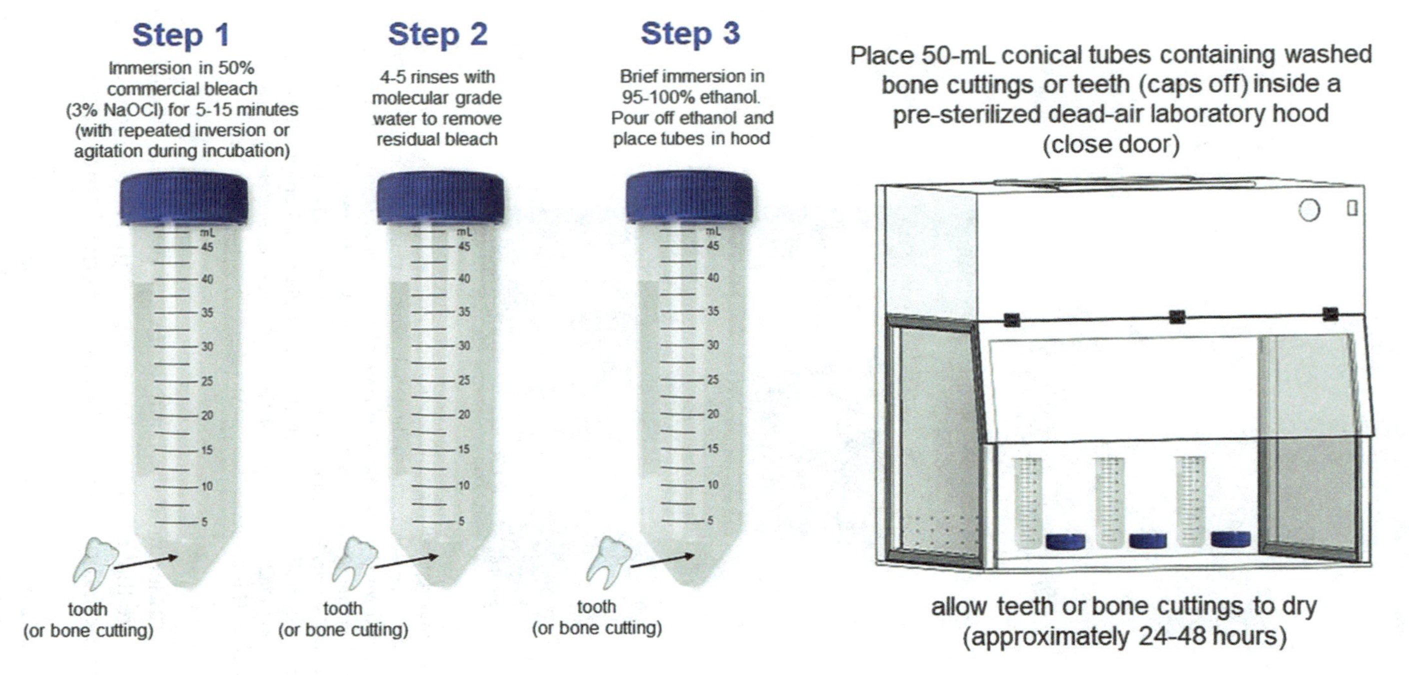

FIG. 21 Schematic of the three-step sequential washing steps for bone cuttings or teeth prior to pulverization in a Freezer/Mill with liquid nitrogen. **Step 1:** immersion and mild agitation in a 50% bleach dilution (3% NaOCl) for 5–15 min (depending on the age, porosity, and condition of the bone); **Step 2:** prompt draining of the bleach solution followed by 4–5 rinses with molecular grade water to remove residual bleach; and **Step 3:** a final, brief immersion in 95%–100% ethanol. After pouring off the ethanol, the 50-mL polypropylene conical tubes (each containing a clean bone cutting or tooth) should be placed in a pre-sterilized, dead-air hood with caps off to allow the bone cuttings and/or teeth to dry (usually for 24–48 h). Note that both the bleach concentration and immersion time should be decreased for old bones, bones of compromised quality, and bones in an advanced state of diagenesis. *Graphic by: Angie Ambers PhD (Henry C. Lee Institute of Forensic Science; Henry C. Lee College of Criminal Justice and Forensic Sciences; Institute for Human Identification, LMU College of Dental Medicine).*

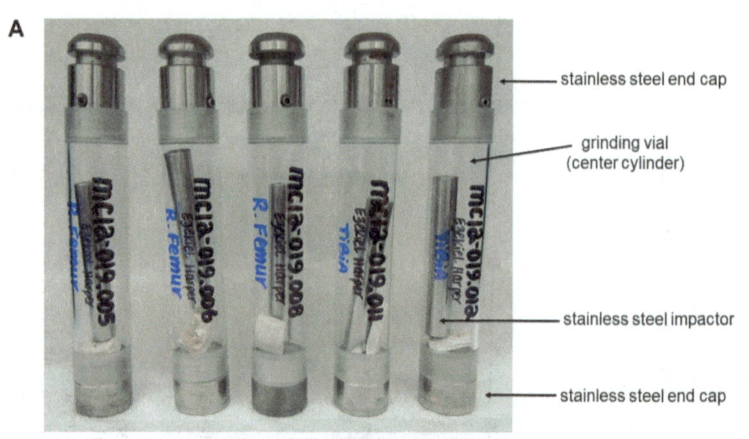

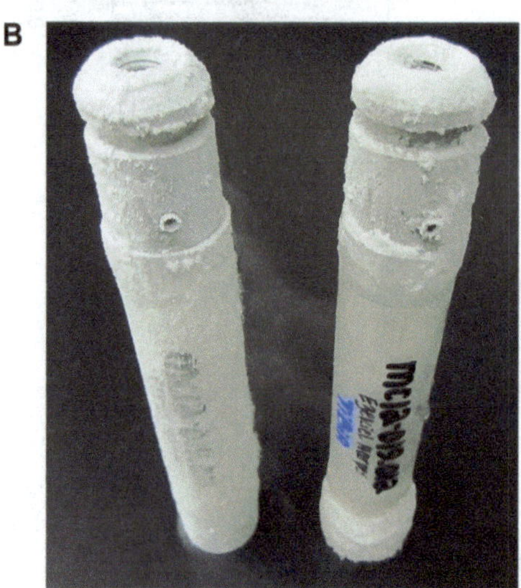

FIG. 22 Example of skeletal remains prepared for pulverization in a Freezer/Mill with liquid nitrogen: (A) a clean bone cutting or tooth is placed inside a pre-sterilized grinding vial (center cylinder) along with a stainless steel impactor, and the center cylinder is sealed with two stainless steel end caps. The entire tube containing the bone cutting or tooth is placed in a Freezer/Mill, submerged in a chamber filled with liquid nitrogen, and pulverized into a fine powder; and (B) post-grinding in the Freezer/Mill, tubes are removed from the instrument and should be allowed to thaw completely prior to removing an end cap to aliquot, weigh, and distribute the bone/tooth powder into tubes for DNA extraction. *Photo by: Angie Ambers PhD (Henry C. Lee Institute of Forensic Science; Henry C. Lee College of Criminal Justice and Forensic Sciences; Institute for Human Identification, LMU College of Dental Medicine).*

imperative that investigators understand the complexities of processing skeletal remains for DNA and the precision and care that must be taken to ensure that any genetic data recovered is endogenous to the decedent and not derived from an external source or due to intralaboratory contamination.

Moreover, DNA recovered from bones and teeth is often damaged and degraded. As discussed in Chapter 2, the molecular chemistry of degraded DNA can be quite complex, and a diverse range of DNA lesions exists, including abasic (AP) sites, deaminated bases, oxidized bases, alkylated bases, intrastrand crosslinks, interstrand crosslinks, single-strand breaks (SSBs), and double-strand breaks (DSBs). In addition to degradation of DNA, the bone microstructural matrix (which is comprised of calcium hydroxyapatite and collagen) is subject to alteration and damage due to environmental exposure and/or intentional insult. This is an important consideration in genetic testing efforts, as a correlation between diagenesis of bone microstructure and DNA quantity and quality has been demonstrated and supported through research and casework experience. Also of importance is an understanding of the non-uniform, heterogenous nature in which bone diagenesis progresses. Awareness of this heterogeneous degradation pattern predicates that sampling multiple regions of the same long bone or multi-site sampling (from different bones within the same skeleton) may be necessary to achieve sufficient results for an identification.

Because DNA in skeletal remains is often of compromised quality and in low quantities, it is imperative that proper pre-processing techniques be employed to maximize recovery and ensure that the DNA recovered is endogenous to the decedent and not from an external source. Fig. 23 summarizes the pre-processing and sterilization procedures for bones/teeth that have been detailed throughout this chapter.

Further, it is important to be reminded that recent advancements in DNA technology have increased the sensitivity of assays and instrumentation, and this increases the likelihood of detecting inadvertent contamination or exogenous DNA deposited through handling or contact

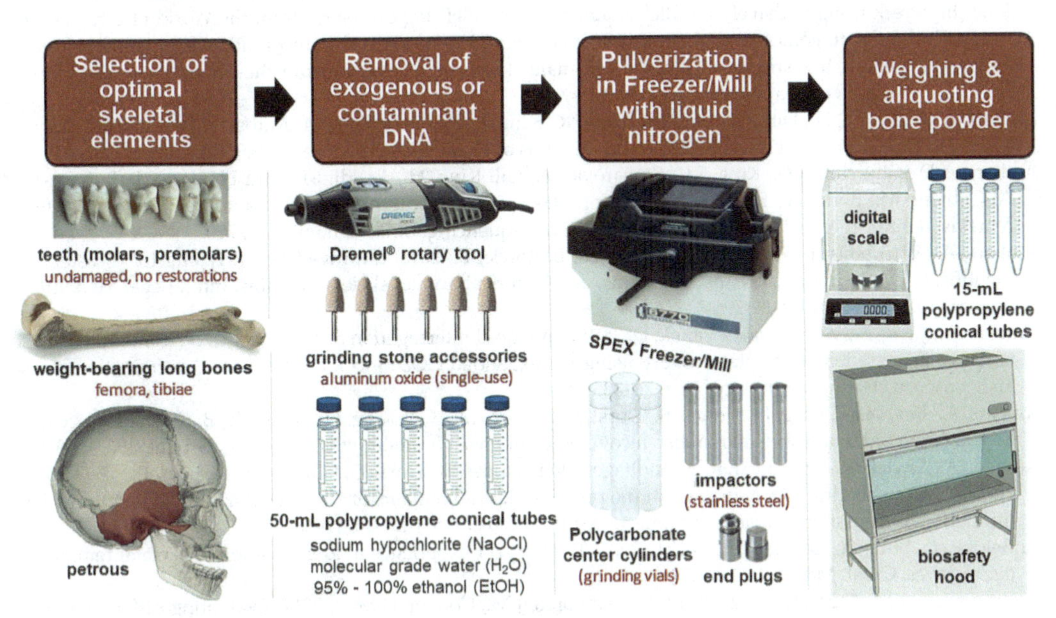

FIG. 23 Overview of optimal skeletal elements for DNA testing, and summary of pre-processing and sterilization procedures prior to DNA extraction. These procedures, along with the use of Forensic DNA Grade products (in accordance with the ISO 18385 Standard), help ensure that genetic data recovered from bone or tooth samples is endogenous to the decedent and not derived from an external source (or due to intralaboratory contamination). *Graphic design by: Angie Ambers PhD (Henry C. Lee Institute of Forensic Science; Henry C. Lee College of Criminal Justice and Forensic Sciences; Institute for Human Identification, LMU College of Dental Medicine).*

with the remains, or that which may be introduced during laboratory processing and testing. In addition to extensive sterilization procedures for the external surfaces of bones and teeth, similar procedures should be performed on all tools/accessories (e.g., saw blades, drill bits, end caps, impactors) that come into contact with skeletal samples. Additionally, laboratories engaged in forensic genetic investigations of unidentified human remains (UHRs) should use Forensic DNA Grade products in accordance with the ISO 18385 Standard, and should incorporate contamination prevention measures that meet or exceed guidelines established by the DAB, SWGDAM, ENFSI, and BSAG.

Acknowledgment

I acknowledge my graduate students in the Henry C. Lee College of Criminal Justice and Forensic Sciences, Paul M. Yount and Amelia Stockman, for their valuable contributions of some of the graphics and schematics used in this chapter.

References

Adler, C.J., Haak, W., Donlon, D., Cooper, A., 2011. Survival and recovery of DNA from ancient bones and teeth. J. Archaeol. Sci. 38, 956–964.

Ambers, A., 2016. Identifying 140-year-old skeletal remains using massively parallel DNA sequencing. In: International Symposium for Human Identification (ISHI) News. https://www.ishinews.com/identifying-140-year-old-remains-using-massively-parallel-dna-sequencing/.

Ambers, A., 2021. Forensic genetic investigation of two adult male skeletons recovered from the 17th-century *La Belle* shipwreck using massively parallel sequencing. The ISHI Report: News from the World of DNA Forensics https://promega.foleon.com/theishireport/november-2020/modern-dna-technology-for-historical-cold-cases-forensic-genetic-investigation-of-two-adult-male-skeletons-recovered-from-the-17th-century-la-belle-shipwreck-using-massively-parallel-sequencing-mps/.

Ambers, A., Gill-King, H., Dirkmaat, D., Benjamin, R., King, J., Budowle, B., 2014. Autosomal and Y-STR analysis of degraded DNA from the 120-year-old skeletal remains of Ezekiel Harper. Forensic Sci. Int. Genet. 9, 33–41.

Ambers, A.D., Churchill, J.D., King, J.L., Stoljarova, M., Gill-King, H., Assidi, M., Abu-Elmagd, M., Buhmeida, A., Al-Qahtani, M., Budowle, B., 2016. More comprehensive forensic genetic marker analyses for accurate human remains identification using massively parallel DNA sequencing. BMC Genomics 17, 750.

Ambers, A., Votrubova, J., Vanek, D., Sajantila, A., Budowle, B., 2018. Improved Y-STR typing for disaster victim identification, missing persons investigations, and historical human skeletal remains. Int. J. Legal Med. https://doi.org/10.1007/s00414-018-1794-8.

Ambers, A., Zeng, X., Votrubova, J., Vanek, D., 2019. Enhanced interrogation of degraded DNA from human skeletal remains: increased genetic data recovery using the expanded CODIS loci, multiple sex determination markers, and consensus testing. Anthropol. Anz. 76, 333–351.

Ambers, A., Bus, M.M., King, J.L., Jones, B., Durst, J., Bruseth, J.E., Gill-King, H., Budowle, B., 2020. Forensic genetic investigation of human skeletal remains recovered from the *La Belle* shipwreck. Forensic Sci. Int. 306, 110050.

Ambers, A., Elwick, K., Cropper, E.R., Brandhagen, M.D., Jones, B., Durst, J., Gilmore, K.K., Bruseth, J.E., Gill-King, H., 2021. Mitochondrial DNA analysis of the putative skeletal remains of Sieur de Marle: genetic support for anthropological assessment of biogeographic ancestry. Forensic Sci. Int. 320, 110682.

Andelovic, S., Sutlovic, D., Erceg, I.I., et al., 2007. Twelve year experience in identification of skeletal remains from mass graves. Croat. Med. J. 46, 530–539.

Andronowski, J.M., Mundorff, A.Z., Pratt, I.V., Davoren, J.M., Cooper, D.M.L., 2017. Evaluating differential nuclear DNA yield rates and osteocyte numbers among human bone tissue types: a synchrotron radiation micro-CT approach. Forensic Sci. Int. Genet. 28, 211–218.

Antinick, T.C., Foran, D.R., 2018. Intra- and Inter-element variability in mitochondrial and nuclear DNA from fresh and environmentally exposed skeletal remains. J. Forensic Sci. https://doi.org/10.1111/1556-4029.13843.

Basset, P., Castella, V., 2019. Positive impact of DNA contamination minimization procedures taken within the laboratory. Forensic Sci. Int. Genet. 38, 232–235.

Behrensmeyer, A.K., 1978. Taphonomic and ecologic information from bone weathering. Paleobiology 4 (2), 150–162.

Buckingham, A.K., Harvey, M.L., van Oorschot, R.A.H., 2016. The origin of unknown source DNA from touched objects. Forensic Sci. Int. Genet. 25, 26–33.

Butler, J., 2012. Forensic DNA advisory groups: DAB, SWGDAM, ENFSI, and BSAG. In: Encyclopedia of Forensic Sciences, second ed.

Casallas, D.A., Moore, K.M., 2012. High soil acidity associated with near complete mineral dissolution of recently buried human remains. Proc. Am. Acad. Forensic Sci. 18, 400–401.

Collins, M.J., Nielson-Marsh, C.M., Hiller, J., Smith, C.J., Roberts, J.P., Prigodich, R.V., Wess, T.J., Csapo, J., Millard, A.R., Turner-Walker, G., 2002. The survival of organic matter in bone: a review. Archaeometry 44, 383–394.

Cooper, A., Poinar, H.N., 2000. Ancient DNA: do it right or not at all. Science 18, 1139–1141.

Corte-Real, A., Andrade, L., Anjos, M.J., Carvalho, M., Vide, M.C., Corte-Real, F., Vieira, D.N., 2006. The DNA extraction from the pulp dentine complex of both with and without carious. Int. Congr. Ser. 1288, 710–712.

Crow, P., 2008. Mineral weathering in forest soils and its relevance to the preservation of the buried archaeological resource. J. Archaeol. Sci. 35, 2262–2273.

Dallas, S.L., Prideaux, M., Bonewald, L.F., 2013. The osteocyte: an endocrine cell … and more. Endocr. Rev. 34 (5), 658–690.

David, T.J., Lewis, J.M., 2018. Forensic Odontology: Principles and Practice. Elsevier Inc, Cambridge, MA.

Draft, D.M., Kasper, K.A., 2019. Dental age assessment in late adolescence. In: Adserias-Garriga, J. (Ed.), Age Estimation: A Multidisciplinary Approach. Elsevier Inc., Academic Press, pp. 107–123.

Edson, S., Ross, J.P., Coble, M.D., Parsons, T.J., Barritt, S.M., 2004. Naming the dead; confronting the realities of rapid DNA identification of degraded skeletal remains. Forensic Sci. Rev. 16, 64–89.

Efremov, J.A., 1940. Taphonomy: new branch of paleontology. Pan.-Am. Geol. 74, 81–93.

Emmons, A.L., Davoren, J., DeBruyn, J.M., Mundorff, A.Z., 2020. Inter and intra-variation in skeletal DNA preservation in buried remains. Forensic Sci. Int. Genet. 44, 102193.

European Network of Forensic Science Institutes (ENFSI), ENFSI DNA Working Group, 2017. DNA Contamination Prevention Guidelines. https://enfsi.eu/wp-content/uploads/2017/09/DNA-contamination-prevention-guidelines-v2.pdf.

Florencio-Silva, R., Sasso, G.R., Sasso-Cerri, E., Simoes, M.J., Cerri, P.S., 2015. Biology of bone tissue: structure, function, and factors that influence bone cells. Biomed. Res. Int. 2015, 1–17.

Fonnelop, A.E., Johannessen, H., Gill, P., 2015. Persistence and secondary transfer of DNA from previous users of equipment. Forensic Sci. Int. Genet. Suppl. Ser. 5, e191–e192.

Foran, D., 2006. Increasing the predictability and success rate of skeletal evidence typing: using physical characteristics of bone as a metric for DNA quality and quantity. National Institute of Justice (NIJ) Report, No. 2002-IJ-CX-K016.

FSR (Forensic Science Regulator), 2020a. The Control and Avoidance of Contamination in Scene Examination Involving DNA Evidence Recovery. FSR-G-206, Issue 2 https://assets.publishing.service.gov.uk/government/uploads/system/uploads/attachment_data/file/915221/FSR_G-206_Issue_2_Final.pdf.

FSR (Forensic Science Regulator), 2020b. The Control and Avoidance of Contamination in Laboratory Activities Involving DNA Evidence Recovery Analysis. FSR-G-208, Issue 2 https://assets.publishing.service.gov.uk/government/uploads/system/uploads/attachment_data/file/914268/208_FSR_lab_anti_contam__V2.pdf.

Gilbert, M.T.P., Hansen, A.J., Willerslev, E., Turner-Walker, G., Collins, M., 2006. Insights into the processes behind the contamination of degraded human teeth and bone samples with exogenous sources of DNA. Int. J. Osteoarchaeol. 16, 156–164.

Gill, P., Rowlands, D., Tully, G., Bastisch, I., Staples, T., Scott, P., 2010. Manufacturer contamination of disposable plasticware and other reagents: an agreed position statement by ENFSI, SWGDAM, and BSAG. Forensic Sci. Int. Genet. 4, 269–270.

Goray, M., Fowler, S., Szkuta, B., van Oorschot, R.A.H., 2016. Shedder status—an analysis of self and non-self DNA in multiple handprints deposited by the same individuals over time. Forensic Sci. Int. Genet. 23, 190–196.

Hansen, H.B., Damgaard, P.B., Margaryan, A., Stenderup, J., Lynnerup, N., Willerslev, E., Allentoft, M.E., 2017. Comparing ancient DNA preservation in petrous bone and tooth cementum. PLoS One 12 (1), e0170940.

Harris, E.F., 2018. Tooth-coding systems in the clinical dental setting. Dental Anthropology 18 (2). https://doi.org/10.26575/daj.v18i2.132. A Publication of the Dental Anthropology Association.

Higgins, D., Kaidonis, J., Austin, J., Townsend, G., James, H., Hughes, T., 2011. Dentine and cementum as sources of nuclear DNA for use in human identification. Austin J. Forensic Sci. 43, 287–295.

Higgins, D., Kaidonis, J., Townsend, G., Hughes, T., Austin, J.J., 2013. Targeted sampling of cementum for recovery of nuclear DNA from human teeth and the impact of common decontamination measures. Investig. Genet. 4, 18.
Hines, D., Vennemeyer, M., Amory, S., Huel, R., Hanson, I., Katzmarzyk, C., Parsons, T., 2014. Prioritized sampling of bone and teeth for DNA analysis in commingled cases. In: Commingled Human Remains Methods Recov. Annal. Identification. Elsevier Inc, pp. 275–305.
Hollund, H.I., Teasdale, M.D., Mattiangeli, V., Sverrisdottir, O.O., Bradley, D.G., O'Connor, T., 2017. Pick the right pocket: sub-sampling of bone sections to investigate diagenesis and DNA preservation. Int. J. Osteoarchaeol. 27, 365–374.
Hopkins, D.W., 2008. The role of soil organisms in terrestrial decomposition. In: Tibbett, M., Carter, D.O. (Eds.), Soil Analysis in Forensic Taphonomy: Chemical and Biological Effects of Buried Human Remains. CRC Press, Boca Raton, FL, pp. 53–66.
Hopkins, D.W., Wiltshire, P.E.J., Turner, D.B., 2000. Microbial characteristics of soils from graves: an investigation at the interface of soil microbiology and forensic science. Appl. Soil Ecol. 14, 283–288.
International Commission on Missing Persons (ICMP), 2015. Standard Operating Procedure for Sampling Bone and Tooth Specimens from Human Remains for DNA Testing at the ICMP. https://www.icmp.int/wp-content/uploads/2016/12/icmp-sop-aa-136-2-W-doc.pdf.
International Geosphere-Biosphere Program—Data and Information System (IGBP-DIS), 1998. SoilData (V.0) A Program for Creating Global Soil-property Databases. IGBP Global Soils Data Task, France.
International Organization for Standardization (ISO), 2016a. ISO3950:2016 Dentistry—Designation System for Teeth and Areas of the Oral Cavity. https://www.iso.org/standard/68292.html.
International Organization for Standardization (ISO), 2016b. ISO 18385:2016—Minimizing the Risk of Human DNA Contamination in Products Used to Collect, Store, and Analyze Biological Material for Forensic Purposes. https://www.iso.org/standard/62341.html.
Janaway, R.C., 2002. Degradation of clothing and other dress materials associated with buried bodies of archaeological and forensic interest. In: Haglund, W.D., Sorg, M.H. (Eds.), Advances in Forensic Taphonomy: Method, Theory, and Archaeological Perspectives. CRC Press, Boca Raton, FL, pp. 379–402.
Janaway, R.C., 2008. The decomposition of materials associated with buried cadavers. In: Tibbett, M., Carter, D.O. (Eds.), Soil Analysis in Forensic Taphonomy. CRC Press, Boca Raton, FL, pp. 153–201.
Jans, M.M.E., 2008. Micro-bioerosion of bone: a review. In: Tapanila, L., Wisshak, M. (Eds.), Current Developments in Bioerosion. Erlangen Earth Science Conferences, Springer Verlag, Berlin, Germany.
Jans, M.M.E., Nielsen-Marsh, C.M., Smith, C.I., Collins, M.J., Kars, H., 2004. Characterization of microbial attack on archaeological bone. J. Archaeol. Sci. 31, 87–95.
Kemp, B., Smith, D.G., 2005. Use of bleach to eliminate contaminating DNA from the surface of bones and teeth. Forensic Sci. Int. 154, 53–61.
Kulstein, G., Hadrys, T., Wiegand, P., 2017. As solid as a rock—comparison of CE- and MPS-based analyses of the petrosal bone as a source of DNA for forensic identification of challenging cranial bones. Int. J. Legal Med. 132, 13–24.
Lehmann, V.J., Mitchell, R.J., Ballantyne, K.N., van Oorschot, R.A.H., 2013. Following the transfer of DNA: how far can it go? Forensic Sci. Int. Genet. 4 (1), e53–e54.
Lowe, A., Murray, C., Whitaker, J., Tully, G., Gill, P., 2002. The propensity of individuals to deposit DNA and secondary transfer of low level DNA from individuals to inert surfaces. Forensic Sci. Int. 129, 25–34.
Lyman, R.L., 1994. Vertebrate Taphonomy. Cambridge University Press, Cambridge.
Malaver, P.C., Yunis, J.J., 2003. Different dental tissues as a source of DNA for human identification in forensic cases. Croat. Med. J. 44, 306–309.
Manjunatha, B.S., 2013. Tooth numbering systems. In: Textbook of Dental Anatomy and Oral Physiology. Jaypee Brothers Medical Publishing, New Delhi, India.
Meakin, G., Jamieson, A., 2013. DNA transfer: review and implications for casework. Forensic Sci. Int. Genet. 7 (4), 434–443.
Micozzi, M.S., 1991. Postmortem Change in Human and Animal Remains: A Systematic Approach. Charles C. Thomas, Springfield, IL.
Milos, A., Selmanovic, A., Smajlovic, L., et al., 2007. Success rates of nuclear short tandem repeat typing from different skeletal elements. Croat. Med. J. 48, 486–493.
Misner, L.M., Halvorson, A.C., Dreier, J.L., Ubelaker, D.H., Foran, D.R., 2009. The correlation between skeletal weathering and DNA quality and quantity. J. Forensic Sci. 54 (4), 822–828.

Mundorff, A.Z., Davoren, J.M., 2014. Examination of DNA yield rates for different skeletal elements at increasing postmortem intervals. Forensic Sci. Int. Genet. 8 (1), 55–63.

Mundorff, A.Z., Bartelink, E.J., Mar-Cash, E., 2009. DNA preservation in skeletal elements from the World Trade Center disaster: recommendations for mass fatality management. J. Forensic Sci. 54 (4), 739–745.

Mundorff, A.Z., Davoren, J.M., Weitz, S., 2013. Developing an Empirically Based Ranking Order for Bone Sampling: Examining the Differential DNA Yield Rates Between Human Skeletal Elements Over Increasing Postmortem Intervals. National Institute of Justice Report No. 2010-DN-BX-K229 241868, Washington, DC.

Nelson Institute for Environmental Studies, Center for Sustainability and the Global Environment (SAGE), University of Wisconsin-Madison. Global Soil pH Map. https://nelson.wisc.edu/sage/data-and-models/atlas/maps/soilph/atl_soilph.jpg. (Accessed 1 August 2021).

Nielsen-Marsh, C.M., Hedges, R.E.M., 2000. Patterns of diagenesis in bone I: the effects of site environments. J. Archaeol. Sci. 27, 1139–1150.

Nielsen-Marsh, C.M., Smith, C.I., Jans, M.M.E., Nord, A., Kars, H., Collins, M.J., 2007. Bone diagenesis in the European holocene II: taphonomic and environmental considerations. J. Archaeol. Sci. 34, 1523–1531.

Obal, M., Zupanic, P.I., Gornjak, P.B., Zupanc, T., 2019. Different skeletal elements as a source of DNA for genetic identification of Second World War victims. Forensic Sci. Int. Genet. Suppl. Ser. 7, 27–29.

Oldoni, F., Castella, V., Hall, D., 2015. Exploring the relative DNA contribution of first and second object's users on mock touch DNA mixtures. Forensic Sci. Int. Genet. Suppl. Ser. 5, e300–e301.

Oleiwi, A.A., Morris, M.R., Schmerer, W.M., Sutton, R., 2015. The relative DNA-shedding propensity of the palm and finger surfaces. Sci. Justice 55 (5), 329–334.

Parker, C., Rohrlac, A.B., Friederich, S., Nagel, S., Meyer, M., Krause, J., Bos, K.I., Haak, W., 2020. A systematic investigation of human DNA preservation in medieval skeletons. Nat. Sci. Rep. 10, 18225.

Parsons, T.J., Weedn, V.W., 1997. Preservation and recovery of DNA in postmortem specimens and trace samples. In: Haglund, W.D., Sorg, M.S. (Eds.), Forensic Taphonomy: The Postmortem Fate of Human Remains. CRC Press, Boca Raton, FL.

Phipps, M., Petricevic, S., 2007. The tendency of individuals to transfer DNA to handled items. Forensic Sci. Int. 168, 162–168.

Pinhasi, R., Fernandes, D., Sirak, K., Novak, M., Connell, S., Alpasian-Roodenberg, S., Gerritsen, F., Moiseyev, V., Gromov, A., Raczky, P., Anders, A., Pietrusewsky, M., Rochefort, G.Y., Pallu, S., Benhamou, C.L., 2010. Osteocyte: the unrecognized side of bone tissue. Osteoporos. Int. 21 (9), 1457–1469.

Pinhasi, R., Fernandes, D., Sirak, K., Novak, M., Connell, S., Alpaslan-Roodenberg, S., Gerritsen, F., Moiseyev, V., Gromov, A., Raczky, P., Anders, A., Pietrusewsky, M., Rollefson, G., Jovanovic, M., Trinhhoang, H., Bar-Oz, G., Oxenham, M., Matsumura, H., Hofreiter, M., 2015. Optimal ancient DNA yields from the inner ear part of the human petrous bone. PLoS One 10 (6). https://doi.org/10.1371/journal.pone.0129102, e0129102.

Poinar, H.N., 2003. Criteria of authenticity for DNA from ancient and forensic samples. Int. Congr. Ser. 1239, 575–579.

Pokines, J.T., 2009. Forensic recoveries of U.S. war dead and the effects of taphonomy and other site-altering processes. In: Steadman, D.W. (Ed.), Hard Evidence: Case Studies in Forensic Anthropology, second ed. Prentice Hall, Upper Saddle River, NJ, pp. 141–154.

Pokines, J.T., Baker, J.E., 2014. Effects of burial environment on osseous remains. In: Pokines, J.T., Symes, S.A. (Eds.), Manual of Forensic Taphonomy. CRC Press, Boca Raton, FL.

Pokines, J.T., Symes, S.A., 2014. Manual of Forensic Taphonomy. CRC Press, Boca Raton, FL.

Rohland, N., Hofreiter, M., 2007. Ancient DNA extraction from bones and teeth. Nat. Protoc. 2, 1756–1762.

Schaffler, M.B., Cheung, W.Y., Majeska, R., Kennedy, O., 2014. Osteocytes: master orchestrators of bone. Calcif. Tissue Int. 94 (1), 5–24.

Scientific Working Group on DNA Analysis Methods (SWGDAM), 2017. Contamination Prevention and Detection Guidelines for Forensic DNA Laboratories. https://media.wix.com/ugd/4344b0_c4d4dbba-84f1400a98eaa2e48f2bf291.pdf.

Siriboonpiputtanaa, T., Rinthachaia, T., Shotivaranona, J., et al., 2018. Forensic genetic analysis of bone remain samples. Forensic Sci. Int. 284, 167–175.

Slessarev, E.W., Lin, Y., Bingham, N.L., Johnson, J.E., Dai, Y., Schimel, P., Chadwick, O.A., 2016. Water balance creates a threshold in soil pH at the global scale. Nature 540, 567–569.

Smith, C.I., Nielsen-Marsh, C.M., Jans, M.M.E., Collins, M.J., 2007. Bone diagenesis in the European holocene I: patterns and mechanisms. J. Archaeol. Sci. 34, 1485–1493.

Szkuta, B., Harvey, M., Ballantyne, K.M., van Oorschot, R.A.H., 2015. DNA transfer by examination tools—a risk for forensic casework? Forensic Sci Int: Genet. 16, 246–254.

Szkuta, B., Ballantyne, K.N., Kokshoorn, B., van Oorschot, R.A.H., 2018. Transfer and persistence of non-self DNA on hands over time: using empirical data to evaluate DNA evidence given activity level propositions. Forensic Sci. Int. Genet. 33, 84–97.

Templeton, E.L., Taylor, D., Handt, O., Linacre, A., 2015. DNA profiles from fingermarks: a mock case study. Forensic Sci. Int. Genet. Suppl. Ser. 5, e154–e155.

Trueman, C.N., Martill, D.M., 2002. The long-term survival of bone: the role of bioerosion. Archaeometry 44, 371–382.

Ubelaker, D.H., Jones, E.B., Landers, D.B., 2003. Human Remains from Voegtly Cemetery, Pittsburgh, Pennsylvania. Smithsonian Contributions to Anthropology Series, vol. 46 Smithsonian Inst Press, Washington, DC.

van Oorschot, R.A.H., Glavich, G., Mitchell, R.J., 2014. Persistence of DNA deposited by the original user on objects after subsequent use by a second person. Forensic Sci. Int. Genet. 8 (1), 219–225.

van Oorschot, R.A.H., Szkuta, B., Meakin, G.E., Kokshoorn, B., Goray, M., 2019. DNA transfer in forensic science: a review. Forensic Sci. Int. Genet. 38, 140–166.

White, T.D., Folkens, P.A., 2005. The Human Bone Manual. Elsevier Academic Press, Burlington, MA.

Yang, D.Y., Watt, K., 2005. Contamination controls when preparing archaeological remains for ancient DNA analysis. J. Archaeol. Sci. 32, 331–336.

Zupanic, P.I., Gornjak, P.B., Balazic, J., 2012. Highly efficient nuclear DNA typing of the World War II skeletal remains using three new autosomal short tandem repeat amplification kits with extended European Standard Set of Loci. Croat. Med. J. 53, 17–23.

Zupanič Pajnič, I., Inkret, J., Zupanc, T., Podovšovnik, E., 2021. Comparison of nuclear DNA yield and STR typing success in Second World War petrous bones and metacarpals III. Forensic Sci. Int. Genet. 55, 102578.

Additional reading—Contamination prevention

Astolphi, R.D., Alves, M.T., Evison, M.P., Francisco, R.A., Guimaraes, M.A., Iwamura, E.S.M., 2019. The impact of burial period on compact bone microstructure: histological analysis of matrix loss and cell integrity in human bones exhumed from tropical soil. Forensic Sci. Int. 298, 384–392.

Brundin, M., Figdor, D., Sundquist, G., Sjorgren, U., 2013. DNA binding to hydroxyapatite: a potential mechanism for preservation of microbial DNA. J. Endod. 39 (2), 211–216.

Cibulskis, K., McKenna, A., Fennell, T., Banks, E., DePristo, M., Getz, G., 2011. ContEst: estimating cross-contamination of human samples in next-generation sequencing data. Bioinformatics 27 (18), 2601–2602.

Dickens, B., Rebolledo-Jaramillo, B., Shu-Wei, S.M., Pau, I.M., Blankenberg, D., Stoler, N., Makova, K.D., Nekrutenko, A., 2014. Controlling for contamination in re-sequencing studies with a reproducible web-based phylogenetic approach. Biotechniques 56 (3), 134–141.

Gill, P., Kirkham, A., 2004. Development of a simulation model to assess the impact of contamination in casework using STRs. J. Forensic Sci. 49 (3), 485–491.

Gill, P., Whitaker, J., Flaxman, C., Brown, N., Buckleton, J., 2000. An investigation of the rigor of interpretation rules for STRs derived from less than 100 pg of DNA. Forensic Sci. Int. 112 (1), 17–40.

Gotherstrom, A., Collins, M.J., Angerbjorn, A., Liden, K., 2002. Bone preservation and DNA amplification. Archaeometry 3, 395–404.

Hedges, R.E.M., Millard, A.R., 1995. Bones and groundwater: towards the modeling of diagenetic processes. J. Archaeol. Sci. 22, 155–164.

Jobling, M.A., Gill, P., 2004. Encoded evidence: DNA in forensic analysis. Nat. Rev. Genet. 5, 739–751.

Jun, G., Flickinger, M., Hetrick, K.N., Romm, J.M., Doheny, K.F., Abecasis, G.R., Boehnke, M., Kang, H.M., 2012. Detecting and estimating contamination of human DNA samples in sequencing and array-based genotype data. Am. J. Hum. Genet. 91, 839–848.

Korlevic, P., Gerber, T., Gansauge, M.T., Hajdinjak, M., Nagel, S., Aximu-Petri, A., Meyer, M., 2015. Reducing microbial and human contamination in DNA extractions from ancient bones and teeth. Biotechniques 58 (2), 87–93.

McCartney, C., Amoako, E., 2018. The UK forensic science regulator: a model for forensic science regulation? Georgia State Univ. Law Rev. 34 (4), 945–982.

Mello, R.B., Silva, M.R.R., Alves, M.T.S., Evison, M.P., Guimaraes, M.A., Francisco, R.A., Astolphi, R.D., Iwamura, E.S.M., 2017. Tissue microarray analysis applied to bone diagenesis. Sci. Rep. 7, 39987. https://doi.org/10.1038/srep39987.

Raymond, K., Setlak, J., Stollberg, C., 2013. Examination of Proposed Manufacturing Standards Using Low Template DNA. http://www.promega.com/resources/profilesin-dna/2013/examination-of-proposed-manufacturing-standards-using-low-template-dna/.

Sajantila, A., 2014. Editors pick: contamination has always been the issue! Investig. Genet. 5, 17.

Scherczinger, C.A., Ladd, C., Bourke, M.T., Adamowicz, M.S., Johannes, P.M., Scherczinger, R., Beesley, T., Lee, H.C., 1999. A systematic analysis of PCR contamination. J. Forensic Sci. 44 (5), 1042–1045.

Sosa, M., Vispe, E., Nunez, C., Baeta, M., Casalod, Y., Bolea, M., Hedges, R.E.M., Martinez-Jarreta, B., 2013. Association between ancient bone preservation and DNA yield: a multi-disciplinary approach. Am. J. Phys. Anthropol. 151, 102–109.

Technical Working Group on Biological Evidence Preservation: National Institute of Standards and Technology, 2013. The Biological Evidence Preservation Handbook: Best Practices for Evidence Handlers. NISTIR. 7928 http://nvlpubs.nist.gov/nistpubs/ir/2013/NIST.IR.7928.pdf.

CHAPTER

6

DNA extraction methods for human skeletal remains

Angie Ambers PhD[a,b,c]

[a]Henry C. Lee Institute of Forensic Science, University of New Haven, West Haven, CT, United States [b]Forensic Science Department, Henry C. Lee College of Criminal Justice and Forensic Sciences, Center for Forensic Investigation of Trafficking in Persons, University of New Haven, West Haven, CT, United States [c]Institute for Human Identification, LMU College of Dental Medicine, Knoxville, TN, United States

"Our DNA does not fade like an ancient parchment; it does not rust in the ground like the sword of the warrior long dead. It is not eroded by wind or rain, nor reduced to ruin by fire and earthquake. It is the traveller from an ancient land who lives within us all"—**Bryan Sykes.**

Introduction

Genetic analyses are highly dependent on the quantity, quality, and purity of DNA obtained during the extraction process. Selection of the appropriate extraction method is crucial for success. There are a variety of different DNA extraction methods available to the forensic geneticist, ranging from organic extraction (the oldest method) to solid-phase (inorganic) extraction to the use of Chelex™ resin or FTA cards. In addition, some DNA extraction methods are performed manually, while other (more recently introduced) protocols involve automated, robotic platforms. The important consideration for skeletal remains cases is that: (1) DNA encompassed within bones and teeth often suffers from the trifecta of degradation/damage, low quantity, and inhibition; (2) the particular DNA extraction method selected—as well as the skill and precision of the laboratory analyst—can have significant impact on whether sufficient quantity and purity of DNA are obtained for downstream analyses; and (3) speed-of-analysis should not be a predominant factor in the selection of extraction method, since there are often sacrifices associated with quicker, less labor-intensive protocols (most importantly, loss of DNA and/or minimal recovery from the substrate).

https://doi.org/10.1016/B978-0-12-815766-4.00006-6

Not all DNA extraction methods are equally effective at maximizing recovery of DNA and concurrently minimizing co-purification of inhibitors. Indeed, the type of extraction method employed should vary based on the estimated age or postmortem interval (PMI) of the sample, the environmental conditions (or intentional damage) that the remains were subjected to, the physical state of the sample, and the amount of sample available for testing. Moreover, it is important to note that protocols employed for high-quality samples (e.g., reference samples or exemplars) will vary significantly from the extraction approaches selected for skeletonized remains, the latter of which are older and often contain DNA that is considerably compromised. Although organic extraction is the oldest and most labor-intensive method available to the forensic geneticist, it has consistently proven successful with skeletal remains across a broad range of PMIs, as well as those from a diverse spectrum of environments, climates, and geographical locations; hence, organic extraction is still the "gold standard" for forensic genetic investigations of skeletonized human remains.

Forensic DNA casework—Standard operating procedure (SOP)

Bone is a complex tissue that consists of two primary components: an inorganic mineral matrix called calcium hydroxyapatite [$Ca_{10}(PO_4)_6(OH)_2$] and an organic structural protein (collagen). Although calcium hydroxyapatite has been demonstrated to play a role in DNA preservation (due to the molecular interactions that occur between the negatively-charged DNA backbone and positively-charged calcium residues in the hydroxyapatite mineral matrix), accessing DNA contained within the bone's microstructure is challenging (Emmons et al., 2020; Astolphi et al., 2019; Antinick and Foran, 2018; Hollund et al., 2017; Brundin et al., 2013; Sosa et al., 2013). Areas of extensive mineralization within bone represent physical barriers to extraction reagents and prevent the release of DNA molecules. Therefore, extensive physical and mechanical pre-processing steps are required, as well as exposure to an array of chemicals, in order to lyse cellular membranes and release DNA for recovery and analysis. Pulverization of bones and teeth into fine powder increases the surface area over which the extraction chemicals can come into contact with DNA-containing cells.

The majority of DNA extraction protocols currently used by the forensic community for recovering DNA from bones and teeth involve: (1) pulverization of skeletal material into fine powder using liquid nitrogen and a Freezer/Mill instrument (or similar apparatus); (2) overnight incubation of bone/tooth powder in a lysis (demineralization) buffer consisting of ethylenediaminetetraacetic acid [$CH_2N(CH_2CO_2H)_2]_2$ (EDTA), sodium N-lauroylsarcosinate ($C_{15}H_{28}NNaO_3$) (also known as Sarkosyl), and Proteinase K; (3) concentration of the supernatant (lysate) into a smaller volume; and (4) an inhibitor removal step to further purify DNA away from environmental inhibitors (e.g., soil acids, vivianite).

Importance of demineralization

Overnight incubation in EDTA-based lysis buffer promotes demineralization of skeletal material, which has been shown to increase DNA yield compared to extraction methods that do not incorporate demineralization. Although the benefits of demineralization were first demonstrated in 1991 (Hagelberg and Clegg, 1991), two formative studies exploring DNA extraction methods for skeletal remains in forensic practice were performed by the Armed Forces DNA Identification Laboratory (AFDIL) (Edson et al., 2004; Loreille et al., 2007). Collectively, these two seminal publications (Edson et al., 2004; Loreille et al., 2007) have provided the foundational knowledge and experience that helped develop the landscape and improve the success of forensic DNA testing of skeletonized human remains.

DNA extraction in designated low copy number (LCN) area

DNA extraction from bones and teeth should always be performed in a laboratory or space that is physically separated from the area where high-quality samples (e.g., reference samples) are processed. In general, "best practices" for extracting DNA from skeletal remains are aligned with the standardized contamination prevention measures used in the ancient DNA studies, including: (1) use of protective suits, hairnets, gloves, and facemasks; (2) bleach decontamination and UV irradiation of work benches, tools, and consumables; (3) extraction of bone samples within a dead-air hood and in a designated low copy number (LCN) laboratory; and (4) preparation and processing of all reference samples in a designated "high copy" laboratory that is physically separate from the space where human remains are handled (Ambers et al., 2014, 2016, 2018, 2019, 2020; Gilbert et al., 2006; Kemp and Smith, 2005; Yang and Watt, 2005; Poinar, 2003; Cooper and Poinar, 2000). To prevent cross-contamination, bones and teeth should never be extracted in the same location as the reference samples that have been submitted for comparison. Additionally, every laboratory should maintain an internal elimination database containing DNA profiles of all laboratory staff and personnel for comparison to DNA results obtained from unidentified human remains (UHR). It is also prudent to collect reference samples from all individuals involved in recovery or exhumation for comparison and for exclusionary purposes in the event of suspected contamination.

Preparation for DNA extraction—Sterilization of tubes, equipment, and work surfaces

Lab coats for DNA extraction protocols serve two primary purposes: (1) to protect the scientist from harsh chemicals and biohazardous materials, and (2) to protect the skeletal casework samples from contamination with the scientist's DNA (e.g., from shed skin cells, sweat, hairs). Disposable (single-use) lab coats with elasticized wrist cuffs are optimal when performing DNA extraction on human skeletal remains. These disposable lab coats can be discarded afterwards and between different cases to prevent cross-contamination between samples. Since all DNA extraction procedures should be performed in a dead-air hood, a more practical and cost-effective option is to incorporate Tyvek™ IsoClean™ sterile sleeves that can be worn on top of the sleeves of disposable lab coats and changed between samples or cases (as opposed to changing the lab coat itself). Traditional cotton or cotton/polyester blend lab coats should *not* be worn in pre-PCR (extraction) areas in order to minimize carryover contamination between cases and during day-to-day operations. In addition to the proper lab coat (and/or Tyvek™ IsoClean™ sterile sleeves), facemasks, gloves, and hair nets should be worn while performing DNA extraction procedures.

All single-use tubes should be UV-irradiated before addition of bone powder, prior to initiation of DNA extraction procedures. This includes all polypropylene conical tubes (50-mL, 15-mL) as well as microcentrifuge tubes (1.7-mL, 1.5-mL). Standard procedure includes placing conical and microcentrifuge tubes inside a UV crosslinker (upright in a tube rack, with caps open) and irradiating the tubes at 1.0 J/cm^2 for 26 min and 13 min, respectively. Although pipettes should not be UV-irradiated because repeated exposure will degrade their plastic casings, it is recommended that the external surface of all pipettes be wiped down with a 10% bleach dilution (0.6% NaOCl) prior to use and between skeletal cases. Note that standard commercial bleach is approximately 6.0% sodium hypochlorite (6.0% NaOCl), so a 10% dilution would result in a 0.6% NaOCl concentration. However, some varieties of commercial bleach are manufactured in concentrated form and therefore contain a much higher initial concentration of NaOCl. It is imperative to pay attention to this information on the label of the bleach container and make proper dilutions accordingly.

Importance of polypropylene tubes and barrier (filter) pipette tips

Polypropylene (not polystyrene) tubes should always be used during DNA extraction to reduce interactions of DNA molecules with the inner walls of tubes. This is especially important for highly degraded (or old) samples, and for DNA extraction protocols that involve multiple transfers between tubes. Reducing adsorption of DNA to the tube wall ensures maximum DNA recovery from the sample and mitigates loss of DNA during the extraction process. DNA LoBind tubes (Eppendorf; San Diego, California USA) are excellent choices for DNA extraction from bone or tooth samples. Using special manufacturing technologies, these tubes are specifically designed for improved assay results by reducing the interaction of nucleic acid molecules with the inner tube surfaces, ensuring nearly 100% recovery of DNA. DNA LoBind tubes are available as 15-mL conical tubes, 50-mL conical tubes, 1.5-mL microcentrifuge tubes, and 2.0-mL microcentrifuge tubes (as well as PCR tubes and plates, for downstream post-extraction assays). Additionally, LoBind polypropylene tubes are independently batch-tested and certified to be free from exogenous DNA, nucleases, and PCR inhibitors (Eppendorf, 2015).

Additionally, barrier filter pipette tips should be used for all pipetting steps throughout the DNA extraction procedure. Barrier tips prevent cross-contamination of samples via incorporation of an aerosol filter inside each disposable tip, and are available in a variety of sizes (10-μL, 20-μL, 50-μL, 100-μL, 200-μL, 1000-μL) to accommodate every step in the extraction procedure. Similar to all tubes used during extraction, pipette tips should also be certified to be sterile and nuclease-free. A variety of sterilization techniques are used by pipette manufacturers for quality assurance (QA). Laboratories performing forensic DNA testing on contemporary, historical, or archaeological skeletal remains should only use pipette tips that have been either gamma-irradiated or electron beam (E-beam) sterilized.

Validated organic DNA extraction method for forensic casework involving skeletonized human remains

One of the most renowned U.S. laboratories that specializes in forensic DNA testing of human skeletal remains, the UNT Center for Human Identification (UNTCHI), has explored a variety of DNA extraction protocols since its inception in 2004. The method that has consistently performed the best in terms of maximizing DNA recovery and effective removal of inhibitors is a modified version of organic extraction. Although organic DNA extraction is one of the oldest methods available—and tends to be even more time-consuming and labor-intensive than some of the more recently introduced methods—the ultimate goal is to make a positive identification of the remains presented to the laboratory. Speed-of-analysis is meaningless in the context of human identification if the extraction method selected does not yield sufficient quantity and purity of DNA. UNTCHI has processed thousands of skeletal remains from diverse environments and across a broad range of geographical locations (nationally and internationally), as well as remains of varying postmortem intervals (PMIs), using this organic extraction approach, with great success (Ambers, 2016, 2021; Ambers et al., 2014, 2016, 2018, 2019, 2020). The steps of this validated DNA extraction protocol are outlined below.

Organic DNA extraction from skeletonized human remains (bones or teeth)

1. Transfer 0.5–1.0 g of bone powder to a pre-labeled 15-mL polypropylene conical tube
   Note: For teeth, up to 1.0 g of tooth powder can be used in the extraction protocol. If pulverization of the entire tooth yields significantly greater than 1.0 g of powder, the excess powder should be transferred to a separate, sterile polypropylene tube. The remaining tooth powder can be retained for additional testing (if initial extraction procedures are unsuccessful), returned to the submitting agency, or placed in long-term storage.
   Note: 1 g of bone powder is preferable for increased DNA recovery. However, inhibited samples may yield better results (i.e., more data) with lower amounts of starting material.
   Important considerations for aliquoting and weighing bone powder:
   Bone powder should be weighed in conical tubes rather than by using weigh boats or weigh paper, since the latter increases the probability of contamination with exogenous DNA. The more items or containers that the bone powder comes in contact with (and as the number of transfers between weighing tools or tubes increases), the greater the chance of contamination. Instead, empty 15-mL conical tubes can be weighed on a digital scale, followed by subsequent weighing after addition of bone or tooth powder. By subtracting the weight of the empty conical tube from the weight of the tube *after* addition of bone or tooth powder, laboratory analysts can calculate the precise amount of bone or tooth powder for DNA extraction while simultaneously minimizing the potential for contamination. Additionally, the process of aliquoting bone or tooth powder into collection tubes should always be performed inside a dead-air hood that has been designated specifically for this purpose and that can be effectively sterilized between samples. PCR hoods equipped with UV light (crosslinking) capability are excellent choices for weighing and aliquoting of powder in preparation for DNA extraction.
2. Add 4.5 mL of *demineralization (lysis) buffer* and 300 μL (μL) of Proteinase K (20 mg/mL) to each conical tube containing bone or tooth powder, as well as to a separate, empty conical tube (i.e., with no powder) to serve as an extraction reagent blank. After addition of demineralization buffer and proteinase K, tightly cap each tube, seal with Parafilm,

and vortex thoroughly to ensure that all bone or tooth powder is completely suspended (Fig. 1).
Note: If only 0.5 g of bone or tooth powder is available for extraction, 3.0 mL of demineralization (lysis) buffer and 200 μL of proteinase K (20 mg/mL) are sufficient.
Preparation of demineralization (lysis) buffer:
Add 1 g of sodium N-lauroylsarcosinate ($C_{15}H_{28}NNaO_3$) (Sarkosyl) to 100 mL of 0.5 M ethylenediaminetetraacetic acid $[CH_2N(CH_2CO_2H)_2]_2$ (EDTA) pH 8.0. Invert and swirl the bottle until the Sarkosyl powder is thoroughly dissolved in the EDTA. Molecular Biology Grade reagents are highly recommended for their purity, sterility, and stability. The 1% sarkosyl demineralization buffer may be stored at room temperature for up to 1 year after preparation. Note that the Proteinase K should *not* be added directly to this stock solution of demineralization buffer. Rather, Proteinase K should be added to each individual conical tube containing a bone sample (i.e., add 300 μL Proteinase K to 1 g of bone powder suspended in 4.5 mL demineralization buffer, or add 200 μL Proteinase K to 0.5 g of bone powder suspended in 3.0 mL demineralization buffer).

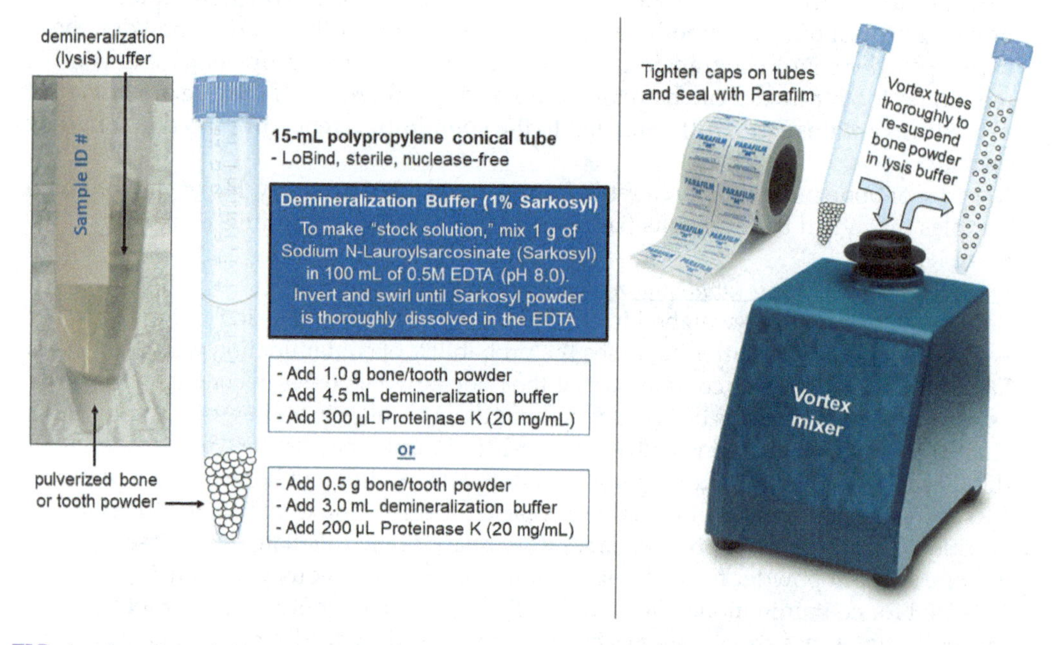

FIG. 1 Overview of preparation of bone/tooth powder for overnight demineralization. Pulverized bone powder (0.5–1.0 g) is mixed with demineralization (lysis) buffer and Proteinase K (20 mg/mL) in sterile, pre-labeled 15-mL polypropylene conical tubes. Each tube should be vortexed thoroughly until all bone powder is completely resuspended in the demineralization (lysis) buffer. Tube caps should then be sealed (wrapped) with Parafilm prior to overnight incubation at 56°C, during which the samples are subjected to constant agitation. Parafilm mitigates leakage of lysate that may occur from repeated inversions throughout the incubation period, thus minimizing sample loss and/or the potential for cross-contamination. *Graphic by: Angie Ambers PhD (Institute for Human Identification—LMU College of Dental Medicine; Henry C. Lee Institute of Forensic Science; Henry C. Lee College of Criminal Justice and Forensic Sciences).*

3. Incubate samples on an orbital shaker at 56°C for 16–24h. Longer incubation times (up to 48h maximum) can be used as necessitated by workflow. However, incubation beyond 24h generally does not improve DNA recovery.
   Note: Conical tubes should be subjected to continuous agitation and inversion to ensure that the bone or tooth powder remains suspended in the lysis buffer for the entirety of the incubation period. A variety of types of orbital shakers will work for this step, including platform rotators and inversion rotators, the latter of which invert tubes during incubation in a rotisserie-like manner (Fig. 2).
4. After the overnight (16–24h) incubation period, remove the tubes from the incubator. Centrifuge tubes briefly to remove any condensation or residual extract from the lids and sides of each tube.
   Note: This protocol was validated using a Thermo Scientific™ Sorvall™ Legend™ Centrifuge equipped with a swinging bucket rotor and adapters to accommodate 15-mL conical tubes. A swinging bucket rotor is preferred over fixed angle rotors for this protocol.

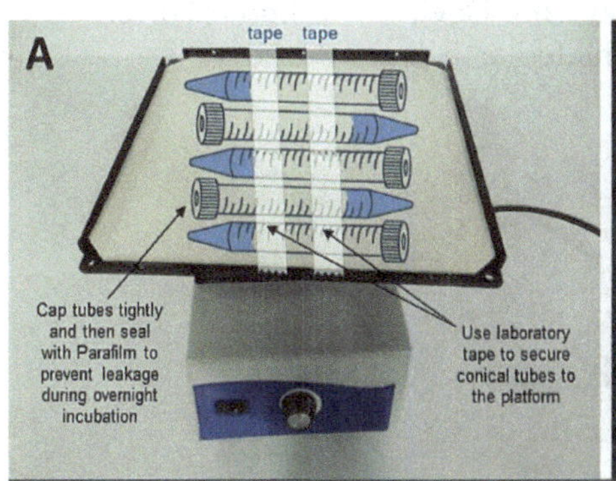
A
tape tape
Cap tubes tightly and then seal with Parafilm to prevent leakage during overnight incubation
Use laboratory tape to secure conical tubes to the platform

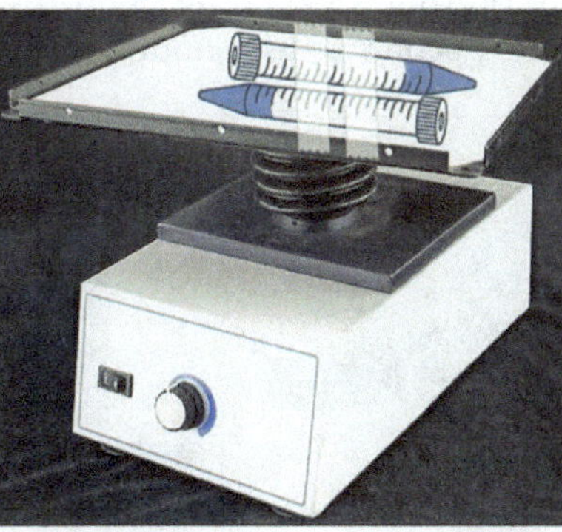

For platform rotators, 15-mL conical tubes should be capped tightly, sealed with Parafilm, placed lying flat on the platform, and secured in place using laboratory tape. The platform rotator then can be placed inside an incubator (oven) preheated to 56°C for overnight demineralization (16-24 hours)

B

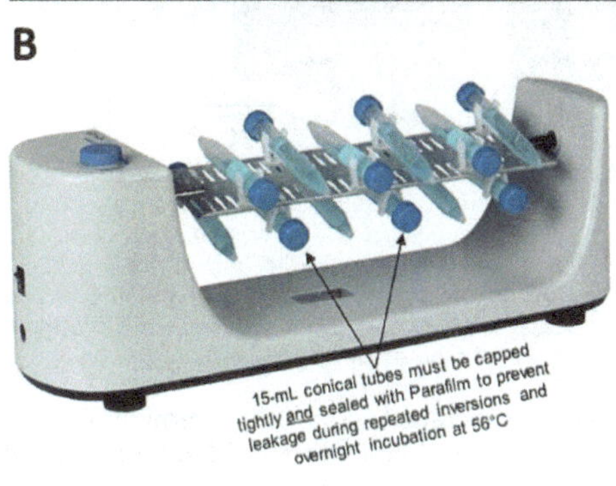
15-mL conical tubes must be capped tightly and sealed with Parafilm to prevent leakage during repeated inversions and overnight incubation at 56°C

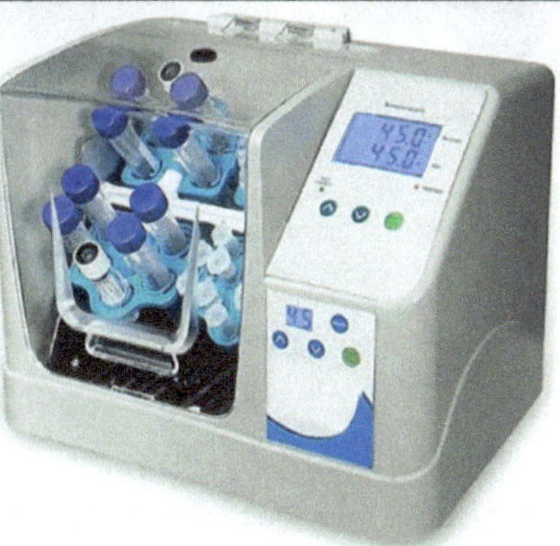

For inversion rotators, 15-mL conical tubes should be capped tightly and sealed with Parafilm. The rotator unit (LEFT) then can be placed inside an incubator (oven) preheated to 56°C for overnight demineralization (16-24 hours). Alternatively, incubators (ovens) equipped with built-in rotating tube holders are available (RIGHT). Rotation speed should be slow and steady, but fast enough to ensure that the bone or tooth powder is continuously re-suspended in the demineralization buffer for the entire incubation period

FIG. 2 Examples of types of orbital shakers that can be used during DNA extraction from bone or tooth powder: (A) platform rotators, which rock back-and-forth and rotate in a tri-directional (nutating) manner for thorough, uniform mixing; and (B) inversion rotators, which spin and invert tubes like a rotisserie. Each 15-mL conical tube containing bone or tooth powder and demineralization buffer should be capped tightly and sealed with Parafilm to prevent leakage during repeated inversions and overnight incubation at 56°C. *Graphic by: Angie Ambers PhD (Institute for Human Identification—LMU College of Dental Medicine; Henry C. Lee Institute of Forensic Science; Henry C. Lee College of Criminal Justice and Forensic Sciences).*

5. Add an *equal volume* of Phenol:Chloroform:Isoamyl Alcohol (25:24:1) (PCIA) to each tube. Vortex on medium speed for 30 s to mix well. Centrifuge tubes at maximum speed for 3 min, in order to separate the solution into the characteristic three layers (i.e., aqueous layer, interface, organic phase) (Fig. 3). After centrifugation, transfer the entire aqueous (top) layer to an Amicon® Ultra-4 Centrifugal Filter Device (30K or 50K, for volumes up to 4 mL), *or* to new, sterile 15-mL polypropylene tube for an additional PCIA cleanup (if needed) (Figs. 4 and 5). Be very careful when removing the aqueous layer for transfer, and do not inadvertently draw up any portion of the interface. The interface contains lipids, which will clog the Amicon® concentrators. Also, do not touch the membrane (filter) of the Amicon® column with the pipette tip.
Note: If 4.5 mL of demineralization buffer were used during Step #2 of this protocol, then 4.5 mL of PCIA (25:24:1) should be added to the lysate in each tube. If 3.0 mL of demineralization buffer were used, add 3.0 mL of PCIA (25:24:1) to the lysate in each tube. This constitutes the "equal volume" described above.
Important considerations for Amicon® Ultra-4 Centrifugal Filter Devices:
This protocol was validated using both the 30K (30,000 NMWL) and 50K (50,000 NMWL) versions of Amicon® Ultra-4 Centrifugal Filter Devices (Millipore, 2018). Which version of Amicon® to use should be evaluated on a case-by-case basis,

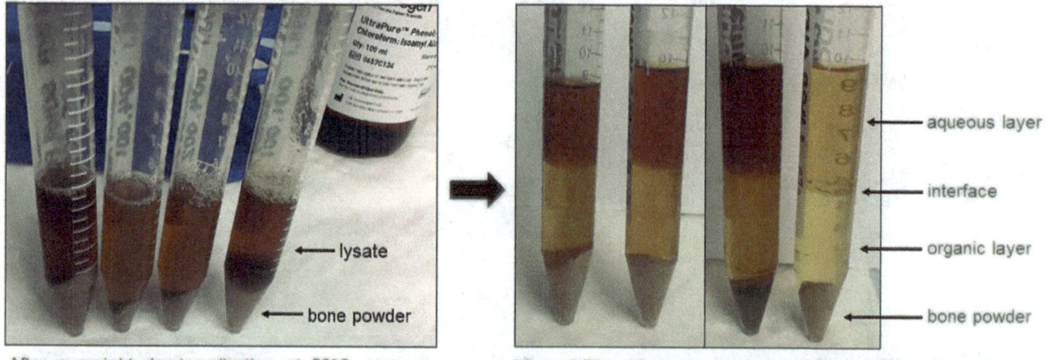

lysate
bone powder
aqueous layer
interface
organic layer
bone powder
After overnight demineralization at 56°C, remove sample tubes from incubator and centrifuge briefly to remove condensation and residual lysate from the lids and sides of the tubes. Bone powder will pellet to the bottom of the tubes.
After addition of an equal volume of Phenol:Chloroform:Isoamyl Alcohol (25:24:1) (PCIA), vortex tubes and mix thoroughly. Once an emulsion is formed, centrifuge the tubes at maximum speed to separate the solution into layers. Denatured proteins and cellular debris will accumulate in the organic (bottom) phase. Lipids collect at the interface between the aqueous and organic layers, and DNA will accumulate in the aqueous (top) layer.

FIG. 3 Photographs depicting: (1) demineralized bone powder and the resultant lysate after overnight incubation at 56°C (*left*), and (2) separation into aqueous and organic phases after addition of Phenol:Chloroform:Isoamyl Alcohol (25:24:1) (PCIA) (*right*). Note that the color of the lysate (post-incubation) may vary significantly from the color of the demineralization buffer added during Step #2 of this extraction protocol, depending on the specific contaminants and inhibitors contained within the skeletal sample as a result of the burial or depositional environment from which it was recovered. Although extensive pre-processing and decontamination steps are performed on bone samples prior to pulverization into powder, these steps (described in Chapter 5) generally only remove surface-level contaminants and do not eliminate those that have permeated or diffused into the bone itself. *Photos by: Angie Ambers PhD (Institute for Human Identification—LMU College of Dental Medicine; Henry C. Lee Institute of Forensic Science; Henry C. Lee College of Criminal Justice and Forensic Sciences).*

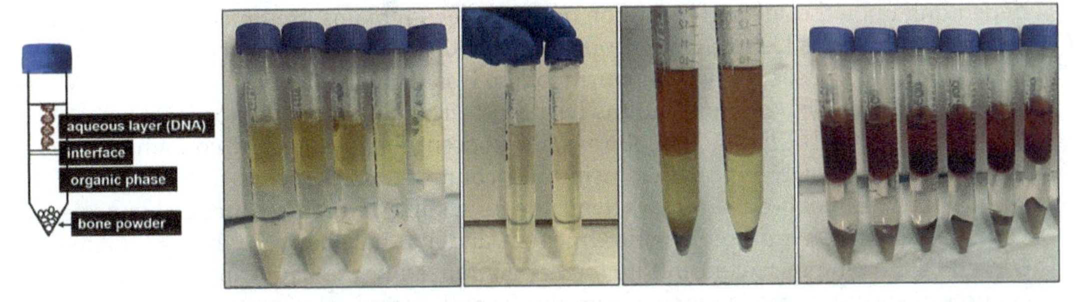

FIG. 4 The color of the aqueous layer (generated during the PCIA cleanup step) may vary significantly on a case-by-case basis. A broad range of aqueous layer colors (from clear to light yellow to dark red) may be observed with skeletal remains samples, depending on the specific inhibitors and contaminants that have diffused into the bone structure during environmental exposure and/or burial. The color of the organic phase also may vary (and may not always be clear, as shown in these photos). The color of the aqueous layer is particularly informative and can be used to make strategic decisions regarding the next step in the extraction procedure [i.e., whether to implement an extra PCIA cleanup step, or to proceed directly with concentration via Amicon® 30K (30,000 NMWL) or 50K (50,000 NMWL) Centrifugal Filter Units]. *Photos by: Angie Ambers PhD (Institute for Human Identification—LMU College of Dental Medicine; Henry C. Lee Institute of Forensic Science; Henry C. Lee College of Criminal Justice and Forensic Sciences).*

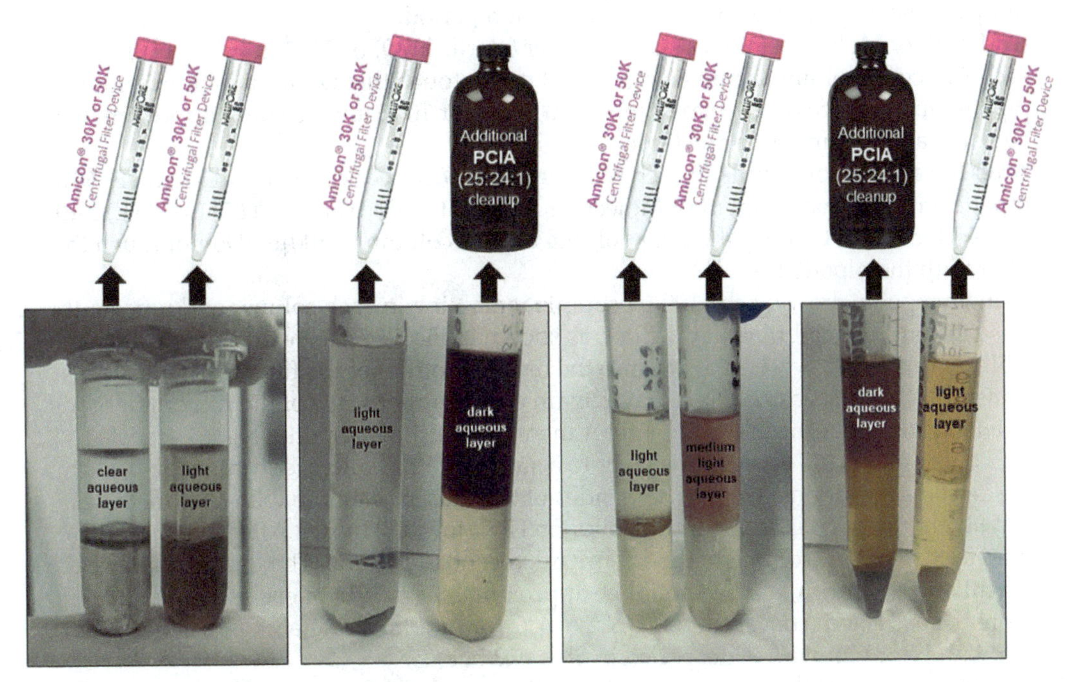

FIG. 5 Examples of the range of colors that may be observed in the aqueous layer during organic extraction of DNA from skeletal remains. The color of the aqueous layer often is an indicator of the purity of the sample and can be used to make strategic decisions regarding the next step in the extraction procedure. For example, for bone samples that produce a clear or lighter colored aqueous layer, analysts can generally proceed directly with concentration using an Amicon® Ultra-4 Centrifugal Filter Device. Darker aqueous layers may benefit from an additional (second) PCIA cleanup step prior to Amicon® concentration. *Photos by: Angie Ambers PhD (Institute for Human Identification—LMU College of Dental Medicine; Henry C. Lee Institute of Forensic Science; Henry C. Lee College of Criminal Justice and Forensic Sciences).*

depending upon the age and condition of the skeletal remains presented for testing. In general, less DNA is lost using Amicon® Ultra-4 30K columns compared to Amicon® Ultra-4 50K columns, although the purity of the lysate will be lower (i.e., more DNA is retained using 30K columns, but the lysate is "dirtier"). Although more DNA is lost using Amicon® Ultra-4 50K columns, the concentrated lysate may contain fewer inhibitors (i.e., less DNA is retained using 50K columns, but the lysate is "cleaner"). If a sufficient amount of bone sample is available for testing, it often is beneficial to perform DNA extractions on multiple powder fractions and alternate between the use of Amicon 30K and 50K columns to concentrate the demineralized lysate. This can maximize chances of success across a range of samples from the same set of remains.

6. Centrifuge the Amicon® columns at a maximum of 4000 × g for 10–20 min, or until the majority of the sample (i.e., the aqueous layer) has passed through the filter. Do not let the filter dry out completely.
   Note: It is very important to stop the centrifuge periodically during the 10–20 min spin period and *visually* check the volume remaining in the filtration section of *each* Amicon® column. Make sure the membrane in the filter does not dry out. Be mindful that the aqueous layer of each individual bone sample may pass through the Amicon® filter at different rates. Some samples may pass through the column in 10 min, while other samples may require a 20–30 min centrifugation period.
7. Add 2 mL of Molecular Biology Grade water (sterile H_2O) or TE^{-4} (low TE) buffer to each Amicon® column, again being careful not to touch the membrane/filter with the pipette tip. Centrifuge at a maximum of 4000 × g for 10–20 min, or until approximately 50 μL of sample remains in the filter.
8. After the amount of sample in each Amicon® has been reduced to approximately 50 μL, pipette an additional amount of Molecular Biology Grade Water or TE^{-4} (low TE) buffer into each column to bring the total volume in each column to 100 μL. Do not touch the filter with the pipette tip.
9. Using a pipette, draw up the 100-μL sample remaining in each column and repeatedly rinse the entire membrane of the filter, as shown in Fig. 6. This ensures that as much DNA as possible is washed down into the narrow bottom of the filter unit prior to transfer. Otherwise, some DNA may remain on the upper walls of the filter membrane.
10. After rinsing the Amicon® Ultra-4 filter membrane using a pipette (Fig. 6), transfer the concentrated 100-μL aqueous layer to a sterile, 1.5-mL microcentrifuge tube. Then, perform one additional rinse of both sides of the filter membrane with an additional 100-μL aliquot of Molecular Grade Water or TE^{-4} (low TE) buffer. After this final rinse, transfer the 100-μL volume to the 1.5-mL microcentrifuge tube. The total volume in the 1.5-mL microcentrifuge tube should be approximately 200 μL for each sample.
11. Store the extracted (concentrated) DNA at 4°C for interim storage, or proceed directly with inhibitor removal, as described in Step #12 below.
12. All bone extracts at this stage should be further purified using either the QIAquick® PCR Purification Kit (Qiagen Science Inc.; Germantown, Maryland USA) or the MinElute® PCR Purification Kit (Qiagen Science Inc.; Germantown, Maryland USA), according to the manufacturer's recommendations. Note that both of these purification kits are effective at removing inhibitors from bone or tooth samples. Which kit to use should be evaluated on a case-by-case basis, depending on the PMI and quality of the

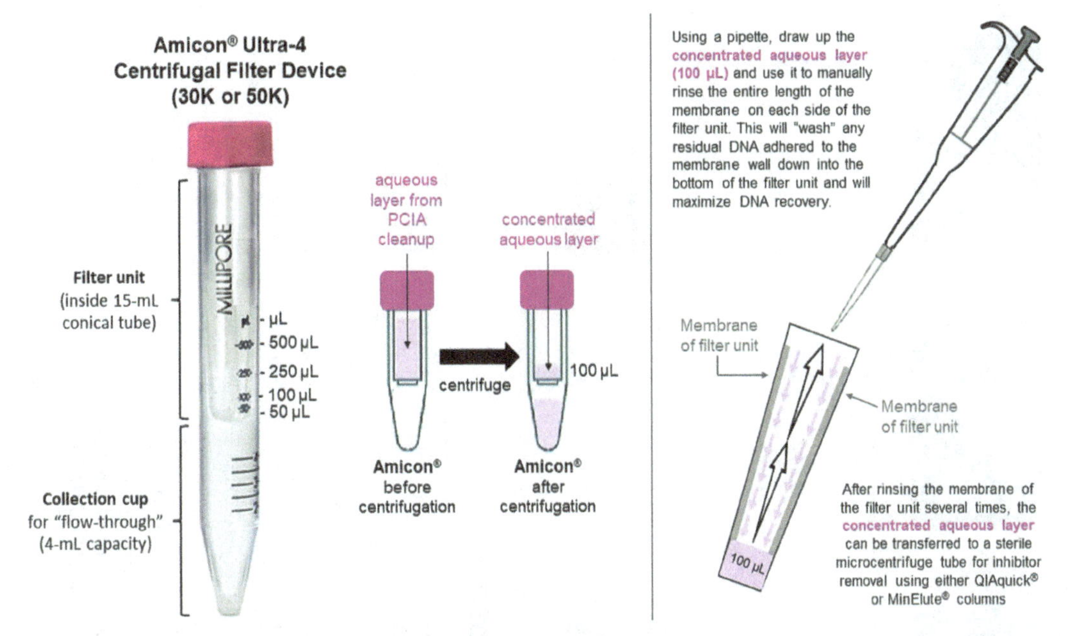
Amicon® Ultra-4
Centrifugal Filter Device
(30K or 50K)
Filter unit
(inside 15-mL
conical tube)
MILLIPORE
- μL
- 500 μL
- 250 μL
- 100 μL
- 50 μL
Collection cup
for "flow-through"
(4-mL capacity)
aqueous
layer from
PCIA
cleanup
concentrated
aqueous layer
centrifuge
100 μL
Amicon®
before
centrifugation
Amicon®
after
centrifugation
Using a pipette, draw up the concentrated aqueous layer (100 μL) and use it to manually rinse the entire length of the membrane on each side of the filter unit. This will "wash" any residual DNA adhered to the membrane wall down into the bottom of the filter unit and will maximize DNA recovery.
Membrane
of filter unit
Membrane
of filter unit
100 μL
After rinsing the membrane of the filter unit several times, the concentrated aqueous layer can be transferred to a sterile microcentrifuge tube for inhibitor removal using either QIAquick® or MinElute® columns

FIG. 6 Illustration of the Amicon® Ultra-4 Centrifugal Filter Device used to concentrate the aqueous layer after cleanup with PCIA. After concentrating the sample (aqueous layer) to 100 μL, a pipette should be used to draw up the liquid and manually "rinse" each side of the membrane in the filter unit. The membrane should be washed several times in order to ensure that any residual DNA adhered to the upper walls of the membrane are washed down into the bottom of the filter unit. The concentrated aqueous layer (100 μL) can then be withdrawn and transferred to a sterile 1.5-μL microcentrifuge tube for downstream inhibitor removal using either QIAquick® or MinElute® columns (which allow for final elution volumes of 30–50 and 10–20 μL, respectively). *Graphic by: Angie Ambers PhD (Institute for Human Identification—LMU College of Dental Medicine; Henry C. Lee Institute of Forensic Science; Henry C. Lee College of Criminal Justice and Forensic Sciences).*

bone samples being processed. QIAquick® columns provide cleanup of DNA fragments between 100 bp and 10 kb (with 90%–95% DNA recovery), and permit elution volumes between 30 μL and 50 μL (Qiagen, 2020a). MinElute® columns provide cleanup of DNA fragments between 70 bp to 7 kb (with 80% DNA recovery), and permit a smaller elution volume than the QIAquick® columns. Using MinElute® columns, the elution volume range is 10–20 μL, producing a more concentrated sample for downstream assays (Qiagen, 2020b).

Note: Although there are other kits available for sample purification, the UNTCHI casework extraction protocol described herein was validated using Qiagen products. If a different kit or protocol is substituted for the Qiagen purification kits, the laboratory should perform internal validation studies to assess and compare the efficacy of the substituted method. This is especially important given the often degraded and low-quantity condition of DNA recovered from skeletal samples, which necessitate minimizing sample loss and manipulation.

13. Store purified samples at 4°C for short-term storage, or at −20°C for long-term storage and preservation. Fig. 7 provides a summary of this DNA extraction protocol.

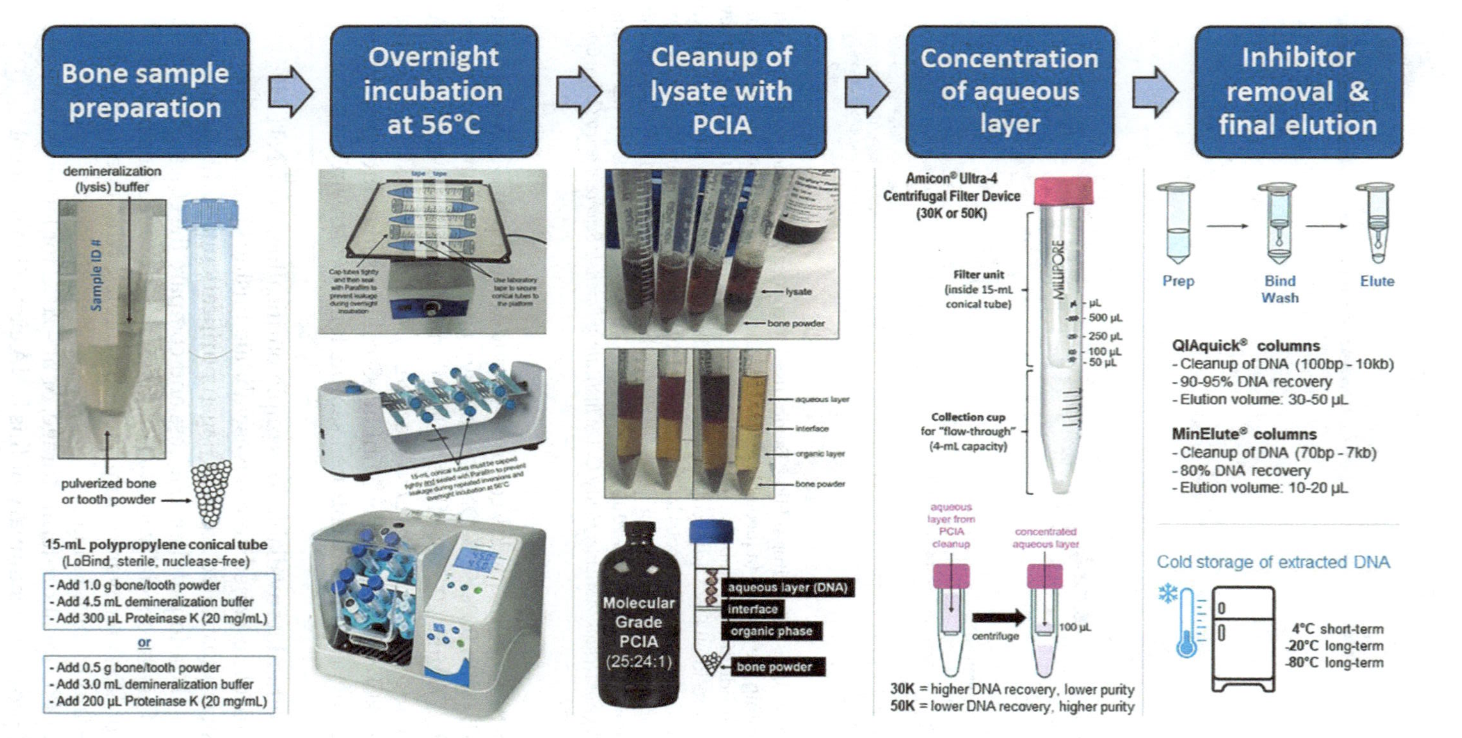

FIG. 7 Overview of a validated forensic casework protocol for DNA extraction from human skeletal remains. One of the most renowned U.S. laboratories that specialize in forensic DNA testing of human skeletal remains, the UNT Center for Human Identification (UNTCHI), has explored a variety of DNA extraction protocols since its inception in 2004. This method—a modified version of organic extraction—has consistently performed the best in terms of maximizing DNA recovery and effective removal of inhibitors. Although organic DNA extraction is one of the oldest methods available—and tends to be even more time- consuming and labor-intensive than some of the more recently introduced methods—the ultimate goal is to make a positive identification of the remains presented to the laboratory. Speed-of-analysis is meaningless in the context of human identification if the extraction method selected does not yield sufficient quantity and purity of DNA. This organic extraction approach has been used to process thousands of skeletal remains from diverse environments and across a broad range of global geographical locations, as well as remains of varying postmortem intervals (PMIs), with great success (and it consistently outperformed nonorganic-based extraction methods in terms of maximizing DNA recovery). *Graphic by: Angie Ambers PhD (Institute for Human Identification—LMU College of Dental Medicine; Henry C. Lee Institute of Forensic Science; Henry C. Lee College of Criminal Justice and Forensic Sciences).*

Field-deployable "Rapid DNA" technology for disaster victim identification (DVI) and mass fatality incidents

Organic-based DNA extraction methods are preferable for fully skeletonized human remains due to the progression of bone diagenesis and its correlation to DNA preservation (refer to Chapter 5 for an overview of diagenesis of bone microstructure). Naturally skeletonized remains that have been exposed to the environment for extended periods of time typically contain DNA that is compromised both in quality and in quantity; hence, selection of a DNA extraction method that maximizes total DNA recovery is prudent (regardless of the time and labor commitments of the extraction protocol). However, in mass casualty scenarios where large numbers of victims need to be identified and the duration of environmental exposure is short, there are other approaches that can be capitalized on to facilitate and speed up the identification process.

There are numerous scenarios encountered in forensic casework that may result in mass fatalities, including bombings/explosions, natural disasters, fires, terrorist attacks, war conflicts, aviation crashes, and other mass transit accidents (e.g., high-speed passenger trains, subways). Human remains recovered in these situations often are severely damaged, dismembered, fragmented, commingled, in varying states of decomposition, or even skeletonized. Although many national-, regional-, and state-level crime laboratories exist throughout the country with extensive expertise and training in forensic DNA identification, these fixed "brick-and-mortar" laboratories are not amenable to field-based operations and testing. Although these laboratories are willing partners in disaster victim identification (DVI) efforts, the logistics of coordinating subcontracts and transporting biological samples long distances for genetic testing is tedious. Additionally, traditional DNA casework methods are both labor-intensive and time-consuming, increasing the amount of time that families of decedents must wait for a positive identification, and this can be a particularly important consideration in mass casualty scenarios.

Rapid DNA technology offers a potential solution to meet the high-throughput and onsite testing demand of mass fatality events. The ANDE® 6C Rapid DNA System (ANDE Corporation; Waltham, Massachusetts USA) is field-deployable and can generate forensic DNA profiles in less than 2h using its patented I-Chip™ and FlexPlex® 27 assay, which includes the 20 core CODIS loci mandated by the FBI for casework (Turingan et al., 2020). The instrument was specifically engineered for field deployment and onsite testing operations. In addition to its relatively small size (dimensions 75cm×45cm×60cm) and weight (117 lbs), it is operational at altitudes as high as 3048m (10,000ft) and was ruggedized to account for vibration and shock during transportation, meeting the strict requirements of the U.S. Military Standard 810G (MIL-STD-810G) [U.S. Department of Defense (DoD), 2008]. Moreover, both the I-Chip™ and accompanying reagents are stable at room temperature for up to 6months (Turingan et al., 2020). Complementing the ANDE® 6C Rapid DNA instrument's compact size, fast run times, and temperature-stable reagents is kinship analysis software. This software, called FAIRS™, has two secure modules which compare DNA profiles obtained from unidentified victims to direct reference samples or family member exemplars: (1) the Claimed Relationships module, and (2) the Familial Search module. When algorithms embedded in the software determine a match, a formal report containing the decedent's genetic profile and the associated match statistic calculations are automatically generated for the case file. Hence, this comprehensive human identification system is particularly well-suited for location-based testing at accident sites, in disaster zones, and/or in temporary buildings or tents constructed by Disaster Mortuary Operational Response Teams (DMORTs).

The Department of Homeland Security (DHS) Science and Technology (S&T) Directorate is currently exploring the possibility of implementing Rapid DNA technology in mass fatality responses, especially since it is already being used to support immigration investigations and to combat transnational human trafficking (Carney et al., 2019; Congress.gov, 2017; Miroff, 2019). DHS S&T has developed numerous drills and mock disaster victim identification exercises in order to: (1) assess if the instruments can be shipped in an expeditious and efficient manner for disaster response operations, and (2) to identify and mitigate potential logistical challenges that might be encountered during complex deployments. Additionally, the American Society of Crime Laboratory Directors (ASCLD) recently formed a Rapid DNA Disaster Victim Identification (DVI) Subcommittee, the goal of which is to determine best practices and develop policies for mass fatality applications (www.ascld.org). Additionally, although the Scientific Working Group on DNA Analysis Methods (SWGDAM) has not yet developed formal guidelines for Rapid DNA analysis of DVI casework samples, the FBI has recently formed a working group to facilitate discussions on this issue.

Recent high-profile (real-world) human identification successes using Rapid DNA technology likely will expedite discussions among the DHS S&T, ASCLD, SWGDAM, and FBI working groups, and may facilitate formal implementation of the technology in mass fatality response efforts. One of the aforementioned high-profile cases involved the 2018 'Camp Fire' in Paradise, California, which claimed the lives of 85 people and is the deadliest wildfire in California history. In collaboration with Butte County's Sheriff's Office and the California Department of Justice (DOJ), ANDE Corporation deployed a team to assist with identification efforts. Remains from 69 victims were recovered; however, due to the intensity and duration of the fire, traditional methods of identification (e.g., fingerprints, odontology, surgical hardware) were only feasible for 22 of the decedents. Using a mobile (recreational) vehicle and three ANDE® 6C Rapid DNA instruments, a variety of sample types from the victims' remains (e.g., bone, muscle, brain tissue, liver) and approximately 300 family reference samples were tested. DNA profiles were obtained from 62 of the 69 victims recovered (89.9% of samples tested). Of the 62 victims for which DNA profiles were successfully generated, 58 victims (93.5%) were able to be positively identified via comparison to exemplars from biological relatives (Gin et al., 2020; Department of Homeland Security, 2019a; Border, 2018; Brown, 2018; Druga, 2019; Gin, 2019; Sernoffsky, 2018; Waitt, 2019; Zimmer, 2018). In 2019, less than a year after the wildfires, another tragedy struck in California. In all, 33 passengers and one crew member aboard the dive boat *Conception* perished when it caught fire off the coast of Santa Cruz Island. All 34 victims were identified within 10 days using ANDE® Rapid DNA technology, much faster than the identification effort would have taken using traditional laboratory-based testing (Department of Homeland Security, 2019b; Woods, 2019).

In addition to the human remains from the California wildfires and the *Conception* boat fire that were successfully processed for DNA, the ANDE® Rapid DNA platform has been subjected to extensive research with a broad spectrum of biological samples. During developmental validation, a total of 101 samples across seven different DVI-relevant tissue types were processed for accuracy and concordance. Included in the study were 18 bones (femora, humeri, ribs), 3 molar teeth, 24 liver samples, 34 skeletal muscles, 2 brain cortices, 11 lung tissues, and 9 kidney samplings (Turingan et al., 2020). In another study, 10 deceased (donated) human subjects were exposed to the environment (surface deposition) at an approved outdoor research facility, and various tissue samplings were collected at regular time intervals over the course of a 1-year study period. In general, successful DNA typing of soft tissues (brain, muscle, liver) was limited to relatively short postmortem time intervals (1–11 days), attributed to accelerated decomposition during environmental exposure compared to hard tissues. The majority of the bone/tooth samples tested yielded useful DNA profiles from the first day of collection through the end of the 1-year experimental period (Turingan et al., 2019). Both of these research studies—as well as the casework performed during the California wildfires and the *Conception* boat fire—demonstrate proof-of-concept for DVI applications and support the use of this technology as a way to expedite human identification and family reunification efforts.

In summary, ANDE Corporation's Rapid DNA platform has already demonstrated robust and reliable forensic genetic testing of human remains with relatively short postmortem intervals (i.e., recent deaths ranging from 1 day to 1 year). However, human remains from mass fatality events are not always located and recovered within such short time frames. Victims of homicide, oppressive regimes, genocide, and war conflicts, for example, often are buried in clandestine graves and may not be discovered for many years (sometimes decades). These types of samples—which have been buried and have naturally skeletonized over time due to environmental insult—may pose additional challenges for Rapid DNA platforms. Buried remains present an entirely different set of challenges compared to 'fresh' bones that are recovered immediately or shortly after a fatality event has occurred. As the structural components of bone (i.e., calcium hydroxyapatite, collagen) decompose in a process called diagenesis, endogenous DNA molecules become far more susceptible to damage. Thus, buried bones typically yield much lower DNA quantities and present templates with a multitude of problematic lesions that can prevent successful profiling. Future research efforts could expand to include buried human remains of prolonged postmortem intervals. Successful application of Rapid DNA technology to naturally skeletonized remains which have been exposed to extended burial conditions or surface deposition across a broad range of climates and environments could dramatically expedite and change the landscape of forensic human identification casework.

Summary

Although a myriad of DNA extraction approaches are available to the forensic geneticist, these methods should not be considered as equally effective across all samples or cases, especially given the often degraded nature of DNA encompassed within skeletonized human remains that have been exposed to environmental insults for extended periods of time. Best practices dictate that laboratories should validate a combination of different extraction methods, providing the forensic geneticist with a range of methodologies from which to select that will maximize the chances of successful DNA recovery. Every skeletal remains case will be different, depending on the specific environmental conditions that the remains were exposed to and the duration of exposure.

Furthermore, when dealing with highly decomposed or skeletonized remains, speed-of-analysis should not be a primary determinant in the selection of extraction method.

While it is understandable to be attracted to methods that take less time and effort to complete, one must consider that certain sacrifices are typical with quicker protocols. Generally, some DNA is lost as the speed of the extraction method increases, and while a loss of 10%–20% DNA during the extraction process may have negligible effects on the genetic typing of other (fresher, less degraded) types of samples, it can have significant impact on DNA typing results from skeletal remains. This seemingly minimal loss of DNA during extraction (due to non-optimal selection of extraction method) can mean the difference between a complete genetic profile and a partial profile, or between obtaining a partial profile or no profile at all. When the ultimate goal is to recover sufficient genetic data to make an identification, the amount of time and labor necessary to obtain this data is unimportant.

Although challenges or barriers to identification may be encountered due to incomplete (or insufficient) DNA databases and/or lack of reference samples for comparison, obtaining DNA of the highest possible quantity, quality, and purity should be the first priority when remains are presented to the laboratory for analysis—and this is largely dependent on the DNA extraction approach employed and the skills/training of the scientist performing the testing. The specific genetic markers and type of downstream testing selected can be tailored to the circumstances and resources available to make the identification.

In addition to selecting optimal skeletal elements (if available) and sampling from dense, cortical bone, the DNA extraction method used is equally as important for downstream profiling success. As advanced as our detection instrumentation are, their efficacy is almost entirely dependent on the quality, quantity, and purity of DNA added into the system. Although the quality of the DNA present within highly decomposed or skeletonized human remains is beyond the control of the analyst, DNA extraction protocols can be selected to maximize the quantity and purity of DNA obtained.

References

Ambers, A., 2016. Identifying 140-year-old skeletal remains using massively parallel DNA sequencing. In: International Symposium for Human Identification (ISHI) News. https://www.ishinews.com/identifying-140-year-old-remains-using-massively-parallel-dna-sequencing/.

Ambers, A., 2021. Forensic genetic investigation of two adult male skeletons recovered from the 17th-century *La Belle* shipwreck using massively parallel sequencing. In: The ISHI Report: News from the World of DNA Forensics. https://promega.foleon.com/theishireport/november-2020/modern-dna-technology-for-historical-cold-cases-forensic-genetic-investigation-of-two-adult-male-skeletons-recovered-from-the-17th-century-la-belle-shipwreck-using-massively-parallel-sequencing-mps/.

Ambers, A., Gill-King, H., Dirkmaat, D., Benjamin, R., King, J., Budowle, B., 2014. Autosomal and Y-STR analysis of degraded DNA from the 120-year-old skeletal remains of Ezekiel Harper. Forensic Sci. Int. Genet. 9, 33–41.

Ambers, A.D., Churchill, J.D., King, J.L., Stoljarova, M., Gill-King, H., Assidi, M., Abu-Elmagd, M., Buhmeida, A., Al-Qahtani, M., Budowle, B., 2016. More comprehensive forensic genetic marker analyses for accurate human remains identification using massively parallel DNA sequencing. BMC Genomics 17, 750.

Ambers, A., Votrubova, J., Vanek, D., Sajantila, A., Budowle, B., 2018. Improved Y-STR typing for disaster victim identification, missing persons investigations, and historical human skeletal remains. Int. J. Leg. Med. https://doi.org/10.1007/s00414-018-1794-8.

Ambers, A., Zeng, X., Votrubova, J., Vanek, D., 2019. Enhanced interrogation of degraded DNA from human skeletal remains: increased genetic data recovery using the expanded CODIS loci, multiple sex determination markers, and consensus testing. Anthropol. Anz. 76, 333–351.

Ambers, A., Bus, M.M., King, J.L., Jones, B., Durst, J., Bruseth, J.E., Gill-King, H., Budowle, B., 2020. Forensic genetic investigation of human skeletal remains recovered from the *La Belle* shipwreck. Forensic Sci. Int. 306, 110050.

Antinick, T.C., Foran, D.R., 2018. Intra- and inter-element variability in mitochondrial and nuclear DNA from fresh and environmentally exposed skeletal remains. J. Forensic Sci. https://doi.org/10.1111/1556-4029.13843.

Astolphi, R.D., Alves, M.T., Evison, M.P., Francisco, R.A., Guimaraes, M.A., Iwamura, E.S.M., 2019. The impact of burial period on compact bone microstructure: histological analysis of matrix loss and cell integrity in human bones exhumed from tropical soil. Forensic Sci. Int. 298, 384–392.

Border, G., 2018. California taps war-zone DNA specialists after wildfire. In: Reuters. https://www.reuters.com/article/us-california-wildfires-remains/california-taps-war-zone-dna-specialists-after-wildfire-idUSKCN1NK1EI.

Brown, K.V., 2018. California turns to war-zone DNA test to ID fire remains. Bloomberg. https://www.bloomberg.com/news/articles/2018-11-19/dna-testing-company-gets-call-to-help-id-california-fire-victims.

Brundin, M., Figdor, D., Sundquist, G., Sjorgren, U., 2013. DNA binding to hydroxyapatite: a potential mechanism for preservation of microbial DNA. J. Endod. 39 (2), 211–216.

Carney, C., Whitney, S., Vaidyanathan, J., Persick, R., Noel, F., Vallone, P.M., Romsos, E.L., Tan, E., Grover, R., Turingan, R.S., French, J.L., Selden, R.F., 2019. Developmental validation of the ANDE® rapid DNA system with Flex-Plex™ assay for arrestee and reference buccal swab processing and database searching. Forensic Sci. Int. Genet. 40, 120–130.

Congress.gov, 2017. H.R.510—Rapid DNA Act of 2017. https://www.congress.gov/bill/115th-congress/house-bill/510.

Cooper, A., Poinar, H.N., 2000. Ancient DNA: do it right or not at all. Science 18, 1139–1141.

Department of Homeland Security, 2019a. Snapshot: S&T's Rapid DNA technology identified victims of California Wildfire. In: DHS Science and Technology News. https://www.dhs.gov/science-and-technology/news/2019/04/23/snapshot-st-rapid-dna-technology-identified-victims.

Department of Homeland Security, 2019b. Snapshot: Rapid DNA identifies Conception boat fire victims. In: DHS Science and Technology News. https://www.dhs.gov/science-and-technology/news/2019/12/10/snapshot-rapid-dna-identifies-conception-boat-fire-victims.

Druga, M., 2019. Rapid DNA technology identified victims of California wildfires. In: Homeland Preparedness News. https://homelandprepnews.com/stories/33583-rapid-dna-technology-identified-victims-of-california-wildfires/.

Edson, S.M., Ross, J.P., Coble, M.D., Parsons, T.J., Barritt, S.M., 2004. Naming the dead: confronting the realities of rapid identification of degraded skeletal remains. Forensic Sci. Rev. 16 (1), 63–90.

Emmons, A.L., Davoren, J., DeBruyn, J.M., Mundorff, A.Z., 2020. Inter and intra-variation in skeletal DNA preservation in buried remains. Forensic Sci. Int. Genet. 44, 102193.

Eppendorf, 2015. Eppendorf LoBind® tubes and plates. https://cdn2.hubspot.net/hubfs/2753166/Products/Eppendorf/Tubes/Instructions%20for%20use%20-%20LoBind%20Tubes%20and%20Plates.pdf.

Gilbert, M.T.P., Hansen, A.J., Willerslev, E., Turner-Walker, G., Collins, M., 2006. Insights into the processes behind the contamination of degraded human teeth and bone samples with exogenous sources of DNA. Int. J. Osteoarchaeol. 16, 156–164.

Gin, K., 2019. California wildfires: rapid DNA and the future of mass fatality identification. In: Forensic Science Executive. www.forensicscienceexecutive.org.

Gin, K., Tovar, J., Bartelink, E.J., Kendell, A., Milligan, C., Willey, P., Wood, J., Tan, E., Turingan, R.S., Selden, R.F., 2020. The 2018 California wildfires: integration of rapid DNA to dramatically accelerate victim identification. J. Forensic Sci. https://doi.org/10.1111/1556-4029.14284.

Hagelberg, E., Clegg, J.B., 1991. Isolation and characterization of DNA from archaeological bone. Proceedings R. Soc. London 244, 45–50.

Hollund, H.I., Teasdale, M.D., Mattiangeli, V., Sverrisdottir, O.O., Bradley, D.G., O'Connor, T., 2017. Pick the right pocket: sub-sampling of bone sections to investigate diagenesis and DNA preservation. Int. J. Osteoarchaeol. 27, 365–374.

Kemp, B., Smith, D.G., 2005. Use of bleach to eliminate contaminating DNA from the surface of bones and teeth. Forensic Sci. Int. 154, 53–61.

Loreille, O.M., Diegoli, T.M., Irwin, J.A., Coble, M.D., Parsons, T.J., 2007. High efficiency DNA extraction from bone by total demineralization. Forensic Sci. Int. Genet. 1, 191–195.

Millipore, 2018. Amicon® Ultra-4 Centrifugal Filter Devices User Manual. PR05034, Rev. 10/18, https://www.emdmillipore.com/US/en/product/Amicon-Ultra-4-Centrifugal-Filter-Units, MM_NF-C7719#documentation.

Miroff, N., 2019. Homeland Security to test DNA of families at border in cases of suspected fraud. In: The Washington Post. https://www.washingtonpost.com/immigration/homeland-security-to-test-dna-of-families-at-border-in-cases-of-suspected-fraud/2019/05/01/8e8c042a-6c46-11e9-a66d-a82d3f3d96d5_story.html?utm_term=.b4611915101f.

Poinar, H.N., 2003. Criteria of authenticity for DNA from ancient and forensic samples. Int. Congr. Ser. 1239, 575–579.
Qiagen, 2020a. QIAquick® PCR Purification Kit User Manual. https://www.qiagen.com/us/products/discovery-and-translational-research/dna-rna-purification/dna-purification/dna-clean-up/qiaquick-pcr-purification-kit/.
Qiagen, 2020b. MinElute® PCR Purification Kit User Manual. https://www.qiagen.com/us/products/discovery-and-translational-research/dna-rna-purification/dna-purification/dna-clean-up/minelute-pcr-purification-kit/.
Sernoffsky, E., 2018. DNA technology helps identify Camp Fire victims; challenge is finding family. In: San Francisco Chronicle. https://www.sfchronicle.com/california-wildfires/article/DNA-technology-helps-identify-Camp-Fire-victims-13406853.php.
Sosa, M., Vispe, E., Nunez, C., Baeta, M., Casalod, Y., Bolea, M., Hedges, R.E.M., Martinez-Jarreta, B., 2013. Association between ancient bone preservation and DNA yield: a multi-disciplinary approach. Am. J. Phys. Anthropol. 151, 102–109.
Turingan, R.S., Brown, J., Kaplun, L., Smith, J., Watson, J., Boyd, D.A., Steadman, D.W., Selden, R.F., 2019. Identification of human remains using rapid DNA analysis. Int. J. Leg. Med. https://doi.org/10.1007/s00414-019-02186-y.
Turingan, R.S., Tan, E., Jiang, H., Brown, J., Estari, Y., Krautz-Peterson, G., Selden, R.F., 2020. Developmental validation of the ANDE 6C system for rapid DNA analysis of forensic casework and DVI samples. J. Forensic Sci. https://doi.org/10.1111/1556-4029.14286.
U.S. Department of Defense (DoD), 2008. Department of Defense test method standard: Environmental engineering considerations and laboratory tests (MIL-STD-810G). https://www.atec.army.mil/publications/Mil-Std-810G/Mil-Std-810G.pdf.
Waitt, T., 2019. Rapid DNA ID's deadly 'Camp Fire' disaster victims. In: American Security Today. https://american-securitytoday.com/rapid-dna-ids-deadly-camp-fire-disaster-victims-see-how-video/.
Woods, A., 2019. Officials used "rapid DNA" tech to identify California boat fire victims. In: The New York Post. https://nypost.com/2019/09/05/officials-used-rapid-dna-tech-to-identify-california-boat-fire-victims/.
Yang, D.Y., Watt, K., 2005. Contamination controls when preparing archaeological remains for ancient DNA analysis. J. Archaeol. Sci. 32, 331–336.
Zimmer, K., 2018. Rapid DNA analysis steps in to identify remains of wildfire victims. In: The Scientist. https://www.the-scientist.com/news-opinion/rapid-dna-analysis-steps-in-to-identify-remains-of-wildfire-victims-65156.

Further reading

Correa, H.S.D., Carvalho, F.T., Brito, M.C.N.C., de O-Dantas, E.S., Junior, D.P., Silva, F.G.S., 2018. Forensic DNA typing from demineralized human bone sections. SM J. Foren. Res. Criminol. 2 (1), 1016.
Correa, H., Cortellini, V., Franceschetti, L., Verzeletti, A., 2021. Large fragment demineralization: an alternative pre-treatment for forensic DNA typing of bones. Int. J. Leg. Med. 135, 1417–1424.
Edson, S.M., 2019. Getting ahead: extraction of DNA from skeletonized cranial material and teeth. J. Forensic Sci. 64, 1646–1657.
Edson, S.M., McMahon, T.P., 2016. Extraction of DNA from skeletal remains. In: Goodwin, W. (Ed.), Forensic DNA Typing Protocols: Methods in Molecular Biology. Springer Science + Business Media, New York.
Gorden, E.M., Sturk-Andreaggi, K., Marshall, C., 2021. Capture enrichment and massively parallel sequencing for human identification. Forensic Sci. Int. Genet. 53, 102496.
Hofreiter, M., Sneberger, J., Pospisek, M., Vanek, D., 2021. Progress in forensic bone DNA analysis: lessons learned from ancient DNA. Forensic Science Int. Genet. 54, 102538.
Silva, R.C.F., Ekert, M.H.F., Mazanek, M.L., Miranda, C.S., Santos, A.L.N., dos Santos, A.R., do Monte, S.M.C., de Castro, S.G., de Souza, A.C., Ramalho-Neto, C.E., 2018. Alternative methodology for extraction of high-quality DNA from ancient bones by demineralizaition without pulverization. Forensic Sci. Criminol. 3 (2), 1–7.
Xavier, C., Eduardoff, M., Bertoglio, B., Amory, C., Berger, C., Casas-Vargas, A., Pallua, J., Parson, W., 2021. Evaluation of DNA extraction methods developed for forensic and ancient DNA applications using bone samples of different age. Genes 12, 146. https://doi.org/10.3390/genes12020146.

CHAPTER

7

Quantitative and qualitative assessment of DNA recovered from human skeletal remains

Jodie Ward PhD[a] *and Jeremy Watherston PhD*[b]

[a]Centre for Forensic Science, University of Technology Sydney, Sydney, NSW, Australia
[b]Forensic and Analytical Science Service, NSW Health Pathology, Lidcombe, NSW, Australia

Overview

Following extraction, the quantity and quality of DNA in a sample are determined. Assessment of the concentration, fragmentation, and inhibition of human DNA extracted from skeletal remains is a critical step in the DNA profiling workflow to provide key sample information to instruct workflow decisions and ensure optimal performance of downstream assays. In addition to permitting an assessment of DNA extraction efficiency, DNA quantification can assist in determining the optimal amount of DNA extract to use in the polymerase chain reaction (PCR) and the most appropriate DNA profiling procedure to use, prior to utilizing any analytical technique that will consume part or all of a limited sample.

The use of an optimal DNA input amount is important for all PCR-based techniques. Amplifying samples with a suboptimal target DNA quantity can result in inefficient amplification of target loci, as well as the observation of a number of amplification and electrophoretic artifacts. If DNA quantification determines a low quantity of DNA template is present, the maximum volume of DNA extract can be placed into the PCR or the DNA extract can be concentrated prior to PCR. Alternatively, if DNA quantification determines a high quantity of DNA is present, the DNA extract could be diluted prior to PCR. Additionally, having a priori knowledge of a sample's condition can prevent or minimize associated PCR complications, such as complete PCR failure, preferential amplification, jumping PCR, and PCR miscoding lesions for degraded DNA samples (Alaeddini et al., 2010). Furthermore, such knowledge can be beneficial for interpreting the DNA profiling results obtained from forensic samples.

https://doi.org/10.1016/B978-0-12-815766-4.00007-8

DNA quantification methods have evolved over time (Lee et al., 2014), with improvements to the sensitivity, accuracy, precision, and sample information content of modern quantification kits, greatly benefiting forensic DNA profiling. The advantages of using a modern DNA quantification assay that can measure the amount of amplifiable DNA in a sample, in addition to flagging those samples that are degraded or inhibited, include improved amplification efficiency and production of interpretable DNA profiles. This will ultimately result in a reduced number of samples requiring re-testing or re-analysis, offering cost- and time savings for the laboratory, and conservation of precious samples.

DNA quantification methods

There are a number of quantification methods available to estimate the DNA concentration of a sample for DNA normalization purposes prior to DNA profiling. Depending on the method used, quantification can determine the amount of amplifiable total human, male, and/or mitochondrial DNA (mtDNA), the presence of co-extracted inhibitors, and degree of degraded DNA in a DNA extract. The use of a human-specific quantification assay is also important for accurately quantifying the amount of endogenous human DNA recovered from skeletal remains.

Historically, non-PCR-based methods such as ultraviolet (UV) spectrophotometry, fluorometry, agarose yield gel electrophoresis, or slot blot hybridization were commonly used in forensic genetics to estimate the quantity of DNA in a biological sample. For a detailed review of the evolution of DNA quantification methods refer to the following publications: Barbisin and Shewale (2010), Butler (2012), Lee et al. (2014), and Nicklas and Buel (2003). In addition to being time-consuming, labor-intensive, subjective, and less sensitive, these traditional quantification assays present a number of limitations for quantifying human DNA recovered from skeletal samples. Non-PCR-based methods cannot: (1) accurately quantify the amount of amplifiable DNA present in a sample, which is important for degraded skeletal samples, (2) detect the presence of PCR inhibitors in a sample, which is important for skeletal samples in general due to their calcium and collagen content, or those which have been recovered from soil (which may contain humic and tannic acids, or heavy metals), or (3) provide a DNA concentration estimate for the total human, human male, and/or human mtDNA component of a sample simultaneously, which is important for skeletal samples with a high microbial DNA content, when DNA-based species and/or sex determination is important, or when DNA extracts are limited.

Most forensic DNA quantification assays in use today are based on real-time quantitative PCR (qPCR) technology, and these assays will be the focus of this chapter. qPCR is considered a sensitive, reliable, reproducible, specific, accurate, and objective quantification method for estimating the concentration of human DNA in forensic samples destined for PCR-based downstream assays. This is because it only measures amplifiable human DNA; therefore short, degraded DNA fragments (which are commonly extracted from compromised skeletal material) or DNA from nonhuman sources (which can be co-extracted with authentic DNA from skeletal remains) will not contribute to the concentration estimate. In addition, many qPCR assays are commercially manufactured, amenable to automation, and occur in a closed-tube system, making them attractive options for high-throughput forensic DNA laboratories.

Real-time quantitative PCR (qPCR)

qPCR was first described by Higuchi and colleagues in the early 1990s (Higuchi et al., 1993). This thermal cycling technique measures the accumulation of PCR product and associated change in fluorescence signal at the completion of every PCR cycle, commonly using nonspecific double-stranded DNA intercalating dyes (e.g., SYBR® Green I), sequence-specific probe-based fluorescent reporters (e.g., TaqMan®, Scorpions®), or primer quenching (e.g., Plexor®). These chemistries differ by how they detect and quantify the accumulated PCR product. Most of the commercial qPCR kits currently available for the forensic market use TaqMan® probes; however, SYBR® Green I, Scorpions®, or Plexor® assays are also mentioned in this chapter.

SYBR® Green I assays

Intercalating dyes were initially used to measure qPCR products (Higuchi et al., 1993). One of the most commonly used DNA intercalating dyes for qPCR today is SYBR® Green I. SYBR® Green I is a nonspecific fluorogenic minor groove DNA-binding dye that intercalates into double-stranded DNA (Arya et al., 2005). The fluorescence of the SYBR® Green I dye molecules increases up to 1000-fold after binding to double-stranded DNA (Morrison et al., 1998; Wittwer et al., 1997). Therefore, the fluorescent signal is proportional to the amount of double-stranded DNA present and will increase as PCR product accumulates. SYBR® Green I assays offer a simple and economical technique for DNA quantification because sequence-specific labeled hybridization probes are not required. The main disadvantages of this method are that the assay is not human or target specific because the dye molecules will bind to any double-stranded DNA present (including nonhuman DNA and nonspecific amplicons), and there multiplexing capability is not possible because fluorescent signals from different amplicons cannot be distinguished. Owing to the lack of specificity, SYBR® Green I assays often require a secondary test such as melt-curve analysis to identify target products from amplification artifacts (Ririe et al., 1997). Despite these limitations, SYBR® Green I has been used to develop custom qPCR assays for forensic applications (e.g., Goodwin et al., 2018).

TaqMan® probe assays

qPCR assays were improved by the introduction of dual-labeled fluorogenic probes for detecting specific PCR products (Bassler et al., 1995; Lee et al., 1993; Livak et al., 1995). TaqMan® probe assays use standard PCR primers to generate an amplicon and an internal TaqMan® probe that is homologous to the sequence between the PCR primers (Fig. 1). TaqMan® probes are labeled with a fluorescent reporter dye at the 5′ end of the probe and a nonfluorescent quencher dye at the 3′ end of the probe (Heid et al., 1996). The addition of a minor groove binder (MGB) to the 3′ end of the probe increases the melting temperature without increasing probe length, facilitating the use of short, sequence-specific probes (Afonina et al., 1997). For multiplex qPCR assays, each assay target consists of locus-specific PCR primers and a different fluorescent dye-labeled TaqMan® probe.

At the start of PCR, all TaqMan® probe molecules are intact, and the close physical proximity of the quencher dye suppresses the fluorescence emitted by the reporter dye (Heid et al., 1996). This occurs due to fluorescence resonance energy transfer (FRET) (Cardullo et al., 1988;

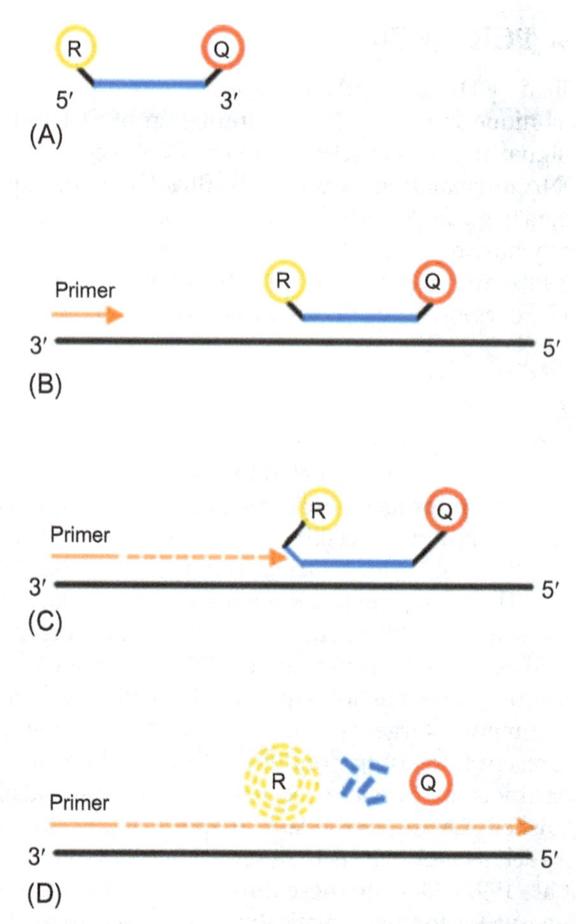

FIG. 1 Overview of the TaqMan® real-time quantitative PCR (qPCR) assay: (A) TaqMan® probes are labeled with a fluorescent reporter dye (R) at the 5′ end of the probe and a nonfluorescent quencher dye (Q) at the 3′ end of the probe, (B) during annealing, the TaqMan® probe binds to the target sequence and no fluorescence is emitted due to the proximity of the R and Q, (C) during extension, Taq DNA polymerase cleaves the hybridized probe and the R and Q separate in distance from each other, and (D) after cleavage, the R emits fluorescence, and primer extension continues to the end of the template strand. *Graphic design by Trent Watherston.*

Clegg, 1992; Förster, 1948). During the annealing phase, the amplicon accumulates, and the probe anneals specifically to a complementary sequence between the forward and reverse primer sites. In the extension phase, Taq DNA polymerase cleaves the hybridized probes with its 5′→3′ exonuclease activity (Holland et al., 1991). The cleavage of the probe separates the fluorescent reporter dye from the quencher dye, thereby increasing the reporter dye signal and allowing primer extension to continue to the end of the template strand (Heid et al., 1996). With each amplification cycle, additional reporter dye molecules are cleaved resulting in an increase in fluorescence intensity proportional to the amount of PCR product produced. All of the recently released forensic DNA quantification kits employ TaqMan® technology (Ewing et al., 2016; Holt et al., 2016; Loftus et al., 2017; QIAGEN, 2018b).

Scorpions® primer assays

Unlike TaqMan® assays which require a set of primers and a separate probe, Scorpions® primer assays employ two primers only, one of which is a tailed primer that serves as the probe. The Scorpions® primer is covalently linked to a hairpin loop structure that includes a probe element, a 5′ fluorophore, and a 3′ quencher (Carters et al., 2008; Thelwell et al., 2000; Whitcombe et al., 1999). The hairpin loop sequence of the Scorpions® probe is complementary to the target sequence, and a PCR stopper between the 3′ quencher and the 5′ end of the PCR primer prevents the Taq DNA polymerase from extending the PCR primer and copying the loop sequence (Arya et al., 2005). During the first PCR cycle, the Scorpions® primer is extended and the sequence complementary to the loop sequence is generated on the same strand as the probe (Whitcombe et al., 1999). During subsequent denaturation and annealing, the probe binds to the amplified target, and then the stem of the hairpin separates and rearranges into a linear structure, resulting in the fluorophore, and the quencher dye present at the ends of the hairpin separating (Whitcombe et al., 1999). This leads to the emission of a fluorescent signal that is proportional to the quantity of PCR product that has accumulated. Although not common in forensics, Scorpions® primer assays were favored for the development of early versions of the Investigator® Quantiplex® kit (QIAGEN) (Frégeau and Laurin, 2015; Thomas et al., 2013).

Plexor® assays

In contrast to the other described assays, Plexor® assays measure reduction in fluorescence. Plexor® chemistry takes advantage of the interaction between two synthetic bases that only pair to each other (Johnson et al., 2004; Moser and Prudent, 2003; Sherrill et al., 2004). The assay uses a set of two primers, one of which has a fluorescently-labeled cytosine moiety, a 5′-methylisocytosine (iso-dC), incorporated at the 5′ end (Sherrill et al., 2004). The PCR mix contains standard deoxynucleotide triphosphates (dNTPs) and an isoguanine (iso-dG) with a covalently linked quencher (Sherrill et al., 2004). During the extension phase of PCR, iso-dG is incorporated on the complementary strand opposite the iso-dC due to the high fidelity of iso-dG/iso-dC (Johnson et al., 2004). Once this occurs, the fluorescent signal is quenched due to the proximity of the fluorophore and quencher. Thus, as the amount of PCR product accumulates, the fluorescent signal decreases (Johnson et al., 2004). The only forensic DNA quantification kit to have used Plexor® chemistry was the Plexor® HY System (Promega Corporation) (Krenke et al., 2008).

Fluorescence detection and measurement

Fluorescence emission produced using any of the qPCR chemistries previously described can be detected and measured in real-time using a qPCR thermal cycling instrument. In the initial cycles of PCR, the fluorescent signal of the reporter dye is below the limit of detection of the instrument (i.e., the baseline) (Arya et al., 2005). As PCR product accumulates, the fluorescent signal reaches a point where the instrument is able to detect it above the baseline (i.e., the threshold) (Arya et al., 2005). The fractional cycle number where PCR product crosses the threshold is termed the threshold cycle (C_T) (Heid et al., 1996).

The C_T will always occur during the exponential phase of the reaction. The exponential phase is considered the optimal place to measure accumulating PCR product in order to extrapolate the starting quantity of DNA template because none of the PCR reagents are limiting (Arya et al., 2005). In the plateau phase, the PCR ceases to generate PCR product at an exponential rate due to factors such as reagent limitation and PCR inhibitors (Arya et al., 2005). An important distinction to note between all other assays and the Plexor® assays is that, for Plexor® assays, the C_T is the cycle at which the fluorescence drops below the threshold.

The analysis software constructs an amplification plot of the fluorescence emission (ΔRn) versus PCR cycle number (Heid et al., 1996) (Fig. 2). The Rn value represents the fluorescence emission of the PCR product divided by the fluorescence of the passive reference dye present in each well (Applied Biosystems™, 2016). This normalization permits the software to account for minor variations in fluorescent signal caused by changes in concentration, volume, or sample effects (Applied Biosystems™, 2017). The ΔRn value represents the normalization of Rn and is obtained by subtracting the fluorescence emission of the baseline from the fluorescence emission of the PCR product (or Rn value) (Heid et al., 1996).

The DNA concentration of a test sample can be estimated from its corresponding C_T value when compared to a standard curve developed from DNA samples of known concentration. A standard curve of C_T versus a logarithm scale of the initial concentration of a set of DNA standards is a straight line (Higuchi et al., 1993). The analysis software automatically constructs the standard curve and quantifies the test samples by comparison to the standard curve formula (refer to section "Assessment of standard curve"). Heid et al. (1996) demonstrated that C_T values decrease linearly with increasing target quantity, such that the lower the C_T value, the higher the starting copy number of DNA.

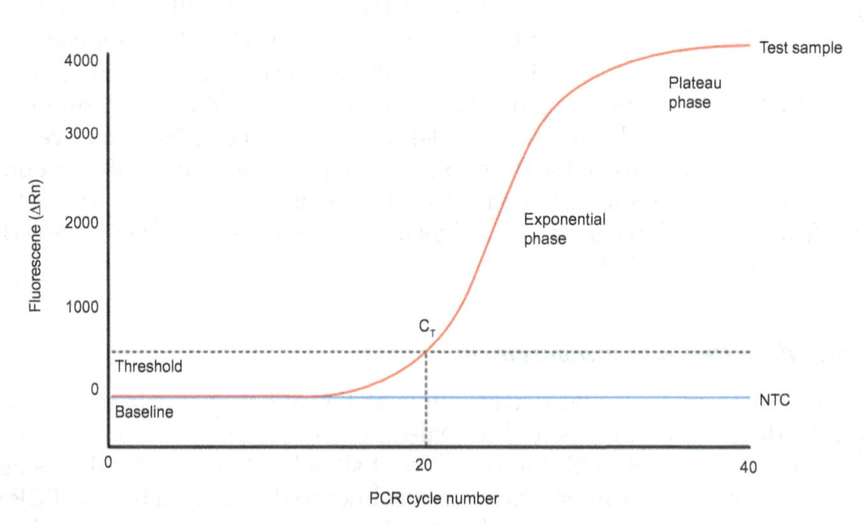

FIG. 2 Example of a real-time quantitative PCR (qPCR) assay amplification plot, with the polymerase chain reaction (PCR) cycle number on the x-axis and the baseline-subtracted fluorescence (ΔRn) on the y-axis. A sample with a threshold cycle (C_T) of 20 and a "no template control" (NTC) are plotted. *Graphic design by Trent Watherston.*

Quality indicators

The incorporation of two different sized autosomal, Y chromosome, or mtDNA targets in an assay allows an assessment of autosomal DNA, male DNA, or mtDNA degradation, respectively. The Degradation Index (DI) is calculated using the following formula:

$$\frac{\text{Small autosomal target DNA concentration}\,(\text{ng}/\mu\text{L})}{\text{Large autosomal target DNA concentration}\,(\text{ng}/\mu\text{L})}$$

An Internal PCR Control or Internal Positive Control (IPC) system can also be incorporated into a multiplex assay to: (1) assess if the reaction components and instrument performed satisfactorily, (2) test for the presence of PCR inhibitors in samples, and (3) identify samples that contain no, or large amounts of, DNA. The IPC system consists of a nonhuman or synthetic DNA template and a pair of unique sequence-specific primers with a dye-labeled TaqMan® probe. A fixed concentration of the IPC is added to each test sample. The primary assay targets and IPC are amplified together in every reaction.

Commercial DNA quantification kits

Currently, there are four manufacturers of forensic DNA quantification kits: Thermo Fisher Scientific (www.thermofisher.com), Promega Corporation (www.promega.com), QIAGEN (www.qiagen.com), and InnoGenomics Technologies (www.innogenomics.com). These kits are used by the majority of forensic laboratories worldwide, and over time have evolved from quantifying the amount of amplifiable (total) human DNA in a sample to quantifying the amount of total human and male DNA separately in a sample. Single copy targets have traditionally been used, but recently released kits contain multicopy targets for greater detection sensitivity and reduced stochastic variability (Holt et al., 2016). Many of these kits include an IPC system and amplification targets of different sizes to provide an objective measure of potential PCR inhibition and DNA degradation, respectively. These kits are routinely developmentally validated by the manufacturer according to the Scientific Working Group on DNA Analysis Methods (SWGDAM) guidelines (SWGDAM, 2016). There are a large number of DNA quantification kits available; however, this chapter will focus on recently released kits that contain features that may aid the quantification of DNA recovered from human skeletal remains.

Thermo Fisher Scientific qPCR kits

Thermo Fisher Scientific has developed a series of Quantifiler® kits to meet the needs of forensic laboratories over the last decade. The features of the different kits are summarized in Table 1.

The Quantifiler® Trio DNA Quantification Kit (Thermo Fisher Scientific) is the latest version in the Quantifiler® quantification kit series. This kit uses TaqMan® assay technology with a fluorescent 5-dye set that allows three genomic targets and an IPC to be multiplexed (Holt et al., 2016). The kit can simultaneously quantify the total amount of amplifiable human DNA and human male DNA in a sample. To estimate the total amount of human DNA present in a sample, the kit uses two different-sized multicopy autosomal targets: (1) a small

TABLE 1 A comparison of the features of each of the Quantifiler® DNA quantification kits manufactured by Thermo Fisher Scientific.

	Quantifiler® Trio	Quantifiler® HP	Quantifiler® Duo	Quantifiler® Y	Quantifiler® Human
Developmental validation	Holt et al. (2016)	Holt et al. (2016)	Barbisin et al. (2009)	Green et al. (2005)	Green et al. (2005)
Target location and size	Human small autosomal target: multiple autosomes—80 bases (multicopy) Human large autosomal target: multiple autosomes—214 bases (multicopy) Male target: Y chromosome—75 bases (multicopy)	Human small autosomal target: multiple autosomes—80 bases (multicopy) Human large autosomal target: multiple autosomes—214 bases (multicopy)	Human target: Ribonuclease P RNA component H1 (RPPH1) locus on chromosome 14-140 bases (single copy) Male target: sex determining region Y (SRY) locus on Y chromosome—130 bases (single copy)	Male target: sex determining region Y (SRY) locus on chromosome Y—64 bases (single copy)	Human target: human telomerase reverse transcriptase (hTERT) locus on chromosome 5-62 bases (single copy)
IPC size	130 bases	130 bases	130 bases	79 bases	79 bases
Sensitivity	< 1 pg	< 1 pg	~ 6 pg	~ 16 pg	~ 16 pg
Degradation Index	Yes	Yes	No	No	No
Inhibitor tolerance	Excellent	Excellent	Good	Good	Good
Cycling time	~ 60 min	~ 60 min	~ 90 min	~ 90 min	~ 90 min
Correlation with current STR kit chemistry	Excellent	Excellent	Poor	Poor	Poor
Correlation of kit targets and genetic marker size range	Excellent	Excellent	Good	Poor	Poor

autosomal (SA) target that is 80 bases in length, and (2) a large autosomal (LA) target that is 214 bases in length. To estimate the amount of human male DNA in a sample, the kit uses a multicopy 75-base long Y-chromosome target. The Quantifiler® Trio DNA Quantification Kit also includes an IPC system, which consists of a 130-base long synthetic DNA sequence, a set of primers, and a TaqMan® probe. The SA and Y targets are used primarily for estimating the total autosomal and male DNA present in a test sample, respectively (Holt et al., 2016). The smaller amplicon sizes of these targets were designed to be aligned with the sizes of typical mini-STR loci (Applied Biosystems™, 2017). The LA target is used primarily as an indicator of DNA degradation.

The DI will increase as DNA degradation selectively depletes the longer LA target template fragments (Holt et al., 2016). The manufacturer indicates that samples with a DI < 1 are not degraded, samples with a DI between 1 and 10 are slightly-to-moderately degraded, and samples with a DI > 10 are significantly degraded (Thermo Fisher Scientific, 2017). The DI can also be evaluated in conjunction with the IPC C_T value to arrive at a sample's Quality Index (Thermo Fisher Scientific, 2017). Vernarecci et al. (2015) investigated the reliability of the Quantifiler® Trio DI in determining the level of degradation in a range of forensic samples. This study supported that the DI was effective in characterizing degraded samples, and four degradation categories were devised: (1) samples with a DI between 0 and 1.5 are not degraded, (2) samples with a DI between 1.5 and 4 are mildly degraded, (3) samples with a DI between 4 and 10 are degraded, and (4) samples with a DI > 10 are severely degraded.

Vernarecci et al. (2015) also proposed operating guidelines for utilizing the qualitative and quantitative information obtained from the Quantifiler® Trio DNA Quantification Kit. For nondegraded or mildly degraded (DI < 4) low-template (3-33.3 pg/μL) samples, a full or partial STR profile is expected when adding the maximum volume of template DNA to the GlobalFiler™ PCR Amplification Kit (or a similar modern STR kit with large DNA input volume capacity). However, for degraded (DI > 4) low-template (3-33.3 pg/μL) samples, a partial or no profile is expected when adding the maximum volume of template DNA. Full profiles are expected when the recommended amount of input DNA (i.e., 0.5 ng) from non or mildly degraded (DI < 4) high-template (> 33.3 pg/μL) samples are added to the GlobalFiler™ PCR Amplification Kit (or a similar contemporary STR kit). However, partial profiles are expected from degraded (DI > 4) high-template (> 33.3 pg/μL) samples, even when the amount of input DNA is doubled (i.e., to 1 ng).

Promega Corporation qPCR kits

The PowerQuant® System (Promega Corporation) is the latest quantification kit manufactured by Promega Corporation. This system uses TaqMan® probes, and replaces previous kits such as the Plexor® HY System, which used Plexor® technology (Krenke et al., 2008). The PowerQuant® System detects targets for quantification of total human and male DNA, in addition to targets to assess if PCR inhibitors or degraded DNA are present in a sample. The short autosomal (Auto) target is a 84-base long sequence from a proprietary multicopy locus, the long degradation (D) target is a 294-base long sequence from a different region of the same proprietary multicopy locus (separated by several kilobases), and there are two different-sized male targets (81 bases and 136 bases) from two different proprietary multicopy loci (Ewing et al., 2016). The use of two male targets increases the sensitivity of detection for male DNA and minimizes the effect that variation in copy number can have on the quantification result. The ratio of [Auto]/[D] constitutes the DI. The IPC, which is generated from a novel 435-base long sequence, was designed to be the longest target in the assay to increase the IPC's susceptibility to PCR inhibitors relative to other targets in the multiplex. This kit has the largest IPC amplicon of all currently available commercial DNA quantification kits, a feature which Pionzio and Mccord (2014) advocate is essential for the increased detection of PCR inhibitors. During developmental validation, Ewing et al. (2016) found the system routinely detected DNA at concentrations too low to yield interpretable STR profiles and accurately identified inhibited versus degraded samples, preventing false identification of a degraded DNA sample due to PCR inhibition. Zupanič Pajnič et al. (2017) advise that STR profiling of aged skeletal remains should only be performed when both the Auto and D targets are detected, regardless of the DI, because bone samples with no detectable DNA did not produce a STR profile. Those samples for which only an Auto target was detected produced partial profiles unsuitable for interpretation.

QIAGEN qPCR kits

The Investigator® Quantiplex® Pro Kit (QIAGEN) was the first kit from QIAGEN to use TaqMan® probes for amplification detection, rather than the Scorpions® primers which were used in earlier kits like the Investigator® Quantiplex® HYres Kit (QIAGEN) (Frégeau and Laurin, 2015). The Investigator® Quantiplex® Pro Kit assay contains two different-sized autosomal targets (91 bases and 353 bases) from the multicopy 4NS1C® locus, an 81-base long male target from a multicopy locus on the Y chromosome, and a 434-base long IPC amplicon (QIAGEN, 2018a). This kit contains the largest autosomal target of all currently available commercial DNA quantification kits. This longer amplicon size is more comparable to the size of a typical STR locus, making this kit suitable for quantifying low quality samples. The IPC is also one of the longest compared to other modern kits, ensuring it is more sensitive to inhibitors than the quantification targets (QIAGEN, 2018c). In addition, incorporation of a DI allows an assessment of the degradation status of the autosomal DNA, a feature which was missing from previous QIAGEN kits.

The latest kit is the Investigator® Quantiplex® Pro RGQ Kit (QIAGEN). This kit contains the same targets as the Investigator® Quantiplex® Pro Kit, with the addition of a large human male target (359 bases) from the same multicopy locus on the Y chromosome as the short human male target (QIAGEN, 2018b). The use of both small and large autosomal and gonosomal targets enables simultaneous measurement of both total human and male-specific DNA degradation. This kit has been designed to be analyzed on the Rotor-Gene Q System (QIAGEN) only, so it will not be suitable for forensic laboratories with other brands of qPCR instruments.

The Investigator® Quantiplex® Pro kits also differ from other qPCR assays by use of a novel QuantiNova® DNA polymerase. The antibody-mediated hot-start mechanism prevents the formation and extension of nonspecific PCR products and primer-dimers during reaction setup and during the first denaturation step. Therefore, this mechanism allows higher PCR specificity and accurate quantification (QIAGEN, 2018c). The reaction buffer also contains the proprietary additive Q-Bond, which allows short cycling times on the Rotor-Gene Q System. Q-Bond increases the affinity of the DNA polymerase for short, single-stranded DNA, reducing the time required for primer/probe annealing to a few seconds. In addition, the unique composition of the buffer supports the melting behavior of DNA, enabling short denaturation and annealing/extension times (QIAGEN, 2018c).

InnoGenomics Technologies qPCR kits

The InnoQuant® HY (InnoGenomics Technologies) kit uses TaqMan® assay technology to amplify and detect four targets in a single reaction. The four DNA targets include: (1) two retro transposable element (RE) targets of different sizes (an 80-base long *Alu* target and a 207-base long SVA target) for the detection of autosomal human DNA and the provision of a DI, (2) a human male DNA-specific target that detects two multicopy 79-base long loci on the Y chromosome, and (3) a synthetic 172-base long IPC for the detection of PCR inhibitors (Loftus et al., 2017). Similar to the PowerQuant® System, targeting two multicopy Y-chromosome loci increases the sensitivity while minimizing the effect that variation in copy number would have on the overall male quantification result (Loftus et al., 2017). This kit has been optimized for use on Thermo Fisher Scientific qPCR instruments, but there is also a different version of this kit available (InnoQuant® HY-R) for laboratories that use qPCR instruments that do not require the use of a reference dye. These kits supersede the InnoQuant® (InnoGenomics Technologies) kit, which only contained the two different-sized RE targets (Pineda et al., 2014).

The developmental validation of the InnoQuant® HY kit demonstrated it was an accurate, sensitive, and reproducible quantification method capable of detecting subpicogram quantities of human and human male DNA (Loftus et al., 2017). The use of high copy number RE targets (i.e., > 1000 copies per genome) has been reported to significantly improve accuracy, sensitivity, and reproducibility of the autosomal targets compared to currently available multicopy commercial qPCR kits (Loftus et al., 2017; Pineda et al., 2014).

Comparison of the latest generation of DNA quantification kits

For those laboratories considering a commercial off-the-shelf quantification kit, published performance studies could assist with determining the most appropriate kit to meet their needs. A comparison of the latest generation of commercially available DNA quantification kits is presented in Table 2.

TABLE 2 A comparison of the features of the latest generation of DNA quantification kits.

	Quantifiler® Trio DNA Quantification Kit	**PowerQuant® System**	**Investigator® Quantiplex® Pro RGQ Kit**	**InnoQuant® HY**
Manufacturer	Thermo Fisher Scientific	Promega Corporation	QIAGEN	InnoGenomics Technologies
Developmental validation	Holt et al. (2016)	Ewing et al. (2016)	QIAGEN (2018b)	Loftus et al. (2017)
Fluorescence detection technology	TaqMan® probes	TaqMan® probes	TaqMan® probes	TaqMan® probes
Target location and size	Small autosomal: 80 bases (multicopy) Large autosomal: 214 bases (multicopy) Male: 75 bases (multicopy)	Small autosomal: 84 bases (multicopy) Large autosomal: 294 bases (multicopy) Male: 81 bases (multicopy) and 136 bases (multicopy)	Small autosomal: 91 bases (multicopy) Large autosomal: 353 bases (multicopy) Small male: 81 bases (multicopy) Large male: 359 bases (multicopy)	Small autosomal: 80 bases (multicopy) Large autosomal: 207 bases (multicopy) Male: 79 bases 2 (multicopy)
IPC size	130 bases	435 bases	434 bases	172 bases
Passive reference	Yes	Yes	No	Yes
Sensitivity	< 1 pg	< 1 pg	< 1 pg	< 1 pg

Continued

TABLE 2 A comparison of the features of the latest generation of DNA quantification kits—cont'd

	Quantifiler® Trio DNA Quantification Kit	**PowerQuant® System**	**Investigator® Quantiplex® Pro RGQ Kit**	**InnoQuant® HY**
Standards range	5pg/μL–50ng/μL	3.2pg/μL–50ng/μL	2.5pg/μL–50ng/μL	5pg/μL–20ng/μL
Degradation Index (DI)	Yes (total human)	Yes (total human)	Yes (total human and male)	Yes (total human)
Inhibitor tolerance	Excellent	Excellent	Excellent	Excellent
Cycling time	~ 60min	~ 60min	~ 60min	~ 60min
Correlation with current STR kit chemistry	Excellent	Excellent	Excellent	Excellent
Correlation of kit targets and genetic marker size range	Good	Excellent	Excellent	Good
Recommended degradation threshold	DI 1-10: slight—moderate degradation DI > 10: significant degradation	DI > 2: degradation flag	DI (human/human degradation) > 10: degradation flag DI (male/male degradation) > 10: degradation flag	*No threshold recommended by manufacturer DI 2.5-20: moderate degradation DI > 20: severe degradation (Van Den Berge et al., 2016) DI > 10: high degradation (Martins et al., 2019)
Recommended inhibition threshold	IPC: plus 2 cycles of average IPC C_T of DNA standards	IPC: plus 0.3 cycle of IPC C_T of the closest DNA standard	IPC: IPC C_T of DNA standards minus sample IPC $C_T < 1$	IPC: plus 2 cycles of average IPC C_T of DNA standards
Thermal cycler compatibility	QuantStudio™ 5/7500 Real-Time PCR System (Thermo Fisher Scientific)	QuantStudio™ 5/7500 Real-Time PCR System (Thermo Fisher Scientific)	Rotor-Gene Q System (5plex/6plex) (QIAGEN)	QuantStudio™ 5/7500 Real-Time PCR System (Thermo Fisher Scientific)

Lin et al. (2018) compared the performance of three newer generation kits: (1) the Quantifiler® Trio DNA Quantification Kit, (2) the Investigator® Quantiplex® HYres Kit, and (3) the PowerQuant® System. Of relevance to the quantification of DNA recovered from skeletal samples, this study concluded that all kits exhibited a higher sensitivity, consistency, and tolerance to PCR inhibition than the Quantifiler® Human DNA Quantification Kit, with the additional benefit of simultaneous detection of autosomal and male DNA in a shorter reaction time. The PowerQuant® System showed better precision for both autosomal and male DNA quantifications. The PowerQuant® System and Quantifiler® Trio DNA Quantification Kit offered better correlations with lower discrepancies between autosomal and male DNA quantifications, and the addition of the DI in the PowerQuant® System and Quantifiler® Trio DNA Quantification Kit provided a detection platform for inhibited and/or degraded DNA template. However, Lin et al. (2018) note that the Quantifiler® Trio DNA Quantification Kit's DI could be a better indicator of inhibition than its IPC amplicon. It is proposed that this could be due to the increased length of the LA target (214 bases versus 130 bases), making it more susceptible to PCR inhibition. In contrast, the 435-base long IPC amplicon of the PowerQuant® System accurately detected inhibitors, and its D amplicon appeared less affected by the presence of inhibitors.

The performance of four commercial qPCR kits was evaluated with inhibited and degraded samples: (1) the Quantifiler® Trio DNA Quantification Kit, (2) the Investigator® Quantiplex® Pro Kit, (3) the PowerQuant® System, and (4) InnoQuant® HY (Holmes et al., 2018). Holmes et al. (2018) found that all kits accurately quantified high-quality DNA samples down to the subpicogram level. However, the PowerQuant® System and Investigator® Quantiplex® Pro Kit produced the least accurate human and male quantification results at lower DNA concentrations, respectively, whilst the InnoQuant® HY kit showed the highest precision across the range of tested DNA concentrations. Kits performed similarly when low concentrations of PCR inhibitors were present. However, the Investigator® Quantiplex® Pro Kit was the most tolerant to a range of forensically relevant inhibitors (i.e., humic acid, calcium) and provided more accurate quantification results (Holmes et al., 2018). Based on these results, Holmes et al. (2018) suggested that the Investigator® Quantiplex® Pro Kit is suitable for DNA quantification of skeletal samples recovered from soil. The kits predicted varying degrees of degradation, with the Investigator® Quantiplex® Pro Kit and Quantifiler® Trio DNA Quantification Kit generating the largest and lowest DIs for degraded samples, respectively (Holmes et al., 2018). Holmes et al. (2018) propose this observation may be a result of the differing sizes of each kit's large autosomal target (353 bases versus 214 bases). When samples were profiled with the qPCR kit's respective STR kit, the Investigator® 24plex QS Kit (QIAGEN) and GlobalFiler™ PCR Amplification Kit (Thermo Fisher Scientific) generated more complete profiles than the PowerPlex® Fusion 6C System (Promega Corporation) when the small autosomal target concentrations were used to calculate DNA input amount (Holmes et al., 2018).

Custom DNA quantification assays

Some forensic laboratories have designed and/or implemented custom qPCR assays for the quantification of mtDNAin a sample—a capability that has not yet been commercialized. In addition, qPCR assays coupled with high-resolution melting (HRM) analysis have been developed for sex determination, which could offer an efficient screening tool to determine the biological sex of skeletal remains by DNA analysis. Some laboratories have even elected to forgo a DNA quantification assay, in lieu of a qualitative screening assay to instruct downstream processing.

Mitochondrial DNA (mtDNA) quantification assays

It is important to estimate the quantity of both nuclear DNA (nDNA) and mitochondrial DNA (mtDNA) recovered from skeletal remains to ensure the finite DNA extract is being used optimally. This includes selection of an optimal profiling method, so that DNA is not being consumed unnecessarily for nDNA profiling if there is not sufficient nDNA present in the sample. For mtDNA profiling, it also aids selection of appropriately sized mtDNA amplicons to target using either large, intermediate, or small primer sets. Additionally, having an accurate indication of the number of mtDNA copies in a sample ensures that an optimal amount of DNA is being used for mtDNA hypervariable region I and II (HVI/II) sequencing, rather than having to substitute the mtDNA concentration with a nDNA quantification assay result. This often results in more template DNA being consumed in each reaction than necessary, leaving less DNA extract available for any additional testing.

A number of qPCR assays have been developed for the quantification of mtDNA in a sample since 2002, either by incorporating amtDNA target into a multiplex qPCR assay, or via development of a single target mtDNA assay (Alonso et al., 2004; Andréasson et al., 2002, 2006; Goodwin et al., 2018; Köchl et al., 2005; Kavlick et al., 2011; Kavlick, 2019; Niederstätter et al., 2007; Timken et al., 2005; Von Wurmb-Schwark et al., 2002; Walker et al., 2005). The mtDNA methods most commonly employed in forensic DNA laboratories, or those that have been recently published, will be described here. For a detailed review of mtDNA quantification methods, refer to the following publications: Alonso and Garcia (2007), Barbisin and Shewale (2010), and Lee et al. (2014).

Andréasson et al. (2002) developed a TaqMan® assay that simultaneously estimates the concentration of nDNA and mtDNA. This duplex assay quantifies a 79-base long nuclear target at the single copy human retinoblastoma susceptibility (RB1) locus located on chromosome 13, and a 143-base long mitochondrial target which spans the junction of the transfer RNA (tRNA) lysine and ATP synthase 8 (ATP8) loci in the mtDNA coding region. Andréasson et al. (2002) noted that inhibitors present in the sample gave rise to an altered curve shape in the quantification analysis, but this assay does not have an incorporated IPC system. A modular approach for DNA quantification was favored by Niederstätter et al. (2007), who extended the Andréasson et al. (2002) assay to include: (1) additional nDNA targets of 156 bases and 246 bases; (2) additional mtDNA targets of 102 bases, 283 bases, and 404 bases; (3) a 156-base long IPC for nDNA; and (4) a 143-base longIPC for mtDNA. The same loci as targeted by Andréasson et al. (2002) were used to design the longer amplicons; however, TaqMan® probes were replaced with shorter MGB probes to maintain PCR efficiency. The availability of only two probes prohibits the assessment of DNA quantity, degradation, and inhibition in the one assay. However, laboratories can elect to use various combinations of the targets in different duplex assays. For example: (1) a nDNA and mtDNA target duplex for nDNA and mtDNA quantification, (2) a nDNA target and nDNA IPC duplex for nDNA quantification and inhibitor detection, or (3) two different-sized nDNA or mtDNA targets for nDNA or mtDNA quantification and assessment of nDNA or mtDNA degradation.

A mtDNA-specific TaqMan® assay was developed by Kavlick et al. (2011) to assess the quantity and quality of mtDNA. This duplex is comprised of a 105-base long target within the mtDNA NADH dehydrogenase subunit 5 (ND5) locus and a commercial IPC. The short target facilitates amplification of degraded DNA, and the IPC identifies inhibited samples requiring repurification. Kavlick et al. (2011) also designed a novel synthetic DNA standard which contained a signature sequence for quality assurance/quality control value, accurate quantification, and to prevent the DNA standard from contaminating any laboratory processes. Sprouse et al. (2014) validated this assay to quantify mtDNA in skeletal samples in their laboratory. It was concluded that this assay was sensitive to 12 mtDNA copies and was useful in identifying inhibited samples that would benefit from repurification. The quantification data obtained from the validation revealed that full HVI and HVII sequence data could be obtained from 20 pg of mtDNA from typical hard tissue samples; however, as little as 0.013 pg could generate a full mtDNA profile when applying an enhanced amplification protocol including smaller amplicon subsets, increased input amounts, and/or increased number of PCR cycles. This method has recently been updated to include an additional target for the assessment of mtDNA degradation and a custom IPC system to replace the commercial IPC used previously (Kavlick, 2019). The larger 316-base long target is found within the 16S ribosomal RNA (rRNA). Kavlick (2019) also described a method for estimating the copy number of mtDNA targets longer than the existing kit targets, so the assay could be used to calculate DNA input volume for mini, midi, maxi, and whole control region amplicons.

Recently, Goodwin et al. (2018) identified three different-sized human-specific mtDNA targets that could be amplified by singleplex qPCR to provide an index of mtDNA degradation from fragment length information. The three targets were selected within highly conserved regions of the 12S rRNA and 16S rRNA loci of the mtDNA coding region. The quantification targets are 452 bases, 190 bases, and 86 bases in size, and are representative of the commonly sequenced large, medium, and small HVI/HVII amplicons, respectively. In addition, the shortest target can be used to represent mtDNA availability for massively parallel sequencing (MPS) enrichment. A 100-base long synthetic IPC was also included to serve as a positive control and detect inhibition. Unlike the other mtDNA quantification methods described here, this method uses SYBR® Green I rather than fluorescent probes, offering a cost-effective quantification option. However, the singleplex nature of the assay will result in more DNA extract being consumed for DNA quantification compared to multiplex methods, if a laboratory is interested in also determining DNA purity and length.

Sex determination DNA quantification assays

HRM is a form of melt curve analysis that permits amplicons to be differentiated according to one or more basepair differences in length, sequence, guanine-cytosine (GC) content, or heterozygosity (Wittwer et al., 2003; Reed et al., 2007). Following qPCR, a double-stranded amplicon is incrementally heated until the DNA denatures, which releases and consequently deactivates the bound fluorescent dye molecules. The melting temperature (T_m) of double-stranded DNA is the temperature at which the normalized fluorescence is 50%, and is characteristic of its length, sequence, and GC content (Reed et al., 2007). The fluorescence can be plotted against temperature to produce an amplicon-specific melt curve.

HRM is a convenient method for high-throughput sex determination of samples because it enables the detection of genetic variation in small PCR products directly following qPCR, without any further sample manipulations or post-PCR processing (Álvarez-Sandoval et al., 2014). To aid the sex determination of highly degraded skeletal remains, Álvarez-Sandoval et al. (2014) proposed a combined qPCR and HRM approach. Two small targets of the third exon of the amelogen in locus covering a 3-base deletion of the AMELX allele are firstly amplified with qPCR, followed by HRM analysis based on the melting temperature difference of the AMELX allele and AMELY allele amplicons (61 and 64 bases, respectively). This rapid and sensitive method demonstrated utility for compromised skeletal remains, and it is expected that the occurrence of random allele drop-out would be minimized due to the small target sizes.

Ginart et al. (2019) designed a triplex qPCR assay (QYDEG HRM) for the simultaneous determination of the sex of a human DNA donor and the extent of DNA degradation. Three target loci are multiplexed in a single PCR and analyzed by HRM: (1) the 84-base Y-chromosome target specific to the transducin (beta)-like 1 Y-linked (TBL1Y) locus, and two different-sized autosomal targets to calculate the DI of a sample, (2) a 152-base short target, and (3) a 244-base long target. After HRM analysis, two or three melting peaks are detected in female and male samples, respectively, and an imbalance between the melting peak heights of the large- and small autosomal target is observed for degraded samples. The assay was shown to be accurate at the picogram level and produced concordant results to those obtained with the PowerQuant® System, with the benefit of being a simple and cost-effective technique that can be performed on any qPCR instrument.

A qualitative PCR assay

Von Wurmb-Schwark et al. (2009) designed a multiplex PCR assay comprised of autosomal STRs, male-specific STRs, and mtDNA to evaluate sample quality in lieu of a qPCR assay for DNA quantification. This assay amplifies: (1) amelogenin and two Y-chromosome STRs (Y-STRs) (i.e., DYS390, DYS391) for sex determination; (2) two autosomal STR loci (TH01, vWA) to assess suitability for STR genotyping; and (3) two different-sized mtDNA amplicons (280 bases and 439 bases) targeting HVI to assess suitability for mtDNA sequencing using intermediate or large amplicons. An additional 271-base long quality sensor was incorporated to detect the presence of inhibitors and to test PCR efficiency. The dual sex marker determination using amelogenin and Y-STRs is important to overcome false results due to variants at the amelogenin locus. The quality sensor assists in differentiating whether a negative PCR result is caused by the presence of PCR inhibitors or a lack of amplifiable DNA. The sensitivity of the assay was demonstrated to be 25 pg, and it was successful for highly degraded samples due to the reduced size of the target STR loci (< 200 bases). This assay provides an easy, reliable, and cost-efficient evaluation of DNA sample quality to aid sample processing decision-making, combined with a preliminary assessment of individualization and sex determination suitable for forensic purposes. Von Wurmb-Schwark et al. (2009) report they routinely apply this method to challenging casework samples to detect nDNA (e.g., single hairs, bones) or PCR inhibitors for samples buried in soil. Benefits over standard qPCR approaches include: (1) no requirement for a specific qPCR instrument, (2) a larger number of amplicons can be multiplexed, and (3) allelic discrimination of target STRs is possible.

DNA quantification instruments

There are a number of commercially available fluorescence-detecting thermal cyclers for real-time PCR applications. All of these instruments are capable of simultaneous amplification and detection of quantification targets. The most widely used qPCR instrument in forensic DNA laboratories is currently the Applied Biosystems™ 7500 Real-Time PCR System (Thermo Fisher Scientific). However, the qPCR kit selected for laboratory use may dictate instrument selection.

Thermo Fisher Scientific real-time PCR systems

Applied Biosystems™ 7500 Real-Time PCR System

The majority of commercial quantification kits have been designed to be performed on the Applied Biosystems™ 7500 Real-Time PCR System. This system can analyze up to 96 samples simultaneously for different qPCR applications. The Applied Biosystems™ 7500 Real-Time PCR System contains a halogen lamp light source, a five-color optical filter set comprised of five excitation and emission filters, and a charge-coupled device (CCD) camera for fluorescence detection (Applied Biosystems™, 2010). This format enables the detection of up to five targets in each well. In addition, this instrument has a high temperature uniformity (± 0.50°C) and a dynamic range of nine orders of magnitude (Applied Biosystems™, 2010).

Applied Biosystems™ QuantStudio™ 5 Real-Time PCR System

The Applied Biosystems™ QuantStudio™ 5 Real-Time PCR System (Thermo Fisher Scientific) is the latest Thermo Fisher Scientific quantification instrument for human identification (HID) applications and has the capacity to analyze 96-384 samples simultaneously (depending on the block configuration) (Fig. 3). The Applied Biosystems™ QuantStudio™

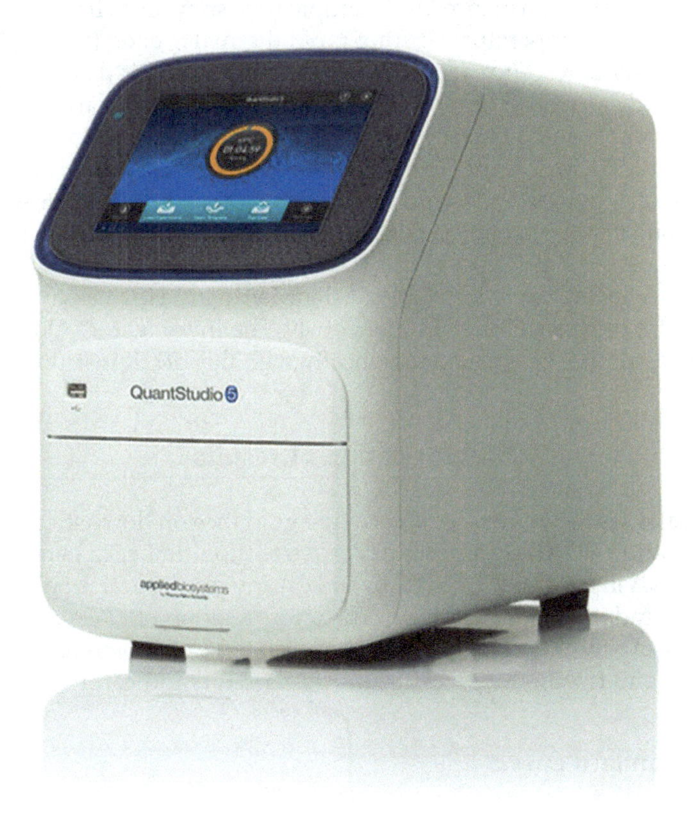

FIG. 3 The Applied Biosystems™ QuantStudio™ 5 Real-time PCR System. © *2019 Thermo Fisher Scientific Inc. Used under permission.*

5 Real-Time PCR System contains a six-color optical filter set (when used in 96-well format) comprised of six excitation and emission filters (450-670 nm excitation, 500-720 nm emission), enabling the detection of six targets in each well (Thermo Fisher Scientific, 2017). Unlike the Applied Biosystems™ 7500 Real-Time PCR System, this instrument has the capacity to accommodate a range of new reporter dyes, as quantification kits expand to include additional amplification targets. Additional improvements with this instrument include better temperature uniformity (± 0.40°C), a dynamic range of 10 orders of magnitude, and an upgraded excitation source from a halogen lamp to the OptiFlex™ System with bright white light-emitting diode (LED) light (Thermo Fisher Scientific, 2017). Many current quantification kits have been designed to be performed on either the Applied Biosystems™ QuantStudio™ 5 Real-Time PCR System or the Applied Biosystems™ 7500 Real-Time PCR System. In addition to HID applications, this system can also accommodate most other qPCR-based chemistries.

Other real-time PCR systems

Rotor-Gene Q System

The Rotor-Gene Q System from QIAGEN has a unique centrifugal rotary design for optimum thermal and optical performance. Each sample tube spins in a chamber of moving air, facilitating uniform sample temperatures during rapid thermal cycling (temperature uniformity: ± 0.02°C) and uniform fluorescence detection (QIAGEN, 2011). This design results in sensitive, precise, and fast qPCR analysis with a dynamic range of up to 10 orders of magnitude and reduces sample-to-sample and edge effects thereby eliminating the need to use a passive reference dye (QIAGEN, 2011). The Rotor-Gene Q System contains a six-color optical filter set comprised of six separate excitation LEDs and emission filters per channel (365-680 nm excitation, 460-750 nm emission), offering a wide optical range to ensure compatibility with reporter dyes used in future multiplexed kits (QIAGEN, 2011). The Rotor-Gene Q is a suitable platform for a range of applications, including SYBR® Green I, TaqMan®, and HRM assays, with the capacity for up to 100 samples to be analyzed simultaneously. The Investigator® Quantiplex® Pro RGQ Kit has been optimized to be performed specifically with this qPCR instrument only.

Interpretation of results

DNA quantification is recommended to provide guidance on downstream sample processing decisions, including: (1) the appropriate type, size, and number of genetic markers to use, (2) the optimum DNA input amount for PCR, and (3) any procedural modifications needed to address sample degradation and inhibition. The interpretation of DNA quantification results first involves assessing the standard curve and associated parameters, then the performance of control samples, and finally the concentration and quality indicators of test samples.

Assessment of standard curve

Standard curve

Absolute quantification involves the use of a standard curve to determine the quantity (or copy number) of a test sample. The standard curve is constructed from a 10-fold dilution

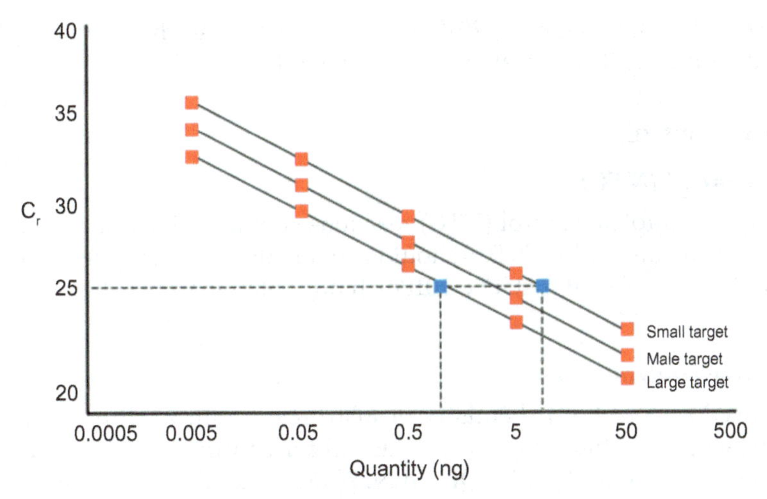

FIG. 4 Example of a real-time quantitative PCR (qPCR) standard curve. The logarithm of the initial copy number of DNA standards is shown on the x-axis, and their respective C_T values are displayed on the y-axis. The regression line is used to determine the copy number of the test sample. A (female) test sample with a threshold cycle (C_T) value of 25 and the respective copy number of the small- and large autosomal targets is plotted. *Graphic design by Trent Watherston.*

series of a DNA standard of known concentration. The logarithm of the initial copy number of DNA standards (log quantity) is on the x-axis and their respective C_T values are on the y-axis (Fig. 4). The software calculates the regression line by calculating the best fit with the quantification standard data points (Applied Biosystems™, 2017). The regression line formula is $y = mx + b$, where y = the cycle threshold (C_T), m = the slope, x = log quantity, and b = the Y-intercept.

Slope

The slope (m) indicates the PCR amplification efficiency for the assay. The amplification efficiency is calculated using the slope of the regression line in the standard curve. A slope of −3.3 indicates 100% amplification efficiency (i.e., meaning the number of copies of PCR product is doubled at each PCR cycle). The acceptable slope range is usually between −3.0 and −3.6 (Applied Biosystems™, 2017; QIAGEN, 2018c).

The R^2 value

The R^2 value is a measure of the closeness-of-fit between the standard curve regression line and the individual C_T data points of the DNA standard reactions. An R^2 value of 1 indicates a perfect fit between the regression line and the data points; therefore, the value of y (C_T) can be used to accurately predict the value of x. An acceptable R^2 value is ≥0.99 (Applied Biosystems™, 2017; QIAGEN, 2018c).

Y-intercept

The Y-intercept (b) indicates the expected C_T value for a sample with a quantity value of 1 (e.g., 1ng/μL). Manufacturers do not normally provide guidance on an acceptable range of Y-intercept values due to variation with kits, standards, amplification targets, and instruments (Applied Biosystems™, 2017; QIAGEN, 2018c). Each laboratory should define Y-intercept values during its internal kit validation study.

Assessment of controls

No template control (NTC)

Replicates of no template control (NTC) reactions should be included in each quantification run to detect contamination. NTCs should contain all of the components of the reaction, except for the DNA template (QIAGEN, 2018c). If no contamination is detected, a C_T value of undetermined is expected.

Internal positive control (IPC)

An IPC is used to detect: (1) chemistry or instrument failure, (2) errors in assay setup, and (3) the presence of inhibitors; and it is useful for confirming the validity of negative results (Applied Biosystems™, 2017; QIAGEN, 2018c). The IPC system should be designed to be more sensitive to inhibition than the human amplification targets, so the quantification will be informative even if inhibitors that do not impact amplification of the nDNA or mtDNA are present in the sample QIAGEN, 2018c). The IPC template DNA is present at a fixed concentration in all wells of the reaction plate such that, in theory, the IPC C_T should be constant in all typical reactions (Applied Biosystems™, 2017). However, the presence of PCR inhibitors and/or higher concentrations of DNA can increase the IPC C_T value of a test sample relative to the IPC C_T value of the quantification standards (Applied Biosystems™, 2017).

For example, if a test sample displays negative amplification of the human-specific targets and positive amplification of the IPC, this indicates the assay was performed correctly but no human DNA is detectable in the sample. An increase in the IPC C_T of a test sample compared to a DNA standard may indicate the presence of an inhibitor in the test sample. If inhibitors are present in high enough concentrations, then the IPC of the test sample may fail to amplify. An increase in C_T could be a result of: (1) the inhibitor (e.g., humic acid) binding to DNA and reducing the total amount of accessible DNA template, (2) disabling the Taq DNA polymerase, resulting in suboptimal performance, (3) deactivation of the enzyme (e.g., as with calcium), or (4) performing both inhibitory mechanisms (e.g., as observed with collagen) (Pionzio and Mccord, 2014; Thompson et al., 2014).

Assessment of DNA quantity

Regardless of the quantification assay selected for use, the laboratory will need to determine which quantification target will primarily be used to estimate the optimal template input volume for amplification of DNA recovered from human skeletal remains. In general, qPCR kits recommend the use of the smallest autosomal target concentration value to accurately estimate autosomal DNA concentration of a sample. However, the largest autosomal target concentration value should be considered when estimating human DNA concentration of compromised skeletal samples to improve decision making on the subsequent DNA profiling approach (Holmes et al., 2018; Loftus et al., 2017;

Pineda et al., 2014). This is because the size of the larger target more accurately reflects the average, or upper, size range of amplicons in current STR multiplexes (approximate size range: 70-500 bases) or routine HVI/HVII sequencing (approximate size range: 100-400 bases). This approach would enable a more accurate calculation of DNA template volumes for the amplification of degraded skeletal samples; assuming targets <100 bases would overestimate the quantity of amplifiable DNA due to fragmentation. Alternatively, Holmes et al. (2018) suggest it may be beneficial to take the average of the smallest and largest target DNA concentrations to reach a balance between minimizing drop-out of larger loci and avoiding introduction of unwanted artifacts (e.g., drop-in, split peaks) due to over-amplification of smaller loci.

In addition to the DNA concentration of a sample, the quantitative information can aid in determining:

1. If a sufficient quantity of amplifiable human DNA is present to proceed with routine autosomal STR profiling and/or Y-STR profiling, or if an alternative genotyping method needs to be employed (e.g., low copy number, mtDNA, or small amplicon approaches);
2. The optimal template DNA volume for PCR;
3. If a sample should be re-extracted, concentrated, or diluted prior to PCR;
4. If a higher number of PCR cycles should be applied;
5. If longer capillary electrophoresis injection times should be applied;
6. The biological sex of the sample donor; and
7. If multiple independent amplifications should be performed to authenticate results.

Assessment of DNA quality

Quantification assays that include quality indicators are advisable for DNA recovered from human skeletal remains. Bones and teeth have a propensity to contain fragmented, damaged, and modified DNA due to the age of the samples at the time of recovery, the variety of environmental insults suffered, and/or a range of endogenous and exogenous inhibitors. The latest quantification kits all offer a Quality Index for estimating the level of sample degradation and/or the presence of PCR inhibitors. This index is derived from consideration of both the DI and IPC C_T values. Manufacturers recommend degradation and inhibition thresholds for quality assessment. However, laboratories are encouraged to evaluate or define these thresholds during their internal kit validation. It is important to determine if a sample is degraded and/or inhibited, as the approach one would take to improve DNA profiling results differs for degraded and inhibited samples (Holmes et al., 2018). Refer to Table 3 for general guidance on interpretation of the Quality Index.

The DI can be affected by the degree of degradation of the large autosomal target DNA and by the presence of inhibitors. When the quantity for one of the autosomal amplification targets is undetermined, the DI cannot be calculated. If the largest autosomal amplicon is undetermined, it can indicate significant degradation and/or inhibition affecting the sample, as extremely high concentrations of inhibitors can inhibit amplification of the large autosomal target (Applied Biosystems™, 2017). Therefore, it is important to note the DI information in conjunction with the IPC data to assess if a sample is degraded and/or inhibited.

TABLE 3 General Quality Index interpretation guidelines.

DI and IPC C_T observations	Interpretation
No amplification of IPC detected No amplification of human targets detected	PCR amplification failure
No IPC shift detected No amplification of human targets detected	No or insufficient DNA present (true negative result)
DI < degradation threshold specified No IPC shift detected	DNA is not degraded or inhibited
DI < degradation threshold specified IPC shift detected above inhibition threshold *or* C_T is undetermined	DNA is not degraded but inhibited
DI > degradation threshold specified *or* DI is undetermined No IPC shift detected	DNA is degraded but not inhibited
DI > degradation threshold specified *or* DI is undetermined IPC shift detected above inhibition threshold *or* C_T is undetermined	DNA *may* be degraded but is inhibited

In addition to detecting inhibited and degraded samples, the qualitative information can aid in determining the following:

1. If a sample should be re-purified or diluted prior to amplification;
2. If a higher concentration of certain PCR components (e.g., Taq DNA polymerase) or amplification enhancers [e.g., bovine serum albumin (BSA)] should be added to the reaction;
3. If a sample should be amplified using multiple template DNA volumes;
4. If sufficient quality of amplifiable human DNA is present to proceed with using a routine STR profiling kit, or a small amplicon profiling system [e.g., mini-STRs, single nucleotide polymorphisms (SNPs), insertions/deletions]; and
5. If a mtDNA sequencing approach should be employed (i.e., with the use of large, intermediate, or small primer sets).

Procedural modifications

The commercialization and automation of qPCR methods has increased the speed at which the quantity and quality of large numbers of DNA extracts can be assessed. However, these kits are substantially more expensive than traditional DNA quantification methods. To reduce the cost-per-reaction, many laboratories have validated the use of reduced reagent volumes. The qPCR data should also be closely evaluated to instruct procedural modifications to the routine DNA profiling workflow under certain conditions, such as: (1) the addition of an optimal inhibitor removal step when inhibitors are detected, or (2) to proceed with PCR irrespective of the quantification result (e.g., when there is a high chance of a "false negative" quantification result).

Reduced reagent volumes

Using DNA quantification kit reaction volumes that are less than those prescribed by the manufacturer has been shown to produce equivalent quantification results whilst offering efficiency gains. Westring et al. (2007) validated reduced-scale (10 μL) Quantifiler™ Human DNA Quantification Kit reaction volumes after demonstrating that this volume displayed equivalent or improved amplification efficiency and sample-to-sample reproducibility compared to full-scale (25 μL) reactions. The practical benefits included increasing the effective number of reactions per kit by 250% and reducing the cost per assay by 60%, while consuming less sample. Similarly, Cho et al. (2018) validated reduced (1/10) reagent volume reactions for the Quantifiler® Trio Quantification Kit to increase the number of reactions per kit by 10-fold and significantly reduce the cost of DNA quantification, without compromising kit performance. Additionally, Frégeau and Laurin (2015) showed that quantification results produced using a full reaction volume (20 μL) were very similar to those of half volume (10 μL) reactions using the Investigator® Quantiplex® HYres Kit.

Removal of PCR inhibitors

Ideally inhibitors, such as metal ions or soil-derived acids, should be removed from DNA extracted from skeletal samples prior to DNA profiling, in order to prevent or minimize artifacts such as peak imbalance or allele drop-out (the latter of which results in an incomplete profile). Importantly, skeletal samples naturally contain calcium, which is reported to inhibit Taq DNA polymerase by preventing magnesium ions interacting with the enzyme during PCR (Combs et al., 2015; Pionzio and Mccord, 2014). Combs et al. (2015) showed that the Quantifiler® Human DNA Quantification Kit was sensitive to certain metals and more resistant to others. For example, the presence of aluminum concentrations (>0.075 mM) resulted in inaccurate quantification estimates, and the ion's presence was not detected by the IPC. Conversely, the presence of calcium (≤18.75 mM) did not negatively affect estimating the concentration of DNA or the amplification of the IPC. Despite this finding, Combs et al. (2015) recommend that DNA extracted from skeletal samples recovered from burial environments should proceed to DNA profiling regardless of the quantification result, due to their susceptibility to produce "false negative" quantification results in the presence of endogenous calcium and other soil-derived metals.

Once DNA is extracted from skeletal remains, methods can be employed to alleviate the effects of known PCR inhibitors and to ensure that accurate quantification results are obtained. Alaeddini (2012) reviewed a number of strategies for overcoming PCR inhibition, including: (1) diluting the DNA sample, (2) spin filtration, (3) adding BSA to the PCR, (4) increasing the concentration of Taq DNA polymerase, (5) increasing the magnesium ion concentration, (6) silica-based spin filtration, and/or (7) the use of smaller (e.g., SNP) or alternative (e.g., mtDNA) amplicon profiling systems. However, the effect these treatments could have on the DNA recovered from compromised samples needs to be considered. For example, it may be preferable to add extra BSA or Taq DNA polymerase to a PCR rather than diluting a LT-DNA sample or using silica-based spin columns to purify degraded DNA and risk DNA loss. Thompson et al. (2014) highlight the importance of using qPCR data to identify the presence of inhibitors, and to therefore select the most appropriate method for their removal without further compromising the amount of DNA.

Summary

DNA identification of skeletal remains can be challenging, due to these samples frequently being degraded, inhibited, and/or contaminated. Selecting a quantification method that facilitates the most effective and efficient assessment of the quantity and quality of DNA recovered is important for triaging purposes, and for informing decisions regarding the DNA profiling approach to be employed. DNA quantification assays that have equivalent, or greater, sensitivity than the downstream DNA profiling systems should be used to identify samples that should not proceed to DNA profiling. However, it is important to remember that quantification assays only provide an estimate of DNA concentration. For skeletal samples that produce a negative quantification result, such as may be the case for LT-DNA samples or those recovered from graves, DNA profiling should still be considered, due to stochastic sampling and amplification effects and/or their susceptibility to produce "false negative" quantification results in the presence of environmental inhibitors.

Determining if a DNA sample is low template, degraded, and/or inhibited prior to DNA profiling can improve first pass success rates, thereby minimizing re-testing or re-analysis of samples, maximizing throughput, and preserving precious samples for specialized or emerging DNA technologies. It is recommended that a quantification kit with target amplicons in the same length range as those targets obtained with routine nDNA and mtDNA profiling approaches be used for quantifying DNA recovered from human skeletal remains, in order to ensure that template DNA volumes are accurately estimated for compromised samples. Furthermore, a kit with an IPC amplicon that is larger than the human amplification targets increases the ability of the IPC system to detect those PCR inhibitors inherent to skeletal samples.

Current commercial qPCR kits offer forensic DNA laboratories a sensitive, reliable, reproducible, specific, accurate, and objective tool for the simultaneous assessment of DNA quantity and purity, while facilitating adherence to quality assurance/quality control requirements. However, until mtDNA quantification assays are commercialized, many laboratories that specialize in DNA profiling of skeletal remains may elect not to validate a custom mtDNA qPCR assay in addition to a commercial assay for total human and male DNA quantification. For those laboratories that do not have access to a mtDNA quantification assay, it is still advisable to analyze human skeletal samples suspected of containing no, or minimal, high quality nDNA using a modern DNA quantification assay. This will assist in determining: (1) the degradation and inhibition state, (2) biological sex, and (3) the optimal DNA profiling approach for the sample; as well as facilitating the interpretation of the resultant nDNA and/or mtDNA profile.

References

Afonina, I., Zivarts, M., Kutyavin, I., Lukhtanov, E., Gamper, H., Meyer, R.B., 1997. Efficient priming of PCR with short oligonucleotides conjugated to a minor groove binder. Nucleic Acids Res. 25, 2657–2660.

Alaeddini, R., 2012. Forensic implications of PCR inhibition—a review. Forensic Sci. Int. Genet. 6, 297–305.

Alaeddini, R., Walsh, S.J., Abbas, A., 2010. Forensic implications of genetic analyses from degraded DNA—a review. Forensic Sci. Int. Genet. 4, 148–157.

Alonso, A., Garcia, O., 2007. Real-time quantitative PCR in forensic science. In: Rapley, R., Whitehouse, D. (Eds.), Molecular Forensics. John Wiley & Sons, Inc., New York.

Alonso, A., Martín, P., Albarrán, C., García, P., García, O., De Simón, L.F., et al., 2004. Real-time PCR designs to estimate nuclear and mitochondrial DNA copy number in forensic and ancient DNA studies. Forensic Sci. Int. 139, 141–149.

Álvarez-Sandoval, B.A., Manzanilla, L.R., Montiel, R., 2014. Sex determination in highly fragmented human DNA by high-resolution melting (HRM) analysis. PLoS ONE 9, e104629.

Andréasson, H., Gyllensten, U., Allen, M., 2002. Real-time DNA quantification of nuclear and mitochondrial DNA in forensic analysis. BioTechniques 33 (402–404), 407–411.

Andréasson, H., Nilsson, M., Budowle, B., Lundberg, H., Allen, M., 2006. Nuclear and mitochondrial DNA quantification of various forensic materials. Forensic Sci. Int. 164, 56–64.

Applied Biosystems™, 2010. Applied Biosystems 7500/7500 Fast Real-Time PCR Systems: System Maintenance. https://assets.thermofisher.com/TFS-Assets/LSG/manuals/4387777d.pdf. (Accessed 03 December 2018).

Applied Biosystems™, 2016. Real-time PCR: Understanding C_T. https://www.thermofisher.com/content/dam/LifeTech/Documents/PDFs/PG1503-PJ9169-CO019879-Re-brand-Real-Time-PCR-Understanding-Ct-Value-Americas-FHR.pdf. (Accessed 03 December 2018).

Applied Biosystems™, 2017. Quantifiler™ HP and Trio DNA Quantification Kits User Guide. http://tools.thermofisher.com/content/sfs/manuals/4485354.pdf. (Accessed 03 December 2018).

Arya, M., Shergill, I.S., Williamson, M., Gommersall, L., Arya, N., Patel, H.R., 2005. Basic principles of real-time quantitative PCR. Expert. Rev. Mol. Diagn. 5, 209–219.

Barbisin, M., Shewale, J.G., 2010. Assessment of DNA extracted from forensic samples prior to genotyping. Forensic Sci. Rev. 22, 199–214.

Barbisin, M., Fang, R., O'shea, C.E., Calandro, L.M., Furtado, M.R., Shewale, J.G., 2009. Developmental validation of the Quantifiler® Duo DNA quantification kit for simultaneous quantification of total human and human male DNA and detection of PCR inhibitors in biological samples. J. Forensic Sci. 54, 305–319.

Bassler, H.A., Flood, S.J., Livak, K.J., Marmaro, J., Knorr, R., Batt, C.A., 1995. Use of a fluorogenic probe in a PCR-based assay for the detection of *Listeria monocytogenes*. Appl. Environ. Microbiol. 61, 3724–3728.

Butler, J.M., 2012. Advanced Topics in Forensic DNA Typing: Methodology. Academic Press, San Diego.

Cardullo, R.A., Agrawal, S., Flores, C., Zamecnik, P.C., Wolf, D.E., 1988. Detection of nucleic acid hybridization by nonradiative fluorescence resonance energy transfer. Proc. Natl. Acad. Sci. U. S. A. 85, 8790–8794.

Carters, R., Ferguson, J., Gaut, R., Ravetto, P., Thelwell, N., Whitcombe, D., 2008. Design and use of Scorpions® fluorescent signaling molecules. Methods Mol. Biol. 429, 99–115.

Cho, Y., Kim, H.S., Kim, M.H., Park, M., Kwon, H., Lee, Y.H., Lee, D.S., 2018. Validation of reduced reagent volumes in the implementation of the Quantifiler® kit. J. Forensic Sci. 63, 517–525.

Clegg, R.M., 1992. Fluorescence resonance energy transfer and nucleic acids. Methods Enzymol. 211, 353–388.

Combs, L.G., Warren, J.E., Huynh, V., Castaneda, J., Golden, T.D., Roby, R.K., 2015. The effects of metal ion PCR inhibitors on results obtained with the Quantifiler® human DNA quantification kit. Forensic Sci. Int. Genet. 19, 180–189.

Ewing, M.M., Thompson, J.M., Mclaren, R.S., Purpero, V.M., Thomas, K.J., Dobrowski, P.A., et al., 2016. Human DNA quantification and sample quality assessment: developmental validation of the PowerQuant® system. Forensic Sci. Int. Genet. 23, 166–177.

Förster, V.T., 1948. Zwischenmolekulare Energiewanderung und Fluoreszenz. Ann. Phys. 6.

Frégeau, C.J., Laurin, N., 2015. The QIAGEN Investigator® Quantiplex® HYres as an alternative kit for DNA quantification. Forensic Sci. Int. Genet. 16, 148–162.

Ginart, S., Caputo, M., Corach, D., Sala, A., 2019. Human DNA degradation assessment and male DNA detection by quantitative-PCR followed by high-resolution melting analysis. Forensic Sci. Int. 295, 1–7.

Goodwin, C., Higgins, D., Tobe, S.S., Austin, J., Wotherspoon, A., Gahan, M.E., Mcnevin, D., 2018. Singleplex quantitative real-time PCR for the assessment of human mitochondrial DNA quantity and quality. Forensic Sci. Med. Pathol. 14, 70–75.

Green, R.L., Roinestad, I.C., Boland, C., Hennessy, L.K., 2005. Developmental validation of the Quantifiler® real-time PCR kits for the quantification of human nuclear DNA samples. J. Forensic Sci. 50, 809–825.

Heid, C.A., Stevens, J., Livak, K.J., Williams, P.M., 1996. Real time quantitative PCR. Genome Res. 6, 986–994.

Higuchi, R., Fockler, C., Dollinger, G., Watson, R., 1993. Kinetic PCR analysis: real-time monitoring of DNA amplification reactions. Biotechnology (NY) 11, 1026–1030.

Holland, P.M., Abramson, R.D., Watson, R., Gelfand, D.H., 1991. Detection of specific polymerase chain reaction product by utilizing the 5′-3′ exonuclease activity of *Thermus aquaticus* DNA polymerase. Proc. Natl. Acad. Sci. U. S. A. 88, 7276–7280.

Holmes, A.S., Houston, R., Elwick, K., Gangitano, D., Hughes-Stamm, S., 2018. Evaluation of four commercial quantitative real-time PCR kits with inhibited and degraded samples. Int. J. Legal Med. 132, 691–701.

Holt, A., Wootton, S.C., Mulero, J.J., Brzoska, P.M., Langit, E., Green, R.L., 2016. Developmental validation of the Quantifiler® HP and Trio Kits for human DNA quantification in forensic samples. Forensic Sci. Int. Genet. 21, 145–157.

Johnson, S.C., Sherrill, C.B., Marshall, D.J., Moser, M.J., Prudent, J.R., 2004. A third base pair for the polymerase chain reaction: inserting isoC and isoG. Nucleic Acids Res. 32, 1937–1941.

Kavlick, M.F., 2019. Development of a triplex mtDNA qPCR assay to assess quantification, degradation, inhibition, and amplification target copy numbers. Mitochondrion 46, 41–50.

Kavlick, M.F., Lawrence, H.S., Merritt, R.T., Fisher, C., Isenberg, A., Robertson, J.M., Budowle, B., 2011. Quantification of human mitochondrial DNA using synthesized DNA standards. J. Forensic Sci. 56, 1457–1463.

Köchl, S., Niederstätter, H., Parson, W., 2005. DNA extraction and quantitation of forensic samples using the phenol-chloroform method and real-time PCR. Methods Mol. Biol. 297, 13–30.

Krenke, B.E., Nassif, N., Sprecher, C.J., Knox, C., Schwandt, M., Storts, D.R., 2008. Developmental validation of a real-time PCR assay for the simultaneous quantification of total human and male DNA. Forensic Sci. Int. Genet. 3, 14–21.

Lee, L.G., Connell, C.R., Bloch, W., 1993. Allelic discrimination by nick-translation PCR with fluorogenic probes. Nucleic Acids Res. 21, 3761–3766.

Lee, S.B., Mccord, B., Buel, E., 2014. Advances in forensic DNA quantification: a review. Electrophoresis 35, 3044–3052.

Lin, S.W., Li, C., Ip, S.C.Y., 2018. A performance study on three qPCR quantification kits and their compatibilities with the 6-dye DNA profiling systems. Forensic Sci. Int. Genet. 33, 72–83.

Livak, K.J., Flood, S.J., Marmaro, J., Giusti, W., Deetz, K., 1995. Oligonucleotides with fluorescent dyes at opposite ends provide a quenched probe system useful for detecting PCR product and nucleic acid hybridization. PCR Methods Appl. 4, 357–362.

Loftus, A., Murphy, G., Brown, H., Montgomery, A., Tabak, J., Baus, J., et al., 2017. Development and validation of InnoQuant® HY, a system for quantitation and quality assessment of total human and male DNA using high copy targets. Forensic Sci. Int. Genet. 29, 205–217.

Martins, C., Ferreira, P.M., Carvalho, R., Costa, S.C., Farinha, C., Azevedo, L., et al., 2019. Evaluation of InnoQuant. Forensic Sci. Int. Genet. 39, 61–65.

Morrison, T.B., Weis, J.J., Wittwer, C.T., 1998. Quantification of low-copy transcripts by continuous SYBR Green I monitoring during amplification. Biotechniques 24. 954–958, 960, 962.

Moser, M., Prudent, J., 2003. Enzymatic repair of an expanded genetic information system. Nucleic Acids Res. 31, 5048–5053.

Nicklas, J.A., Buel, E., 2003. Quantification of DNA in forensic samples. Anal. Bioanal. Chem. 376, 1160–1167.

Niederstätter, H., Köchl, S., Grubwieser, P., Pavlic, M., Steinlechner, M., Parson, W., 2007. A modular real-time PCR concept for determining the quantity and quality of human nuclear and mitochondrial DNA. Forensic Sci. Int. Genet. 1, 29–34.

Pineda, G.M., Montgomery, A.H., Thompson, R., Indest, B., Carroll, M., Sinha, S.K., 2014. Development and validation of InnoQuant™, a sensitive human DNA quantitation and degradation assessment method for forensic samples using high copy number mobile elements Alu and SVA. Forensic Sci. Int. Genet. 13, 224–235.

Pionzio, A.M., Mccord, B.R., 2014. The effect of internal control sequence and length on the response to PCR inhibition in real-time PCR quantitation. Forensic Sci. Int. Genet. 9, 55–60.

QIAGEN, 2011. Rotor-Gene Q – Pure Detection. https://www.qiagen.com/au/resources/download.aspx?id=315ba5fb-ccf9-4a3e-b0b7-aa1558050848&lang=en. (Accessed 07 January 2019).

QIAGEN, 2018a. Developmental Validation of the Investigator® Quantiplex® Pro Kit. https://www.qiagen.com/au/resources/download.aspx?id=e80c5169-8cd1-4906-9359-de26fa0afdcf&lang=en. (Accessed 07 January 2019).

QIAGEN, 2018b. Developmental Validation of the Investigator® Quantiplex® Pro RGQ Kit. https://www.qiagen.com/au/resources/download.aspx?id=60d7e387-febb-4c69-a1e2-def65aed8b15&lang=en. (Accessed 07 January 2019).

QIAGEN, 2018c. Investigator® Quantiplex® Pro RGQ Kit Handbook. https://www.qiagen.com/au/resources/download.aspx?id=57497d59-7a43-4eaf-8c94-086e88742e86&lang=en. (Accessed 07 January 2019).

Reed, G.H., Kent, J.O., Wittwer, C.T., 2007. High-resolution DNA melting analysis for simple and efficient molecular diagnostics. Pharmacogenomics 8, 597–608.

Ririe, K.M., Rasmussen, R.P., Wittwer, C.T., 1997. Product differentiation by analysis of DNA melting curves during the polymerase chain reaction. Anal. Biochem. 245, 154–160.

Sherrill, C.B., Marshall, D.J., Moser, M.J., Larsen, C.A., Daudé-Snow, L., Jurczyk, S., et al., 2004. Nucleic acid analysis using an expanded genetic alphabet to quench fluorescence. J. Am. Chem. Soc. 126, 4550–4556.

Sprouse, M.L., Phillips, N.R., Kavlick, M.F., Roby, R.K., 2014. Internal validation of human mitochondrial DNA quantification using real-time PCR. J. Forensic Sci. 59, 1049–1056.

SWGDAM, 2016. SWGDAM Validation Guidelines for DNA Analysis Methods. https://docs.wixstatic.com/ugd/4344b0_813b241e8944497e99b9c45b163b76bd.pdf. (Accessed 25 February 2019).

Thelwell, N., Millington, S., Solinas, A., Booth, J., Brown, T., 2000. Mode of action and application of scorpion primers to mutation detection. Nucleic Acids Res. 28, 3752–3761.

Thermo Fisher Scientific, 2017. QuantStudio™ Real-Time PCR and Digital PCR Systems. https://www.thermofisher.com/content/dam/LifeTech/Documents/PDFs/QuantStudio-Family-Broch-Q117-Global-FINAL-FLR.pdf. (Accessed 03 December 2018).

Thomas, J.T., Berlin, R.M., Barker, J.M., Dawson Cruz, T., 2013. Qiagen's Investigator™ Quantiplex Kit as a predictor of STR amplification success from low-yield DNA samples. J. Forensic Sci. 58 (5), 1306–1309.

Thompson, R.E., Duncan, G., Mccord, B.R., 2014. An investigation of PCR inhibition using Plexor®-based quantitative PCR and short tandem repeat amplification. J. Forensic Sci. 59, 1517–1529.

Timken, M.D., Swango, K.L., Orrego, C., Buoncristiani, M.R., 2005. A duplex real-time qPCR assay for the quantification of human nuclear and mitochondrial DNA in forensic samples: implications for quantifying DNA in degraded samples. J. Forensic Sci. 50, 1044–1060.

Van Den Berge, M., Wiskerke, D., Gerretsen, R.R., Tabak, J., Sijen, T., 2016. DNA and RNA profiling of excavated human remains with varying postmortem intervals. Int. J. Legal Med. 130, 1471–1480.

Vernarecci, S., Ottaviani, E., Agostino, A., Mei, E., Calandro, L., Montagna, P., 2015. Quantifiler® Trio Kit and forensic samples management: a matter of degradation. Forensic Sci. Int. Genet. 16, 77–85.

Von Wurmb-Schwark, N., Higuchi, R., Fenech, A.P., Elfstroem, C., Meissner, C., Oehmichen, M., Cortopassi, G.A., 2002. Quantification of human mitochondrial DNA in a real time PCR. Forensic Sci. Int. 126, 34–39.

Von Wurmb-Schwark, N., Preusse-Prange, A., Heinrich, A., Simeoni, E., Bosch, T., Schwark, T., 2009. A new multiplex-PCR comprising autosomal and y-specific STRs and mitochondrial DNA to analyze highly degraded material. Forensic Sci. Int. Genet. 3, 96–103.

Walker, J.A., Hedges, D.J., Perodeau, B.P., Landry, K.E., Stoilova, N., Laborde, M.E., et al., 2005. Multiplex polymerase chain reaction for simultaneous quantitation of human nuclear, mitochondrial, and male Y-chromosome DNA: application in human identification. Anal. Biochem. 337, 89–97.

Westring, C.G., Kristinsson, R., Gilbert, D.M., Danielson, P.B., 2007. Validation of reduced-scale reactions for the Quantifiler Human DNA kit. J. Forensic Sci. 52, 1035–1043.

Whitcombe, D., Theaker, J., Guy, S.P., Brown, T., Little, S., 1999. Detection of PCR products using self-probing amplicons and fluorescence. Nat. Biotechnol. 17, 804–807.

Wittwer, C.T., Herrmann, M.G., Moss, A.A., Rasmussen, R.P., 1997. Continuous fluorescence monitoring of rapid cycle DNA amplification. BioTechniques 22 (130–131), 134–138.

Wittwer, C.T., Reed, G.H., Gundry, C.N., Vandersteen, J.G., Pryor, R.J., 2003. High-resolution genotyping by amplicon melting analysis using LCGreen. Clin. Chem. 49, 853–860.

Zupanič Pajnič, I., Zupanc, T., Balažic, J., Geršak, Ž., Stojković, O., Skadrić, I., Črešnar, M., 2017. Prediction of autosomal STR typing success in ancient and second world war bone samples. Forensic Sci. Int. Genet. 27, 17–26.

SECTION II

Types of DNA markers and applications for identification

CHAPTER

8

Autosomal short tandem repeat (STR) profiling of human skeletal remains

Jeremy Watherston PhD[a] *and Jodie Ward PhD*[b]

[a]Forensic and Analytical Science Service, NSW Health Pathology, Lidcombe, NSW, Australia
[b]Centre for Forensic Science, University of Technology Sydney, Sydney, NSW, Australia

Overview

The genotyping of autosomal short tandem repeats (STRs) is the current standard method used in forensic DNA analysis for human identification. STR profiling involves amplifying a core set of human-specific STR loci present in the nuclear genome using the polymerase chain reaction (PCR), and then separating and detecting those DNA fragments by capillary electrophoresis (CE). In addition to being a cost- and time-efficient technique, the polymorphic information content (Ziętkiewicz et al., 2012) and power of discrimination (Chakraborty and Kidd, 1991) of STRs make them both suitable and valuable for forensic use to determine the source of biological evidence. The tetranucleotide (and to a lesser extent tri-, penta-, and hexanucleotide) STR markers used in modern commercially available multiplex kits have been widely characterized in the literature for the individual discrimination they offer and for their ability to type degraded and/or inhibited DNA. An added benefit offered by STRs is that standardized sets have been described in both the United States and Europe (Hares, 2015; Schneider, 2009), facilitating the establishment of centralized DNA databases in most countries.

Human skeletal remains can present some of the most challenging biological samples for successfully recovering genetic information, due to the presence of a large number and variety of PCR inhibitors and the often severely degraded nature of these samples when recovered. However, there are a number of STR-based approaches available to combat these challenges, including: (1) the application of STR profiling of skeletal samples following a highly efficient DNA extraction (Loreille et al., 2007); (2) the use of a large number of different

https://doi.org/10.1016/B978-0-12-815766-4.00008-X

sized STR markers as are present in modern multiplexes; (3) the use of mini-STRs for highly degraded DNA (Fondevila et al., 2008b); and (4) optimized amplification parameters (e.g., increased Taq DNA polymerase, increased cycle number, and/or decreased reaction volume). Laboratories may elect to combine a number of these approaches to type severely degraded skeletal samples (Irwin et al., 2007; Zar et al., 2013). Alternatively, other genetic markers such as Y-chromosome STRs (Y-STRs), X-chromosome STRs (X-STRs), single nucleotide polymorphisms (SNPs), insertions/deletions (indels), or any combination of these in conjunction with STRs can be applied (Fondevila et al., 2008b; Marjanović et al., 2009; Romanini et al., 2012).

Given their extensive use within the forensic community as the preferred and most widely used marker for human identification, improvements to STR profiling methodologies are constantly explored to reduce the turnaround times associated with DNA profiling and analysis. STR multiplexes have been extensively applied to automated platforms for high-throughput genotyping; and protocols are easily modified to alter the number of cycles, reaction volumes, and input amount of DNA, dependent upon the equipment and type of samples tested (Edson et al., 2013; McNevin et al., 2015; Ottens et al., 2013a). McCord et al. (2018) describe several approaches to reduce the processing time of STR profiling including rapid DNA and direct PCR methods, or a combination of the two. While many of these approaches are currently limited to high-yielding forensic and antemortem (AM) samples, their use in large-scale identification efforts for rapid intelligence may still be valuable. Moreover, these technologies are starting to be applied to postmortem (PM) samples (Habib et al., 2017).

Short tandem repeats (STRs)

Short tandem repeats (STRs) are DNA regions with core repeated units that are commonly 2–6 bases in length (Jeffreys et al., 1985; Litt and Luty, 1989; Weller et al., 1984; Wyman and White, 1980). The number of repeats in STR markers can be highly variable among individuals. The size of an STR allele is dependent on the number of repeating units present (as well as the position of the PCR primers). Most alleles differ in size by the length of a repeat unit; however, microvariants are observed at some loci (e.g., TH01), defined as alleles that contain incomplete repeat units (Puers et al., 1993). Additionally, STR sequences not only vary in the length of the repeat unit and the number of repeats, but also in the repeat pattern itself (Jeffreys et al., 1985). These repeat patterns are divided into the categories of simple, compound, and complex repeats. Simple repeats contain identical length and sequence units; compound repeats comprise two or more adjacent simple repeats; and complex repeats comprise multiple repeat blocks of variable length, with variable intervening sequences (Urquhart et al., 1994).

Tetranucleotide repeats are the most commonly targeted STR loci for modern multiplexes. Tetranucleotides contain a 4-base repeat structure, allowing closely spaced heterozygous allelic peaks to be resolved easily by CE. Butler (2005) summarizes the advantages of tetranucleotide STR loci as: (1) a narrow allelic size range for multiplexing and reduced drop-out from preferential amplification of small alleles, (2) reduced stutter formation, and (3) the possibility of generating small PCR product sizes for recovering degraded samples. The relatively small size of STR alleles (<500 bases) facilitates the amplification of degraded DNA,

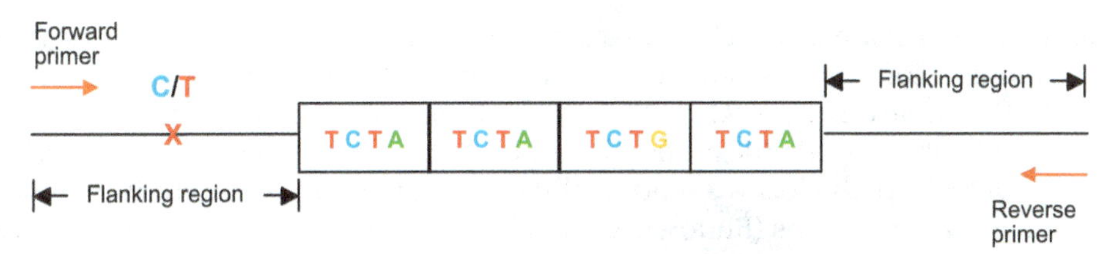

FIG. 1 A complex short tandem repeat (STR) containing a single nucleotide polymorphism (SNP) in the flanking region. The two adjacent simple repeats have different sequences (TCTA and TCTG), and a SNP (C/T) is present in the flanking area between the forward primer and the repeat region. *Graphic design by Trent Watherston.*

with even smaller amplicon sizes having been specifically designed for the STR profiling of more highly degraded DNA samples (Coble and Butler, 2005).

The emerging use of massively parallel sequencing (MPS) for STR profiling has revealed sequence variation within STR alleles of the same length, as well as polymorphisms (e.g., SNPs or indels) in the flanking areas of the repeat region (Gettings et al., 2015). These newly discovered genetic polymorphisms could therefore increase discrimination power substantially (Novroski et al., 2016; Parson et al., 2016). To account for both length and sequence variation, van der Gaag et al. (2016) present four new classes of STRs: (1) simple STRs—loci that only show variation in the number of repeats (e.g., CSF1PO), (2) complex STRs—loci where the repeat motif consists of several repeating blocks with a different sequence (e.g., D19S433), (3) simple STRs with SNPs in the flanking region (e.g., TPOX), and (4) complex STRs containing SNPs in the flanking region (e.g., vWA) (Fig. 1).

For the purposes of inclusion in an STR panel, loci are selected based on their unique characteristics such as the number of alleles present, the type of repeat sequence, the presence of microvariants, and their chromosomal location (Butler and Hill, 2012). Ideal autosomal STR markers for multiplexes reside on separate chromosomes, or are sufficiently separated on a chromosome, so that loci remain independent (which means linkage disequilibrium problems are reduced and the product rule can be applied to exponentially increase discrimination during statistical interpretation) (Butler and Hill, 2012). A high discriminatory power, robustness and reproducibility when multiplexed, low stutter rate, and low mutation rate are also characteristics favored for forensic STR loci (Carracedo and Lareu, 1998; Gill et al., 1996).

Modern STR multiplexes

Butler et al. (2012) purports that the availability of large STR multiplexes has revolutionized forensic genetics because they are relatively affordable, results are obtained quickly, and matching probabilities that easily exceed one-in-a-billion are possible from one assay using a maximum of 1ng of DNA. In addition, many STR multiplexes are commercially manufactured, quality-controlled, amenable to automation, and have been designed to overcome the effects of DNA degradation and inhibition, making them attractive options for accredited, high-throughput forensic DNA laboratories. These kits contain all the necessary components including: (1) allelic ladders, (2) internal sizing standards, (3) positive control samples, (4) primers, and (5) a reaction mix.

A common set of standardized STR markers has been nominated and implemented globally. The standard loci are derived from the United States and Europe in the form of the Combined DNA Index System (CODIS) and the European Standard Set (ESS) of loci, respectively. While there is significant overlap (and all ESS loci are now included within the expanded CODIS loci), the loci were originally selected to offer increased discrimination to the relevant, local populations (Budowle et al., 1998; Ensenberger et al., 2014). The recent expansion of these core loci sets offers a means by which to reduce the chances of adventitious matching, increase international compatibility with regard to data sharing, and increase the discrimination power to assist in database searching for criminal cases, missing persons investigations, and disaster victim identification (DVI) (Butler and Hill, 2012).

Sex-determining markers

Amelogenin is the most common sex-determining marker used in forensic DNA analysis and was first described in 1993 (Sullivan et al., 1993). Amelogenin is a gene that encodes proteins in tooth enamel (Couwenhoven et al., 1993; Diekwisch et al., 1993). There are two forms of the amelogenin gene, one on the X chromosome and the other on the Y chromosome (Sasaki and Shimokawa, 1995). Due to the presence of a 6-base deletion in the amelogenin gene on the X homologue, amplification of the amelogenin locus results in a 106- and 112-base amplicon from the X and Y chromosomes, respectively (Butler, 2012). The amelogenin marker is contained in all modern STR multiplex kits and is a required locus in both the CODIS and ESS core loci set.

While an effective sex-determining marker for most samples, there are certain limitations associated with the amelogenin marker. A rare deletion on the Y chromosome can cause Y allele drop-out in some populations, so only the X amplicon would be present in males (Santos et al., 1998). Conversely, amelogenin X allele drop-out has been observed in males, likely due to a primer binding site mutation, where only the Y amplicon is present (Alves et al., 2006; Maciejewska and Pawłowski, 2009; Shewale et al., 2000). However, the sex of an individual can be verified by typing additional Y-chromosome markers, such as amplification of the sex-determining region Y (SRY) or Y-STRs (Thangaraj et al., 2002). The latest generation of STR multiplexes now includes one or more Y-chromosome markers to ensure an accurate sex determination can be made in the event of amelogenin X or Y allele drop-out. These include a Y-STR(s), a Y chromosome indel (Y-indel), or both.

STR markers

In November 1997, in the United States, the Federal Bureau of Investigation (FBI) named a core set of 13 STR loci in an effort to standardize STR markers for uploading to national DNA databases (Budowle et al., 1998). In 2012, plans to expand this core set to 20 loci began (Hares, 2012). The selection and implementation plan of the expanded CODIS core loci was announced in early 2015 (Hares, 2015), with an additional seven loci being implemented to the National DNA Index System (NDIS) on January 1, 2017. The 20 CODIS core loci include: CSF1PO, D3S1358, D5S818, D7S820, D8S1179, D13S317, D16S539, D18S51, D21S11, FGA, TH01, TPOX, vWA, D1S1656, D2S441, D2S1338, D10S1248, D12S391, D19S433, and D22S1045. Amelogenin is also included as the sex-determining locus.

The seven core ESS loci were originally selected in 1999 (Martin et al., 2001). In April 2009, the European Network of Forensic Science Institute (ENFSI) adopted an additional five loci for a total of 12 core loci (Gill et al., 2006; Schneider, 2009). The 12 ESS loci include: D3S1358, D8S1179, D18S51, D21S11, FGA, TH01, vWA, D1S1656, D2S441, D10S1248, D12S391, and D22S1045. Like with the CODIS core loci, amelogenin is included as the sex-determining locus.

Discrimination power

The increased number of core loci in both the United States and Europe has resulted in an increase in discrimination power (Moretti et al., 2016). In addition to these core loci, commercially available STR kits often contain extra loci that may offer more discrimination for specific population groups. One of these loci is SE33, which has been described as the most polymorphic locus in modern STR multiplexes (Coticone et al., 2004). This is due to its high mutation rate (0.64%), which is more than three times higher than the average mutation rate of most tetranucleotide STR loci (Butler et al., 2012). However, Butler et al. (2012) do emphasize that while the large numbers of alleles make these polymorphic markers well-suited for certain forensic applications such as identity testing and mixture detection, the high mutation rate might present challenges for kinship testing. This would be similar for the inclusion of rapidly mutating Y-STRs in STR multiplexes.

The release of modern STR kits with an excess of 20 STR loci has seen the Probability of Identity (PI) values being achieved increase exponentially. For example, the PowerPlex® 21 System (Promega Corporation), which is a 20 STR multiplex, can provide statistical discrimination in the order of 10^{-25} depending on local population databases used and in conjunction with θ values to account for relatedness within populations (Bright et al., 2014). In comparison, two of the latest generation kits (which have at least 23 loci) have been reported to provide even more discriminatory statistical power: (1) a PI of 7.73×10^{-28} for the GlobalFiler™ PCR Amplification Kit (Thermo Fisher Scientific), and (2) a PI of 6.58×10^{-29} for the PowerPlex® Fusion System (Promega Corporation) (Butler et al., 2012). In terms of STR loci, these kits only differ by inclusion of the SE33 locus in the GlobalFiler™ PCR Amplification Kit and two pentanucleotides (Penta D, Penta E) in the PowerPlex® Fusion System. It should be noted that some loci, such as Penta D and Penta E, are susceptible to drop out of profiles from degraded skeletal samples due to their larger amplicon size. Moreover, a kit that has a large number of smaller amplicon sizes, such as the GlobalFiler™ PCR Amplification Kit (i.e., amplicons are mostly <400 bases), can be an important consideration when selecting an appropriate STR kit for typing skeletal remains.

Quality indicators

The most recently released STR multiplexes now include quality markers that can indicate (and differentiate between) degraded and/or inhibited DNA. Consequently, these markers can assist in decision making for DNA profiling following a prediction of the extent of DNA degradation and/or PCR inhibition. For example, the quality markers facilitate differentiation between failed PCR amplification due to a lack of DNA and failed PCR amplification due to the presence of inhibitors (Kraemer et al., 2017). It therefore becomes important to determine if a sample is degraded and/or inhibited, in order to devise the most appropriate improvement strategy for DNA profiling (Holmes et al., 2018). Furthermore, sample quality information can be enhanced by using an STR kit with quality markers in conjunction with a forensic DNA quantification kit that includes an Internal Positive Control (IPC) system and amplification targets of different sizes to provide an objective measure of PCR inhibition and DNA degradation, respectively. The latest quantification kits all offer a Quality Index (QI) for estimating the level of sample degradation and/or the presence of PCR inhibitors. This index is derived from consideration of both the Degradation Index (DI) and IPC threshold cycle (C_T) values.

Commercial STR kits

Forensic DNA laboratories are unlikely to develop and use custom STR assays when a plethora of extensively validated, quality-controlled STR multiplexes are already available for purchase (Butler and Hill, 2012). Several modern multiplexes are currently used throughout the world for STR profiling. The optimum efficiency of a standard STR multiplex system is usually associated with an input DNA amount of ~1 ng and 28–30 PCR amplification cycles.

Autosomal STR kits

Currently, there are three main manufacturers of forensic STR multiplex kits: Thermo Fisher Scientific (www.thermofisher.com), Promega Corporation (www.promega.com), and QIAGEN (www.qiagen.com). These kits are used by the majority of forensic laboratories worldwide and have evolved over time to include: (1) a larger number of autosomal STR markers to increase discrimination power of the kit, (2) gonosomal STR markers to also increase discrimination and/or aid amelogenin result verification, and (3) quality markers to assist in decision making for subsequent DNA profiling following a prediction of DNA degradation and PCR inhibition. These kits are routinely developmentally validated by the manufacturer according to the Scientific Working Group on DNA Analysis Methods (SWGDAM) guidelines (SWGDAM, 2016). There are a large number of STR multiplex kits available; however, this chapter will focus on recently released kits which amplify all of the CODIS and ESS core STR loci, and which contain additional features that may aid in amplification of DNA recovered from human skeletal remains.

Thermo Fisher Scientific STR kits

The GlobalFiler™ PCR Amplification Kit is a 6-dye STR multiplex that amplifies 21 autosomal STR loci (including 10 autosomal mini-STRs <220 bases), one Y-STR, one Y-indel, and amelogenin. The GlobalFiler™ PCR Amplification Kit encompasses the ESS- and CODIS-recommended composite set of loci (Ludeman et al., 2018) and the highly discriminating SE33 locus (Thermo Fisher Scientific, 2016a). The Master Mix configuration has been designed to maximize the input sample volume (Ludeman et al., 2018). Additionally, the Master Mix has been developed for increased sensitivity and tolerance to inhibitors, while also being optimized for a fast thermal cycling time of approximately 80 min (Thermo Fisher Scientific, 2016a).

The NGM Detect™ PCR Amplification Kit (Thermo Fisher Scientific) is a six-dye STR multiplex designed specifically for laboratories using the ESS loci. The kit amplifies 20 autosomal STR loci (including 7 autosomal mini-STRs <230 bases), two internal quality control (IQC) markers, and two sex-determining markers (Y-indel and amelogenin). The kit offers a reduced amplification time and a higher input volume, but has also been shown to display high allelic imbalance (Burch et al., 2017). Other advantages of the kit include a short amplification time of approximately 60 min, a rearrangement of certain loci (e.g., D2S1338 and SE33) to position them in the low molecular weight range so they are recovered more often, and the inclusion of an IQC. The IQC is comprised of two markers: (1) the IQCS, which is a 70-base amplicon, and (2) the IQCL, which is a 456-base amplicon (Thermo Fisher Scientific, 2019). The IQC markers offer the ability to confirm the validity of negative results and predict if samples are degraded and/or inhibited. However, these IQC marker peaks have also been found to distort one of the dye channels because of their much higher peak height compared to sample peaks (Burch et al., 2017). This kit has been designed to be used in conjunction with the AmpF*l*STR™ NGM SElect™ PCR Amplification Kit (Thermo Fisher Scientific). The kits contain the same loci, but they are amplified using different primers and some markers are repositioned, an approach that can confirm genotypes obtained from other kits and maximize information recovery, especially from the 12 independent mini-STRs included across the kits (Thermo Fisher Scientific, 2016b).

Promega Corporation STR kits

The PowerPlex® Fusion System is a 5-dye 24-locus multiplex that contains 22 autosomal STR loci, including all CODIS and ESS core loci, as well as a Y-STR locus and amelogenin for sex determination (Oostdik et al., 2014). The PowerPlex® Fusion 6C System (Promega Corporation) is a 6-dye 27-locus STR multiplex that was expanded to include SE33 and two rapidly mutating Y-STRs. The inclusion of SE33 in this kit, in addition to Penta D and Penta E, provides additional discrimination, while 8 autosomal mini-STRs (<220 bases) aid amplification of degraded samples (Ensenberger et al., 2016). The kit has also been reported to provide robust performance for changes in PCR conditions or exposure to PCR inhibitors, and the inclusion of the three Y-STR loci is beneficial for sex determination or recognizing mixed DNA profiles (Feng et al., 2017). It has also been determined that applying low-template DNA (LT-DNA) modifications to this kit (such as increasing cycle numbers or CE injection voltage) resulted in limited extra genetic information, but introduced more elevated stutter peaks and background noise (Duijs et al., 2018). Both PowerPlex® Fusion Systems are capable of direct amplification of reference and casework (i.e., swab) samples, thus supporting database workflows (Ensenberger et al., 2016; Oostdik et al., 2014).

QIAGEN STR kits

The Investigator® 24plex QS Kit (QIAGEN) is a 6-dye multiplex that co-amplifies 22 STR loci, plus amelogenin and an internal Quality Sensor for analyzing challenging samples (Kraemer et al., 2017). The Investigator® 24plex QS Kit contains all 20 autosomal STR loci from the core ESS and CODIS set, the SE33 locus, and DYS391 to complement the amelogenin sex-determination marker (Kraemer et al., 2017). Of the autosomal STR loci, there are 12 mini-STRs (<250 bases). The Quality Sensor is contained in the Primer Mix and is amplified with the sample simultaneously. It is comprised of one small amplicon (74 bases) sensor (QS1) and one large amplicon (435 bases) sensor (QS2) (Kraemer et al., 2017). The inclusion of the Quality Sensor enables failed PCR amplification due to a lack of DNA to be distinguished from failed PCR amplification due to the presence of inhibitors, and inhibited DNA from degraded DNA (Kraemer et al., 2017). This feature has been reported to improve laboratory workflow due to the extra information obtained earlier in the genotyping process (Zgonjanin et al., 2017).

Comparison of the latest generation of STR multiplex kits

This latest generation of STR multiplex kits amplifies all CODIS and ESS loci (except for the NGM Detect™ PCR Amplification Kit which does not amplify all CODIS loci), the highly discriminating SE33 locus, amelogenin, and one or more Y chromosome markers to enhance discrimination power, sex-determination, and detection of male DNA in mixtures. The GlobalFiler™ PCR Amplification Kit includes a Y-indel and DYS391, Investigator® 24plex QS Kit includes DYS391, and the PowerPlex® Fusion 6C System includes DYS391, DYS576, and DY570. Refer to Table 1 for a list of loci contained in the commonly used or recently released commercial STR multiplex kits.

The use of 6-dye technology enables the amplicon lengths to be shortened and minimizes overlap of STR markers. Additionally, 7–12 mini-STRs are included in the kits for profiling highly degraded DNA. Chemistry improvements have resulted in reduced amplification times (60–80 min) and have enhanced the profiling success from inhibited and LT-DNA samples, with full STR profiles being generated from as little as 100–250 pg of template DNA (Lin et al., 2017; Tan et al., 2017). This improved DNA recovery could also be attributed to now having the ability to add a greater volume of template (up to 15 μL) to the PCR for these kits. Finally, inclusion of an internal quality control system in the NGM Detect™ PCR Amplification Kit and Investigator® 24plex QS Kit will greatly aid in the detection of sample inhibition and/or degradation and determining if sample repurification or increasing DNA input volume would be the most appropriate action for retesting.

Tan et al. (2017) evaluated the GlobalFiler™ PCR Amplification Kit, Investigator® 24plex QS Kit, and PowerPlex® Fusion 6C System in terms of sensitivity, profile recovery from degraded DNA samples, tolerance to PCR inhibitors, and detection of minor components in DNA mixtures. The kits performed similarly. However, the PowerPlex® Fusion 6C System and the Investigator® 24plex QS Kit were shown to tolerate PCR inhibitors such as hematin and tannic acid better, while the GlobalFiler™ PCR Amplification Kit had a higher average percentage recovery of alleles when less than 100 pg of starting DNA template was amplified (Tan et al., 2017). The study was able to demonstrate the utility of the Investigator® 24plex QS Kit Quality Sensors, by showing that the heights of QS1 and QS2 remained similar for degraded samples, while showing that the height of QS2 decreases more compared to QS1 when inhibitors were introduced. However, total drop-out of both sensors was observed when inhibitor concentrations were very high.

Similar to Tan et al. (2017), the performance of the GlobalFiler™ PCR Amplification Kit, Investigator® 24plex QS Kit, and PowerPlex® Fusion 6C System has been reviewed by Lin et al. (2017). This study demonstrated that the Investigator® 24plex QS Kit showed a higher tolerance to common PCR inhibitors (including humic acid and collagen), and the GlobalFiler™ PCR Amplification Kit had the highest sensitivity (yielding nearly full profiles using 100 pg of template DNA and higher peak heights at all DNA concentrations). Also, the

TABLE 1 Comparison of loci amplified by a range of commercially available short tandem repeat (STR) multiplex kits. Loci listed in blue are core CODIS loci, loci listed in red are both CODIS and ESS loci, and loci listed in black are neither CODIS nor ESS loci.

	NGM™	NGM SElect™	NGM Detect™	GlobalFiler™	Investigator® 24plex QS	PowerPlex® 21	PowerPlex® Fusion	PowerPlex® Fusion 6C
1	Amelogenin	Amelogenin	Amelogenin	Amelogenin	Amelogenin	Amelogenin	Amelogenin	Amelogenin
2	D3S1358	D3S1358	D3S1358	CSF1PO	CSF1PO	CSF1PO	CSF1PO	CSF1PO
3	D8S1179	D8S1179	D8S1179	D3S1358	D3S1358	D3S1358	D3S1358	D3S1358
4	D16S539	D16S539	D16S539	D5S818	D5S818	D5S818	D5S818	D5S818
5	D18S51	D18S51	D18S51	D7S820	D7S820	D7S820	D7S820	D7S820
6	D21S11	D21S11	D21S11	D8S1179	D8S1179	D8S1179	D8S1179	D8S1179
7	FGA	FGA	FGA	D13S317	D13S317	D13S317	D13S317	D13S317
8	TH01	TH01	TH01	D16S539	D16S539	D16S539	D16S539	D16S539
9	vWA	vWA	vWA	D18S51	D18S51	D18S51	D18S51	D18S51
10	D1S1656	D1S1656	D1S1656	D21S11	D21S11	D21S11	D21S11	D21S11
11	D2S441	D2S441	D2S441	FGA	FGA	FGA	FGA	FGA
12	D2S1338	D2S1338	D2S1338	TH01	TH01	TH01	TH01	TH01
13	D10S1248	D10S1248	D10S1248	TPOX	TPOX	TPOX	TPOX	TPOX
14	D12S391	D12S391	D12S391	vWA	vWA	vWA	vWA	vWA
15	D19S433	D19S433	D19S433	D1S1656	D1S1656	D1S1656	D1S1656	D1S1656
16	D22S1045	D22S1045	D22S1045	D2S441	D2S441	D2S1338	D2S441	D2S441
17		SE33	SE33	D2S1338	D2S1338	D12S391	D2S1338	D2S1338
18			Y-indel	D10S1248	D10S1248	D19S433	D10S1248	D10S1248
19				D12S391	D12S391	Penta D	D12S391	D12S391

Continued

TABLE 1 Comparison of loci amplified by a range of commercially available short tandem repeat (STR) multiplex kits. Loci listed in blue are core CODIS loci, loci listed in red are both CODIS and ESS loci, and loci listed in black are neither CODIS nor ESS loci—cont'd

	NGM™	NGM SElect™	NGM Detect™	GlobalFiler™	Investigator® 24plex QS	PowerPlex® 21	PowerPlex® Fusion	PowerPlex® Fusion 6C
20				D19S433	D19S433	Penta E	D19S433	D19S433
21				D22S1045	D22S1045	D6S1043	D22S1045	D22S1045
22				SE33	SE33		Penta D	SE33
23				DYS391	DYS391		Penta E	Penta D
24				Y-indel	QS1/QS2		DYS391	Penta E
25								DYS391
26								DYS570
27								DYS576

Quality Sensors of the Investigator® 24plex QS Kit were found to be useful, with the ratio of sensors being affected in the presence of PCR inhibitors, but rather stable with a range of DNA concentrations, or with degraded DNA ranging in size from 150 to 500 bases.

One study using LT-DNA, highly inhibited, and challenging samples compared the GlobalFiler™ PCR Amplification Kit and the Investigator® 24plex QS Kit (Elwick et al., 2018). This study reported that the GlobalFiler™ PCR Amplification Kit was slightly more sensitive than the Investigator® 24plex QS Kit with more balanced and higher peak heights. However, the Investigator® 24plex QS Kit was shown to be more tolerant to common PCR inhibitors characteristic of human remains samples when both 1 ng and 100 pg of DNA were amplified (Elwick et al., 2018). In addition to confirming that the Quality Sensor system in the Investigator® 24plex QS Kit operates effectively to indicate inhibition with high levels of inhibitors, Elwick et al. (2018) also innovatively used the sensors to quantitatively assess sample quality. The Q/S ratio was calculated by dividing the peak height of the QS1 peak by the height of the QS2 peak.

Direct amplification for genotyping reference samples

While the Promega Corporation multiplexes (e.g., PowerPlex® 21 and PowerPlex® Fusion 6C) also incorporate a direct amplification methodology for a database workflow, the other above-mentioned commercially available kits have complementary kits for the direct amplification of reference samples for databasing purposes. These include the GlobalFiler™ Express PCR Amplification Kit (Thermo Fisher Scientific), the AmpF*l*STR™ Identifiler™ Direct PCR Amplification Kit (Thermo Fisher Scientific), the AmpF*l*STR™ NGM SElect™ Express Kit (Thermo Fisher Scientific), and the Investigator® 24plex GO! Kit (QIAGEN). These kits support direct amplification of a swab or FTA® card sample (Kraemer et al., 2017; Wang et al., 2011, 2015).

Mini short tandem repeat (mini-STR) kits

Mini-STRs are achieved by redesigning the primers of existing STRs to reduce the length of STR marker amplicons (Butler et al., 2003; Wiegand and Kleiber, 2001). PCR primers are repositioned as close as possible to the STR repeat region, reducing the flanking region and, consequently, the overall size of the STR marker (Luce et al., 2009). It is recommended that mini-STR products should be <150 bases to ensure optimal sensitivity (Dixon et al., 2006). This narrow size range of loci also limits the number of mini-STRs able to be multiplexed using CE technology. The CE conditions for mini-STR assays are often the same as those used for standard STR profiling (Fondevila et al., 2008b).

The benefits of mini-STRs for degraded DNA samples have been well-characterized in the literature (Butler et al., 2003; Chung et al., 2004; Coble and Butler, 2005; Dixon et al., 2006; Grubwieser et al., 2006). It has been shown that this approach results in an increased DNA recovery success rate for degraded DNA samples (Chung et al., 2004; Dixon et al., 2006; Grubwieser et al., 2006; Opel et al., 2006). This is because, in compromised samples, the highest molecular weight loci most often fail to amplify (Mulero et al., 2008). Their value has also been highlighted in skeletal remains casework (Barrot et al., 2011; Fondevila et al., 2008b; Pajnic, 2013; Parsons et al., 2007; Zar et al., 2015) due to their ability to type degraded DNA at a relatively low cost and with a high success rate (Parsons et al., 2007).

Thermo Fisher Scientific mini-STR kit

The AmpFlSTR™ MiniFiler™ PCR Amplification Kit (Thermo Fisher Scientific) is the only commercially available mini-STR kit to date. This kit amplifies eight reduced-size STR markers (D7S820, D13S317, D16S539, D21S11, D2S1338, D18S51, CSF1PO, and FGA) and amelogenin (Mulero et al., 2008). All of these loci are now represented in the core set of CODIS loci. Originally designed to type mini-STRs in conjunction with STR loci from the AmpFlSTR™ Identifiler™ and AmpFlSTR™ SGM Plus™ PCR Amplification Kits (Thermo Fisher Scientific), MiniFiler™ has been shown to produce robust and reliable profiles from samples exhibiting both degradation and PCR inhibition (Andrade et al., 2008; Luce et al., 2009; Mulero et al., 2008). A number of the STR loci in the MiniFiler™ PCR Amplification Kit correspond to some of the largest loci in older STR multiplex kits, demonstrating the value of having this kit available to supplement routine STR profiling of compromised samples; this approach is used by the International Commission on Missing Persons (ICMP) on occasion for determining kinship associations (Parsons et al., 2019). The small amplicon sizes of the MiniFiler™ PCR Amplification Kit, combined with improved PCR amplification conditions and an optimized buffer system, provide increased sensitivity (Andrade et al., 2008). Additionally, this kit has a lower DNA input amount compared to standard STR kits. During developmental validation, it was determined that the optimum template range was 0.2ng - 0.6ng, with 0.3ng yielding the best results (Luce et al., 2009).

Custom mini-STR kits

The National Institute of Standards and Technology (NIST) has developed two short-amplicon mini-STR sets, the Mini-NC01 (Coble and Butler, 2005) and Mini-SGM (Hill et al., 2007). However, when compared to the performance of the MiniFiler™ PCR Amplification Kit for typing severely degraded and burnt bone samples, the Mini-SGM set has been reported to perform poorly and the Mini-NC01 set, while amplifying degraded DNA well, also amplified nonspecific peaks (Fondevila et al., 2008b). The ICMP also developed their own mini-STR multiplexes as a less expensive and more sensitive screening test for reassociation of skeletal elements (specifically, a 5-plex, 6-plex, and 7-plex) (Parsons et al., 2019). These multiplexes targeted loci from large commercial multiplexes, with an average decrease in amplicon size of 144 bases (Parsons et al., 2007). However, for operational reasons, the ICMP moved away from using these small multiplexes, which often did not produce sufficient direct match statistics; hence, they reverted to employing a commercial STR multiplex (Parsons et al., 2019).

Y-chromosome short tandem repeat (Y-STR) kits

While often exploited for analysis and interpretation of male/female mixtures in sexual assault casework (Johns et al., 2006), Y-STRs are useful for establishing paternal lineages because Y-STR profiles are expected to remain the same along a patrilineage (Gill et al., 2001). Supplementing the analysis of autosomal STRs with Y-STRs can help achieve extra discrimination, especially when using male-to-male siblings as reference samples for DNA identification or in motherless paternity cases (Diegoli, 2015). Commercial Y-STR kits have long been shown to perform well with regard to sensitivity, reproducibility, and ability to distinguish mixtures (Johns et al., 2006). The current Yfiler™ Plus PCR Amplification Kit (Thermo

Fisher Scientific) and PowerPlex® Y23 System (Promega Corporation) are much more sensitive than older Y-STR kits and are able to type samples with a male quantity less than 0.5 ng (Ambers et al., 2018; Ferreira-Silva et al., 2018). The Yfiler™ Plus PCR Amplification Kit currently has the most loci, with 27 Y-STR loci in a 6-dye fluorescent system format.

X-chromosome short tandem repeat (X-STR) kits

Like Y-STRs, supplementing analysis of autosomal STRs with X-STRs can help achieve extra discrimination, especially when using siblings as reference samples for DNA identification or for confirming any parent-child relationship that involves at least one female (e.g., father-daughter, mother-son, or mother-daughter) (Diegoli, 2015). X-STRs are highly polymorphic and can meet Hardy Weinberg and linkage equilibrium expectations, if not within the same linkage group (Alzate et al., 2015; Diegoli, 2015; Szibor et al., 2003). Currently, the only commercially available X-STR kit is the Investigator® Argus X-12 QS Kit (QIAGEN); however, the ForenSeq™ DNA Signature Prep Kit (Illumina) types seven X-STR markers (Illumina, 2015). The Investigator® Argus X-12 QS Kit co-amplifies 12 X-STRs, D21S11, and amelogenin, along with a Quality Sensor for predicting sample inhibition and degradation. The autosomal marker D21S11 is included to align the kit with other autosomal STR kits for the purpose of detecting sample mix-ups (Scherer et al., 2015). Degenerate primers have also been added to the kit to overcome allelic drop-out due to known variants in primer binding sites (Elakkary et al., 2014). The kit reaction mix allows short PCR cycling protocols with higher inhibitor tolerance, meaning X-STR profiling can be achieved within 70 min (Scherer et al., 2015).

Dual amplification strategies

For challenging samples, current SWGDAM and International Society of Forensic Genetics (ISFG) guidelines recommend using replicate amplification with the same STR multiplex kit to improve DNA profiling results of samples with incomplete STR profiles (Prinz et al., 2007; SWGDAM, 2014). However, another strategy is to amplify the sample using at least two complementary kits (Prinz et al., 2007). This is beneficial because different kits may use different primer sets for STR markers (resulting in different amplicon lengths for the same marker) or different primer sequences (in order to avoid primer binding site variants). Frequent drop-out in larger amplicons, and false homozygous patterns due to primer binding variations (Delamoye et al., 2004), can be overcome by the use of at least two kits with different primer sequences (Harder et al., 2012).

A number of dual amplification strategies with the MiniFiler™ PCR Amplification Kit and different STR multiplex kits have been reported. The MiniFiler™ PCR Amplification Kit was used to type mini-STRs that overlap with the AmpF*l*STR™ Identifiler™ Plus PCR Amplification Kit to offer an extremely sensitive multiplex STR amplification system for old skeletal remains (Barrot et al., 2011). Similarly, the MiniFiler™ and Identifiler™ PCR Amplification Kits were amplified together to type compromised samples (Andrade et al., 2008). These same kits were also used to obtain a consensus DNA profile from 100- to 1000-year old skeletal remains from mass graves in Pakistan using both kits under optimized PCR amplification conditions (i.e., reduced reaction volume and 33 PCR cycles), even from ≤10 pg/μL of input DNA (Zar et al., 2015).

Combining Y-STRs with other genetic markers allows the complementary addition of a lineage marker to contribute to the identification of human remains. Marjanović et al. (2009) report used a combination of STRs, mini-STRs, and Y-STRs to provide sufficient genetic information for the identification of World War II (WWII) victims. For those samples that were too compromised to generate a DNA profile, the authors suggested that refining DNA extraction procedures might provide more success than applying additional typing methodologies (Marjanović et al., 2009).

SNPs and indels have also been described as additional tools to complement routine STR and mini-STR typing. The combination of mini-STRs and SNPs has been applied to bone samples from decomposed and burnt remains (Fondevila et al., 2008b). The study compared three mini-STR sets and two SNP multiplexes. Mini-STR profiling offered improved success; however, SNP genotyping displayed the best performance. Due to the low discrimination power of SNPs, it was suggested that they offer most value when combined with other genetic markers (Fondevila et al., 2008b). Furthermore, Fondevila et al. (2008a) concluded that standard STR typing methods should be sufficient for most degraded samples if inhibition is properly managed; however, for severely degraded samples, a combination of small amplicon approaches and optimized extraction protocols would offer a better approach. When two multiplexes of small amplicon markers comprising 50 SNPs and 28 indels were compared with standard STR and mini-STR kits for the typing of skeletal remains recovered 35 years after burial, Romanini et al. (2012) concluded that the additional 78 binary markers significantly increased the power of discrimination obtained with the commercial STR kits alone. Furthermore, STR and SNP markers will be able to be readily multiplexed to add DNA intelligence information concerning the appearance and ancestry of an individual to the identification information gleaned from STR markers when typed by massively parallel sequencing (MPS) (Kidd et al., 2013, 2014, 2015; Mehta et al., 2017).

While mini-STRs have been shown to offer much greater DNA recovery from compromised samples, it has been suggested that their advantages may not outweigh the advantages of employing a single modern STR multiplex for the majority of cases (Parsons et al., 2007). Indeed, most of the new commercial multiplexes include up to 10 mini-STRs, so laboratories might deem it more pertinent to validate one of the newer-generation STR kits rather than validating a separate mini-STR kit for profiling DNA from old or degraded remains. In this instance, the dual amplification approach could apply to two complementary modern STR multiplex kits. Parys-Proszek et al. (2018) assessed the value of including the NGM Detect™ PCR Amplification Kit in a dual amplification strategy with the GlobalFiler™ PCR Amplification Kit given its additional features (i.e., different primer sequences, shorter amplicon sizes, and inclusion of the IQC). The NGM Detect™ PCR Amplification Kit was useful for increasing the number of positively typed alleles for degraded bone and tooth samples that generated partial profiles with the GlobalFiler™ PCR Amplification Kit. Despite the associated increase in cost and time, Parys-Proszek et al. (2018) advocated a dual amplification strategy with two compatible kits to counteract the stochastic effects often observed with low template DNA (LT-DNA), degraded, or inhibited samples (e.g., allelic or locus drop-out).

PCR thermal cycling instruments

There are a number of commercially available thermal cyclers for PCR. These instruments are capable of amplifying sections of DNA by subjecting samples to multiple rounds of optimal denaturation, annealing, and extension temperatures during PCR. The Applied Biosystems™ thermal cyclers manufactured by Thermo Fisher Scientific are used by most forensic laboratories throughout the world.

Thermo Fisher Scientific PCR thermal cyclers

Applied Biosystems™ GeneAmp® PCR System 9700

The majority of commercial STR kits have been designed to be amplified on the Applied Biosystems™ GeneAmp® PCR System 9700 (Thermo Fisher Scientific). The Applied Biosystems™ GeneAmp® PCR System 9700 was first introduced in 1997 and is one of the most common thermal cyclers in forensic DNA laboratories (Applied Biosystems™, 2015). However, this thermal cycler was discontinued in December 2015 and, hence, forensic laboratories will soon need to upgrade to a newer model. The system is an automated instrument offering a full numeric keypad control panel with a graphical display screen showing the time and temperature profile for each run (Applied Biosystems™, 2003). The Applied Biosystems™ GeneAmp® PCR System 9700 is usually used with a silver 96-well block, or with the gold-plated silver 96-well block.

Applied Biosystems™ Veriti® 96-Well Thermal Cycler

The Applied Biosystems™ Veriti® 96-Well Thermal Cycler (Thermo Fisher Scientific) uses VeriFlex™ Block technology, which allows six independently regulated thermal blocks to assist in designing a primer set and run method for optimal PCR conditions (Applied Biosystems™, 2008). Setup is offered by a color touch screen and interface. The instrument also has a run recovery feature in the event of power failure. The Applied Biosystems™ Veriti® 96-Well Fast Thermal Cycler includes optional setups for a fast or standard PCR method with an option to shorten PCR cycling times (Applied Biosystems™, 2008).

Applied Biosystems™ ProFlex™ PCR System

The Applied Biosystems™ ProFlex™ PCR System (Thermo Fisher Scientific) (Fig. 2) is the latest Thermo Fisher Scientific thermal cycler for human identification (HID) applications. The user interface includes a touchscreen with a graphical display that shows the time, status, and temperature for each run. The system also offers a flexible configuration and interchangeable block formats to allow a high-throughput or focused approach. For example, it can operate using dual 96-well or 384-well blocks for high-throughput laboratories, or a 3×32-well block for running three separate thermal cycling programs simultaneously (Applied Biosystems™, 2016). Many current STR kits have been designed to be amplified on either the Applied Biosystems™ ProFlex™ PCR System or the Applied Biosystems™ GeneAmp® PCR System 9700.

Other PCR thermal cyclers

The Mastercycler™ series (Eppendorf™), C1000 Touch™ 96-Well Thermal Cycler (Bio-Rad™), and the MJ Research DNA Engine® PTC-200 Peltier Thermal Cycler (Bio-Rad™) are examples of other thermal cyclers on the market. Currently, these thermal cyclers are much less characterized in the literature; however, during developmental validation of the

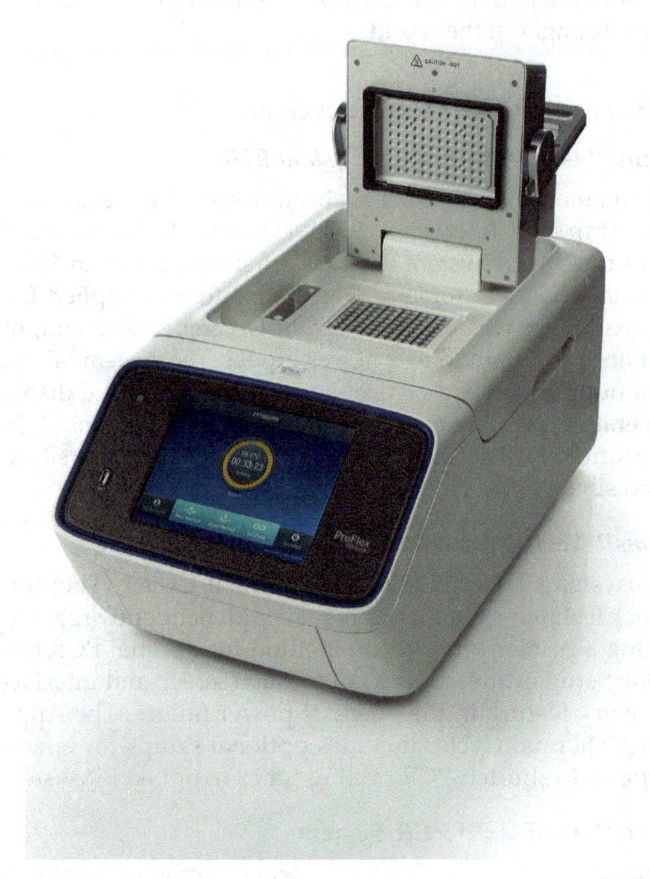

FIG. 2 The Applied Biosystems™ ProFlex™ PCR System. © 2019 Thermo Fisher Scientific. Used under permission.

Investigator® 24plex QS Kit, the Mastercycler™ ep and MJ Research DNA Engine® PTC-200 Peltier Thermal Cycler all gave comparable mean peak heights to those obtained with the Applied Biosystems™ GeneAmp® PCR System 9700 and the Applied Biosystems™ Veriti® 96-Well Thermal Cycler. The C1000 Touch™ 96-Well Thermal Cycler and the Philisa® Thermal Cycler (Streck) have also been reported in rapid PCR protocols (with improved Taq DNA polymerases), alongside the Applied Biosystems™ thermal cyclers. Refer to Romsos and Vallone (2015) for a detailed review.

Optimization of STR profiling of skeletal samples

Standard manufacturer methods for STR profiling may not be appropriate for laboratories that process (a large number of) compromised samples, such as skeletal elements. STR multiplex systems normally operate at 28–30 amplification cycles, but these systems assume an input amount of ~1 ng of DNA. At approximately 250 pg, STR multiplex systems are traditionally within the lower limits of sensitivity as recommended by manufacturers (Gill, 2001). For bone samples, specifically compromised or aged bone, the ideal 1 ng of template DNA may not be recoverable, or even present, within the sample. Alternatively, if a sufficient amount of DNA is recovered, PCR inhibition can be a common cause of PCR failure (Alaeddini, 2012). Optimization of a number of PCR and CE parameters may assist to improve STR allele recovery in these circumstances.

Options for optimizing amplification of STR markers include: (1) reduced reaction volume, (2) dilution of the sample to reduce the concentration of inhibitors, (3) increased DNA input amount, (4) increased number of cycles, (5) increased concentration of Taq DNA polymerase or Bovine Serum Albumin (BSA), or (6) post-PCR purification. Many forensic laboratories elect to use a combination of these approaches. Budowle et al. (2009) reviewed a number of these modifications in the context of low copy number (LCN) typing. Some of these PCR and post-PCR options specifically for optimizing the amplification of skeletal samples are discussed and should be guided by both manufacturer guidelines and internal validation data.

Manufacturer guidelines

Published manufacturer guidelines provide a baseline for optimizing the STR profiling process and will be dependent on: (1) the equipment available, (2) the STR kit selected, and (3) the type and number of samples analyzed in the laboratory. Manufacturers of commercial STR multiplex kits will perform and publish a developmental validation for each kit intended for HID applications. Similarly, each forensic laboratory will also perform and document an internal validation for each kit selected for use in their laboratory. A laboratory's internal validation can be guided by documents prepared by the FBI/National Standards or SWGDAM (FBI/National Standards, 2000; SWGDAM, 2016). Some of these validations, and the accompanying standard operating procedures, are published in scientific journals, available on websites, and/or presented at conferences. STR kit manuals will also outline the required input amounts of DNA, reagent volumes, and thermal cycling parameters for optimum kit performance. An optimal input amount of DNA, reaction volume, and number of PCR cycles are usually determined by the laboratory's internal validation study using forensic type samples that may simply confirm manufacturer's guidelines or demonstrate that modifications are beneficial.

DNA input amount

Commercial kits tend to be optimized for a specific range of DNA concentrations. However, some forensic samples (specifically skeletal remains) tend to yield quantities that fall below the manufacturer's minimum and optimum concentrations (Roeder et al., 2009). This can occur because there is simply not enough DNA in the sample or extract, or because the extract is so dilute and, consequently, enough cannot be added to the PCR. In any event, an increased input amount of DNA can improve chances of successful DNA recovery if samples do not

TABLE 2 Manufacturer-recommended DNA input quantities and polymerase chain reaction (PCR) cycle number options for a range of commercial short tandem repeat (STR) multiplex kits (including maximum volume capacity for each reaction).

	NGM™	NGM SElect™	NGM Detect™	GlobalFiler™	Investigator® 24plex QS	PowerPlex® 21	PowerPlex® Fusion	PowerPlex® Fusion 6C
29 cycles	1.0 ng; 10 μL	1.0 ng; 10 μL	–	1.0 ng; 15 μL	–	–	–	1.0 ng; 15 μL
30 cycles	–	–	0.5 ng; 15 μL	0.5 ng; 15 μL	0.5 ng; 15 μL	0.5 ng; 15 μL	0.5 ng; 15 μL	–

contain a large amount of PCR inhibitors. Note that different STR multiplex kits offer different sample input amounts (i.e., 10 or 15 μL), so this may be an important consideration for laboratories that commonly process skeletal remains. Refer to Table 2 for the manufacturer-recommended DNA input quantities for a range of commercial STR multiplex kits. However, it is important to understand that the number of attempts that can be made to genotype such degraded samples is limited by the volume of DNA extract (Roeder et al., 2009).

Reaction volume

Many forensic laboratories commonly employ reduced volume reactions of commercial STR multiplex kits. Some manufacturers state in accompanying user guides that their kit can be run with half-reaction mix volumes (albeit with the caveat that the highest success rates are expected to be obtained using full reaction volumes) (e.g., QIAGEN, 2018). However, half-reactions have been validated extensively in the literature on a range of multiplex kits. These validation studies report either comparable or increased quality profiles, as well as significant savings on the cost of genotyping due to less reagent used per reaction (Bessekri et al., 2013; Gaines et al., 2002; Hoffman and Fenger, 2010). Another advantage is that less sample is consumed per reaction. This approach facilitates duplicate or dual amplification approaches, reanalysis using low copy number (LCN) approaches, or preserving DNA extract for specialized genotyping approaches.

It has been shown that lowering the reaction volume increases the sensitivity of the amplification reaction. Gaines et al. (2002) reported that, for samples less than 250 pg, up to four-fold increases in sensitivity could be gained with the AmpF*l*STR® Profiler Plus® PCR Amplification Kit (Thermo Fisher Scientific) by drying the template DNA solution directly in the PCR tube and amplifying in 5 μL of reaction master mix. Brito et al. (2017) showed that half-reactions using the GlobalFiler™ PCR Amplification Kit often produced higher quality profiles with LT-DNA samples within a range of 0.004–0.03 ng/μL. An application of a reduced volume reaction to aged skeletal remains has also been applied in conjunction with a total demineralization extraction method and increased PCR cycle number (Zar et al., 2013).

Cycle number

Increasing the number of PCR amplification cycles is the simplest way to increase the number of amplicons and therefore the sensitivity of testing (Gill, 2001). The LCN method for examining samples with less than 100 pg is traditionally characterized by the use of 34 cycles (Gill et al., 2000; Gill, 2001; Whitaker et al., 2001). Above 34 cycles, artifact production has been reported to be enhanced with no increase in sensitivity (Gill, 2001). Of benefit to those laboratories that do not want to validate LCN methodology, some manufacturers have validated a standard and extra sensitivity option for amplification of routine and compromised forensic samples, respectively. Refer to Table 2 for manufacturer-recommended PCR cycle numbers for a range of commercial STR multiplex kits. For example, the GlobalFiler™ PCR Amplification Kit User Guide offers two PCR options, which vary by the recommended DNA input amount and PCR cycle number (Thermo Fisher Scientific, 2016a). Furthermore, the standard thermal cycling conditions for all of the newly released commercial STR kits use at least 29 cycles (compared to 28 cycles for many of the predecessor kits). Alternatively, laboratories can employ increased cycle numbers for routine processing of old and degraded skeletal samples, like the ICMP does (i.e., 31–32 cycles depending on kit used) (Parsons et al., 2019), or on a case-by-case basis as Zar et al. (2013) did when they increased the cycle number from 28 to 33 cycles to improve STR typing from 200 to 500 year old skeletal remains.

The Armed Forces DNA Identification Laboratory (AFDIL) increased PCR cycles in conjunction with an increase in Taq DNA polymerase concentration for its casework using the PowerPlex® 16 System (Promega Corporation) (Irwin et al., 2007, 2012). The protocol used twice the manufacturer's recommended concentration of Taq DNA polymerase and an additional 6 cycles, for a total of 36 cycles (Irwin et al., 2012). In terms of allelic recovery, this modified amplification protocol resulted in more than twice as much data across several DNA inputs. This approach was also found to produce authentic data from an input amount of DNA of <0.5 pg, with only a minimal increase in artifacts; also, the modified parameters actually appeared to mitigate stochastic effects observed under manufacturer-recommended parameters at lower DNA inputs (Irwin et al., 2012). It was also determined that (on average) the modified parameters required four times less input DNA to reproduce profiles under standard conditions (Irwin et al., 2012). This is an important consideration in the pursuit of generating a complete consensus STR profile.

An even higher PCR cycle number has been applied to ancient bone samples. Gill et al. (1994) used 38–43 cycles to analyze STRs from 70 year old Romanov family bone samples, while Burger et al. (1999) used 50 cycles and Schmerer et al. (1999) used 60 cycles, to analyze STRs from bone samples that were reported to be thousands of years old. Strom and Strom and Rechitsky (1998) applied a nested primer PCR protocol using an initial 40-cycle amplification with subsequent 20–30 cycle amplification for charred human remains. As a consequence of LCN methodology, there are implications of allelic drop-out and the possibility of contamination, which mean special considerations are needed to interpret results (Gill, 2001).

PCR reagent modifications

Manufacturers have made efforts to increase tolerance to PCR inhibitors, and optimization of the buffer system and Taq DNA polymerase has been shown to relieve inhibition (Abu Al-Soud and Râdström, 1998; Hedman et al., 2010). PCR inhibition can also be successfully minimized by increasing the Taq DNA polymerase concentration, deoxynucleotide triphosphate (dNTP) concentration, and PCR elongation time (Dietrich et al., 2013). As mentioned previously, AFDIL developed a protocol that uses increased Taq DNA polymerase concentration and cycle number (Irwin et al., 2007, 2012). Additionally, it has been shown that the combination of reduced reaction volumes and increased Taq DNA polymerase concentration can improve STR profiling for LT-DNA samples, even more so than increasing the number of amplification cycles (McNevin et al., 2015).

BSA studies have demonstrated its beneficial effect (in the absence of any other additive) on the yield of amplified DNA (Tarrand et al., 1978; Woide et al., 2010). It has been proposed that BSA prevents inhibitors from interacting with Taq DNA polymerase (Woide et al., 2010), and PCR inhibition has been reported to be overcome by the addition of BSA (Hagelberg et al., 1989). The addition of BSA is thought to bind a soluble inhibitory factor for PCR that copurifies with DNA (Rogan and Salvo, 1990) to the PCR (Hagelberg et al., 1989), making amplification of inhibited bone samples possible (Eilert and Foran, 2009; Hochmeister et al., 1991; Pagan et al., 2012). Moreover, when amplifying DNA in the absence of PCR inhibitors, BSA has been shown to have no effect on amplification yield (Kramer and Coen, 2001). Refer to Farell and Alexandre (2012) for a detailed review of the effects of BSA on PCR.

In some circumstances, it may be beneficial to increase the concentration of Taq DNA polymerase and add BSA to PCR. Seo et al. (2013) investigated multiple methods for obtaining STR profiles from high humic acid-content samples. To overcome inhibition, it was concluded that modification to PCR reagents was more effective than applying post-PCR purification methods designed to remove PCR inhibitors, especially since the latter causes loss of DNA. The use of Ex Taq™ DNA Polymerase Hot Start Version (TaKaRa) with BSA addition improved allele recovery due to a higher resistance to humic acid compared to AmpliTaq Gold® DNA Polymerase (Thermo Fisher Scientific) (possibly due to the inclusion of Nonidet P-40). Furthermore, it was suggested that BSA addition is a more effective and efficient option to overcome humic acid inhibition than increasing the concentration of Taq DNA polymerase.

Post-PCR purification

Post-PCR purification methods involve purifying PCR products using a range of filtration and silica spin column methods in order to remove salts, ions, unused dNTPs, and primers from the PCR (van Oorschot et al., 2010). These ions can interfere with downstream processes by competing with DNA during electrokinetic injection (Budowle et al., 2001; Forster et al., 2008; Smith and Ballantyne, 2007). Several commercial post-PCR clean-up kits are available, such as: 1) the MinElute® PCR Purification Kit (QIAGEN), 2) Microcon® Centrifugal Filters (Millipore), and 3) the NucleoSpin® PCR Clean-Up Kit (Clontech). The selection of which kit to use is usually based on PCR product size range and elution volume. For example, the MinElute® PCR Purification Kit is used for direct purification of double-stranded PCR products ranging in size between 70 bases and 4 kilobases, with a final elution volume of 10 μL (QIAGEN, 2008). This reduction in PCR product volume (i.e., 10 μL) compared to a standard PCR reaction volume (e.g., 25 μL) concentrates DNA. This serves to allow more of the PCR product to be injected during CE (van Oorschot et al., 2010).

Post-PCR purification, and subsequent concentration, can substantially increase sensitivity and performance of STR kits (Forster et al., 2008; Mayntz-Press et al., 2008; Smith and Ballantyne, 2007). Smith and Ballantyne (2007) used the MinElute® PCR Purification Kit to increase the sensitivity of standard 28-cycle PCR amplification. This method produced a four-fold increase in fluorescent signal intensity, with an expectation of up to a 19-fold increase if the entire concentrated product was used. Full Identifiler™ PCR Amplification Kit profiles were obtained with as little as 20 pg of input DNA template. Forster et al. (2008) combined standard PCR with the MinElute® PCR Purification Kit as an alternative approach to increasing PCR cycle number, comparing a post-28 cycle PCR purification with modified CE method against a 34-cycle LCN method. It was found that the 28-cycle approach with the MinElute® PCR Purification Kit clean-up step recovered equivalent or improved results compared to the LCN method, while reducing the complications of elevated stutter peak ratios and allelic drop-in which are usually associated with LCN (Forster et al., 2008).

High-throughput and Rapid STR profiling for large-scale identification efforts

There are now several technologies and methods available that can significantly increase output and reduce turnaround times for STR profiling; hence, DNA-based identification can be achieved in a rapid, streamlined, and high-throughput manner.

Automation

The entire STR profiling process is amenable to automation. Automated platforms offer the advantages of increased throughput, reduced turnaround time, and enhanced sample tracking (Frégeau et al., 2010). Modern STR multiplexes have been extensively applied to automated platforms and this workflow can be modified to alter the number of cycles, reaction volumes, and input amount of DNA (Edson et al., 2013; McNevin et al., 2015; Ottens et al., 2013a).

Rapid amplification

McCord et al. (2018) have reviewed several approaches to reduce the processing time of STR profiling. This can include reducing incubation times (Laurin et al., 2015) or a more rapid amplification and separation (Aboud et al., 2015; Wang and Hennessy, 2017). Rapid PCR protocols reduce PCR amplification time by the use of a faster Taq DNA polymerase combined with the use of rapid thermal cyclers (Bahlmann et al., 2014; Butts and Vallone, 2014; Gibson-Daw et al., 2017). While commercial kit protocols can be modified, multiplexes have been designed for fast genotyping (Aboud et al., 2015; Iyavoo et al., 2015). Finally, amplification time can be reduced by lower total reaction volumes, which consequently reduces the heating and cooling times of the sample (Connon et al., 2016).

Rapid DNA systems

Rapid DNA analysis systems offer a fully automated sample-to-profile workflow for STR profiling and are considered suitable for mobile and nonexpert operation (Verheij et al., 2013). Despite being designed primarily for the rapid processing of reference samples, with further optimization, these instruments may facilitate rapid DNA-based identifications during mass

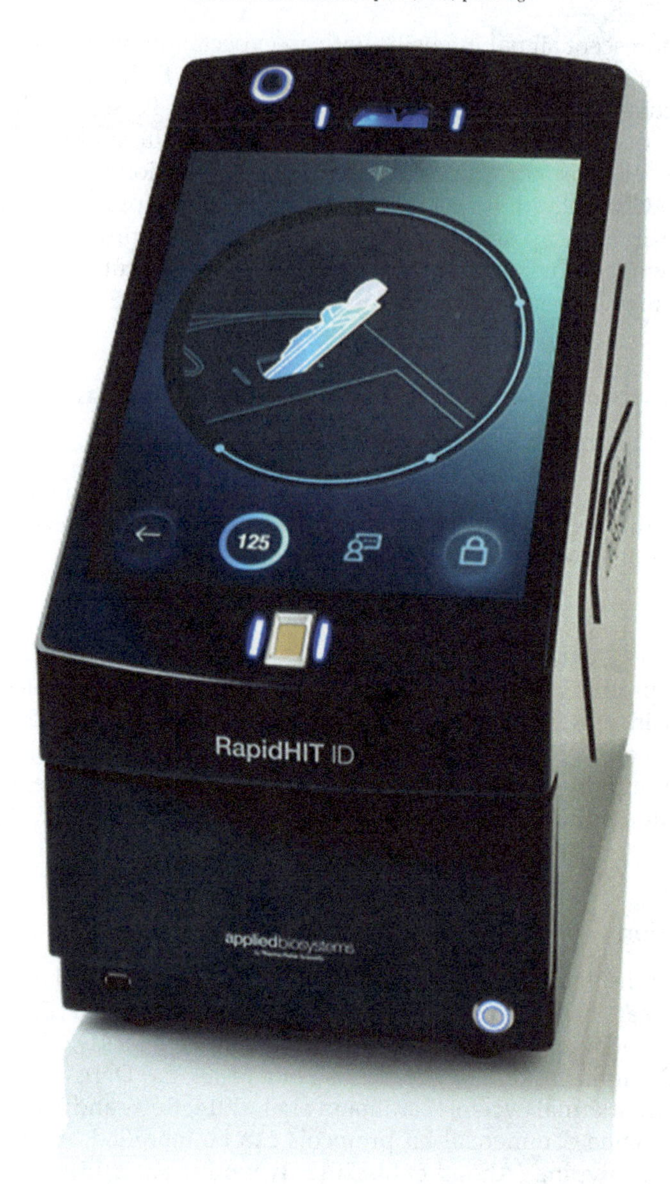

FIG. 3 Applied Biosytems™ RapidHIT™ ID System. © 2019 Thermo Fisher Scientific. Used under permission.

fatality operations. Current technologies include the Applied Biosystems™ RapidHIT™ ID System (Thermo Fisher Scientific) (Fig. 3) and the ANDE™ Rapid DNA Identification System (ANDE Corporation).

The Applied Biosystems™ RapidHIT™ ID System (Salceda et al., 2017) is a fully integrated system capable of generating a CODIS-compatible STR profile from a single high quality sample in approximately 90 min, with minimal hands on time (less than 1 min). After sample collection, a buccal swab is placed in the single-use ACE sample cartridge for processing on the instrument. For casework samples, an EXT sample cartridge was developed for the processing of previously extracted and quantified DNA (Amick and Swiger, 2019). The Applied Biosystems™ RapidHIT™ ID System uses either the GlobalFiler™ Express PCR Amplification Kit or the AmpF*l*STR™ NGM SElect™ Express Kit chemistry, and profiles from multiple instruments can be centralized and reviewed using the RapidLINK™ software. To date, there are no published studies demonstrating the use of this instrument for the rapid profiling of skeletal samples.

The ANDE™ Rapid DNA Identification System (Carney et al., 2019; Grover et al., 2017) is comprised of three components: the fully automated ANDE™ 6C instrument, a single-use microfluidic chip (i.e., an A-Chip or I-Chip for reference and casework samples, respectively), and fully integrated Expert System Software. This system is capable of performing fully automated STR profiling on 4–5 samples simultaneously in approximately 90 min (using the custom FlexPlex™ assay). This assay is based on the PowerPlex® Fusion 6C System and contains 23 autosomal STR loci, 3 Y-STR loci, and amelogenin. Turingan et al. (2016) assessed the ANDE™ 4C instrument for STR profiling of low DNA content samples in a field setting. The system successfully produced PowerPlex® 16 HS System (Promega Corporation) STR profiles from pulverized bone fragments up to 6 days postmortem. More recently, Turingan et al. (2020) showed that the ANDE™ 6C instrument could successfully profile skeletal remains exposed above ground for up to 1 year postmortem, generating useable STR profiles from 5 to 500 mg of bone (i.e., femur, distal foot phalanx, and rib) and tooth (root only) samples that had been cleaned, fragmented with a hammer, and incubated in ANDE™ Bone Solution prior to a swab soaked in the lysate being loaded on the instrument. These studies demonstrate that this rapid DNA testing platform has the potential for in-field identification of human remains for disaster victim identification (DVI).

Direct PCR

Direct PCR is an approach that involves elimination of the extraction and quantification steps in routine DNA processing. The approach has gained interest in the forensic community due to the potential gains in time efficiency (Ottens et al., 2013b; Verheij et al., 2012), an increase in sensitivity (Templeton et al., 2013), and minimization of steps to reduce potential for error or contamination (Linacre et al., 2010). Additionally, there is less opportunity for loss of DNA due to the extraction process (Balogh et al., 2003; Barbaro et al., 2004; Sorensen et al., 2016; van Oorschot et al., 2003; Vandenberg et al., 1997).

Direct PCR has also been found to generate a significant increase in the height of electropherogram peaks (as a result of more available DNA template) compared to traditional extraction methodologies (Templeton et al., 2015). The success and quality of the DNA profiles recovered using the direct PCR approach is dependent on the nature of the material sampled and the presence of PCR inhibitors, which are usually removed during the DNA extraction process (Templeton et al., 2013). Habib et al. (2017) and Watherston et al. (2019) both report successful applications to antemortem (AM) samples such as blood, buccal cells, and hair, with the latter utilizing an automated workflow. Although at this stage direct PCR is mostly only applicable to AM samples, Habib et al. (2017) report the successful direct PCR application to postmortem (PM) samples including bone shavings (following a Proteinase K lysis step).

Summary

DNA-based identification of skeletal remains can be challenging, due to these samples frequently being severely degraded and/or inhibited. While the larger STR multiplex kits, currently commercially available, have a buffering capacity that is largely inhibitor-tolerant, contain a number of mini-STRs, and can distinguish degradation from inhibition via quality sensors, successful STR profiling of skeletal samples will still be dependent on ensuring that a DNA extraction protocol optimized for skeletal samples is employed prior to any profiling attempts. For highly degraded skeletal samples, a number of LCN-type approaches have been described, and their use (often in combination) has been shown to improve DNA profiling outcomes. These approaches range from using specialized markers (e.g., mini-STRs, SNPs, and/or indels), a combination of markers (e.g., STRs and mini-STRs, or STRs and SNPs), or optimizing amplification parameters (e.g., reduced reaction volume, increased Taq DNA polymerase concentration, BSA addition, and/or increased cycle number).

If this arsenal of genetic tools is not practical given the time and cost constraints involved in validating and implementing them, laboratories may be better positioned to optimize their existing STR kit or upgrade to one of the modern multiplexes (i.e., PowerPlex® Fusion 6C System, Investigator® 24plex QS, or GlobalFiler™). It is also encouraged that laboratories validate, or have access to, at least two different STR multiplexes for recovering more genetic information, confirming genotypes, and/or generating a complete consensus profile where possible. Subsequently, it may then be more appropriate to modify amplification parameters for an individual sample, based on whether it is degraded or inhibited (or both). However, as massively parallel sequencing (MPS) for forensic applications becomes more mainstream, many of these limitations may be overcome, with the technology affording laboratories the capability to amplify large numbers of nuclear and mitochondrial DNA loci from multiple skeletal samples simultaneously. In addition, rapid DNA technology is showing the potential to offer investigators an on-site DNA testing option for real-time identification of skeletal remains at mass casualty scenes and in mortuary settings.

References

Aboud, M.J., Gassmann, M., McCord, B., 2015. Ultrafast STR separations on short-channel microfluidic systems for forensic screening and genotyping. J. Forensic Sci. 60 (5), 1164–1170.

Abu Al-Soud, W., Râdström, P., 1998. Capacity of nine thermostable DNA polymerases to mediate DNA amplification in the presence of PCR-inhibiting samples. Appl. Environ. Microbiol. 64 (10), 3748–3753.

Alaeddini, R., 2012. Forensic implications of PCR inhibition—a review. Forensic Sci. Int. Genet. 6 (3), 297–305.

Alves, C., Coelho, M., Rocha, J., Amorim, A., 2006. The amelogenin locus displays a high frequency of X homologue failures in Sao Tome Island (West Africa). Int. Congr. Ser. 1288, 271–273.

Alzate, L., Agudelo, N., Builes, J., 2015. X-STRs as a tool for missing persons identification using only siblings as reference. Forensic Sci. Int. Suppl. Ser. 5, e636–e637.

Ambers, A., Votrubova, J., Vanek, D., Sajantila, A., Budowle, B., 2018. Improved Y-STR typing for disaster victim identification, missing persons investigations, and historical human skeletal remains. Int. J. Legal Med. 132 (6), 1545–1553.

Amick, G.D., Swiger, R.R., 2019. Internal validation of RapidHIT® ID ACE sample cartridge and assessment of the EXT sample cartridge. J. Forensic Sci. 64 (3), 857–868.

Andrade, L., Bento, A., Serra, A., Carvalho, M., Gamero, J., Oliveira, C., Anjos, M., 2008. AmplSTR® MiniFiler™ PCR amplification kit: the new miniSTR multiplex kit. Forensic Sci. Int. Genet. Suppl. Ser., 189–191.

Applied Biosystems™, 2003. GeneAmp® PCR System 9700 User Manual. https://assets.thermofisher.com/TFS-Assets/LSG/manuals/cms_040970.pdf. (Accessed 23 February 2019).

Applied Biosystems™, 2008. Applied Biosystems™ Veriti™ Thermal Cycler User Guide. https://assets.thermofisher.com/TFS-Assets/LSG/manuals/cms_042832.pdf. (Accessed 23 February 2019).

Applied Biosystems™, 2015. Thermal cycler comparison: the ProFlex 96-Well PCR System demonstrates consistent, reliable performance. https://www.thermofisher.com/content/dam/LifeTech/global/life-sciences/PCR/pdfs/ProFlexvs9700C1000MasterCyclerPro_AppNote.pdf. (Accessed 23 February 2019).

Applied Biosystems™, 2016. ProFlex™ PCR System User Guide. http://tools.thermofisher.com/content/sfs/manuals/MAN0007697.pdf. (Accessed 23 February 2019).

Bahlmann, S., Hughes-Stamm, S., Gangitano, D., 2014. Development and evaluation of a rapid PCR method for the PowerPlex® S5 system for forensic DNA profiling. Legal Med. 16 (4), 227–233.

Balogh, M.K., Burger, J., Bender, K., Schneider, P.M., Alt, K.W., 2003. STR genotyping and mtDNA sequencing of latent fingerprint on paper. Forensic Sci. Int. 137 (2–3), 188–195.

Barbaro, A., Staiti, N., Cormaci, P., Saravo, L., 2004. DNA profiling by different extraction methods. Int. Congr. Ser. 1261, 562–564.

Barrot, C., Moreno, C., Sanchez, C., Rodriguez, M., Ortega, M., Huguet, E., Gene, M., 2011. Comparison of Identifiler®, Identifiler Plus® and Minifiler® performance in an initial paternity testing study on old skeletal remains at the forensic and legal medicine area of the Government of Andorra (Pyrenees). Forensic Sci. Int. Genet. Suppl. Ser. 3, e15–e16.

Bessekri, M., Aggoune, A., Lazreg, S., Bucht, R., Fuller, V., 2013. Comparative study on the effects of reduced PCR reaction volumes and increased cycle number, on the sensitivity and the stochastic threshold of the AmpFlSTR Identifiler® Plus kit. Forensic Sci. Int. Genet. Suppl. Ser. 4 (1), e306–e307.

Bright, J.A., Allen, C., Fountain, S., Gray, K., Grover, D., Neville, S., Wilson-Wilde, L., 2014. Australian population data for the twenty Promega PowerPlex 21 short tandem repeat loci. Aust. J. Forensic Sci. 46 (4), 442–446.

Brito, P., Gouveia, N., Bogas, V., Bento, A., Balsa, F., Lopes, V., Porto, M., 2017. Evaluation and comparative analysis on reduction of Globalfiler™ reaction volume in low template samples. Forensic Sci. Int. Genet. Suppl. Ser. 6, e362–e364.

Budowle, B., Moretti, T., Niezgoda, S., Brown, B., 1998. CODIS and PCR-based short tandem repeat loci: law enforcement tools. In: Proceedings of the 2nd European Symposium on Human Identification, Innsbruck, Austria, 10-12 June, pp. 73–88.

Budowle, B., Hobson, D., Smerick, J., Smith, J., 2001. Low copy number—consideration and caution. In: Proceedings of the 12th International Symposium on Human Identification, Biloxi, Mississippi, 9-12 October.

Budowle, B., Eisenberg, A.J., van Daal, A., 2009. Validity of low copy number typing and applications to forensic science. Croat. Med. J. 50 (3), 207–217.

Burch, S., Sulzer, A., Voegeli, P., Morf, N., Gysi, M., Kratzer, A., 2017. The applied biosystems™ NGM detect™ PCR amplification kit—as promising as promised? Forensic Sci. Int. Genet. Suppl. Ser. 6, e504–e506.

Burger, J., Hummel, S., Hermann, B., Henke, W., 1999. DNA preservation: a microsatellite-DNA study on ancient skeletal remains. Electrophoresis 20 (8), 1722–1728.

Butler, J., 2005. Forensic DNA Typing: Biology, Technology and Genetics of STR Markers. Elsevier Academic Press, New York.

Butler, J.M., 2012. Advanced Topics in Forensic DNA Typing: Interpretation. Academic Press, San Diego.

Butler, J.M., Hill, C.R., 2012. Biology and genetics of new autosomal STR loci useful for forensic DNA analysis. Forensic Sci. Rev. 24 (1), 15–26.

Butler, J.M., Shen, Y., McCord, B.R., 2003. The development of reduced size STR amplicons as tools for analysis of degraded DNA. J. Forensic Sci. 48 (5), 1054–1064.

Butler, J.M., Hill, C.R., Coble, M.D., 2012. Variability of New STR Loci and Kits in U.S. Population Groups. https://www.promega.com/resources/profiles-in-dna/2012/variability-of-new-str-loci-and-kits-in-us-population-groups/. (Accessed 26 February 2019).

Butts, E.L., Vallone, P.M., 2014. Rapid PCR protocols for forensic DNA typing on six thermal cycling platforms. Electrophoresis 35 (21 – 22), 3053–3061.

Carney, C., Whitney, S., Vaidyanathan, J., Persic, R., Noel, F., Vallone, P.M., Selden, R.F., 2019. Developmental validation of the ANDE© rapid DNA system with FlexPlex™ assay for arrestee and reference buccal swab processing and database searching. Forensic Sci. Int. Genet. 40, 120–130.

Carracedo, A., Lareu, M., 1998. Development of new STRs for forensic casework: criteria for selection, sequencing & population data and forensic validation. In: Proceedings of the 9th International Symposium on Human Identification, Orlando, Florida, 7-10 October, pp. 89–107.

Chakraborty, R., Kidd, K.K., 1991. The utility of DNA typing in forensic work. Science 254 (5039), 1735–1739.
Chung, D.T., Drábek, J., Opel, K.L., Butler, J.M., McCord, B.R., 2004. A study on the effects of degradation and template concentration on the amplification efficiency of the STR Miniplex primer sets. J. Forensic Sci. 49 (4), 733–740.
Coble, M.D., Butler, J.M., 2005. Characterization of new miniSTR loci to aid analysis of degraded DNA. J. Forensic Sci. 50 (1), 43–53.
Connon, C.C., LeFebvre, A.K., Benjamin, R.C., 2016. Validation of low volume, fast PCR amplification of STR loci for DNA reference samples. J. Forensic Leg. Investig. Sci. 2 (008).
Coticone, S.R., Oldroyd, N., Philips, H., Foxall, P., 2004. Development of the AmpFlSTR SEfiler PCR amplification kit: a new multiplex containing the highly discriminating ACTBP2 (SE33) locus. Int. J. Legal Med. 118 (4), 224–234.
Couwenhoven, R.I., Schwartz, S.A., Snead, M.L., 1993. Arrest of amelogenin transcriptional activation in bromodeoxyuridine-treated developing mouse molars in vitro. J. Craniofac. Genet. Dev. Biol. 13 (4), 259–269.
Delamoye, M., Duverneuil, C., Riva, K., Leterreux, M., Taieb, S., De Mazancourt, P., 2004. False homozygosities at various loci revealed by discrepancies between commercial kits: implications for genetic databases. Forensic Sci. Int. 143 (1), 47–52.
Diegoli, T.M., 2015. Forensic typing of short tandem repeat markers on the X and Y chromosomes. Forensic Sci. Int. Genet. 18, 140–151.
Diekwisch, T., David, S., Bringas, P., Santos, V., Slavkin, H.C., 1993. Antisense inhibition of AMEL translation demonstrates supramolecular controls for enamel HAP crystal growth during embryonic mouse molar development. Development 117 (2), 471–482.
Dietrich, D., Uhl, B., Sailer, V., Holmes, E.E., Jung, M., Meller, S., Kristiansen, G., 2013. Improved PCR performance using template DNA from formalin-fixed and paraffin-embedded tissues by overcoming PCR inhibition. PLoS One 8 (10), e77771.
Dixon, L.A., Dobbins, A.E., Pulker, H.K., Butler, J.M., Vallone, P.M., Coble, M.D., Gill, P., 2006. Analysis of artificially degraded DNA using STRs and SNPs-results of a collaborative European (EDNAP) exercise. Forensic Sci. Int. 164 (1), 33–44.
Duijs, F., van de Merwe, L., Sijen, T., Benschop, C.C.G., 2018. Low-template methods yield limited extra information for PowerPlex® Fusion 6C profiling. Leg. Med. (Tokyo) 3, 362–365.
Edson, J., Brooks, E.M., McLaren, C., Robertson, J., McNevin, D., Cooper, A., Austin, J.J., 2013. A quantitative assessment of a reliable screening technique for the STR analysis of telogen hair roots. Forensic Sci. Int. Genet. 7 (1), 180–188.
Eilert, K.D., Foran, D.R., 2009. Polymerase resistance to polymerase chain reaction inhibitors in bone. J. Forensic Sci. 54 (5), 1001–1007.
Elakkary, S., Hoffmeister-Ullerich, S., Schulze, C., Seif, E., Sheta, A., Hering, S., Augustin, C., 2014. Genetic polymorphisms of twelve X-STRs of the investigator Argus X-12 kit and additional six X-STR centromere region loci in an Egyptian population sample. Forensic Sci. Int. Genet. 1, 126–130.
Elwick, K., Mayes, C., Hughes-Stamm, S., 2018. Comparative sensitivity and inhibitor tolerance of GlobalFiler® PCR Amplification and Investigator® 24plex QS kits for challenging samples. Leg. Med. (Tokyo) 3, 231–236.
Ensenberger, M.G., Hill, C.R., McLaren, R.S., Sprecher, C.J., Storts, D.R., 2014. Developmental validation of the PowerPlex® 21 System. Forensic Sci. Int. Genet. 9, 169–178.
Ensenberger, M.G., Lenz, K.A., Matthies, L.K., Hadinoto, G.M., Schienman, J.E., Przech, A.J., Storts, D.R., 2016. Developmental validation of the PowerPlex® Fusion 6C System. Forensic Sci. Int. Genet. 21, 134–144.
Farell, E.M., Alexandre, G., 2012. Bovine serum albumin further enhances the effects of organic solvents on increased yield of polymerase chain reaction of GC-rich templates. BMC Res. Notes 5, 257.
FBI/National Standards, 2000. Quality assurance standards for forensic DNA testing laboratories. Forensic Sci. Commun. 2 (3), 1–29.
Feng, X., Li, S., Liu, H., Liu, C., Zhang, X., Wang, H., Chen, L., 2017. Validation of the PowerPlex Fusion 6C system: a six-dye STR system for forensic case applications. Aust. J. Forensic Sci. 51 (3), 280–308.
Ferreira-Silva, B., Fonseca-Cardoso, M., Porto, M.J., Magalhães, T., Cainé, L., 2018. A comparison among three multiplex Y-STR profiling kits for sexual assault cases. J. Forensic Sci. 63 (6), 1836–1840.
Fondevila, M., Phillips, C., Naveran, N., Cerezo, M., Rodriguez, A., Calvo, R., Lareu, M.V., 2008a. Challenging DNA: assessment of a range of genotyping approaches for highly degraded forensic samples. Forensic Sci. Int. Genet. Suppl. Ser. 1 (1), 26–28.
Fondevila, M., Phillips, C., Naveran, N., Fernandez, L., Cerezo, M., Salas, A., Lareu, M.V., 2008b. Case report: identification of skeletal remains using short-amplicon marker analysis of severely degraded DNA extracted from a decomposed and charred femur. Forensic Sci. Int. Genet. 2 (3), 212–218.

Forster, L., Thomson, J., Kutranov, S., 2008. Direct comparison of post-28-cycle PCR purification and modified capillary electrophoresis methods with the 34-cycle "low copy number" (LCN) method for analysis of trace forensic DNA samples. Forensic Sci. Int. Genet. 2 (4), 318–328.

Frégeau, C.J., Lett, C.M., Fourney, R.M., 2010. Validation of a DNA IQ-based extraction method for TECAN robotic liquid handling workstations for processing casework. Forensic Sci. Int. Genet. 4 (5), 292–304.

Gaines, M.L., Wojtkiewicz, P.W., Valentine, J.A., Brown, C.L., 2002. Reduced volume PCR amplification reactions using the AmpFlSTR Profiler Plus kit. J. Forensic Sci. 47 (6), 1224–1237.

Gettings, K.B., Aponte, R.A., Vallone, P.M., Butler, J.M., 2015. STR allele sequence variation: current knowledge and future issues. Forensic Sci. Int. Genet. 18, 118–130.

Gibson-Daw, G., Albani, P., Gassmann, M., McCord, B., 2017. Rapid microfluidic analysis of a Y-STR multiplex for screening of forensic samples. Anal. Bioanal. Chem. 409 (4), 939–947.

Gill, P., 2001. Application of low copy number DNA profiling. Croat. Med. J. 42 (3), 229–232.

Gill, P., Ivanov, P.L., Kimpton, C., Piercy, R., Benson, N., Tully, G., Sullivan, K., 1994. Identification of the remains of the Romanov family by DNA analysis. Nat. Genet. 6 (2), 130–135.

Gill, P., Urquhart, E., Millican, E., Oldroyd, N., Watson, S., Sparkes, R., Kimpton, C., 1996. Criminal intelligence databases and interpretation of STRs. In: Proceedings of the 16th Congress of the International Society for Forensic Haemogenetics, Santiago de Compostela, Spain, pp. 235–242.

Gill, P., Whitaker, J., Flaxman, C., Brown, N., Buckleton, J., 2000. An investigation of the rigor of interpretation rules for STRs derived from less than 100 pg of DNA. Forensic Sci. Int. 112 (1), 17–40.

Gill, P., Brenner, C., Brinkmann, B., Budowle, B., Carracedo, A., Jobling, M.A., Tyler-Smith, C., 2001. DNA Commission of the International Society of Forensic Genetics: recommendations on forensic analysis using Y-chromosome STRs. Forensic Sci. Int. 124 (1), 5–10.

Gill, P., Fereday, L., Morling, N., Schneider, P.M., 2006. The evolution of DNA databases-recommendations for new European STR loci. Forensic Sci. Int. 156 (2–3), 242–244.

Grover, R., Jiang, H., Turingan, R.S., French, J.L., Tan, E., Selden, R.F., 2017. FlexPlex27—highly multiplexed rapid DNA identification for law enforcement, kinship, and military applications. Int. J. Legal Med. 131 (6), 1489–1501.

Grubwieser, P., Mühlmann, R., Berger, B., Niederstätter, H., Pavlic, M., Parson, W., 2006. A new "miniSTR-multiplex" displaying reduced amplicon lengths for the analysis of degraded DNA. Int. J. Legal Med. 120 (2), 115–120.

Habib, M., Pierre-Noel, A., Fogt, F., Budimlija, Z., Prinz, M., 2017. Direct amplification of biological evidence and DVI samples using the Qiagen investigator 24plex GO! kit. Forensic Sci. Int. Genet. Suppl. Ser. 6, e208–e210.

Hagelberg, E., Sykes, B., Hedges, R., 1989. Ancient bone DNA amplified. Nature 342 (6249), 485.

Harder, M., Renneberg, R., Meyer, P., Krause-Kyora, B., von Wurmb-Schwark, N., 2012. STR-typing of ancient skeletal remains: which multiplex-PCR kit is the best? Croat. Med. J. 53 (5), 416–422.

Hares, D.R., 2012. Expanding the CODIS core loci in the United States. Forensic Sci. Int. Genet. 6 (1), e52–e54.

Hares, D.R., 2015. Selection and implementation of expanded CODIS core loci in the United States. Forensic Sci. Int. Genet., 1733–1734.

Hedman, J., Nordgaard, A., Dufva, C., Rasmusson, B., Ansell, R., Rådström, P., 2010. Synergy between DNA polymerases increases polymerase chain reaction inhibitor tolerance in forensic DNA analysis. Anal. Biochem. 405 (2), 192–200.

Hill, C.R., Kline, M.C., Mulero, J.J., Lagacé, R.E., Chang, C.W., Hennessy, L.K., Butler, J.M., 2007. Concordance study between the AmpFlSTR MiniFiler PCR amplification kit and conventional STR typing kits. J. Forensic Sci. 52 (4), 870–873.

Hochmeister, M.N., Budowle, B., Borer, U.V., Eggmann, U., Comey, C.T., Dirnhofer, R., 1991. Typing of deoxyribonucleic acid (DNA) extracted from compact bone from human remains. J. Forensic Sci. 36 (6), 1649–1661.

Hoffman, N.H., Fenger, T., 2010. Validation of half-reaction amplification using Promega PowerPlex 16. J. Forensic Sci. 55 (4), 1044–1049.

Holmes, A.S., Houston, R., Elwick, K., Gangitano, D., Hughes-Stamm, S., 2018. Evaluation of four commercial quantitative real-time PCR kits with inhibited and degraded samples. Int. J. Legal Med. 132 (3), 691–701.

Illumina, 2015. ForenSeq™ DNA Signature Prep Reference Guide. https://support.illumina.com/content/dam/illumina-support/documents/documentation/chemistry_documentation/forenseq/forenseq-dna-signature-prep-guide-15049528-01.pdf. (Accessed 26 February 2019).

Irwin, J.A., Leney, M.D., Loreille, O., Barritt, S.M., Christensen, A.F., Holland, T.D., Parsons, T.J., 2007. Application of low copy number STR typing to the identification of aged, degraded skeletal remains. J. Forensic Sci. 52 (6), 1322–1327.

Irwin, J.A., Just, R.S., Loreille, O.M., Parsons, T.J., 2012. Characterization of a modified amplification approach for improved STR recovery from severely degraded skeletal elements. Forensic Sci. Int. Genet. 6 (5), 578–587.

Iyavoo, S., Wolejko, A., Furmanczyk, D., Graham, R., Myers, R., Haizel, T., 2015. Reduced PCR cycling time amplification using AmpFℓSTR® Identifiler® kit. Forensic Sci. Int. Genet. Suppl. Ser. 5, e286–e288.

Jeffreys, A.J., Wilson, V., Thein, S.L., 1985. Hypervariable 'minisatellite' regions in human DNA. Nature 314 (6006), 67–73.

Johns, L., Burton, R., Thomson, J., 2006. Study to compare three commercial Y-STR testing kits. Int. Congr. Ser. 1288, 192–194.

Kidd, K., Pakstis, A., Speed, W., Lagace, R., Chang, J., Wootton, S., Ihuegbu, N., 2013. Microhaplotype loci are a powerful new type of forensic marker. Forensic Sci. Int. Genet. Suppl. Ser. 4, 123–124.

Kidd, K.K., Pakstis, A.J., Speed, W.C., Lagacé, R., Chang, J., Wootton, S., Kidd, J.R., 2014. Current sequencing technology makes microhaplotypes a powerful new type of genetic marker for forensics. Forensic Sci. Int. Genet. 12, 215–224.

Kidd, K., Speed, W., Wootton, S., Lagace, R., Langit, R., Haigh, E., Pakstis, A., 2015. Genetic markers for massively parallel sequencing in forensics. Forensic Sci. Int. Genet. Suppl. Ser. 5, 677–679.

Kraemer, M., Prochnow, A., Bussmann, M., Scherer, M., Peist, R., Steffen, C., 2017. Developmental validation of QIAGEN Investigator® 24plex QS kit and Investigator® 24plex GO! kit: two 6-dye multiplex assays for the extended CODIS core loci. Forensic Sci. Int. Genet. 29, 9–20.

Kramer, M., Coen, D., 2001. Enzymatic amplification of DNA by PCR: standard procedures and optimization. In: Bonifacino Juan, S. (Ed.), Current Protocols in Cell Biology. John Wiley & Sons.

Laurin, N., Célestin, F., Clark, M., Wilkinson, D., Yamashita, B., Frégeau, C., 2015. New incompatibilities uncovered using the Promega DNA IQ™ chemistry. Forensic Sci. Int. 257, 134–141.

Lin, S.W., Li, C., Ip, S.C.Y., 2017. A selection guide for the new generation 6-dye DNA profiling systems. Forensic Sci. Int. Genet. 30, 34–42.

Linacre, A., Pekarek, V., Swaran, Y.C., Tobe, S.S., 2010. Generation of DNA profiles from fabrics without DNA extraction. Forensic Sci. Int. Genet. 4 (2), 137–141.

Litt, M., Luty, J.A., 1989. A hypervariable microsatellite revealed by in vitro amplification of a dinucleotide repeat within the cardiac muscle actin gene. Am. J. Hum. Genet. 44 (3), 397–401.

Loreille, O.M., Diegoli, T.M., Irwin, J.A., Coble, M.D., Parsons, T.J., 2007. High efficiency DNA extraction from bone by total demineralization. Forensic Sci. Int. Genet. 1 (2), 191–195.

Luce, C., Montpetit, S., Gangitano, D., O'Donnell, P., 2009. Validation of the AmpFlSTR MiniFiler PCR amplification kit for use in forensic casework. J. Forensic Sci. 54 (5), 1046–1054.

Ludeman, M.J., Zhong, C., Mulero, J.J., Lagacé, R.E., Hennessy, L.K., Short, M.L., Wang, D.Y., 2018. Developmental validation of GlobalFiler™ PCR amplification kit: a 6-dye multiplex assay designed for amplification of casework samples. Int. J. Legal Med. 132 (6), 1555–1573.

Maciejewska, A., Pawłowski, R., 2009. A rare mutation in the primer binding region of the Amelogenin X homologue gene. Forensic Sci. Int. Genet. 3 (4), 265–267.

Marjanović, D., Durmić-Pasić, A., Kovacević, L., Avdić, J., Dzehverović, M., Haverić, S., Primorac, D., 2009. Identification of skeletal remains of Communist Armed Forces victims during and after World War II: combined Y-chromosome (STR) and MiniSTR approach. Croat. Med. J. 50 (3), 296–304.

Martin, P.D., Schmitter, H., Schneider, P.M., 2001. A brief history of the formation of DNA databases in forensic science within Europe. Forensic Sci. Int. 119 (2), 225–231.

Mayntz-Press, K.A., Sims, L.M., Hall, A., Ballantyne, J., 2008. Y-STR profiling in extended interval (> or = 3 days) postcoital cervicovaginal samples. J. Forensic Sci. 53 (2), 342–348.

McCord, B., Gauthier, Q., Cho, S., Roig, M., Gibson-Daw, G., Young, B., Duncan, G., 2018. Forensic DNA analysis. Anal. Chem. 91 (1), 673–688.

McNevin, D., Edson, J., Robertson, J., Austin, J.J., 2015. Reduced reaction volumes and increased Taq DNA polymerase concentration improve STR profiling outcomes from a real-world low template DNA source: telogen hairs. Forensic Sci. Med. Pathol. 11 (3), 326–338.

Mehta, B., Daniel, R., Phillips, C., McNevin, D., 2017. Forensically relevant SNaPshot. Int. J. Legal Med. 131 (1), 21–37.

Moretti, T.R., Moreno, L.I., Smerick, J.B., Pignone, M.L., Hizon, R., Buckleton, J.S., Onorato, A.J., 2016. Population data on the expanded CODIS core STR loci for eleven populations of significance for forensic DNA analyses in the United States. Forensic Sci. Int. Genet. 25, 175–181.

Mulero, J.J., Chang, C.W., Lagacé, R.E., Wang, D.Y., Bas, J.L., McMahon, T.P., Hennessy, L.K., 2008. Development and validation of the AmpFlSTR MiniFiler PCR Amplification Kit: a MiniSTR multiplex for the analysis of degraded and/or PCR inhibited DNA. J. Forensic Sci. 53 (4), 838–852.

Novroski, N.M.M., King, J.L., Churchill, J.D., Seah, L.H., Budowle, B., 2016. Characterization of genetic sequence variation of 58 STR loci in four major population groups. Forensic Sci. Int. Genet. 25, 214–226.

Oostdik, K., Lenz, K., Nye, J., Schelling, K., Yet, D., Bruski, S., Storts, D.R., 2014. Developmental validation of the PowerPlex® Fusion System for analysis of casework and reference samples: a 24-locus multiplex for new database standards. Forensic Sci. Int. Genet. 12, 69–76.

Opel, K.L., Chung, D.T., Drábek, J., Tatarek, N.E., Jantz, L.M., McCord, B.R., 2006. The application of miniplex primer sets in the analysis of degraded DNA from human skeletal remains. J. Forensic Sci. 51 (2), 351–356.

Ottens, R., Taylor, D., Abarno, D., Linacre, A., 2013a. Optimising direct PCR from anagen hair samples. Forensic Sci. Int. Genet. Suppl. Ser. 4 (1), e109–e110.

Ottens, R., Templeton, J., Paradiso, V., Taylor, D., Abarno, D., Linacre, A., 2013b. Application of direct PCR in forensic casework. Forensic Sci. Int. Genet. Suppl. Ser. 4, e47–e48.

Pagan, F., Keglovic, M., McNevin, D., 2012. Comparison of DNA extraction methods for identification of human remains. Aust. J. Forensic Sci. 44 (2), 117–127.

Pajnic, I., 2013. A comparative analysis of the AmpFlSTR Identifiler and PowerPlex 16 autosomal short tandem repeat (STR) amplification kits on the skeletal remains excavated from second world war mass graves in Slovenia. Rom. J. Leg. Med. 21 (1), 73–78.

Parson, W., Ballard, D., Budowle, B., Butler, J.M., Gettings, K.B., Gill, P., Phillips, C., 2016. Massively parallel sequencing of forensic STRs: considerations of the DNA commission of the International Society for Forensic Genetics (ISFG) on minimal nomenclature requirements. Forensic Sci. Int. Genet. 22, 54–63.

Parsons, T.J., Huel, R., Davoren, J., Katzmarzyk, C., Milos, A., Selmanović, A., Rizvić, A., 2007. Application of novel "mini-amplicon" STR multiplexes to high volume casework on degraded skeletal remains. Forensic Sci. Int. Genet. 1 (2), 175–179.

Parsons, T.J., Huel, R.M.L., Bajunović, Z., Rizvić, A., 2019. Large scale DNA identification: the ICMP experience. Forensic Sci. Int. Genet. 38, 236–244.

Parys-Proszek, A., Wróbel, M., Marcińska, M., Kupiec, T., 2018. Dual amplification strategy for improved efficiency of forensic DNA analysis using NGM Detect™, NGM™ or Globalfiler™ kits. Forensic Sci. Int. Genet. 35, 46–49.

Prinz, M., Carracedo, A., Mayr, W.R., Morling, N., Parsons, T.J., Sajantila, A., Schneider, P.M., 2007. DNA Commission of the International Society for Forensic Genetics (ISFG): recommendations regarding the role of forensic genetics for disaster victim identification (DVI). Forensic Sci. Int. Genet. 1 (1), 3–12.

Puers, C., Hammond, H.A., Jin, L., Caskey, C.T., Schumm, J.W., 1993. Identification of repeat sequence heterogeneity at the polymorphic short tandem repeat locus HUMTH01[AATG]n and reassignment of alleles in population analysis by using a locus-specific allelic ladder. Am. J. Hum. Genet. 53 (4), 953–958.

QIAGEN, 2008. MinElute® Handbook. https://www.qiagen.com/us/resources/resourcedetail?id=-fa2ed17d-a5e8-4843-80c1-3d0ea6c2287d&lang=en. (Accessed 2 March 2019).

QIAGEN, 2018. Investigator 24plex QS Handbook. https://www.qiagen.com/au/resources/download.aspx?id=de-be09ab-5483-478b-aeb3-e5c128e78a92&lang=en. (Accessed 26 February 2019).

Roeder, A.D., Elsmore, P., Greenhalgh, M., McDonald, A., 2009. Maximizing DNA profiling success from sub-optimal quantities of DNA: a staged approach. Forensic Sci. Int. Genet. 3 (2), 128–137.

Rogan, P., Salvo, J., 1990. Study of nucleic acids isolated from ancient remains: yearbook of physical anthropology. Am. J. Phys. Anthropol. 33 (S11), 195–214.

Romanini, C., Catelli, M.L., Borosky, A., Pereira, R., Romero, M., Salado Puerto, M., Vullo, C.M., 2012. Typing short amplicon binary polymorphisms: supplementary SNP and Indel genetic information in the analysis of highly degraded skeletal remains. Forensic Sci. Int. Genet. 6 (4), 469–476.

Romsos, E.L., Vallone, P.M., 2015. Rapid PCR of STR markers: applications to human identification. Forensic Sci. Int. Genet. 18, 90–99.

Salceda, S., Barican, A., Buscaino, J., Goldman, B., Klevenberg, J., Kuhn, M., King, D., 2017. Validation of a rapid DNA process with the RapidHIT® ID system using GlobalFiler® Express chemistry, a platform optimized for decentralized testing environments. Forensic Sci. Int. Genet. 28, 21–34.

Santos, F.R., Pandya, A., Tyler-Smith, C., 1998. Reliability of DNA-based sex tests. Nat. Genet. 18 (2), 103.

Sasaki, S., Shimokawa, H., 1995. The amelogenin gene. Int. J. Dev. Biol. 39 (1), 127–133.

Seo, S.B., Jin, H.X., Lee, H.Y., Ge, J., King, J.L., Lyoo, S.H., Shin, D.H., Lee, S.D., 2013. Improvement of short tandem repeat analysis of samples highly contaminated by humic acid. J. Forensic Leg. Med. 20 (7), 922–928.

Scherer, M., Konig, M., Bussmann, M., Prochnow, A., Peist, R., 2015. Development and validation of the new Investigator® Argus X-12 QS kit. Forensic Sci. Int. Genet. Suppl. Ser. 5, e256–e257.

Schmerer, W.M., Hummel, S., Herrmann, B., 1999. Optimized DNA extraction to improve reproducibility of short tandem repeat genotyping with highly degraded DNA as target. Electrophoresis 20 (8), 1712–1716.

Schneider, P.M., 2009. Expansion of the European Standard Set of DNA database loci-the current situation. Profiles DNA 12 (1), 6–7.

Shewale, J., Richey, S., Sinha, S., 2000. Anomalous amplification of the amelogenin locus typed by AmpFlSTR Profiler Plus amplification kit. Forensic Sci. Commun. 2 (4).

Smith, P.J., Ballantyne, J., 2007. Simplified low-copy-number DNA analysis by post-PCR purification. J. Forensic Sci. 52 (4), 820–829.

Sorensen, A., Berry, C., Bruce, D., Gahan, M.E., Hughes-Stamm, S., McNevin, D., 2016. Direct-to-PCR tissue preservation for DNA profiling. Int. J. Legal Med. 130 (3), 607–613.

Strom, C.M., Rechitsky, S., 1998. Use of nested PCR to identify charred human remains and minute amounts of blood. J. Forensic Sci. 43 (3), 696–700.

Sullivan, K.M., Mannucci, A., Kimpton, C.P., Gill, P., 1993. A rapid and quantitative DNA sex test: fluorescence-based PCR analysis of X-Y homologous gene amelogenin. BioTechniques 15 (4), 636–638. 640–641.

SWGDAM, 2014. Guidelines for STR Enhanced Detection Methods. http://media.wix.com/ugd/4344b0_29feed748e3742a5a7112467cccec8dd.pdf. (Accessed 25 February 2019).

SWGDAM, 2016. SWGDAM Validation Guidelines for DNA Analysis Methods. https://docs.wixstatic.com/ugd/4344b0_813b241e8944497e99b9c45b163b76bd.pdf. (Accessed 25 February 2019).

Szibor, R., Krawczak, M., Hering, S., Edelmann, J., Kuhlisch, E., Krause, D., 2003. Use of X-linked markers for forensic purposes. Int. J. Legal Med. 117 (2), 67–74.

Tan, J.Y.Y., Tan, Y.P., Ng, S., Tay, A.S., Phua, Y.H., Tan, W.J., Syn, C.K.C., 2017. A preliminary evaluation study of new generation multiplex STR kits comprising of the CODIS core loci and the European Standard Set loci. J. Forensic Legal Med., 5216–5223.

Tarrand, J.J., Krieg, N.R., Döbereiner, J., 1978. A taxonomic study of the Spirillum lipoferum group, with descriptions of a new genus, Azospirillum gen. nov. and two species, Azospirillum lipoferum (Beijerinck) comb. nov. and Azospirillum brasilense sp. nov. can. J. Microbiol. 24 (8), 967–980.

Templeton, J., Ottens, R., Paradiso, V., Handt, O., Taylor, D., Linacre, A., 2013. Genetic profiling from challenging samples: direct PCR of touch DNA. Forensic Sci. Int. Genet. Suppl. Ser. 4, e224–e225.

Templeton, J.E., Taylor, D., Handt, O., Skuza, P., Linacre, A., 2015. Direct PCR improves the recovery of DNA from various substrates. J. Forensic Sci. 60 (6), 1558–1562.

Thangaraj, K., Reddy, A., Singh, L., 2002. Is the amelogenin gene reliable for gender identification in forensic casework and pre-natal diagnosis? Int. J. Legal Med. 116, 121–123.

Thermo Fisher Scientific, 2016a. GlobalFiler™ PCR Amplification Kit User Guide. https://tools.thermofisher.com/content/sfs/manuals/4477604.pdf. (Accessed 25 February 2019).

Thermo Fisher Scientific, 2016b. The New NGM Detect Kit. https://www.thermofisher.com/content/dam/LifeTech/Documents/PDFs/NGM-Detect-Brochure-Global.pdf. (Accessed 28 February 2019).

Thermo Fisher Scientific, 2019. NGM Detect™ PCR Amplification Kit User Guide. https://assets.thermofisher.com/TFS-Assets/LSG/manuals/100044085_NGMD_UG_EN.pdf. (Accessed 28 February 2019).

Turingan, R.S., Vasantgadkar, S., Palombo, L., Hogan, C., Jiang, H., Tan, E., Selden, R.F., 2016. Rapid DNA analysis for automated processing and interpretation of low DNA content samples. Investig. Genet. 7 (2), 1–12.

Turingan, R.S., Brown, J., Kaplun, L., Smith, J., Watson, J., Boyd, D.A., Selden, R.F., 2020. Identification of human remains using rapid DNA analysis. Int. J. Legal Med. 134, 863–872.

Urquhart, A., Kimpton, C.P., Downes, T.J., Gill, P., 1994. Variation in short tandem repeat sequences-a survey of twelve microsatellite loci for use as forensic identification markers. Int. J. Legal Med. 107 (1), 13–20.

van der Gaag, K.J., de Leeuw, R.H., Hoogenboom, J., Patel, J., Storts, D.R., Laros, J.F.J., de Knijff, P., 2016. Massively parallel sequencing of short tandem repeats-population data and mixture analysis results for the PowerSeq™ system. Forensic Sci. Int. Genet. 24, 86–96.

van Oorschot, R., Phelan, D., Furlong, S., Scarfo, G., Holding, N., Cummins, M., 2003. Are you collecting all the available DNA from touched objects? Int. Congr. Ser. 1239, 803–807.

van Oorschot, R.A., Ballantyne, K.N., Mitchell, R.J., 2010. Forensic trace DNA: a review. Investig. Genet. 1 (1), 14.

Vandenberg, N., van Oorschot, R., Mitchell, R., 1997. An evaluation of selected DNA extraction strategies for short randem repeat typing. Electrophoresis 18, 1624–1626.

Verheij, S., Harteveld, J., Sijen, T., 2012. A protocol for direct and rapid multiplex PCR amplification on forensically relevant samples. Forensic Sci. Int. Genet. 6 (2), 167–175.

Verheij, S., Clarisse, L., van den Berge, M., Sijen, T., 2013. RapidHIT™ 200, a promising system for rapid DNA analysis. Forensic Sci. Int. Genet. Suppl. Ser. 4 (1), e254–e255.

Wang, D., Hennessy, L., 2017. Fast PCR for STR Genotyping. USA Patent Application US 2017/0081715 A1.

Wang, D.Y., Chang, C.W., Lagacé, R.E., Oldroyd, N.J., Hennessy, L.K., 2011. Development and validation of the AmpFlSTR® Identifiler® direct PCR amplification kit: a multiplex assay for the direct amplification of single-source samples. J. Forensic Sci. 56 (4), 835–845.

Wang, D.Y., Gopinath, S., Lagacé, R.E., Norona, W., Hennessy, L.K., Short, M.L., Mulero, J.J., 2015. Developmental validation of the GlobalFiler® express PCR amplification kit: a 6-dye multiplex assay for the direct amplification of reference samples. Forensic Sci. Int. Genet. 19, 148–155.

Watherston, J., Bruce, D., Ward, J., Gahan, M.E., McNevin, D., 2019. Automating direct-to-PCR for disaster victim identification. Aust. J. Forensic Sci. 51 (1), S39–S43.

Weller, P., Jeffreys, A.J., Wilson, V., Blanchetot, A., 1984. Organization of the human myoglobin gene. EMBO J. 3 (2), 439–446.

Whitaker, J.P., Cotton, E., Gill, P., 2001. A comparison of the characteristics of profiles produced with the AmpFlSTR SGM plus multiplex system for both standard and low copy number (LCN) STR DNA analysis. Forensic Sci. Int. 123 (2–3), 215–223.

Wiegand, P., Kleiber, M., 2001. Less is more-length reduction of STR amplicons using redesigned primers. Int. J. Legal Med. 114 (4–5), 285–287.

Woide, D., Zink, A., Thalhammer, S., 2010. Technical note: PCR analysis of minimum target amount of ancient DNA. Am. J. Phys. Anthropol. 142 (2), 321–327.

Wyman, A.R., White, R., 1980. A highly polymorphic locus in human DNA. Proc. Natl. Acad. Sci. U. S. A. 77 (11), 6754–6758.

Zar, M., Shahid, A., Shahzad, M., Shin, K.-J., Lee, H., Israr, M., Husnain, T., 2013. Forensic DNA typing of old skeletal remains using AmpFlSTR® Identifiler® PCR amplification kit. J. Forensic Res. 5 (1), 1–6.

Zar, M., Shahid, A., Shahzad, M., Shin, K.-J., Lee, H., Israr, M., Husnain, T., 2015. Comparative study of STR loci for typing old skeletal remains with modified protocols of AmpFlSTR Identifiler and AmpFlSTR MiniFiler STR kits. Aust. J. Forensic Sci. 47 (2), 200–223.

Zgonjanin, D., Soler, M., Antov, M., Redhead, P., Stojiljkovic, G., Milic, A., 2017. Validation and implementation of the Investigator® 24plex QS kit for forensic casework. Forensic Sci. Int. Genet. Suppl. Ser., e77–e79.

Ziętkiewicz, E., Witt, M., Daca, P., Zebracka-Gala, J., Goniewicz, M., Jarząb, B., 2012. Current genetic methodologies in the identification of disaster victims and in forensic analysis. J. Appl. Genet. 53 (1), 41–60.

CHAPTER

9

Y-chromosome analysis for unidentified human remains (UHR) investigations

Angie Ambers PhD[a,b,c]

[a]Henry C. Lee Institute of Forensic Science, University of New Haven, West Haven, CT, United States [b]Forensic Science Department, Henry C. Lee College of Criminal Justice and Forensic Sciences, Center for Forensic Investigation of Trafficking in Persons, University of New Haven, West Haven, CT, United States [c]Institute for Human Identification, LMU College of Dental Medicine, Knoxville, TN, United States

Introduction

Testing of autosomal short tandem repeat (auSTR) loci is the "gold standard" in forensic DNA casework. The FBI recently expanded the original set of 13 core CODIS loci (adopted in 1997) to 20 STR core loci, increasing the discriminatory power of multiplex assays and improving global compatibility (Hares, 2015). However, autosomal STRs undergo genetic recombination each generation, and therefore the informativeness of such analyses decreases as the number of familial generations increases between the evidentiary sample and the reference sample(s) available for comparison. Distant relatives are difficult, if not impossible, to be traced using autosomal STRs because genetic recombination events produce dissimilarities between genotypes with every subsequent generation. In such scenarios, lineage markers such as mitochondrial DNA (mtDNA) and the Y-chromosome can be used to assist in the investigation.

Y-chromosome testing has advanced significantly over the past decade, with numerous commercially available multiplex testing kits now accessible for both Y-chromosome short tandem repeat (Y-STR) and Y-chromosome single nucleotide polymorphism (Y-SNP) analysis. Today, Y-chromosome typing is used for a broad range of forensic investigations, including missing persons cases, disaster victim identification (DVI), commingled (mass) graves, male lineage (kinship or paternity) testing, and in predicting the biogeographic

https://doi.org/10.1016/B978-0-12-815766-4.00009-1

ancestry of unidentified human remains (UHR) (Kayser, 2017). In addition to its value in contemporary forensic casework, Y-chromosome analysis has been applied in a number of historical and archaeological human remains investigations (Ambers, 2021, 2016; Ambers et al., 2014, 2016, 2018, 2020; Olasz et al., 2019; Parsons et al., 2019; King et al., 2014; Haas et al., 2013; Lee et al., 2010; Loreille et al., 2010; Zupanič Pajnič et al., 2010; Crubézy et al., 2010; Coble et al., 2009; Marjanović et al., 2009; Vanek et al., 2009; Gojanovic and Sutlovic, 2007; Irwin et al., 2007; Kemp et al., 2007). One of the most famous historical case studies using Y-chromosome analysis investigated the assertion that former U.S. President Thomas Jefferson fathered a son with his African-American slave, Sally Hemings. Y-STR and Y-SNP analysis demonstrated that several living male relatives of Thomas Jefferson share the same Y haplotype as a living descendant of Eston Hemings Jefferson (Sally Hemings' son), except for one repeat difference at a single Y-STR locus, which could be easily explained by a mutation. This analysis supported that President Jefferson had indeed sired Eston Hemings Jefferson—or alternatively, his brother Randolph did (since Thomas and Randolph share the same Y haplotype) (Foster et al., 1998, 1999). Y-chromosome analysis has also been used extensively in forensic genetic genealogy investigations (Calafell and Larmuseau, 2017; Fitzpatrick, 2016, 2011; Butler et al., 2008).

Recommendations regarding forensic analysis of Y-STRs have been established and published extensively by the DNA Commission of the International Society of Forensic Genetics (ISFG) (Roewer et al., 2020; Gusmao et al., 2006; Gill et al., 2001).

Basic structure and inheritance patterns of the Y chromosome

Barring the presence of a chromosomal abnormality, humans have 46 chromosomes in each cell, or 23 pairs of homologous chromosomes, as shown in the karyotype depicted in Fig. 1A. The first 22 pairs are called autosomes, while the 23rd pair comprises the sex chromosomes (XX = female, XY = male). The Y chromosome is unique to males, and it is the third smallest human chromosome (only slightly larger than chromosome 21 and chromosome 22). It contains approximately 60 million base pairs (60 Mb) and is divided into two primary regions: (1) the pseudoautosomal regions (PAR1, PAR2), located at the telomeres (ends) of the p arm (short arm) and q arm (long arm), respectively; and (2) the male-specific Y (MSY) region, which comprises the rest of the chromosome (Fig. 1B). The MSY region was formerly referred to as the non-combining region of the Y chromosome, or NRY region. During meiosis in males, PAR1 and PAR2 undergo recombination with homologous regions on the X chromosome; however, the rest of the Y chromosome (i.e., MSY region) does not participate in homologous recombination. Therefore, it is inherited intact (unchanged) from the biological father and is passed down in the same manner to all biological sons and throughout subsequent generations of male offspring (Skaletsky et al., 2003). The X-homologous regions of the Y chromosome are typically not used for forensic applications. The collection of alleles in a Y-STR profile comprises what is referred to as a *haplotype* (instead of a genotype) because it is inherited *en bloc* from a single chromosome. The word *haplotype* is derived from the word "haploid" (which describes cells with only one set of chromosomes) and from the word "genotype" (which refers to the genetic makeup of an organism).

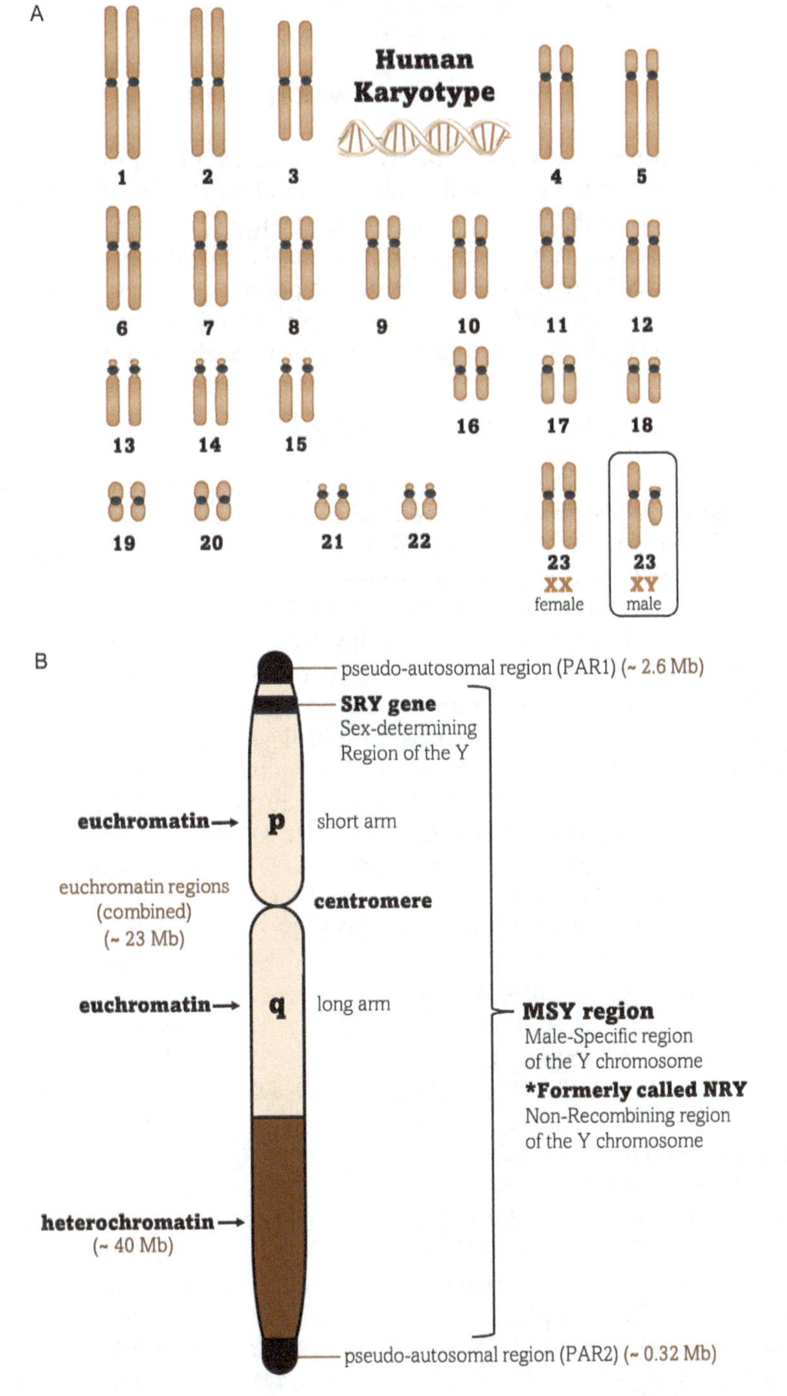

FIG. 1 (A) Basic schematic of a normal human karyotype, which consists of 46 total chromosomes (or 23 pairs of homologous chromosomes). The first 22 pairs of chromosomes are referred to as autosomes. The 23rd pair represents the sex chromosomes and determines an individual's biological sex (XX = female; XY = male). (B) The Y chromosome is one of the smallest human chromosomes, comprising approximately 60 Mb of DNA, and is divided into two major parts: (1) the pseudoautosomal regions (PAR 1, PAR2), which undergo recombination with homologous regions of the X chromosome during meiosis, and (2) the male-specific region (MSY), which does not participate in genetic recombination and is therefore passed intact (unchanged) from biological fathers to sons, as well as to subsequent generations of male offspring. *Graphic by: Angie Ambers PhD (Institute for Human Identification—LMU College of Dental Medicine; Henry C. Lee Institute of Forensic Science; Henry C. Lee College of Criminal Justice and Forensic Sciences).*

Y-chromosome polymorphisms

The male Y chromosome contains a variety of different types of polymorphisms (variations) in its genome, including: (1) variable number tandem repeats (VNTRs), or minisatellites; (2) short tandem repeats (Y-STRs), or microsatellites; (3) insertion-deletions (InDels); and (4) single nucleotide polymorphisms (Y-SNPs). Similar to the variation that exists within the mitochondrial (mtDNA) genome, Y-chromosome polymorphisms can be used to classify Y haplotypes into a series of haplogroups. The two most commonly used Y-chromosome polymorphisms in forensic, historical, and archaeological casework are Y-STRs and Y-SNPs.

Y-chromosome short tandem repeats (Y-STRs)

In 1997, a core set of Y-STR loci—referred to as the European Minimal Haplotype (EMH)—was recommended by the International Y-STR User Group for forensic applications (Kayser et al., 1997). The EMH consists of nine Y-STR loci: DYS19, DYS385a/b, DYS389I, DYS389II, DYS390, DYS391, DYS392, and DYS393. Subsequently, in 2003, the U.S. haplotype loci were recommended by the Scientific Working Group on DNA Analysis Methods (SWGDAM) (2014), and included the EMH loci plus two additional Y-STRs (DYS438 and DYS439). More than 400 Y-STR loci have been identified in the male Y-chromosome genome, and the precise locations of each locus have been sequentially mapped using human genome sequencing data.

A number of commercially available Y-STR multiplexes exist for Y-chromosome testing in contemporary, historical, and archaeological investigations. The Yfiler® Plus kit (Thermo Fisher Scientific, Waltham, Massachusetts, USA) is a multiplex of 27 Y-STR loci, including DYS19, DYS385a/b, DYS387S1a/b, DYS389I, DYS389II, DYS390, DYS391, DYS392, DYS393, DYS437, DYS438, DYS439, DYS448, DYS449, DYS456, DYS458, DYS460, DYS481, DYS518, DYS533, DYS570, DYS576, DYS627, DYS635 (Y GATA C4), and Y GATA H4. Seven of the Y-STR loci in the Yfiler® Plus kit (DYF387S1a/b, DYS449, DYS518, DYS570, DYS576, DYS627) are Rapidly Mutating Y-STRs (RM Y-STRs), which provide improved resolution for close paternal relatives (Ambers et al., 2018; Andersen et al., 2017; Gopinath et al., 2016; Ottaviani et al., 2015; Pickrahn et al., 2016; Rapone et al., 2016). The PowerPlex® Y23 kit (Promega Corporation, Madison, Wisconsin, USA) co-amplifies 23 Y-STR loci: DYS576, DYS389I, DYS389II, DYS448, DYS19, DYS391, DYS481, DYS549, DYS533, DYS438, DYS437, DYS570, DYS635, DYS390, DYS439, DYS392, DYS643, DYS393, DYS458, DYS385a/b, DYS456, and Y-GATA-H4 (Thompson et al., 2012, 2013).

Lastly, the Investigator® Argus Y-28 QS kit (Qiagen Corporation, Germantown, Maryland, USA) co-amplifies a total of 27 Y-STRs (including 6 RM Y-STRs), and the assay incorporates a Quality Sensor (QS) system (QS1 and QS2) to provide information about PCR efficiency and the presence of inhibitors. The Y-STR markers included in the Investigator® Argus Y-28 multiplex are DYS19, DYS385a/b, DYS389I, DYS389II, DYS390, DYS391, DYS392, DYS393, DYS437, DYS438, DYS439, DYS448, DYS449, DYS456, DYS458, DYS460, DYS481, DYS518, DYS533, DYS549, DYS570, DYS576, DYS627, DYS635, DYS643, and YGATAH4. The six RM Y-STRs in this assay are DYS449, DYS481, DYS570, DYS576, DYS518, and DYS627 (Qiagen, 2021).

In the massively parallel sequencing (MPS) arena—previously referred to as next generation sequencing (NGS)—there are a number of assays available that incorporate Y-STRs into their multiplexes. One such assay is the *ForenSeq™ DNA Signature Prep Kit* (Verogen, San Diego, California, USA), which co-amplifies over 200 DNA markers in a single reaction, thus providing investigators with a substantial amount of genetic information that can be used to make associations or identifications in highly decomposed or skeletonized human remains cases. More specifically, the *ForenSeq™ DNA Signature Prep Kit* co-amplifies the following DNA markers in a single reaction: 27 global autosomal short tandem repeats (STRs), 24 Y-chromosome STRs (Y-STRs), 7 X-chromosome STRs (X-STRs), 94 human identity single nucleotide polymorphisms (iSNPs), 22 phenotype-informative SNPs (piSNPs), and 56 biogeographical ancestry informative SNPs (aiSNPs) (Verogen, 2018).

Another MPS multiplex kit that incorporates Y-STR DNA markers is the *ForenSeq™ MainstAY Kit* (Verogen, San Diego, California, USA). The *ForenSeq™ MainstAY* assay offers the largest combination of STRs in a single amplification reaction, and its design meets the European/SWGDAM minimal Y-haplotype and European Standard Set (ESS) loci requirements. The *ForenSeq™ MainstAY Kit* contains 53 standard STR loci for use in forensic analysis, relationship testing, and research. These established DNA markers co-amplified in the *ForenSeq™ MainstAY* panel are broadly accepted as informative by the forensic community and by globally recognized databases (including the FBI's CODIS database and Interpol). (Verogen, 2021a). In total, twenty-seven (27) autosomal STRs (auSTRs) and twenty-six (26) Y-chromosome STRs (Y-STRs)—plus the amelogenin sex-determining locus—are co-amplified by the *ForenSeq™ MainstAY* panel in a single reaction. The 27 autosomal STR (auSTR) loci included in the assay are D1S1656, TPOX, D2S441, D2S1338, D3S1358, D4S2408, FGA, D5S818, CSF1PO, D6S1043, D7S820, D8S1179, D9S1122, D10S1248, vWA, D12S391, D13S317, Penta E, D16S539, D17S1301, D18S51, D19S433, D20S482, D21S11, Penta D, D22S1045, and THO1. The 26 Y-chromosome STRs (Y-STRs) included in the *ForenSeq™ MainstAY* panel are DYF387S1, DYS19, DYS385a/b, DYS389I, DYS389II, DYS390, DYS391, DYS392, DYS393, DYS437, DYS438, DYS635, DYS643, DYS439, DYS448, DYS460, DYS481, DYS505, DYS522, DYS533, DYS549, DYS570, DYS576, DYS612, and Y-GATA-H4 (Verogen, 2021a).

Y-chromosome single nucleotide polymorphisms (Y-SNPs)

Y-SNP testing has not yet gained widespread use in forensics due to limitations in multiplexing capabilities using CE-based systems. However, massively parallel sequencing (MPS) technology has the potential to allow investigators to take full advantage of the large number of Y-SNPs needed for inferring biogeographic ancestry on a detailed, discriminatory level. One recent proof-of-concept study demonstrated that 530 Y-SNPs could be analyzed simultaneously in a single targeted MPS run (Ralf et al., 2015). However, in order to achieve accurate biogeographic ancestry prediction for an unknown individual, ancestry-informative SNPs (aiSNPs) from the Y chromosome for paternal ancestry inference should be combined with those from mitochondrial DNA (mtDNA) for maternal ancestry inference, and from autosomal DNA for bi-parental ancestry inference. This combined genetic data analysis approach will consider and account for the potential existence of population admixture in the unknown individual. Admixed individuals possess the DNA of biological ancestors that are from very different geographic regions, and this admixture cannot be assessed using Y-chromosome DNA or mtDNA alone (Kayser, 2017).

One of the most respected and cutting-edge manufacturers of forensic DNA assays for challenging samples, Verogen, recently introduced a massively parallel sequencing (MPS) multiplex SNP assay that includes 85 Y-SNPs for forensic genetic genealogy investigations. More specifically, this multiplex assay—called the ForenSeq Kintelligence Kit—co-amplifies a total of 10,230 SNP markers, including 56 ancestry-informative SNPs (aiSNPs), 94 human identity SNPs, 9867 kinship SNPs, 22 phenotype-informative SNPs (piSNPs), 106 X-chromosome SNPs (X-SNPs), and 85 Y-chromosome SNPs (Y-SNPs) (Verogen, 2021b).

Forensic genetic genealogy (FGG)—discussed in Chapter 19 of this book—is a powerful tool for lead generation in a variety of different types of investigations, including missing persons and unidentified human remains (UHR) cases. FGG typically combines microarray genotyping or whole-genome sequencing (WGS) methods for DNA profile generation with comparisons to genetic relatives in genealogy databases such as GEDmatch®, the largest database of voluntarily submitted DNA profiles for forensic comparisons. Comparisons made in the database allow genealogists to construct family trees using census records, vital records, obituaries, and newspaper archives, and then trace the source of the DNA through the family tree. Short tandem repeat (STR) typing using massively parallel sequencing (MPS) or capillary electrophoresis (CE) then can confirm the identity of the DNA source (Tillman et al., 2020). However, microarray and WGS-based methods do not align with the full spectrum of biological samples associated with forensic cases, including unidentified human remains (UHR) that are highly decomposed or completely skeletonized.

UHR samples are often degraded or inhibited due to environmental exposure over time, resulting in low-quality data that limits FGG applications. Additionally, microarrays and WGS produce large amounts of medically-relevant SNP data that have no (or limited) forensic applications, but also which raise serious concerns about genetic data privacy and usage. To help forensic laboratories address these DNA degradation and privacy challenges, Verogen developed the ForenSeq Kintelligence Kit—the only FGG assay appropriate for forensic casework purposes. This kit was designed for sequencing on the National DNA Index System (NDIS)-approved MiSeq FGx Sequencing System, data interpretation with Verogen's Universal Analysis Software (UAS), and long-range kinship analysis in GEDmatch PRO (Jager et al., 2017).

Y-chromosome databases

Decades of extensive study have resulted in a detailed understanding of the diversity and distribution patterns of Y-chromosome haplotypes, and this knowledge elucidated the value of creating an accessible population database consisting of Y-chromosome data from different geographical regions around the world. Currently, the largest and most widely used population database for Y-chromosome data is the Y-chromosome haplotype reference database (YHRD) (Roewer, 2019a,b; Willuweit and Roewer, 2015). The YHRD currently contains more than a hundred national databases and continues to grow. In contrast to a law enforcement database that contains DNA profiles of known criminal offenders or unmatched profiles from crime scenes, the YHRD is a reference database of anonymous Y-STR profiles that have been obtained primarily through academic research. Hence, the YHRD does not provide personal information on an individual match, but rather utilizes bulk population data to calculate haplotype frequencies and other parameters that can provide statistical assessment of the value of a particular match (Roewer, 2019a,b).

Y-STR haplotype frequencies and matching: Identical by Descent (IBD) vs. Identical by State (IBS); Rapidly Mutating Y-STRs (RM Y-STRs)

Y-STR loci are genetically linked and are therefore inherited *en bloc* (together as a unit) throughout generations of a paternal lineage. Hence, statistical methods relying on the independence of loci, such as the "multiplication rule" used for autosomal STRs, cannot be applied. All members of a patrilineage are expected to share the same Y haplotype. However, it's important to consider that unrelated males may also share the same Y haplotype. When legitimate male relatives possess identical Y haplotypes, this is referred to as "identical by descent (IBD)," whereas identical Y haplotypes that arise from random mutation (not relatedness) are called "identical by state (IBS)" (de Kniff, 2000). The number of IBD and IBS haplotypes that will match in a database is influenced by a variety of factors, including the number of loci in the haplotype, the mutation rates of the loci, the size of the database, and the extent of family/clan structures in the sample population(s). The higher the number of loci typed per haplotype and the higher the mutation rates of each locus, the greater the probability is that any resultant matching haplotypes are IBD instead of IBS (Roewer, 2019b).

Rapidly Mutating Y-STR markers (RM Y-STRs) offer a potential solution to resolving identical IBD and IBS haplotypes. Elevated mutation rates make haplotypes consisting of RM Y-STRs more discriminatory than low-resolution (conventional) Y-haplotypes, and therefore have the potential to differentiate even close paternal relatives (Liu et al., 2016; Ballantyne et al., 2010, 2012, 2014). In all, 13 RM Y-STRs in particular have been identified as being extremely useful for paternal lineage differentiation and identification: DYF387S1a/b, DYF399S1, DYF403S1a/b, DYF404S1, DYS449, DYS518, DYS526a/b, DYS547, DYS570, DYS576, DYS612, DYS626, and DYS627 (Kayser, 2017; Ballantyne et al., 2012). Empirical evidence of male relative differentiation using these 13 RM Y-STRs has been explored extensively. Discrimination rates supported by currently available data are: (1) 27% for fathers-sons based on 2378 father-son pairs (Ballantyne et al., 2014); (2) 44% for brothers and grandfathers-grandsons separated by two meioses, based on 480 pairs (Adnan et al., 2016); and (3) 61% for male relatives separated by four meioses, based on 277 pairs (Adnan et al., 2016). Two commercially available Y-STR kits incorporate RM Y-STR loci into their multiplexes. Two RM Y-STR loci are included in the PowerPlex® Y23 kit (Promega Corporation, Madison, Wisconsin, USA) and seven RM Y-STRs are included in the Yfiler® Plus assay (Thermo Fisher Scientific, Waltham, Massachusetts, USA). Neither of these kits delivers the full power of male relative differentiation that would be provided using the complete 13 RM Y-STR panel, and a commercially available 13 RM Y-STR multiplex kit does not currently exist. However, non-commercial, RM Y-STR multiplex amplification protocols have been published (Alghafri et al., 2015; Ballantyne et al., 2012).

Predicting biogeographic ancestry of unidentified human remains (UHR)

The YHRD is a useful resource to infer the ancestry of an unknown male, primarily because Y haplotypes are highly geographically structured (Roewer, 2019a,b). It should be noted that Y-SNPs have approximately 100,000 times lower mutation rates than most Y-STRs;

as such, geographic ancestry signatures remain stable for much longer at Y-SNPs compared to Y-STRs. Hence, Y-SNPs are generally more suitable for paternal biogeographic ancestry inference compared to Y-STRs (Kayser, 2017; Ballantyne et al., 2010; Xue et al., 2009).

Ultimately, combining both Y-STR and Y-SNP data—in addition to data from ancestry-informative single nucleotide polymorphisms (aiSNPs) included in some of the current massively parallel sequencing (MPS) assays—can substantially improve the precision of the ancestry assessment and biogeographical origin prediction for an unidentified male. In one archaeological case, a combination of Y-STRs and aiSNPs was used to predict the biogeographic ancestry of two adult male skeletons recovered from a 17th-century French shipwreck, resulting in an unexpected finding (i.e., Native American ancestry) for one of the UHR samples, which was later corroborated by historical archives and diaries (Ambers, 2021; Ambers et al., 2020).

Y haplogroup prediction tools

A single Y haplogroup represents a group or a family of Y chromosomes related by descent or ancestry. A variety of Y haplogroup predictors are available online. However, several recent studies have indicated that Whit Athey's Haplogroup Predictor (http://www.hprg.com/hapest5/index.html) provides the most accurate haplogroup assignment using Y-STR data (Babic Jordamović et al., 2021; Dogan et al., 2016). Another program, the NevGen Y-DNA Haplogroup Predictor (YHP) (https://www.nevgen.org/), is an open-access software program that integrates previously published datasets with Y-STR haplotypes and high-resolution haplogroups to predict the Y haplogroup for an unknown sample (Cetkovic Gentula and Nevski, 2015).

Genetic genealogy applications for Y-chromosome DNA

Genetic genealogists suggest additional forensic applications of the human Y-chromosome, including prediction of an unknown male's surname from his Y-DNA (Calafell and Larmuseau, 2017; King and Jobling, 2009). In patrilineal societies (which comprise most societies), surnames are transferred from a biological father to all of his offspring, and the father's Y-chromosome haplotype is passed down to all of his sons. Hence, in patrilineal societies, co-ancestry of surnames and Y haplotypes are expected. However, caution should be applied because, in reality, men from different paternal families with different Y-chromosomes can have the same surname. Regardless, the combined use of Y-chromosome data and surname information can be valuable and highly informative in many types of forensic, historical, and archaeological investigations (Kayser, 2017).

Summary

Y-chromosome analysis has been used to assist in contemporary forensic, historical, and archaeological investigations for more than two decades. With regard to missing persons and unidentified human remains (UHR) cases, the predominant applications for generating

Y-haplotypes include: (1) to differentiate individuals (i.e., males) for human identification purposes; (2) to trace the male lineage of a profile for kinship associations; and (3) to infer the biogeographic ancestry and ethnicity of an unknown male. In 1998, a minimal Y-STR haplotype was first established in Europe and included nine Y-STR loci (Kayser et al., 1997). Since then, additional Y-STR loci have been incorporated into Y-STR multiplexes to increase the discriminatory power of the generated Y-haplotypes. Among the commercially available Y-STR multiplexes include: (1) the *Yfiler® Plus kit* (Thermo Fisher Scientific, Waltham, Massachusetts, USA), which co-amplifies a set of **27 Y-STR loci**, including 7 Rapidly Mutating Y-STRs **(7 RM Y-STRs)** (Ambers et al., 2018; Andersen et al., 2017; Gopinath et al., 2016; Ottaviani et al., 2015; Pickrahn et al., 2016; Rapone et al., 2016); (2) the *PowerPlex® Y23 kit* (Promega Corporation, Madison, Wisconsin, USA), which co-amplifies a panel of **23 Y-STRs** (Thompson et al., 2012; Thompson et al., 2013); and (3) the *Investigator® Argus Y-28 QS kit* (Qiagen Corporation, Germantown, Maryland, USA), which co-amplifies a total of **27 Y-STRs** (including **6 RM Y-STRs**), plus incorporates a Quality Sensor (QS) system to assess PCR efficiency and the presence of inhibitors (Qiagen, 2021). Although all males of the same patrilineage are expected to share the same Y-STR haplotype, Rapidly Mutating Y-STRs (RM Y-STRs) offer the increased ability to distinguish between close paternal relatives. In addition, Y-STR, Y-SNP, and aiSNP data can be considered in concert for more accurate biogeographic ancestry assessment and to provide higher resolution in ancestry prediction for unknown male human remains.

In addition to DNA testing kits specifically designed for Y-chromosome analysis, several recently developed massively parallel sequencing (MPS) multiplex assays have been introduced to maximize the amount of genetic data that can be generated from a biological sample in a single reaction. Each of these MPS multiplexes includes a set of Y-chromosome DNA markers (i.e., either Y-STRs or Y-SNPs). Among the commercially available MPS multiplex assays are: (1) the *ForenSeq™ DNA Signature Prep Kit* (Verogen, San Diego, California, USA), which co-amplifies **24 Y-chromosome STRs (Y-STRs)**, 27 global autosomal short tandem repeats (auSTRs), 7 X-chromosome STRs (X-STRs), 94 human identity single nucleotide polymorphisms (iSNPs), 22 phenotype-informative SNPs (piSNPs), and 56 biogeographical ancestry informative SNPs (aiSNPs) (Verogen, 2018); (2) the *ForenSeq™ MainstAY Kit*, which co-amplifies **26 Y-chromosome STRs (Y-STRs)** and 27 autosomal STRs (auSTRs) (Verogen, 2021a); (3) the *ForenSeq™ Kintelligence Kit*, which co-amplifies **85 Y-chromosome SNPs (Y-SNPs)** for forensic genetic genealogy investigations, as well as 56 ancestry-informative SNPs (aiSNPs), 94 human identity SNPs, 9867 kinship SNPs, 22 phenotype-informative SNPs (piSNPs), and 106 X-chromosome SNPs (X-SNPs) (Verogen, 2021b).

After Y-chromosome genetic data is obtained from highly decomposed or skeletonized human remains, several Y-DNA-specific databases can be used to assess the informativeness of the Y-profile generated. The Y-chromosome haplotype reference database (YHRD) is the largest and most widely used database for Y-haplotype matching and population frequency estimates. In addition, two open-access online haplogroup prediction tools—Whit Athey's Haplogroup Predictor (http://www.hprg.com/hapest5/index.html) and the NevGen Y-Haplogroup Predictor (YHP) (https://www.nevgen.org/)—can be used to predict the Y haplogroup for an unknown sample (Babic Jordamović et al., 2021; Dogan et al., 2016; Cetkovic Gentula and Nevski, 2015).

References

Adnan, A., Ralf, A., Rakha, A., Kousouri, N., Kayser, M., 2016. Improving empirical evidence on differentiating closely related men with RM Y-STRs: a comprehensive pedigree study from Pakistan. Forensic Sci. Int. Genet. 25, 45–51.

Alghafri, R., Goodwin, W., Ralf, A., Kayser, M., Hadi, S., 2015. A novel multiplex assay for simultaneously analyzing 13 rapidly mutating Y-STRs. Forensic Sci. Int. Genet. 17, 91–98.

Ambers, A., 2016. Identifying 140-year-old skeletal remains using massively parallel DNA sequencing. In: International Symposium for Human Identification (ISHI) News. https://www.ishinews.com/identifying-140-year-old-remains-using-massively-parallel-dna-sequencing/.

Ambers, A., 2021. Forensic genetic investigation of two adult male skeletons recovered from the 17th-century *La Belle* shipwreck using massively parallel sequencing. News from the World of DNA Forensics, The ISHI Report. https://promega.foleon.com/theishireport/november-2020/modern-dna-technology-for-historical-cold-cases-forensic-genetic-investigation-of-two-adult-male-skeletons-recovered-from-the-17th-century-la-belle-shipwreck-using-massively-parallel-sequencing-mps/.

Ambers, A., Gill-King, H., Dirkmaat, D., Benjamin, R., King, J., Budowle, B., 2014. Autosomal and Y-STR analysis of degraded DNA from the 120-year-old skeletal remains of Ezekiel Harper. Forensic Sci. Int. Genet. 9, 33–41.

Ambers, A.D., Churchill, J.D., King, J.L., Stoljarova, M., Gill-King, H., Assidi, M., Abu-Elmagd, M., Buhmeida, A., Al-Qahtani, M., Budowle, B., 2016. More comprehensive forensic genetic marker analyses for accurate human remains identification using massively parallel DNA sequencing. BMC Genomics 17, 750.

Ambers, A., Votrubova, J., Vanek, D., Sajantila, A., Budowle, B., 2018. Improved Y-STR typing for disaster victim identification, missing persons investigations, and historical human skeletal remains. Int. J. Leg. Med. https://doi.org/10.1007/s00414-018-1794-8.

Ambers, A., Bus, M.M., King, J.L., Jones, B., Durst, J., Bruseth, J.E., Gill-King, H., Budowle, B., 2020. Forensic genetic investigation of human skeletal remains recovered from the *La Belle* shipwreck. Forensic Sci. Int. 306, 110050.

Andersen, M.M., Mogensen, H.S., Eriksen, P.S., Morling, N., 2017. Yfiler® plus population samples and dilution series: stutters, analytical thresholds, and drop-out probabilities. Int. J. Leg. Med. 131, 1503–1511.

Babic Jordamović, N., Kojović, T., Dogan, S., Bešić, L., Salihefendić, L., Konjhodžić, R., Škaro, V., Projić, P., Hadžiavdić, V., Ašić, A., Marjanović, D., 2021. Haplogroup prediction using Y-chromosomal short tandem repeats in the general population of Bosnia and Herzegovina. Front. Genet. 12. https://doi.org/10.3389/fgene.2021.671467, 671467.

Ballantyne, K.N., Goedbloed, M., Fang, R., Schaap, O., Lao, O., et al., 2010. Mutability of Y-chromosome microsatellites: rates, characteristics, molecular bases, and forensic implications. Am. J. Hum. Genet. 87, 341–353.

Ballantyne, K.N., Keerl, V., Wollstein, A., Choi, Y., Zuniga, S.B., Ralf, A., et al., 2012. A new future of forensic Y-chromosome analysis: rapidly mutating STRs for differentiating male relatives and paternal lineages. Forensic Sci. Int. Genet. 6 (2), 208–218.

Ballantyne, K.N., Ralf, A., Aboukhalid, R., et al., 2014. Toward male individualization with rapidly mutating Y-chromosomal short tandem repeats. Hum. Mutat. 35 (8), 1021–1032.

Butler, J.M., Kline, M.C., Decker, A.E., 2008. Addressing Y-chromosome short tandem repeat allele nomenclature. J. Genetic Geneal. 4 (2), 125–148.

Calafell, F., Larmuseau, M.H.D., 2017. The Y chromosome as the most popular genetic marker in genetic genealogy benefits interdisciplinary research. Hum. Genet. 136, 559–573.

Cetkovic Gentula M., Nevski A., 2015. NevGen Y-DNA Haplogroup predictor. Serbian DNA Project. https://dnk.poreklo.rs/DNK-projekat/.

Coble, M.D., Loreille, O.M., Wadhams, M.J., Edson, S.M., Maynard, K., Meyer, C.E., Niederstätter, H., Berger, C., Berger, B., Falsetti, A.B., Gill, P., Parson, W., Finelli, L.N., 2009. Mystery solved: the identification of the two missing Romanov children using DNA analysis. PLoS One 4 (3), e4838.

Crubézy, E., Amory, S., Keyser, C., Bouakaze, C., Bodner, M., Gibert, M., Röck, A., Parson, W., Alexeev, A., Ludes, B., 2010. Human evolution in Siberia: from frozen bodies to ancient DNA. BMC Evol. Biol. 10, 25.

de Kniff, P., 2000. Y-chromosomes shared by descent or state. In: Renfrew, C., Boyle, K. (Eds.), Archaeogenetics: DNA and the Population Pre-History of Europe. Cambridge, England, The McDonald Institute.

Dogan, S., Babic, N., Gurkan, C., Goksu, A., Marjanović, D., Hadziavdic, V., 2016. Y-chromosomal haplogroup distribution in the Tuzla Canton of Bosnia and Herzegovina: a concordance study using four different in silico assignment algorithms based on Y-STR data. Homo 67, 471–483.

Fitzpatrick C, 2016. Identifinders International used forensic genetic genealogy to solve the 1992–1993 Phoenix Canal Murders. https://identifinders.com/case-study-phoenix-canal-killer/.

Fitzpatrick C, 2011. The 1991 Sarah Yarborough homicide: the first case ever to use genetic genealogy to generate investigative leads in a cold case. https://identifinders.com/first-genetic-genealogy-leads-cold-case/.

Foster, E.A., Jobling, M.A., Taylor, P.G., Donnelly, P., de Kniff, P., Mieremet, R., et al., 1998. Jefferson fathered slave's last child. Nature 396, 27–28.

Foster, E.A., Jobling, M.A., Taylor, P.G., Donnelly, P., de Kniff, P., Mieremet, R., et al., 1999. The Thomas Jefferson paternity case—reply. Nature 397, 32.

Gill, P., Brenner, C., Brinkmann, B., Budowle, B., Carracedo, A., Jobling, M.A., de Kniff, P., Kayser, M., Krawczak, M., Mayr, W.R., Morling, N., Olaisen, B., Pascali, V., Prinz, M., Roewer, L., Schneider, P.M., Sajantila, A., Tyler-Smith, C., 2001. DNA Commission of the International Society of forensic genetics: recommendations on forensic analysis using Y-chromosome STRs. Forensic Sci. Int. 124, 5–10.

Gojanovic, M.D., Sutlovic, D., 2007. Skeletal remains from world war II mass grave: from discovery to identification. Croat. Med. J. 48 (4), 520–527.

Gopinath, S., Zhong, C., Nguyen, V., Ge, J., Lagace, R., Short, M.L., Mulero, J.J., 2016. Developmental Validation of the Yfiler® Plus PCR Amplification Kit: An Enhanced Y-STR Multiplex for Casework and Database Applications.

Gusmao, L., Butler, J.M., Carracedo, A., Gill, P., Kayser, M., Mayr, W.R., Morling, N., Prinz, M., Roewer, L., Tyler-Smith, C., Schneider, P.M., 2006. DNA Commission of the International Society of forensic genetics (ISFG): an update of the recommendations on the use of Y-STRs in forensic analysis. Forensic Sci. Int. 157, 187–197.

Haas, C., Shved, N., Rühli, F.J., Papageorgopoulou, C., Purps, J., Geppert, M., Willuweit, S., Roewer, L., Krawczake, M., 2013. Y-chromosomal analysis identifies the skeletal remains of Swiss national hero Jörg Jenatsch (1596–1639). Forensic Sci. Int. Genet. 7 (6), 610–617.

Hares, D., 2015. Selection and implementation of expanded CODIS core loci in the United States. Forensic Sci. Int. Genet. 17, 33–34.

Irwin, J.A., Edson, S.M., Loreille, O., Just, R.S., Barritt, S.M., Lee, D.A., Holland, T.D., Parsons, T.J., Leney, M.D., 2007. DNA identification of "earthquake McGoon" 50 years postmortem. J. Forensic Sci. 52 (5), 1115–1118.

Jäger, A.C., Alvarez, M.L., Davis, C.P., Guzmán, E., Han, Y., Way, L., Walichiewicz, P., et al., 2017. Developmental validation of the MiSeq FGx Forensic Genomics System for targeted next generation sequencing in forensic DNA casework and database laboratories. Forensic Sci. Int. Genet. 28, 52–70. https://doi.org/10.1016/j.fsigen.2017.01.011.

Kayser, M., 2017. Forensic use of Y-chromosome DNA: a general overview. Hum. Genet. 136, 621–635.

Kayser, M., Caglia, A., Corach, D., Fretwell, N., Gehrig, C., Graziosi, G., Heidorn, F., Hermann, S., Herzog, B., Hidding, M., Honda, K., Jobling, M., Krawczak, M., Leim, K., et al., 1997. Evaluation of Y-chromosomal STRs: a multicenter study. Int. J. Leg. Med. 110, 125–133.

Kemp, B.M., Malhi, R.S., McDonough, J., Bolnick, D.A., Eshleman, J.A., Rickards, O., Martinez-Labarga, C., Johnson, J.R., Lorenz, J.G., Dixon, E.J., Fifield, T.E., Heaton, T.H., Worl, R., Smith, D.G., 2007. Genetic analysis of early Holocene skeletal remains from Alaska and its implications for the settlement of the Americas. Am. J. Biol. Anthropol. 132 (4), 605–621.

King, T.E., Jobling, M.A., 2009. What's in a name? Y chromosomes, surnames, and the genetic genealogy revolution. Trends Genet. 25, 351–360.

King, T., Fortes, G., Balaresque, P., Thomas, M.G., Balding, D., Delser, P.M., Neumann, R., Parson, W., Knapp, M., Walsh, S., Tonasso, L., Holt, J., Kayser, M., Appleby, J., Forster, P., Ekserdjian, D., Hofreiter, M., Schürer, K., 2014. Identification of the remains of King Richard III. Nat. Commun. 5, 5631.

Lee, H.Y., Kim, N.Y., Park, M.J., Sim, J.E., Yang, W.I., Shin, K., 2010. DNA typing for the identification of old skeletal remains from Korean war victims. J. Forensic Sci. 55 (6), 1422–1429.

Liu, H., Li, X., Mulero, J., Carbanaro, A., Short, M., Ge, J., 2016. A convenient guideline to determine if two Y-STR profiles are from the same lineage. Electrophoresis 37, 1659–1668.

Loreille, O.M., Parr, R.L., McGregor, K.A., Fitzpatrick, C.M., Lyon, C., Yang, D.Y., Speller, C.F., Grimm, M.R., Grimm, M.J., Irwin, J.A., Robinson, E.M., 2010. Integrated DNA and fingerprint analyses in the identification of 60-year-old mummified human remains discovered in an Alaskan glacier. J. Forensic Sci. 55 (3), 813–818.

Marjanović, D., Durmić-Pašić, A., Jasna Avdić, L., Džehverović, M., Haverić, S., Ramić, J., Bilela, L.L., Škaro, V., Projić, P., Bajrović, K., Drobnič, K., Davoren, J., Primorac, D., 2009. Identification of skeletal remains of communist armed forces victims during and after World War II: combined Y-chromosome short tandem repeat (STR) and Mini-STR approach. Croat. Med. J. 50 (3), 296–304.

Olasz, J., Seidenberg, V., Hummel, S., Szentirmay, Z., Szabados, G., Melegh, B., Kásler, M., 2019. DNA profiling of Hungarian King Béla III and other skeletal remains originating from the Royal Basilica of Székesfehérvár. Archaeol. Anthropol. Sci. 11, 1345–1357.

Ottaviani, E., Vernarecci, S., Asili, P., Agostino, A., Montagna, P., 2015. Preliminary assessment of the prototype Yfiler® plus kit in a population study of northern Italian males. Int. J. Leg. Med. 129, 729–730.

Parsons, T.J., Huel, R.M.L., Bajunović, Z., Rizvić, A., 2019. Large scale DNA identification: the ICMP experience. Forensic Sci. Int. Genet. 38, 236–244.

Pickrahn, I., Muller, E., Zahrer, W., Dunkelmann, B., Cemper-Kiesslich, J., Kreindi, G., Neuhuber, 2016. Yfiler® plus amplification kit validation and calculation of forensic parameters for two Austrian populations. Forensic Sci. Int. Genet. 21, 90–94.

Qiagen, 2021. Qiagen Investigator® Argus Y-28 QS Kit Technical Data. https://www.qiagen.com/us/products/human-id-and-forensics/investigator-solutions/investigator-argus-y-28-qs-kit/.

Ralf, A., van Oven, M., Zhong, K., Kayser, M., 2015. Simultaneous analysis of hundreds of Y-chromosomal SNPs for high-resolution paternal lineage classification using targeted semiconductor sequencing. Hum. Mutat. 36, 151–159.

Rapone, C., D'Atanasio, E., Agostino, A., Mariano, M., Papaluca, M.T., Cruciani, F., Berti, A., 2016. Forensic genetic value of a 27 Y-STR loci multiplex (Yfiler® plus kit) in an Italian population sample. Forensic Sci. Int. Genet. 21, e1–e5.

Roewer, L., 2019a. Using the YHRD Database for Casework Analysis. The ISHI Report: News from the World of DNA Forensics. https://promega.foleon.com/theishireport/april-2019-final/using-the-yhrd-database-for-casework-analysis/.

Roewer, L., 2019b. Y-chromosome short tandem repeats in forensics—sexing, profiling, and matching male DNA. WIREs. Forensic Sci. e1336. https://doi.org/10.1002/wfs2.1336.

Roewer, L., Andersen, M.M., Ballantyne, J., Butler, J.M., Caliebe, A., Corach, D., D'Amato, M.E., Gusmao, L., Hou, Y., de Kniff, P., Parson, W., Prinz, M., Schneider, P.M., Taylor, D., Venneman, M., Willuweit, S., 2020. DNA commission of the International Society of Forensic Genetics (ISFG): recommendations on the interpretation of Y-STR results in forensic analysis. Forensic Sci. Int. Genet. 48, 102308.

Scientific Working Group on DNA Analysis Methods (SWGDAM), 2014. SWGDAM interpretation guidelines for Y-chromosome STR typing by forensic DNA laboratories. https://www.swgdam.org/publications.

Skaletsky, H., Kuroda-Kawaguchi, T., Minx, P.J., Cordum, H.S., Hillier, L., Brown, L.G., et al., 2003. The male-specific region of the human Y chromosome is a mosaic of discrete sequence classes. Nature 423, 825–837.

Thompson, J.M., et al., 2012. The PowerPlex® Y23 system: a new Y-STR multiplex for casework and database applications. https://www.promega.com/resources/profiles-in-dna/2012/the-powerplex-y23-system-a-new-y-str-multiplex-for-casework-and-database-applications/.

Thompson, J.M., et al., 2013. Developmental validation of the PowerPlex® Y23 system: a single multiplex Y-STR analysis system for casework and database samples. Forensic Sci. Int. Genet. 7 (2), 240–250.

Tillman, A., Sjölund, P., Lundqvist, B., Klippmark, T., Älgenäs, C., Green, H., 2020. Whole genome sequencing of human remains to enable genealogy DNA database searches—a case report. Forensic Sci. Int. Genet. 46, 102233. https://doi.org/10.1016/j.fsigen.2020.102233.

Vanek, D., Saskova, L., Koch, H., 2009. Kinship and Y-chromosome analysis of 7th century human remains: novel DNA extraction and typing procedure for ancient material. Croat. Med. J. 50 (3), 286–295.

Verogen, 2018. ForenSeq™ DNA Signature Prep Kit Datasheet, Publication #VD2018002. https://verogen.com/wp-content/uploads/2018/08/ForenSeq-prep-kit-data-sheet-VD2018002.pdf.

Verogen, 2021a. ForenSeq™ MainstAY Kit Datasheet, Document #VD2021055. https://verogen.com/wp-content/uploads/2021/07/ForenSeq-MainstAY-Kit-Datasheet-Document-VD2020055.pdf.

Verogen, 2021b. ForenSeq Kintelligence Kit Datasheet, Document #VD2020054 Rev. A. https://verogen.com/wp-content/uploads/2021/03/forenseq-kintelligence-datasheet-vd2020054-a.pdf.

Willuweit, S., Roewer, L., 2015. The new Y-chromosome haplotype reference database. Forensic Sci. Int. Genet. 15, 43–48.

Xue, Y., Wang, Q., Long, Q., Ng, B.L., Swerdlow, H., Burton, J., Skuce, C., Taylor, R., Abdellah, Z., Zhao, Y., Asan MacArthur, D.G., Quail, M.A., Carter, N.P., Yang, H., Tyler-Smith, C., 2009. Human Y chromosome base substitution mutation rate measured by direct sequencing in a deep-rooting pedigree. Curr. Biol. 19, 1453–1457.

Zupanič Pajnič, I., Gornjak Pogorelc, B., Balažic, J., 2010. Molecular genetic identification of skeletal remains from the second world war Konfin I mass grave in Slovenia. Int. J. Leg. Med. 124, 307–317.

Further reading

Alghafri, R., Zupanic Pajnic, I., Zupanc, T., Balazic, J., Shrivastava, P., 2018. Rapidly mutating Y-STR analyses of compromised forensic samples. Int. J. Leg. Med. 132, 397–403.

Butler, J.M., 2003. Recent developments in Y short tandem repeat and Y single nucleotide polymorphism analysis. Forensic Sci. Rev. 15 (2), 91–111.

Gayden, T., Bukhari, A., Chennakrishnaiah, S., Stojkovic, O., Herrera, R.J., 2012. Y-chromosomal microsatellite diversity in three culturally defined regions of historical Tibet. Forensic Sci. Int. Genet. 6, 437–446.

Khan, K., Siddiqi, M.H., Abbas, M., Almas, M., Idrees, M., 2017. Forensic applications of Y chromosomal properties. Leg. Med., 86–91.

Purps, J., Siegert, S., Willuweit, S., Nagy, M., Alves, C., et al., 2014. A global analysis of Y-chromosomal haplotype diversity for 23 Y-STR loci. Forensic Sci. Int. Genet. 12, 12–23.

Roewer, L., 2009. Y-chromosome STR typing in crime casework. Forensic Sci. Med. Pathol. 5, 77–84.

Song, M., Song, F., Zhao, C., Hou, Y., 2021. YHP: Y-chromosome Haplogroup predictor for predicting male lineages based on Y-STRs. Cold Spring Harbor Laboratory bioRxiv. https://doi.org/10.1101/2021.01.11.426186.

CHAPTER

10

Mitochondrial DNA and its use in the forensic analysis of skeletal material

Brandon Letts PhD

Federal Bureau of Investigation (FBI) Laboratory, Quantico, VA, United States

Introduction

Skeletal material from unidentified human remains can be poorly preserved and quite degraded, making nuclear DNA analysis difficult. For this reason, the analysis of mitochondrial DNA (mtDNA), which is found in a much higher concentration than its nuclear DNA (nDNA) counterpart, has become commonplace when working with calcified tissue. mtDNA is often available for extraction and analysis when nuclear DNA has degraded beyond the reach of current forensic STR (short tandem repeat) approaches, the latter of which generally utilize polymerase chain reaction (PCR) technology and require extracted DNA fragments of 150 bases or more. While not unique to an individual like nuclear DNA, most mtDNA sequences (i.e., haplotypes) are still uncommon in a given population, and therefore, can provide valuable evidence to assist in identifying skeletal remains.

Cellular location of mtDNA

mtDNA is a relatively short, circular DNA molecule located within the mitochondria of eukaryotic cells. Mitochondria are the cytoplasmic organelles responsible for energy production, and their inner membrane is the site of the electron transport chain, which produces adenosine triphosphate (ATP) molecules required by the body for numerous energetic processes, including muscle contraction.

The central compartment of a mitochondrion is called the lumen, and it is here that the mtDNA molecule is found in numbers ranging from tens to hundreds, typically depending on the energy requirements of the cell. In addition, cells requiring high levels of energy typically possess numerous mitochondria, multiplying the number of mtDNA genomes present even further. For example, cardiac muscle cells (which have massive energy requirements) possess

Forensic Genetic Approaches for Identification of Human Skeletal Remains
https://doi.org/10.1016/B978-0-12-815766-4.00010-8

on the order of 7000–10,000 mtDNA copies per cell, while skin cells (with few requirements in the way of energy) contain only about 120 mtDNA molecules per cell (Naviaux, 2008; Pohjoismäki et al., 2010). Erythrocytes (red blood cells) are the only cells that lack mitochondria, though blood is still an excellent source of mtDNA due to the presence of leukocytes (white blood cells).

When viewed under a microscope, mtDNA is found in concentrated regions called nucleoids, which are complexes consisting of mtDNA molecules and the proteins associated with their replication and transcription. High resolution fluorescence microscopy suggests that each nucleoid contains a single mtDNA molecule, which aligns with the observation that numerous nucleoids are present in a cell (Kukat et al., 2011).

Endosymbiont theory and the origins of mtDNA

Research suggests that eukaryotic cells have not always contained mitochondria. In fact, phylogenetic analyses of mitochondrial rRNA suggest that the mitochondrion is most closely related to α-proteobacteria (Yang et al., 1985; Gray et al., 1999). When cells first developed early in Earth's history, oxygen levels were low and anaerobic respiration was sufficient for energy production. As the oxygen content of the planet's atmosphere increased, some early bacteria developed aerobic respiration through oxidative phosphorylation. According to the generally accepted *endosymbiont origin theory*, approximately 1.5 billion to 2 billion years ago an archebacterial cell (i.e., proto-eukaryote) ingested or was predatorily invaded by an aerobic eubacterium, and subsequently, benefited from the energy it produced through respiration (Gray et al., 1999; Davidov et al., 2006). This association became a symbiotic relationship, in which the archebacterial proto-eukaryote provided nutrients and the aerobic eubacterium provided energy through respiration in the increasingly oxygen-rich environment. Over time, the endosymbiont became increasingly dependent on the proto-eukaryotic cell for redundant functions such as cellular repair and gene regulation and began to lose parts of its own genome, including through transfer to the (nuclear) genome of the host cell. Eventually, enough genetic material was lost and the bacterium ceased to be a separate organism and became what it is today—a cytoplasmic organelle with remnants of its original, circular bacterial genome.

Substantial evidence exists to support endosymbiotic theory. First and foremost, genetic analysis has established the close relationship between the mitochondrial genome and the α-proteobacterial genome (particularly that of *Rickettsia*), showing that they shared a common ancestor in the distant past (Andersson et al., 1998). Due in part to its bacterial origin, cell-free mtDNA circulating in the blood has been shown to trigger an immune response, adding to the body's defense after a physical injury (Zhang et al., 2010). Interestingly, many insects alive today contain endosymbiotic bacteria that confer advantages to the host. For example, weevil cells contain mitochondria; however, they also contain a bacterial endosymbiont termed *Sitophilus oryzae* principal endosymbiont (SOPE), which supplies high levels of pantothenic acid and riboflavin, increasing rates of oxidative phosphorylation in the mitochondria and potentially making flight more sustainable (Heddi et al., 1999). These endosymbionts are confined to the cells, divide through binary fission, and are transferred maternally, reflecting the early history of the organism that became the mitochondrion.

Structure of mtDNA

While the majority of the genes required to build a mitochondrion have been transferred over time to the nuclear genome, some are still found within a circular DNA molecule within the mitochondrion itself. These genes typically are the ones which code for proteins important to oxidative phosphorylation (particularly the electron transport chain). These proteins are under constant regulation and must be readily supplied to the inner membrane of the mitochondrion when energetic stress is experienced. This rapid response and its associated regulation necessitate a close proximity of the DNA coding for the involved proteins and maintain the evolutionary pressure that prevents the mtDNA genome from being lost completely.

The mtDNA molecule, or mitochondrial genome, varies greatly in size and structure between species. In humans, the mtDNA genome is ~ 16,569 bases long, is double-stranded, and includes 37 genes (Fig. 1). Fourteen of these genes code for proteins, including 13 found within the electron transport chain. The remainder code for various functional RNAs, such as ribosomal RNA (rRNA) and transfer RNA (tRNA), which are needed to build the encoded proteins. Interestingly, some genes in the mitochondrial genome overlap and are found on

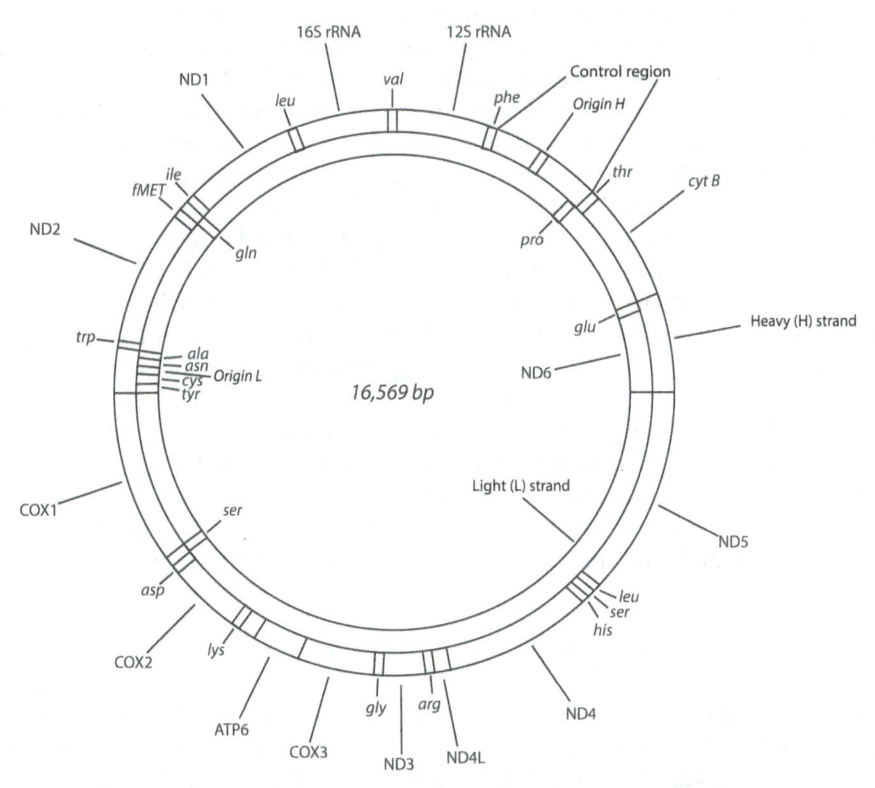

FIG. 1 The human mitochondrial genome is approximately 16,569 bases long and contains 37 genes dispersed across a heavy (H) strand and a light (L) strand. Fourteen of the genes code for proteins, 13 of which are found within the electron transport chain. The rest of the genes code for transfer RNA (tRNA) and ribosomal RNA (rRNA). The control region (CR), which surrounds the origin of replication on the H strand, is targeted for forensic analysis.

opposite DNA strands, known as the light (L) strand and heavy (H) strand (named according to their differential base compositions).

Forensic analysis currently focuses on a region of the mtDNA genome called the control region (CR). This region, which is approximately 1100 bases long, contains the sequence required for replication of the genome, including the "origin of replication" and its associated promoters. Because no proteins are coded for by the control region, it is less constrained by evolutionary pressure and mutates on average at a rate about 10 times higher than the rest of the mtDNA genome. There is, however, evidence that the binding sites for regulatory proteins are maintained (Gilbert et al., 2003). This may explain why, within the readily mutating control region, there are areas which mutate more quickly than others. The most quickly mutating regions are referred to as hypervariable (HV) regions and are of particular forensic interest because they are the most useful for differentiating between individuals. There are two main hypervariable regions, termed hypervariable region 1 (HV1), which consists of bases 16,024–16,383 of the CR, and hypervariable region 2 (HV2), which consists of bases 57–372. Additionally, there is a smaller region called hypervariable region 3 (HV3) at base positions 438–574 (van Oven and Kayser, 2009). The amplification strategy employed in any given case centers around these hypervariable regions (in order to obtain as much discrimination power as possible) and is discussed in detail in the next few paragraphs.

mtDNA can recombine with the nuclear genome, and such past recombination events have resulted in what are called pseudogenes, or nuclear mitochondrial insertions (NUMTs). These (theoretically) useless segments of DNA are pieces of the mitochondrial genome that have been copied into the nuclear genome and are found scattered across all chromosomes (Dayama et al., 2014). Because NUMTs are no longer subjected to the proofreading mechanisms of the mtDNA genome and are not maintained as genes by nDNA, they have a higher mutation rate and can often be identified in this way. For example, unexpected insertions and deletions (indels) in a generated sequence are a good clue that a NUMT has been inadvertently amplified.

Because they are part of the nuclear genome, NUMTs are more likely to be encountered in samples with high levels of high molecular weight (HMW) nuclear DNA. Degraded skeletal material is not typically a good source of high molecular weight nuclear DNA, and therefore, NUMTs are not as commonly observed when analyzing bones compared to blood samples or other body fluids or tissues. However, fresh or well-preserved bones can certainly contain high enough levels of high-quality nuclear DNA to yield NUMTs during analysis. The best way to avoid inadvertent amplification of a NUMT is through proper primer design during developmental validation and by targeting the largest amplicon possible for any particular sample. This helps to select for the more readily available and (on average) longer fragments of mtDNA during PCR.

Inheritance patterns of mtDNA

Mitochondria are directly transferred from the mother during cell division, beginning with female gametogenesis and ovum formation. mtDNA, therefore, follows a maternally related mode of inheritance, meaning that mtDNA is passed directly from a mother to her children, and only her female children will pass this mtDNA to their children. Because sons cannot pass their mtDNA to their children, a mtDNA lineage is broken when a woman has no daughters.

The biological mechanism for maternal inheritance of the mitochondria is incredibly efficient, with only one confirmed example of paternal inheritance of mtDNA in humans. In this example, the individual (who displayed severe muscle myopathy) possessed paternal mtDNA in his muscle cells and maternal mtDNA in all other tissues (Schwartz and Vissing, 2002). Three other potential examples of suspected biparental inheritance were uncovered through deep sequencing of the mtDNA of individuals suspected of possessing mitochondrial disease. In these instances, the pattern was multigenerational and appeared to originate from females who had inherited mtDNA from both parents and then passed the heteroplasmic mtDNA to their offspring (see section later in this chapter for a detailed description and definition of heteroplasmy) (Luo et al., 2018). However, the conclusion of paternal inheritance in these three examples has proved controversial. Lutz-Bonengel and Parson (2019) have suggested that the results may have instead been due to the unintentional co-amplification and sequencing of NUMTs.

In all but the rarest cases, when a spermatozoa (sperm cell) penetrates an ovum (egg cell), its own mitochondria are immediately targeted for destruction by a chemical modification called ubiquitination. Research has shown that male mitochondria are eliminated within 48 hours of fertilization (or two cell divisions), and that the paternal mtDNA is destroyed even before the mitochondria themselves (Nishimura et al., 2006). The purpose of this targeted destruction remains a topic of research.

Because paternal mtDNA is destroyed upon introduction to the ovum, recombination between maternal and paternal mtDNA does not occur. While recombination between mtDNA molecules may occur, any signal of such an event would be effectively undetectable, as all molecules share the same DNA sequence (with the exception of heteroplasmy) (Kraytsberg et al., 2004). The lack of recombination and uniparental mode of inheritance means that mtDNA can be traced in an unbroken line across numerous generations, with evidence suggesting an average rate of mutation of one control region (CR) mutation every 33 generations (Parsons et al., 1997). This allows the use of a reference sample from a maternally related individual to be used in place of a direct reference sample when such a sample is not available, as may often be the case for missing persons and unidentified human remains.

The relatively slow rate of mutation of mtDNA (compared to nuclear short tandem repeats, or STRs) permits even a distant relative to serve as a reference or exemplar, as long as that relative is maternally related to the decedent. For example, when the purported remains of England's King Richard III were found under a parking lot in 2012, researchers compared mtDNA recovered from the skeleton to sequences developed from two modern descendants of King Richard's sister who were 19 and 22 generations distant from him, respectively. In one case, the sequence matched, and in the other descendant, only a single nucleotide difference was observed (across the entire mtDNA genome). Given the distance of the relationships, the mtDNA sequences confirmed the hypothesis that the skeleton was that of King Richard (King et al., 2014).

Heteroplasmy in the mtDNA genome

When the cells of a single individual contain two or more different mtDNA sequences, the individual is referred to as *heteroplasmic*. While heteroplasmy was once thought rare, increasingly sensitive detection methods are demonstrating that most, if not all, individuals are heteroplasmic to some degree (Dayama et al., 2014). There are two types of heteroplasmy: (1) point heteroplasmy and (2) length heteroplasmy.

Point heteroplasmy refers to a situation in which a person possesses mitochondrial genomes differing in sequence at a particular base. In practice, such a position will appear in the electropherogram as a mixed base (Fig. 2). Irwin et al. (2009) analyzed the control region of 5015 individuals and found point heteroplasmy in 310 samples (approximately 6% of the individuals tested). Of these, 302 individuals showed heteroplasmy at one position, 7 individuals possessed heteroplasmy at 2 positions, and 1 person was heteroplasmic at three positions. The frequency of point heteroplasmy across the entire mtDNA genome is not as well-established; however, Li et al. (2010) showed that when using a 10% variant threshold with Next-Generation Sequencing (NGS) methods, 25% of the 131 European individuals tested in their study were heteroplasmic, with five of the individuals possessing two points of heteroplasmy each. Furthermore, 13 of the 34 observed heteroplasmic positions were located in the control region, demonstrating the more frequent occurrence in the control region compared to the much larger (and more regulated) coding region. A more recent study of 1526 mother-offspring pairs found that 45.1% of individuals were heteroplasmic above a 1% detection threshold (Wei et al., 2019). The transmission of point heteroplasmy is typically reduced between generations by the mtDNA bottleneck that occurs during oogenesis and early embryonic development (Lee et al., 2012). This ensures that the embryo contains a single predominant mtDNA sequence. High levels of heteroplasmy are deleterious to the health of an individual; hence, the bottleneck is an important step in development.

Length heteroplasmy refers to the presence of two or more mtDNA genomes differing in length within an individual and results from polymerase slippage during the replication of homopolymeric regions such as the HV1 and HV2 C-stretches (Fig. 3). It can also occur at the AC repeat beginning at position 515. While length heteroplasmy occurs naturally, it also results in vitro from slippage during PCR, complicating its utility in forensics (Seo et al., 2010).

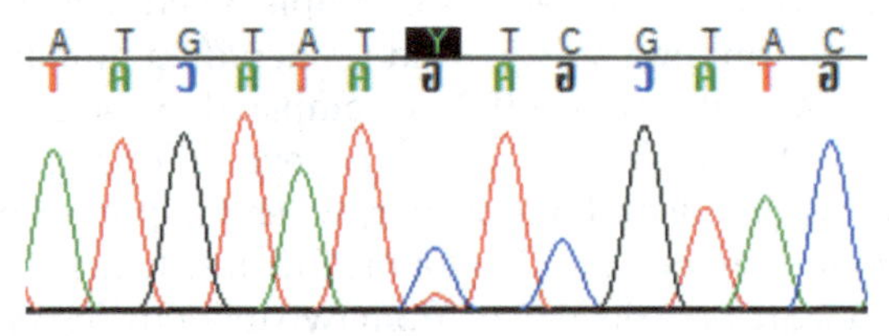

FIG. 2 Point heteroplasmy appears in sequencing data as a mixed base. In this case, at the highlighted position, the individual possesses both a cytosine (C) and a thymine (T) base.

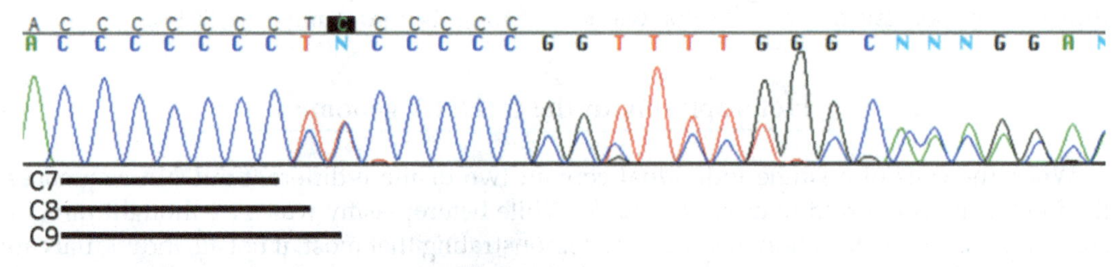

FIG. 3 Length heteroplasmy appears in sequencing data as overlapping sequences that are offset by one or more bases following a homonucleotide tract. In this case, which illustrates a forward sequencing reaction of the HV2 C-stretch, three length variants differing by one nucleotide each (C7, C8, C9) are present in the first tract of cytosines. In this example, the C9 variant is barely detectable.

The use of mtDNA in forensic casework

Sample processing in the laboratory

mtDNA processing is extremely sensitive due to the high copy number of mtDNA in a sample, especially when compared to nDNA. For example, if an errant epithelial (skin) cell finds its way into a PCR tube during setup, it will not only introduce 2 copies of nuclear DNA, but also approximately 120 copies of mtDNA (Naviaux, 2008). For this reason, extra precautions above and beyond those typically used for nuclear DNA processing are needed when working with mtDNA.

The most important safeguard when processing mtDNA is unidirectional movement. As a rule of thumb, a mtDNA technician should always move from areas of low quantity DNA (e.g., extraction areas) to areas of high quantity DNA (e.g., postamplification rooms). PCR product on skin and clothes can easily find its way into an extract and overwhelm the endogenous, often degraded, template. If a technician must return to a preamplification space after spending time in a postamplification area, a shower and change of clothes are necessary.

To reduce the risk of contaminating a sample, all individuals who enter a mtDNA preamplification laboratory should wear appropriate personal protective equipment (PPE), including gloves, a mask, and a clean lab coat (even when not directly involved in sample processing). For exceptionally degraded specimens, disposable paper sleeves are an extra layer of security against contamination. During bone cutting and powdering, a disposable lab coat can be used to prevent accidental wearing of a dirty lab coat that was used in processing a previous specimen. All bone processing should be conducted in a hood equipped with HEPA filtration to trap fine powder. A 1:10 bleach dilution (made fresh daily) should be used to decontaminate all surfaces. When possible, it is good practice to UV-irradiate plasticware and any solutions that do not contain enzymes. This crosslinks any contaminant or exogenous DNA, preventing its downstream amplification. For additional information regarding facility and workflow considerations, refer to Chapter 3 of this book.

Bone surfaces should be decontaminated to the fullest possible extent before powdering, in order to eliminate any exogenous human DNA and to eliminate bacterial or fungal DNA from the soil in which the bone was preserved (the latter of which can often amplify and appear as underlying sequence, even with well-designed primers) (Fig. 4). In order to decontaminate a bone, at a minimum the surface should be sanded off with a Dremel tool equipped with a sanding stone/disk before cutting. The bone cutting can then be placed in a 1:10 bleach solution or 5% Tergazyme, and then rinsed with distilled (molecular grade) water. Both steps can be performed in a sonicator for 10–15minutes, if one is available. The cutting is then dried and powdered. Because contamination is difficult to avoid when working with mtDNA, evidence samples should be extracted individually to avoid potential contact, even with aerosolized DNA. Each sample should be extracted with an accompanying reagent blank to monitor the extraction reagents for contamination.

When choosing a specimen for extraction, any bone will yield sufficient DNA for analysis when the skeletal material is well-preserved. Conventional wisdom suggests that when working with weathered material, however, the long bones such as the femur and humerus are ideal (Alonso et al., 2001; Parsons and Weedn, 1997). Several recent studies cast doubt on this (Edson et al., 2004; Mundorff et al., 2009; Mundorff and Davoren, 2014). These studies, which

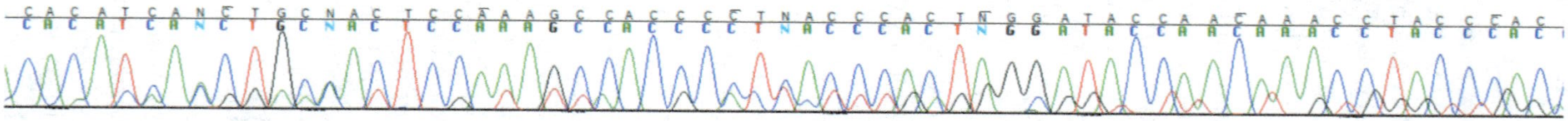

FIG. 4 Example of sequencing data from calcified tissue, showing contamination from fungal DNA as an underlying sequence. Note that, in a mixture of individuals, few positions will be mixed; however, when fungal or bacterial DNA is co-amplified, most positions will appear mixed. Sometimes, as in this case, the contaminating sequence may be clear enough over short stretches that it is possible to manually transcribe and search 20–30 nucleotides using BLAST (Altschul et al., 1990; https://blast.ncbi.nlm.nih.gov/Blast.cgi). Such a search can confirm the contamination to be nonhuman in origin.

compared skeletal elements originating from the same skeletons to reduce sample-dependent variation, found that while femurs provided high levels of DNA, hand and foot bones (the metatarsals in particular) provided the highest DNA yields. If a skull is present, teeth are easy to work with and will typically yield high levels of useable DNA as well. Several studies from the field of ancient DNA have demonstrated that the highest quality DNA in poorly preserved skeletal remains is typically obtained from the petrous area of the temporal bone, which includes and surrounds the ear canal (Gamba et al., 2014; Hansen et al., 2017). The petrous bone is extremely dense, and it is hypothesized that the high density protects the endogenous DNA from degradation. While the petrous bone may contain the highest quality DNA, its removal is labor-intensive, making other skeletal elements such as the metatarsals a better first choice.

mtDNA amplification strategy

Because the analysis of mtDNA in most forensic laboratories involves PCR, the average DNA fragment size in a sample is a critical limiting factor to amplification success. Skeletal material can range in quality from pristine to extremely degraded, depending on age and preservation conditions. Crime laboratories have developed multiple amplification strategies to address this (Wilson et al., 1995a,b; Gabriel et al., 2001; Eichmann and Parson, 2008; Berger and Parson, 2009), with the goal of targeting the largest possible amplicon for: (1) increased processing efficiency and (2) to avoid NUMTs and contamination, both of which are more likely to be a problem with smaller amplicons. Fig. 5 outlines the amplification strategy employed by the FBI Laboratory.

The largest amplicon targeted in forensics is known as the whole control region (WCR) and is approximately 1200 bases in length. Typically, WCR amplifications are used for high-quality reference samples such as blood or buccal swabs. It is also possible to successfully amplify the WCR from well-preserved bones. Depending on the primers used, the WCR generally spans from position 15,998 to position 616. This encompasses HV1 (16,024–16,365) and HV2 (73–340), as well as the smaller HV3 region. The inclusion of additional genetic information from HV3, combined with the intervening region between HV1 and HV2, increases resolution beyond that achieved by the two-region amplification strategy (known as the HV1/HV2 amp).

The HV1/HV2 strategy focuses on the hypervariable regions and uses two separate amplifications of approximately 375 bases in length. This amplification strategy targets the most variable regions of the mtDNA genome, while reducing the amplicon size to a more reasonable length for degraded samples.

For many remains, the DNA is too degraded to target amplicons of even 375 bases. In these cases, an amplification strategy known as HV1A/B, HV2A/B, or the four-region amp, is used. In this strategy, 5 separate amplifications of about 250 bases in length target hypervariable regions 1 and 2, with each of the hypervariable regions covered by 2 overlapping amplicons. The two amplicons for each hypervariable region are designed to overlap so that, when aligned to a reference sequence, the overlapping bases can be compared. This allows the examiner to confirm that both amplicons originated from the same source, and thus, helps to prevent sample switches and assists in recognizing contamination.

The most degraded skeletal samples require even smaller amplification targets. The mini-primer strategy achieves this goal by tiling 5 amplicons of about 90 bases each across hypervariable region 1, which is slightly more variable than hypervariable region 2 (Weitz et al., 1999).

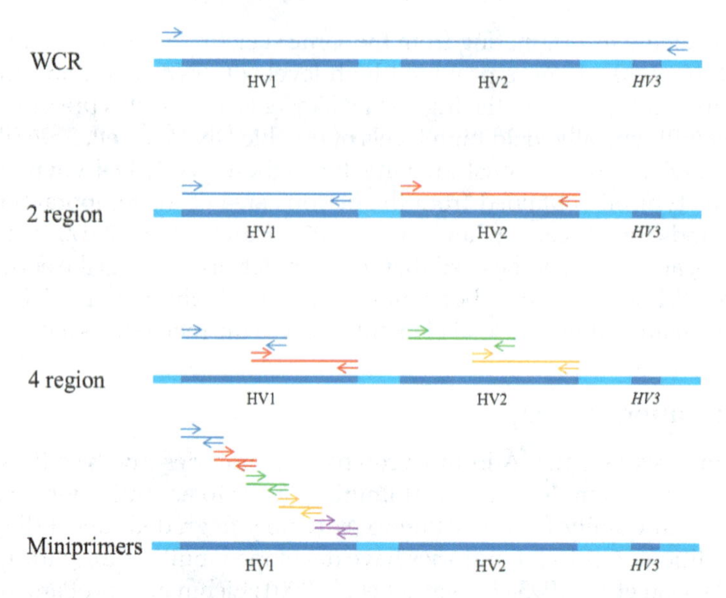

FIG. 5 The four amplification strategies used for mtDNA analysis by the FBI. Whole control region (WCR) amplification targets the entire control region in one amplification, but requires high-quality DNA. As the level of degradation increases, strategies targeting increasingly smaller targets are employed.

Like the four-region amplification strategy, the mini-primer strategy allows the examiner to use overlapping bases to check for contamination.

Although less common with modern dye terminator cycle sequencing kits, sequencing artifacts can occur which may obscure a sequenced nucleotide in an electropherogram (Brandstätter et al., 2005). Furthermore, especially in degraded samples, a noisy baseline can hide low-level heteroplasmy and mixtures, or can make a clear base call difficult. Because of these challenges, it is common practice to sequence both strands of an amplicon. This allows a direct comparison between two different sequencing reactions covering the same nucleotide position and ensures an accurate call.

Data interpretation and comparison

Before a comparison can be made, generated mtDNA sequences must be aligned to a reference sequence. In forensic mtDNA analysis, this reference sequence is the first fully sequenced human mitochondrial genome and is called the Cambridge Reference Sequence (CRS; Anderson et al., 1981). The CRS was revised in 1999, using updated sequencing technology to address several areas outside of the control region in which ambiguous nucleotides had been elucidated based on Hela cell and bovine sequences (Andrews et al., 1999). In total, 11 incorrect bases were identified and corrected. This updated sequence, known as the revised Cambridge Reference Sequence (rCRS), is the current forensic reference sequence for mtDNA. After alignment to the rCRS, the examiner notes differences from the rCRS. In this way, different examiners and even different labs can communicate the complete sequence without having to report all of the sequenced bases, since the vast majority of the sequenced positions will be the same as the rCRS.

Because multiple individuals may share a DNA sequence, a match is treated differently in mtDNA analysis compared to a match in nDNA analysis. When an mtDNA match occurs between the evidence and known reference samples across their common sequence range, a conclusion of *cannot exclude* is appropriate. This means that the source of the reference sample cannot be excluded as the source of the evidence sample. This is equivalent to a statement that the reference can be "included" as a possible source of the evidence. This is because, at best, in the absence of other evidence, any of a group of maternally related individuals could be included as possible sources of the sample.

A single difference between a reference sample and an evidence sample is grounds for an *inconclusive* interpretation. This is because heteroplasmy can result in differences between two samples taken from the same individual. For example, Stewart et al. (2001) showed that separate hairs from a single individual can yield sequences in which heteroplasmic positions (which normally appear as mixed bases) have become fixed for one base or the other, giving each hair a different sequence. Furthermore, because the mitochondrial bottleneck acts to reduce heteroplasmy between generations (by fixing variants), a reference buccal sample obtained from a maternally related individual may contain a single-base difference (or even two) from a sample taken from unidentified human remains, even when the relationship is legitimately that of mother/offspring (Sekiguchi et al., 2004).

While interpretation can vary between laboratories, typically two or more differences between a reference sample and an evidence sample will result in an *exclusion*, where the donor of the reference sample (and his or her maternal relatives) can be excluded as the source of the evidence. This is based on known rates of mutation where it would not be expected for more than one difference to exist between two samples from a single individual (or a missing person) and his or her maternally related family member who donated a reference sample. Some laboratories that deal with closed populations, such as those working to identify victims of a plane crash, may allow two differences when making comparisons to account for differences due to heteroplasmy.

When multiple mixed bases are observed in mtDNA sequence data, the length of the sequence should be considered in order to differentiate between a mixture and multiple sites of heteroplasmy. For example, three heteroplasmic positions across a WCR sequence are rare, but possible. However, three (or even two) heteroplasmic sites within an HV1B sequence (which is much shorter) would be far less likely, and therefore, a mixture should be strongly considered as a possibility. Typically, if a WCR sequence is developed from an evidentiary sample and three mixed bases are observed, the sequence is called a mixture out of an abundance of caution and is re-extracted. Because length heteroplasmy can result from PCR, few (if any) labs consider the absence of shared length variants between evidence and reference samples to be grounds for an exclusion.

A final consideration when working with skeletal material is DNA damage. Cytosine deamination (in which cytosine is deaminated to uracil) is also known as Type II deamination and results in a C to T conversion (or G to A, depending on the strand sequenced) when compared to the original sequence because the uracil is copied as thymine during PCR. The difficulty lies in recognizing a damaged position in sequencing data. Typically, if DNA damage is present, affected positions in a sample will be mixed, with most DNA fragments containing the original cytosine and a few containing the deaminated base (Hofreiter et al., 2001; Dabney et al., 2013) (Fig. 6).

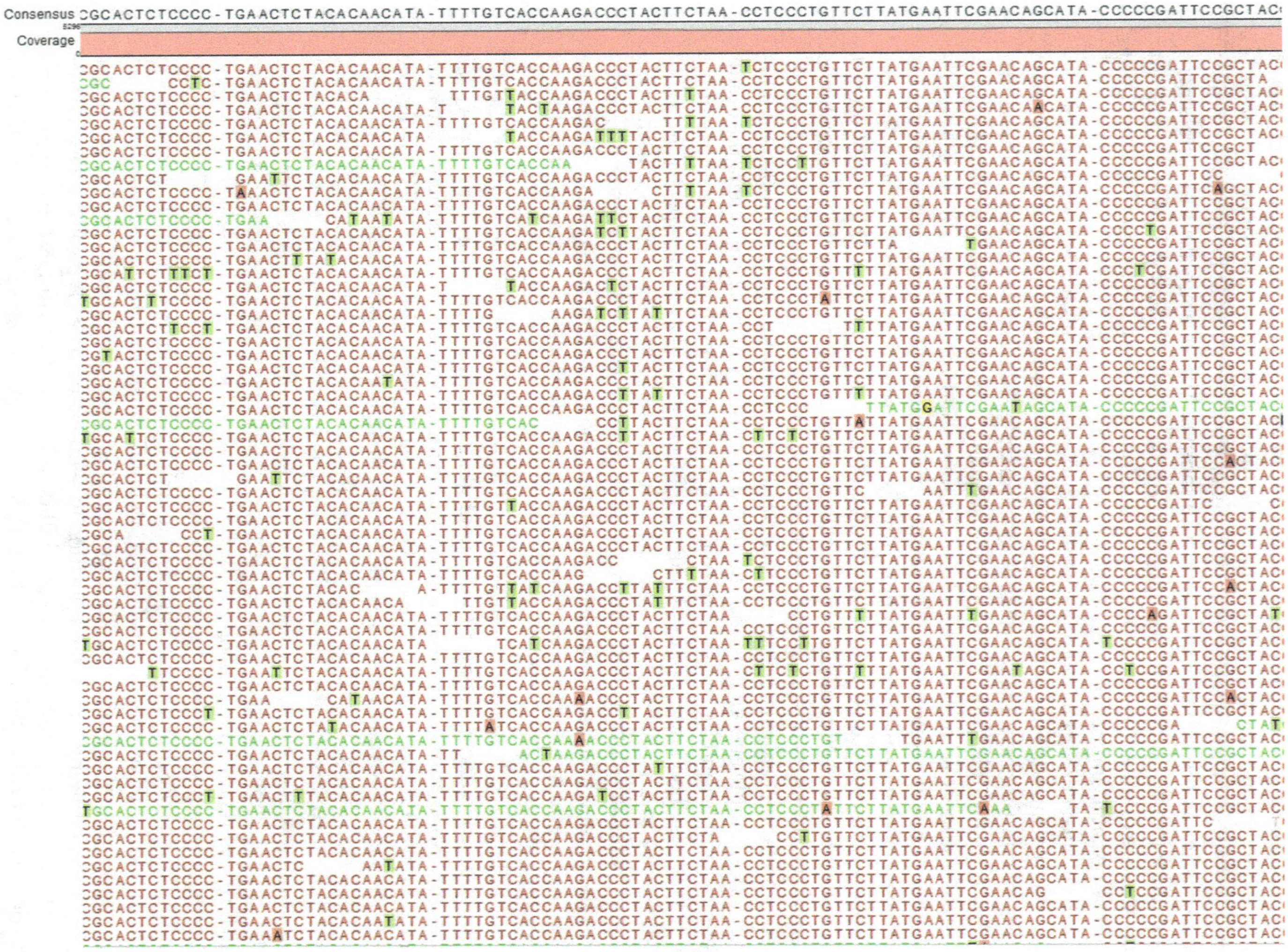

FIG. 6 Example of Type II DNA damage (specifically cytosine deamination). In this screenshot of Next-Generation Sequencing (NGS) data, nucleotides differing from the consensus sequence (top) are highlighted. Particularly noticeable are the thymine (T) bases marked in *green*. While the majority of DNA fragments remain undamaged at any given position, these highlighted thymine (T) bases indicate sites of cytosine deamination. When a specific position has been "hit" by deamination in a large number of extracted DNA fragments, it is possible to see a mixture of thymine and cytosine at this position.

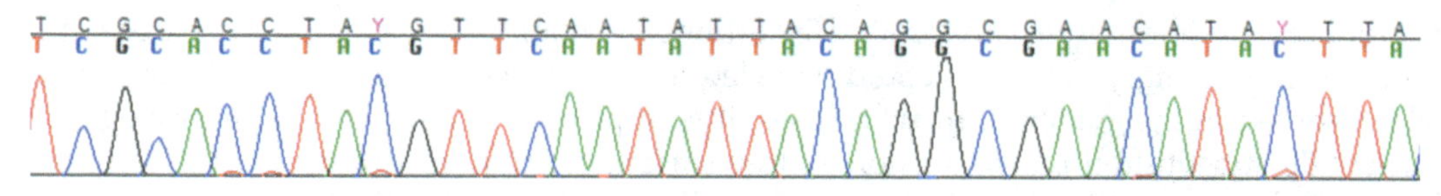

FIG. 7 In Sanger sequencing data, Type II DNA damage appears as a low-level mixture of numerous mixed bases, with a major cytosine (C) component and a minor thymine (T) component.

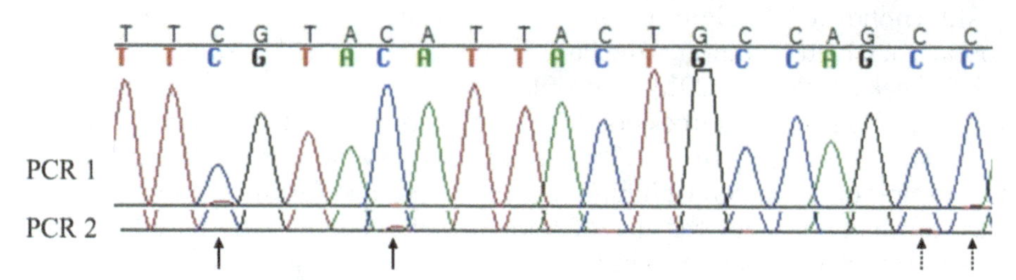

FIG. 8 With Type II DNA damage, the positions of mixed bases will change when the sample is subjected to reamplification.

If the starting copy number during PCR is low enough, damaged positions across an amplicon may appear as mixed C/T bases in an electropherogram, giving the appearance of a mixture of DNA from multiple individuals (Fig. 7). However, performing a second PCR (which begins from different DNA fragments damaged at different positions from the first) will result in a different "mixture," and always of C → T/G → A bases (Fig. 8). On the other hand, performing a second PCR reaction on a true (legitimate) mixed sample will yield the same mixture as that obtained in the first amplification. In this way, an examiner can differentiate between a true mixture of multiple individuals and DNA that has been damaged over time due to exposure to environmental conditions.

mtDNA nomenclature

The collection of nucleotide differences (i.e., polymorphisms) by which an evidence or reference sample differs from the rCRS is called its profile (or *haplotype*), and it is important to note that this collection of polymorphisms is not entirely random. Instead, as mutations are gathered and passed down through lineages over time, patterns are developed and maintained due to the fact that the mitochondrial genome is inherited as a single, nonrecombining unit. These inherited patterns are what we refer to as haplogroups, i.e., sets of shared alleles that groups of related individuals have in common. The evolutionary pathway over which mutations are accumulated is called a *phylogeny*. By sequencing thousands of individuals and creating phylogenetic trees, researchers have been able to assess phylogenetic patterns, and as a result, have identified specific haplogroups and their diagnostic allelic patterns. Typically, a *haplogroup* is defined by one or more "private" mutations which are shared by all individuals with a common ancestry.

The nomenclature of a sample refers to the way its differences from the rCRS are called. For example, if at position 16,162, the rCRS has an adenine (A) and the evidence has a guanine (G) at the same position, the nomenclature for the evidence would be 16162G at that particular position. Sometimes, especially around insertions/deletions (indels), biology discerns that multiple evolutionary pathways could have led to differences between the rCRS and the allele observed in the evidence. In such cases, it is important to consult known phylogenetic nucleotide substitution patterns in order to ensure that the nomenclature of the sample is interpreted correctly (Bandelt and Parson, 2008; Parson et al., 2014). By checking a developed profile against known phylogenetic patterns, the forensic scientist has a method of double-checking their interpretation prior to comparison. The European DNA Profiling (EDNAP) Group's Mitochondrial DNA Population Database (EMPOP) (Parson and Dur, 2007; https://empop.online/), which determines the haplogroup of a searched sample, can be used to perform such a check (Röck et al., 2013). A profile developed from an evidence sample should fit into a known haplogroup. If it does not, there is a good chance that an error was made either during laboratory processing (e.g., in the case of HV1/HV2 amplicons being combined from separate samples) or when determining the sample's nomenclature.

Further information about the use of known phylogenetic patterns when determining a sample's nomenclature can be found in (1) Section 2.3.2 of the Scientific Working Group on DNA Analysis Methods (SWGDAM) Interpretation Guidelines for Mitochondrial DNA Analysis by Forensic DNA Testing Laboratories (2019) and (2) the associated SWGDAM Mitochondrial DNA Nomenclature Examples Document (2014).

Statistical calculations

Unlike nDNA, a mtDNA match can never be used on its own to make an identification. This is because (as mentioned previously) maternal relatives, even distant ones, share an mtDNA sequence with the source of the evidentiary profile. Furthermore, while most haplotypes are rare in any given population, some can be quite common. For example, 6.9% of U.S. Caucasians in the EMPOP database in 2014 shared the same control region sequence (FBI, internal communication, unpublished). For these reasons, it is especially important to add a measure of significance to the conclusion when the sequences developed from evidence and known reference samples match.

Frequency statistics estimate how often the evidentiary profile would be observed in a given population if everyone in that population was sampled. Because this is not possible, databases are designed to create a representative (and random) sampling of a population from which estimates can be made. For mtDNA haplotypes, the most common profiles are likely to be captured without extensive sampling; however, as a database grows, it captures rarer profiles as well. The larger the database becomes, the better the estimate for those rare profiles will be. EMPOP is the leading forensic database for searches of worldwide populations. The Combined DNA Index System (CODIS) software used by forensic laboratories in the United States captures a subset of the EMPOP database and presents the data in a format that is useful to investigators within the United States. Both databases use the Clopper-Pearson equation to provide an upper bound frequency estimate.

Currently, the statistic used in the United States and Europe to estimate the weight of a match in a mtDNA comparison is the upper bound frequency estimate calculated via the Clopper-Pearson equation. This equation is a cumulative binomial distribution formula that solves for p (proportion) through serial iterations and describes an exact interval, as opposed to the approximation described by a normal distribution (Clopper and Pearson, 1934). It is favored in forensic mtDNA analysis because it is both highly conservative and powerful when the number of observations is small. In forensics, the number of observations of a searched mtDNA profile in a database is typically low, and a conservative statistic (lower proportion) will favor the defendant (Melton et al., 2012).

Given the frequency at which an evidentiary profile is observed in a representative database, the Clopper-Pearson equation calculates a confidence interval (CI)—typically a one-tailed 95% CI by convention—to account for sampling bias. The upper bound of this confidence interval gives a conservative estimate for the highest frequency at which the profile would be expected to occur in the population from which the sampling occurred.

Emerging mtDNA methods

As described in this chapter, currently the standard mtDNA methodology employed in forensic labs involves amplification using PCR and sequencing by Sanger terminator cycle sequencing. While these methods are robust and reliable, they are limited by the requirement that extracted DNA fragments must still be large enough to accommodate primer binding sites. In addition, the chemistry behind Sanger sequencing means that peak heights in an electropherogram do not correlate to the amount of DNA present (as observed in the case of STR analysis). For this reason, mixture deconvolution is not possible in mtDNA analysis (Melton et al., 2012).

Next-Generation Sequencing (NGS), also known as high-throughput sequencing and massively parallel sequencing (MPS), is quickly becoming a reality in the field (Børsting and Morling, 2015). This technology carries with it two major advantages over current methodology. First, by incorporating primer binding sites into the adaptors, fragments as small as 30 bases in length can be sequenced. Second, the chemistry which generates the sequences (or reads) works in such a way that the number of reads is proportional to the amount of DNA present; hence, mixture deconvolution is possible. In addition to these two advantages, NGS makes it possible to sequence the entire mitochondrial genome in a single reaction, either through multiplexing or via shotgun sequencing. In short, NGS will allow the forensic laboratory to obtain far more data from less sample.

Another technique gaining traction is known as hybridization capture. This technique uses purposely designed RNA baits to "fish" DNA-of-interest (e.g., mtDNA) out of an extract (Horn, 2012; Carpenter et al., 2013). This approach works by hybridizing single-stranded RNA baits with complementary denatured DNA. The baits are attached to magnetic beads, so that any DNA not captured can be washed away before degrading the RNA and removing the beads. Hybridization capture techniques are game-changing for the analysis of unidentified human remains, particularly because this type of material is often exposed to the environment for long periods of time and can become heavily contaminated with bacterial and fungal DNA. By concentrating the mtDNA in an extract and washing away both bacterial and fungal DNA, the proportion of reads in a shotgun sequencing library representing the DNA-of-interest becomes much higher, which can potentially mean the difference between an unusable sample and high-quality, probative results.

Conclusion

mtDNA analysis does not offer the same power of discrimination that nuclear DNA can provide. However, in forensic casework, skeletonized material is often highly weathered, sometimes resulting in degraded DNA with fragments too short and in too low quantities to allow for nuclear DNA amplification. In such cases, mtDNA provides vital information to the investigator who often has little (if any) other evidence to consider.

Acknowledgments

The author would like to thank Odile Loreille and Lara Adams for their helpful comments in the preparation of this book chapter.

Disclaimer

This is publication number 19-19 of the Laboratory Division of the Federal Bureau of Investigation (FBI). Names of commercial manufacturers are provided for identification only and inclusion does not imply endorsement of the manufacturer, or its products or services, by the FBI or the U.S. Government. The views expressed are those of the author and do not necessarily reflect the official policy or position of the FBI or the U.S. Government.

References

Alonso, A., Andelinovic, S., Martin, P., Sutlovic, D., Erceg, I., Huffine, E., de Simon, L.F., Albarran, C., Definis-Gojanovic, M., Fernandez-Rodriguez, A., Garcia, P., Drmic, I., Rezic, B., Kuret, S., Sancho, M., Primorac, D., 2001. DNA typing from skeletal remains: evaluation of multiplex and megaplex STR systems on DNA isolated from bone and teeth samples. Croat. Med. J. 42 (3), 260–266.

Altschul, S.F., Gish, W., Miller, W., Myers, E.W., Lipman, D.J., 1990. Basic local alignment search tool. J. Mol. Biol. 215 (3), 403–410.

Anderson, S., Bankier, A.T., Barrell, B.G., de Bruijn, M.H.L., Coulson, A.R., Drouin, J., Eperon, I.C., Nierlich, D.P., Roe, B.A., Sanger, F., Schreier, P.H., Smith, A.J.H., Staden, R., Young, I.G., 1981. Sequence and organization of the human mitochondrial genome. Nature 290, 457–465.

Andersson, S.G.E., Zomorodipour, A., Andersson, J.O., Sicheritz-Pontén, T., Alsmark, U.C.M., Podowski, R.M., Näslund, A.K., Eriksson, A., Winkler, H.H., Kurland, C.G., 1998. The genome sequence of *Rickettsia prowazekii* and the origin of mitochondria. Nature 396, 133–140.

Andrews, R.M., Kubacka, I., Chinnery, P.F., Lightowlers, R.N., Turnbull, D.M., Howell, N., 1999. Reanalysis and revision of the Cambridge reference sequence for human mitochondrial DNA. Nat. Genet. 23, 147.

Bandelt, H.J., Parson, W., 2008. Consistent treatment of length variants in the human mtDNA control region: a reappraisal. Int. J. Legal Med. 122 (1), 11–21.

Berger, C., Parson, W., 2009. Mini-midi-mito: adapting the amplification and sequencing strategy of mtDNA to the degradation state of crime scene samples. Forensic Sci. Int. Genet. 3, 149–153.

Børsting, C., Morling, N., 2015. Next generation sequencing and its applications in forensic genetics. Forensic Sci. Int. Genet. 18, 78–89.

Brandstätter, A., Sänger, T., Lutz-Bonengel, S., Parson, W., Béraud-Colomb, E., Wen, B., Kong, Q.P., Bravi, C.M., Bandelt, H.J., 2005. Phantom mutation hotspots in human mitochondrial DNA. Electrophoresis 26 (18), 3414–3429.

Carpenter, M.L., Buenrostro, J.D., Valdiosera, C., Schroeder, H., Allentoft, M.E., Sikora, M., Rasmussen, M., Gravel, S., Guillé, S., Nekhrizov, G., Leshtakov, K., Dimitrova, D., Theodossiev, N., Pettener, D., Luiselli, D., Sandoval, K., Moreno-Estrada, A., Li, Y., Wang, J., Gilbert, M.T.P., Willersle, E., Greenleaf, W.J., Bustamante, C.D., 2013. Pulling out the 1%: whole-genome capture for the targeted enrichment of ancient DNA sequencing libraries. Am. J. Hum. Genet. 93 (5), 852–864.

Clopper, C.J., Pearson, E.S., 1934. The use of confidence or fiducial limits illustrated in the case of the binomial. Biometrika 26, 404–413.

Dabney, J., Meyer, M., Pääbo, S., 2013. Ancient DNA damage. Cold Spring Harb. Perspect. Biol. 5, a012567.

Davidov, Y., Huchon, D., Koval, S.F., Jurkevitch, E., 2006. A new α-proteobacterial clade of Bdellovibrio-like predators: implications for the mitochondrial endosymbiotic theory. Environ. Biol. 8 (12), 2179–2188.

Dayama, G., Emery, S.B., Kidd, J.M., Mills, R.E., 2014. The genomic landscape of polymorphic human nuclear mitochondrial insertions. Nucleic Acids Res. 42 (20), 12640–12649.

Edson, S., Ross, J., Coble, M., Parsons, T.J., Barritt, S., 2004. Naming the dead—confronting the realities of rapid identification of degraded skeletal remains. Forensic Sci. Rev. 16 (1), 63–90.

Eichmann, C., Parson, W., 2008. 'Mitominis': multiplex PCR analysis of reduced size amplicons for compound sequence analysis of the entire mtDNA control region in highly degraded samples. Int. J. Legal Med. 122 (5), 385–388.

Gabriel, M., Huffine, E., Ryan, J., Holland, M., Parsons, T., 2001. Improved mtDNA sequence analysis of forensic remains using a "mini-primer set" amplification strategy. J. Forensic Sci. 46 (2), 247–253.

Gamba, C., Jones, E.R., Teasdale, M.D., McLaughlin, R.L., Gonzalez-Fortes, G., Mattiangeli, V., Domboróczki, L., Kővári, I., Pap, I., Anders, A., Whittle, A., Dani, J., Raczky, P., Higham, T.F.G., Hofreiter, M., Bradley, D.G., Pinhasi, R., 2014. Genome flux and stasis in a five millennium transect of European prehistory. Nat. Commun. 5 (5257), 1–9.

Gilbert, M.T.P., Willerslev, E., Hansen, A.J., Barnes, I., Rudbeck, L., Lynnerup, N., Cooper, A., 2003. Distribution patterns of postmortem damage in human mitochondrial DNA. Am. J. Hum. Genet. 72 (1), 32–47.

Gray, M.W., Berger, G., Lang, B.F., 1999. Mitochondrial evolution. Science 283 (5407), 1476–1481.

Hansen, H.B., Damgaard, P.B., Margaryan, A., Stenderup, J., Lynnerup, N., Willerslev, E., Allentoft, M.E., 2017. Comparing ancient DNA preservation in petrous bone and tooth cementum. PLoS One 12 (1), e0170940.

Heddi, A., Grenier, A., Khatchadourian, C., Charles, H., Nardon, P., 1999. Four intracellular genomes direct weevil biology: nuclear, mitochondrial, principal endosymbiont, and Wolbachia. Proc. Natl. Acad. Sci. U. S. A. 96 (12), 6814–6819.

Hofreiter, M., Jaenicke, V., Serre, D., von Haeseler, A., Pääbo, S., 2001. DNA sequences from multiple amplifications reveal artifacts induced by cytosine deamination in ancient DNA. Nucleic Acids Res. 29 (23), 4793–4799.

Horn, S., 2012. Target enrichment via DNA hybridization capture. In: Shapiro, B., Hofreiter, M. (Eds.), Ancient DNA: Methods in Molecular Biology (Methods and Protocols). Humana Press.

Irwin, J.A., Saunier, J.L., Niederstätter, H., Strouss, K.M., Sturk, K.A., Diegoli, T.M., Brandstätter, A., Parson, W., Parsons, T.J., 2009. Investigation of heteroplasmy in the human mitochondrial DNA control region: a synthesis of observations from more than 5000 global population samples. J. Mol. Evol. 68 (5), 516–527.

King, T.E., Fortes, G.G., Balaresque, P., Thomas, M.G., Balding, D., Delser, P.M., Neumann, R., Parson, W., Knapp, M., Walsh, S., Tonasso, L., Holt, J., Kayser, M., Appleby, J., Forster, P., Ekserdjian, D., Hofreiter, M., Schürer, K., 2014. Identification of the remains of King Richard III. Nat. Commun. 5 (5631), 1–8.

Kraytsberg, Y., Schwartz, M., Brown, T.A., Ebralidse, K., Kunz, W.S., Clayton, D.A., Vissing, J., Khrapko, K., 2004. Recombination of human mitochondrial DNA. Science 304 (5673), 981.

Kukat, C., Wurm, C.A., Spåhr, H., Falkenberg, M., Larsson, N., Jakobs, S., 2011. Super-resolution microscopy reveals that mammalian mitochondrial nucleoids have a uniform size and frequently contain a single copy of mtDNA. Proc. Natl. Acad. Sci. U. S. A. 108 (33), 13534–13539.

Lee, H., Ma, H., Juanes, R.C., Tachibana, M., Sparman, M., Woodward, J., Ramsey, C., Xu, J., Kang, E., Amato, P., Mair, G., Steinborn, R., Mitalipov, S., 2012. Rapid mitochondrial DNA segregation in primate preimplantation embryos precedes somatic and germline bottleneck. Cell Rep. 1 (5), 506–515.

Li, M., Schönberg, A., Schaefer, M., Schroeder, R., Nasidze, I., Stoneking, M., 2010. Detecting heteroplasmy from high-throughput sequencing of complete human mitochondrial DNA genomes. Am. J. Hum. Genet. 87 (2), 237–249.

Luo, S., Valenciaa, A., Zhang, J., Leed, N., Slone, J., Guia, B., Wang, X., Lia, Z., Della, S., Brown, J., Chen, S.M., Chien, Y., Hwu, W., Fan, P., Wong, L., Atwal, P.S., Huanga, T., 2018. Biparental inheritance of mitochondrial DNA in humans. Proc. Natl. Acad. Sci. U. S. A. 115 (51), 13039–13044.

Lutz-Bonengel, S., Parson, W., 2019. No further evidence for paternal leakage of mitochondrial DNA in humans yet. Proc. Natl. Acad. Sci. U. S. A. 116 (6), 1821–1822.

Melton, T., Holland, C., Holland, M., 2012. Forensic mitochondrial DNA: current practice and future potential. Forensic Sci. Rev. 24 (2), 101–122.

Mundorff, A., Davoren, J.M., 2014. Examination of DNA yield rates for different skeletal elements at increasing postmortem intervals. Forensic Sci. Int. Genet. 8 (1), 55–63.

Mundorff, A.Z., Bartelink, E.J., Mar-Cash, E., 2009. DNA preservation in skeletal elements from the World Trade Center Disaster: recommendations for mass fatality management. J. Forensic Sci. 54 (4), 739–745.

Naviaux, R.K., 2008. Mitochondrial control of epigenetics. Cancer Biol. Ther. 7 (8), 1191–1193.

Nishimura, Y., Yoshinari, T., Naruse, K., Yamada, T., Sumi, K., Mitani, H., Higashiyama, T., Kuroiwa, T., 2006. Active digestion of sperm mitochondrial DNA in single living sperm revealed by optical tweezers. Proc. Natl. Acad. Sci. U. S. A. 103 (5), 1382–1387.

Parson, W., Dur, A., 2007. EMPOP—a forensic mtDNA database. Forensic Sci. Int. Genet. 1 (2), 88–92.

Parson, W., Gusmão, L., Hares, D.R., Irwin, J.A., Mayr, W.R., Morling, N., Pokorak, E., Prinz, M., Salas, A., Schneider, P.M., Parsons, T.J., 2014. DNA Commission of the International Society for Forensic Genetics: revised and extended guidelines for mitochondrial DNA typing. Forensic Sci. Int. Genet. 13, 134–142.

Parsons, T.J., Weedn, V.W., 1997. Preservation and recovery of DNA in postmortem specimens and trace samples. In: Haglund, W., Sorg, M. (Eds.), Forensic Taphonomy: The Postmortem Fate of Human Remains. CRC Press, pp. 109–138.

Parsons, T.J., Muniec, D.S., Sullivan, K., Woodyatt, N., Alliston-Greiner, R., Wilson, M.R., Berry, D.L., Holland, K.A., Weedn, V.W., Gill, P., Holland, M.M., 1997. A high observed substitution rate in the human mitochondrial DNA control region. Nat. Genet. 15, 363–368.

Pohjoismäki, J.L.O., Goffart, S., Taylor, R.W., Turnbull, D.M., Suomalainen, A., Jacobs, H.T., Karhunen, P.J., 2010. Developmental and pathological changes in the human cardiac muscle mitochondrial DNA organization, replication, and copy number. PLoS One 5 (5), e10426.

Röck, A.W., Dür, A., van Oven, M., Parson, W., 2013. Concept for estimating mitochondrial DNA haplogroups using a maximum likelihood approach (EMMA). Forensic Sci. Int. Genet. 7 (6), 601–609.

Schwartz, M., Vissing, J., 2002. Paternal inheritance of mitochondrial DNA. N. Engl. J. Med. 347, 576–580.

Scientific Working Group on DNA Analysis Methods (SWGDAM), 2014. Mitochondrial DNA nomenclature examples document. Available from: https://docs.wixstatic.com/ugd/4344b0_2044739c57574dbea97f2f85b6f73c9d.pdf.

Scientific Working Group on DNA Analysis Methods (SWGDAM), 2019. Interpretation Guidelines for Mitochondrial DNA Analysis by Forensic DNA Testing Laboratories. Available from: https://docs.wixstatic.com/ugd/4344b0_f61de6abf3b94c52b28139bff600ae98.pdf.

Sekiguchi, K., Sato, H., Kasai, K., 2004. Mitochondrial DNA heteroplasmy among hairs from single individuals. J. Forensic Sci. 49 (5), JFS2003216–3.

Seo, S.B., Jang, B.S., Zhang, A., Yi, J.A., Kim, H.Y., Yoo, S.H., Lee, Y.S., Lee, S.D., 2010. Alterations of length heteroplasmy in mitochondrial DNA under various amplification conditions. J. Forensic Sci. 55 (3), 719–722.

Stewart, J., Fisher, C., Aagaard, P., Wilson, M., Isenberg, A., Polanskey, D., Pokorak, E., DiZinno, J., Budowle, B., 2001. Length variation in HV2 of the human mitochondrial DNA control region. J. Forensic Sci. 46 (4), 862–870.

Oven, M., Kayser, M., 2009. Updated comprehensive phylogenetic tree of global human mitochondrial DNA variation. Hum. Mutat. 30 (2), E386–E394. https://doi.org/10.1002/humu.20921. http://www.phylotree.org.

van Wei, W., Tuna, S., Keogh, M.J., Smith, K.R., Aitman, T.J., Beales, P.L., Bennett, D.L., Gale, D.P., Bitner-Glindzicz, M.A.K., Black, G.C., Brennan, P., Elliott, P., Flinter, F.A., Floto, R.A., Houlden, H., Irving, M., Koziell, A., Maher, E.R., Markus, H.S., Morrell, N., Newman, W.G., Roberts, I., Sayer, J.A., Smith, K.G.C., Taylor, J.C., Watkins, H., Webster, A.R., Wilkie, A.O., Williamson, C., NIHR BioResource-Rare Diseases, 100,000 Genomes Project - Rare Diseases Pilot, Ashford, S., Penkett, C.J., Stirrups, K.E., Rendon, A., Ouwehand, W.H., Bradley, J.R., Raymond, F.L., Caulfield, M., Turro, E., Chinnery, P.F., 2019. Germline selection shapes the landscape of human mitochondrial DNA. Science 364 (6442), eaau6520.

Weitz, B.S., Meyer, S., Weiss, G., Haeseler, A., 1999. Pattern of nucleotide substitution and rate heterogeneity in the hypervariable regions I and II of human mtDNA. Genetics 152 (3), 1103–1110.

Wilson, M.R., DiZinno, J.A., Polanskey, D., Replogle, J., Budowle, B., 1995a. Validation of mitochondrial DNA sequencing for forensic casework analysis. Int. J. Legal Med. 108, 68–74.

Wilson, M.R., Polanskey, D., Butler, J., DiZinno, J.A., Replogle, J., Budowle, B., 1995b. Extraction, PCR amplification, and sequencing of mitochondrial DNA from human hair shafts. BioTechniques 18, 662–669.

Yang, D., Oyaizu, Y., Oyaizu, H., Olsen, G.J., Woese, C.R., 1985. Mitochondrial origins. Proc. Natl. Acad. Sci. U. S. A. 82 (13), 4443–4447.

Zhang, Q., Raoof, M., Chen, Y., Sumi, Y., Sursal, T., Junger, W., Brohi, K., Itagaki, K., Hauser, C.J., 2010. Circulating mitochondrial DAMPs cause inflammatory responses to injury. Nature 464, 104–107.

CHAPTER

11

X-chromosome short tandem repeats (X-STRs): Applications for human remains identification

Vivek Sahajpal PhD[a] *and Angie Ambers PhD*[b,c,d]

[a]State Forensic Science Laboratory, Directorate of Forensics Services, Shimla, Himachal Pradesh, India [b]Henry C. Lee Institute of Forensic Science, University of New Haven, West Haven, CT, United States [c]Forensic Science Department, Henry C. Lee College of Criminal Justice and Forensic Sciences, Center for Forensic Investigation of Trafficking in Persons, University of New Haven, West Haven, CT, United States [d]Institute for Human Identification, LMU College of Dental Medicine, Knoxville, TN, United States

Introduction

The nucleus of human cells contains 23 pairs of chromosomes, thread-like structures which package the tightly coiled DNA molecule. The 23rd pair of chromosomes are sex-determining chromosomes and therefore differ between males and females. Females typically have two X chromosomes, while males possess one X chromosome and one Y chromosome. Although autosomal short tandem repeats (STRs), Y-chromosome STRs (Y-STRs), and mitochondrial DNA (mtDNA) typically dominate DNA-based human identification efforts, in recent years STR markers on the X chromosome (X-STRs) have been an intensive area of research for potential forensic applications. X-chromosomal loci (specifically X-STRs) can be used to complement the types of markers traditionally used in forensic genetic investigations and can be of great utility in complex kinship analyses, with particular relevance to mass death scenarios (e.g., natural disasters, mass graves, aviation accidents, forced disappearances, terrorist attacks) (Szibor et al., 2003; Szibor, 2007; Tillmar et al., 2017).

Identification of skeletal remains is a challenging sphere in human identification due to long-term exposure to environmental insults and advanced degradation of endogenous DNA. X-chromosome analysis offers an additional tool in the forensic geneticist's repertoire to assist in identification efforts and may yield useful inference, especially when direct reference samples are not available for comparison; when familial reference samples are limited; and/or when autosomal STR analysis, Y-STR testing, and/or mtDNA results are inconclusive.

https://doi.org/10.1016/B978-0-12-815766-4.00011-X

The X chromosome: Discovery and pattern of inheritance

The X chromosome was first discovered by cytologist Hermann Henking in the late 19th century (c.1890–91) during observation of cellular meiotic events. During these studies, Henking noticed that one chromosome did not take part in meiosis in males. At the time (although it did adhere staining in a similar manner to other chromosomes), questions arose regarding whether this structure was indeed an actual chromosome or something else entirely; as such, it was originally referred to as the X *element*. It was later established (in 1901 by Clarence Erwin McClung) that the X *element* was actually a chromosome and henceforth it was referred to as the X chromosome (Schwartz, 2008; Bainbridge, 2003).

The X chromosome is the eighth largest chromosome in humans, is composed of approximately 155 million base pairs (155 Mb), and represents ~5% of the total human genome (Ross et al., 2005; Online Mendelian Inheritance in Man, n.d.). Normal human males have one copy of the X chromosome (hemizygous) and one Y chromosome; normal human females possess two X chromosomes (dizygous). In addition, chromosomal abnormalities involving the X chromosome do exist, e.g., female karyotype XO (Turner syndrome), male karyotype XXY (Klinefelter syndrome), and female karyotype XXX (Triple X syndrome). Interestingly, in human females one of the two X chromosomes gets inactivated during early embryonic development, a phenomenon referred to as *X-inactivation* or *Lyonization* (Lyon, 1961).

Differences in the karyotypes of males and females result in interesting inheritance patterns of the X chromosome. In males, the X chromosome is passed on in its entirety to daughters since recombination cannot occur during gametogenesis (because males possess only a single copy of the X chromosome). Furthermore, a male cannot (and does not) transmit an X chromosome to his biological son. Since females have two X chromosomes, genetic recombination occurs during gametogenesis; hence, offspring (both daughters and sons) inherit a combination of alleles from the mother's two X chromosomes. These unique patterns of inheritance make the X chromosome an interesting genetic target that can be used to supplement traditional forensic analyses based on autosomal STRs, Y-STRs, and mitochondrial DNA. An overview of the inheritance patterns of genetic material contained within the X chromosome is summarized in Table 1.

In addition, there are linkage groups present on the X chromosome. Szibor et al. (2003) described four such linkage groups for X-STR markers used in forensic analysis. Due to linkage, the set of alleles on linked markers are considered as a haplotype and therefore cosegregate as blocks during meiosis (Diegoli, 2014). Hence, linked X-STRs are inherited as haplotypes by offspring. This behavior adds another interesting dimension to the analysis of the X-chromosome for forensic purposes, given that the product rule cannot be applied to alleles for loci located within a linkage group; instead, the product rule is applied to haplotypes to estimate the rarity of an X-STR profile (Butler, 2012). Hence, haplotype frequency data is also very important in forensic analyses based on X-STRs.

Forensic applications of X-chromosome analysis

X-chromosome typing is useful in kinship analyses for any parent-offspring relationship that involves at least one female (Krawczak, 2007). X-chromosome analysis can exhibit

TABLE 1 Inheritance patterns of human X-chromosome DNA.

Biological relationship	Percentage of X-chromosome DNA inherited	Genetic inheritance pattern
Mother to son	50	Sons inherit a combination of the two X chromosomes from the mother
Mother to daughter	50	Daughters inherit a combination of the two X chromosomes from the mother
Father to son	0	No transmission (sons do not inherit the father's X chromosome)
Father to daughter	100	Daughters inherit the father's X chromosome in totality (sans recombination)
Paternal grandmother to granddaughter	50	Combination of the two X chromosomes inherited by the son is passed on in entirety to granddaughter
Paternal grandmother to grandson	0	X chromosome not transmitted by the son to the grandson
Maternal grandmother to grandchild	25	Inheritance pattern is similar to the inheritance of autosomal DNA markers
Paternal grandfather to grandchild	0	No transmission (because the X chromosome was not transmitted to the son)
Maternal grandfather to grandchild	25	Inheritance pattern is similar to the inheritance of autosomal DNA markers

particular utility in missing person cases and skeletal remains identification because human remains in such scenarios are often recovered after a considerable period of time has elapsed. Direct reference samples and/or familial exemplars from biological parents or offspring may not be available for comparison. In these cases, routine methods of DNA identification may not be informative. Identification of human remains frequently requires data recovered from a combination of sophisticated methods and, in some cases, X-chromosome markers may be more efficient and informative than autosomal STR markers (Szibor, 2007). When only distant relatives are available for comparison testing, X-chromosome markers can be very useful because they can assign pedigree members over long distances (Szibor, 2007). Furthermore, the small amplicons representative of X-chromosome (X-STR) markers make them particularly suitable for the degraded DNA templates typically encountered with skeletal remains. Indeed, the suitability of X-chromosome analysis for degraded samples has been previously demonstrated (Asamura et al., 2006).

One of the important applications of X-chromosome analysis involves genetically linking grandparents and grandchildren when the child's biological parents are not available for testing. For example, as a general rule, a granddaughter will only inherit 25% of autosomal DNA from her paternal grandmother; hence, establishing a relationship can be very difficult in the absence of reference DNA samples from the child's biological parents. Table 2 and Fig. 1 display autosomal STR data from a paternal grandmother and her granddaughter, when reference

TABLE 2 Autosomal STR profiles of a **Paternal Grandmother** and her biological **Granddaughter**.

STR locus	Grandmother		Granddaughter	
Amelogenin	X	X	X	X
D3S1358	15	**17**	**17**	18
D1S1656	13	14	11	12
D6S1043	**17**	19	12	**17**
D13S317	8	8	11	12
Penta E	**14**	15	**14**	14
D16S539	9	14	12	13
D18S51	13	18	14	16
D2S1338	19	23	17	24
CSF1PO	11	**12**	11	**12**
Penta D	**9**	13	**9**	13
TH01	9	9.3	7	7
vWA	**16**	17	14	**16**
D21S11	**29**	29	**29**	30
D7S820	11	12	10	10
D5S818	**11**	12	**11**	11
TPOX	8	**11**	9	**11**
D8S1179	13	16	14	14
D12S391	18	20	18.3	21
D19S433	13	15.2	12.2	15
FGA	**23**	24	21	**23**

Autosomal STR data were generated using the PowerPlex 21 System (Promega Corporation). Common (inherited) alleles are displayed in red boldface font.

samples from the biological parent(s) of the granddaughter were not available for comparison. Autosomal STR data were generated using the PowerPlex 21 System (Promega Corporation), a multiplex of 20 autosomal STR loci plus amelogenin. In this scenario, a grandparent-grandchild relationship can be inferred; however, additional lineage marker testing would strengthen this assessment and increase the potential for making a positive identification.

In this particular type of case, X-chromosome analysis is more informative than traditional autosomal STR typing. A grandmother's son inherits a combination of her two X chromosomes. In terms of X-STRs, he inherits a combination of X-STR alleles present on the two X chromosomes of his mother. For clarification on the utility of X-chromosome testing, in this case, the X-STR profiles of the paternal grandmother, her son, and her granddaughter are displayed in Fig. 2. X-STR data was generated using the Investigator Argus X-12

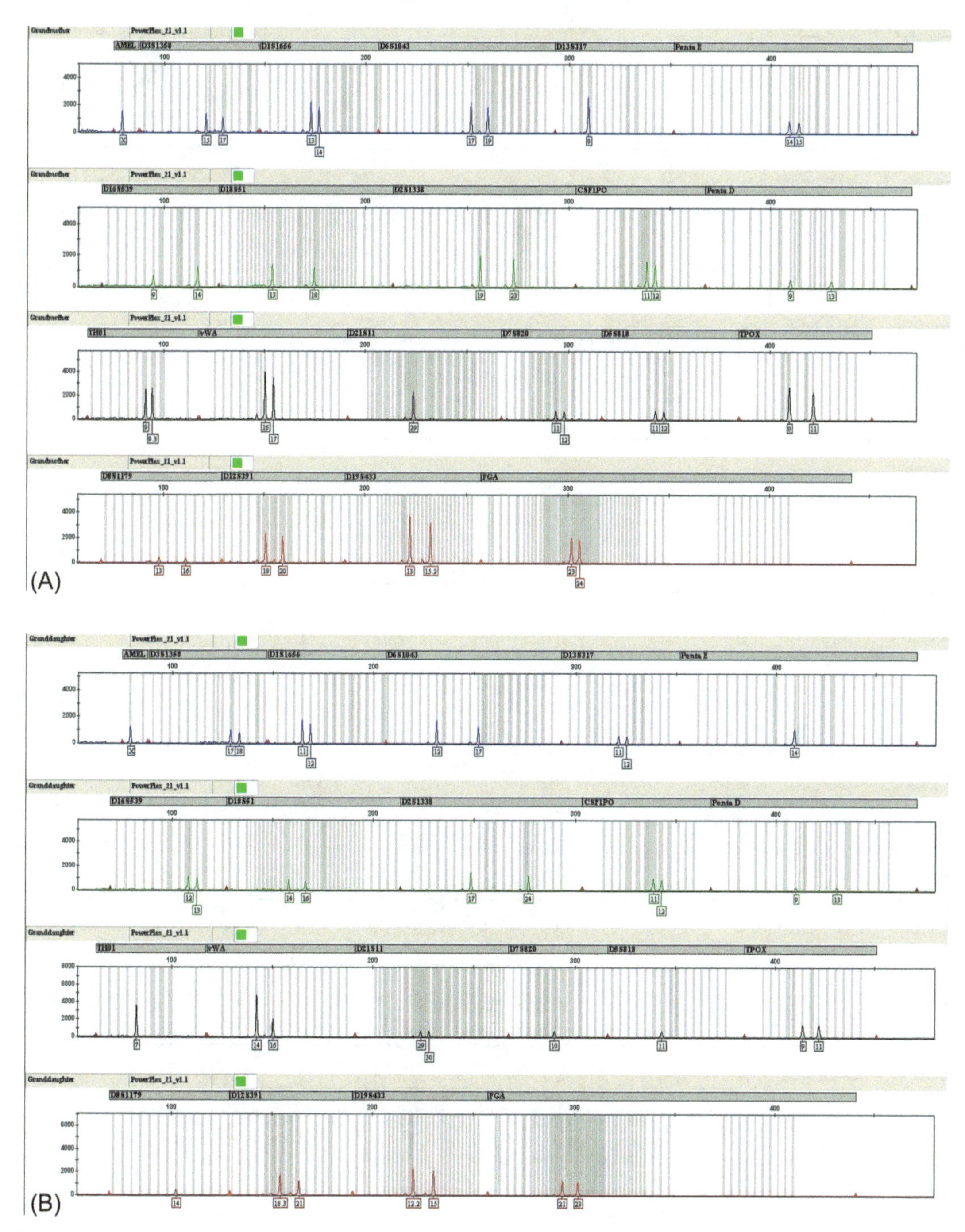

FIG. 1 Electropherograms displaying the autosomal STR profiles of (A) a **Paternal Grandmother**, and (B) her biological **Granddaughter**. Autosomal STR data were generated using the PowerPlex 21 System (Promega Corporation).

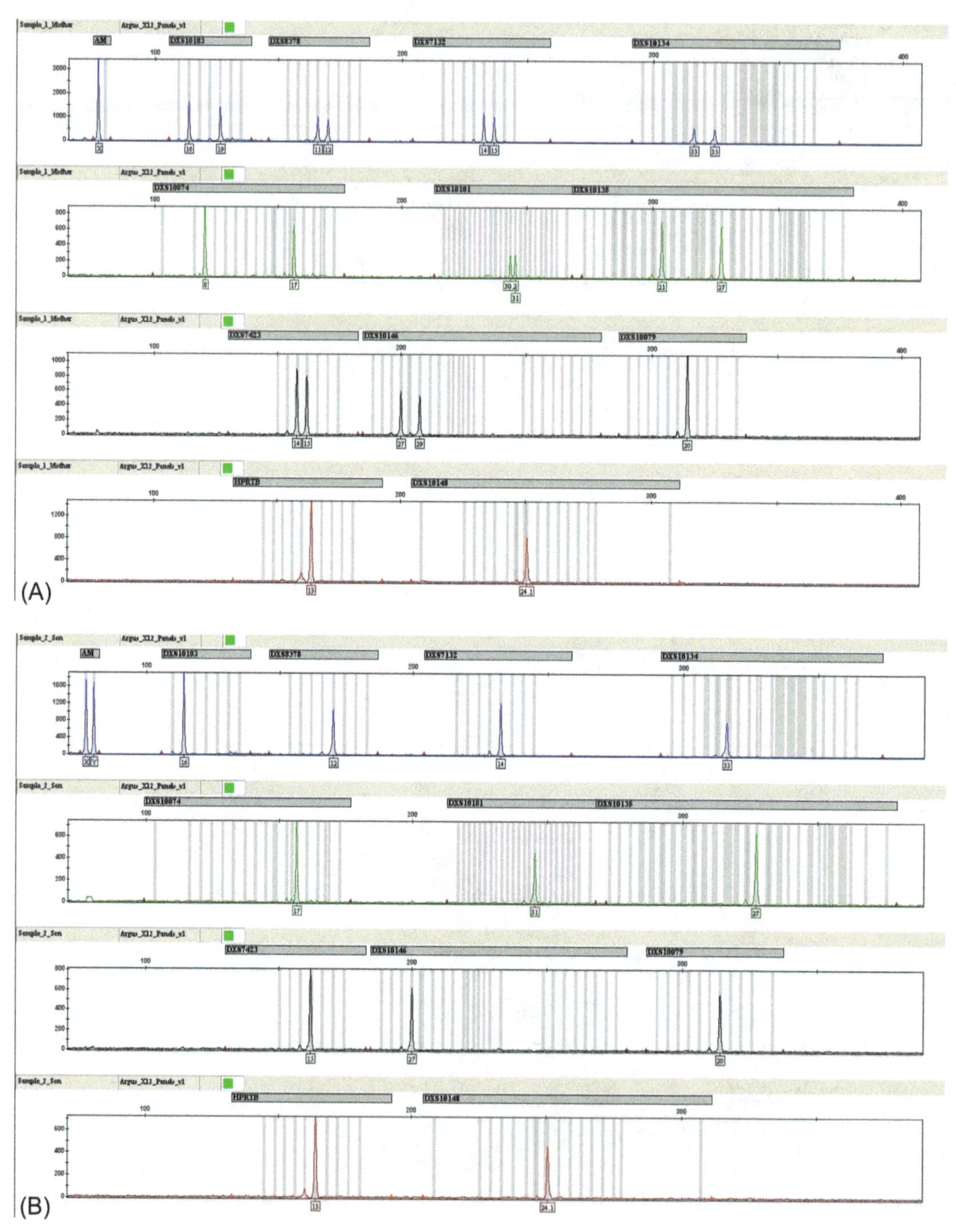

FIG. 2 Electropherograms displaying the X-STR profiles of (A) the **Paternal Grandmother**, (B) the **Grandmother's son**,

(Continued)

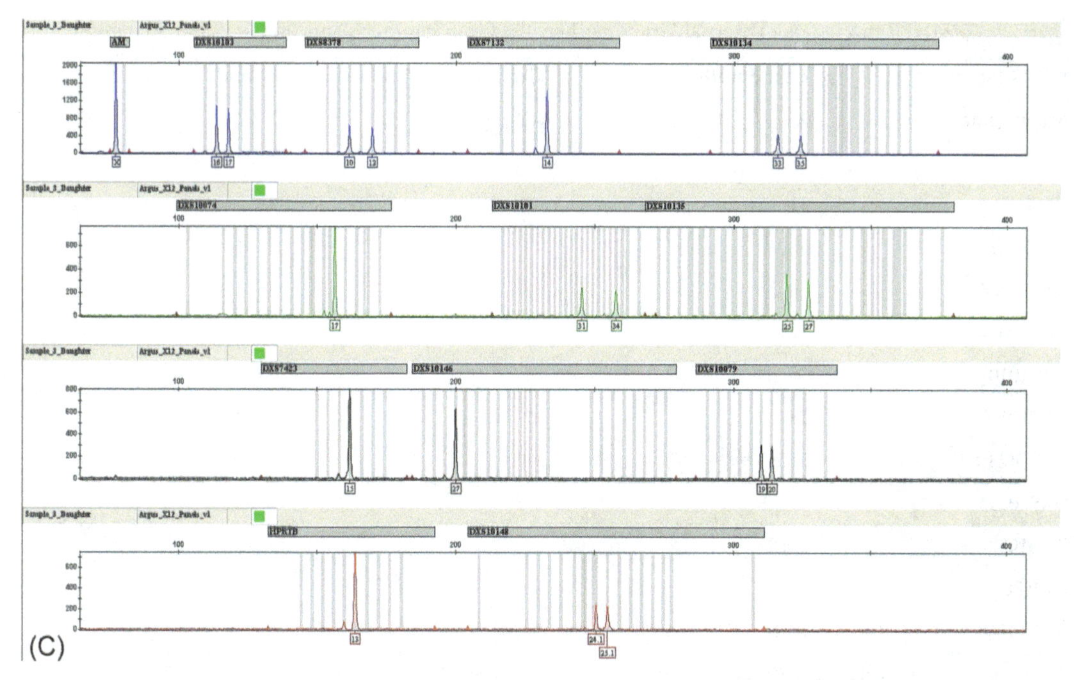

FIG. 2, CONT'D and (C) the **Granddaughter**. X-STR data were generated using the Investigator Argus X-12 QS Kit (Qiagen Corporation).

QS Kit (Qiagen Corporation), a multiplex of 12 X-STRs, the autosomal STR locus D21S11, and amelogenin. Table 3 depicts the X-STR profiles of these three individuals, with common (inherited) alleles displayed in red boldface font. During meiosis, the granddaughter inherits her father's X chromosome (intact and unchanged), as a haplotype. All STR alleles present on her father's X-chromosome were inherited directly from his mother (i.e., the child's grandmother). In this scenario, analysis of X-STR markers could also confirm the mother-son relationship and can further complement autosomal STR typing and mtDNA sequencing results. In a similar manner, the X-STR genotype of the putative grandmother could also be constructed based on X-STR data of her grandchildren and biological children (i.e., sons and daughters).

In addition to establishing grandparent-grandchild relationships, X-STRs can be used for confirming female half-sibling relationships. By definition, half-siblings share one biological parent. In the case of half-siblings with a common biological father but different mothers, half-sisters will inherit the same X chromosome from the putative father (which means they will share an allele at each X-STR locus tested). Half-sisters share 50% of their X-STR alleles. This can help to establish the relationship between the duos or can exclude paternity even if the biological parents are not available for testing (Szibor, 2007).

Consider another type of case in which a biological son goes missing and a reference sample from the biological mother is the only sample available for comparison (i.e., the biological father of the missing son is not available to provide an exemplar). Autosomal STR

TABLE 3 X-STR profiles of a **Paternal Grandmother**, the **Grandmother's son**, and her **Granddaughter**.

X-STR locus	Grandmother	Grandmother's son	Granddaughter
Amelogenin	XX	XY	XX
DXS10103	**16**, 19	**16**	**16**, 17
DXS8378	11, **12**	**12**	10, **12**
DXS7132	**14**, 15	**14**	**14**, 14
DXS10134	**33**, 35	**33**	**33**, 35
DXS10074	8, **17**	**17**	17, **17**
DXS10101	30.2, **31**	**31**	**31**, 34
DXS10135	21, **27**	**27**	25, **27**
DXS7423	14, **15**	**15**	15, **15**
DXS10146	**27**, 29	**27**	27, **27**
DXS10079	20, **20**	**20**	19, **20**
HPRTB	13, **13**	**13**	13, **13**
DXS10148	24.1, **24.1**	**24.1**	**24.1**, 25.1

X-STR data was generated using the Investigator Argus X-12 QS Kit (Qiagen Corporation). Common (inherited) alleles are displayed in red boldface font.

comparison between a biological mother and the putative remains of her missing son can be complicated by the presence of a mutation at a particular locus. In such a case, the findings of autosomal STR analysis may benefit from additional (complementary) X-STR or mtDNA testing. This type of scenario is further challenged if allele frequency data for the autosomal markers tested are not available in that jurisdiction (as is the case in some countries which lack the infrastructure and resources to build an allele frequency database representative of their respective populations). Further complicating a case like this is the fact that not all laboratories conduct mitochondrial DNA analysis. STR markers remain the current mainstay of the majority of forensic DNA testing labs. Mitochondrial DNA analysis also lacks the degree of statistical certainty required for positive identification of human remains (Szibor et al., 2003). An example is illustrated in Table 4, in which a mother's autosomal STRs were compared to the putative remains of her missing son. However, a mutation existed at the FGA locus, so there was a nonmatch at one of the markers tested. X-STR analysis was performed to supplement the autosomal STR data and further confirm the potential mother-son association between these samples. The results of X-STR testing are shown in Table 5, revealing that the X-STR profile obtained from the skeletal remains shares one allele at each X-STR locus with the reference sample obtained from the biological mother (i.e., there is 50% sharing of X-STR alleles between the bone samples and the maternal exemplar). Thus, X-STR data were beneficial in complementing the findings of autosomal STR analysis and in further supporting the mother-son relationship (and in the subsequent positive identification of the remains of her missing son).

TABLE 4 Autosomal STR profiles of a biological mother and the putative remains of her missing son.

STR locus	Biological mother		Putative remains of missing son	
Amelogenin	**X**	X	**X**	Y
D3S1358	16	**17**	17	**17**
D1S1656	15	**17.3**	12	**17.3**
D6S1043	11	**11**	**11**	18
D13S317	8	**11**	11	**11**
Penta E	12	**17**	16	**17**
D16S539	**9**	11	**9**	11
D18S51	18	**18**	14	**18**
D2S1338	**19**	23	**19**	24
CSF1PO	**10**	12	**10**	11
Penta D	9	**10**	10	**10**
TH01	**6**	7	**6**	7
vWA	14	**15**	**15**	18
D21S11	30	**30**	28	**30**
D7S820	**8**	11	**8**	12
D5S818	12	**13**	10	**13**
TPOX	**11**	12	9	**11**
D8S1179	12	**15**	14	**15**
D12S391	18	**21**	17	**21**
D19S433	**13**	15	**13**	13
FGA	19	25	24	24

Common (shared) alleles are displayed in red boldface font. A mutation is present at the FGA locus (highlighted in gray). Autosomal STR data were generated using the PowerPlex 21 System (Promega Corporation).

Forensically relevant X-chromosome markers, commercial amplification kits, and other multiplex systems available for X-STR analysis

A large number of STR markers located on the human X chromosome have been identified for their potential utility in forensic DNA analysis (Tillmar et al., 2017; Nothnagel et al., 2012). A total of 26 trinucleotide- and 90 tetranucleotide repeat polymorphisms on the X chromosome have been described (Szibor et al., 2003). Among these existing polymorphisms, 30 X-STR markers are currently being applied in forensic genetic investigative applications (Butler, 2012). A complete list of X-STR markers (and associated information) is maintained by the Forensic ChrX Research Group in Germany. This research

TABLE 5 X-STR profiles of a biological mother and the putative remains of her missing son.

X-STR locus	Biological mother	Putative remains of missing son
Amelogenin	X, X	X, Y
DXS10103	18, **19**	**19**
DXS8378	**11**, 12	**11**
DXS7132	**12**, 16	**12**
DXS10134	**37**, 37	**37**
DXS10074	16, **17**	**17**
DXS10101	29, **30.2**	**30.2**
DXS10135	27, **29**	**29**
DXS7423	**14**, 15	**14**
DXS10146	29, **31**	**31**
DXS10079	**19**, 21	**19**
HPRTB	**12**, 14	**12**
DXS10148	20, **24.1**	**24.1**

This is an example of how X-STR data can complement autosomal STR testing (shown in Table 4). Results were generated using the Investigator Argus X-12 QS Kit (Qiagen Corporation). Common (shared) alleles are displayed in red boldface font.

consortium maintains an online X-chromosome database called ChrX-STR.org 2.0 (http://www.chrx-str.org) that calculates population genetic data on the basis of X-chromosomal allele frequencies (Szibor et al., 2006).

Currently, there are a variety of commercially available X-STR amplification kits for forensic casework. These kits—multiplexes of X-STR loci—have been developmentally validated with forensically relevant sample types. In addition to commercial manufacturers, several "in-house" X-STR multiplexes have been created by independent research groups around the world. An overview of existing X-STR testing systems is provided below.

Investigator Argus X-12 QS (Qiagen Corporation)

One of the most popular X-STR kits for forensic casework is the Investigator Argus X-12 QS Kit, manufactured by Qiagen Corporation. This kit supplements the Investigator Argus Y-12 QS Kit for kinship and paternity testing. The Investigator Argus X-12 multiplex is based on 5-dye chemistry and coamplifies 12 X-STR loci (DXS7132, DXS7423, DXS8378, DXS10074, DXS10079, DXS10101, DXS10103, DXS10134, DXS10135, DXS10146, DXS10148, HPRTB) plus amelogenin. These 12 X-chromosomal markers are clustered into four linkage groups (three markers per group), and therefore each set of three markers is handled as a haplotype for genotyping (Fig. 3). Coamplification of these 12 X-STRs provides heightened discrimination power and meets the criteria for complex kinship and paternity testing. This X-STR panel also assists in fulfilling the demands of complicated deficiency cases involving at least one female.

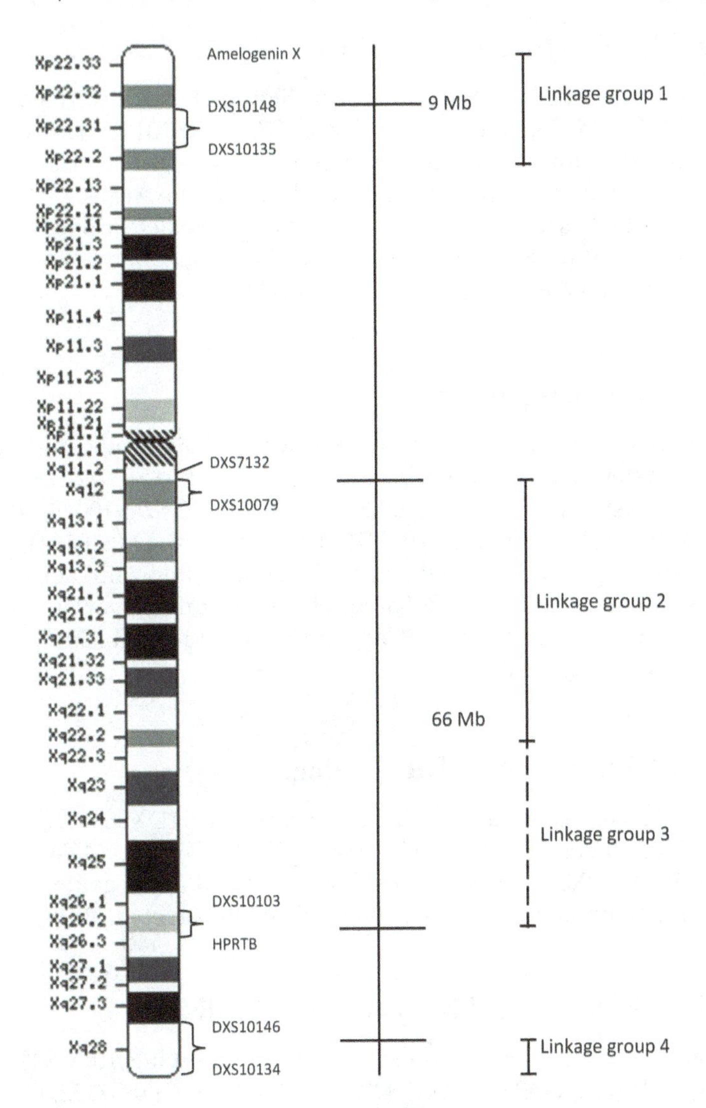

FIG. 3 Ideogram of the human X chromosome, displaying the physical location of X-STR loci assayed by the Investigator Argus X-12 QS Kit (Qiagen Corporation). Distances from the p-telomere are shown in megabases (Mb) (http://www.ncbi.nlm.nih.gov/genome/guide/human as of 11/2014).

In addition to high discriminatory power, the Investigator Argus X-12 QS Kit was developed as a high-sensitivity and inhibitor-tolerant assay to account for challenging conditions typically encountered in forensic casework samples. The detection limit is 100 picograms (pg) of genomic DNA, with an optimal input range of 0.2–0.5 nanograms (ng) DNA (under standard thermal cycling conditions). The kit also contains an internal positive control (Quality Sensor QS1) to provide the user with feedback on the efficiency of the PCR reaction (Qiagen Inc., 2015).

Mentype Argus X-8 Kit (Biotype Forensics)

The Mentype Argus X-8 PCR Amplification Kit (Biotype Forensics) is a multiplex of eight X-STR loci (DXS7132, DXS7423, DXS8378, DXS10074, DXS10101, DXS10134, DXS10135, HPRTB) plus amelogenin for gender determination. These eight X-STRs are clustered into four linkage groups (two markers per group) and, similar to the Argus X-12 assay, each set of linked markers is handled as a haplotype for genotyping. The detection limit of the Mentype Argus X-8 PCR Amplification Kit is approximately 100 pg genomic DNA, although the recommended input range is 0.1–1.0 ng DNA (Biotype, 2009).

AGCU X19 STR Kit (Jiangsu, China)

The AGCU X19 STR Kit (AGCU ScienTech Incorporation, Jiangsu, China) is a recently developed multiplex system which coamplifies 19 X-STR loci (DXS8378, DXS7423, DXS10148, DXS10159, DXS6809, DXS7424, DXS10164, DXS10162, DXS7132, DXS10079, DXS6789, DXS101, DXS10103, DXS10101, HPRTB, DXS10075, DXS10074, DXS10135, DXS10134). Published literature indicates that these specific 19 X-STR loci can be divided into 7 linkage groups. Extensive research on the use of the AGCU X19 STR Kit for kinship testing on various ethnic groups has been undertaken by researchers in China (Chen et al., 2018; Liu et al., 2017; Meng et al., 2017; Yang et al., 2016, 2017; Zhang et al., 2016).

Goldeneye DNA ID System 17X Kit (Beijing, China)

The Goldeneye DNA ID System 17X Kit assays 16 X-STR loci (DXS6795, DXS9902, DXS8378, HPRTB, GATA165B12, DXS7132, DXS7424, DXS6807, DXS6803, GATA172D05, DXS6800, DXS10134, GATA31E08, DXS10159, DXS6789, DXS6810) and amelogenin. Studies using this X-STR assay in Chinese populations have been published (Gao et al., 2017).

X-STR Decaplex: System A and System B (GHEP-ISFG)

The X-STR Decaplex System A is a multiplex of the following 10 X-STR loci (DXS8378, DXS9898, DXS8377, HPRTB, GATA172D05, DXS7423, DXS6809, DXS7132, DXS101, DXS6789) (Gomes et al., 2007). Also a 10-locus multiplex, the X-STR Decaplex System B assays the following X-STRs: DXS8378, DXS9898, DXS7133, GATA172D05, GATA31E08, DXS7423, DXS6809, DXS7132, DXS9902, and DXS6789. The System A and System B Decaplexes differ from each other with respect to three loci (i.e., DXS8377, HPRTB, and DXS101 in System A; DXS7133, GATA31E08, and DXS9902 in System B).

System B was optimized during a collaborative study between 17 laboratories affiliated with the Spanish and Portuguese-Speaking Working Group of the International Society for Forensic Genetics (GHEP-ISFG). In this collaborative effort, 15 Iberian and Latin American populations from Argentina, Brazil, Costa Rica, Colombia, Portugal, and Spain were studied (Gusmão et al., 2009). Notably, this system is listed as an "in-house" kit by the Forensic ChrX Research Group (http://www.chrx-str.org).

12X-Plex System (Turrina et al., 2007)

The 12X-Plex System coamplifies 12 X-STR loci (DXS7132, DXS8378, DXS6809, DXS7133, DXS6789, DXS7424, GATA172D05, HPRTB, DXS7423, GATA31E08, DXS101, DXS6807) plus amelogenin. Initial studies were conducted on 200 unrelated individuals (100 males, 100 females) in Northern Italy (Turrina et al., 2007).

11X-Plex System (Rodrigues et al., 2008)

The 11X-Plex System coamplifies 11 X-STRs (DXS9895, DXS7132, DXS6800, DXS9898, DXS6789, DXS7133, DXS7130, HPRTB, GATA31E08, DXS7423, DXS10011). This system was developed for use in paternity assessment and for complex kinship analyses in forensic cases (Rodrigues et al., 2008).

Two X-chromosomal multiplexes—15 X-STR loci (Diegoli et al., 2013)

An independently developed X-STR system exists that is comprised of 2 multiplexes that collectively assay 15 X-STR markers (DXS6789, DXS7130, DXS9902, GATA165B12, DXS101, GATA31E08, DXS7424, DXS6795, GATA172D05, DXS10147, DXS8378, DXS7132, DXS6803, HPRTB, DXS7423) (Diegoli et al., 2011, 2013, 2014). These assays have been used to study X-STRs in the U.S. population and in populations within Bosnia and Herzegovina.

17 X-Plex System (Prieto-Fernández et al., 2015)

The 17 X-Plex multiplex system assays 11 X-STRs included in the decaplex of the Spanish and Portuguese-Speaking Working Group of the International Society for Forensic Genetics (GHEP-ISFG) and the Investigator Argus X-12 kit, as well as an additional 6 X-STR loci. The 11 X-STR markers included in the decaplex of the GHEP-ISFG and the Investigator Argus X-12 kit are DXS8378, DXS9902, DXS7132, DXS9898, DXS6809, DXS6789, DXS7133, GATA172D05, GATA31E08, DXS10079, and DXS7423. The additional 6 X-STR markers included in the 17 X-Plex System are DXS6800, DXS6801, DXS6803, DXS6807, DXS6799, and DXS10075. Sensitivity studies indicate that the 17 X-Plex System requires a minimum of 100 pg DNA to generate complete profiles. Validation studies involving population samples from four different continents have been conducted, including studies on Asian populations (Thailand), European Caucasoids (Spain), African populations (Malawi, Equatorial Guinea), and Hispanics (Colombia) (Prieto-Fernández et al., 2015).

ForenSeq™ DNA Signature Prep Kit (Verogen)

In addition to traditional capillary electrophoresis (CE) based testing, massively parallel sequencing (MPS) assays which include X-STRs are now available to the forensic community. The ForenSeq™ DNA Signature Prep Kit (Verogen), a multiplex of more than 200 DNA markers, is designed for the MiSeq™ FGx Forensic Genomics System (Illumina). The ForenSeq™ panel simultaneously analyzes 27 autosomal STRs, 7 X-STRs, 24 Y-STRs, amelogenin, and 94 human identity(HID) single nucleotide polymorphisms (SNPs). In addition, the system analyzes 56 ancestry-informative SNPs (aiSNPs) and 24 phenotype-informative SNPs (piSNPs), the latter of which can be useful in the prediction of eye color and hair color (Hussing et al., 2018; Ambers et al., 2016; Ballantyne et al., 2010; Rodig et al., 2010; Becker et al., 2008; Hering et al., 2006; Kidd et al., 2006; Sanchez et al., 2006; Edelmann et al., 2002; Mertens et al., 1999). The seven X-STRs assayed by the ForenSeq™ panel are DXS10074, DXS10103, DXS10135, DXS7132, DXS7423, DXS8378, and HPRTB.

X-STR population data

For statistical interpretation of X-STR results, it is imperative to maintain a reference population database of both allele frequencies and haplotype frequencies. Haplotype frequencies are very important in X-STR analyses due to the linkage exhibited by the markers used in various X-STR typing systems. For the purpose of statistical analysis, data from approximately 100–250 unrelated individuals within a population should be available, according to recommendations of the International Society for Forensic Genetics (ISFG). Moreover, any database generated and maintained should be representative of the population in question, as allele- and haplotype frequencies (and the degree of X-STR polymorphism) will vary between different ethnic and geographic populations (which ultimately directly affects the statistical discriminatory power of any X-STR panel).

In the past decade, X-STR data from numerous population studies have been reported, and the forensic application of X-STR markers in dealing with complex cases is increasing. An important resource on X-STR data is the ChrX-STR.org 2.0 database maintained by the Forensic ChrX Research Group. In addition to providing access to allele and haplotype data, the site also provides online tools for the analysis of X-STR data (similar to the Y-chromosome STR Haplotype Reference Database at YHRD.org). Furthermore, the ChrX-STR site also provides statistical parameters for the evaluation of X-STR profiles and has a data submission tool that allows X-STR data to be submitted for inclusion in the database. Currently, the ChrX-STR database contains haplotype data from studies conducted in Germany (Becker et al., 2008; Hering et al., 2006; Edelmann et al., 2012), in Ghana (Becker et al., 2008), in Japan (Becker et al., 2008), and in China (Zeng et al., 2011). As X-STR research continues to be conducted across various global populations, the amount of data contained within this database will increase exponentially, expanding the utility of X-STRs in forensic genetic investigations.

Conclusions

X-STR analysis can serve as a powerful supplemental (and complementary) tool in unidentified human remains (UHR) cases and for disaster victim identification (DVI). Forensic utilization of X-STR multiplex systems is increasing worldwide, and extensive research on X-STR loci continues to emerge from different parts of the globe. The importance and utility of X-STR analysis will undoubtedly increase as it continues to prove itself to be an effective tool for resolving complex kinship analyses.

References

Ambers, A.D., Churchill, J.D., King, J.L., Stoljarova, M., Gill-King, H., Assidi, M., Abu-Elmagd, M., Buhmeida, A., Budowle, B., 2016. More comprehensive forensic genetic marker analyses for accurate human remains identification using massively parallel DNA sequencing. BMC Genomics 17 (9), 750.

Asamura, H., Sakai, H., Kobayashi, K., Ota, M., Fukushima, H., 2006. MiniX-STR multiplex system population study in Japan and application to degraded DNA analysis. Int. J. Legal Med. 120 (3), 174–181.

ainbridge, D., 2003. The X in Sex: How the X Chromosome Controls Our Lives. Harvard University Press, ISBN: 0674016211.

Ballantyne, K.N., Goedbloed, M., Fang, R., Sehaap, O., Lao, O., Wollstein, A., Choi, Y., van Duijn, K., Vereulen, M., Brauer, S., Docorte, R., Poetsch, M., von Wurmb-Schwark, N., de Knijff, P., Labuda, D., Vezina, H., Knoblauch, H., Lessig, R., Roewer, L., Ploski, R., Dobosz, T., Henke, L., Furtado, M.R., Kayser, M., 2010. Mutability of Y-chromosomal microsatellites: rates, characteristics, molecular bases, and forensic implications. Am. J. Hum. Genet. 87, 341–353.

Becker, D., Rodig, H., Augustin, C., Edelmann, J., Gotz, F., Hering, S., Szibor, R., Brabetz, W., 2008. Population genetic evaluation of eight X-chromosomal short tandem repeat loci using Mentype Argus X-8 PCR amplification kit. Forensic Sci. Int. Genet. 2, 69–74.

Biotype, 2009. Mentype® Argus X-8 PCR Amplification Kit. www.biotype.de.

Butler, J.M., 2012. X-chromosome analysis. In: Advanced Topics in Forensic DNA Typing: Methodology. Academic Press, pp. 457–472.

Chen, L., Guo, Y., Xiao, C., Wu, W., Lan, Q., Fang, Y., Chen, J., Zhu, B., 2018. Genetic polymorphisms and forensic efficiency of 19 X-chromosomal STR loci for Xinjiang Mongolian population. PeerJ 6, e5117. https://doi.org/10.7717/peerj.5117.

Diegoli, T.M., 2014. Forensic application of X chromosome STRs. In: Primorac, D., Schanfield, M. (Eds.), Forensic DNA Applications and Interdisciplinary Perspective. CRC Press, pp. 135–169.

Diegoli, T.M., Kovacevic, L., Pojskic, N., Coble, M.D., Marjanovic, D., 2011. Population study of fourteen X chromosomal short tandem repeat loci in a population from Bosnia and Herzegovina. Forensic Sci. Int. Genet. 5, 350–351.

Diegoli, T.M., Linacre, A., Coble, M.D., 2013. Developmental validation of 15 X chromosomal short tandem repeat markers. Forensic Sci. Int. Genet. Suppl. Ser. 4, e142–e143.

Diegoli, T.M., Linacre, A., Coble, M.D., 2014. Population genetic data for 15 X chromosomal short tandem repeat markers in three U.S. populations. Forensic Sci. Int. Genet. 8, 64–67.

Edelmann, J., Deichsel, D., Hering, S., Plate, I., Szibor, R., 2002. Sequence variation and allele nomenclature for the X-linked STRs DXS9895, DXS8378, DXS7132, DXS6800, DXS7133,GATA172D05, DXS7423 and DXS8377. Forensic Sci. Int. 129, 99–103.

Edelmann, J., Lutz-Bonengel, S., Naue, J., Hering, S., 2012. X-chromosomal haplotype frequencies of four linkage groups using the Investigator Argus X-12 Kit. Forensic Sci. Int. Gen. 6 (1), 24–34.

Gao, M., Wang, C., Han, S.Y., Sun, S.H., Xiao, D.J., Wang, Y.S., Li, C.T., Zhang, M.X., 2017. Analysis of the 19 Y-STR and 16 X-STR loci system in the Han population of Shandong province, China. Genet. Mol. Res. 16 (1). https://doi.org/10.4238/gmr16019573. gmr16019573.

Gomes, I., Prinz, M., Pereira, R., Meyers, C., Mikulasovich, R.S., Amorim, A., Carracedo, A., Gusmão, L., 2007. Genetic analysis of three U.S. population groups using an X-chromosomal STR decaplex. Int. J. Legal Med. 121, 198–203.

Gusmão, L., Sánchez-Diz, P., Alves, C., Gomes, I., Zarrabeitia, M.T., Abovich, M., Atmetlla, I., Bobillo, C., Bravo, L., Builes, J., 2009. A GEP-ISFG collaborative study on the optimization of an X-STR Decaplex: data on 15 Iberian and Latin American populations. Int. J. Legal Med. 12, 227–234.

Hering, S., Augustin, C., Edelmann, J., Heidel, M., Dressler, J., Rodig, H., Kuhlisch, E., Szibor, R., 2006. DXS10079, DXS10074 and DXS10075 are STRs located within a 280-kb region of Xq12 and provide stable haplotypes useful for complex kinship cases. Int. J. Legal Med. 120, 337–345.

Hussing, C., Huber, C., Bytyci, R., Mogensen, H.S., Morling, N., Børsting, C., 2018. Sequencing of 231 forensic genetic markers using the MiSeq FGxTM forensic genomics system—an evaluation of the assay and software. Forensic Sci. Res. 3 (2), 111–123.

Kidd, K.K., Pakstis, A.J., Speed, W.C., Grigorenko, E.L., Kajuna, S.L., Karoma, N.J., Kungulilo, S., Kim, J.J., Lu, R.B., Odunsi, A., Okonofua, F., Parnas, J., Schulz, L.O., Zhukova, O.V., Kidd, J.R., 2006. Developing a SNP panel for forensic identification of individuals. Forensic Sci. Int. 164, 20–32.

Krawczak, M., 2007. Kinship testing with X-chromosomal markers: mathematical and statistical issues. Forensic Sci. Int. Genet. 1 (2), 111–114.

Liu, Y.S., Meng, H.T., Mei, T., Zhang, L.P., Chen, J.G., Zhang, Y.D., Chen, J., Guo, Y.X., Dong, Q., Yan, J.W., Zhu, B.F., 2017. Genetic diversity and haplotypic structure of Chinese Kazak ethnic group revealed by 19 STRs on the X chromosome. Gene 600, 64–69.

Lyon, M.F., 1961. Gene action in the X-chromosome of the mouse. Nature 190, 372–373.

Meng, H.T., Shen, C.M., Zhang, Y.D., Dong, Q., Guo, Y.X., Yang, G., Yan, J.W., Liu, Y.S., Mei, T., Shi, J.F., Zhu, B.F., 2017. Chinese Xibe population genetic composition according to linkage groups of X-chromosomal STRs: population genetic variability and interpopulation comparisons. Ann. Hum. Biol. 44 (6), 546–553.

Mertens, G., Gielis, M., Mommers, N., Mularoni, A., Lamartine, J., Heylen, H., Muylle, L., Vandenberghe, A., 1999. Mutation of the repeat number of the HPRTB locus and structure of rare intermediate alleles. Int. J. Legal Med. 112, 192–194.

Nothnagel, M., Szibor, R., Vollrath, O., Augustin, C., Edelmann, J., Geppert, M., Alves, C., Gusmão, L., Vennemann, M., Hou, Y., Immel, U.D., Inturri, S., Luo, H., Lutz-Bonengel, S., Robino, C., Roewer, L., Rolf, B., Sanft, J., Shin, K.J., Sim, J.E., Wiegand, P., Winkler, C., Krawczak, M., Hering, S., 2012. Collaborative genetic mapping of 12 forensic short tandem repeat (STR) loci on the human X chromosome. Forensic Sci. Int. Genet. 6, 778–784.

Online Mendelian Inheritance in Man, n.d. OMIM® McKusick-Nathans Institute of Genetic Medicine, Johns Hopkins University, Baltimore, MD. http://omim.org/.

Prieto-Fernández, E., Baeta, M., Núñez, C., Jiménez-Moreno, S., de Pancorbo, M.M., 2015. A new 17 X-STR multiplex for forensic purposes. Forensic Sci. Int. Genet. Suppl. Ser. 5, e283–e285.

Qiagen Inc., 2015. Investigator Argus X-12 Handbook. www.Qiagen.com.

Rodig, H., Kloep, F., Weissbach, L., Augustin, C., Edelmann, J., Hering, S., Szibor, R., Gotz, F., Brabetz, W., 2010. Evaluation of seven X-chromosomal short tandem repeat loci located within the Xq26 region. Forensic Sci. Int. Genet. 4, 194–199.

Rodrigues, R.E.M., Leite, F.P., Hutz, M.H., Palha, T.J., dos Santos, A.K.R., dos Santos, S.E., 2008. A multiplex PCR for 11 X chromosome STR markers and population data from a Brazilian Amazon Region. Forensic Sci. Int. Genet. 2, 154–158.

Ross, M.T., Grafham, D., Coffey, A., et al., 2005. The DNA sequence of the human X chromosome. Nature 434, 325–337. https://doi.org/10.1038/nature03440.

Sanchez, J.J., Phillips, C., Borsting, C., Balogh, K., Bogus, M., Fondevila, M., Harrison, C.D., Musgrave-Brown, E., Salas, A., Syndercombe-Court, D., Schneider, P.M., Carracedo, A., Morling, N., 2006. A multiplex assay with 52 single nucleotide polymorphisms for human identification. Electrophoresis 27, 1713–1724.

Schwartz, J., 2008. Pursuit of the gene: From Darwin to DNA. Harvard University Press, ISBN: 0674034910, pp. 145–183.

Szibor, R., 2007. X-chromosomal markers: past, present, and future. Forensic Sci. Int. Genet. 1, 93–99.

Szibor, R., Krawczak, M., Hering, S., Edelmann, J., Kuhlisch, E., Krause, D., 2003. Use of X-linked markers for forensic purposes. Int. J. Legal Med. 117, 67–74.

Szibor, R., Hering, S., Edelmann, J., 2006. A new web site compiling forensic chromosome X research is now online. Int. J. Legal Med. 120, 252.

Tillmar, A.O., Kling, D., Butler, J.M., Parson, W., Prinz, M., Schneider, P.M., Egeland, T., Gusmao, L., 2017. DNA Commission of the International Society for Forensic Genetics (ISFG): guidelines on the use of X-STRs in kinship analysis. Forensic Sci. Int. Genet. 29, 269–275.

Turrina, S., Atzei, R., Filippini, G., De Leo, D., 2007. Development and forensic validation of a new multiplex PCR assay with 12 X-chromosomal short tandem repeats. Forensic Sci. Int. Genet. 1, 201–204.

Yang, X., Wu, W., Chen, L., Liu, C., Zhang, X., Chen, L., Feng, X., Wang, H., Liu, C., 2016. Development of the 19 X-STR loci multiplex system and genetic analysis of a Zhejiang Han population in China. Electrophoresis 37 (15–16), 2260–2272.

Yang, X., Zhang, X., Zhu, J., Chen, L., Liu, C., Feng, X., Chen, L., Wang, H., Liu, C., 2017. Genetic analysis of 19 X chromosome STR loci for forensic purposes in four Chinese ethnic groups. Sci. Rep. 7, 42782.

Zeng, X.P., Ren, Z., Chen, J.D., Lv, D.J., Tong, D.Y., Chen, H., Sun, H.Y., 2011. Genetic polymorphisms of twelve X-chromosomal STR loci in Chinese Han population from Guangdong Province. Forensic Sci. Int. Genet. 5 (4), 114–116.

Zhang, Y.D., Shen, C.M., Meng, H.T., Guo, Y.X., Dong, Q., Yang, G., Yan, J.W., Liu, Y.S., Mei, T., Huang, R.Z., Zhu, B.F., 2016. Allele and haplotype diversity of new multiplex of 19 ChrX-STR loci in Han population from Guanzhong region (China). Electrophoresis 37 (12), 1669–1675.

CHAPTER

12

Single nucleotide polymorphisms (SNPs): Ancestry-, phenotype-, and identity-informative SNPs

Nicole Novroski PhD

Department of Anthropology, Forensic Science Program, University of Toronto Mississauga, Toronto, ON, Canada

Introduction

Although short tandem repeats (STRs) are traditionally the marker of choice for forensic DNA typing, single nucleotide polymorphisms (SNPs; pronounced "snips") are an additional genetic marker class than can be utilized for identification of human skeletal remains. SNPs provide an abundant form of human genetic variation, with approximately one SNP located every 1000 nucleotides throughout the genome (Brookes, 1999). Given the diversity of genetic variation that is present in the genome, there are certain criteria that polymorphisms must meet in order to be considered a SNP. For example, the least abundant allele at the SNP locus must have a frequency of ≥1% in the sampled population, and any allele present at less than 1% in the genome is typically considered a *rare variant* (Brookes, 1999). In instances in which rare variants (i.e., private polymorphisms) are identified, it is not uncommon that nucleotide differences are represented by only a small number of individuals (or potentially only a single individual).

SNPs present themselves in unique ways within the human genome. First, a *transition mutation* can arise, in which a purine replaces a purine (A→G or G→A) or a pyrimidine replaces a pyrimidine (C→T or T→C). Transitional SNPs tend to exist as silent substitutions in the genome. Fig. 1 illustrates an example of a SNP between two individuals. The other three possibilities can be accounted for by *transversion mutations* (A→T or T→A; A→C or C→A; C→G or G→C). Two thirds of all SNPs occur due to transition mutations, which result in bi-allelic loci. The remaining one-third are comprised of equal frequencies of the three possible transversion mutations (Brookes, 1999). In principle, SNPs can exist as bi-, tri-, or tetra-allelic polymorphisms at any given locus (Fig. 2). In humans, however, tri- and tetra-allelic

Forensic Genetic Approaches for Identification of Human Skeletal Remains
https://doi.org/10.1016/B978-0-12-815766-4.00012-1

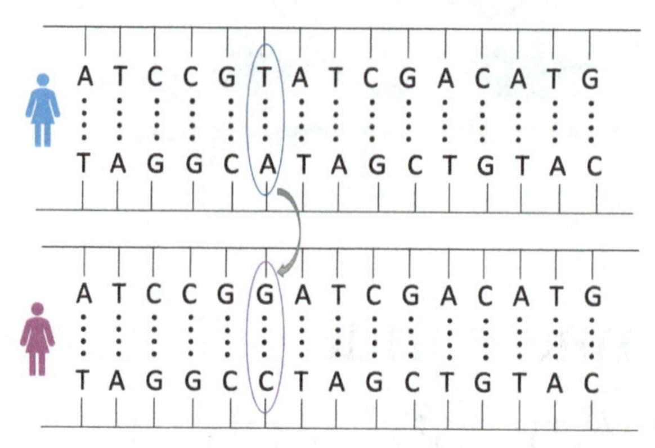

FIG. 1 A single nucleotide polymorphism (SNP) observed at a random nucleotide in the human genome.

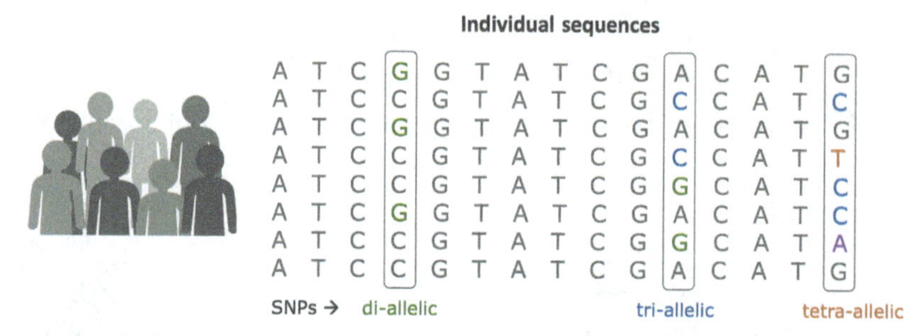

FIG. 2 SNPs can be di-allelic (two possible alleles/nucleotides), tri-allelic (three possible nucleotides), or tetra-allelic (all four nucleotides A, C, T, G) at a locus.

SNPs are rare. Bi-allelic SNPs (typically referred to as di-allelic) are the most common form of SNP present throughout the human genome.

When considering the number of possible genotypes at a SNP locus, one must consider the number of possible allelic combinations. In the simplest case of a di-allelic locus, there are three possible genotypes (Butler et al., 2007) (Fig. 3). For example, if the alleles for a SNP locus are A and B [in which "A" could represent any of the four nucleotides (C, T, A, or G) and "B" could also be a C, T, A or G], the three possible genotypes would be AA, AB, or BB.

SNPs are commonplace in human populations. As a consequence of the high frequency of SNPs throughout the genome, it is not surprising to identify up to several million SNP differences between any two individual DNA sequences. However, from an evolutionary perspective, many SNPs observed in the human genome have been present since before the human migrations out of Africa (Brookes, 1999). Approximately 85% of all SNPs found in the human genome are common to all global populations; only 15% of SNP differences between individuals can be accounted for by private or population-specific frequencies (Brookes, 1999; Holden, 2002). Therefore, given the random nature of mutation coupled with a relatively low rate of nucleotide substitution per generation (approximately 10^{-8} changes per nucleotide per generation), SNPs can be considered a relatively stable form of variation in the human genome.

Alleles	Genotypes
A	AA
G	GG
	AG

FIG. 3 When a SNP presents as a di-allelic locus, three different genotypes are possible. In this example, the di-allelic locus presents as either an A or G, giving rise to either an AA, GG, or AG genotype.

SNPs are valuable human identity and inference markers from a forensic genetic perspective, but one cannot underestimate the significant role that SNPs may play in human development and function. SNPs ultimately contribute to diversity among individuals, genome evolution, and govern externally visible characteristics (EVCs) such as eye and hair color. SNPs also confer interindividual phenotypic variation that impacts drug metabolism and response pathways as well as complex diseases (e.g., diabetes, obesity, hypertension, psychiatric disorders) (Brookes, 1999). When SNPs present themselves within the noncoding regions of the human genome, they are often considered *silent mutations*, meaning that there is no immediate consequence to the functionality of the organism as a result of the newly incorporated polymorphism. However, when SNPs are introduced in the coding regions of some genes, they may change the encoded amino acids (i.e., *non-synonymous mutations*) or remain silent (i.e., *synonymous mutations*). Finally, SNPs have the potential to influence promoter activity in regulatory regions of the genome, which may affect gene expression or alter messenger RNA (mRNA) conformation and stability (Brookes, 1999). The consequence of each SNP is dependent on both the type of polymorphism that occurs and its location within the genome.

History of SNP use in forensic genetics

With the advent of the polymerase chain reaction (PCR) in 1985, forensic approaches for genetic marker testing were revolutionized, as the analysis of picogram (pg) levels of DNA became possible (Mullis et al., 1986). One of the first post-PCR applications that were introduced to the forensic genetics community was the use of SNP-based marker systems. Allele-specific oligonucleotide (ASO) hybridization probes using a reverse dot-blot typing method were implemented to evaluate sequence polymorphisms at the HLA-DQA1 locus and the PolyMarker (PM) loci (i.e., LDLR, GYPA, HBGG, D7S8, Gc) (Budowle et al., 1995; Comey and Budowle, 1991; Gyllensten and Erlich, 1988; Walsh et al., 1991). Although these SNP-based assays were cost-effective and offered high sensitivity of detection, they lacked the power of discrimination that was possible using variable number of tandem repeat (VNTR) loci coupled with restriction fragment length polymorphism (RFLP) techniques (Budowle et al., 1991). Further, ASO reverse dot-blot methods were ineffective at DNA mixture deconvolution due to the increased allele-sharing between individuals at the loci being typed.

Since the early 1990s, PCR-based short tandem repeat (STR) typing replaced VNTR-RFLP methods to become the gold standard for DNA analysis in forensic human identification (Edwards et al., 1991; Whitaker et al., 1995; Budowle et al., 1998). However, many sample types (e.g., ancient specimens or poorly preserved human skeletal remains) are not particularly amenable to traditional STR typing methods. Typically, these types of samples contain too little autosomal template DNA or have undergone substantial DNA degradation. Alternate approaches, such as characterization of the mitochondrial DNA (mtDNA) genome, have been attempted as a means to circumvent challenges of low quantity and low-quality DNA; however, the power of discrimination using lineage-based markers (e.g., mtDNA) can be limiting (Ballantyne, 1997; Wilson et al., 1995; Budowle et al., 2003). Autosomal SNPs are a potential ideal marker for degraded and inhibited human remains samples due to their small target size, i.e., one nucleotide. Primers designed to capture individual SNPs may have amplicon sizes less than 100 nucleotides, making SNP loci an ideal marker choice for degraded samples. DNA fragment sizes recovered from human remains samples can be small, and therefore, are often not amenable to traditional STR analysis using PCR and capillary electrophoresis (CE) methods. Because SNP-based approaches interrogate very small amplicons, they are particularly suited for degraded samples, such as those that might be encountered in missing persons, mass disaster, and ancient remains cases.

SNP typing methodologies

Due to the inherent small size of SNP loci, there exist many suitable methods for SNP genotyping. Selection of an appropriate technique should depend on the user requirements and the facility capabilities. Typically, when applying SNP typing methods for characterization of human skeletal remains, the focus often becomes high accuracy in the genotyping results coupled with increased multiplexing capabilities of the instrumentation or technology. Described below are several common methods utilized in forensic genetic laboratories to characterize SNP loci. While this list is extensive, new genotyping methods, chemistries, and platforms are continuously being developed, and referring to the literature will always be the most efficient way to remain current with respect to the totality of options available for SNP typing.

Allele-specific oligonucleotide hybridization

Allele-specific oligonucleotide (ASO) hybridization (also referred to simply as allele-specific hybridization) relies on distinguishing between two alleles at a locus that differ at one nucleotide position (Wallace et al., 1979; Budowle et al., 1995; Comey and Budowle, 1991). Using a probe-based design, unique ASOs are created that are complementary for each allele at each locus being typed, usually with the polymorphic nucleotide in a central position in the probe sequence. Each ASO probe is applied directly to a nylon membrane strip, and PCR-amplified alleles will hybridize to any complementary immobilized probes present. Under optimized assay conditions, only the perfectly matched probe-target hybrids remain stable; hybrids containing a nucleotide mismatch become unstable. The use of ASO hybridization was one of the original forensic SNP typing methods introduced for human identification purposes and targeted two loci: (1) the HLA-DQA1 locus and (2) Polymarker (PM) loci (Budowle et al., 1995; Comey and Budowle, 1991). Although still used by some laboratories worldwide, this method has since been replaced by approaches with increased accuracy, power of discrimination, and sensitivity. Thus, ASO hybridization is no longer considered a common methodology for forensic SNP typing.

Taqman 5′ nuclease assays

Similar to real-time quantification methods, fluorescent probes consisting of a reporter dye and a quencher dye are added to each sample in a PCR reaction. This method relies upon 5′ nuclease activity of *Taq* polymerase, which ultimately will displace and cleave an oligonucleotide probe that has hybridized to the target DNA (Holland et al., 1991). An important feature of this method is the use of individual probes for each allele at the locus. Typically, in a two-probe model (e.g., used with loci that are di-allelic), one probe will be complementary to a predefined wild-type allele and a second probe is designed to be complementary to the variant allele. Each probe has a unique fluorophore (i.e., reporter dye) attached to its 5′ end and another fluorophore (i.e., a "quencher" dye) attached to the 3′ end. Amplification results in cleavage of the probe by *Taq* polymerase, cleaving and releasing the 5′ reporter dye from the substrate (which ultimately results in detectable fluorescent signal due to physical separation from the quencher dye, which suppresses fluorescence when the probe is intact). Determination of the sample genotype is possible by measuring the signal (fluorescence) intensity captured by the instrument (Holland et al., 1991; Sobrino et al., 2005).

Pyrosequencing

Pyrosequencing uses sequencing-by-synthesis, which involves an enzyme cascade system. This technology is typically capable of providing real-time determination of a string of 20–30 nucleotides of the target DNA strand (Ronaghi et al., 1996; Ronaghi et al., 1998; Sobrino et al., 2005). In order to determine the exact nucleotide sequence of the template DNA strand, deoxyribonucleotide triphosphates (dNTPs; synthetic nucleotides) are added in a specific order to the reaction, and incorporation of a particular dNTP to the growing DNA sequence of the complementary (synthetic) strand will result in release of pyrophosphate and light (Ronaghi et al., 1996; Ronaghi et al., 1998). This light signal is detected, and the instrument registers the dNTP that was incorporated. This process is repeated for the entirety of the DNA strand being sequenced, until all data are detected and recorded (Sobrino et al., 2005). Pyrosequencing is still a viable technology for SNP typing; however, massively parallel sequencing (MPS) offers higher throughput and greater multiplexing capabilities (as discussed in the MPS section of this chapter, and in Chapter 16).

Minisequencing (SNaPshot assays)

Considered one of the most common SNP typing assays, the minisequencing reaction involves an allele-specific primer that anneals to its target DNA immediately adjacent to the SNP of interest (Sanchez et al., 2003; Daniel et al., 2015; Mehta et al., 2017). This method, often referred to as single base extension (SBE), relies on the high accuracy of nucleotide incorporation by DNA polymerases (Sokolov, 1990; Syvanen et al., 1990; Kuppuswamy et al., 1991). Prior to SBE, the DNA template is subjected to multiplex PCR (i.e., copying of multiple targets simultaneously) to generate millions of copies of amplicons containing the SNPs of interest. Purification of PCR product is performed via exonuclease activity, in order to degrade any unbound primers and unincorporated dNTPs from the PCR reaction. Minisequencing products are often analyzed using spatial separation via electrophoresis coupled with fluorescence detection. Thus, during minisequencing, the 3′ end of the oligonucleotide (detection) primer binds immediately adjacent to the SNP of interest and is extended by the polymerase to incorporate a fluorescently labeled di-deoxynucleotide triphosphate (ddNTP; chain-terminating synthetic nucleotide) complementary to the base adjacent the SNP (Sokolov, 1990; Syvanen et al., 1990; Kuppuswamy et al., 1991). Further, incorporation of an oligonucleotide tail can be added to the 5′ end of each primer to assist with spatial separation of each fragment during capillary electrophoresis (CE). The biggest limitation with minisequencing technology is that the multiplexing capability is frequently maximized at ten SNP loci (Sanchez et al., 2003; Mehta et al., 2017).

Massively parallel sequencing

Massively parallel sequencing (MPS), often referred to as next-generation sequencing (NGS), is a high-throughput alternative to the previously mentioned SNP typing methods. MPS allows for increased multiplexing capabilities, in which hundreds of targets (or thousands of targets, if using a chip-based assay) can be simultaneously sequenced in a single sequencing reaction (Berglund et al., 2011). Multiple manufacturers have designed MPS instruments that utilize unique chemistries in order to sequence their genomic targets (refer to Chapter 16). However, what remains consistent across all approaches is the ability to target multiple amplicons (both small and large) with high accuracy and resolution. Furthermore, MPS-based methods allow for simultaneous multiplexing capabilities of multiple marker types; that is, SNPs can be sequenced simultaneously with other length and sequence-based polymorphisms, therefore maximizing the data output from a single (often very limited) forensic DNA extract.

SNP databases for forensic genetic applications

SNP databases have now become commonplace in the global forensic genetic research community. Public access to SNP records for thousands of human samples is freely available. The SNP Consortium (TSC) is considered one of the first global SNP databases and was established in 1999 with the goal of creating a high-density SNP map of the human genome (Thorisson and Stein, 2003). The massive effort from the TSC resulted in several million mapped and characterized human SNPs that have since been shared with multiple public databases, including the renowned dbSNP database, which is housed at the National Institutes of Health's (NIH) National Center for Biotechnology Information (NCBI) (http://www.ncbi.nlm.nih.gov/SNP). The original compendium of SNP records generated by the TSC contributed to the launch of the HapMap project in 2002. The HapMap project was an international collaboration between researchers from academic institutions, private companies, and nonprofit biomedical research companies from Canada, China, Japan, Nigeria, the United Kingdom, and the United States (The International HapMap Consortium, 2003). The HapMap dataset includes thousands of SNPs in 270 individuals from African, European, and Asian populations (https://www.genome.gov/10001688/international-hapmap-project).

The dbSNP database is currently one of the most comprehensive SNP repositories used worldwide. As a public domain, dbSNP includes records relating to single nucleotide variations, microsatellites, and small-scale insertions and deletions in the human genome. The online database also includes publication and contributor information, population frequency, molecular consequence(s) of the polymorphism as well as genomic mapping information for both common variants and clinical mutations (Sherry et al., 2001).

The GWAS Catalog, founded in 2008 by the NIH National Human Genome Research Institute (NHGRI), is a collaborative database in partnership with the European Bioinformatics Institute (EMBL-EBI). It offers a variety of search tools to mine the human genome for SNPs based on gene and genomic location, trait information, and associated publications (https://www.ebi.ac.uk/gwas/). Additionally, the forensic community often relies on the National Institute of Standards and Technology (NIST) for forensic genetic marker resources. NIST has created a website that maintains a compendium of genetic marker database information and relevant resources for both practitioners and researchers (https://strbase.nist.gov/SNP.htm).

The Kidd Lab at Yale University has established ALFRED (ALlele FREquency Database), which acts as a repository of global allele frequency data (Osier et al., 2001). While dbSNP and STRbase (which characterizes forensically relevant STRs) focus specifically on forensic loci, ALFRED was designed for both research and education in a variety of areas in human genetic diversity and molecular anthropology. This repository contains extensive information on human SNPs and INDELS (insertions/deletions), including identity-informative SNPs (iiSNPs) and ancestry-informative SNPs (aiSNPs) routinely utilized for forensic applications. The ALFRED database is consistently updated and is freely accessible through the web interface https://alfred.med.yale.edu.

As researchers continue to share their data via global platforms, the availability of SNP loci and accompanying population data will continue to improve. Free access to large SNP databases allows for a more comprehensive understanding of the frequency of SNPs in various human populations, and subsequent exploitation of SNPs for forensic applications. SNPs have a very straightforward nomenclature, which has resulted in a substantially more straightforward approach to information exchange, data handling, and general reporting of SNP genotypes.

SNPs in forensic DNA mixtures

DNA mixtures have become commonplace in forensic biological casework. Whether as a consequence of contamination, poor preservation of forensic evidence, or potential technical error(s) made by the analyst, DNA mixtures can destroy the potential informativeness of DNA evidence if a reliable interpretation and accounting of all contributors is not possible. Additionally, when compared to STRs for inferring identity, SNPs have an overall lower power of discrimination on a "per-locus" basis. The literature suggests that approximately 2.6-to-4 times the number of SNPs are necessary to offer the same level of informativeness as an average forensic STR locus (Brookes, 1999; Gill, 2001; Sobrino et al., 2005). Thus, SNP multiplexes often tend to include a greater number of genomic targets (i.e., >40) when utilized for inferring human identity. However, SNPs offer two important benefits over STRs in challenging forensic scenarios: (1) the design of much shorter amplicons for PCR is possible, which is especially important when DNA in the sample is highly degraded or in low quantity; and (2) the lower mutation rate (compared to STRs) can be useful for kinship and lineage inferences in cases where STR genotyping yields an ambiguous result (Gill, 2001; Budowle, 2004).

A greater amount of allele-sharing is the primary outcome of typing di-allelic SNPs for human identity purposes. However, as multiplexing capabilities for genetic marker genotyping continue to expand (especially in the era of MPS), improved discrimination power will be possible using di-allelic SNP-based assays. In addition, not all SNPs are binary, meaning that not all SNPs present as di-allelic loci (Brookes, 1999). Nonbinary SNPs have increased power of discrimination due to the increased number of alleles (and therefore genotypes) that are possible at each locus. Hence, when thinking about mixture deconvolution applications using SNP-based approaches, SNPs with the highest possible discrimination power should be included in multiplexes. Phillips et al. (2015) mined the 1000 genomes data for tetra-allelic SNPs and their overall value in forensic typing panels (1000 Genomes Project Consortium et al., 2005). Their study targeted markers that met a minimum overall heterozygosity of 25% and an overall minimum allele frequency (MAF) of 0.008 for the least frequent of the four alleles (A, C, T, or G) in the tetra-allelic loci. Although the authors found only a few tri- and tetra-allelic SNP loci suitable for forensic analysis (i.e., 961 in total, with only 160 loci meeting the minimum heterozygosity thresholds), their overall findings suggested that highly polymorphic SNPs can play a significant role in mixture deconvolution efforts when working with low-quality and low-quantity DNA samples (Phillips et al., 2015).

Overall, SNPs may present unique challenges when attempting to interpret complex DNA mixtures, especially in instances where the integrity of the forensic sample may be compromised. Although improved statistical capabilities have been developed (Gill et al., 2015), challenges may still prevent the analyst from determining the difference between a true heterozygote SNP genotype and a mixture containing two homozygote genotypes, or a DNA profile that contains both a heterozygote and a homozygote. Therefore, it will be important to assess the quantitative information from SNP allele calls when attempting to decipher complex DNA mixtures (although mixtures should not be an issue with skeletonized remains if proper contamination prevention measures are taken during processing).

SNP classes for forensic investigative and intelligence purposes

Di-allelic SNPs present with a very specific set of challenges as follows: (1) di-allelic SNPs are not as informative on a per-locus basis as their STR counterparts; (2) SNPs are oftentimes incapable of deconvoluting DNA mixtures; and (3) there currently does not exist a forensic SNP database that can compare SNPs to the CODIS-style forensic STR databases that exist worldwide. However, SNPs continue to be useful in a diversity of forensic investigative and intelligence applications. Relatively low mutation rates in SNPs allow them to function as stable genetic markers in the human genome, which increases their utility in lineage-based casework, especially in situations where a direct reference sample is not available (Romanini et al., 2012).

SNPs for forensic investigative and intelligence analyses can typically be classified into four main categories:

1. *Identity-informative SNPs (iiSNPs)*

   These types of SNPs are selected for individualization (similar to the use of STRs for human identification). Criteria used to select SNPs in this category include high heterozygosity (as observed with tri- and tetra-allelic SNPs) and a low coefficient of inbreeding (F_{st}) (i.e., population heterogeneity is minimized).

2. *Lineage-informative SNPs (liSNPs)*
   These types of SNPs are often used for genealogy and kinship analyses and are often tightly linked (and therefore function more like haplotype markers).
3. *Ancestry-informative SNPs (aiSNPs)*
   aiSNPs assist to predict the biogeographical ancestry (BGA) of an individual and may be used to infer some phenotypic information for investigative lead/intelligence value. Criteria used to select SNPs in this category include low heterozygosity and high Fst values (i.e., heterogeneity between populations is maximized).
4. *Phenotype-informative SNPs (piSNPs)*
   piSNPs estimate phenotypic variation as it relates to externally visible characteristics (EVCs) (e.g., hair color, eye color, skin pigmentation, freckling) for investigative lead/intelligence value.

Identity-informative SNPs

Identity-informative SNP (iiSNP) profile determination is an alternate method (to STR genotyping) that forensic geneticists can use to individualize a DNA sample. Like forensically relevant STRs, iiSNPs are typed to generate a unique DNA profile that can be used to differentiate between individuals. SNP targets can be found in the autosome, the Y chromosome, the X chromosome, and within the mitochondrial genome. For human skeletal remains cases, an iiSNP profile can be used to exclude individuals as being the source of the evidentiary sample or to determine the inclusion or exclusion of a putative family member. Ideally, when selecting iiSNPs for individualization, high heterozygosity (i.e., 50% when a locus is di-allelic) and low F_{st} are desirable characteristics, in order to minimize the total number of SNPs needed to reach high levels of discrimination and, in turn, rely on fewer reference population datasets for statistical assessments in forensic casework (Sanchez et al., 2006). Despite the fact that SNPs tend to be di-allelic and may be limited in their power-of-discrimination when compared locus-by-locus to STRs, SNPs are useful for establishing identity by descent (IBD) due to their smaller mutation rate ($\approx 10^{-8}$) (compared to average mutation rate of STRs at $\approx 10^{-3}$).

Although typical STR amplicons range from 100 to 450 nucleotides (base pairs), amplification of forensically relevant SNPs involves much smaller targets, i.e., one nucleotide (Brookes, 1999; Sobrino et al., 2005). Therefore, it is possible to design PCR primers for SNP amplification to target substantially smaller genomic fragments, typically in the range of 100 base pairs or less. With the advent of MPS, the possibility of constructing large multiplex assays (coupled with automation) is now possible, which increases overall throughput and quality control. MPS also offers the absence of extraneous artifacts (e.g., stutter), which are routinely encountered when genotyping forensically relevant STRs on CE-based platforms. Hence, the resulting sequence data using SNP-based methods requires less complex interpretation approaches.

Commercially available kits for iiSNP genotyping

The forensic community is constantly developing novel approaches and/or multiplexes to type challenging samples for human identity purposes. The 21-locus autosomal SNP panel by Dixon et al. (2005) and a multiplex assay comprising 52 SNP targets (Sanchez et al., 2006) were the first major identity SNP panels to be developed in European forensic laboratories.

In 2006, a team at Yale University focused on developing a fundamental set of criteria needed to develop robust and useful iiSNP panels for forensic applications. Starting with a candidate set of 90,000 SNPs, they refined the panel using key criteria (i.e., high heterozygosity, low F_{st}) and reported that only nineteen iiSNPs met the heterozygosity and F_{st} parameters in all forty population groups tested. This study demonstrated that selection of iiSNPs can be challenging, and that inclusion of non-robust iiSNP markers in a human identity panel has many limitations if the loci are not sufficiently discriminatory (Kidd et al., 2006). The nineteen loci that were established to be of highest identification value have been adopted into a variety of multiplexes from other research groups, including commercial entities that market iiSNP panels to the forensic community.

Commercial entities have also focused on characterizing iiSNPs for forensic genetic applications. Verogen (formerly Illumina; San Diego, California USA) manufactures the ForenSeq™ DNA Signature Prep Kit, an MPS-based multiplex which comprises a variety of diverse and forensically relevant genomic targets, and recommends one nanogram (ng) of DNA input. Within the primer mix, 94 iiSNPs (ranging in amplicon size from 63 to 231 nucleotides) are included; and iiSNP loci from Sanchez et al. (2006) and Kidd et al. (2006) are incorporated into the panel. The specific iiSNPs used, chromosomal coordinates, and reference allele information are available on the manufacturer website: (https://support.illumina.com/content/dam/illumina-support/documents/documentation/chemistry_documentation/forenseq/forenseq-dna-signature-prep-guide-15049528-01.pdf).

Applied Biosystems™, a subsidiary of Thermo Fisher Scientific (Waltham, Massachusetts, USA), has designed an MPS-based multiplex known as the Precision ID Identity Panel. This multiplex assay advertises discrimination power comparable to STR genotyping assays (i.e., between 1×10^{-31} and $6\text{x}10^{-35}$). The panel encompasses iiSNPs from preexisting multiplexes (e.g., Pakstis et al., 2010; Phillips et al., 2007), for a total of 124 iiSNPs that can be assayed in a single reaction. The Applied Biosystems™ Precision ID Identity Panel was optimized for use with degraded or challenging forensic samples and requires only 1 nanogram (ng) of input DNA (a common desired input amount for many forensic applications). Although skeletonized or highly decomposed human remains often yield DNA quantities well below the optimal 1-ng target, this panel has been tested and demonstrated acceptable performance using input quantities as low as 100 pg. Amplicon sizes range from 132 to 141 nucleotides in length. The specific iiSNPs included chromosomal coordinates, and reference allele information are available on the manufacturer website (https://assets.thermofisher.com/TFS-Assets/LSG/manuals/MAN0017767_PrecisionID_SNP_Panels_S5_UG.pdf).

In collaboration with QIAGEN Corporation (and designed for use on the QIAGEN GeneReader NGS System), the International Commission on Missing Persons (ICMP) and a number of respected forensic genetic leaders (e.g., at the University of Santiago de Compostela, Linkoping University, the Swedish National Board of Forensic Medicine) have developed one of the largest forensic SNP-based assays to be used for forensic purposes (Peck et al., 2018). In its current state, the multiplex includes ~1200 SNP loci (including autosomal and X-chromosomal targets) designed specifically for missing persons applications. The ICMP panel targets many tri-allelic SNPs with high heterozygosity, coupled with a subset of carefully selected microhaplotype loci adapted from Kidd et al. (2014a). Furthermore, the amplicons were designed to ensure that the multiplex is effective when typing heavily degraded samples. Approximately 95% of the primer pairs are less than 100 nucleotides from their respective target, and at least 75% of those primer pairs are less than fifty nucleotides from their respective target, ensuring the smallest possible amplicons in order to have the greatest possible typing success. From the current published simulated data, this multiplex has demonstrated significant utility in kinship analysis. The panel has demonstrated a robust ability to produce high-certainty matches using single first cousin(s) as reference samples (Peck et al., 2018).

An important consideration when targeting iiSNPs (or any genetic markers) in forensic assays is the potential characterization of off-target, or ancillary, phenotypic information. Wendt and Novroski (2019) demonstrated the potential for associations between iiSNPs in current forensic multiplexes with phenotypic traits or characteristics (sometimes EVCs) found in GWAS data that is freely available online. Hence, as forensic experts continue to push the envelope with selection of novel forensic markers in human identity multiplexes, it will be important for the community to consider the genetic proximity to phenotypically informative loci, especially when selecting novel *phenotypically uninformative* loci is the goal. Additionally, if phenotypic information can be derived, investigators should ensure that the totality of individualizing genetic information is disclosed.

Lineage-informative SNPs

Lineage-informative SNPs (liSNPs) are loci that typically reside in mitochondrial DNA (mtDNA) and on the Y-chromosome. liSNPs have been well-documented in the literature due to their lack of recombination and a low mutation rate, and because they offer substantial degrees of informativeness for evolutionary studies and kinship analyses (Daly et al., 2001; Gabriel et al., 2002; Budowle et al., 2004; Coble et al., 2004; Ge et al., 2010). The usefulness of liSNPs is especially evident in investigations in which the reference sample(s) and the evidence sample(s) are separated by several generations. To date, very few validated liSNP assays have been developed for forensic applications. Coble et al. (2004) developed a set of eight multiplex panels which targeted a total of fifty-nine SNPs that defined Caucasian mtDNA haplotypes from hypervariable regions I and II (HVI/HVII) into eighteen specific groups.

The use of liSNPs will be most useful for missing person cases and/or mass disaster identifications. Although liSNPs from mtDNA or the Y-chromosome have limited power of discrimination due to uniparental inheritance, haploblocks (i.e., several neighboring, tightly linked SNPs that are inherited together) may serve as useful lineage-based markers and could play a role in forensic genealogy applications (Daly et al., 2001; Gabriel et al., 2002; Coble et al., 2004).

Ancestry-informative SNPs (aiSNPs) for prediction of biogeographic ancestry

In some forensic cases, especially those involving skeletonized or highly decomposed human remains, the inference of biogeographic ancestry (BGA) from the recovered DNA sample may indirectly provide information about the general appearance of a person, which could provide some investigative lead or intelligence value in the case. Although forensic STRs have proven to be powerful identity markers, they are poorly suited for inferring the biogeographical ancestry of an individual due to the high degree of locus-specific allele-sharing within populations (Shriver et al., 1997; Pritchard et al., 2000a, b; Rosenberg et al., 2002; Shriver et al., 2003). Additionally, while liSNPs (i.e., Y-STR and mtDNA haplotypes) are used for recreating the evolutionary history of human populations and may infer general predictions of the genetic ancestry of an individual (Daly et al., 2001; Gabriel et al., 2002; Coble et al., 2004), uniparental inheritance and limited representation of the human genome prevent these genetic markers from being useful predictors of an individual's unique ancestral composition.

In recent years, an influx of forensic genetic research has explored the use of genetic targets as a tool for reconstructing population histories, primarily by exploring how genetic markers infer population origin, evolution, migration, and admixture (Shriver et al., 1997; Pritchard et al., 2000a, b; Rosenberg et al., 2002; Shriver et al., 2003). From a forensic perspective, determination of biogeographic ancestry and characteristics of physical appearance from limited evidence can offer probative information not achievable using routine STR genotyping, especially in the absence of a viable reference sample. Population genetics research has revealed that gene frequencies vary between populations, offering the potential for localizing a specific source population from any particular ancestry-informative SNP (aiSNP) profile. In the simplest of terms, ancestry-informative markers (AIMs) are SNPs that are distributed throughout the genome and occur at very different frequencies among different world populations (Shriver et al., 1997; Pritchard et al., 2000a, b; Rosenberg et al., 2002; Shriver et al., 2003; Yang et al., 2005; Cheung et al., 2018). Private polymorphisms have been identified, appear to be restricted to individual populations, and aid in the refinement of placing individuals in uniquely diverged populations.

However, a number of difficulties can be encountered when attempting to estimate the BGA of an individual. For example, private polymorphisms (or rare variants) have the potential to be so rare that they are not useful from a forensic standpoint (i.e., meaning that, because so few individuals possess the rare variant, the polymorphism may not aid in BGA determination for the individual). Alternatively, when a frequently occurring polymorphism is present, it may lack sufficient resolution to any particular population, creating another uninformative BGA assignment. It should be noted, however, that sufficient researches into population-specific AIMs have revealed large SNP datasets that are useful for BGA inference using aiSNPs.

The number of critical aiSNPs necessary for correct BGA inference is always shifting. Halder et al. (2008) developed a megaplex of 176 autosomal AIMs for estimating individual biogeographical ancestry and admixture from four continental population groups. In 2009, Kosoy et al. developed a 128-locus AIM panel for determining continental origin and admixture proportions for seven common populations in America. The SNPforID Consortium is another pioneering group to infer ancestral origin using a panel of AIMs. The consortium's original design involved a single-tube 34-Plex SNP assay capable of classifying individuals into three ancestral population groups: African, European, and East Asian (Phillips et al., 2007). Nievergelt et al. (2013) used informativeness measures to select 41 AIMs to distinguish populations from seven continental regions: Africa, the Middle East, Europe, Central/South Asia, East Asia, the Americas, and Oceania. In 2014, groups from around the world were actively developing novel aiSNP panels for BGA determinations in forensic casework (Gettings et al., 2014). Kidd et al. (2014b) developed an efficient and globally useful panel of 55 AIMs and tested 73 populations from around the world. Phillips et al. (2014) set out to identify the most divergent markers by focusing on differentiating between five global population groups: Africans, Europeans, East Asians, Native Americans, and Oceanians. This study investigated the use of a total set of 128 aiSNPs, and refined multiplexes of 88, 55, 28, 20, and 12 AIMs, so that each multiplex could be customized for different population datasets and for different forensic chemistries/technologies. In 2015, Zeng et al. developed an aiSNP panel which focused on loci that had one allele fixed in a specific population, whereas that same allele would be completely absent in other populations. The multiplex comprised 23 aiSNPs, with the intent to use a minimum number of aiSNPs to characterize individuals into one of four major American populations: African American, East Asian, European American, and Hispanic American (Zeng et al., 2016). The findings demonstrated that individuals were assigned to their *expected* groups, therefore supporting the use of the 23 aiSNPs for U.S. population samples.

Commercial entities have created multiplexes that allow for BGA inferences to be made using well-validated aiSNPs. There are currently two ancestry kits developed for forensic BGA estimation, designed for MPS platforms: (1) the ForenSeq™ DNA Signature Prep Kit (Verogen), which became commercially available in 2015, and (2) the Applied Biosystems™ Precision ID Ancestry Panel (Thermo Fisher Scientific), which was originally launched in 2014. The ForenSeq™ DNA Signature Prep Kit prepares samples for sequencing of 56 aiSNPs, ranging in size from 67 to 200 nucleotides (nt). Verogen's Universal Analysis Software (UAS) incorporates principle component analysis (PCA) to estimate the BGA of each sample using the European, East Asian, and African (excepting ASW) super populations data from the 1000 Genomes Project dataset (1000 Genomes Project Consortium et al., 2005). The result is a classification of each sample into one of four population groups: Admixed American, African, East Asian, or European. Fig. 4 demonstrates a simple prediction of ancestry and phenotype for the author and illustrates how the output data of the ForenSeq™ DNA Signature Prep Kit is presented to the user.

The Precision ID Ancestry Panel is much larger and evaluates 165 autosomal aiSNPs, with 55 aiSNPs selected from Kidd et al. (2012) and the remaining 123 aiSNPs selected from the extensive number of AIMs characterized by Kosoy et al. (2009). Although the Precision ID Ancestry Panel has demonstrated high discriminative power among seven continental level

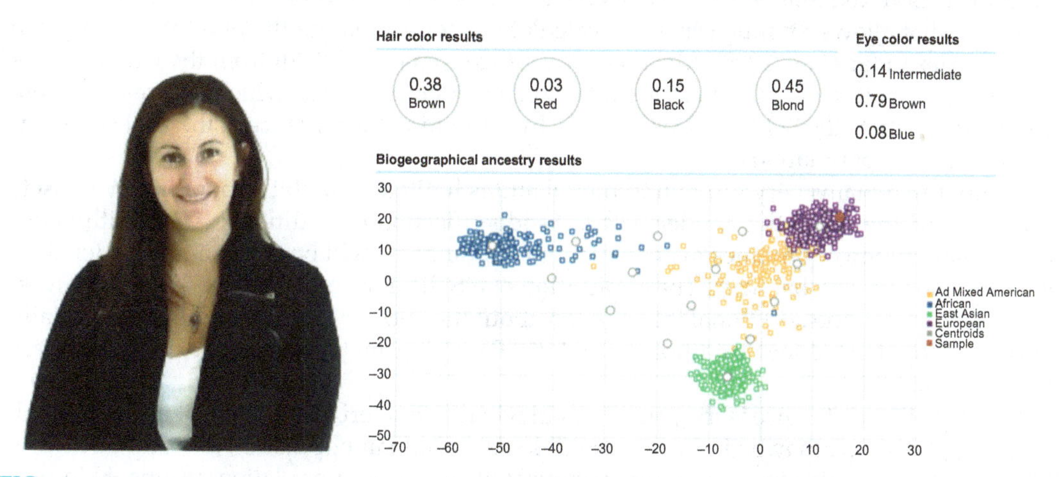

FIG. 4 Phenotypic and ancestry predictions for the author, Dr. Nicole Novroski, using the ForenSeq™ DNA Signature Library Prep Kit. Dr. Novroski identifies as European, with natural brown hair and brown eyes.

populations (i.e., Europe, Oceania, East Asia, Sub-Saharan Africa, South Asia, North America, Middle East/North Africa) for forensic ancestry inference, the panel has limitations when estimating BGA of individuals in closely related subpopulations. This limitation remains one of the greatest constraints for all panels that utilize aiSNPs to infer BGA in forensic identification cases.

In response to growing datasets of BGA SNP information, several databases have been developed. ALFRED (https://alfred.med.yale.edu), the SNIPPER Bayesian classification system (http://mathgene.usc.es/snipper), and the Estonian Biocentre (http://evolbio.ut.ee) represent only a few of the many tools available for users exploring BGA determination using AIM and aiSNP data. Biostatistical predictive tools have also been developed to assist in BGA inferences for forensic genetic applications. The primary approach to BGA determination incorporates maximum likelihood estimates of the AIM profile (multi-locus genotype) in various populations (defined by the software and dependent on reference databases), in which the population with the highest likelihood is assigned as the most likely population-of-origin of the sample being assessed (Pritchard et al., 2000a, b). This is achieved by evaluating similar patterns of genetic variation observed between individuals in different population groups. Individuals who share similar patterns of genetic variation are grouped together, in order to give rise to various groups, or clumps of individuals, based on their genetic differences. Furthermore, BGA tools attempt to infer different degrees of admixture to each sample by evaluating the presence of distinct populations and assigning samples within these admixed groups. The program STRUCTURE, developed at Stanford University, is an example of a software tool that functions under these parameters. STRUCTURE is designed to analyze human population samples, infer the presence of distinct populations, and assign individuals to populations (Pritchard et al., 2000a, b). STRUCTURE also studies hybrid zones and attempts to identify migrants and admixed individuals (as well as estimate population allele frequencies in situations where many individuals are migrants or admixed) (Pritchard et al., 2000a, b; Hubisz et al., 2009). ADMIXTURE, another biostatistical predictive tool that works similarly to STRUCTURE, utilizes a maximum likelihood estimation of individual ancestries using multi-locus SNP datasets. Unlike STRUCTURE, however, ADMIXTURE utilizes a different algorithm that allows for much speedier calculations for estimating the BGA of an unknown sample (Alexander et al., 2009). Most recently, Mogensen et al. (2020) from the University of Copenhagen created *GenoGeographer* (http://genogeographer.org), which utilizes a z-score method to evaluate the BGA of AIMs markers by estimating the variance of the allele frequencies in reference populations.

Admixture remains very difficult to model and is limited by both the population datasets available and the predictive models used. As identification of additional AIMs with better discrimination power becomes available, prediction models will be better able to infer BGA of individuals on a global scale. The application of aiSNPs for human skeletal remains investigations may provide additional information about the ancestry of the individual, especially when anthropological and traditional DNA (STR) typing methods are not possible or are unsuccessful.

Although BGA inference is the primary goal when characterizing aiSNPs of an individual, assessing phenotype may be indirectly possible using human population ancestry structure. For example, skin pigmentation is often associated to specific population groups. Skin pigmentation is typically lighter in individuals of Northern European descent, so assessment of admixture and genetic ancestry can infer potential externally visible characteristics (EVCs) (Shriver et al., 1997; Pritchard et al., 2000a, b; Rosenberg et al., 2002; Shriver et al., 2003; Yang et al., 2005).

The number of aiSNPs that allow for better discrimination between populations are constantly being rectified. Reflecting on the past decade, forensic genetic researchers have significantly refined the number of population classifications for individual BGA and admixture. As we move forward with increasingly powerful technologies and machine learning models, the constant refinement of aiSNPs best suited for human identity, ancestry prediction, and external characteristic determination will allow for increased capabilities

Phenotype-informative SNPs for prediction of externally visible characteristics

Phenotype-informative SNPs (piSNPs) are a unique group of polymorphisms which—unlike iiSNPs, liSNPs, and aiSNPs—are found within genes (i.e., coding regions) of the genome. Consequently, variation in piSNP genotypes can also be impacted by the environment, giving rise to a continuum of phenotypic variation at these markers. Despite the increased variability between individuals at these sites, externally visible characteristics (EVCs) lend well to forensic applications because they allow for the prediction of appearance traits (i.e., non-private genetic traits) of an individual. The use of piSNPs in the identification of unidentified human remains is known as forensic DNA phenotyping (FDP). Using traditional forensic samples, or DNA recovered from identified human remains, the goal of FDP is to develop a *biological witness*, or a prediction of the physical characteristics of an individual, notably in cases in which there are no viable investigative leads or in instances when recovered remains are unidentifiable using current comparative DNA profiling. The use of FDP for estimating EVCs differs greatly from the traditional forensic DNA typing methods used for human identification in the criminal justice system.

The forensic use of phenotyping began in the early 2000s, but has progressed relatively slowly compared to other forensic DNA typing methods. The primary challenge of using piSNPs to predict EVCs lies in the overall limited body of knowledge surrounding the genotype-phenotype relationship of many EVCs. In the simplest genetic scenario, a given genotype (i.e., both copies of the genetic makeup of an individual) will correspond with a particular phenotype (i.e., physical characteristic of the individual). However, human genetics is complicated, and several genes may contribute to a particular phenotype. Combined with environmental factors that also impact phenotype expression, this gives rise to a spectrum of phenotypic variation for each characteristic being investigated. Therefore, prediction of EVCs in humans continues to remain challenging, as environmental parameters cannot be measured from a person's DNA sample, and scientists have not yet fully elucidated the complex gene interactions involved in physical trait determination.

Current research into piSNPs has revealed that, in general, human pigmentation traits seem to be less genetically complex than other EVCs. With pigmentation, a small number of genes are involved in phenotype determination, at least when evaluating the pigmentation phenotype on a broad categorical level. Human pigmentation studies focus on variation in the coloration of the human iris (eye), head hair, and skin (including the presence or absence of freckling) (Sulem et al., 2007; Frudakis et al., 2007; Walsh et al., 2011; Breslin et al., 2019). However, FDP has also attempted to infer estimates related to age, height, and face shape.

Eye (iris) color

As one of the most thoroughly investigated EVCs, eye (iris) color is routinely predicted using piSNPs in human identity and unidentified remains investigations. Two studies in 2007 provided seminal work in DNA-based eye color prediction. Frudakis et al. (2007) targeted 33 SNPs from the human OCA2 gene, which allowed for classification of approximately 8% of the eye colors observed using a sample size exceeding 1000 individuals. In contrast, Sulem et al. (2007) used a refined set of 9 SNPs from 6 regions within the human genome (i.e., SLC24A4, KITLG, 6p25.3, TYR, OCA2–HERC2, MC1R) in order to infer categorical eye color prediction. The combined findings from these studies led to an influx of research efforts dedicated to identifying the key candidate SNP loci involved in iris pigmentation. Manfred Kayser and his team made significant contributions to phenotype estimation, by developing and publishing a comprehensive predictive model for assessing the utility of specific genomic targets and their viability in phenotype determination (Liu et al., 2009). Subsequently, Walsh et al. (2011) published the first DNA-based eye color prediction system for use in a forensic setting. This panel, named the IrisPlex, incorporated both the recommended genomic targets and the analysis pipeline developed by Liu et al. (2009), which focused on a refined set of six piSNPs most capable of predicting eye color (i.e., HERC2 rs12913832, OCA2 rs1800407, SLC24A4 rs12896399, SLC45A2 rs16891982, TYR rs1393350, IRF4 rs12203592). Results were imported into an interactive and easy-to-use Excel spreadsheet to provide categorical eye color probabilities. The IrisPlex assay was well-received by the forensic community. Validation studies were conducted, and the methodology is fully compatible with SWGDAM guidelines. Furthermore, the IrisPlex assay offers remarkable sensitivity, generating complete 6-SNP profiles from samples containing as low as 30 picograms (pg) of input DNA. Overall, the IrisPlex assay is easy to implement in the forensic laboratory and is highly reliable, allowing for useful estimates of eye color from potentially heavily degraded forensic samples.

The IrisPlex is not the only multiplex used to predict eye color in humans. Spichenok et al. (2011) proposed a different SNP panel including three novel SNPs and three SNPs found in the IrisPlex (HERC2 rs12193832, IRF4 rs12203592, SLC45A2 rs16891982, OCA2 rs1545397, ASIP rs6119471, MC1R rs885479). Both IrisPlex and the multiplex proposed by Spichenok et al. (2011) are able to accurately predict blue and brown eye color; however, intermediate eye color determination remains a limitation.

Prediction of eye color using SNP-based approaches typically incorporates statistical modeling to aid in the prediction determination. Eye color is one of the human traits exhibiting the most color variability, ranging from light shades of blue to dark shades of brown or black, through intermediate colors such as gray, hazel, yellow, and green. These color differences follow a pattern similar to the pigmentation of skin and hair. Eye color is defined by the amount of melanin and number of melanosomes in the outer layer of the iris. For example, blue eyes have less melanin and fewer melanosomes than brown eyes.

Two main approaches that attempt to infer eye color involve: (1) multinomial logistic regression (MLR), and (2) Bayes analysis (https://hirisplex.erasmusmc.nl/ and http://mathgene.usc.es/snipper/eyeclassifier.html) (McNevin et al., 2013). Each prediction model has its own strengths and weaknesses. For example, the IrisPlex SNPs have a high success rate for correctly predicting blue/brown phenotypes, but not for green-hazel or intermediate dark phenotypes (Walsh et al., 2011). In contrast, although the Snipper tool was better able to determine intermediate phenotypes, it increased the number of unclassified individuals given the prediction probability threshold applied (McNevin et al., 2013). Overall, the forensic community agrees that intermediate eye color prediction is still a challenge, requiring additional research to identify new genetic variants. Currently the accuracy of predicting intermediate eye color is significantly lower when compared with blue eye and brown eye color. Finally, gender may act as a possible influencing factor in determination of eye pigmentation. The literature has demonstrated that women tend to have darker eyes (i.e., predominantly brown and green) than men (i.e., predominantly blue and gray) in some European countries (Sulem et al., 2007; Martinez-Cadenas et al., 2013). However, the genetic causes for these differences in eye color have not yet been elucidated and more studies will be needed to evaluate the potential correlation.

Hair color

Hair color exhibits a wide range of phenotypes, similar to the variation that exists with eye and skin color pigmentation. Hair color is usually divided into two main categories of melanin: brown/black eumelanin and red/yellow pheomelanin. Although several genes are involved in the melanogenesis process, the MC1R gene was determined to be one of the first to demonstrate strong discrimination power for red hair, fair skin, and freckles (Branicki et al., 2007). Additional research has elucidated associations with other genes, including SLC45A2, SLC24A5, and HERC2. In 2013, a new panel to characterize hair color was developed, adding eighteen new loci to the six preexisting SNPs in the IrisPlex panel. This expanded panel, called the HIrisplex System, encompassed loci from the MC1R, HERC2, OCA2, SLC45A2, KITLG, EXOC2, TYR, SLC24A4, IRF4, ASIP, and TYRP1 genes (Walsh et al., 2013; Walsh and Kayser, 2016). However, current hair prediction models continue to be challenged with accurate prediction of hair colors from individuals who have had changes to their natural hair color throughout life (e.g., individuals whose hair color has naturally changed to a darker tone after childhood) (Söchtig et al., 2015). Thus, a limiting factor in many hair phenotype prediction models is that they do not currently take into account age-dependent phenotypes, especially in instances in which individuals transitioned from blonde to darker shades during their adolescent years.

Skin color

Skin pigmentation traits are currently the best predicted (i.e., highest accuracy) EVCs that can be typed using an FDP approach. Skin pigmentation variability is likely to have emerged as an evolutionary response to the intensity of ultraviolet (UV) radiation among different geographical regions on Earth. Regions that are closer to the equator, where higher UV light levels are likely, result in a higher frequency of individuals with darker skin, whereas regions with less UV intensity seem to exhibit less selective pressure, which has resulted in the appearance of lighter skin tones in those populations. Although skin pigmentation seems to be driven by environmental (location) factors, the genotype/phenotype associations of skin pigmentation in different populations remain difficult, and some specific population groups do not seem to exhibit consistent trends in pigmentation. Furthermore, pigmentation prediction in admixed populations does not have the same discriminatory power as with more homogeneous populations; however, researchers have developed multiplexes that are capable of differentiating skin colors between discrete groups of Asian, African, and Native American descent. With the HIrisPlex-S panel, three (light, dark, dark-black) to five (very pale, pale, intermediate, dark, dark-black) skin tones can be predicted with at least 72% accuracy (Walsh et al., 2017; Breslin et al., 2019).

Currently, results from the IrisPlex, HIrisPlex, and HIrisPlex-S systems have been compiled into a publicly available interactive tool that can be used to predict eye, hair, and skin color from genotyped DNA data. Using the website https://hirisplex.erasmusmc.nl/, anyone can utilize the tool by inserting genotype data from up to 41 markers available in order to predict EVC probabilities for 3 eye-, 4 hair-, and 5 skin color categories (Chaitanya et al., 2018; Breslin et al., 2019). As refinement of high value FDP loci continues to be explored, a substantially more detailed description of an unknown person's appearance from DNA may be possible (e.g., facial features, height, age), delivering increased value for investigations in criminal and missing person cases.

Forensic DNA phenotyping as a private forensic service

Characterization of phenotypic traits has been commercialized. The Identitas V1 Forensic Chip was the first of its kind to incorporate six eye-color piSNPS into a chip-based assay (Keating et al., 2013). The Identitas V1 Forensic Chip uses a genome-wide SNP array to predict eye color with high accuracy, but it also can be used to infer BGA, rough estimates of appearance, relatedness to other individuals, and the biological sex of an unidentified individual. This tool (http://identitascorp.com/) follows the IrisPlex system of EVC prediction. However, as with all SNP microarrays, the underlying hybridization technology is challenged with low-quantity and/or low-quality input DNA (Keating et al., 2013); hence, the chip is not well-suited for human remains cases, as the input DNA for this service typically needs to be of high quantity and quality.

Parabon Nanolabs is a for-hire company that specializes in FDP for forensic investigative purposes. Using large SNP datasets, Parabon utilizes machine learning algorithms and bioinformatics to identify genetic markers that are associated with particular EVCs. Using a large enough number of individuals in the modeling program has allowed for mathematical predictions to be made that can allow for individual forensic samples to be predicted and excluded for EVCs such as skin color, eye color, pigmentation, and freckling (Wiley et al., 2016). Parabon reports the results as "percent confidence" in the prediction, so that the user can infer the strength of the prediction.

Microhaplotypes

Microhaplotypes (also referred to as microhaps) are regions of the genome, typically 200 base pairs or less, that contain two or more SNPs. Fig. 5 illustrates the difference between a SNP and a microhaplotype. The concept behind using microhaplotypes in forensic genetics is that di-allelic SNPs can be converted into a multi-allelic system (Ge et al., 2010; Kidd et al., 2014a; Oldoni et al., 2019). Each multi-allelic microhaplotype is clearly more informative than an individual di-allelic SNP and is suitable for forensic applications with degraded DNA. Hence, this application lends itself well to the typing of human skeletal remains, in which the quality and quantity of DNA available for human identification can be quite limited.

Microhaplotypes, like SNPs, offer several advantages over traditional forensic DNA typing. Unlike traditional typing of microsatellite (STR) loci, sequencing of SNPs and/or microhaplotypes does not generate stutter peaks (Kidd et al., 2014a; Van der Gaag et al., 2018;

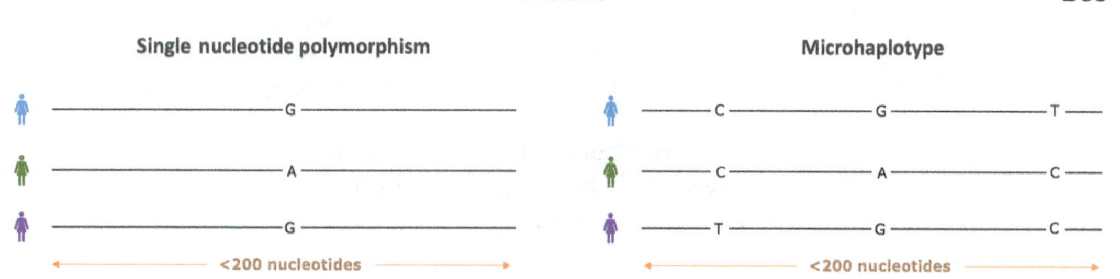

FIG. 5 An example of a single nucleotide polymorphism (SNP) versus a microhaplotype. A SNP captures only a single variant in an amplified product, while a microhaplotype captures multiple loci (2+) within the same sized amplicon (≤200 nucleotides).

Oldoni et al., 2019). Furthermore, there is a great deal of amplicon size variability when typing forensic microsatellite loci. Often when a sample is degraded, larger loci and larger alleles in an STR multiplex will experience reduced amplification efficiency, and this may result in allele dropout or entire locus dropout, yielding little or no data for the sample (Van der Gaag et al., 2018). Additionally, the decreased mutation rate in SNPs/microhaplotypes is advantageous for kinship determinations.

However, similar to independent SNPs, microhaplotypes generally have fewer alleles, and consequently, a larger panel of markers is required to obtain a comparable power of discrimination. Allele frequencies have been shown to vary significantly across populations, creating a challenge for forensic statistical calculations (i.e., random match probabilities [RMPs] or likelihood ratios [LRs]) that rely on allele frequency datasets (Kidd et al., 2011, 2014a; Oldoni et al., 2019).

A substantial benefit of using microhaplotypes in human remains investigations relies upon the concept that microhaplotypes seem to be unaffected by recombination. As such, microhaplotypes have been advocated in recent years as the optimal type of forensically useful DNA marker for family or lineage inference (Pakstis et al., 2012; Oldoni et al., 2019). The multiple alleles of these haplotypes have the potential to be more informative than simple two-allele SNPs for many types of forensic analyses, including (1) identification of biological relatives, (2) individual identification, and (3) ancestry prediction of an individual or group of biological relatives (Kidd et al., 2014a,b). Hence, on a "per-locus" basis, sequencing of microhaplotypes can yield more information than sequencing a single SNP. Finally, by sequencing microhaplotypes (either by traditional Sanger sequencing or via massively parallel sequencing), rare variants can be observed and detected when they fall within a microhaplotype region, but can be missed when the same SNPs are typed individually, depending on primer placement.

Conclusion

Currently, SNPs and microhaplotypes present a reliable alternative DNA typing approach to traditional STR typing in skeletal remains cases. With poor quality and low-quantity DNA samples, a SNP-based approach could be invaluable in providing investigators information about a person's ancestry and physical characteristics and could provide a viable human identification profile for comparison (King et al., 2018). As more information is uncovered about the nature and content of the human genome, we will be better able to identify specific genetic variants that code for phenotypic characteristics. Perhaps, in the future, SNP sites can be identified that will reliably correlate to facial features, chronological age approximation, and other physically identifying characteristics that will aid investigators and humanitarian agencies in unidentified remains investigations (Bulbul et al., 2011). However, due to the complexity of multigenic traits and external factors such as aging and environmental influence, it is unlikely that a few carefully chosen SNPs will present a foolproof picture of a sample's source.

References

1000 Genomes Project Consortium, Auton, A., Brooks, L.D., Durbin, R.M., Garrison, E.P., Kang, H.M., Korbel, J.O., Marchini, J.L., McCarthy, S., McVean, G.A., Abecasis, G.R., 2005. A global reference for human genetic variation. Nature 526, 68–74.

Alexander, D.H., Novembre, J., Lange, K., 2009. Fast model-based estimation of ancestry in unrelated individuals. Genome Res. 19, 1655–1664.

Ballantyne, J., 1997. Mass disaster genetics. Nat. Genet. 15, 329–331.

Berglund, E.C., Kiialainen, A., Syvänen, A.C., 2011. Next-generation sequencing technologies and applications for human genetic history and forensics. BMC Invest. Genet. 2 (23).

Branicki, W., Brudnik, U., Kupiec, T., Wolańska-Nowak, P., Wojas-Pelc, A., 2007. Determination of phenotype associated SNPs in the MC1R gene. J. Forensic Sci. 52 (2), 349–354.

Breslin, K., Wills, B., Ralf, A., Garcia, M.V., Kukla-Bartoszek, M., Pospiech, E., Freire-Aradas, A., Xavier, C., Ingold, S., de La Puente, M., van der Gaag, K.J., Herrick, N., Haas, C., Parson, W., Phillips, C., Sijen, T., Branicki, W., Walsh, S., Kayser, M., 2019. HIrisPlex-S system for eye, hair, and skin color prediction from DNA: massively parallel sequencing solutions for two common forensically used platforms. Forensic Sci. Int. Genet. 43, 102152.

Brookes, A.J., 1999. The essence of SNPs. Gene 234, 177–186.

Budowle, B., 2004. SNP typing strategies. Forensic Sci. Int. 146, S139–S142.

Budowle, B., Chakraborty, R., Giusti, A.M., Eisenberg, A.J., Allen, R.C., 1991. Analysis of the variable number of tandem repeats locus D1S80 by the polymerase chain reaction followed by high resolution polyacrylamide gel electrophoresis. Am. J. Hum. Genet. 48, 137–144.

Budowle, B., Lindsey, J.A., DeCou, J.A., Koons, B.W., Giusti, A.M., Comey, C.T., 1995. Validation and population studies of the loci LDLR, GYPA, HBGG, D7S8, and Gc (PM loci), and HLA-DQα using a multiplex amplification and typing procedure. J. Forensic Sci. 40, 45–54.

Budowle, B., Moretti, T.R., Niezgoda, S.J., Brown, B.L., 1998. CODIS and PCR-based short tandem repeat loci: law enforcement tools. In: Second European Symposium on Human Identification, Promega Corporation, Madison, WI, pp. 73–88.

Budowle, B., Allard, M.W., Wilson, M.R., Chakraborty, R., 2003. Forensics and mitochondrial DNA: applications, debates, and foundations. Ann. Rev. Genom. Human Genet. 4, 119–141.

Budowle, B., Planz, J., Campbell, R., Eisenberg, A., 2004. SNPs and microarray technology in forensic genetics: development and application to mitochondrial DNA. Forensic Sci. Rev. 16, 22–36.

Bulbul, O., Filoglu, G., Altuncul, H., Freire Aradas, A., Ruiz, Y., Fondevila, M., Phillips, C., Carracedo, A., Kriegel, A.K., Schneider, P.M., 2011. A SNP multiplex for the simultaneous prediction of biogeographic ancestry and pigmentation type. Forensic Sci. Int. Genet. Suppl. Ser. 3, e500–e501.

Butler, J.M., Coble, M.D., Vallone, P.M., 2007. STRs vs. SNPs: thoughts on the future of forensic DNA typing. Forensic Sci. Med. Path. 3, 200–205.

Chaitanya, L., Breslin, K., Zuñiga, S., Wirken, L., Popiech, E., Kukla-Bartoszek, M., Sijen, T., De Knijff, P., Liu, F., Branicki, W., Kayser, M., Walsh, S., 2018. The HIrisPlex-S system for eye, hair and skin colour prediction from DNA: introduction and forensic developmental validation. Forensic Sci. Int. Genet. 35, 123–135.

Cheung, E.C.C., Gahan, M.E., McNevin, D., 2018. Prediction of biogeographical ancestry in admixed individuals. Forensic Sci. Int. Genet. 36, 104–111.

Coble, M.D., Just, R.S., O'Callaghan, J.E., Letmanyi, I.H., Peterson, C.T., Irwin, J.A., Parsons, T.J., 2004. Single nucleotide polymorphisms over the entire mtDNA genome that increase the power of forensic testing in Caucasians. Int. J. Legal Med. 118, 137–146.

Comey, C.T., Budowle, B., 1991. Validation studies on the analysis of the HLA-DQ alpha locus using the polymerase chain reaction. J. Forensic Sci. 36, 1633–1648.

Daly, M.J., Rioux, J.D., Schaffner, S.F., Hudson, T.J., Lander, E.S., 2001. High-resolution haplotype structure in the human genome. Nat. Genet. 29, 229–232.

Daniel, R., Santos, C., Phillips, C., Fondevila, M., van Oorschot, R.A.H., Carracedo, A., Lareu, M.V., McNevin, D., 2015. A SNaPshot of next generation sequencing for forensic SNP analysis. Forensic Sci. Int. Genet. 14, 50–60.

Dixon, L.A., Murray, C.M., Archer, E.J., Dobbins, A.E., Koumi, P., Gill, P., 2005. Validation of a locus autosomal SNP multiplex for forensic identification purposes. Forensic Sci. Int. 154, 62–77.

Edwards, A., Civitello, A., Hammond, H.A., Caskey, C.T., 1991. DNA typing and genetic mapping with trimeric and tetrameric tandem repeats. Am. J. Hum. Genet. 49, 746–756.

Frudakis, T., Terravainen, T., Thomas, M., 2007. Multi-locus OCA2 genotypes specify human iris colors. Hum. Genet. 122, 311–326.

Gabriel, S.B., Schaffner, S.F., Nguyen, H., Moore, J.M., Roy, J., Blumenstiel, B., Higgins, J., DeFelice, M., et al., 2002. The structure of haplotype blocks in the human genome. Science 296, 2225–2229.

Ge, J., Budowle, B., Planz, J.V., Chakraborty, R., 2010. Haplotype block: a new type of forensic DNA markers. Int. J. Legal Med. 124, 353–361.

Gettings, K.B., Lai, R., Johnson, J.L., Peck, M.A., Hart, J.A., Gordish-Dressman, H., Schanfield, M.S., Podini, D.S., 2014. A 50-SNP assay for biogeographic ancestry and phenotype prediction in the U.S. population. Forensic Sci. Int. Genet. 8, 101–108.

Gill, P., 2001. An assessment of the utility of single nucleotide polymorphisms (SNPs) for forensic purposes. Int. J. Legal Med. 114, 204–210.

Gill, P., Haned, H., Eduarardoff, M., Santos, C., Phillips, C., Parson, W., 2015. The open-source software LRmix can be used to analyse SNP mixtures. Forensic Sci. Int. Genet. Suppl. Ser. 5, e50–e51.

Gyllensten, U.B., Erlich, H.A., 1988. Generation of single-stranded DNA by the polymerase chain reaction and its application to direct sequencing of the HLA-DQ alpha locus. Proc. Natl. Acad. Sci. 85, 7652–7656.

Halder, I., Shriver, M., Thomas, M., Fernandez, J.R., Frudakis, T., 2008. A panel of ancestry informative markers for estimating individual biogeographical ancestry and admixture from four continents: utility and applications. Human Mut. 29 (5), 648–658.

Holden, A.L., 2002. The SNP consortium: summary of a private consortium effort to develop an applied map of the human genome. BioTechniques 32, S22–S26.

Holland, P.M., Abramson, R.D., Watson, R., Gelfand, G.H., 1991. Detection of specific polymerase chain reaction product by utilizing the 5′-3′ exonuclease activity of Thermus aquaticus DNA polymerase. Proc. Natl. Acad. Sci. 88, 7276–7280.

Hubisz, M.J., Falush, D., Stephens, M., Pritchard, J.K., 2009. Inferring weak population structure with the assistance of sample group information. Mol. Ecol. Res. 9, 1322–1332.

Keating, B., Bansal, A.T., Walsh, S., Millman, J., Newman, J., Kidd, K., Budowle, B., Eisenberg, A., Donfack, J., Gasparini, P., Budimlija, Z., Henders, A.K., Chandrupatla, H., Duffy, D.L., Gordon, S.D., Hysi, P., Liu, F., Medland, S.E., Rubin, L., Martin, N.G., Spector, T.D., Kayser, M., International Visible Trait Genetics (VisiGen) Consortium, 2013. First all-in-one diagnostic tool for DNA intelligence: genome-wide inference of biogeographic ancestry, appearance, relatedness, and sex with the Identitas v1 Forensic Chip. Int J Legal Med 127 (3), 559–572.

Kidd, K.K., Pakstis, A.J., Speed, W.C., Grigorenko, E.L., Kajuna, S.L., Karoma, N.J., Kungulilo, S., Kim, J.J., Lu, R.B., Odunsi, A., Okonofua, F., Parnas, J., Schulz, L.O., Zhukova, O.V., Kidd, J.R., 2006. Developing a SNP panel for forensic identification of individuals. Forensic Sci. Int. 164 (1), 20–32.

Kidd, J.R., Friedlaender, F., Pakstis, A.J., Furtado, M., Fang, R., Wang, X., Nievergelt, C.M., Kidd, K.K., 2011. Single nucleotide polymorphisms and haplotypes in Native American populations. Am. J. Phys. Anthropol. 146, 495–502.

Kidd, K.K., et al., 2012. Better SNPs for better forensics: ancestry, phenotype, and family identification. In: NIJ Annual Meeting, Arlington, VA.

Kidd, K.K., Pakstis, A.J., Speed, W.C., Lagacé, R., Chang, J., Wootton, S., Haigh, E., Kidd, J.R., 2014a. Current sequencing technology makes microhaplotypes a powerful new type of genetic marker for forensics. Forensic Sci. Int. Genet. 12, 215–224.

Kidd, K.K., Speed, W.C., Pakstis, A.J., Furtado, M.R., Fang, R., Madbouly, A., Maiers, M., Middha, M., Friedlaender, F.R., Kidd, J.R., 2014b. Progress toward an efficient panel of SNPs for ancestry inference. Forensic Sci. Int. Genet. 10, 23–32.

King, J.L., Churchill, J.D., Novroski, N.M.M., Zeng, X., Warshauer, D.H., Seah, L.-H., Budowle, B., 2018. Increasing the discrimination power of ancestry- and identity-informative SNP loci within the ForenSeq™ DNA Signature Prep Kit. Forensic Sci. Int. Genet. 36, 60–76.

Kosoy, R., Nassir, R., Tian, C., White, P.A., Butler, L.M., Silva, G., Kittles, R., Alarcon-Riquelme, M.E., Gregersen, P.K., Belmont, J.W., De La Vega, F.M., Seldin, M.F., 2009. Ancestry informative marker sets for determining continental origin and admixture proportions in common populations in America. Human Mut. 30 (1), 69–78.

Kuppuswamy, M.N., Hoffmann, J.W., Kasper, C.K., Spitzer, S.G., Groce, S.L., Bajaj, S.P., 1991. Single nucleotide primer extension to detect genetic diseases: experimental application to hemophilia B (factor IX) and cystic fibrosis genes. Proc. Natl. Acad. Sci. 88, 1143–1147.

Liu, F., van Duijn, K., Vingerling, J.R., Hofman, A., Uitterlindenm, A.G., Janssens, A.C.J.W., Kayser, M., 2009. Eye color and the prediction of complex phenotypes from genotypes. Curr. Biol. 19 (5), R192–R193.

Martinez-Cadenas, C., Peña-Chilet, M., Ibarrola-Villava, M., Ribas, G., 2013. Gender is a major factor explaining discrepancies in eye colour prediction based on *HERC2/OCA2* genotype and the IrisPlex model. Forensic Sci. Int. Genet. 7 (4), 453–460.

McNevin, D., Santos, C., Gómez-Tato, A., Alvarez-Dios, J., Casares de Cal, M., Daniel, R., Phillips, C., Lareu, M.V., 2013. An assessment of Bayesian and multinomial logistic regression classification systems to analyze admixed individuals. Forensic Sci. Int. Genet. Suppl. Ser. 4 (1), e63–e64.

Mehta, B., Daniel, R., Phillips, C., McNevin, D., 2017. Forensically relevant SNaPshot assays for human DNA SNP analysis: a review. Int. J. Legal Med. 131, 21–37.

Mogensen, H.S., Tvedebrink, T., Børsting, C., Pereira, V., Morling, N., 2020. Ancestry prediction efficiency of the software *GenoGeographer* using a z-score method and the ancestry informative markers in the Precision ID Ancestry Panel. Forensic Sci. Int. Genet. 44, 102154.

Mullis, K., Faloona, F., Scharf, S., Saiki, R., Horn, G., Erlich, H., 1986. Specific enzymatic amplification of DNA in vitro: the polymerase chain reaction. Cold Spring Harb Symp Quant Biol. 51 (1), 263–273.

Nievergelt, C.M., Maihofer, A.X., Shekhtman, T., Libiger, O., Wang, X., Kidd, K.K., Kidd, J.R., 2013. Inference of human continental origin and admixture proportions using a highly discriminative ancestry informative 41-SNP panel. Investig. Genet. 4, 13.

Oldoni, F., Kidd, K.K., Podini, D., 2019. Microhaplotypes in forensic genetics. Forensic Sci. Int. Genet. 38, 54–69.

Osier, M.V., Cheung, K.H., Kidd, J.R., Pakstis, A.J., Miller, P.L., Kidd, K.K., 2001. ALFRED: an allele frequency database for diverse populations and DNA polymorphisms—an update. Nucleic Acids Res. 29 (1), 317–319.

Pakstis, A.J., Speed, W.C., Fang, R., Hyland, F.C., Furtado, M.R., Kidd, J.R., Kidd, K.K., 2010. SNPs for a universal individual identification panel. Hum. Genet. 127 (3), 315–324.

Pakstis, A.J., Fang, R., Furtado, M.R., Kidd, J.R., Kidd, K.K., 2012. Mini haplotypes as lineage informative SNPs and ancestry inference SNPs. Eur. J. Hum. Genet. 20, 1148–1154.

Peck, M., Idrizbegovic, S., Bittner, F., Parsons, T., 2018. Optimization and performance of a very large MPS SNP panel for missing persons. In: Presentation at the 7th QIAGEN Investigator Forum, San Antonio, TX, USA. https://www.slideshare.net/QIAGENscience/icmp-mps-snp-panel-for-missing-persons-michelle-peck-et-al.

Phillips, C., Salas, A., Sánchez, J.J., Fondevila, M., Gómez-Tato, A., Álvarez-Dios, J., Calaza, M., Casares de Cal, M., Ballard, D., Lareu, M.V., Carracedo, A., The SNPforID Consortium, 2007. Inferring ancestral origin using a single multiplex assay of ancestry-informative marker SNPs. Forensic Sci. Int. Genet. 1, 273–280.

Phillips, C., Parson, W., Lundsberg, B., Santos, C., Freire-Aradas, A., Torres, M., Eduardoff, M., Børsting, C., Johansen, C., Fondevila, M., Morling, N., Schneider, P., EUROFORGEN-NoE Consortium, Carracedo, A., Lareu, M.V., 2014. Building a forensic ancestry panel from the ground up: the EUROFORGEN Global AIM-SNP set. Forensic Sci. Int. Genet. 11, 13–25.

Phillips, C., Amigo, J., Carracedo, A., Lareu, M.V., 2015. Tetra-allelic SNPs: informative forensic markers compiled from public whole-genome sequence data. Forensic Sci. Int. Genet. 19, 100–106.

Pritchard, J.K., Stephens, M., Donnelly, P., 2000a. Inference of population structure using multi-locus genotype data. Genetics 155, 945–959.

Pritchard, J.K., Stephens, M., Rosenberg, N.A., Donnelly, P., 2000b. Association mapping in structured populations. Am. J. Hum. Genet. 67, 170–181.

Romanini, C., Catelli, M.L., Borosky, A., Pereira, R., Romero, M., Salado Puerto, M., Phillips, C., Fondevila, M., Freire, A., Santos, C., et al., 2012. Typing short amplicon binary polymorphisms: supplementary SNP and Indel genetic information in the analysis of highly degraded skeletal remains. Forensic Sci. Int. Genet. 6, 469–476.

Ronaghi, M., Karamohamed, S., Pettersson, B., Uhlen, M., Nyren, P., 1996. Real-time DNA sequencing using detection of pyrophosphate release. Anal. Biochem. 242, 84–89.

Ronaghi, M., Uhlen, M., Nyren, P., 1998. A sequencing method based on real-time pyrophosphate. Science 281, 363–365.

Rosenberg, N.A., Pritchard, J.K., Weber, J.L., Cann, H.M., Kidd, K.K., Zhivotovsky, L.A., Feldman, M.W., 2002. Genetic structure of human populations. Science 298, 2381–2385.

Sanchez, J., Børsting, C., Hallenberg, C., Buchard, A., Hernendez, N., Morling, N., 2003. Multiplex PCR and minisequencing of SNPs—a model with 35 Y chromosome SNPs. Forensic Sci. Int. 137, 74–84.

Sanchez, J.J., Phillips, C., Børsting, C., Balogh, K., Bogus, M., Fondevila, M., Harrison, C.D., Musgrave-Brown, E., Salas, A., Syndercombe-Court, D., Schneider, P.M., Carracedo, A., Morling, N., 2006. A multiplex assay with 52 single nucleotide polymorphisms for human identification. Electrophoresis 27, 1713–1724.

Sherry, S.T., Ward, M.H., Kholodov, M., Baker, J., Phan, L., Smigielski, E.M., Sirotkin, K., 2001. dbSNP: the NCBI database of genetic variation. Nucleic Acids Res. 29 (1), 308–311.

Shriver, M.D., Smith, M.W., Jin, L., Marcini, A., Akey, J.M., Deka, R., Ferrell, R.E., 1997. Ethnic affiliation estimation by use of population-specific DNA markers. Am. J. Hum. Genet. 60, 957–964.

Shriver, M.D., Parra, E.J., Dios, S., Bonilla, C., Norton, H., Jovel, C., Pfaff, C., Jones, C., et al., 2003. Skin pigmentation, biogeographical ancestry and admixture mapping. Hum. Genet. 112, 387–399.

Sobrino, B., Brión, M., Carracedo, A., 2005. SNPs in forensic genetics: a review on SNP technologies. Forensic Sci. Int. 154, 181–194.

Söchtig, J., Phillips, C., Maroñas, O., Gómez-Tato, A., Cruz, R., Alvarez-Dios, J., Casares de Cal, M.-A., Ruiz, Y., Reich, K., Fondevila, M., Carracedo, A., Lareu, M., 2015. Exploration of SNP variants affecting hair color prediction in Europeans. Int. J. Legal Med. 129, 963–975.

Sokolov, B.P., 1990. Primer extension technique for the detection of single nucleotide in genomic DNA. Nucleic Acids Res. 18, 3671.

Spichenok, O., Budimlija, Z.M., Mitchell, A.A., Jenny, A., Kovacevic, L., Marjanovic, D., Caragine, T., Prinz, M., Wurmbach, E., 2011. Prediction of eye and skin color in diverse populations using seven SNPs. Forensic Sci. Int. Genet. 5 (5), 472–478.

Sulem, P., Gudbjartsson, D.F., Stacey, S.N., Helgason, A., Rafnar, T., Magnusson, K.P., Manolescu, A., Karason, A., et al., 2007. Genetic determinants of hair, eye and skin pigmentation in Europeans. Nat. Genet. 39, 1443–1452.

Syvanen, A.C., Aalto-Setala, K., Harju, L., Kontula, K., Soderlund, H., 1990. A primer-guided nucleotide incorporation assay in the genotyping of apolipoprotein E. Genomics 8, 684–692.

The International HapMap Consortium, 2003. The International HapMap Project. Nature 426, 789–796.

Thorisson, G.A., Stein, L.D., 2003. The SNP Consortium website: past, present and future. Nucleic Acids Res. 31 (1), 124–127.

Van der Gaag, K.J., de Leeuw, R.H., Laros, J.F.J., den Dunnen, J.T., 2018. Short hypervariable microhaplotypes: a novel set of very short high discriminating power loci without stutter artefacts. Forensic Sci. Int. Genet. 35, 169–175.

Wallace, R.B., Shaffer, J., Murphy, R.F., Bonner, J., Hirose, T., Itakura, K., 1979. Hybridization of synthetic oligodeoxyribonucleotides to phi 174 DNA: the effect of single base pair mismatch. Nucleic Acids Res. 6, 3543–3557.

Walsh, S., Kayser, M., 2016. A practical guide to the HIrisPlex system: simultaneous prediction of eye and hair color from DNA. Forensic DNA Typing Protocols. pp. 213–231.

Walsh, P.S., Fildes, N., Louie, A.S., Higuchi, R., 1991. Report of the blind trial of the Cetus AmpliType HLA DQα forensic deoxyribonucleic acid (DNA) forensic deoxyribonucleic acid (DNA) amplification and typing kit. J. Forensic Sci. 36, 1551–1556.

Walsh, S., Liu, F., Ballantyne, K.N., van Oven, M., Lao, O., Kayser, M., 2011. IrisPlex: a sensitive DNA tool for accurate prediction of blue and brown eye colour in the absence of ancestry information. Forensic Sci. Int. Genet. 5, 170–180.

Walsh, S., Liu, F., Wollstein, A., Kovatsi, L., Ralf, A., Kosiniak-Kamysz, A., Branicki, W., Kayser, M., 2013. The HIrisPlex system for simultaneous prediction of hair and eye colour from DNA. Forensic Sci. Int. Genet. 7 (1), 98–115.

Walsh, S., Chaitanya, L., Breslin, K., Muradlidharan, C., Bronikowska, A., Pospiech, E., Koller, J., Kovatsi, L., Wollstein, A., Branicki, W., Liu, F., Kayser, M., 2017. Global skin colour prediction from DNA. Hum. Genet. 136 (7), 847–863.

Wendt, F.R., Novroski, N.M.M., 2019. Identity informative SNP associations in the UK Biobank. Forensic Sci. Int. Genet. 42, 45–48.

Whitaker, J.P., Clayton, T.M., Urquhart, A.J., Millican, E.S., Downes, T.J., Kimpton, C.P., Gill, P., 1995. Short tandem repeat typing of bodies from a mass disaster: high success rate and characteristic amplification patterns in highly degraded samples. BioTechniques 18, 670–677.

Wiley, R., Zeng, X., LaRue, B., Greytak, E.M., Armentrout, S., Budowle, B., 2016. Blind testing and evaluation of a comprehensive DNA phenotyping system. In: Presentation at the 27th International Symposium for Human Identification, Minneapolis, Minnesota, USA.

Wilson, M.R., DiZinno, J.A., Polansky, D., Replogle, J., Budowle, B., 1995. Validation of mitochondrial DNA sequencing for forensic casework analysis. Int. J. Legal Med. 108, 68–74.

Yang, N., Li, H., Criswell, L.A., Gregersen, P.K., Alarcon-Riquelme, M.E., Kittles, R., Shigeta, R., Silva, G., Patel, P.I., Belmont, J.W., Seldin, M.F., 2005. Examination of ancestry and ethnic affiliation using highly informative di-allelic DNA markers: application to diverse and admixed populations and implications for clinical epidemiology and forensic medicine. Hum. Genet. 118, 382–392.

Zeng, X., Chakraborty, R., King, J.L., LaRue, B., Moura-Neto, R.S., Budowle, B., 2016. Selection of highly informative SNP markers for population affiliation of major U.S. populations. Int. J. Legal Med. 130 (2), 341–352.

CHAPTER

13

Diallelic Markers: INDELs and INNULs

Bobby L. LaRue PhD

Department of Forensic Science, Sam Houston State University, Huntsville, TX, United States; Verogen Inc., San Diego, CA, United States

Introduction

Short tandem repeat markers (STRs) are the basis of modern human identity testing. Contemporary STR assays are capable of reliably and robustly genotyping trace amounts of biological materials, with very high powers of discrimination. The evolution of these assays has allowed STR kits to be developed that robustly perform with inhibited samples from a variety of types of biological materials (Ensenberger et al., 2016; Wang et al., 2015; Kraemer et al., 2017). However, there are few instances where the technology can be improved upon. STR assays sometimes fall short when the template DNA is highly degraded and fragmented.

STR assays amplify short repeated DNA segments (typically tetranucleotide, or 4-bp, repeat units) and their associated flanking regions using the polymerase chain reaction (PCR). During PCR, a prokaryotic DNA polymerase is used in conjunction with oligonucleotide primers that bind to the flanking regions of each STR locus to amplify the existing template exponentially. This reaction occurs for all alleles and markers in the multiplex simultaneously. PCR products (amplicons) are then separated via capillary electrophoresis (CE) in a size-dependent fashion to determine the specific alleles present in the sample (based on the number of repeats in each STR locus). For this molecular copying process to work effectively, however, the template DNA molecule must be intact and nondegraded (Butler, 2007).

Severe environmental degradation of DNA can lead to fragmentation of template DNA molecules in the sample to a size that is smaller than the complete length of the amplicon in the STR assay. Without the complete (intact) fragment, the polymerase chain reaction cannot amplify that particular STR marker. The result is a gradual loss of STR alleles in the profile (i.e., in the form of locus- or allele dropout) from the longest amplicons to the shortest as the degree of degradation becomes more severe (Butler, 2007; Hughes-Stamm et al., 2011; Prinz and Schmitt, 1994).

Environmental degradation of DNA leads to small fragments of amplifiable DNA that are approximately 150 base pairs (bp), which is roughly the size of nucleic acids bound to

Forensic Genetic Approaches for Identification of Human Skeletal Remains
https://doi.org/10.1016/B978-0-12-815766-4.00013-3

histone proteins in the nucleosome (Bina et al., 1980). These small DNA fragment sizes do not allow for successful amplification of the majority of markers in any existing standard STR or mini-STR kit (Mulero et al., 2008). However, diallelic markers—such as insertion-deletion (INDEL) polymorphisms and single nucleotide polymorphisms (SNPs)—allow for design of very small amplicon sizes and multiplexing into an assay with all target loci being less than 150bp (Dixon et al., 2005; Børsting et al., 2009).

Single nucleotide polymorphisms (SNPs) are quite plentiful in the human genome and, if selected based on the proper criteria, can be used for a number of forensically relevant tasks such as human identity, ancestral determination, maternal lineage, paternal lineage, and phenotype prediction. These polymorphisms involve substitution of one nucleotide (and its complement on the opposite strand) by another (different) nucleotide. The distribution of SNP alleles in global populations, as well as their locations within the genome, determines their utility (Budowle and Van Daal, 2008). SNPs with high amounts of heterozygosity (minor allele frequencies approaching 50%), low population substructure (F_{ST} less than 0.06), and those which do not exhibit linkage disequilibrium with other markers in an assay are good candidates for human identity applications (Kidd et al., 2006). If the desired goal is biogeographic ancestry prediction, one should basically look for opposite values in the categories of heterozygosity (e.g., rare variants in the majority of populations that are the major alleles in a single population) and population substructure (e.g., F_{ST} greater than 0.20) (Phillips et al., 2007).

Multiple SNP-based assays exist for human identification purposes; however, unfortunately, since there is no length difference between SNP variants (but rather a base substitution in the sequence), SNP detection is not amenable to the standard workflows used in modern crime laboratories, and in general, is cumbersome when conducted on CE instrumentation. SNP detection requires either single-base extension or chain termination sequencing, both of which are more labor-intensive and require instrument modification compared to traditional STR-based fragment analysis.

In contrast to SNPs but similar to STRs, INDEL alleles [also described as Deletion-Insertion Polymorphisms (DIPs)] are differentiated by the length of the polymorphism, so they are quite amenable to the fragment size separation that is a standard part of CE-based testing (Weber et al., 2002). Other than the need to perform a spectral calibration for the dye set being used in an INDEL multiplex, there are no changes to instrumentation or workflow required when switching from STR to INDEL workflows on a CE instrument. Likewise, INDEL data analysis uses the same standard genetic analysis software pipelines.

The formation of insertion/deletion polymorphisms could occur through a myriad of mechanisms, but in most cases INDELs result from three primary methods. First, there could be an uneven crossing over that occurs during meiosis where a small/short extra fragment of DNA gets transferred. Alternatively, slippage of the polymerase during replication at repeating sequences could expand or reduce a proto-STR segment. A third mechanism occurs during nonhomologous end repair. When a double-stranded break (DSB) occurs in the DNA molecule, repair enzymes will often trim back nucleotides around the site of the break during repair, leaving a short section of DNA that gets deleted with regard to the original sequence (Cooper and Krawczak, 1991; Krawczak and Cooper, 1991).

A special case of "insertion only" polymorphisms is related to retrotransposable elements. These polymorphisms result in the length difference associated with insertion of large pieces of mobile DNA elements that have a basis in RNA. The most common of these mobile DNA elements are primate-specific *alu* sequences, which make up a large portion of the human genome by mass (Batzer et al., 1994; Batzer and Deininger, 2002). These *alu* elements and other small mobile elements are collectively referred to as short interspersed nuclear elements (SINEs) and, along with much larger interspersed nuclear elements (LINEs), can also be treated as diallelic markers. One major difference between these markers and INDELs is size (i.e., from 400 to 600 bases for small SINEs like *alu* elements) to the much larger LINEs (10 kb or more for some L1 LINEs). Another major difference is that it is often unknown if the ancestral state of an INDEL is the insertion or deletion allele; however, with mobile elements, either the insertion occurred or it did not. For this reason, when developing the first commercially available kit for these types of markers, they were termed insertion-null markers, or INNULs (also known as restriction-insertion polymorphisms, or RIPs).

The origins of these polymorphisms involve insertion of a new copy of the mobile element into the new receptor site by an enzyme called transposase. In the case of *alu* elements, the inserted fragment has the molecular signature of a corrupted messenger RNA (mRNA) species that was converted to DNA by a rogue reverse transcriptase. The original event happened sometime before primates differentiated, and all primates carry *alu* sequences. There is a branch of *alu* elements that are specific to humans. The mechanism for transposition of other SINEs and LINEs is similar to *alu* elements, with the only difference being the size and nature of the sequence inserted. A comprehensive database of INNULs with limited population-specific allele distributions makes it possible to select markers for an INNUL-based human identification (HID) panel in silica and has greatly advanced the field in this area (Wang et al., 2006).

INDELs and INNULs, while different in the nature of their structure, are also diallelic like SNPs. The alleles of the markers are designated by either the presence or absence of a fragment of DNA, rather than one nucleotide substituted for another. As such, the same selection criteria mentioned earlier for SNPs can be used to develop panels of INDEL or INNUL markers for either human identity purposes or biogeographic ancestry prediction, depending on the distribution of alleles in global populations.

Existing INDEL panels in the literature

While great attention has been given to STRs and SNPs in the literature, INDELs have been often neglected. Two studies by Krawczak and Cooper describe a compendium of diseases caused by insertions and deletions, and the lengths of these INDELs are catalogued. As might be expected, smaller INDELs were more common than larger insertions, with a steep decline in frequency for INDELs larger than 6 bp (Cooper and Krawczak, 1991; Krawczak and Cooper, 1991).

Another study utilized newly available genomic data to characterize INDELs in multiple populations across the genome. This work, which was deposited in GenBank along with the allelic frequencies in the sampled population, was fundamental in the development of INDEL panels for forensic investigation purposes. The only limitation of the study was the small population size of some global population groups, notably East Asian and African (Weber et al., 2002).

In 2009, using information from the 2002 Weber et al. and 2006 Mills et al. studies, a 38-marker INDEL multiplex panel was developed that could generate random match probabilities (RMPs) in the magnitudes of 10^{-14} to 10^{-15}, depending on the ethnic group (Pereira et al., 2009a,b). All of the amplicons in this assay were 160-bp or less, which would enhance success with degraded DNA templates. This included successful genotyping of bones and teeth that were 35 years postmortem. The only drawback to the panel is that there was some evidence of population substructure in African populations when compared to Caucasian and East Asian populations and substructure between Caucasian and East Asian populations as measured by F_{ST}. Multiple population studies in other niche populations have followed using this panel (Pereira et al., 2009a,b; Manta et al., 2012; Martínez-Cortés et al., 2015).

In a separate effort, a private-sector German company called Mentype developed the first commercially available INDEL assay that included 30 INDEL markers (Friis et al., 2012). Mentype was purchased by Qiagen (Hilden, Germany) shortly after this study, and the name of the assay was changed to Investigator DIPplex®. The Investigator DIPplex® kit was marketed as a supplemental application for use in paternity and human identity cases. The assay was shown in multiple studies to be extremely sensitive, tolerant of degraded samples, and resistant to inhibitors. One study compared the Investigator DIPplex® kit to the 38-plex assay developed by Pereira et al. and found that both assays work well with low-quantity and low-quality DNA (Fondevila et al., 2012).

Population studies demonstrate that the Investigator DIPplex® assay works extremely well with Caucasian populations (RMP = 10^{-14}) and quite well with both African and East Asian populations (RMP = 10^{-11}). The only shortcoming of this assay is that many of the markers were selected based on the studies of Weber et al. (2002); hence, for non-Caucasian populations, some markers had minor allele frequencies that were less than 10% (LaRue et al., 2012a,b). This is not to suggest that the assay is not informative in these populations, but a few of the markers (2–3, depending on the population) might be less informative because they are approaching fixation in that population. Note that this assay works with samples that would not be typeable using standard STR assays, so the other alternative would be mitochondrial haplotypes—which would provide multiple orders of magnitude less statistical power than the DIPplex® assay would return even in its worst performing populations. The Investigator DIPplex® assay has been used to genotype multiple population groups globally, and an extensive catalog of microvariants (and even tri-alleles) has been described (Liang et al., 2013; Neuvonen et al., 2012; Nunotani et al., 2015; Seong et al., 2014; Tomas et al., 2016; Wang et al., 2014, 2016; Ferreira Palha et al., 2015; Jian et al., 2019).

Potentially in response to the minor shortcomings that other assays have had with East Asian populations, there is also a Chinese-developed, Asian-specific 30-INDEL multiplex that generates similar results to the DIPplex® and 38-plex assay described by Pereira et al. This multiplex has been the subject of multiple population studies, using populations within China and across Asia in general. For identification of degraded remains that are of known or suspected Asian ancestry, this kit—referred to as INDEL Typer 30—would be the assay of choice for human identity testing (Zhang et al., 2018).

Improvements in marker selection have been suggested, and larger population studies have been completed to empirically test markers for suitability in a global INDEL panel for human identity. One hundred fourteen INDELs were tested in 8 major populations; after testing, 38 markers were selected (including 12 markers from the Pereira et al. 38-plex) for additional population studies. Syntenic loci in this global panel were at least 40 Mbp distant from each other; have random match probabilities approaching 10^{-18}; were in Hardy-Weinberg Equilibrium (HWE); displayed no evidence of linkage disequilibrium; and had F_{ST} values less than 0.06 in four major U.S. populations (LaRue et al., 2014; Moura-Neto et al., 2015). These markers have since been developed into a multiplex for both capillary electrophoresis (CE) and massively parallel sequencing (MPS) platforms (LaRue, 2018; Wendt et al., 2016). Due to small amplicon sizes (less than 180 bp), the assays are capable of generating complete profiles with samples as diverse as cadaver bone, formalin-fixed paraffin-embedded (FFPE) tissue, and burned cadaver tissue (LaRue, 2018).

Existing INNUL panels in the literature

Initial efforts to use INNULs or RIPs for human identification (HID) purposes faced difficulties. These early approaches, primarily focused on *alu* sequences, attempted to amplify the insertion site with a common set forward and reverse primers. With such a large size difference in amplicon lengths for heterozygotes, preferential amplification of the shorter amplicon resulted in a dramatic decrease in the amount of longer amplicon fragment (Mamedov et al., 2010). Hence, this approach is less than ideal.

Two subsequent groups independently developed similar strategies for overcoming the preferential amplification issue (LaRue et al., 2012a,b; Asari et al., 2012). The overall strategy, illustrated in Fig. 1, involves use of a common forward primer and a set of reverse primers that overlap the insertion site size differences dependent on a smaller discrepancy in amplicon sizes (i.e., a few base pairs vs a few hundred base pairs). Average random match probabilities (RMP) range from 10^{-13} for a 30-marker 3-primer pool panel (Asari et al., 2012) and 10^{-9} for the single tube InnoTyper® 21 (IT21) assay (InnoGenomics Technologies LLC, New Orleans, Louisiana, United States) described in Brown et al. (2017). At first glance, these RMPs might seem low compared to modern STR kits, but the reduced amplicon size (i.e., all amplicons in the InnoTyper® 21 assay are less than 120 bp) allows for genotyping of samples that would ordinarily only be suitable for mitochondrial haplotype analysis.

InnoTyper® 21 is the only commercially available INNUL-based assay for HID purposes. It has been shown to generate full profiles with highly degraded samples, including skeletal remains, recovered postblast bomb fragments, and rootless hair shafts (Li et al., 2017; Grisedale et al., 2018; Martins et al., 2019; Tasker et al., 2017). Optimization of extraction chemistries to retain small fragments of template DNA (as would be found in highly degraded samples)

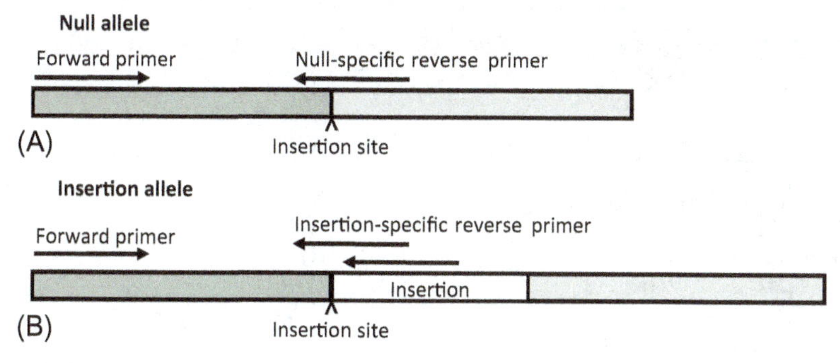

FIG. 1 Primer design strategy for INNUL-based assays to avoid preferential amplification of the smaller null allele: (A) Null allele, (B) insertion allele.

allows an even greater recovery of alleles less than 150 bp in length. This improved extraction method facilitates recovery of degraded (< 150 bp) but amplifiable nuclear DNA in approximately 20% of rootless hair shafts that have been vigorously washed to remove exterior contaminating DNA (Grisedale et al., 2018). Electropherograms for the same DNA extract obtained from rootless hair shafts, but amplified separately using GlobalFiler™ (Thermo Fisher Scientific, Waltham, Massachusetts, United States) and InnoTyper® 21 (InnoGenomics Technologies LLC, New Orleans, Louisiana, United States), are shown in Fig. 2.

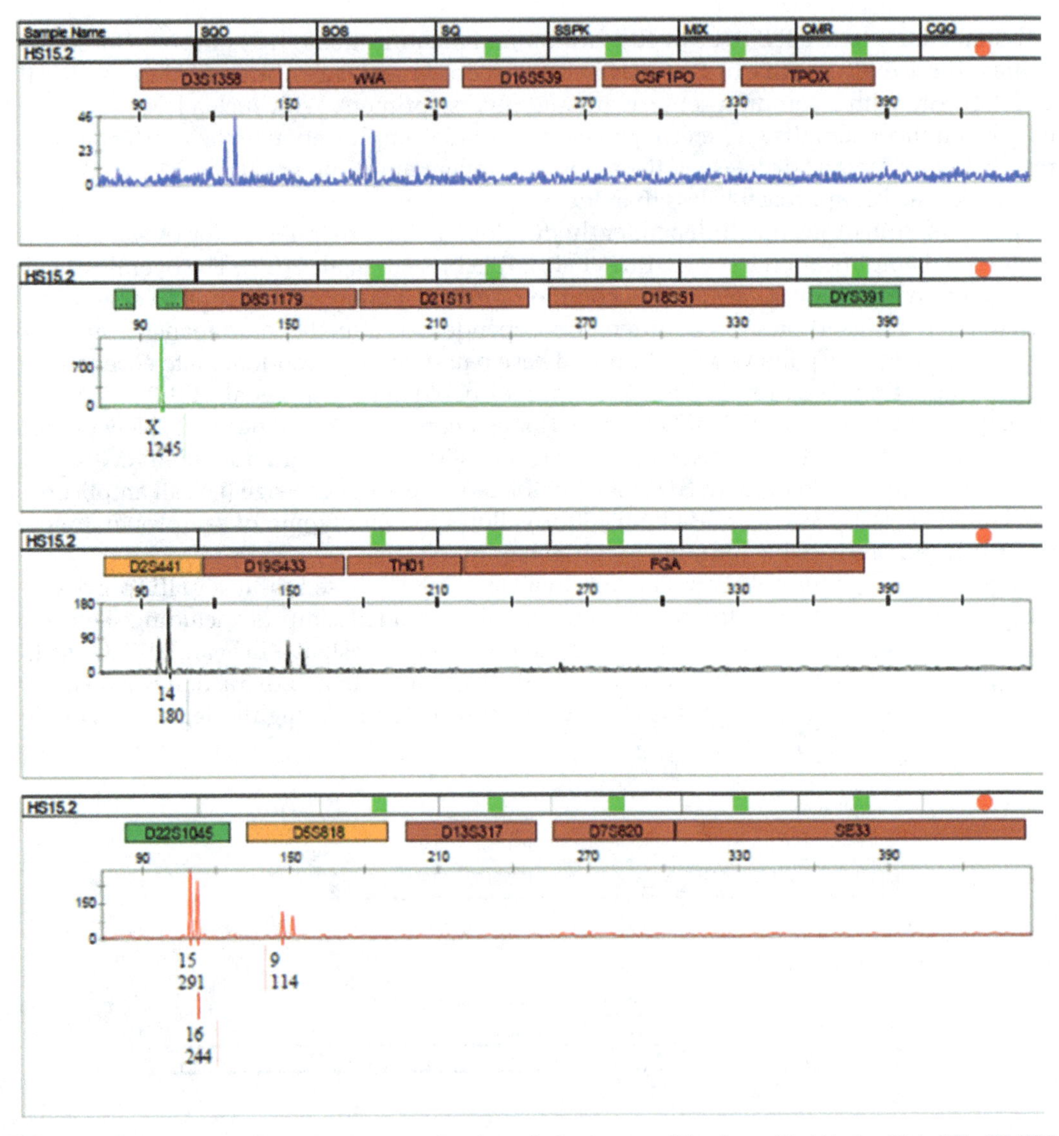

FIG. 2 DNA typing results for a 2-cm rootless hair shaft after amplification with (A) the GlobalFiler™ PCR Amplification Kit (Thermo Fisher Scientific, Waltham, Massachusetts, United States) and

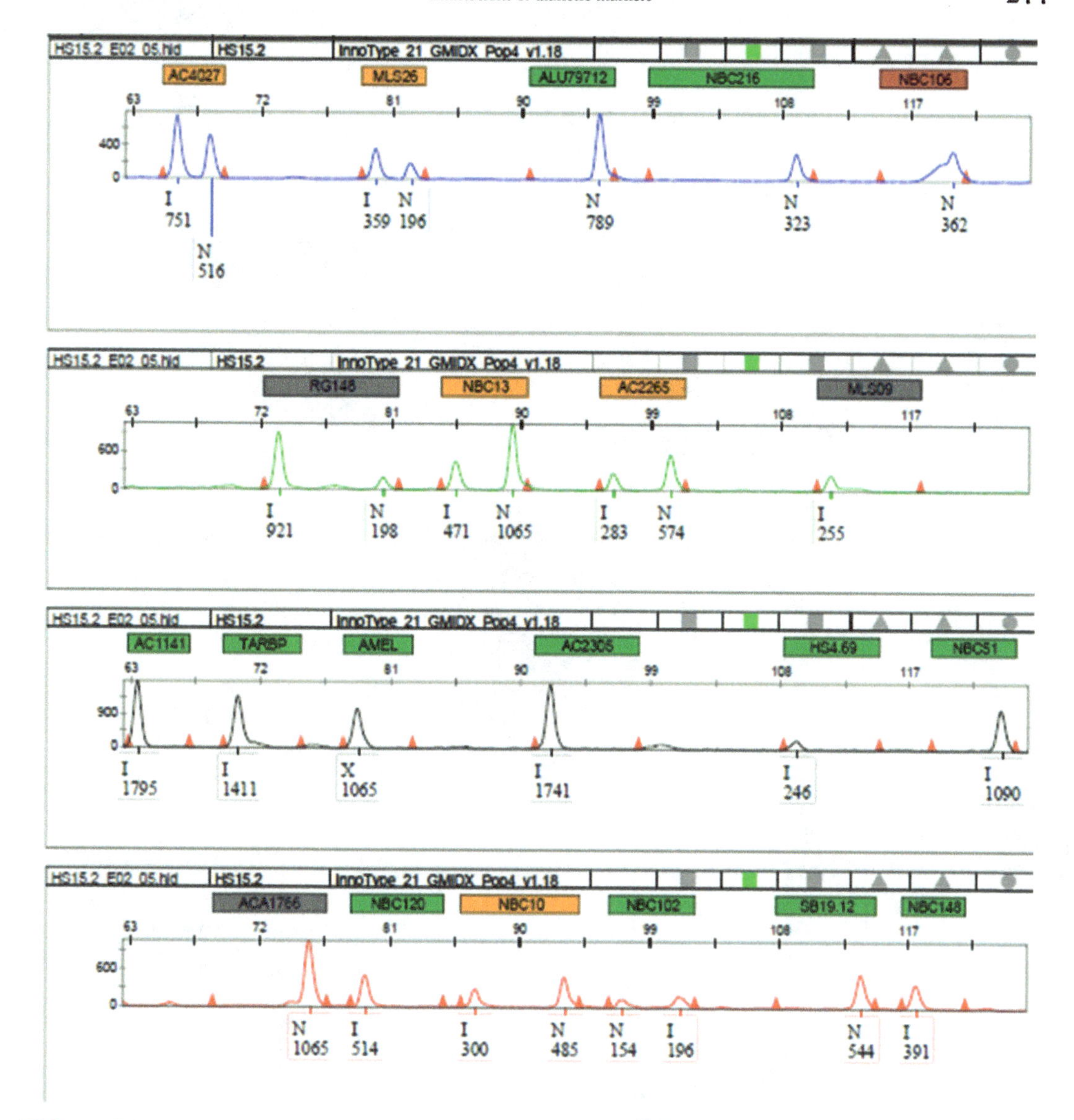

FIG. 2, CONT'D (B) the InnoTyper® 21 assay (InnoGenomics Technologies LLC, New Orleans, Louisiana, United States).

Limitations of diallelic markers

Due to the binary nature of diallelic markers, two possible alleles exist, which yield a total of three potential genotypes (aside from rare microvariants and tri-alleles). These genotypes include heterozygous insertion-deletion/null, or homozygous for the two different allele types (respectively). Since diallelic markers are not as polymorphic as STR loci, the detection of mixtures is more difficult. While it is possible to detect peak height imbalance, anything

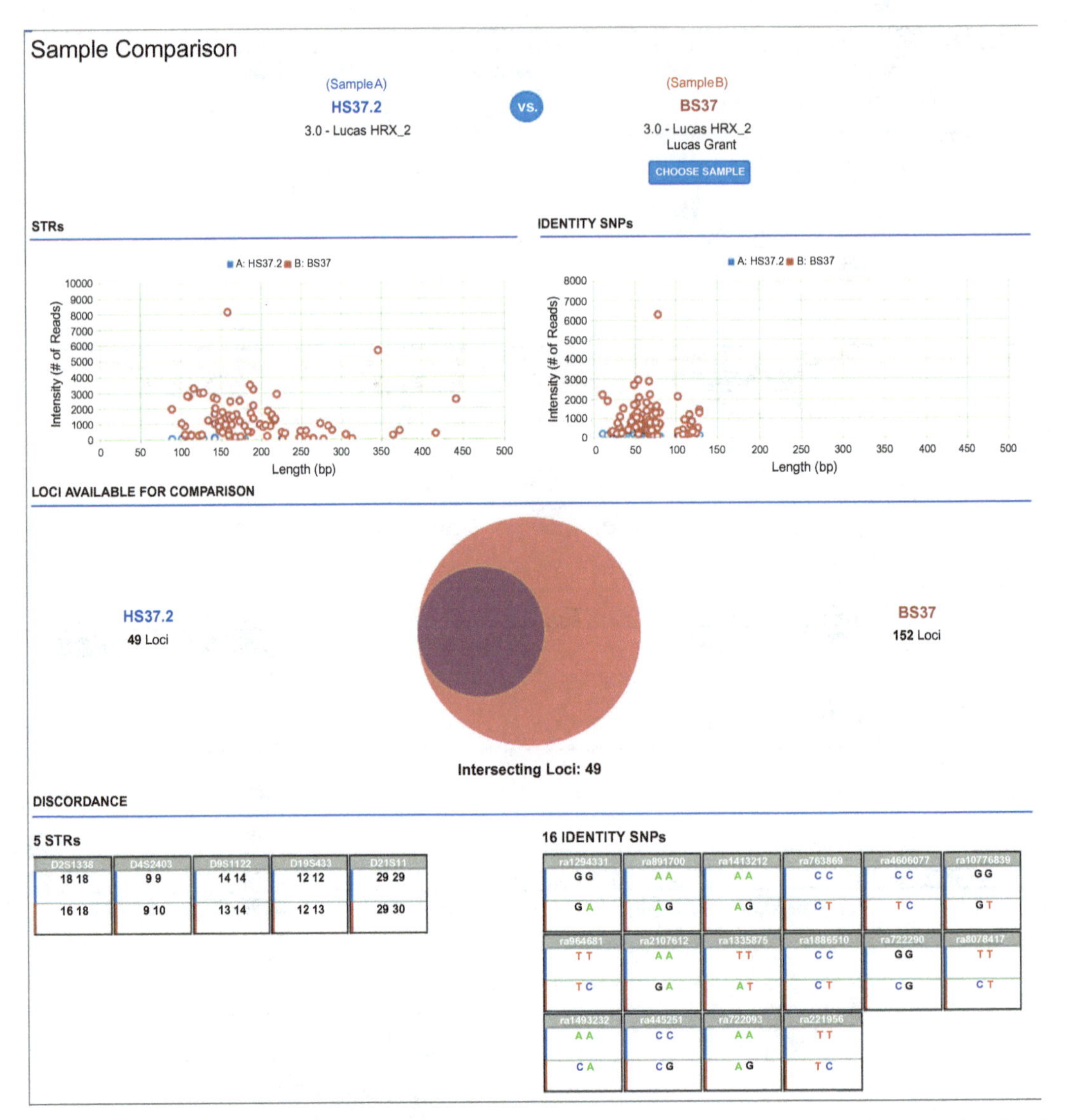

FIG. 3 Massively Parallel Sequencing (MPS) results for the same rootless hair shaft sample reported in Fig. 2, using the ForenSeq™ Signature Prep Kit and the MiSeq FGx™ Forensic Genomics System (Verogen, San Diego, California United States).

more than a two-person mixture is difficult (or virtually impossible) to identify with any sort of certainty and could potentially pass unnoticed. There have been discussions in the small community that works with these markers about including a few highly polymorphic STR loci to INDEL or INNUL panels for the purpose of mixture detection; however, doing so with amplicons less than 150 bp has proven to be difficult, limiting this approach with highly degraded samples.

While length-based diallelic markers are much more suited as an alternative marker for capillary electrophoresis-based genotyping systems than SNPs, the same cannot be said for MPS platforms. While the chemistry of MPS platforms with regard to INDELs vs SNPs possesses no real bias, bioinformatic workflows co-opted from the biomedical genomics field are more heavily targeted toward SNP detection than INDELs. Additionally, since there is no need to compete for amplicon size "real estate" in dye channels (i.e., all markers can be designed as small as possible), all STR markers can be designed as a "mini-primer" amplicon which gives a higher chance for allele generation. With extracts that would fail with GlobalFiler™ but would generate successful profiles using InnoTyper® 21, these MPS platforms are capable of generating profiles that have RMPs with 7–10 orders of magnitude higher than InnoTyper® 21. Additionally, a profile (and its corresponding reference profile) generated with the same extract shown in Fig. 2 yields many more markers with the ForenSeq™ Signature Prep assay (Verogen, San Diego, California United States) (Fig. 3). The recovery of more alleles—including some mini-amplicon STRs—generates substantially more powerful RMPs; additionally, co-amplification of STR markers makes this assay more amenable to mixture detection. As MPS platforms gain a broader foothold in forensic DNA laboratories, the utility of length-based diallelics will become less important because MPS platforms offer superior characteristics with regard to discriminatory power and mixture detection. Regardless, INDEL and INNUL marker systems provide a powerful set of supplementary alternative markers for analysis of degraded samples on CE-based platforms.

References

Asari, M., Omura, T., Oka, K., Maseda, C., Tasaki, Y., Shiono, H., Matsubara, K., Matsuda, M., Shimizu, K., 2012. Multiplex PCR-based Alu insertion polymorphisms genotyping for identifying individuals of Japanese ethnicity. Genomics 99 (4), 227–232.

Batzer, M.A., Deininger, P.L., 2002. Alu repeats and human genomic diversity. Nat. Rev. Genet. 3 (5), 370–379.

Batzer, M.A., Stoneking, M., Alegria-Hartman, M., Bazan, H., Kass, D.H., Shaikh, T.H., Novick, G.E., Ioannou, P.A., Scheer, W.D., Herrera, R.J., 1994. African origin of human-specific polymorphic Alu insertions. Proc. Natl. Acad. Sci. 91 (25), 12288–12292.

Bina, M., Sturtevant, J.M., Stein, A., 1980. Stability of DNA in nucleosomes. Proc. Natl. Acad. Sci. 77 (7), 4044–4047.

Børsting, C., Rockenbauer, E., Morling, N., 2009. Validation of a single nucleotide polymorphism (SNP) typing assay with 49 SNPs for forensic genetic testing in a laboratory accredited according to the ISO 17025 standard. Forensic Sci. Int. Genet. 4 (1), 34–42.

Brown, H., Thompson, R., Murphy, G., Peters, D., LaRue, B., King, J., Montgomery, A.H., Carroll, M., Baus, J., Sinha, S., 2017. Development and validation of a novel multiplexed DNA analysis system, InnoTyper® 21. Forensic Sci. Int. Genet. 29, 80–99.

Budowle, B., Van Daal, A., 2008. Forensically relevant SNP classes. BioTechniques 44 (5), 603–610.

Butler, J.M., 2007. Short tandem repeat typing technologies used in human identity testing. Biotechniques 43 (4), Sii–Sv.

Cooper, D.N., Krawczak, M., 1991. Mechanisms of insertional mutagenesis in human genes causing genetic disease. Hum. Genet. 87 (4), 409–415.

Dixon, L., Murray, C., Archer, E., Dobbins, A., Koumi, P., Gill, P., 2005. Validation of a 21-locus autosomal SNP multiplex for forensic identification purposes. Forensic Sci. Int. 154 (1), 62–77.

Ensenberger, M.G., Lenz, K.A., Matthies, L.K., Hadinoto, G.M., Schienman, J.E., Przech, A.J., Morganti, M.W., Renstrom, D.T., Baker, V.M., Gawrys, K.M., 2016. Developmental validation of the PowerPlex® Fusion 6C system. Forensic Sci. Int. Genet. 21, 134–144.

Ferreira Palha, T.J.B., Ribeiro Rodrigues, E.M., Cavalcante, G.C., Marrero, A., de Souza, I.R., Seki Uehara, C.J., Silveira da Motta, C.H.A., Koshikene, D., da Silva, D.A., de Carvalho, E.F., Chemale, G., Freitas, J.M., Alexandre, L., Paranaiba, R.T.F., Soler, M.P., Santos, S., 2015. Population genetic analysis of insertion–deletion polymorphisms in a Brazilian population using the Investigator DIPplex kit. Forensic Sci. Int. Genet. 19, 10–14.

Fondevila, M., Phillips, C., Santos, C., Pereira, R., Gusmao, L., Carracedo, A., Butler, J., Lareu, M., Vallone, P., 2012. Forensic performance of two insertion–deletion marker assays. Int. J. Legal Med. 126 (5), 725–737.

Friis, S., Børsting, C., Rockenbauer, E., Poulsen, L., Fredslund, S., Tomas, C., Morling, N., 2012. Typing of 30 insertion/deletions in Danes using the first commercial indel kit–Mentype® DIPplex. Forensic Sci. Int. Genet. 6 (2), e72.

Grisedale, K.S., Murphy, G.M., Brown, H., Wilson, M.R., Sinha, S.K., 2018. Successful nuclear DNA profiling of rootless hair shafts: a novel approach. Int. J. Legal Med. 132 (1), 107–115.

Hughes-Stamm, S.R., Ashton, K.J., Van Daal, A., 2011. Assessment of DNA degradation and the genotyping success of highly degraded samples. Int. J. Legal Med. 125 (3), 341–348.

Jian, H., Wang, L., Wang, H., Bai, X., Lv, M., Liang, W., 2019. Population genetic analysis of 30 insertion–deletion (INDEL) loci in a Qinghai Tibetan group using the Investigator DIPplex Kit. Int. J. Legal Med. 133 (4), 1039–1041.

Kidd, K.K., Pakstis, A.J., Speed, W.C., Grigorenko, E.L., Kajuna, S.L., Karoma, N.J., Kungulilo, S., Kim, J.J., Lu, R.B., Odunsi, A., 2006. Developing a SNP panel for forensic identification of individuals. Forensic Sci. Int. 164 (1), 20–32.

Kraemer, M., Prochnow, A., Bussmann, M., Scherer, M., Peist, R., Steffen, C., 2017. Developmental validation of QIAGEN Investigator® 24plex QS Kit and Investigator® 24plex GO! Kit: two 6-dye multiplex assays for the extended CODIS core loci. Forensic Sci. Int. Genet. 29, 9–20.

Krawczak, M., Cooper, D.N., 1991. Gene deletions causing human genetic disease: mechanisms of mutagenesis and the role of the local DNA sequence environment. Hum. Genet. 86 (5), 425–441.

LaRue, B.L., 2018. Development of Improved Insertion-Deletion Assays for Human and Ancestral identifications From Degraded Samples. Office of Justice Programs' National Criminal Justice Reference Service, Document number 251818.

LaRue, B.L., Ge, J., King, J.L., Budowle, B., 2012a. A validation study of the Qiagen Investigator DIPplex® kit: an INDEL-based assay for human identification. Int. J. Legal Med. 126 (4), 533–540.

LaRue, B.L., Sinha, S.K., Montgomery, A.H., Thompson, R., Klaskala, L., Ge, J., King, J., Turnbough, M., Budowle, B., 2012b. INNULs: a novel design amplification strategy for retrotransposable elements for studying population variation. Hum. Hered. 74 (1), 27–35.

LaRue, B.L., Lagacé, R., Chang, C.W., Holt, A., Hennessy, L., Ge, J., King, J.L., Chakraborty, R., Budowle, B., 2014. Characterization of 114 insertion/deletion (INDEL) polymorphisms, and selection for a global INDEL panel for human identification. Legal Med. 16 (1), 26–32.

Li, F., Zhang, M., Wang, Y., Shui, J., Yan, M., Jin, X., Zhu, X., 2017. Determination of hair shafts by InnoTyper® 21 kit. Fa Yi Xue Za Zhi 33 (6), 615–618.

Liang, W., Zaumsegel, D., Rothschild, M., Lu, M., Zhang, L., Li, J., Liu, F., Xiang, J., Schneider, P., 2013. Genetic data for 30 insertion/deletion polymorphisms in six Chinese populations with Qiagen Investigator DIPplex Kit. Forensic Sci. Int. Genet. Suppl. Ser. 4 (1), e268–e269.

Mamedov, I.Z., Shagina, I.A., Kurnikova, M.A., Novozhilov, S.N., Shagin, D.A., Lebedev, Y.B., 2010. A new set of markers for human identification based on 32 polymorphic Alu insertions. Eur. J. Hum. Genet. 18 (7), 808–814.

Manta, F., Caiafa, A., Pereira, R., Silva, D., Amorim, A., Carvalho, E.F., Gusmão, L., 2012. Indel markers: genetic diversity of 38 polymorphisms in Brazilian populations and application in a paternity investigation with post mortem material. Forensic Sci. Int. Genet. 6 (5), 658–661.

Martínez-Cortés, G., Gusmao, L., Pereira, R., Salcido, V., Favela-Mendoza, A., Muñoz-Valle, J., Inclán-Sánchez, A., López-Hernández, L., Rangel-Villalobos, H., 2015. Genetic structure and forensic parameters of 38 Indels for human identification purposes in eight Mexican populations. Forensic Sci. Int. Genet. 17, 149–152.

Martins, C., Ferreira, P.M., Carvalho, R., Costa, S.C., Farinha, C., Azevedo, L., Amorim, A., Oliveira, M., 2019. Evaluation of InnoQuant® HY and InnoTyper® 21 kits in the DNA analysis of rootless hair samples. Forensic Sci. Int. Genet. 39, 61–65.

Mills, R.E., Luttig, C.T., Larkins, C.E., Beauchamp, A., Tsui, C., Pittard, W.S., Devine, S.E., 2006. An initial map of insertion and deletion (INDEL) variation in the human genome. Genome Res. 16 (9), 1182–1190.

Moura-Neto, R., Silva, R., Mello, I., Nogueira, T., Al-Deib, A., LaRue, B., King, J., Budowle, B., 2015. Evaluation of a 49 InDel Marker HID panel in two specific populations of South America and one population of Northern Africa. Int. J. Legal Med. 129 (2), 245–249.

Mulero, J.J., Chang, C.W., Lagace, R.E., Wang, D.Y., Bas, J.L., McMahon, T.P., Hennessy, L.K., 2008. Development and validation of the AmpFℓSTR® MiniFilerTM PCR Amplification Kit: a miniSTR multiplex for the analysis of degraded and/or PCR inhibited DNA. J. Forensic Sci. 53 (4), 838–852.

Neuvonen, A.M., Palo, J.U., Hedman, M., Sajantila, A., 2012. Discrimination power of Investigator DIPplex loci in Finnish and Somali populations. Forensic Sci. Int. Genet. 6 (4), e99–e102.

Nunotani, M., Shiozaki, T., Sato, N., Kamei, S., Takatsu, K., Hayashi, T., Ota, M., Asamura, H., 2015. Analysis of 30 insertion–deletion polymorphisms in the Japanese population using the Investigator DIPplex® kit. Legal Med. 17 (6), 467–470.

Pereira, R., Phillips, C., Alves, C., Amorim, A., Carracedo, A., Gusmão, L., 2009a. A new multiplex for human identification using insertion/deletion polymorphisms. Electrophoresis 30 (21), 3682–3690.

Pereira, R., Phillips, C., Alves, C., Amorim, A., Carracedo, A., Gusmão, L., 2009b. Insertion/deletion polymorphisms: a multiplex assay and forensic applications. Forensic Sci. Int. Genet. Suppl. Ser. 2 (1), 513–515.

Phillips, C., Salas, A., Sanchez, J., Fondevila, M., Gomez-Tato, A., Alvarez-Dios, J., Calaza, M., de Cal, M.C., Ballard, B., Lareu, M., 2007. Inferring ancestral origin using a single multiplex assay of ancestry-informative marker SNPs. Forensic Sci. Int. Genet. 1 (3–4), 273–280.

Prinz, M., Schmitt, C., 1994. Effect of degradation on PCR based DNA typing. In: Advances in Forensic Haemogenetics. Springer, pp. 375–378.

Seong, K.M., Park, J.H., Hyun, Y.S., Kang, P.W., Choi, D.H., Han, M.S., Park, K.W., Chung, K.W., 2014. Population genetics of insertion–deletion polymorphisms in South Koreans using Investigator DIPplex kit. Forensic Sci. Int. Genet. 8 (1), 80–83.

Tasker, E., LaRue, B., Beherec, C., Gangitano, D., Hughes-Stamm, S., 2017. Analysis of DNA from post-blast pipe bomb fragments for identification and determination of ancestry. Forensic Sci. Int. Genet. 28, 195–202.

Tomas, C., Poulsen, L., Drobnič, K., Ivanova, V., Jankauskiene, J., Bunokiene, D., Børsting, C., Morling, N., 2016. Thirty autosomal insertion-deletion polymorphisms analyzed using the Investigator® DIPplex Kit in populations from Iraq, Lithuania, Slovenia, and Turkey. Forensic Sci. Int. Genet. 25, 142–144.

Wang, J., Song, L., Grover, D., Azrak, S., Batzer, M.A., Liang, P., 2006. dbRIP: a highly integrated database of retrotransposon insertion polymorphisms in humans. Hum. Mutat. 27 (4), 323–329.

Wang, Z., Zhang, S., Zhao, S., Hu, Z., Sun, K., Li, C., 2014. Population genetics of 30 insertion–deletion polymorphisms in two Chinese populations using Qiagen investigator® DIPplex kit. Forensic Sci. Int. Genet. 11, e12–e14.

Wang, D.Y., Gopinath, S., Lagacé, R.E., Norona, W., Hennessy, L.K., Short, M.L., Mulero, J.J., 2015. Developmental validation of the GlobalFiler® Express PCR Amplification Kit: a 6-dye multiplex assay for the direct amplification of reference samples. Forensic Sci. Int. Genet. 19, 148–155.

Wang, L., Lv, M., Zaumsegel, D., Zhang, L., Liu, F., Xiang, J., Li, J., Schneider, P.M., Liang, W., Zhang, L., 2016. A comparative study of insertion/deletion polymorphisms applied among Southwest, South and Northwest Chinese populations using Investigator® DIPplex. Forensic Sci. Int. Genet. 21, 10–14.

Weber, J.L., David, D., Heil, J., Fan, Y., Zhao, C., Marth, G., 2002. Human diallelic insertion/deletion polymorphisms. Am. J. Hum. Genet. 71 (4), 854–862.

Wendt, F.R., Warshauer, D.H., Zeng, X., Churchill, J.D., Novroski, N.M., Song, B., King, J.L., LaRue, B.L., Budowle, B., 2016. Massively parallel sequencing of 68 insertion/deletion markers identifies novel microhaplotypes for utility in human identity testing. Forensic Sci. Int. Genet. 25, 198–209.

Zhang, S., Zhu, Q., Chen, X., Zhao, Y., Zhao, X., Yang, Y., Gao, Z., Fang, T., Wang, Y., Zhang, J., 2018. Forensic applicability of multi-allelic InDels with mononucleotide homopolymer structures. Electrophoresis 39 (16), 2136–2143.

SECTION III

Traditional platforms, alternative strategies, and emerging technologies for DNA analysis of human skeletal remains

CHAPTER

14

Genotyping and sequencing of DNA recovered from human skeletal remains using capillary electrophoresis (CE)

Jodie Ward PhD[a], Jeremy Watherston PhD[b], Irene Kahline MSFS[c], Timothy P. McMahon PhD[d], and Suni M. Edson PhD[c]

[a]Centre for Forensic Science, University of Technology Sydney, Sydney, NSW, Australia [b]Forensic and Analytical Science Service, NSW Health Pathology, Lidcombe, NSW, Australia [c]Past Accounting Section, Armed Forces DNA Identification Laboratory (AFDIL), Armed Forces Medical Examiner System, Dover, DE, United States [d]Department of Defense (DoD) DNA Operations, Armed Forces Medical Examiner System, Defense Health Agency, Dover, DE, United States

Overview

Capillary electrophoresis (CE) is the preferred method for forensic DNA profiling. CE instruments (i.e., Genetic Analyzers) are the modern platform of choice for forensic DNA laboratories performing routine fragment and/or sequence analysis. Current Genetic Analyzers offer a fully automated workflow with the capacity to analyze up to 96 samples simultaneously. The injection, separation, detection, and primary data analysis are all automated. Prior to CE, target regions of DNA are amplified using fluorescently labeled primers (in the case of fragment analysis) or fluorescently labeled chain-terminating nucleotides are incorporated during the polymerase chain reaction (PCR) (in the case of DNA sequencing). Both techniques result in fluorescent fragments of different lengths, which are subsequently separated and detected using CE.

CE can be used for both genotyping and sequencing applications for profiling DNA recovered from human skeletal remains, including genotyping of autosomal short tandem repeats

Forensic Genetic Approaches for Identification of Human Skeletal Remains
https://doi.org/10.1016/B978-0-12-815766-4.00014-5

(STRs), Y-chromosome STRs (Y-STRs), single nucleotide polymorphisms (SNPs), insertions/deletions (indels), and Sanger sequencing of mitochondrial DNA (mtDNA). This chapter will focus on STR genotyping and mtDNA sequencing, and the platform, chemistry, and analysis software requirements for each application. Furthermore, due to the often-compromised nature of human skeletal remains, procedural modifications to improve the success of STR genotyping and mtDNA sequencing will be described. CE artifacts associated with the DNA profiling of skeletal samples will be detailed, including those attributable to DNA degradation, inhibition, and damage, and the co-extraction and amplification of nonhuman DNA.

Principles of capillary electrophoresis (CE)

CE was first used to separate and detect STRs in the early 1990s (Butler et al., 1994; McCord et al., 1993). The mid-1990s saw the introduction of the first commercially available CE platform for human identification (HID) applications, with the release of a single-capillary CE instrument by Applied Biosystems™ in 1995. Prior to the introduction of CE instruments, amplified STR products were separated using an agarose or polyacrylamide gel-based apparatus. Since then, the general workflow for CE instruments has remained the same across all CE platforms (Fig. 1); however, advancements include increasing the throughput by introducing multiple capillary arrays, as well as efficiencies gained through maintenance, reagent tracking, and data collection and analysis software. For a detailed review of the evolution of CE technology and its application to forensic STR profiling, refer to the following publications: Butler et al. (2004), Butler (2012b), and Shewale et al. (2012).

Sample injection

Unlike slab-gel electrophoresis systems, in which STR or sequencing products are manually loaded onto the gel, CE relies on electrokinetic injections and the inherent negative charge of DNA molecules. To accomplish this, the CE instrument must create a closed current circuit that will allow the negatively charged DNA products to enter the capillary and then provide a continuous current, enabling the separation of DNA products according to size. CE instruments provide a closed current circuit by connecting the anode buffer container to the cathode buffer container via a capillary array, which is connected to the polymer delivery pump. The polymer delivery pump is also connected to the Performance Optimized Polymer (POP) pouch and the waste trap container. The anode and cathode buffer containers located on opposite ends of the circuit contain 1 × ethylenediaminetetraacetic acid (EDTA) running buffer, which is an electrical conducting buffer that allows a negative current to pass from the anode buffer container through the capillary to the cathode buffer container.

DNA products are prepared in highly deionized (Hi-Di™) formamide (a strong chemical denaturant) and then heat-denatured at 95°C. During the denaturation step, double-stranded DNA is separated into single strands and hydrogen ions (H^+) are donated by the Hi-Di™ formamide, which keeps the DNA in a single-stranded conformation. The DNA products are then loaded onto the instrument. The instrument will move the capillaries to the waste container and the polymer delivery pump will inject new buffer from the polymer pouch through the capillary and into the waste container. This step removes previously used polymer and

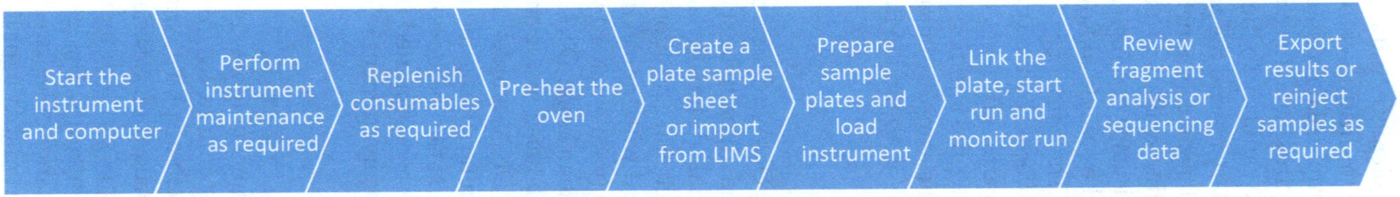

Start the instrument and computer

Perform instrument maintenance as required

Replenish consumables as required

Pre-heat the oven

Create a plate sample sheet or import from LIMS

Prepare sample plates and load instrument

Link the plate, start run and monitor run

Review fragment analysis or sequencing data

Export results or reinject samples as required

FIG. 1 A typical capillary electrophoresis (CE) workflow for automated CE platforms. After the instrument and computer are turned on, scheduled maintenance tasks, verification of consumable levels and expiration dates, removal of polymer bubbles, and preheating the oven should be performed prior to beginning a run. Sample data sheets are either manually entered or imported from a Laboratory Information Management System (LIMS) and subsequently linked to the prepared sample plates once loaded in the instrument. After the run is complete, analysts can review the CE data and export, or reinject or retype samples if not of requisite quality. *Graphic design by: Timothy P. McMahon.*

reconditions the capillary. The capillary is moved to the cathode buffer and a pre-run is initiated to test the closed circuit by a voltage tolerance. If the voltage tolerance fails, the run is aborted and an error message will appear on the instrument screen. After the pre-run, the capillaries are washed in water, moved to the sample plate containing the DNA product, and then electrokinetically injected at 1.2–3kV. In electrokinetic injection, a negative charge is passed from the anode buffer container through the capillary to the sample well and the negatively charged DNA molecules move into the capillary toward the positively charged anode, with no loss in liquid from the sample well. Low voltage electrokinetic injections allow for a uniform amount of DNA (regardless of size) to enter the capillary in a tight bundle, commonly referred to as stacking. Stacking occurs due to the existence of a higher salt content in the buffer compared to in the sample well, which causes the sample to flow quickly into the capillary, but then stall or "stack" just inside the capillary. The efficiency of electrokinetic injection is highly dependent on salt concentration and can be affected by the quality of the formamide used. The use of a nondeionized formamide will decrease the efficiency of the injection, as the formamide will compete with the DNA fragments during the injection process; for this reason, Hi-Di™ formamide is recommended for use. After the sample is injected, the capillaries are washed, placed back into the cathode buffer, and 15kV is applied to the samples in the capillary.

Sample separation

Upon stacking and entering the capillary, DNA products are separated by size using POP containing either 4% linear noncross-linked dimethylacrylamide (i.e., POP-4™) for STR fragment separation or 6% linear noncross-linked dimethylacrylamide (i.e., POP-6™) for sequencing product separation, along with high concentrations of urea to help keep DNA in a denatured form. Application of 15kV, coupled with the sieving activity of the dimethylacrylamide, allows DNA to be separated by size (with smaller DNA products migrating toward the detection cell heater block more quickly than larger DNA products). Central to this whole process is the fact that the capillary array is housed inside an oven that keeps both the array and polymer at a constant 60°C to maintain DNA samples in a single-stranded state by preventing DNA secondary structure from forming during the run. At a set time (based on the defined run module being used), the shutter on the charge-coupled device (CCD) camera will open and close every second to collect fluorescence emitted from the DNA fragments as they pass through the detection cell window in the instrument.

Sample detection

The CE optical system is made up of many components (Fig. 2), with sample detection occurring within the detection cell. The laser emits a 488/514.5 or 505nm beam (dependent on the instrument model) that passes through a splitter, which divides the beam into two separate beams. The two beams are then focused off a series of mirrors to the bottom and top of the detection cell. By splitting the beams, the laser light is accumulated across all capillaries providing a more uniform, balanced intensity. The laser crossing the capillaries in the detection cell window excites the fluorescent molecules attached to the DNA fragments, and light of a longer wavelength is emitted. The emitted light is passed through a transmission grating that divides the light into their respective emission spectra, which are collected by the

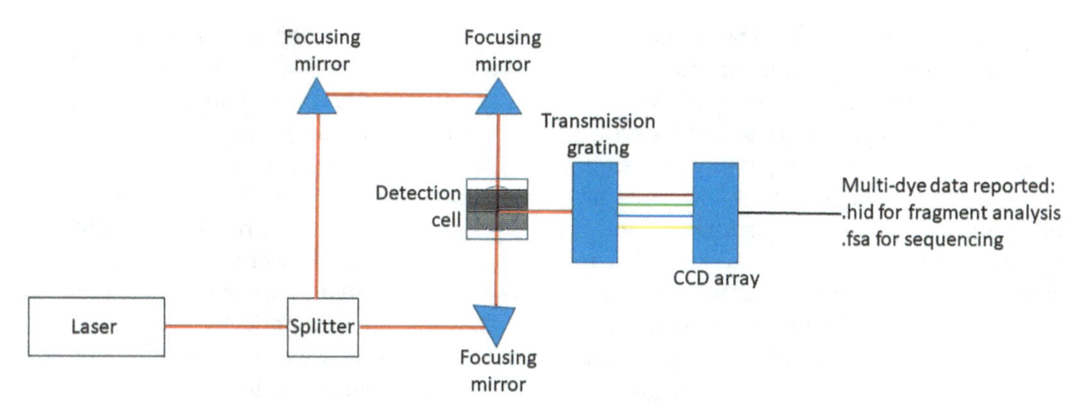

FIG. 2 The capillary electrophoresis (CE) optical system components in a typical 31XX instrument. A split laser beam is focused across all capillaries of the array in the detection cell. Once the DNA fragments enter the detection cell, the attached dye labels are excited by the laser causing them to fluoresce. The emitted light passes through a transmission grating to disperse the light by wavelength and the emission spectra data are detected and recorded by the charge-coupled device (CCD) camera. The raw data are then processed by the Data Collection Software, which applies an appropriate spectral calibration file and creates a .hid file for fragment analysis and a .fsa file for sequencing. *Graphic design by: Timothy P. McMahon.*

CCD camera. The CCD camera has 20 collection zones assigned to each capillary and sends the collected raw data to the Data Collection Software, which applies an appropriate spectral calibration file. The Data Collection Software creates a .hid file for fragment analysis and a .fsa file for sequencing.

Spatial and spectral calibration

In order for the CCD camera to know where to collect emitted data from fluorescently labeled STR fragments or sequencing products, it needs to be aligned with the capillaries in the detection cell. This is accomplished by performing a spatial calibration, which establishes a spatial relationship between the capillary array and the CCD camera. A spatial array is only run if the capillary is replaced, the service engineer updates the optics, or the detection cell is opened. For the Applied Biosystems™ 3500 Series Genetic Analyzers (Thermo Fisher Scientific), the software will look for a single sharp peak in each capillary. Peak heights should be uniform and of approximately the same height, between 13 and 16 units apart, and one (+) symbol is above the apex of each peak. The user will need to verify these thresholds and accept or reject the spatial calibration. Once this is completed, an appropriate spectral calibration dye set will need to be run.

Spectral calibrations are performed to create a deconvolution matrix. The deconvolution matrix corrects for the spectral emission overlaps in the 4-, 5-, or 6-dye systems used for forensic DNA profiling applications in order to: (1) reduce the raw data from the instrument, (2) prevent pull-up peaks in overlapping color channels, and (3) enable normalization of the baseline to assist with data analysis. Spectral calibrations need to be performed when a capillary is replaced, when the type of polymer being used is changed, if a new dye set is being applied (and has not previously been calibrated), or when any changes are made to the optics system. The spectral calibration is completely automated with predefined modules on the

Applied Biosystems™ 3500 Series Genetic Analyzers (Thermo Fisher Scientific). The module automatically sets up three injections of the spectral dye in case of a capillary failure during a single run. The software does allow the user to specify whether borrowing of spectral data from neighboring capillary is either permitted or disallowed. If borrowing is disallowed, then all capillaries must meet passing criteria within the three injections. If all capillaries pass in injection one, then runs two and three are not performed. If one capillary fails during the first run, a second run is performed, and if all capillaries pass, a third run is not performed. If a third run is performed and a single-capillary does not pass, the user will need to perform another (new) spectral calibration. If borrowing is allowed from an adjacent capillary, all capillaries must meet defined borrowing specifications. Runs are automatically performed if the previous run does not meet passing specifications. Once the spectral calibration has been completed, a spectral calibration report can be created and exported for review and application.

CE instrumentation

Since 1995, Thermo Fisher Scientific (www.thermofisher.com) has introduced a number of Applied Biosystems™ Genetic Analyzer platforms to the forensic market and, until recently, has primarily been the sole vendor of the technology. The technology has evolved from laser-induced 5-dye fluorescence detection on single-capillary instruments to using 6-dye fluorescence detection on capillary array instruments. The most widely used instruments in forensic DNA laboratories are currently the Applied Biosystems™ 3130 Series Genetic Analyzers (Thermo Fisher Scientific) and Applied Biosystems™ 3500 Series Genetic Analyzers. However, having been released in 2010, the Applied Biosystems™ 3500 Genetic Analyzer is the newest Thermo Fisher Scientific instrument available and offers forensic laboratories a number of new features designed specifically for HID applications. Soon Promega Corporation (www.promega.com) will release two new CE instruments to the forensic market, offering forensic DNA laboratories a choice of manufacturer for the first time. A comparison of the features of these modern CE instruments is presented in Table 1.

Thermo Fisher Scientific genetic analyzers

Applied Biosystems™ 31XX Series Genetic Analyzer

The 16-capillary Applied Biosystems™ 3100 Genetic Analyzer (Thermo Fisher Scientific) was introduced in 2000 as a higher throughput alternative to the single-capillary Applied Biosystems™ 310 Genetic Analyzer (Thermo Fisher Scientific). The Applied Biosystems™ 3100 Genetic Analyzer was similar to the Applied Biosystems™ 310 Genetic Analyzer in that it used a syringe-filled polymer to fill the capillaries and had cathode and anode buffer chambers, a laser, CCD camera, and an oven for stabilization. However, like the Applied Biosystems™ 310 Genetic Analyzer, the Applied Biosystems™ 3100 Genetic Analyzer was prone to air bubbles being introduced into the system after refilling the polymer syringe, resulting in run failures. This was corrected for in the Applied Biosystems™ 3130 and 3130*xl* Genetic Analyzers, with the introduction of the polymer delivery pump and a polymer bottle. The Applied Biosystems™ 3130 Series Genetic Analyzers were released in 2003; however, these can no longer be purchased as a new instrument. Sales of the Applied Biosystems™

TABLE 1 A comparison of the features of modern capillary electrophoresis (CE) instruments manufactured by Thermo Fisher Scientific and Promega Corporation.

	Applied Biosystems™ 3130 Genetic Analyzer	Applied Biosystems™ 3130*xl* Genetic Analyzer	Applied Biosystems™ 3500 Genetic Analyzer	Applied Biosystems™ 3500xL Genetic Analyzer	Spectrum Compact CE System	Spectrum CE System
Manufacturer	Thermo Fisher Scientific	Thermo Fisher Scientific	Thermo Fisher Scientific	Thermo Fisher Scientific	Promega Corporation	Promega Corporation
Years of manufacture	2003–2011	2003–2011	2010–	2010–	Expected launch 2020	Expected launch 2020
Number of capillaries	4	16	8	24	4	8, 24
Capillary length (cm)	22, 36, 50, 80	22, 36, 50, 80	36, 50	36, 50	36	36
Polymer type	POP-4™, POP-6™, POP-7™	POP-4™, POP-6™, POP-7™	POP-4™, POP-6™, POP-7™	POP-4™, POP-6™, POP-7™	Polymer4, Polymer7	Polymer4
Dye channels	5	5	6	6	6	8
Laser	Argon ion	Argon ion	Solid state	Solid state	Proprietary	Proprietary
Excitation wavelength/s	488 nm, 514.5 nm	488 nm, 514.5 nm	505 nm	505 nm	Not disclosed	Not disclosed
Tracking	None	None	RFID	RFID	2D barcoding	RFID
Sample capacity	96 × 2	96 × 2	96 × 2	96 × 2	32	96 × 4
Application/s	Fragment analysis, sequencing	Fragment analysis, sequencing	Fragment analysis, sequencing	Fragment analysis, sequencing	Fragment analysis, sequencing	Fragment analysis

31XX Series Genetic Analyzers ceased in 2011, so an Applied Biosystems™ 3130 Series Genetic Analyzer is now only available as an upgrade to an existing Applied Biosystems™ 3100 instrument.

Applied Biosystems™ 3500 Series Genetic Analyzer

The Applied Biosystems™ 3500 Series Genetic Analyzers are the newest Thermo Fisher Scientific CE platforms on the market (Fig. 3). They were modeled after the Applied Biosystems™ 3130 Series Genetic Analyzers, but incorporated changes specifically for HID applications. The basic workflow for operating and running samples on the Applied Biosystems™ 3130 or 3500 Series Genetic Analyzers is similar; however, the Applied Biosystems™ 3500 Series Genetic Analyzers incorporate numerous modifications at the software, maintenance, and optics level. The Applied Biosystems™ 3500 Series Genetic Analyzers use a single excitation (505 nm) line solid state laser instead of an argon laser, which has reduced the heat output of the instruments and has allowed the instruments to run on standard 110 V outlets, instead of 220 V outlets needed for the Applied Biosystems™ 3130 Series

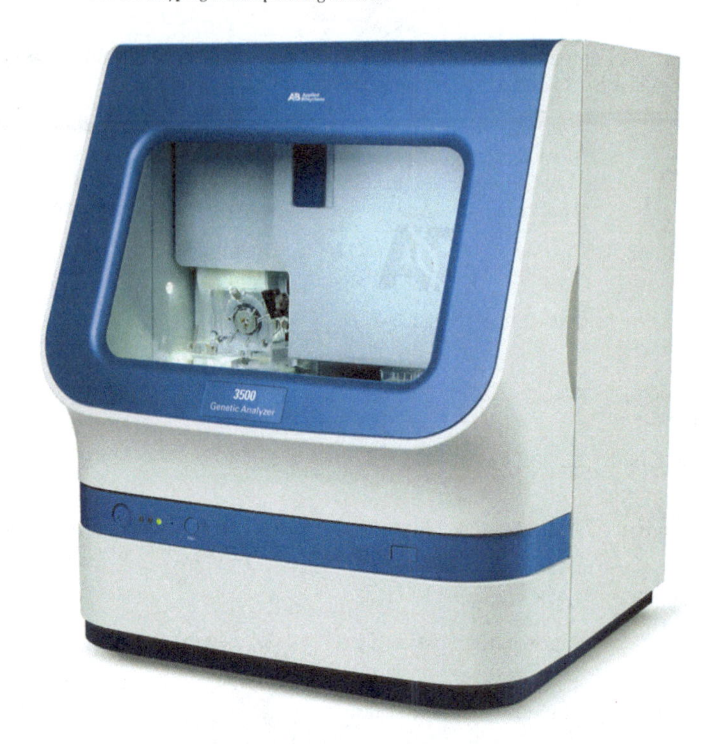

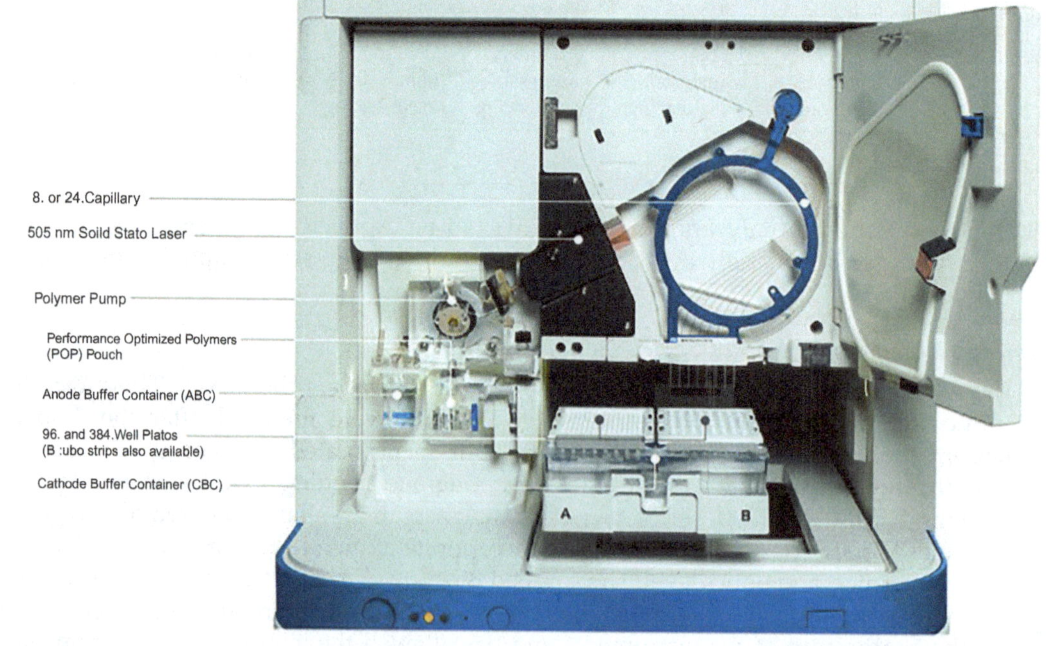

FIG. 3 Applied Biosystems™ 3500 Genetic Analyzer: (A) External view, and (B) Internal components. *© 2019 Thermo Fisher Scientific Inc. Used under permission.*

Genetic Analyzers. A reduced heat output creates efficient temperature control in and around the instruments and so multiple instruments may be colocated in smaller laboratory spaces. Preventing temperature shifts throughout the run increases the quality of the base resolution and removes the need to run multiple STR sizing ladders per plate. The remaining portions of the optics system are consistent with that of the Applied Biosystems™ 3130 Series Genetic Analyzers.

Other major improvements of the Applied Biosystems™ 3500 Series Genetic Analyzers compared to the earlier generation Applied Biosystems™ 3130 Series Genetic Analyzers include:

1. Reduced size;
2. Improved temperature control;
3. Improved polymer delivery pump design;
4. Increased throughput with 8-capillary or 24-capillary instrument options;
5. Increased STR multiplexing capacity with 6-dye detection;
6. Capillary arrays are constructed in a solid support unit for ease of replacement;
7. Ready-to-use reagents and consumables;
8. RFID technology for monitoring of consumable lot numbers, usage, and expiry dates;
9. A normalization feature to reduce peak height variation between multiple instruments and injections;
10. Higher optimal signal intensity;
11. Integrated data collection and primary analysis software for real-time assessment of data quality; and
12. Additional security and maintenance software features.

Promega Corporation genetic analyzers

While Thermo Fisher Scientific monopolizes much of the market for CE instrumentation, Promega Corporation will soon offer two alternative instrument choices: the Spectrum CE System and the Spectrum Compact CE System. The general theory behind the chemistry of the Spectrum instruments are the same as other CE instruments and will thus not be discussed further. Rather, some of the unique features of the Spectrum CE systems are presented in the next two sections.

Spectrum CE System

The Spectrum CE System (Promega Corporation) is of a similar construction and size as the Applied Biosystems™ 3500 Series Genetic Analyzer. The Spectrum CE System (Promega Corporation, 2018) uses Spectrum Polymer4 for the optimized dimethylacrylamide polymer and is capable of running any of the commercial STR kits that are currently available, including kits using 4-, 5-, or 6-dye chemistries. At present, this is the only available instrument also capable of running and analyzing new 8-dye STR kits, such as the soon to be released PowerPlex® 35GY 8C System (Promega Corporation). In addition, the Spectrum CE System can be simultaneously loaded with four 96-well plates, which can be accessed continuously and replaced with fresh plates while the instrument is running. However, the Spectrum CE System is unable to perform any fragment separation other than that of STR analysis, so Sanger-sequenced DNA cannot be separated using this instrument.

Spectrum Compact CE System

The Spectrum Compact CE System (Promega Corporation) is a smaller benchtop system that has an integrated computer and is operated using a touch screen. The Spectrum Compact CE System (Promega Corporation, 2019) uses Spectrum Polymer4 or Polymer7 to run 4-, 5-, or 6-dye chemistry STR kits, or Sanger-sequenced DNA, respectively. While having expanded capabilities for chemistry analysis, the Spectrum Compact CE System has been designed for low-throughput laboratories in mind, as it is only capable of running 32 samples in four 8-well strips.

Short tandem repeat (STR) genotyping

The current gold standard for forensic DNA profiling is considered STR genotyping with commercially available multiplex kits using PCR and CE.

Polymerase chain reaction (PCR)

Commercial STR kits are comprised of a multiplex of autosomal and Y-chromosome markers, with the latest kits containing up to 26 STR loci plus the sex-determining locus amelogenin (Ensenberger et al., 2016). These kits typically contain a master mix of PCR reagents and a primer mix; hence, an analyst only has to combine template DNA with the master mix and primer mix reagents. The reactions are then placed on a thermal cycler such as the GeneAmp® PCR System 9700 or ProFlex™ PCR System (Thermo Fisher Scientific) for PCR, according to the thermal cycling program specified by the manufacturer. Some laboratories may elect to validate a higher number of PCR cycle numbers to improve allele recovery from compromised and low template DNA (LT-DNA) samples. At the completion of the PCR, PCR products are diluted approximately 1/10 in Hi-Di™ formamide and then heated to 95°C followed by rapid cooling. These processes assist to denature the double-stranded DNA, with the formamide also acting to reduce salt levels in order to facilitate electrokinetic injection (Butler, 2012b). Additionally, diluted PCR products are mixed with an internal size standard for accurate sizing of the DNA fragments after electrophoresis.

Each STR locus is amplified by PCR using a forward and reverse PCR primer that anneals on either side of the target repeat region. One of the primers has a fluorescent dye label on the 5′ end that enables detection of the PCR product following PCR (Butler et al., 2004). The primers used to amplify a given STR locus may vary between manufacturers, or even between two STR kits produced by the same manufacturer. The ability to multiplex multiple STR loci in the one assay is reliant on the use of different fluorescent dye labels and nonoverlapping PCR product lengths for STRs amplified with the same fluorescent dye label. The length of the PCR product is determined by the position of the primers and the number of repeats in the STR region (Butler et al., 2004). The CE instrument's capability to resolve spectral overlap of the dyes, physiochemical properties of the dyes, and sensitivity and robustness of the multiplex PCR determines the number of loci that can be analyzed simultaneously (Shewale et al., 2012).

Appropriate controls should be initiated with each amplification and processed through all post-amplification procedures in parallel with the sample. These include an amplification negative control, an amplification positive control, and a reagent blank initiated during the extraction step. The amplification negative control monitors for the presence of exogenous DNA in amplification reagents and/or consumables. This amplification negative control consists of all the reagents used in setting up the PCR amplification reaction and the appropriate amount of sterile water (based on kit or manufacturer specifications). The amplification positive control monitors the success of amplification. An amplification positive control must be initiated with each amplification and subsequently processed through all post-amplification procedures in parallel with the sample. The positive control should be from a known source and can be purchased commercially or made in-house. Often a positive control (i.e., known cell line DNA) sample is included as a component of commercial STR multiplex kits.

CE requirements

PCR products are separated by size and dye color using CE on a Genetic Analyzer instrument, with current technology detecting up to 27 STRs in a single CE injection. The separation of PCR products can be influenced by the polymer used, the capillary diameter and length, the electrophoresis buffer, temperature, and the voltage applied (Butler, 2012b; Shewale et al., 2012). The internal size standard and an allelic ladder are also critical components for determining the size of the PCR products and genotyping of alleles, respectively.

Polymer type and capillary length

Generally, a combination of POP-4™ polymer and a 36-cm capillary array is used to achieve the desired resolution for STR profiling (Butler, 2012b; Connon et al., 2016); this combination of polymer type and capillary length has been validated for HID applications on the Applied Biosystems™ 3500 Series Genetic Analyzers (Thermo Fisher Scientific, 2018). POP-4™ polymer consists of 4% linear polydimethylacrylamide and a high concentration (8 M) of urea to maintain the denatured state of DNA fragments within the capillaries (Butler, 2012b). A 36-cm capillary is suitable for adequate fragment separation, while ensuring shorter run times compared to longer capillaries.

This polymer/array combination can resolve alleles that differ in size by four bases, which is important given the majority of STR markers used in forensic casework are tetranucleotide repeats. However, the presence of microvariants requires single base resolution for those alleles that differ by one to three bases (Butler et al., 2004). Complete separation of alleles differing in size by a single base cannot be consistently achieved for all observed microvariants using the current HID conditions recommended by Thermo Fisher Scientific (Connon et al., 2016), so other polymer/array combinations have been explored. For example, Connon et al. (2016) validated the use of the more viscous POP-6™ polymer with a shorter 22-cm capillary array on the Applied Biosystems™ 3130*xl* Genetic Analyzer to improve resolution, while significantly reducing CE detection time. However, single base resolution could only be achieved for fragment sizes less than 200 bases.

Run modules

A run module contains electrophoresis parameters, such as oven temperature, detector cell temperature, ramp rate, injection time and voltage, and run time and voltage (Thermo Fisher Scientific, 2014). A number of run modules have been developed by the manufacturer and are provided with the Data Collection Software for each Genetic Analyzer instrument, based on the configuration of polymer type and capillary length. For example, the Applied Biosystems™ 3500 Series Genetic Analyzers include specific run modules for HID applications. Alternatively, laboratories can optimize run module settings for each genotyping application, including the ability to vary run times, injection times, and/or injection voltage strength. The HID-validated run module for fragment analysis requires POP-4™ polymer and a 36-cm capillary. The length of each run is approximately 35 minutes, enabling up to 312 or 936 samples to be run each day on the Applied Biosystems™ 3500 or 3500xL Genetic Analyzer, respectively (Thermo Fisher Scientific, 2018).

Internal size standard

Each test sample is mixed with size standard prior to the CE run, so the standard is electrophoresed in the same capillary under the same conditions as the sample. The size standard is labeled with a unique fluorescent dye and so it can be spectrally resolved and detected separately from the target loci. Internal size standards allow sizing of all sample peaks and correct for injection-to-injection variations that result in differences when comparing the same fragments from different capillaries, runs, and instruments (Thermo Fisher Scientific, 2014). The size standard is comprised of a number of uniformly spaced fragments of known size and is used to generate a sizing curve for each sample. The sizes of each sample peak can then be determined through a relative comparison of the migration speeds during electrophoresis (Thermo Fisher Scientific, 2014). Commercial STR kits include an internal size standard; however, the recommended size standard will vary between kits due to differing fragment size ranges, precision requirements, and fluorescent dye label compatibility. A size standard for one kit cannot be substituted for the size standard created by a different manufacturer. Signal intensity of the size standard peaks should be 30%–100% of the signal intensity of the sample peaks (Thermo Fisher Scientific, 2014).

Allelic ladder

Commercial STR kits contain an amplified allelic ladder. An allelic ladder is a collection of DNA fragments corresponding to common human alleles observed in the general population for a given STR kit. The alleles are generated with the same primers used to amplify test samples (Butler, 2012b). It is recommended that multiple allelic ladders are included with every batch of test samples on a 96-well plate (ideally, one per injection). This is because minor variations in electrical current, temperature, buffer, and/or polymer variation can alter the electrophoretic mobility of fragments and contribute to differences in the sizing of alleles in the allelic ladder across the plate and during subsequent genotyping of test samples (Shewale et al., 2012).

Quality thresholds

Genetic Analyzers can convert a limited range of fluorescence into digital values (Thermo Fisher Scientific, 2014). Different instruments have different fluorescence saturation thresholds, and therefore, different recommended signal intensity ranges. For example, the Applied Biosystems™ 3500 Series Genetic Analyzers have a fluorescence saturation of 30,000 relative fluorescent units (RFU), with a recommended signal intensity range of 175–10,000 RFUs (Thermo Fisher Scientific, 2014). Despite being provided with manufacturer-recommended peak height thresholds, it is important for laboratories to establish their own validation-defined quality thresholds for allele calling due to sensitivity variation between instruments. The revised Scientific Working Group on DNA Analysis Methods (SWGDAM) Validation Guidelines (SWGDAM, 2004) state that a laboratory's internal validation should include reproducibility, precision, sensitivity, and stochastic studies. These studies will ultimately establish thresholds to confidently identify true allelic peaks in a sample. Additionally, laboratories may elect to apply different thresholds for different kits based on the results of the individual kit validations.

Analytical thresholds

SWGDAM (2017) defines an analytical threshold as the minimum peak height above which recovered peaks can be reliably distinguished from background noise. This limit of detection is based on signal-to-noise analyses during the laboratory's internal validation. An allelic peak is considered to be a true peak if it is above this empirically determined peak height. However, under this threshold, artifact labelling should not necessarily be avoided as it may result in allelic data loss (SWGDAM, 2017). After assessing different methods to determine analytical thresholds, Bregu et al. (2013) concluded that analytical thresholds derived from negatives should only be applied to LT-DNA samples and a new analytical threshold should be validated for each post-PCR procedure employed in the laboratory.

Stochastic thresholds

A laboratory's internal validation should also address stochastic effects associated with PCR-based assays (SWGDAM, 2004). The stochastic threshold is the peak height value at which it is reasonable to assume that allelic drop-out has not occurred within a single-source sample (Bieber et al., 2016; SWGDAM, 2017). It is established by assessing allelic drop-out and peak height ratios at each locus for a dilution series of DNA. Its application to allelic data allows the distinction between drop-out, heterozygous balance at a locus, and the determination of a DNA profile as a mixture.

In LT-DNA and degraded DNA samples, imbalance between heterozygous peaks is expected to increase due to stochastic effects. Optimal peak balance among heterozygous peaks at a locus is usually characterized as a percentage. For example, a peak balance of 60% specifies that the smaller peak is expected to be at least 60% of the height of the paired peak. In a suboptimal DNA sample (such as may be encountered with skeletal remains), it is likely that a lower percentage (i.e., an increased imbalance) may be observed. Internal validation will determine this balance percentage as well as the peak height where increased imbalance may commonly be observed.

A homozygote threshold also needs to be determined. This is the peak height value where it is assumed that allelic drop-out of the partner allele is unlikely to have occurred. Another consideration for a true homozygous peak will be observation of the peak multiple times in the formation of a consensus profile (i.e., reproduced alleles from two or more replicate analyses). Probabilistic genotyping software will also consider the locus-specific amplification efficiency when modeling paired peaks (Bright et al., 2013b).

Stutter thresholds

Stutter peaks are commonly observed as a smaller peak, one repeat unit less than the corresponding parent allele peak (Walsh et al., 1996). This occurs due to slipped-strand mispairing (Hauge and Litt, 1993; Walsh et al., 1996). In addition, when genotyping on modern CE equipment such as the Applied Biosystems™ 3500 or 3500xL Genetic Analyzer, stutter of

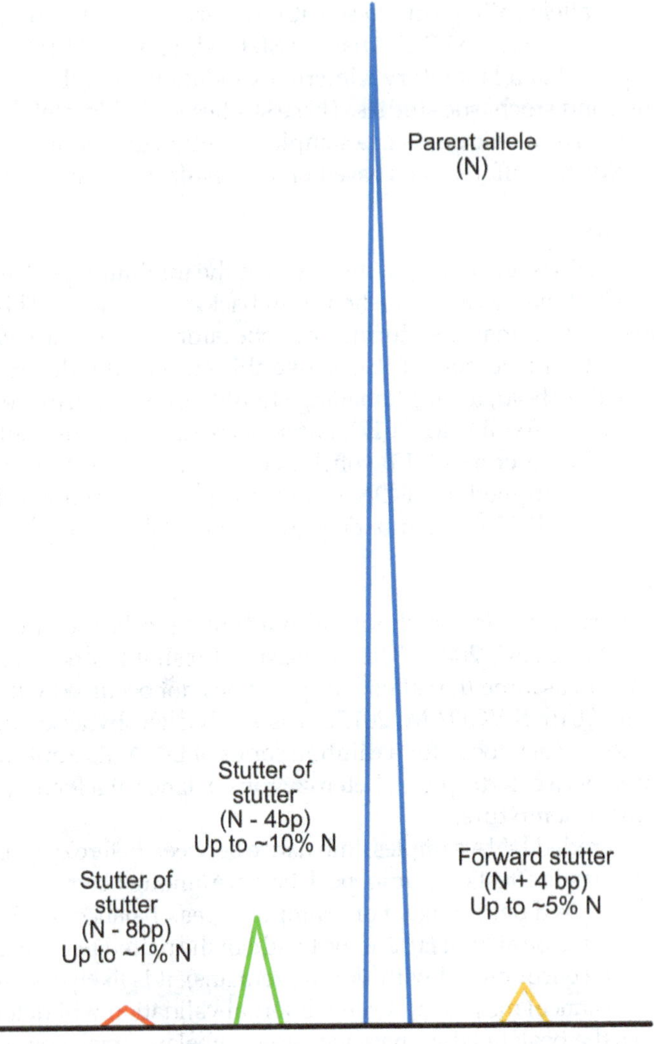

FIG. 4 A parent allele peak exhibiting stutter, stutter of stutter, and forward stutter. Stutter peak heights tend to be approximately 10%, 1%, and up to 5% of the parent peak, respectively. *Graphic design by: Trent Watherston.*

stutter and forward (or over or post) stutter can also be observed, two repeat units less and one repeat unit more than the parent peak, respectively (Bright et al., 2013b, 2014a) (Fig. 4). This is a consequence of DNA typing technologies becoming more sensitive.

Manufacturers will determine stutter values during the developmental validation of an STR kit. When determining stutter ratio thresholds, the laboratory can designate a single average stutter ratio, a range of stutter percentages, or a combination of the two. However, internal validations now routinely determine locus-specific stutter ratio means and ranges. This data is input into data analysis software programs such as GeneMapper® ID-X (GMID-X) (Thermo Fisher Scientific). Furthermore, locus-specific stutter ratios will also need to be determined for probabilistic genotyping software for the deconvolution and statistical interpretation of STR profiles.

The longest uninterrupted stretch (LUS) is the longest continuous stretch of tandem repeats. Increased rates of stutter are directly related to an increased length of LUS alleles (Vilsen et al., 2017; Brookes et al., 2012; Walsh et al., 1996; Bright et al., 2014b, 2013b). Ultimately, a different stutter ratio is associated with every allele at every locus. It has been suggested that the LUS is the best predictor of stutter ratio (Brookes et al., 2012; Walsh et al., 1996; Vilsen et al., 2017; Bright et al., 2014b). However, it has also been shown that there are examples where parental allele length is a better predictor in commonly assayed forensic markers (Woerner et al., 2017). As probabilistic genotyping software such as STRmix™ starts using LUS as its predictor of stutter ratios, this will prove an important implication for the accurate modeling of STR profiles and mixed DNA samples.

Data analysis

The data interpretation workflow typically involves:

1. Assessing internal size standards, allelic ladders, and positive and negative controls;
2. Assessing each sample for the presence of artifacts or extraneous peaks, and determining if these peaks interfere with the interpretation process; and
3. Analyzing the data for each test sample; this should be conducted independently by two different analysts.

Two Thermo Fisher Scientific software programs are commonly used to process data from Applied Biosystems™ Genetic Analyzers and produce an STR genotype: Applied Biosystems™ Data Collection Software and GMID-X. Additional probabilistic genotyping software, such as STRmix™ and TrueAllele®, can be used to interpret degraded STR profiles. Using a fully continuous model, the software will determine a likely genotype combination at each locus associated with a probabilistic certainty for subsequent database searching and calculation of a likelihood ratio (LR) for identification. These expert systems also help to remove subjectivity in DNA profile interpretation (Gill et al., 2006, 2012) and standardize interpretation in the laboratory.

Applied Biosystems™ data collection software

The Data Collection Software (Thermo Fisher Scientific) controls, monitors, and collects data generated from the Genetic Analyzer. The software performs three primary functions: (1) controls electrophoresis run conditions, (2) controls which wavelengths of light will be examined on the CCD camera through the use of "virtual filters," and (3) enables sample sheets and injection lists to be created (Butler et al., 2004). For Applied Biosystems™ 3500 Series Genetic Analyzers, the Data Collection Software is preconfigured for AmpF*l*STR® kits and enables real-time assessment of data quality due to the integration of peak detection, sizing, and quality value assignment functions. Electrophoresis data (.hid files) contain information on samples, electrophoresis run conditions, consumables, reinjection information, and analysis parameters (Shewale et al., 2012). Additionally, the software has a normalization feature, which can be applied to the data to normalize the observed intensity of amplified fragments with reference to the internal size standard. Finally, analyzed sample data are displayed as an electropherogram (with data arranged for each dye set used). Electropherograms are plotted as RFUs on the y-axis and the number of data points on the x-axis (Butler et al., 2004).

Applied Biosystems™ genotyping software

GeneMapper® ID-X (GMID-X) is a fragment analysis software package that provides DNA sizing and allele-calling functions. Electrophoresis data are analyzed using GMID-X to convert it into a peak representative of an allele and to generate STR genotyping information (Fig. 5). The sizing process involves the software first identifying an allele based on validation-defined threshold values, separating the alleles into their respective dye colors after application of an appropriate matrix file, and sizing the allele by comparison to a sizing or calibration curve generated from the internal size standard that is co-electrophoresed with each sample (Butler et al., 2004). The software produces a sizing curve by plotting the actual data points of the size standard against the expected size of each standard peak (Thermo Fisher Scientific, 2014). The Local Southern method (Elder and Southern, 1983) is the most commonly applied sizing method for forensic DNA profiling (Butler et al., 2004). It determines the sizes of fragments by selecting four size standard data points that are closest in size to the unknown fragment to derive a "best fit" line value (Thermo Fisher Scientific, 2014).

A genotype is then assigned by comparing the size of the alleles obtained for the test sample to the averaged sizes obtained for the alleles in the allelic ladder. Automated allele calling is performed with the use of specifically designed panels, bins, and stutter text profiles for the relevant STR kit. If an allele does not match an allele represented in the allelic ladder, it is labeled as an off-ladder allele. The reinjection and re-analysis of a sample can confirm if the observed off-ladder allele is a true microvariant or caused by a migration anomaly, as these

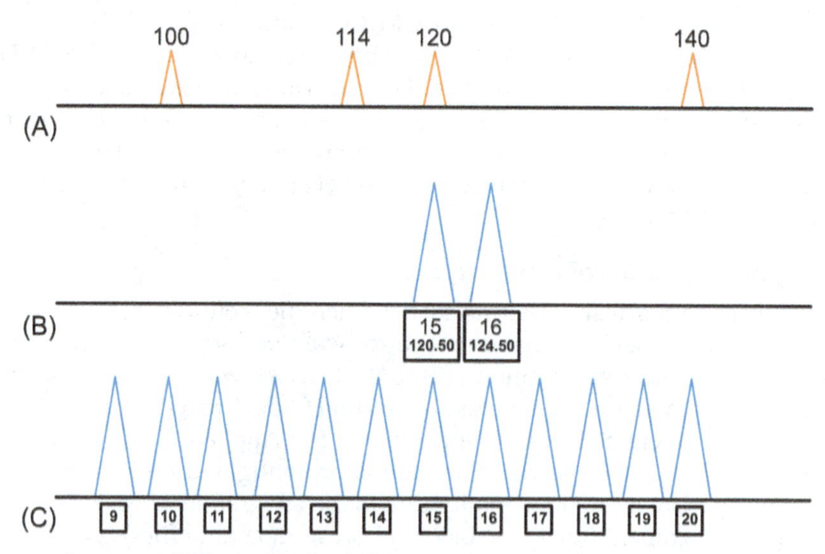

FIG. 5 Short tandem repeat (STR) genotyping of alleles 15 and 16 at the D3S1358 locus: (A) The internal size standard is used to generate a sizing curve to size the sample peaks (i.e., 120.50 and 124.50 bases); (B) DNA amplified with an STR kit produced a pair of heterozygote alleles at D3S1358; and (C) Sized sample peaks are compared to known alleles in the STR kit's allelic ladder to identify the sample alleles and determine the number of STRs (i.e., 15, 16). *Graphic design by: Trent Watherston.*

should not be reproducible. For the accurate typing of microvariant alleles, Shewale et al. (2012) advise that the size of these rarer alleles should be no more than 0.5 bases apart from the measured allele in the allelic ladder.

GMID-X software has a system of Process Quality Values (PQVs) that automatically assigns quality values to the fragment analysis data. The PQVs can be used to identify and sort samples or genotypes based on validation-defined sizing and allele-calling quality thresholds and alert the user to samples or genotypes with potential problems (Applied Biosystems™, 2009). The PQVs will flag the presence of a number of artifacts including (but not limited to) split peaks, pull-up, off-scale peaks, broad peaks, and low or imbalanced peak heights.

Probabilistic genotyping software

The possibility of drop-out and stochastic peak imbalance can complicate profile interpretation, particularly when observing STR profiles from compromised or degraded samples. Binary DNA interpretation models have since been superseded by continuous, and fully continuous, models which make better use of the observed profile data (Cowell et al., 2008; Perlin et al., 2011). Fully continuous models applied in expert systems such as STRmix™ and TrueAllele® use biological modeling of the data parameters, which can be used to generate expected peak heights of alleles for individual STR loci.

For example, STRmix™ describes a biological model whereby allelic and stutter peak height, as well as degradation of STR profiles, are modeled. There are a number of publications outlining the biological and mathematical models used by STRmix™ (Bright et al., 2013a, b, 2016; Taylor et al., 2013, 2016). Additionally, validation studies of TrueAllele® have also been published (Greenspoon et al., 2015; Perlin et al., 2011, 2013). Such software programs help remove subjectivity in DNA profile interpretation (particularly mixed DNA profiles) and will also standardize the analysis of profiles within the laboratory (Gill et al., 2006, 2012).

These expert systems require a range of data from the laboratory's internal validation including the analytical threshold, saturation limit, stutter ratios, and drop-in parameters. These laboratory-specific values are required to be able to validate expert systems and implement them into routine casework. Local population data and any population-specific F_{ST} (or θ) values to account for interrelatedness will also need to be included for calculation of a likelihood ratio (LR).

While largely focused on mixed DNA profiles, the biological modeling and fully continuous models used in these expert systems are able to objectively model degraded STR profiles commonly observed from skeletal remains. Modeling of such profiles will also allow the determination of a genotype combination for upload to a database for profile searching purposes. This genotype combination will also be associated with a level of certainty (e.g., the laboratory will often upload genotype combinations at a weight of 99%). Consequently, it is possible to calculate a LR using relevant population data for identification purposes.

Genotyping considerations for human skeletal remains

The genotyping of DNA recovered from human skeletal remains can present a number of challenges, due to these samples frequently being degraded, inhibited, and contaminated, or containing scarce amounts of nuclear DNA (nDNA). Additionally, forensic DNA laboratories that routinely perform STR profiling on DNA recovered from skeletal samples commonly observe spurious off-ladder peaks, so consideration of these artifacts and how they may interfere with the interpretation of human nDNA profiles is important. In an effort to improve the quality of STR profiles from more compromised skeletal samples, laboratories have implemented a number of post-PCR manipulations and/or modifications to standard CE procedures.

Common artifacts

In addition to common amplification and electrophoretic artifacts, skeletal samples can exhibit artifacts associated with low template, degraded, or inhibited DNA and have been reported to exhibit a number of sample-specific artifacts (Figs. 6 and 7). Refer to Butler (2012a) and SWGDAM (2017) for a detailed review of amplification and electrophoretic artifacts commonly observed for STR profiling. STR multiplex kit user manuals and developmental validation documents also contain kit-specific information regarding artifacts observed in their respective kits.

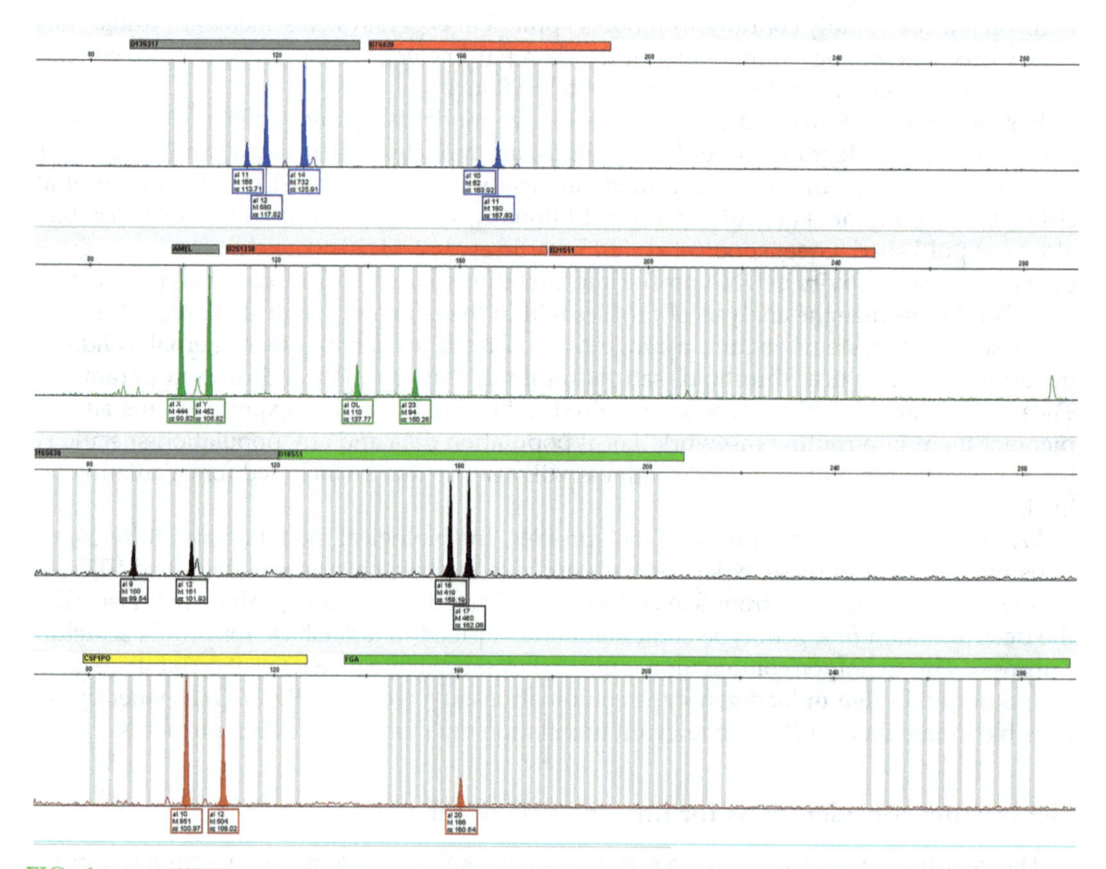

FIG. 6 AmpF*l*STR™ MiniFiler™ PCR Amplification Kit DNA profile recovered from a bone sample exhibiting a number of artifacts commonly associated with low template DNA (LT-DNA). For example, elevated stutter, peak height imbalance, amplification failure, and an off-ladder artifact peak can be observed at D13S317, CSF1PO, D21S11, and D2S1338, respectively. Note that the analytical threshold applied was 40 RFU.

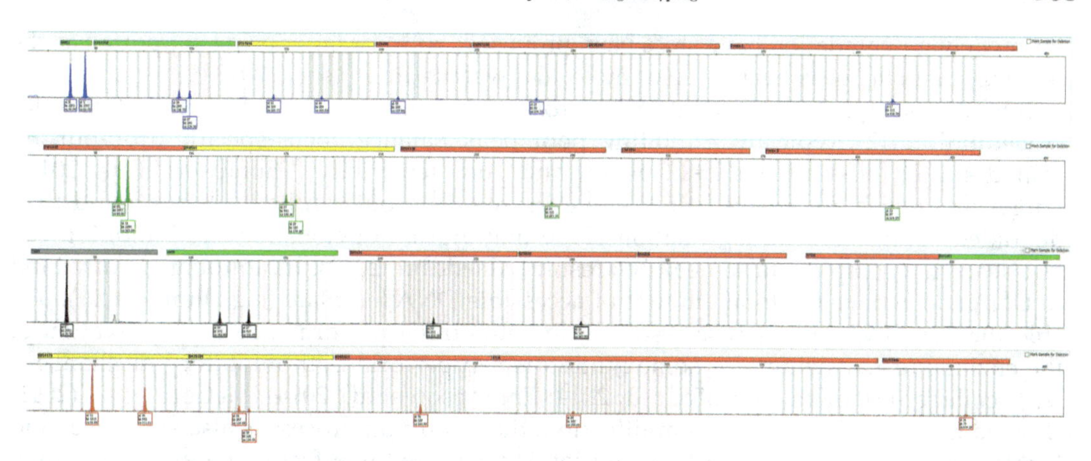

FIG. 7 PowerPlex® Fusion System DNA profile recovered from a bone sample exhibiting moderate DNA degradation and a confirmed nonspecific peak. A ski slope pattern can be observed across the profile from left to right and an off-ladder artifact peak can be observed at D16S539. The peak morphology of the artifact peak is similar to the authentic allele (allele 11), so this locus could be misinterpreted as heterozygous instead of homozygous for allele 11. Note that the analytical threshold applied was 70 RFU.

Due to the effective low copy number of DNA templates, amplification of low yielding skeletal samples may cause a number of stochastic effects including: (1) allelic drop-out due to the preferential amplification of one allele at one or more heterozygous loci, (2) peak height imbalance within and between loci, (3) allele drop-in due to amplification artifacts such as stutter, and (4) allelic drop-in caused by sporadic contamination from the environment (van Oorschot et al., 2010). Reporting guidelines instructed by the laboratory's internal validation will assist with the interpretation of such DNA profiles. Re-amplifying the sample, either using the same STR kit to generate a consensus profile or an alternative kit designed for lower quality samples such as the AmpF*l*STR™ MiniFiler™ PCR Amplification Kit (Thermo Fisher Scientific), can also assist in distinguishing true allelic and artifactual peaks. Finally, while largely focused on mixed DNA samples in forensic casework, deconvolution of the profile and statistical interpretation using probabilistic genotyping software will allow the objective interpretation of LT-DNA and degraded profiles, while ensuring the laboratory standardizes its interpretation of challenging samples.

The three main complications observed from amplifying degraded DNA samples are amplification failure, miscoding lesions, and preferential amplification of shorter alleles at a locus (Alaeddini et al., 2010). Degraded sample profiles are typically identifiable by a downward slope of STR allele peak heights with increasing molecular weight (i.e., left to right) in the electropherogram and is often referred to as the degradation slope, or a "ski slope" pattern (Nicklas et al., 2012; Chung et al., 2004; McCord et al., 2011). In highly degraded samples, it is common for the largest molecular weight loci to completely fail to amplify (Mulero et al., 2008). While the slope has been considered as both linear (Bright et al., 2013b) and exponential (Tvedebrink et al., 2012), the downward trend is easily observed and identified. The linear model suggests expected peak height decreases as a result of increased molecular weight (Bright et al., 2013b); however, it is now acknowledged that degradation is a result of a complex combination of biological and environmental events. Linear and exponential equations for modeling degradation have since been proposed, with the latter being supported as the best way to model peak height versus molecular weight (Bright et al., 2013a).

Inhibited samples can generate STR profiles that mimic those of degraded samples (Alaeddini, 2012), typically exhibiting poor peak balance, locus-specific drop-out, elevated stutter, off-scale and split peaks, and poor sensitivity (Thompson et al., 2014). Larger amplicons are affected first; however, the drop-out can be more profound than degraded profiles and a degradation slope may not be obvious. Severe inhibition can also lead to a complete false negative result (Alaeddini, 2012). Inhibitors typically bind to DNA and reduce the total amount of accessible DNA template (e.g., humic acid), disable the Taq DNA polymerase resulting in suboptimal performance or deactivation of the enzyme (e.g., calcium), or some inhibitors can perform both inhibitory mechanisms (e.g., collagen) (Pionzio and McCord, 2014; Thompson et al., 2014). McCord et al. (2011) showed that in STR multiplex amplification, the inhibition process is also sequence- and length-dependent. Consequently, inhibition is purported to affect larger amplicons first, although inhibitors that bind DNA may have additional sequence-specific effects in addition to these generic length effects.

A number of studies have attempted to confirm the origin of the sample-specific artifacts observed in bone DNA profiles. Dembinski and Picard (2017) observed an extraneous allele (allele 5) at the TPOX locus when two different microbial species associated with decomposition were amplified, with and without human DNA, using the PowerPlex® 16 HS System (Promega Corporation). Interestingly, this extra allele was not amplified with the PowerPlex® Fusion System (Promega Corporation), likely due to the different locus-specific PCR primers used in the kits. Dembinski and Picard (2017) urge caution in DNA profile interpretation for samples collected from decomposing and/or buried remains where the co-extraction and amplification of microbial DNA are realistic.

Similarly, Koehn (2013) observed nonspecific PCR products in bone STR profiles amplified with the PowerPlex® 16 HS System and the Identifiler® Plus Amplification Kit (Thermo Fisher Scientific); however, bone samples did not produce the same artifacts across kits. Koehn (2013) was able to demonstrate that similar, reproducible artifacts were observed in STR profiles for different soil samples amplified with the PowerPlex® 16 HS System (but not with the Identifiler® Plus Amplification Kit), supporting that nonhuman DNA found in soil might be the source of artifact peaks observed in the skeletal sample profiles. In addition, some kit manufacturers have acknowledged cross-reactivity with certain nonhuman species [e.g., pig DNA amplification with the PowerPlex® 21 System (Promega Corporation)]. It is important to be cognisant of artifactual peaks when profiling skeletal remains, especially where the origin of the sample may be unknown.

Optimizing electrophoresis parameters

A number of post-PCR and CE parameters can be optimized for genotyping DNA recovered from human skeletal remains that contain low quantities of DNA, in order to improve allele recovery for STR profiling. When signal intensity of an allele is too low, the signal-to-noise ratio is also low and can make it difficult for an analyst to discriminate between sample peaks and background noise. To increase the signal intensity at the electrophoresis stage, either one or more post-PCR processing steps can be employed, or the sample injection time and/or injection voltage could be increased.

In addition to PCR modifications, van Oorschot et al. (2010) reviewed a number of post-PCR manipulations that have been investigated to improve the detection of amplified PCR products from trace DNA samples. These modifications include: (1) introducing a post-PCR purification step such as filtration, silica gel membranes, or enzyme-mediated hydrolysis, (2) concentrating the PCR product prior to injection, or (3) increasing injection time or voltage. There is no consensus on the optimal injection profile to use for LT-DNA samples to date; however, Thermo Fisher Scientific (2014) reports that the maximum recommended injection time for current Genetic Analyzer instruments is 30 seconds and the maximum possible injection voltage is 15 kV. Hedell et al. (2015) found that adding one PCR cycle or doubling the CE injection time resulted in similar peak height increases. For example, increasing the PCR cycle number from 30 to 32 cycles with a 5-second injection produced similar improvement in allele detection and peak height compared to using 30 cycles with a 20-second injection time. After evaluating various combinations of cycle numbers and CE injection times, 32 and 33 PCR cycles with a 10-second injection was determined to be the optimal conditions for profiling LT-DNA using the PowerPlex® ESX 16 System (Promega Corporation). Notably, 34–35 cycles showed no further improvement in allele recovery and did complicate interpretation through the introduction of artifact peaks and CE saturation problems.

Higher CE injection settings were favored by Westen et al. (2009) to improve incomplete AmpF*l*STR® SGM Plus® PCR Amplification Kit (Thermo Fisher Scientific) STR profiles generated from low-level DNA samples under standard PCR conditions (28 cycles). The injection settings on an Applied Biosystems™ 3130*xl* Genetic Analyzer were increased from 3 kV/10 seconds to 9 kV/15 seconds, resulting in a six-fold peak height increase for the same amplified 28-cycle PCR product. If the amplified 28-cycle PCR product was then subjected to an additional six PCR cycles (28 + 6 cycles) prior to the boosted CE conditions, the peak heights increased 35-fold on average. Artifacts were observed for all approaches due to stochastic effects, including heterozygous peak imbalance and allele drop-out. However, samples subjected to elevated cycling were more susceptible to allele drop-ins, increased stutter peaks, and pull-up. An important consideration for the DNA profiling of skeletal samples is that the standard injection, boosted injection, and 28 + 6 PCR cycle approaches can be performed on the same amplified product, thereby preserving DNA extract.

Alternatively, some laboratories have validated a combination of post-PCR strategies to improve allele detection. A similar staged approach to low-level sample processing was reported by Forster et al. (2008), with a focus on post-PCR enhancements. They compared the SGM Plus® PCR Amplification Kit STR profiles from samples amplified using a 34-cycle low copy number (LCN) approach with the same samples processed using a 28-cycle PCR, concentrated using a MinElute® PCR Purification Kit (QIAGEN) with increased sample loading, and increasing injection conditions from 3 kV/10 seconds to 4 kV/30 seconds. This combination of PCR product cleanup, concentration, increased sample loading, and increased injection parameters produced STR profiles with a similar or better quality and sensitivity as those generated with a 34-cycle PCR (with more allele drop-ins and higher stutter peak ratios being observed for the elevated cycling conditions). The advantages of employing this staged post-PCR processing method compared to a 34-cycle LCN approach include reduced sample consumption, as only one further 28-cycle PCR is required to be performed to generate a consensus profile (as recommended for LT-DNA samples) rather than two new 34-cycle PCR reactions.

Mitochondrial DNA (mtDNA) sequencing

Sanger sequencing of mtDNA provides an alternative DNA profiling method for human skeletal remains that may contain little to no nDNA and are compromised and/or highly commingled. Sanger sequencing, or the chain termination method, was developed by Frederick Sanger and colleagues in the 1970s (Sanger et al., 1977). Sequence differences from a reference sequence, known as the revised Cambridge Reference Sequence (rCRS) (Andrews et al., 1999), are reported as a mtDNA profile (Anderson et al., 1981). For forensic DNA profiling purposes, the Control Region (CR) of the mtDNA genome is typically targeted using PCR and CE because it is highly polymorphic between individuals and can be used for putative identification within a maternal line. For a detailed review of the evolution of mtDNA sequencing, and its application for forensic DNA profiling, refer to the following publications: Amorim et al. (2019), Holland and Parsons (1999), and Stewart et al. (2003).

DNA template preparation

PCR

The CR contains two regions that are the primary sequence targets: Hypervariable Region I (HVI: 16024–16,365) and Hypervariable Region II (HVII: 73–340). Other portions of the CR can be amplified as needed, the most common being Hypervariable Region III (HVIII: 424–576). Amplification of the CR and hypervariable regions may be undertaken in different strategies. If the sample being tested is of high quality, the entire CR may be amplified in a single reaction with a single primer pair (e.g., F15971 and R599). Samples of lower quality can be amplified using smaller amplicon sizes. These range from amplicons that encompass an entire hypervariable region to amplicons as small as 89 bases (Edson et al., 2004; Edson, 2019; Gabriel et al., 2001; Loreille and Irwin, 2012). For the smaller amplicons, amplifications will need to be performed in a set of independent reactions using different combinations of primer pairs. The results will be a series of overlapping amplicons from which the polymorphisms can be confirmed and reported.

The PCR typically contains a master mix of: 10 × PCR Buffer (100 mM Tris–HCl, pH 8.3, 500 mM KCl, 15 mM $MgCl_2$), 2.5 mM deoxynucleotide triphosphates (dNTPs), 6.25 μg/μL nonacetylated bovine serum albumin (NA-BSA), 10 μM each of paired primers, 5 units/μL Taq DNA polymerase, and sterile water. The components of the master mix can be purchased and added individually or as a composed reaction [e.g., MyTaq™ HS Red Mix (Bioline Meridian Bioscience)]. Primers can be made in-house using equipment such as the MerMade Oligonucleotide Synthesizer (LGC BioAutomation) or purchased. Primers are not typically multiplexed, but can be if the DNA is of high enough quality and the desired amplicon is larger (Ballard, 2016). NA-BSA is best used for degraded DNA amplification as it increases the specificity of binding of the small amplicon primers. This element can be removed for high-quality sample reactions, such as those for the entire CR or HVI/II, or inhibition may occur. NA-BSA is also not recommended for use in amplification of nonhuman DNA as it may cause false results, indicating the origin of the sample is bovine in nature. A Taq DNA polymerase that has no inherent repair or proof-reading capabilities, such as AmpliTaq Gold®

(Thermo Fisher Scientific), is well-suited for use with degraded DNA since proof-reading the DNA template may introduce false polymorphisms into the template.

Similar to STR genotyping, PCR amplification of the mtDNA CR is conducted using a thermal cycler such as the GeneAmp® PCR System 9700 or ProFlex™ PCR System. A thermal cycling program commonly used for the amplification of medium-sized amplicons is: a 10 minute soak at 96°C, followed by 38 cycles of 20 seconds at 94°C, 20 seconds at 56°C, and 30 seconds at 72°C, and ending with a hold of 4°C upon completion. Modifications to this program should be made based on the primer combinations used and the laboratory validation (Edson et al., 2004; Gabriel et al., 2001). Additionally, appropriate negative and positive amplification controls, and an extraction reagent blank, should be included in the amplification setup and all subsequent post-PCR procedures.

Quantification of PCR products

After PCR, the product needs to be both qualified and quantified. This allows the analyst to determine whether the PCR was successful and to optimize the DNA template quantity for the sequencing reaction. There are several methods by which PCR product may be quantified, but perhaps the most common are agarose gel electrophoresis and microchip CE (Montesino and Prieto, 2012).

Agarose gels are a cost-effective method to quantify PCR products. The percentage of agarose gel (*v*/v) required depends on the laboratory needs. A 2% agarose gel is fairly common (Edson et al., 2004; Loreille and Irwin, 2012) and provides sufficient separation of PCR products. There are a variety of dyes by which the DNA may be visualized on a gel. Ethidium bromide (EtBr) is an intercalating dye that can be added directly to the liquid gel prior to casting. However, EtBr is a strong mutagen. Safer, but more expensive, alternatives include SYBR® Green (Thermo Fisher Scientific) and GelRed® or GelGreen® (Biotium). All of the intercalating dyes are visualized using a UV transilluminator. An alternative to pouring and casting agarose gels are FlashGel® Systems (VWR). These are fully contained and pre-stained gels that can be run on the FlashGel® System. No UV irradiation or running buffer is needed; however, the proprietary system and software is required. A mass ladder should be electrophoresed in parallel with the samples in order to evaluate the quantity of DNA present. Commonly used mass ladders are DNA Ladder II (APExBIO Research) and HyperLadder™ 100 bp (Bioline). The brightness of the DNA band is compared to that of the accompanying ladder to determine the input into the subsequent sequencing reaction, which is rather subjective and can be prone to user error.

Microchip CE coupled with a laser-induced fluorescence DNA detection method is a more accurate manner by which to determine the quantity of DNA present (Alonso et al., 2006; Fernández and Alonso, 2012; Kline et al., 2005). The Agilent 2100 Bioanalyzer (Agilent) uses a DNA chip and microcapillary gel electrophoresis to visualize PCR product and provide the concentration of all DNA present. There are multiple kits available for use with the instrument; however, the most commonly used for mtDNA sequencing are the Agilent DNA 1000 Kit and the Agilent High Sensitivity DNA Kit. The Agilent DNA 1000 Kit provides higher resolution of smaller fragments (25–1000 bases), and the High Sensitivity DNA Kit is available for samples that may have a much lower concentration of DNA and has an enlarged sizing range (50–7000 bases).

The DNA chip contains an interconnected set of 15-mm microchannels. The chip is primed with a gel-dye mix (comprised of a sieving polymer and fluorescent dye), the PCR product is loaded onto the instrument with two internal size standards (upper and lower marker), and an external ladder is added in a separate well of the chip. The software establishes a standard curve by plotting the ladder against the migration time. The internal standards balance the samples to the ladder and correct for any electrophoretic drift effects. In order to quantify the DNA fragments, the area under the upper marker peak is compared with the area under the sample peaks. Fluorescence data is converted into a concentration (ng/μL), and either an electropherogram or a gel image that resembles those generated from an agarose gel can be exported from the instrument.

Purification of PCR products

Prior to sequencing, the PCR product needs to be purified to remove excess components from the PCR mixture, such as residual single-stranded primers, single-stranded DNA, and excess unincorporated dNTPs that may interfere with sequencing. There are a number of different techniques that may be used for this process. Some laboratories may also choose to purify the PCR products prior to quantification.

Membrane-based filtration systems are often the simplest and least expensive. PCR product is added directly to a column filled with a set amount of solution. The solution depends on the column used and the protocol developed by the laboratory and can simply be sterile H_2O, a detergent such as sodium dodecyl sulphate (SDS), or Tris-ethylenediaminetetraacetic acid (TE) buffer. Once the PCR product is added to the column, it is spun in a centrifuge until the liquid spins through. The waste materials are carried through the membrane with the solution, and purified DNA is retained on the membrane. The PCR product can then be recovered using a pipette or by flipping the column and centrifuging the materials into a separate tube. The columns need to be chosen based on the pore size of the membrane to ensure the PCR product size of interest is retained and not washed away. Commonly used columns include Microcon® 100 Centrifugal Filters and Amicon Ultra-0.5 mL Centrifugal Filters (Millipore).

Another column-based purification system is based on a series of washes using a silica-based column coupled with chemistry to remove the waste elements from the PCR product. The target product binds with the silica column, which is then treated with a series of chemical washes to remove waste materials and then finally to recover the target. Commonly used kits include the Wizard® SV Gel and PCR Cleanup System (Promega Corporation) and MinElute® PCR Purification Kit.

The final manner of purification does not use a column at all, but rather an enzymatic process. ExoSAP-IT® (Thermo Fisher Scientific) combines the action of two hydrolytic enzymes: Exonuclease I and Shrimp Alkaline Phosphatase (SAP). These enzymes collectively remove components from the PCR mixture that may interfere with downstream processing. The components are stored separately under refrigeration and combined immediately prior to use. The reagent is directly added to the PCR product and incubated at 37°C for 15 minutes. Following treatment, the ExoSAP-IT® reagent is inactivated by heating to 80°C for 15 minutes, and the PCR product is then purified and suitable for mtDNA sequencing. No further treatment of the sample is necessary. While this process necessarily dilutes the PCR product, no adjustment to the input for sequencing is required.

Cycle sequencing

Unlike the cycle sequencing method first reported by Sanger et al. (1977) which used radioactive material, fluorescence-based cycle sequencing involves the use of fluorescently labeled dideoxynucleotide triphosphates (ddNTPs) and unlabelled dNTPs. When an unlabelled dNTP is incorporated into the 3′ end of a growing nucleotide strand, chain extension can continue. However, when a labeled ddNTP is incorporated into a growing nucleotide strand, chain extension terminates for that strand. This is because ddNTPs do not have a 3′ hydroxyl group, so after their addition, no further dNTPs can be added since no phosphodiester bond can be created. Despite the increasing use of massively parallel sequencing (MPS), this cycle sequencing approach remains the most common method by which a DNA sequence may be determined.

Process

Fluorescence-based cycle sequencing requires a single-stranded DNA template, a sequencing primer, modified DNA polymerase, all four dNTPs, all four dye-labeled ddNTPs, and sequencing buffer. Fluorescent dyes are used to label the extension products and the components are combined in a reaction that is subjected to cycles of denaturation, annealing, and extension in a thermal cycler. Thermal cycling the sequencing reactions creates and amplifies extension products that are terminated by one of the four ddNTPs. The ratio of dNTPs to ddNTPs is optimized to produce a balanced population of long and short extension products. Each of the four ddNTPs is tagged with a different fluorescent dye. A CE instrument separates the different sized extension products and detects these fluorescent labels as they migrate passed the CCD camera. Because each dye emits a unique wavelength when excited by light, the fluorescent dye on the extension product identifies the 3′ terminal ddNTP as adenine (A), cytosine (C), guanine (G), or thymine (T). Therefore, the sequence of nucleotide bases in a DNA strand can be determined.

Sequencing chemistries

Sanger sequencing can be performed using one of the Applied Biosystems™ Cycle Sequencing Kits. The commonly used sequencing chemistries include: the BigDye® Terminator v1.1 Cycle Sequencing Kit, the BigDye® Terminator v3.1 Cycle Sequencing Kit, and the dGTP BigDye® Terminator Cycle Sequencing Kit. Many laboratories have validated the use of these kits using a quarter strength reaction (for high-quality samples amplified using the CR amplicon) or a half strength reaction (for lower quality samples amplified with all other amplicons) in order to reduce sequencing costs.

The BigDye® Terminator v1.1/3.1 Cycle Sequencing Ready Reaction Kits are comprised of AmpliTaq® DNA Polymerase, sequencing dilution buffer (400 mM Tris, 10 mM $MgCl_2$, pH 9.0), and a mixture of four dNTPs. In addition, the kit contains the four ddNTPs that are each tagged with a different fluorescent dye. The desired primers are added to each reaction mix and can be the same as those used for amplification or primers that are set slightly internally to the original binding sites. BigDye® Terminator v1.1 is optimized for shorter reads and is therefore better suited for use with mtDNA fragments. BigDye® Terminator v3.1 is optimized for longer reads and can be used for sequencing entire CR amplicons, although v1.1 can also be used.

For sequencing of degraded or difficult templates, such as those containing homopolymer regions (commonly referred to as C stretches), dGTP BigDye® Terminator Cycle Sequencing Kit may be added to reactions using BigDye® Terminator v1.1 Cycle Sequencing Ready Reaction Kit to improve sequence quality. The dGTP BigDye® Terminator Cycle Sequencing Kit is optimized for sequencing GT-rich templates or sequence motifs comprised of G and C nucleotide bases and generally improves the quality of the reverse (heavy strand) sequencing reactions. However, when used alone, the dGTP BigDye® Terminator Cycle Sequencing Kit can cause peak compression, thus obscuring the authentic data (Loreille and Irwin, 2012).

Thermal cycling conditions

The manufacturer recommends specific DNA template quantities for sequencing different PCR product lengths. However, the optimal amount of DNA template should be determined for each target amplicon during a laboratory's internal validation. For example, for amplicons 200–500 bases in length, up to 10ng of template is often used to generate quality sequence data.

Sequencing reactions are placed on a thermal cycler such as the GeneAmp® PCR System 9700 or ProFlex™ PCR System and typically amplified according to manufacturer guidelines. The thermal cycling conditions are the same for all sequencing reactions, regardless of the region to be sequenced. A typical program is as follows: 96°C for 1 minute, followed by 25 cycles at 96°C for 15 seconds, 50°C for 5 seconds, and 60°C for 4 minutes, with a final hold at 4°C. Some laboratories may choose to validate a shorter extension time of 2 minutes instead of 4 minutes due to the shorter sequencing reads associated with routine mtDNA HVI/II sequencing.

Purification of sequencing products

Following cycle sequencing, the reaction product needs to be purified of any waste dyes and extraneous materials. Failure to adequately clean the product will result in low-quality sequencing data containing elevated background or dye blobs that can obscure authentic data peaks. There are a number of different methods to achieve a purified sequencing product, including chemical purification, spin columns, and ethanol precipitation. The use of both spin columns and ethanol precipitation requires an additional step of drying the product and then resuspension in Hi-Di™ formamide prior to loading on the CE instrument.

A common form of chemical purification is using BigDye® XTerminator™ Purification Kit (Thermo Fisher Scientific). The XTerminator™ Kit is comprised of two components: BigDye® XTerminator™ Solution and BigDye® SAM™ Solution. The XTerminator™ Solution captures excess salt ions and unincorporated dNTPs/ddNTPs remaining from the sequencing reaction, to prevent their coinjection with dye-labeled extension products. The SAM™ Solution improves the performance of the XTerminator™ Solution and stabilizes the sample after purification. Both solutions are added directly to the completed sequencing reactions and vortexed. It is during vortexing that the XTerminator™ Solution captures and immobilizes the unwanted components. After vortexing, the reactions are briefly centrifuged to separate the insoluble XTerminator™ mix from the purified dye-labeled extension products in the upper supernatant layer. The purified samples can then be loaded onto a CE instrument, where they are injected directly from the supernatant layer and analyzed using specialized BigDye® XTerminator™ (BDX) run modules.

Spin columns can be used in either a single column or 96-well plate format. Performa® DTR Gel Filtration Cartridges or Performa DTR Ultra 96-Well Plate Kits (Edge BioSystems) use a gel filtration matrix to remove waste dye terminators, dNTPs, and salts that would impact the quality of the downstream sequencing product. The sequencing reactions are placed individually in the center of each column and centrifuged. After purification, the product is dried in a CentriVap Centrifugal Concentrator (Labconco) that uses low heat and a vacuum to completely dry the product prior to resuspension in Hi-Di™ formamide.

Ethanol precipitation is a cost-effective alternative to either spin columns or chemical purification. This method requires 125mM EDTA to be added directly to the sequencing product, followed by 100% ethanol. The samples should be vortexed and then left at room temperature for at least 15minutes. Samples should be centrifuged until the product is pelleted, at which point the plate can be flipped and the ethanol poured off. A second wash is performed with 70% ethanol, followed by a second drying step. The plate is left at room temperature until completely dry, after which point the product can be stored at − 20°C or resuspended in Hi-Di™ formamide.

CE requirements

A 50-cm capillary array filled with POP-6™ (6% dimethylacrylimide, 8M Urea, 5% 2-pyrrolidinone) is typically used for the separation of Sanger-sequenced fragments. This allows for a more even separation of the fragments. The shorter capillary array (36-cm) used for STR analysis is not used for sequencing because it can cause a loss of resolution, especially in samples containing homopolymeric stretches. The Applied Biosystems™ 3130, 3130*xl*, and 3500 Series Genetic Analyzers are recommended for electrophoresis of Sanger-sequenced mtDNA. The Applied Biosystems™ 3730 Genetic Analyzer (Thermo Fisher Scientific) uses POP-7™, which is optimized for longer runs and may remove the first 20–30 bases of any sequencing data. While this may not be an issue for amplicons of longer size, removal of 30 bases from an 89-base fragment would impede data analysis.

Instrument run modules are largely the same for any size of mtDNA fragment being sequenced; however, injection and run times can be modified to suit the target sequence length. The following run parameters are the same for all amplicon sizes: oven temperature = 60°C, run voltage = 18kV, pre-run voltage = 19.5kV, and injection voltage = 1.6kV. The larger CR or hypervariable region fragments can be injected for up to 24seconds, while the smaller fragments can be injected for a shorter length of time. An entire plate of larger fragments can be run in 4.5hours, with a plate of smaller fragments taking 3.5hours. There are standard run modules available on instruments (e.g., RapidSeq50_POP6); however, standard modules may need to be modified for shorter reads to allow for a read slightly outside of the primer region, in an attempt to generate as much data as possible. Laboratories should determine optimal injection times during the internal validation, as injection times may vary widely depending on purification strategy, instrument, or even the temperature of the laboratory.

Data analysis

The data interpretation workflow typically involves:

1. Assessing the positive and negative controls;
2. Assessing each sample for the presence of artifacts and determining if they interfere with the interpretation process; and
3. Analyzing the data for each test sample—conducted independently by two analysts.

Different software programs are commonly used to process data from Applied Biosystems™ Genetic Analyzers and produce a mtDNA profile. Applied Biosystems™ Data Collection Software and Sequence Analysis Software (Thermo Fisher Scientific) are used to automatically call bases and trim data. Additional sequence analysis software, such as Sequencher™ (Gene Codes Corporation) or SeqScape™ (Thermo Fisher Scientific), are required to assemble, edit, and analyze sequence data and confirm the mtDNA profile. Base calls should be made according to the recommendations published by the International Society for Forensic Genetics (Parson et al., 2014). In addition, any potential misalignments or false polymorphisms can be detected if the sequence generated is checked using phylogenetic approaches (Ballard, 2016); for example EMPOP (www.empop.com) or Phylotree (www.phylotree.org).

Applied Biosystems™ data collection software

The Data Collection Software automatically performs data analysis on the electrophoresis data (.ab1 files) using various algorithms. The Data Collection Software for Applied Biosystems™ 3500 Series Genetic Analyzers applies the following analysis settings to the sequencing results: multicomponent analysis, mobility shift correction, base calling using the KB™ Basecaller, and quality value assignment (Thermo Fisher Scientific, 2016). The software allows sequences to be viewed immediately following run completion so that data quality can be assessed and samples requiring retesting identified. A number of reports can also be generated to document this information, such as the Quality Control Report. Analyzed sample data are displayed as an electropherogram, where different colored peaks represent one of four nucleotide bases. It is important to note that the KB™ Basecaller is not integrated into the Data Collection Software for other Applied Biosystems™ Genetic Analyzers.

It is also relevant to note that the automated analysis on any of the Genetic Analyzer systems may remove a large number of bases from the beginning of the generated sequence. This is based on the algorithm inherent in the system that automatically removes the first 20–30 bases from a sequence under the presumption that the data present will be of low quality. However, this is generally not so. Data can be exported from the instrument without analysis and manually analyzed using the Applied Biosystems™ Sequencing Analysis Software, thus retaining data that would otherwise be removed.

Applied Biosystems™ sequence analysis software

If an Applied Biosystems™ 3500 Series Genetic Analyzer is not used, the sequencing analysis (.ab1) files generated are required to be analyzed using an additional program such as the Applied Biosystems™ Sequencing Analysis Software. This program uses the KB™ Basecaller to perform base calling, assign quality values for each base, trim low-quality bases, and generate quality reports. If necessary, the sample's spacing, the first peak, and the Start/Stop values can be adjusted, and the sample can then be reanalyzed. In addition, electropherograms can be printed through this software.

Other sequence analysis software

Additional sequence analysis software is necessary to align the sequence data to a reference sequence and combine multiple amplicons into a consensus profile after data normalization. Sequencher™ is an example of a software program that can be used to assemble, edit, and analyze mtDNA sequence data to produce a contiguous (contig) consensus sequence. Once the forward and reverse sequences are assembled for a sample, the electropherograms should be reviewed base-by-base, in order to confirm the software base calls and identify any mixed nucleotide bases (contamination/heteroplasmy), or ambiguous results caused by sequencing artifacts/errors or other phenomena (e.g., deamination). Sequence information at each nucleotide position should be confirmed by data from both forward and reverse DNA strands when possible. Single-stranded regions present due to length heteroplasmy can be confirmed by additional sequencing of the same strand in the same direction. The software complementary strand alignment will flag conflicts between the two sequencing directions for all strands imported into the contig. While the program will flag bases that are questionable based on overall peak quality or spacing issues, it is not infallible and should not be considered an expert system. An analyst can modify the base calls if the underlying data support it.

Common artifacts observed for skeletal samples

Artifacts are commonly observed in the sequence data of human skeletal samples. Environmental factors and conditions that the skeletal sample is exposed to can greatly impact DNA preservation, as well as the quality and quantity of recoverable DNA (Latham and Miller, 2019). When processing degraded skeletal samples, DNA is expected to be smaller in length, amount, and of poor quality. Some common artifacts encountered with skeletal samples include postmortem changes, cross-contamination from modern or exogenous DNA sources, coextracted microbial DNA, and the presence of nuclear mitochondrial pseudogenes (NUMTs). Consideration of these artifacts and how they may interfere with the interpretation of human mtDNA sequences is important. In an effort to improve the quality of mtDNA profiles from more compromised skeletal samples, some laboratories have implemented other CE-based technologies for targeting mtDNA.

Common artifacts

DNA damage is a major concern for attempting to recover DNA from human skeletal remains. Within 4–5 minutes of death, cellular death begins to occur and degradation of DNA can be rapid to the extent that, after weeks or months, there could be little-to-no amplifiable DNA remaining (Gilbert, 2006). The damage that is occurring to the DNA includes double-strand breakage and oxidative dinucleotide modification (Gilbert et al., 2003). DNA damage will often result in only mtDNA being recoverable, as it is typically in higher copy number than nDNA. Additionally, since the template DNA is often in such low copy number, there is a greater chance of sample contamination with modern DNA, resulting in either amplifying only the modern DNA or coamplifying both the modern DNA and the skeletal sample DNA (Gilbert, 2006).

In older, degraded samples cytosine deamination is a common artifact. Deamination occurs when a cytosine nucleotide base is converted to uracil, resulting in C→T and G→A transitions (Hofreiter et al., 2001). The data can appear as secondary T peaks under C peaks or A peaks under G peaks (Fig. 8). The sequencing data may appear to be mixed; however, when the sample is amplified and sequenced again in the same region, the

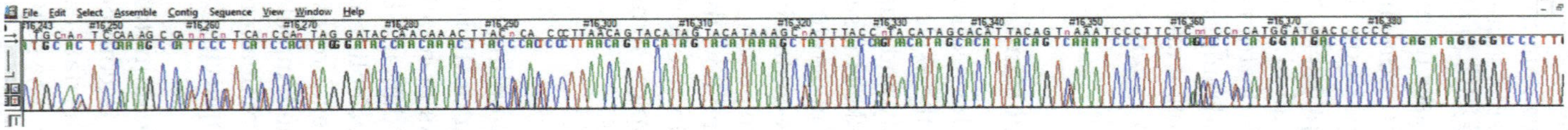

FIG. 8 DNA sequence exhibiting cytosine deamination. Secondary *(red)* thymine (T) peaks can be observed underneath or overlapping *(blue)* cytosine (C) peaks at numerous nucleotide positions.

occurrence of deamination will commonly shift and appear at different base positions. This is occurring due to the limited amount of template molecules and the stochastic amplification of either the original or modified template during the early PCR cycles (Loreille and Irwin, 2012). This can result in: (1) the number of original templates in the reaction being significantly outnumbered by the modified templates, yielding the incorrect sequence; (2) the number of original templates in the reaction significantly outnumbering the modified templates, yielding the correct sequence, or (3) a mixture of the two being observed (Gilbert, 2006).

During decomposition, skeletal samples are prone to exposure and contamination by microorganisms (Korlević et al., 2015). Microbial DNA can make up more than 95% of recoverable DNA in older, skeletal samples (Korlević et al., 2015). Treatment prior to DNA extraction or amplification/library preparation of the sample, as well as the type of extraction method performed, can assist in removing most of the microbial DNA present, leaving primarily the endogenous DNA for testing.

NUMTs are another type of artifact that may be observed in DNA sequences recovered from skeletal samples. A NUMT occurs due to mtDNA fragments inserting into the human nuclear genome. In degraded samples, there is an increase of possible contamination by the NUMT (especially if it is present in high copy) because it can be co-amplified with the target mtDNA or amplified in the place of the target mtDNA sequence (Goios et al., 2009). The presence of a high number of polymorphisms can be an indication of the presence of a NUMT (Fig. 9). False polymorphisms can also be introduced by other means. Phantom mutations are artifacts that are created during the sequencing process which tend to have a different pattern compared to natural mutations (Bandelt et al., 2002). Phantom mutations can also be of concern because they could be inaccurately reported as part of a skeletal sample's true sequence.

Alternative sequencing technologies

SNP genotyping is an alternative sequencing technology that can be employed to aid the profiling of compromised mtDNA. SNP profiling can be conducted on a CE platform, targeting specific positions in either the noncoding or coding regions of the mtDNA genome. This approach allows small amplicon sizes to be designed for each target SNP, making it ideal for sequencing degraded and fragmented DNA. It is most typically used for biogeographical analysis and may allow for samples to be rapidly sorted; however, it is not frequently used for identification purposes per se as the profile may be shared among many members of a population and are not typically specific to a maternal lineage. This is due to the design of the assays, which target population level differences (Ballantyne et al., 2012; Chaitanya et al., 2014; Chemale et al., 2013). A simple system for analyzing SNPs involves the use of the SNaPshot™ assay (Thermo Fisher Scientific). SNaPshot™ is a multiplex system, which allows for multiple SNP markers to be interrogated in a single assay. Data generated from this assay resembles STR electropherograms and may be analyzed using the same software. At present, there is a move to take these multiplexes and combine large numbers into panels designed for MPS platforms.

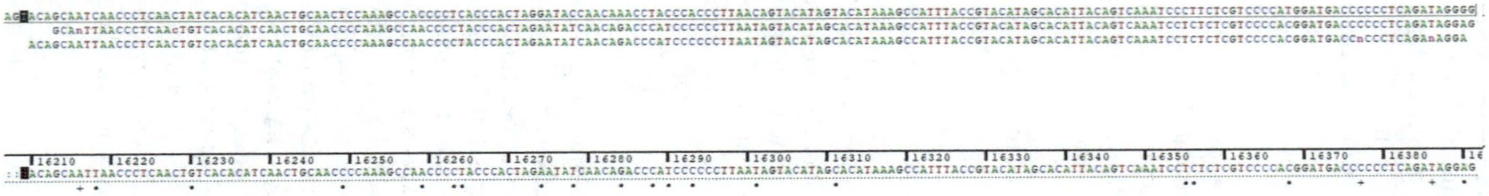

```
AGTACAGCAATCAACCCTCAACTATCACACATCAACTGCAACTCCAAAGCCACCCCTCACCCACTAGGATACCAACAAACCTACCCACCCTTAACAGTACATAGTACATAAAGCCATTTACCGTACATAGCACATTACAGTCAAATCCCTTCTCGTCCCCATGGATGACCCCCCTCAGATAGGGG
      GCAnTTAACCCTCAAcTGTCACACATCAACTGCAACCCCAAAGCCAACCCCTACCCACTAGAATATCAACAGACCCATCCCCCCTTAATAGTACATAGCACATAAAGCCATTTACCGTACATAGCACATTACAGTCAAATCCTCTCTCGTCCCCACGGATGACCCCCCTCAGATAGGAG
   ACAGCAATTAACCCTCAACTGTCACACATCAACTGCAACCCCAAAGCCAACCCCTACCCACTAGAATATCAACAGACCCATCCCCCCTTAATAGTACATAGCACATAAAGCCATTTACCGTACATAGCACATTACAGTCAAATCCTCTCTCGTCCCCACGGATGACCnCCCTCAGAnAGGA

16210 16220 16230 16240 16250 16260 16270 16280 16290 16300 16310 16320 16330 16340 16350 16360 16370 16380 16
ACAGCAATTAACCCTCAACTGTCACACATCAACTGCAACCCCAAAGCCAACCCCTACCCACTAGAATATCAACAGACCCATCCCCCCTTAATAGTACATAGCACATAAAGCCATTTACCGTACATAGCACATTACAGTCAAATCCTCTCTCGTCCCCACGGATGACCCCCCTCAGATAGGAG
```

FIG. 9 DNA sequence of a nuclear mitochondrial pseudogene (NUMT). A high number of polymorphisms can be observed when compared to the revised Cambridge Reference Sequence (rCRS) (top row of sequence).

Limitations of CE

The introduction of MPS has highlighted some of the limitations of traditional CE platforms. Unlike CE, MPS technology enables the large-scale multiplexing of both samples and genetic markers, to drastically improve the capacity and capabilities of forensic DNA laboratories. The advantages for STR profiling include high-throughput, low cost, simultaneous detection of large numbers of STRs on both autosomal and sex chromosomes, and the ability to distinguish alleles of identical length due to detection of intra-STR SNPs (Sobiah et al., 2018). In their recent review, McCord et al. (2019) summarize some of the improvements of MPS technology over CE for STR and mtDNA profiling applications, including the ability to: (1) multiplex 10-fold more genetic markers from the nDNA and mtDNA genome, reducing sample consumption and total assay time; (2) multiplex up to 96 samples per analysis using 1 ng or less of DNA increasing sample throughput; (3) detect sequence variants of STR alleles of the same size, or in the repeat flanking regions, providing higher discrimination power; (4) recover more genetic information from degraded and inhibited samples due to increased sensitivity and availability of small amplicon approaches; (5) generate full mtDNA genomes from compromised samples; (6) resolve mtDNA mixtures; and (7) improve the detection of mtDNA heteroplasmy.

Sequencing the whole mtDNA genome has not been practical for forensic applications until the introduction of MPS. Forensic mtDNA profiling has traditionally focused on sequencing HVI and HVII; however, more than 70% of the total variation within the whole mtDNA genome has been reported to exist outside of HVI and HVII (Brotherton et al., 2013). Therefore, sequencing the whole mtDNA genome can provide far greater resolving power for human identification and facilitate haplogroup assignment (Melton et al., 2001; King et al., 2014). HVI/HVII Sanger sequencing approaches typically amplify 2–8 overlapping fragments of approximately 126–400 bases in length (Ginther et al., 1992; Gill et al., 1994; Boles et al., 1995; Allen et al., 1998). However, hundreds of short overlapping amplicons would be required to sequence the whole mtDNA genome of skeletal samples in this manner, presenting a laborious, expensive, and time-consuming exercise that consumes large amounts of DNA extract. A commercial Precision ID mtDNA Whole Genome Panel (Thermo Fisher Scientific) is now available for sequencing the whole mtDNA genome of compromised samples using 162 short overlapping amplicons (average size of 163 bases) and is currently undergoing developmental validation (Churchill et al., 2017). Furthermore, the development of customized hybridization-based DNA capture techniques has been shown to recover whole mtDNA genome sequences from highly degraded and fragmented DNA samples (Templeton et al., 2013; Marshall et al., 2017). Other advantages of whole mtDNA genome sequencing using MPS include enhanced detection and/or resolution of heteroplasmy (Just et al., 2015), damage-induced lesions (Rathbun et al., 2017), length heteroplasmy (Holland et al., 2011; Davis et al., 2015; Lin et al., 2017), and mixture components (Holland et al., 2011, 2017; Vohr et al., 2017).

Of relevance to human skeletal remains, MPS can also be used to determine phenotypic characteristics of an individual, including appearance, ancestry, and age (Bruijns et al., 2018; Tillmar et al., 2018), which is important information for scenarios where DNA reference samples are not known or available for comparison, or to aid the craniofacial reconstruction of a skull. Other specific advantages of MPS technology for the DNA profiling of human skeletal remains include improved allele recovery of autosomal STRs using the ForenSeq™ DNA Signature Prep Kit (Illumina) compared to CE-based STR multiplexes, plus the additional benefit of gleaning DNA intelligence information without further consumption of limited sample (Almohammed et al., 2017). However, for challenging samples, this will only be observable if the recommended DNA input amount for the MPS assay is met (Votrubova et al., 2017). Additionally, it has been shown that the PowerPlex® Fusion System was able to tolerate at least 200 times more humic acid compared with the ForenSeq™ DNA Signature Prep Kit, demonstrating the need for further optimization of MPS STR kits for challenging forensic samples (Sidstedt et al., 2019).

In regard to STR profiling, de Knijff (2019) presents a number of practical consequences of transitioning from CE to MPS platforms. First, in addition to stutter alleles, MPS can reveal other previously unseen artifacts, including erroneous base pair substitutions caused by DNA editing errors or sequencing miscalls. Second, the full spectrum of technological errors, and how they may influence the interpretation of MPS-based STR genotyping, is yet to be thoroughly investigated. Third, there is no standardized STR workflow for the forensic community, in terms of platform, commercial STR panels, or software; nor is there consensus on nomenclature, analysis, and interpretation and reporting guidelines. However, various international initiatives are currently in progress to facilitate application of MPS technology for STR profiling (Alonso et al., 2018), including the development of a consistent, platform-independent, database-compatible nomenclature system (Parson et al., 2016). Finally, a collaborative effort is needed to generate sufficient population frequency data for sequence variation observed within or adjacent to forensically relevant STR loci to support statistical calculations for forensic casework, including kinship analysis (Staadig and Tillmar, 2019).

Summary

A CE-based STR profiling workflow is considered the gold standard approach for human identification. This type of fragment analysis is relatively simple, low cost, and high throughput, especially when processed using robotic platforms. In comparison, Sanger sequencing is labor-intensive, time-consuming, and relatively expensive. However, mtDNA sequencing is an invaluable application for the identification of human remains, especially for compromised skeletal samples or when only distant maternal relatives are available for DNA comparisons. The introduction of MPS into forensic laboratories will likely make mtDNA sequencing a more accessible application.

Specific CE platforms, chemistries, and analysis software are favored by forensic laboratories specializing in the identification of human skeletal remains. The Applied Biosystems™ 3500 or 3500xL Genetic Analyzer is the most modern CE platform used, depending on the processing throughput of the laboratory. For STR profiling, a typical workflow involves: (1) amplifying large commercial STR multiplexes, (2) electrophoresing PCR products using POP-4™ and a 36-cm capillary array according to a predefined run module, and (3) analyzing the data using a combination of Data Collection Software and GMID-X. For mtDNA sequencing, a typical workflow involves: (1) amplifying different sized target amplicons using custom primer sets; (2) quantifying PCR products using the DNA 1000 Kit on the 2100 Bioanalyzer; (3) purifying PCR products using ExoSAP-IT®; (4) sequencing PCR products using the BigDye® Terminator v1.1 Cycle Sequencing Kit; (5) purifying sequencing products using the BigDye®

XTerminator™ Purification Kit; (6) electrophoresing sequencing products using POP-6™ and a 50-cm capillary array according to a customized run module; and (7) analyzing the data using a combination of Data Collection Software and Sequencher™.

The compromised nature of human skeletal remains often requires laboratories to validate a number of CE-based procedural modifications to improve the success of STR genotyping and mtDNA sequencing. These may include modifying injection voltage, injection time, and/or run time. The condition of the samples may also increase the likelihood of observing CE artifacts attributable to DNA degradation, inhibition and damage, and the co-extraction and amplification of nonhuman DNA. The detection and interpretation of these artifacts will improve as a laboratory gains more experience with profiling of human skeletal remains. Additionally, artifacts can be verified, or their impact reduced, by using alternative STR kits or mtDNA primer sets; hence, it is advantageous for a laboratory to have access to multiple validated kits or primers for this type of casework.

References

Alaeddini, R., 2012. Forensic implications of PCR inhibition—a review. Forensic Sci. Int. Genet. 6 (3), 297–305.

Alaeddini, R., Walsh, S.J., Abbas, A., 2010. Forensic implications of genetic analyses from degraded DNA—a review. Forensic Sci. Int. Genet. 4 (3), 148–157.

Allen, M., Engström, A.S., Meyers, S., Handt, O., Saldeen, T., von Haeseler, A., Gyllensten, U., 1998. Mitochondrial DNA sequencing of shed hairs and saliva on robbery caps: sensitivity and matching probabilities. J. Forensic Sci. 43 (3), 453–464.

Almohammed, E., Zgonjanin, D., Iyengar, A., Ballard, D., Devesse, L., Sibte, H., 2017. A study of degraded skeletal samples using ForenSeq DNA signature™ kit. Forensic Sci. Int. Genet. Suppl. Ser. 6, e410–e412.

Alonso, A., Albarran, C., Martín, P., García, P., Capilla, J., García, O., Sancho, M., 2006. Usefulness of microchip electrophoresis for the analysis of mitochondrial DNA in forensic and ancient DNA studies. Electrophoresis 27 (24), 5101–5109.

Alonso, A., Barrio, P.A., Müller, P., Köcher, S., Berger, B., Martin, P., Budowle, B., 2018. Current state-of-art of STR sequencing in forensic genetics. Electrophoresis 39 (21), 2655–2668.

Amorim, A., Fernandes, T., Taveira, N., 2019. Mitochondrial DNA in human identification: a review. PeerJ 7, e7314.

Anderson, S., Bankier, A.T., Barrell, B.G., de Bruijn, M.H., Coulson, A.R., Drouin, J., Young, I.G., 1981. Sequence and organization of the human mitochondrial genome. Nature 290 (5806), 457–465.

Andrews, R.M., Kubacka, I., Chinnery, P.F., Lightowlers, R.N., Turnbull, D.M., Howell, N., 1999. Reanalysis and revision of the Cambridge reference sequence for human mitochondrial DNA. Nat. Genet. 23 (2), 147.

Applied Biosystems™, 2009. GeneMapper® Software Version 4.1: Reference and Troubleshooting Guide. http://tools.thermofisher.com/content/sfs/manuals/cms_070162.pdf. (Accessed 18 February 2019).

Ballantyne, K.N., van Oven, M., Ralf, A., Stoneking, M., Mitchell, R.J., van Oorschot, R.A., Kayser, M., 2012. MtDNA SNP multiplexes for efficient inference of matrilineal genetic ancestry within Oceania. Forensic Sci. Int. Genet. 6 (4), 425–436.

Ballard, D., 2016. Analysis of mitochondrial control region using Sanger sequencing. Methods Mol. Biol. 1420, 143–155.

Bandelt, H.J., Quintana-Murci, L., Salas, A., Macaulay, V., 2002. The fingerprint of phantom mutations in mitochondrial DNA data. Am. J. Hum. Genet. 71 (5), 1150–1160.

Bieber, F.R., Buckleton, J.S., Budowle, B., Butler, J.M., Coble, M.D., 2016. Evaluation of forensic DNA mixture evidence: protocol for evaluation, interpretation, and statistical calculations using the combined probability of inclusion. BMC Genet. 17 (1), 125.

Boles, T.C., Snow, C.C., Stover, E., 1995. Forensic DNA testing on skeletal remains from mass graves: a pilot project in Guatemala. J. Forensic Sci. 40 (3), 349–355.

Bregu, J., Conklin, D., Coronado, E., Terrill, M., Cotton, R.W., Grgicak, C.M., 2013. Analytical thresholds and sensitivity: establishing RFU thresholds for forensic DNA analysis. J. Forensic Sci. 58 (1), 120–129.

Bright, J.A., Taylor, D., Curran, J., Buckleton, J., 2013a. Degradation of forensic DNA profiles. Aust. J. Forensic Sci. 45 (4), 445–449.

Bright, J.A., Taylor, D., Curran, J.M., Buckleton, J.S., 2013b. Developing allelic and stutter peak height models for a continuous method of DNA interpretation. Forensic Sci. Int. Genet. 7 (2), 296–304.

Bright, J.A., Buckleton, J.S., Taylor, D., Fernando, M.A., Curran, J.M., 2014a. Modeling forward stutter: toward increased objectivity in forensic DNA interpretation. Electrophoresis 35 (21 – 22), 3152–3157.

Bright, J.A., Stevenson, K.E., Coble, M.D., Hill, C.R., Curran, J.M., Buckleton, J.S., 2014b. Characterising the STR locus D6S1043 and examination of its effect on stutter rates. Forensic Sci. Int. Genet. 8 (1), 20–23.

Bright, J.A., Taylor, D., McGovern, C., Cooper, S., Russell, L., Abarno, D., Buckleton, J., 2016. Developmental validation of STRmix™ expert software for the interpretation of forensic DNA profiles. Forensic Sci. Int. Genet. 23, 226–239.

Brookes, C., Bright, J.A., Harbison, S., Buckleton, J., 2012. Characterising stutter in forensic STR multiplexes. Forensic Sci. Int. Genet. 6 (1), 58–63.

Brotherton, P., Haak, W., Templeton, J., Brandt, G., Soubrier, J., Jane Adler, C., et al., Consortium, G., 2013. Neolithic mitochondrial haplogroup H genomes and the genetic origins of Europeans. Nat. Commun. 4, 1764.

Bruijns, B., Tiggelaar, R., Gardeniers, H., 2018. Massively parallel sequencing techniques for forensics: a review. Electrophoresis 39 (21), 2642–2654.

Butler, J.M., 2012a. Advanced Topics in Forensic DNA Typing: Interpretation. Academic Press, San Diego.

Butler, J.M., 2012b. Advanced Topics in Forensic DNA Typing: Methodology. Academic Press, San Diego.

Butler, J.M., McCord, B.R., Jung, J.M., Allen, R.O., 1994. Rapid analysis of the short tandem repeat HUMTH01 by capillary electrophoresis. Biotechniques 17 (6), 1062–1070.

Butler, J.M., Buel, E., Crivellente, F., McCord, B.R., 2004. Forensic DNA typing by capillary electrophoresis using the ABI prism 310 and 3100 genetic analyzers for STR analysis. Electrophoresis 25 (10 – 11), 1397–1412.

Chaitanya, L., van Oven, M., Weiler, N., Harteveld, J., Wirken, L., Sijen, T., Kayser, M., 2014. Developmental validation of mitochondrial DNA genotyping assays for adept matrilineal inference of biogeographic ancestry at a continental level. Forensic Sci. Int. Genet. 11, 39–51.

Chemale, G., Paneto, G.G., Menezes, M.A., de Freitas, J.M., Jacques, G.S., Cicarelli, R.M., Fagundes, P.R., 2013. Development and validation of a D-loop mtDNA SNP assay for the screening of specimens in forensic casework. Forensic Sci. Int. Genet. 7 (3), 353–358.

Chung, D.T., Drábek, J., Opel, K.L., Butler, J.M., McCord, B.R., 2004. A study on the effects of degradation and template concentration on the amplification efficiency of the STR Miniplex primer sets. J. Forensic Sci. 49 (4), 733–740.

Churchill, J., Peters, D., Capt, C., Strobl, C., Parson, W., Budowle, B., 2017. Working towards implementation of whole genome mitochondrial DNA sequencing into routine casework. Forensic Sci. Int. Genet. Suppl. Ser. 6, e388–e389.

Connon, C.C., LeFebvre, A.K., Benjamin, R.C., 2016. Validation of alternative capillary electrophoresis detection of STRs using POP-6 polymer and a 22cm array on a 3130xl genetic analyzer. Forensic Sci. Int. Genet. 22, 113–127.

Cowell, R., Lauritzen, S., Mortera, J., 2008. Probabilistic modelling for DNA mixture analysis. Forensic Sci. Int. Genet. Suppl. Ser. 1 (1), 640–642.

Davis, C., Peters, D., Warshauer, D., King, J., Budowle, B., 2015. Sequencing the hypervariable regions of human mitochondrial DNA using massively parallel sequencing: enhanced data acquisition for DNA samples encountered in forensic testing. Legal Med. 17 (2), 123–127.

de Knijff, P., 2019. From next generation sequencing to now generation sequencing in forensics. Forensic Sci. Int. Genet. 38, 175–180.

Dembinski, G.M., Picard, C.J., 2017. Effects of microbial DNA on human DNA profiles generated using the PowerPlex. J. Forensic Legal Med. 52, 208–214.

Edson, S.M., 2019. Extraction of DNA from skeletonized postcranial remains: a discussion of protocols and testing modalities. J. Forensic Sci. 64 (5), 1312–1323.

Edson, S.M., Ross, J.P., Coble, M.D., Parsons, T.J., Barritt, S.M., 2004. Naming the dead—confronting the realities of rapid identification of degraded skeletal remains. Forensic Sci. Rev. 16 (1), 63–90.

Elder, J.K., Southern, E.M., 1983. Measurement of DNA length by gel electrophoresis II: comparison of methods for relating mobility to fragment length. Anal. Biochem. 128 (1), 227–231.

Ensenberger, M.G., Lenz, K.A., Matthies, L.K., Hadinoto, G.M., Schienman, J.E., Przech, A.J., Storts, D.R., 2016. Developmental validation of the PowerPlex® fusion 6C system. Forensic Sci. Int. Genet. 21, 134–144.

Fernández, C., Alonso, A., 2012. Microchip capillary electrophoresis protocol to evaluate quality and quantity of mtDNA amplified fragments for DNA sequencing in forensic genetics. Methods Mol. Biol. 830, 367–379.

Forster, L., Thomson, J., Kutranov, S., 2008. Direct comparison of post-28-cycle PCR purification and modified capillary electrophoresis methods with the 34-cycle "low copy number" (LCN) method for analysis of trace forensic DNA samples. Forensic Sci. Int. Genet. 2 (4), 318–328.

Gabriel, M.N., Huffine, E.F., Ryan, J.H., Holland, M.M., Parsons, T.J., 2001. Improved mtDNA sequence analysis of forensic remains using a "mini-primer set" amplification strategy. J. Forensic Sci. 46 (2), 247–253.

Gilbert, M., 2006. Postmortem damage of mitochondrial DNA. In: Bandelt, H., Macaulay, V., Richards, M. (Eds.), Human Mitochondrial DNA and the Evolution of *Homo sapiens*. Springer, Berlin, Germany.

Gilbert, M.T., Willerslev, E., Hansen, A.J., Barnes, I., Rudbeck, L., Lynnerup, N., Cooper, A., 2003. Distribution patterns of postmortem damage in human mitochondrial DNA. Am. J. Hum. Genet. 72 (1), 32–47.

Gill, P., Ivanov, P.L., Kimpton, C., Piercy, R., Benson, N., Tully, G., Sullivan, K., 1994. Identification of the remains of the Romanov family by DNA analysis. Nat. Genet. 6 (2), 130–135.

Gill, P., Brenner, C.H., Buckleton, J.S., Carracedo, A., Krawczak, M., Mayr, W.R., Weir, B.S., 2006. DNA commission of the International Society of Forensic Genetics: recommendations on the interpretation of mixtures. Forensic Sci. Int. 160 (2–3), 90–101.

Gill, P., Gusmão, L., Haned, H., Mayr, W.R., Morling, N., Parson, W., Weir, B.S., 2012. DNA commission of the International Society of Forensic Genetics: recommendations on the evaluation of STR typing results that may include drop-out and/or drop-in using probabilistic methods. Forensic Sci. Int. Genet. 6 (6), 679–688.

Ginther, C., Issel-Tarver, L., King, M.C., 1992. Identifying individuals by sequencing mitochondrial DNA from teeth. Nat. Genet. 2 (2), 135–138.

Goios, A., Carvalho, A., Amorim, A., 2009. Identifying NUMT contamination in mtDNA analyses. Forensic Sci. Int. Genet. Suppl. Ser. 2, 278–280.

Greenspoon, S.A., Schiermeier-Wood, L., Jenkins, B.C., 2015. Establishing the limits of TrueAllele® casework: a validation study. J. Forensic Sci. 60 (5), 1263–1276.

Hauge, X.Y., Litt, M., 1993. A study of the origin of 'shadow bands' seen when typing dinucleotide repeat polymorphisms by the PCR. Hum. Mol. Genet. 2 (4), 411–415.

Hedell, R., Dufva, C., Ansell, R., Mostad, P., Hedman, J., 2015. Enhanced low-template DNA analysis conditions and investigation of allele dropout patterns. Forensic Sci. Int. Genet. 14, 61–75.

Hofreiter, M., Jaenicke, V., Serre, D., von Haeseler, A., Pääbo, S., 2001. DNA sequences from multiple amplifications reveal artifacts induced by cytosine deamination in ancient DNA. Nucleic Acids Res. 29 (23), 4793–4799.

Holland, M.M., Parsons, T.J., 1999. Mitochondrial DNA sequence analysis—validation and use for forensic casework. Forensic Sci. Rev. 11 (1), 21–50.

Holland, M.M., McQuillan, M.R., O'Hanlon, K.A., 2011. Second generation sequencing allows for mtDNA mixture deconvolution and high resolution detection of heteroplasmy. Croat. Med. J. 52 (3), 299–313.

Holland, M.M., Wilson, L.A., Copeland, S., Dimick, G., Holland, C.A., Bever, R., McElhoe, J.A., 2017. MPS analysis of the mtDNA hypervariable regions on the MiSeq with improved enrichment. Int. J. Legal Med. 131 (4), 919–931.

Just, R., Scheible, M., Fast, S., Sturk-Andreaggi, K., Rock, A., Bush, J., Irwin, J., 2015. Full mtGenome reference data: development and characterization of 588 forensic-quality haplotypes representing three US populations. Forensic Sci. Int. Genet. 14, 141–155.

King, J.L., LaRue, B.L., Novroski, N.M., Stoljarova, M., Seo, S.B., Zeng, X., Budowle, B., 2014. High-quality and high-throughput massively parallel sequencing of the human mitochondrial genome using the Illumina MiSeq. Forensic Sci. Int. Genet. 12, 128–135.

Kline, M.C., Vallone, P.M., Redman, J.W., Duewer, D.L., Calloway, C.D., Butler, J.M., 2005. Mitochondrial DNA typing screens with control region and coding region SNPs. J. Forensic Sci. 50 (2), 377–385.

Koehn, A., 2013. Identification of Unknown PCR Products Generated During STR Analysis of Bone Samples. Masters Thesis, University of North Texas Health Science Center.

Korlević, P., Gerber, T., Gansauge, M.T., Hajdinjak, M., Nagel, S., Aximu-Petri, A., Meyer, M., 2015. Reducing microbial and human contamination in DNA extractions from ancient bones and teeth. Biotechniques 59 (2), 87–93.

Latham, K.E., Miller, J.J., 2019. DNA recovery and analysis from skeletal material in modern forensic contexts. Forensic Sci. Res. 4 (1), 51–59.

Lin, C.Y., Tsai, L.C., Hsieh, H.M., Huang, C.H., Yu, Y.J., Tseng, B., Lee, J.C., 2017. Investigation of length heteroplasmy in mitochondrial DNA control region by massively parallel sequencing. Forensic Sci. Int. Genet. 30, 127–133.

Loreille, O.M., Irwin, J.A., 2012. Capillary electrophoresis of human mtDNA control region sequences from highly degraded samples using short mtDNA amplicons. Methods Mol. Biol. 830, 283–299.

Marshall, C., Sturk-Andreaggi, K., Daniels-Higginbotham, J., Oliver, R.S., Barritt-Ross, S., McMahon, T.P., 2017. Performance evaluation of a mitogenome capture and Illumina sequencing protocol using non-probative, case-type skeletal samples: implications for the use of a positive control in a next-generation sequencing procedure. Forensic Sci. Int. Genet. 31, 198–206.

McCord, B., Jung, J., Holleran, E., 1993. High resolution capillary electrophoresis of forensic DNA using a non-gel sieving buffer. J. Liq. Chromatogr. 16 (9–10), 1963–1981.

McCord, B., Opel, K., Funes, M., Zoppis, S., Jantz, L., 2011. An Investigation of the Effect of DNA Degradation and Inhibition on PCR Amplification of Single Source and Mixed Forensic Samples. https://www.ncjrs.gov/pdffiles1/nij/grants/236692.pdf. (Accessed 25 February 2019).

McCord, B.R., Gauthier, Q., Cho, S., Roig, M.N., Gibson-Daw, G.C., Young, B., Duncan, G., 2019. Forensic DNA analysis. Anal. Chem. 91 (1), 673–688.

Melton, T., Clifford, S., Kayser, M., Nasidze, I., Batzer, M., Stoneking, M., 2001. Diversity and heterogeneity in mitochondrial DNA of north American populations. J. Forensic Sci. 46 (1), 46–52.

Montesino, M., Prieto, L., 2012. Capillary electrophoresis of big-dye terminator sequencing reactions for human mtDNA control region haplotyping in the identification of human remains. Methods Mol. Biol. 830, 267–281.

Mulero, J.J., Chang, C.W., Lagacé, R.E., Wang, D.Y., Bas, J.L., McMahon, T.P., Hennessy, L.K., 2008. Development and validation of the AmpFlSTR MiniFiler PCR amplification kit: a MiniSTR multiplex for the analysis of degraded and/or PCR inhibited DNA. J. Forensic Sci. 53 (4), 838–852.

Nicklas, J.A., Noreault-Conti, T., Buel, E., 2012. Development of a real-time method to detect DNA degradation in forensic samples. J. Forensic Sci. 57 (2), 466–471.

Parson, W., Gusmão, L., Hares, D.R., Irwin, J.A., Mayr, W.R., Morling, N., Parsons, T.J., 2014. DNA Commission of the International Society for forensic genetics: revised and extended guidelines for mitochondrial DNA typing. Forensic Sci. Int. Genet. 13, 134–142.

Parson, W., Ballard, D., Budowle, B., Butler, J.M., Gettings, K.B., Gill, P., Phillips, C., 2016. Massively parallel sequencing of forensic STRs: considerations of the DNA commission of the International Society for Forensic Genetics (ISFG) on minimal nomenclature requirements. Forensic Sci. Int. Genet. 22, 54–63.

Perlin, M.W., Legler, M.M., Spencer, C.E., Smith, J.L., Allan, W.P., Belrose, J.L., Duceman, B.W., 2011. Validating TrueAllele® DNA mixture interpretation. J. Forensic Sci. 56 (6), 1430–1447.

Perlin, M.W., Belrose, J.L., Duceman, B.W., 2013. New York state TrueAllele® casework validation study. J. Forensic Sci. 58 (6), 1458–1466.

Pionzio, A.M., McCord, B.R., 2014. The effect of internal control sequence and length on the response to PCR inhibition in real-time PCR quantitation. Forensic Sci. Int. Genet. 9, 55–60.

Promega Corporation, 2018. Spectrum CE System. https://www.promega.com.au/-/media/files/products-and-services/genetic-identity/analysis/36383631-brgispectrum-br284-digital.pdf. (Accessed 1 November 2019).

Promega Corporation, 2019. Spectrum Compact CE System. https://promega.media/-/media/files/resources/brochures/genomics/49402999-brgnspectrumcce-br292-digital.pdf. (Accessed 1 November 2019).

Rathbun, M.M., McElhoe, J.A., Parson, W., Holland, M.M., 2017. Considering DNA damage when interpreting mtDNA heteroplasmy in deep sequencing data. Forensic Sci. Int. Genet. 26, 1–11.

Sanger, F., Nicklen, S., Coulson, A.R., 1977. DNA sequencing with chain-terminating inhibitors. Proc. Natl. Acad. Sci. U. S. A. 74 (12), 5463–5467.

Shewale, J.G., Qi, L., Calandro, L.M., 2012. Principles, practice, and evolution of capillary electrophoresis as a tool for forensic DNA analysis. Forensic Sci. Rev. 24 (2), 79–100.

Sidstedt, M., Steffen, C.R., Kiesler, K.M., Vallone, P.M., Rådström, P., Hedman, J., 2019. The impact of common PCR inhibitors on forensic MPS analysis. Forensic Sci. Int. Genet. 40, 182–191.

Sobiah, R., Syeda, R., Zunaira, E., Nageen, Z., Maria, K., Syeda, A., Muhammad, R., 2018. Implications of targeted next generation sequencing in forensic science. J. Forensic Res. 9 (1), 1–8.

Staadig, A., Tillmar, A., 2019. An overall limited effect on the weight-of-evidence when taking STR DNA sequence polymorphism into account in kinship analysis. Forensic Sci. Int. Genet. 39, 44–49.

Stewart, J.E., Aagaard, P.J., Pokorak, E.G., Polanskey, D., Budowle, B., 2003. Evaluation of a multicapillary electrophoresis instrument for mitochondrial DNA typing. J. Forensic Sci. 48 (3), 571–580.

SWGDAM, 2004. Revised Validation Guidelines-Scientific Working Group on DNA Analysis Methods (SWGDAM). https://docs.wixstatic.com/ugd/4344b0_813b241e8944497e99b9c45b163b76bd.pdf. (Accessed 25 February 2019).

SWGDAM, 2017. SWGDAM Interpretation Guidelines for Autosomal STR Typing by Forensic DNA Testing Laboratories. https://docs.wixstatic.com/ugd/4344b0_50e2749756a242528e6285a5bb478f4c.pdf. (Accessed 25 February 2019).

Taylor, D., Bright, J.A., Buckleton, J., 2013. The interpretation of single source and mixed DNA profiles. Forensic Sci. Int. Genet. 7 (5), 516–528.

Taylor, D., Bright, J.A., McGoven, C., Hefford, C., Kalafut, T., Buckleton, J., 2016. Validating multiplexes for use in conjunction with modern interpretation strategies. Forensic Sci. Int. Genet. 20, 6–19.

Templeton, J.E., Brotherton, P.M., Llamas, B., Soubrier, J., Haak, W., Cooper, A., Austin, J.J., 2013. DNA capture and next-generation sequencing can recover whole mitochondrial genomes from highly degraded samples for human identification. Investig. Genet. 4 (1), 26.

Thermo Fisher Scientific, 2014. DNA Fragment Analysis by Capillary Electrophoresis. https://www.thermofisher.com/content/dam/LifeTech/global/Forms/PDF/fragment-analysis-chemistry-guide.pdf. (Accessed 5 February 2019).

Thermo Fisher Scientific, 2016. DNA Sequencing by Capillary Electrophoresis. https://www.thermofisher.com/content/dam/LifeTech/Documents/PDFs/sequencing_handbook_FLR.pdf. (Accessed 20 January 2019).

Thermo Fisher Scientific, 2018. 3500/3500xL Genetic Analyzer User Guide—Data Collection Software v3.1. https://assets.thermofisher.com/TFS-Assets/LSG/manuals/100031809_3500_3500xL_Software_v3_1_UG.pdf. (Accessed 5 February 2019).

Thompson, R.E., Duncan, G., McCord, B.R., 2014. An investigation of PCR inhibition using Plexor®-based quantitative PCR and short tandem repeat amplification. J. Forensic Sci. 59 (6), 1517–1529.

Tillmar, A., Grandell, I., Montelius, K., 2018. DNA identification of compromised samples with massive parallel sequencing. Forensic Sci. Res. 4, 1–7.

Tvedebrink, T., Eriksen, P.S., Mogensen, H.S., Morling, N., 2012. Statistical model for degraded DNA samples and adjusted probabilities for allelic drop-out. Forensic Sci. Int. Genet. 6 (1), 97–101.

van Oorschot, R.A., Ballantyne, K.N., Mitchell, R.J., 2010. Forensic trace DNA: a review. Investig. Genet. 1 (14), 1–17.

Vilsen, S.B., Tvedebrink, T., Mogensen, H.S., Morling, N., 2017. Statistical modelling of ion PGM HID STR 10-plex MPS data. Forensic Sci. Int. Genet. 28, 82–89.

Vohr, S.H., Gordon, R., Eizenga, J.M., Erlich, H.A., Calloway, C.D., Green, R.E., 2017. A phylogenetic approach for haplotype analysis of sequence data from complex mitochondrial mixtures. Forensic Sci. Int. Genet. 30, 93–105.

Votrubova, J., Ambers, A., Budowle, B., Vanek, D., 2017. Comparison of standard capillary electrophoresis based genotyping method and ForenSeq DNA signature prep kit (Illumina) on a set of challenging samples. Forensic Sci. Int. Genet. Suppl. Ser. 6, e140–e142.

Walsh, P.S., Fildes, N.J., Reynolds, R., 1996. Sequence analysis and characterization of stutter products at the tetranucleotide repeat locus vWA. Nucleic Acids Res. 24 (14), 2807–2812.

Westen, A.A., Nagel, J.H., Benschop, C.C., Weiler, N.E., de Jong, B.J., Sijen, T., 2009. Higher capillary electrophoresis injection settings as an efficient approach to increase the sensitivity of STR typing. J. Forensic Sci. 54 (3), 591–598.

Woerner, A.E., King, J.L., Budowle, B., 2017. Flanking variation influences rates of stutter in simple repeats. Gene 8 (11), 329.

CHAPTER

15

Rapid DNA identification of human skeletal remains

Rosemary Turingan Witkowski PhD, Ranjana Grover PhD, Eugene Tan PhD, and Richard F. Selden MD, PhD

ANDE Corporation, Waltham, MA, United States

Introduction and historical context of skeletal samples in DNA-based forensic human identification

With the seminal observation in the 1980s that human minisatellite repeats could be utilized to identify individuals (Jeffreys et al., 1985), it might be tempting to assume that the techniques of the burgeoning new field of forensic DNA analysis were quickly and routinely adapted and applied to bone samples. In comparison to the relatively crude discrimination provided by the immunochemistry techniques of 50 years ago, the exquisite specificity characteristic of nucleic acid hybridization demonstrated that even early forensic DNA experiments could generate useful data. For example, in 1989, human Alu-family and Y-chromosomal repetitive DNA sequences were used to reliably determine gender in teeth and bone (Yokoi et al., 1989). However, studies such as this were the exceptions to the rule. The primary use of bone DNA was not initially related to forensic identification, but instead focused on human paleogenetics. A 1997 study purified nucleic acids from Neandertal bones and sequenced the resulting mitochondrial DNA (mtDNA), ushering in the modern study of ancient human populations (Krings et al., 1997).

During the 20 years that followed Jeffrey's work, an armamentarium of forensic DNA analysis tools and an international infrastructure to apply those tools to law enforcement were created. The application of the polymerase chain reaction (PCR) to amplification of variable repeat loci enabled DNA fingerprinting to be performed on smaller quantities of DNA template, and the analysis on polyacrylamide gels (as opposed to agarose gels) improved resolution of the resulting amplicons (Weber and May, 1989; Litt and Luty, 1989). As the analysis of variable repeats improved, the selection of loci appropriate for human identification evolved in parallel. The use of trimeric and tetrameric short tandem repeats (STRs) was pioneered in 1991 (Rassmann et al., 1991; Edwards et al., 1991), ultimately leading to the selection of a

Forensic Genetic Approaches for Identification of Human Skeletal Remains
https://doi.org/10.1016/B978-0-12-815766-4.00015-7

standard set of 13 tetrameric STR loci to be used in DNA fingerprinting in the United States (Budowle, 1997; Moretti and Budowle, 1998). The sensitivity and reliability of STR technology led law enforcement and military agencies to develop databases of DNA IDs (i.e., STR profiles) from individuals and sample materials of interest.

Relatively little of the law enforcement forensic DNA work during this period involved the use of bone samples. It may on the surface appear puzzling that techniques sensitive enough to trace the migration of human populations over tens of thousands of years were not being applied to the identification of bones from present day humans impacted by crime and mass casualty events. After all, the fundamental approach to generating STR information from bone has been in place since the late 1990s, as exemplified by the analysis of skeletal remains from mass graves in Croatia and Bosnia and Herzegovina (Alonso et al., 2001). Following the painstaking identification of large numbers of samples from World Trade Center attack victims (Holland et al., 2003) and plane crashes (Hsu et al., 1999; Leclair et al., 2004), STR analysis of bone and tooth samples has now become a standard tool for disaster victim identification (DVI) following mass casualty events (Deng et al., 2005; Sudoyo et al., 2008; Donkervoort et al., 2008; Hartman et al., 2011; McEntire et al., 2012).

The reasons that bone samples have lagged behind other forensic sample types in law enforcement are likely due to the fact that purifying DNA from bone is more cumbersome and time-consuming and requires more equipment than from most other casework sample types (e.g., blood, soft tissue, handled objects); additionally, bone DNA often is highly degraded and may be contaminated with microbial DNA (endogenous or environmental) or human DNA from those who have handled or processed the sample. Even today in forensic laboratories, bone samples are typically the "sample of last resort" for casework processing. As a result, tens of thousands of sets of human remains languish unidentified in coroners' and medical examiners' offices; it can take years to identify some crime- and mass casualty event victims. The nearly institutionalized reluctance to process bone and tooth samples and the need to bring closure to families inspired ANDE Corporation to adapt its Rapid DNA Identification System for these sample types (Turingan et al., 2019, 2020; Gin et al., 2020).

Overview of the ANDE Rapid DNA Identification System

Rapid DNA Identification may be defined as the fully automated process of generating a DNA ID from a forensic sample, typically performed outside the laboratory by nontechnical operators with results available in less than 2 hours (Carney et al., 2019; Turingan et al., 2019; Grover et al., 2017; Moreno et al., 2017; Della Manna et al., 2016; Turingan et al., 2016; Palmbach et al., 2014; Tan et al., 2013). The ANDE Rapid DNA System has been applied to bone and tooth samples in a variety of settings, including identification of intact and highly degraded human remains. In disaster victim identification, the system has been utilized in a wide range of mass casualty events, notably in the 2018 Butte County wildfires, the deadliest in California history (Gin et al., 2020).

The ANDE system consists of four components: the swab, the consumable chip, the ANDE 6C instrument, and the automated Expert System. There are two types of chips: (1) the A-Chip, and (2) the I-Chip. The A-Chip processes up to five buccal samples and, along with the ANDE instrument and Expert System, received NDIS approval for searching and uploading to

CODIS following processing of buccal samples in accredited laboratories (Carney et al., 2019; Della Manna et al., 2016).

The I-Chip was designed for the processing of casework or DVI samples and is a single-use, disposable consumable which includes all reagents, materials, and waste containment required to perform fully automated STR analysis on up to four samples. DNA purification reagents, PCR reagents, buffers, and separation polymer are all preloaded on the chip. There is no direct contact between the instrument and the sample or the reagents; all liquids within the chip are driven by pneumatic pressure. This closed system design, coupled with swabs that lock and seal into the chips, minimizes the potential for cross-contamination. The major difference between the A-Chip and the I-Chip is that the I-Chip (Fig. 1) incorporates a sample concentration module positioned downstream of the purification module. The purified DNA solution is directed to a semipermeable ultrafiltration membrane by application of pneumatic pressure. As a result, DNA present in the retentate is concentrated, critical to ensure an optimal limit-of-detection for the System.

The FlexPlex assay (Fig. 2) was developed to support compatibility with DNA databases around the world (Grover et al., 2017). FlexPlex contains 23 autosomal STR loci (D1S1656, D2S1338, D2S441, D3S1358, D5S81, D6S1043, D7S820, D8S1179, D10S1248, D12S391, D13S317, D16S539, D18S51, D19S433, D21S11, D22S1045, FGA, CSF1PO, Penta E, TH01, vWA, TPOX, SE33), three Y-chromosomal STR loci (DYS391, DYS570, DYS576), and amelogenin. FlexPlex generates STR data compatible with databases around the world, including the CODIS core 20 loci, ENFSI/EDNAP Expanded European Standard Set, Australia's National Criminal Investigation DNA Database, Canada's National DNA Data Bank, China's National DNA Database, Germany's DNA-Analyze-Datei, New Zealand's National DNA Profile Databank, and United Kingdom's National DNA Database.

The ANDE instrument is comprised of several subsystems, including (1) a pneumatic subsystem for driving fluids throughout the chip; (2) a thermal subsystem for performing multiplexed amplification; (3) a high voltage subsystem for electrophoresis; (4) a 6-color optical subsystem for exciting and detecting fluorescently labeled STR fragments during electrophoresis; and (5) a ruggedization subsystem to allow transport and field-forward operation

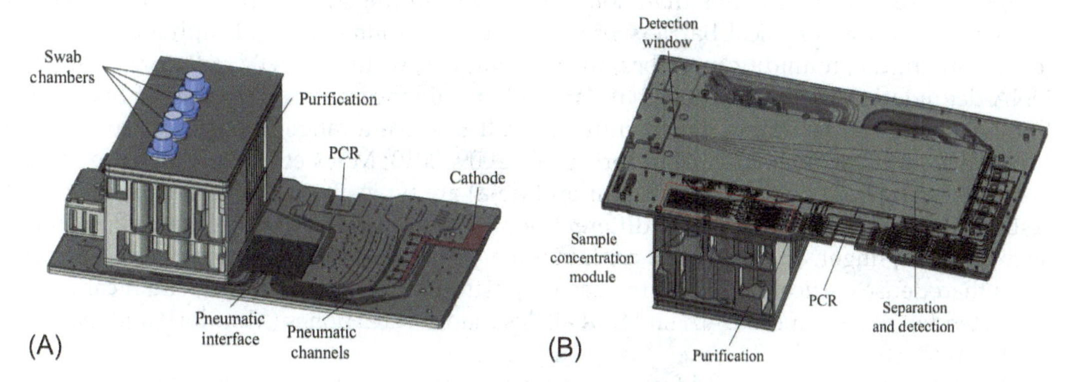

FIG. 1 Schematic diagram of the ANDE I-Chip: (A) top view, and (B) bottom view (showing the ultrafiltration modules). The dimensions of the I-Chip are 295 mm × 165 mm × 93 mm.

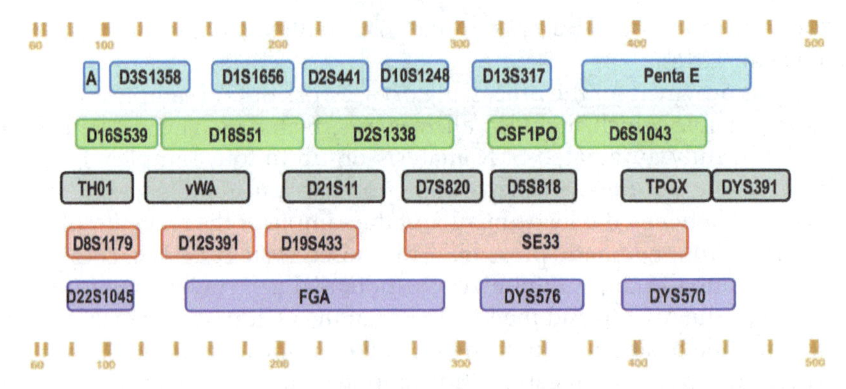

FIG. 2 Configuration of the FlexPlex STR assay. Flexplex is a 6-color assay (including an Internal Lane Standard, or ILS) that interrogates 23 autosomal STR loci, three Y-chromosomal STRs, and amelogenin. The dye labels in the FlexPlex assay are: FAM *(blue)*, JOE *(green)*, TMR *(yellow)*, CXR *(red)*, TOM *(purple)*, and WEN *(orange)*.

without recalibration or optical realignment. The System includes several integrated software packages for instrument control, data collection, and Expert System interpretation of STR profiles. The instrument's integrated touchscreen features a graphical user interface with workflow-driven instructions. After the I-Chip containing the swab samples has been inserted into the instrument and the door closed, sample processing starts automatically. Following electrophoretic separation and laser-based detection of amplified STR fragments, the Expert System software automatically analyzes and interprets the data and provides rapid feedback on the usability of the STR profiles for database enrollment and searching. The output files are available for optional review by a qualified DNA analyst. All output files are encrypted and can be exported and decrypted by FAIRS, a multifunction, multitiered user privilege software package. FAIRS integrates database generation and management, search and match, and kinship determination functionalities.

Selection of bone samples for Rapid DNA Analysis

Bone preserves DNA better than soft tissues due to the structural elements in skeletal remains that act as physical barriers from various environmental and taphonomic factors (e.g., temperature, humidity, microbes/insects/animals, water, fire, UV radiation) that cause DNA degradation. Several studies have provided guidance on the selection of bone types that will optimize the chance of obtaining DNA IDs across a range of postmortem intervals (Mundorff and Davoren, 2014; Mundorff et al., 2009, 2013; Miloš et al., 2007). Dense cortical weight-bearing bones (such as the femur and tibia) are typically preferred for forensic DNA testing. However, some studies on different bone types suggest that small cancellous bones such as ankle, finger, and toe bones contain (on average) greater quantities of DNA per unit mass than dense cortical bones. At increasing postmortem intervals (PMI), small cancellous bones tended to yield more DNA and STR alleles than cortical bones (Mundorff and Davoren, 2014; Mundorff et al., 2013).

Data from remains excavated from mass graves originating from conflicts that occurred in the former Yugoslavia during the 1990s suggest that the highest success rates for DNA typing were observed with samples from dense cortical weight-bearing bones (e.g., femur), whereas long bones of the arms showed significantly lower success (Miloš et al., 2007). Reported success rates were higher for the lower body (femur, tibia, fibula) than those for the upper body (humerus, radius, ulna). This study also indicated that bones which performed poorly tended to be less dense or have a greater proportion of spongy or diploic bone (e.g., vertebrae, ilium, cranium). Other studies have confirmed that DNA of reasonable quality and quantity can be more reliably obtained from compact/long (cortical) bones than from spongy (trabecular) bones such as rib (Alonso et al., 2001; Parsons and Weedn, 1996). It is likely that the high moisture content in spongy bones influences the rate of DNA degradation. Case studies involving DNA analyses of older, historical skeletal remains (100+ years old) have proven successful using femora, tibiae, fibulae, ribs, and molar teeth (Ambers et al., 2014, 2016, 2019a). In historical cases (including mass graves from oppressive regimes) and in cases in which remains have been exposed to the environment for extended periods of time, smaller cancellous bones of the feet and hands are often not recovered (and therefore are not available for testing). As soft tissue decomposes and the skeleton disarticulates, animal scavenging and dispersal by weather and gravity contribute to the loss of these smaller bones from the original site of deposition or burial.

In addition to STR analysis, mitochondrial DNA (mtDNA) typing is another widely used method for identification of human remains, particularly severely degraded skeletal elements. One study on mtDNA typing success rates in degraded skeletal specimens reported that weight-bearing long bones (e.g., femora, tibiae) were the optimal sample type (Edson et al., 2004); however, metatarsals also generally produced reportable results, despite their relatively small size. Ribs were reported to be highly successful, but required a larger sample size and the evaluation was limited to a single case. The spiny projections of the vertebrae that provide an area for muscle attachment were reported to generate superior results as compared to the porous body. As with nuclear DNA typing methods, cranial fragments were also found to be the most challenging of all samples tested, presumably because of the large porous areas of diploic bone which are exposed to environmental elements and contaminants.

Samples submitted for Rapid DNA Identification may range from those present in intact or fragmented remains (allowing the operator to select the bone sample type) to those preselected onsite by first responders or by external agencies. Preferences for bone types for Rapid DNA Identification using the ANDE system generally conform to the findings from the research previously cited in this chapter. When possible, it is worthwhile to select a sample with sufficient mass to allow duplicate analyses to be performed. Furthermore, as Rapid DNA processing will frequently be performed outside the laboratory, the ease of sample collection should also be considered. Specifically, the phalanges of the foot are readily obtained in the field. Multirooted teeth are also ideal for Rapid DNA analysis; however, depending on the overall state of the remains, it may be important to preserve teeth for odontologic evaluation. The mid-shaft (diaphysis) of femora is an excellent sample type when bone saws or drills are available. If other skeletal elements are presented for analysis, methods for sample processing can be adjusted to increase the likelihood of successful typing (e.g., by increasing the sample mass input, using a longer incubation step, etc.).

Bones may be described as either "fresh" or "old," although suitability for Rapid DNA Identification is not as simple as merely knowing the PMI. The quantity and quality of DNA within bones are affected by a variety of environmental exposures, including varying temperature and humidity, sunlight, microbes, insects, and animals (particularly carnivores). Excessive heat exposure during a fire, as well as the deposition state of the remains (e.g., buried, above-ground, submerged), can also have profound impact on DNA preservation. Accordingly, bones exposed to temperate climates may have longer PMIs but higher DNA typing success rates compared to bones recovered from tropical climates with shorter PMIs. Remains stored in refrigerated morgues can often still be processed as "fresh" samples for many years after death. Burned remains may still have intact marrow and suffer minimal degradation, whereas bones subjected to extreme heat for only several minutes may become calcined with poor quality DNA. Another important consideration in evaluating bone samples for processing is the presence of contaminants such as fuels, accelerants, explosives, plant oils, by-products of decomposition, and microbial DNAs that may co-purify with endogenous human DNA and inhibit PCR reactions (Edson and Roberts, 2019; Alonso et al., 2001). The Rapid DNA methods described below have been developed to minimize the potential for PCR inhibition.

For Rapid DNA Identification, the physical appearance of the bone should be assessed to inform the optimal processing approach. Questions to consider include:

- Does the bone contain marrow or adhered soft tissue(s)?
- Can the specific bone type be identified?
- Is the approximate postmortem interval (PMI) known?
- How much material is available for testing?
- Is the bone dense or is it spongy?
- Are potential PCR inhibitors (e.g., jet fuel) present?
- Is the bone dry?
- Is the bone fragile or crumbling even with minimal pressure?

Answers to these questions provide guidance as to which of the methods described in the following sections should be utilized for processing. In general, bones that are thought to contain relatively intact DNA are subjected to a brief preprocessing protocol, whereas bones which likely contain highly degraded DNA require a longer preprocessing method (typically overnight) and may require a greater quantity of starting material. Although beyond the scope of this chapter, it is also critical to assess whether or not the bone(s) presented for testing are commingled with other remains; these and related observations may require the expertise of a forensic anthropologist.

Bone preprocessing protocols for Rapid DNA Identification

Note: The use of personal protective equipment (PPE) during sample processing is required both to protect the operator from biohazardous materials and to minimize operator contamination of the samples.

Bone cleaning

Specimens should be thoroughly cleaned of organic and inorganic contaminants including adhered tissues, soil, debris, and chemical residues. A variety of methods have been utilized, including irradiation with UV light, chemical cleaning (e.g., with alcohol, acid, bleach), and combinations of these methods (Ambers et al., 2014; Lee et al., 2010a,b; Kemp and Smith, 2005). Careful handling of the materials minimizes the risk of endogenous DNA being subjected to further degradation and human or microbial contamination (Whiteman et al., 2002; Hayatsu et al., 1971). Cortical or compact bones are typically easier to clean than trabecular/porous (i.e., cancellous or spongy) bones.

The cleaning method may be selected based on the condition of the bone, and a standard approach is described here. Adhered tissue, if any, is first removed using a scalpel. Surface debris on external surfaces is removed by thorough brushing with a toothbrush under running tap water or sanded down using a grinding stone attached to a Dremel rotary tool. The bone fragment is further cleaned by fully immersing the specimen in a conical tube (or other specimen container) with 10% bleach solution; the tube is then vigorously shaken for approximately 30 seconds. The specimen is then transferred to a second clean tube containing sterile water and is washed by vigorously shaking for approximately 30 seconds. The water-wash step is then repeated. The cleaned bone is transferred into a tube containing absolute ethanol and washed with shaking one final time. Finally, the bone is removed and placed on a sterile plate (or weigh boat) and allowed to dry at room temperature. Depending on the size and density of the specimen, drying can be completed within a few hours and can be expedited by placing the sample in a heated oven at 35–50°C.

Certain bone samples (particularly degraded trabecular or cancellous bones) may fracture very easily. Depending on the specific skeletal element, such bones may be cleaned as described above for cortical or compact bones. However, some samples may be too fragile to subject to standard cleaning methods. Fragile specimens may be cleaned by simply rubbing the external surfaces with 10% bleach wipes, followed by sterile water and alcohol wipes. The intensity of cleaning is based on evaluating the trade-off between losing DNA during preprocessing and removing potential contaminants that may affect downstream analysis.

Bone fragmentation

Bone and tooth samples are often the best samples for DNA analysis because extensive mineralization within the bone provides physical barriers to DNA degradation. However, the very same feature that protects DNA also prevents the ready release of DNA from bone using standard purification procedures. Accordingly, bone and tooth samples are conventionally converted to homogenous powder using a liquid nitrogen and a grinding mill, followed by bone demineralization to maximize DNA purification from bone. Techniques for drilling and pulverizing bone using freezing mills followed by demineralization with EDTA and other reagents have been successfully used to obtain sufficient amounts of DNA for typing (Ambers et al., 2014, 2016, 2018, 2019a,b; Uzair et al., 2017; Pajnič, 2016; Amory et al., 2012; Lee et al., 2010a,b; Loreille et al., 2007). DNA purification from bone powder has been performed using lengthy incubations in demineralization and lysis solutions followed by organic extraction, Chelex, or silica-based purifications. When using large volumes for demineralization, sample concentration steps have been added to the protocols (Lee et al., 2010b; Loreille et al., 2007) prior to DNA quantification and amplification. These techniques are laborious and time-consuming and require extensive equipment, a carefully controlled laboratory environment, and highly skilled technical operators. This combination of drawbacks is reasonable justification for the typical reluctance of forensic laboratories to process bone and tooth samples.

These time-, labor-, and equipment-intensive procedures have been eliminated or minimized in Rapid DNA processing of bone without compromising purification efficiency, and consequently, DNA typing success. After cleaning and drying of bone specimens, the use

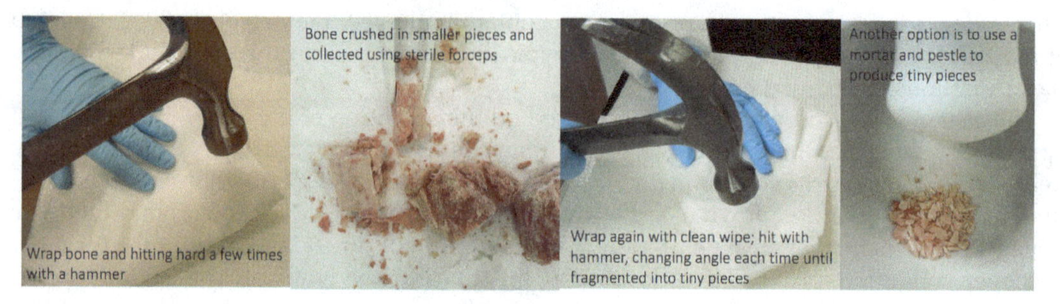

FIG. 3 Fragmenting bone into smaller pieces via hammering or use of a mortar-and-pestle.

of a cryogenic grinder or freezer/mill to pulverize the sample into powder can be avoided; instead, the sample can simply be pummeled into smaller fragments using a hammer or a mortar-and-pestle (Fig. 3). ANDE has even had success by scraping the bone sample with a heavy-duty nail clipper. Improvised mini blender cups (and similar apparatuses) can also be used to pulverize the bone while avoiding the generation of airborne particles. The bone can first be wrapped in DurX 770 wipe, which is a clean sheet of nonwoven polyester/cellulose material, or its equivalent. The use of plastic to cover the bone is not recommended since bone penetrates through the plastic and it can be difficult to separate plastic from fragmented bone. Striking the bone at different angles with a hammer will easily break the bone into smaller fragments. Use of a mortar-and-pestle to further break bone fragments into smaller pieces is optional but recommended, as it is easy to contain and collect the final pulverized sample. The ideal size for Rapid DNA processing is similar to a finely ground peppercorn, with thin shards of less than 1/8 inch in length or diameter.

Processing intact bone

Bone can be considered intact (or "fresh") if marrow or blood is present or if tissue is still attached (Fig. 4). For these samples, after cleaning, place approximately 5–10 mg crushed bone in a 2-mL microcentrifuge tube and add 120 μL of ANDE Bone Solution (a proprietary reagent designed to demineralize the bone matrix to facilitate DNA release). Mix the sample by vortexing for 5–10 seconds using a benchtop vortexer set at maximum speed and then incubate for

FIG. 4 Photographs of intact ("fresh") bones processed by Rapid DNA Analysis.

1–2 minutes at room temperature. Pipette 15 μL of the solution onto an ANDE swab for Rapid DNA processing. In general, weight-bearing bones such as femora or foot phalanges will generate complete (or near complete) DNA IDs using this approach. For other bone types, sample input may be increased by pipetting the entire solution onto an ANDE swab.

Processing burned bone

Bone fragments can range from moderately burned samples to fully calcinated remains. Figs. 5 and 6 show transformations and color changes in bone (respectively) as a function of heat (Ellingham, 2018).

FIG. 5 Unburnt bone to 1100°C burnt bones in 100°C increments. Dehydration occurs between 100°C and 600°C; decomposition (loss of organic components) between 300°C and 800°C; inversion (changes in the inorganic phase) between 500°C and 1100°C; and fusion (coalescence of the crystal structure) at temperatures above 700°C. *Courtesy of Ellingham, S., 2018. Fire Destruction of the Body and Biochemical and Structural Changes to the Bone Matrix. American Academy of Forensic Sciences (AAFS), Seattle, WA.*

FIG. 6 Changes in bone color from unburnt bone (top left) to 1100°C burnt bones in 100°C increments. *Courtesy of Ellingham, S., 2018. Fire Destruction of the Body and Biochemical and Structural Changes to the Bone Matrix. American Academy of Forensic Sciences (AAFS), Seattle, WA.*

Bone fragments are crushed by hammering until a sandy texture is achieved and, based on the initial condition of the samples, one of four protocols can be applied, as follows:

- **Protocol #1** (for bone fragments with minimal thermal damage, i.e., with an appearance similar to that shown in Fig. 6 when burned at temperatures between 100°C and 200°C). Place approximately 5–10 mg crushed bone in a 2-mL microcentrifuge tube and add 120 μL of ANDE Bone Solution. Mix the sample by vortexing for 5–10 seconds using a benchtop vortexer set at maximum speed and then incubate for 1–2 minutes at room temperature. Pipette the entire solution onto an ANDE swab for Rapid DNA processing.
- **Protocol #2** (for bone fragments with moderate thermal damage, i.e., with an appearance similar to that shown in Fig. 6 when burned at temperatures of 300°C). Place approximately 50–100 mg crushed bone in a 2-mL microcentrifuge tube and add 120 μL of ANDE Bone Solution. Mix the sample by vortexing for 5–10 seconds using a benchtop vortexer set at maximum speed and then incubate at 56°C for 3 hours. Pipette the entire solution or an aliquot onto an ANDE swab for Rapid DNA processing.
- **Protocol #3** (for bone fragments with severe thermal damage, essentially ashes or fused bone; i.e., with an appearance similar to that shown in Fig. 6 when burned at temperatures above 400°C). Place approximately 500–700 mg crushed bone in a 2-mL microcentrifuge tube and add 1.4 mL of ANDE Bone Solution and 70 μL of 20 mg/mL Proteinase K (Qiagen, Inc.). Place the tube in a thermomixer at 56°C and agitate for 90 minutes before incubating the sample overnight. Centrifuge the sample at maximum speed (at least 16,000 rcf) for 1 minute to pellet any remaining bone particulates. Concentrate the supernatant to approximately 50–100 μL using a 10K Amicon Ultra-0.5-mL Centrifugal Filter Unit (Millipore Sigma, Burlington, MA). Pipette the entire concentrated sample onto an ANDE swab for Rapid DNA processing. *Note*: ANDE Corporation typically utilizes this protocol for processing samples from burned bone.
- **Protocol #4** (for bone fragments with severe thermal damage, essentially ashes or fused bone; i.e., with an appearance similar to that shown in Fig. 6 when burned at temperatures above 400°C). Place approximately 1–2 g crushed bone in a 15-mL conical tube. Add 15 mL of ANDE Bone Solution and 200 μL of 20 mg/mL Proteinase K. Place the tube in an incubated rotator (e.g., a Benchmark Scientific Roto-Therm) set to 56°C and 12 rpm overnight. Centrifuge the tube (at least at 3000 rcf) for 1 minute to pellet any remaining bone particulates. Concentrate the supernatant to approximately 50–100 μL using a 10K Amicon Ultra-0.5-mL Centrifugal Filter Unit (MilliporeSigma, Burlington, MA). Pipette the entire concentrated sample onto an ANDE swab for Rapid DNA processing.

Processing old and degraded bone

Bones that appear to be extensively degraded (regardless of apparent PMI) or those that have been exhumed from mass graves should be processed using the same protocols previously described for burned bones (specifically Protocol #3 or Protocol #4). It is noted that the DNA content within a given bone sample is unlikely to be homogenous and, accordingly, it is recommended to prepare and process highly degraded samples in duplicate (when possible) to increase likelihood of success.

Processing demineralized bone and tooth samples in the ANDE Rapid DNA Identification System

Fig. 7 summarizes the overall sample processing workflow for bone and tooth samples for Rapid DNA Identification. Once the bone has been demineralized, undissolved bone particulates can be separated from the solution via a brief centrifugation. The solution is then directly pipetted onto an ANDE swab. Swabs are stored in a desiccant-containing protective tube until processing, and each swab cap contains a radio-frequency identification (RFID) tag for sample tracking. To perform a run, up to four ANDE swabs with the loaded solution are inserted into the swab chambers of an I-Chip. Note that when fewer than four samples are processed, blank swabs are inserted into the unused chambers. The I-Chip is then inserted into the ANDE instrument. All processes are performed within the I-Chip and the instrument without human intervention. Lastly, the Expert System software automatically processes the raw data, designates STR alleles, and employs a set of analytical rules to interpret the STR profiles. Automated processing and interpretation lead to DNA ID generation in 106 minutes (Turingan et al., 2019, 2020).

Representative results from Rapid DNA processing of degraded bones and teeth

Above-ground exposed and refrigerated remains

A collaborative study between ANDE and the Forensic Anthropology Center (FAC) at the University of Tennessee evaluated the use of Rapid DNA identification from exposed human

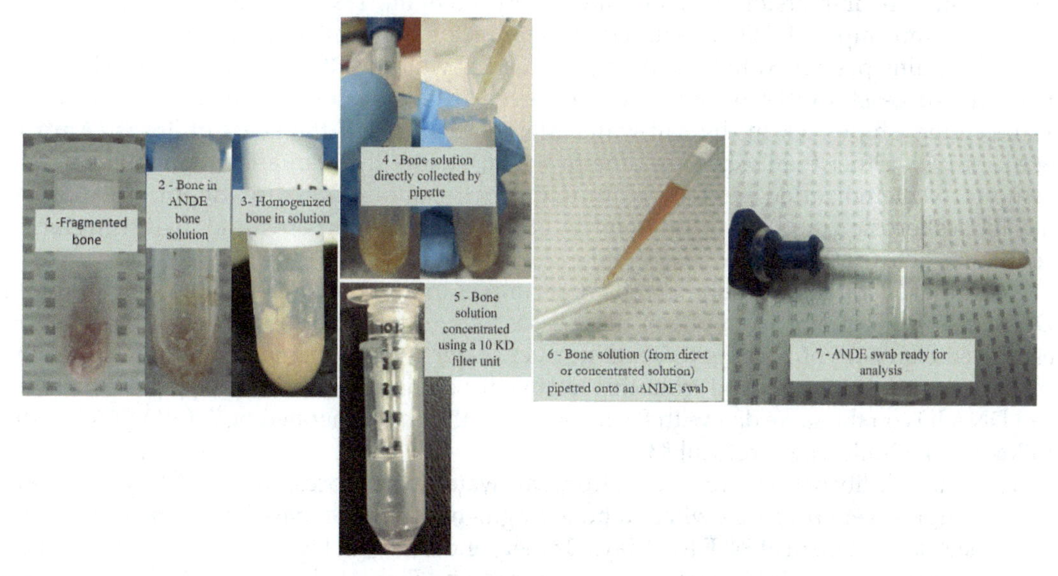

FIG. 7 Simplified workflow for processing bone samples on the ANDE Rapid DNA Identification System (after clean-up and fragmentation). Note that for some specimens (e.g., those that are fresh or which exhibit minimal degradation), sample concentration is not necessary.

bodies placed above-ground or stored in a refrigerated morgue; two scenarios commonly encountered following mass disasters (Turingan et al., 2019). Deceased human subjects were placed above-ground at the Anthropological Research Facility (a.k.a. "body farm"), a natural outdoor laboratory on a forested bluff above the Tennessee River. One set of cadavers were exposed outdoors for 1 year and another set refrigerated for 3 months, with longitudinal sampling throughout the decomposition process. The goal of this work was to provide guidance to first responders as to which sample types (e.g., buccal, muscle, brain, tooth, bone) represent the best combination of ease-of-collection and highest likelihood of generating useful DNA IDs using the ANDE Rapid DNA System.

Results demonstrated that for exposed remains, buccal swabs are the sample of choice for up to 11 days postexposure; bone and tooth samples generated excellent DNA IDs for the entire 1-year duration of the study. Bones submitted for Rapid DNA analysis included femora, ribs, and phalanges. Phalanges were superior to femur samples in terms of quality of DNA IDs generated; ribs were the least preferred bone type for successful typing. For the refrigerated remains, all sample types generated excellent DNA IDs. Figs. 8 and 9 display DNA IDs generated from femur and phalanx samples (respectively) from the same cadaver; DNA IDs were generated from samples collected at Day 1 and Month 12 of body placement.

Submerged remains

Human remains can be recovered in freshwater or salt-water following plane crashes, boats capsizing, tsunamis, accidental drownings, and criminal acts. Histological changes in bone structures when submerged in water have been previously described (Vacchiano and Vyshka, 2015). Successful DNA typing has been reported for bones recovered after up to 18 months immersion in river water (D'Errico et al., 1998; Hochmeister et al., 1991), and use of an improved DNA extraction method for bone led to a successful identification of remains presumed to be submerged in river water for 3 years (Crainic et al., 2002). Recently, successful DNA recovery was achieved from skeletal remains recovered from a 17th-century shipwreck in the salt-water Matagorda Bay off the coast of Texas (Ambers et al., 2019a). Other than exposure time in water, factors that can affect bone density and complete skeletonization are temperature and pH, osmotic pressure, depth of exposure, mechanical erosion, and the diluting effect of the water that favors rapid disintegration of cellular structures (Vacchiano and Vyshka, 2015).

Recently, several cases of submerged bone samples in water have been submitted for Rapid DNA analysis. Remains from one of these cases were reported to have been found after an estimated exposure of months-to-years in river water. Fig. 10 includes photographs of the phalanges and tibia recovered from this freshwater environment. The phalanges generated full DNA IDs on the same day with Protocol #2 and the tibia generated near-full DNA IDs the following day following Protocol #3.

To test the ability of bones recovered from salt-water to be processed by the ANDE System, mock samples were prepared wherein bone fragments were incubated in a sample of water from the Atlantic Ocean at 84°F for 4 days, 1 week, and 1 month. Fig. 11A shows a photograph of a bone fragment soaked in seawater for 1 month, and Fig. 11B displays the same bone fragment after it had been cleaned, dried, and prepared for fragmentation and analysis. Protocol #2 was utilized, and all samples generated full DNA IDs.

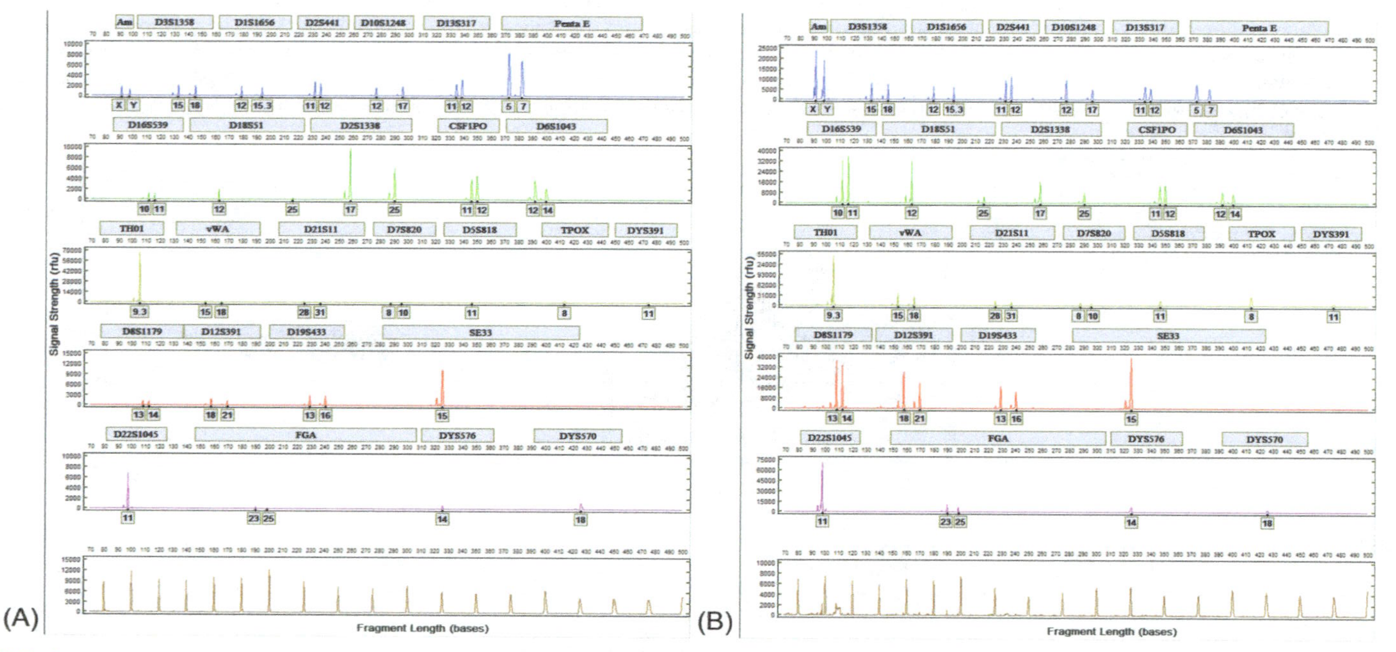

FIG. 8 Full DNA IDs resulting from processing a femoral section collected (A) at Day 1 and (B) after 12-months' exposure. For the Day 1 sample, 5 mg of hammered bone was processed by adding 120 μL of ANDE Bone Solution, mixing well, incubating for 1 minute at room temperature, and then transferring 15 μL of the solution onto an ANDE swab for Rapid DNA Identification. For the 12-month exposure sample, 500 mg of hammered bone was processed by adding 1.4 mL of ANDE Bone Solution and 70 μL of 20 mg/mL Proteinase K, agitating for 90 minutes at 56°C, and incubating overnight for demineralization. The solution was then concentrated and transferred onto an ANDE swab for Rapid DNA Identification.

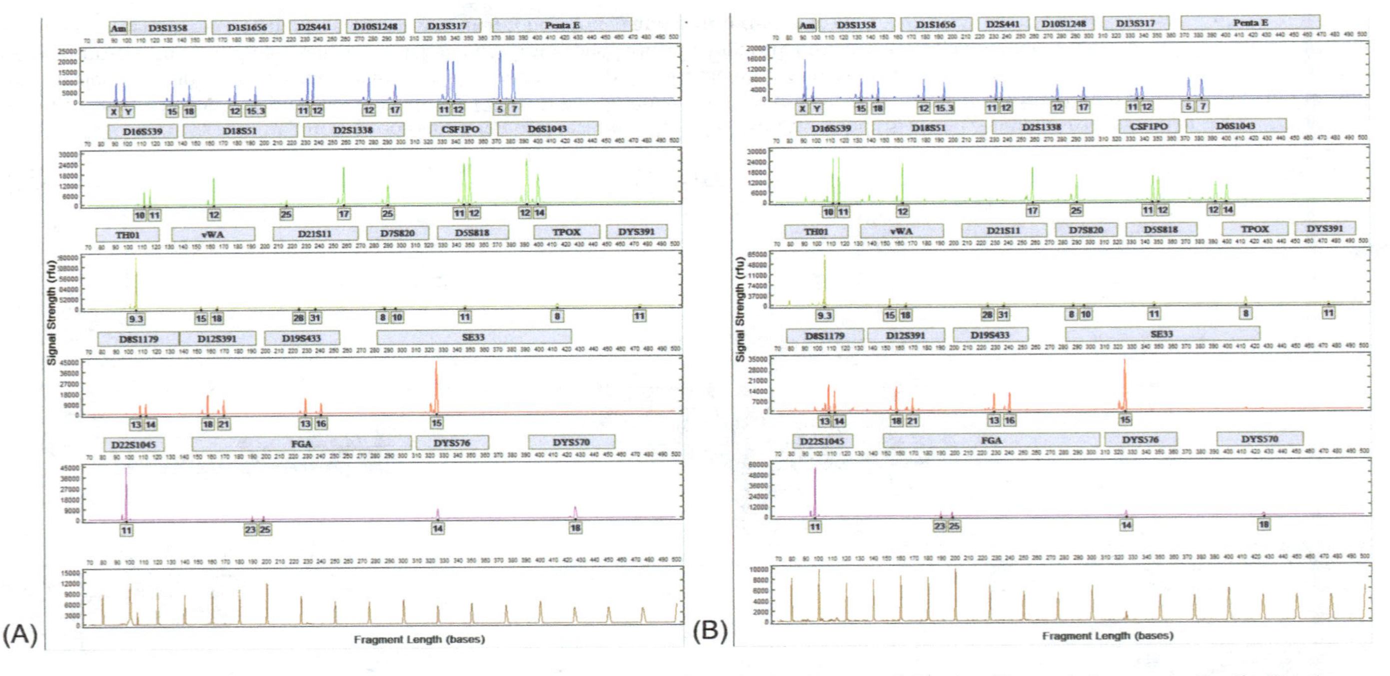

FIG. 9 Full DNA IDs resulting from processing a distal foot phalanx collected (A) at Day 1 and (B) after 12-months' exposure. For the Day 1 sample, 5 mg of hammered bone was processed by adding 120 μL of ANDE Bone Solution, mixing well, incubating for 1 minute at room temperature, and then transferring 15 μL of the solution onto an ANDE swab. For the 12-month exposure sample, 500 mg of hammered bone was processed by adding 1.4 mL of ANDE Bone Solution and 70 μL of 20 mg/mL Proteinase K, agitating for 90 minutes at 56°C, and then incubating overnight for demineralization. The solution was then concentrated and transferred onto an ANDE swab for Rapid DNA Identification.

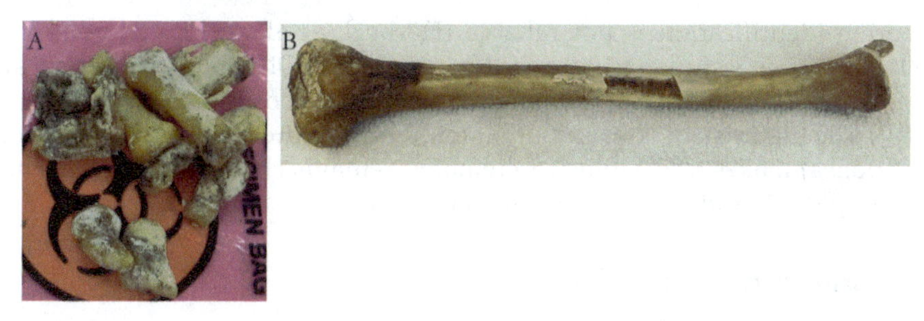

FIG. 10 Sample intake photographs of bones recovered from a freshwater river: (A) phalanges, and (B) tibia.

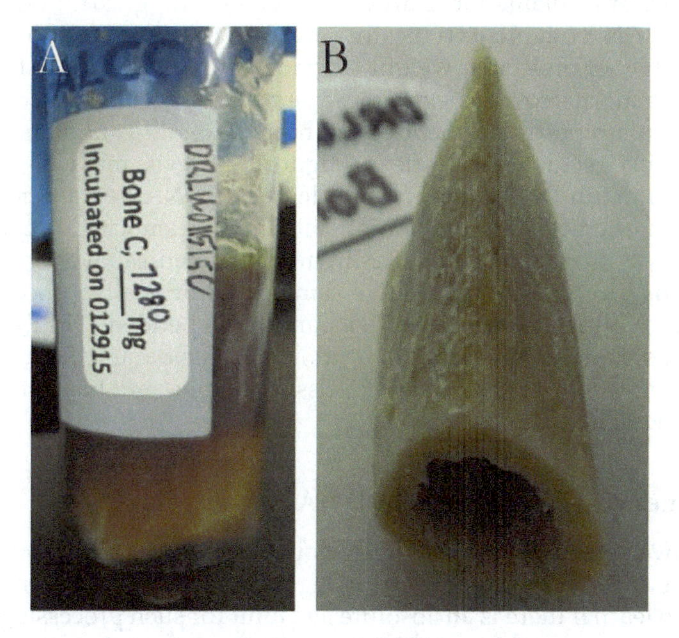

FIG. 11 (A) Mock decomposition of bone by soaking in seawater from the Atlantic Ocean for 1 month at 84°C; and (B) the same bone after cleaning in preparation for Rapid DNA analysis.

Cremated remains (cremains)

Cremation usually takes place in a chamber subjected to over 1000°C for at least an hour until the remains are reduced to ashes (i.e., until the remains are comprised of severely charred, brittle, and calcified bones and teeth). To resolve criminal cases and establish kinship, there is an occasional need to perform genetic analysis of cremated remains. Studies of DNA preservation and stable isotope values at varying heat exposure have been performed (Harbeck et al., 2011). Amplifiable DNA was reported to be obtained even after exposure to temperatures of 600°C and 700°C. The duration of exposure plays an important role on the integrity of biological elements inside bone structures, and although genetic analyses (e.g., STR typing) of cremains indicate the presence of amplifiable DNA, postcremation identification is not reliable (Harbeck et al., 2011; von Wurmb-Schwark et al., 2004). In fact, in a study of 10 corpses, precremation DNA IDs from buccal swabbings of cadavers did not match the postcremation STR peaks observed (von Wurmb-Schwark et al., 2004). The high risk of contamination (either from direct handling or during preprocessing of bones) makes STR typing of cremated remains challenging. ANDE has not yet attempted Rapid DNA Identification using cremains.

Formalin/formaldehyde-treated remains

DNA typing of embalmed cadavers is particularly challenging as embalming chemicals (aqueous formaldehyde and glutaraldehyde) cause DNA fragmentation and also reduce the quantity of DNA available for analysis (Gielda and Rigg, 2017; Wheeler et al., 2017). After embalming, DNA fragments typically range between 50 and 300 bp in length, making amplification of higher molecular weight fragments impossible. One study reported that, postembalming, bone marrow is the preferred sample type for STR success (Wheeler et al., 2017). Other sample types recommended for human identification are samples obtained from gastrocnemius, rectus femoris, flexor digitorum brevis, masseter, and brachioradialis muscles. Variation in DNA quantity and degree of DNA degradation can result from differential and unpredictable perfusion of embalming chemicals. Accordingly, sampling marrow and muscle from a range of sites in an embalmed cadaver should be attempted to maximize efforts to find an informative sample. Additionally, procedures for improving DNA purification from formalin-fixed paraffin-embedded (FFPE) biological tissues or embalmed bodies do exist (Sarnecka et al., 2019; Gielda and Rigg, 2017). ANDE has successfully generated DNA IDs from FFPE tissue samples, but has not yet attempted this approach on fixed bone samples.

The oldest bones processed by Rapid DNA

As noted above, age is not the only factor or even the dominant factor in determining if a bone sample can generate a DNA ID using Rapid DNA processing. However, ANDE has attempted to learn if there is an absolute age limit for such processing, beyond which a DNA ID cannot be generated. The ANDE Rapid DNA platform has successfully generated useful DNA IDs from decades-old bones in skeletal teaching collections, as well as from bones in medical office cold case files. Additionally, success has been obtained from bones and teeth from remains from the Vietnam War (>50 years old). Most recently, the ANDE 6C Rapid DNA system was used to successfully generate DNA IDs from skeletal remains recovered from the *La Belle* shipwreck (from the French explorer La Salle's last expedition); *La Belle* sank 315 years ago in the Gulf of Mexico (Matagorda Bay) on the Texas coast. Massively parallel sequencing (MPS) was previously successfully conducted on skeletonized remains recovered from *La Belle*, using Verogen's ForenSeq panel and a MiSeq FGx Desktop Sequencer (Illumina) (Ambers et al., 2019a). Data for autosomal STR loci common between the ForenSeq panel and ANDE's FlexPlex assay were concordant (Ambers et al., manuscript in preparation). All of these samples were processed on the ANDE System following Protocol #3 and Protocol #4. Accordingly, the maximum age limit or PMI for Rapid DNA Identification is not known.

Developmental validation of Rapid DNA Identification for bones and teeth

ANDE has previously published a comprehensive developmental validation for the processing of casework and disaster victim identification (DVI) samples using I-Chips on the ANDE Rapid DNA Identification system (Turingan et al., 2020). A total of 1705 samples were evaluated, including (1) blood, (2) oral epithelial samples from drinking containers, (3) samples on FTA and untreated paper, (4) semen, (5) bones/teeth, and (6) soft tissues. The results of this study demonstrated that the automated Expert System performs at least as well as conventional laboratory data analysis; additionally, Rapid DNA analysis demonstrated accuracy, precision, resolution, concordance, and reproducibility that were comparable or better than conventional processing (along with appropriate testing involving species specificity, limits of detection, performance in the presence of inhibitors). No lane-to-lane or run-to-run contamination was observed, and the system correctly identified the presence of mixtures. Collectively, the ANDE instrument, I-Chip consumable, FlexPlex chemistry, and automated Expert System successfully processed and interpreted more than 1700 samples with over 99.99% concordant CODIS alleles. Among this sample set, 18 bones and 3 tooth samples were processed. Recently, the number of bone and tooth samples was expanded to 100.

Reproducibility

Several 5 mg samples from a fresh femur (less than 1-week PMI, stored frozen) were processed. Two 15 μL aliquots (1 and 2) obtained from two different tubes (A and B) of stock 5 mg sample were analyzed (i.e., four samples equivalent to one full I-Chip run) per instrument; this work was performed in triplicate on three independent instruments to monitor intra-chip (Fig. 12) and intra-run (Fig. 13) reproducibility. Fig. 13 shows representative profiles from two different 5 mg preparations. All samples generated full and concordant profiles. Signal strength (peak heights) across all loci and peak height ratios (PHR) for heterozygote loci were generally comparable in all samples, demonstrating the reproducibility of the system.

Accuracy and concordance

A total of 100 independent samples (fresh and aged samples) from 28 different donors were processed and analyzed. Concordance was determined by comparing Rapid DNA results to those generated by conventional laboratory processing. Conventional processing utilized the PowerPlex Fusion 6C and PowerPlex 21 assays (Note: the Fusion 6C panel does not include the D6S1043 locus). All Rapid DNA IDs generated were concordant with DNA IDs generated from conventional assay.

The accuracy of the ANDE system was evaluated by generating profiles for these 100 bone samples. A sample was classified as "passing" if it generated calls for 12 or more CODIS core or FlexPlex 27 loci, all processed using the ANDE Expert System. Of the 100 samples tested, 96 samples gave DNA IDs with 12-27 CODIS core loci on the first pass, for an overall 96% first pass success rate.

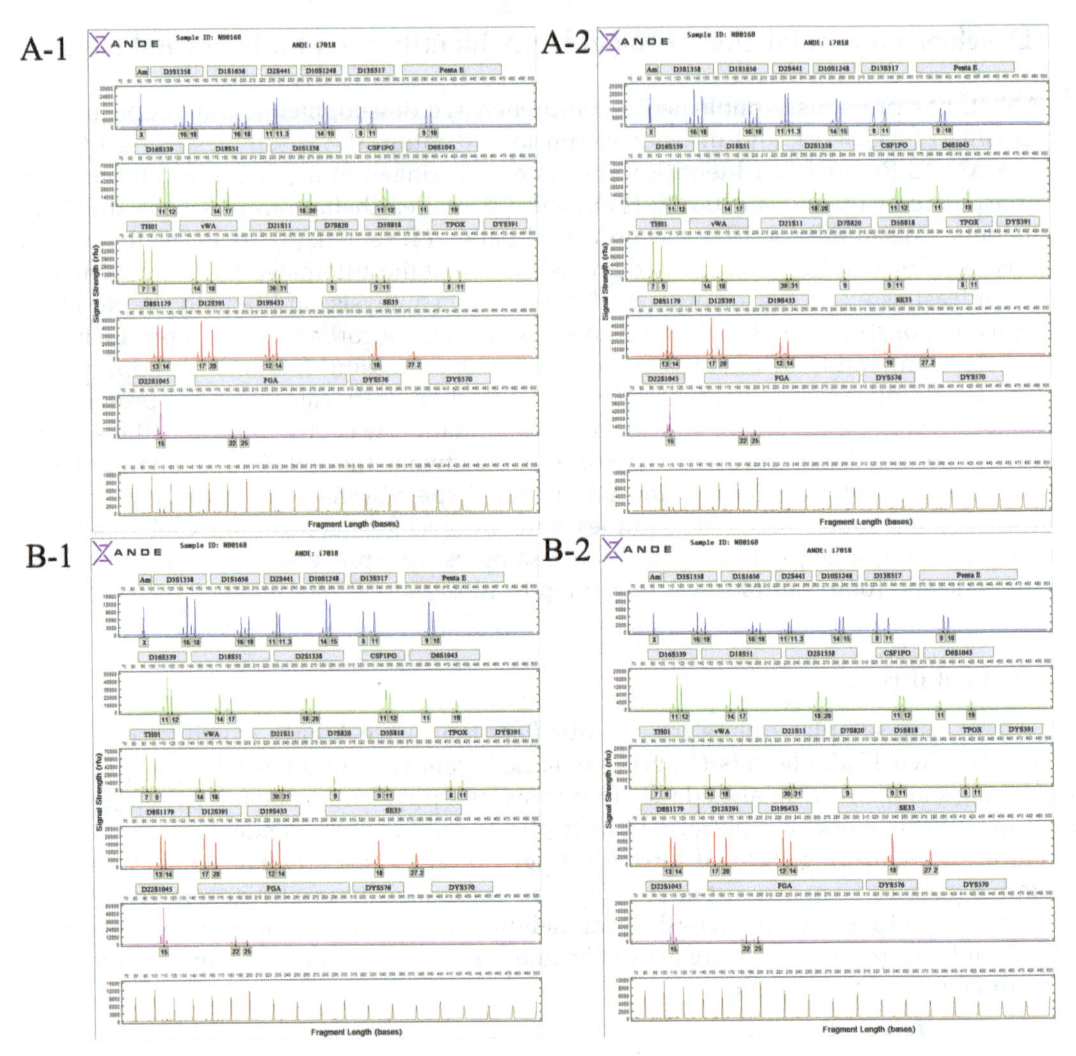

FIG. 12 Full DNA IDs generated from processing 5mg of hammered fresh femur, after brief demineralization in 120μL of ANDE Bone Solution. 15μL of the solution was added onto each ANDE swab for Rapid DNA Identification to assess intra-run reproducibility. Samples A-1 and A-2 were from same tube of 5mg hammered femur; Samples B-1 and B-2 were from a different tube of 5mg hammered femur.

Peak height and Peak Height Ratios (PHRs)

The heterozygous peak height for each locus was calculated for the 100 bone DNA IDs generated by summing the peak height of all called alleles within the locus and dividing by two. The average heterozygous peak height ranged from 2018 RFU at DYS570 to 68,567 RFU at TH01 (Fig. 14). The heterozygous peak height ratio (PHR) for each locus was calculated for each of the 100 DNA IDs by dividing the height of the smaller peak by the height of the larger peak. The average peak height ratio ranged from 0.661 at SE33 to 0.839 at TH01 (Fig. 15).

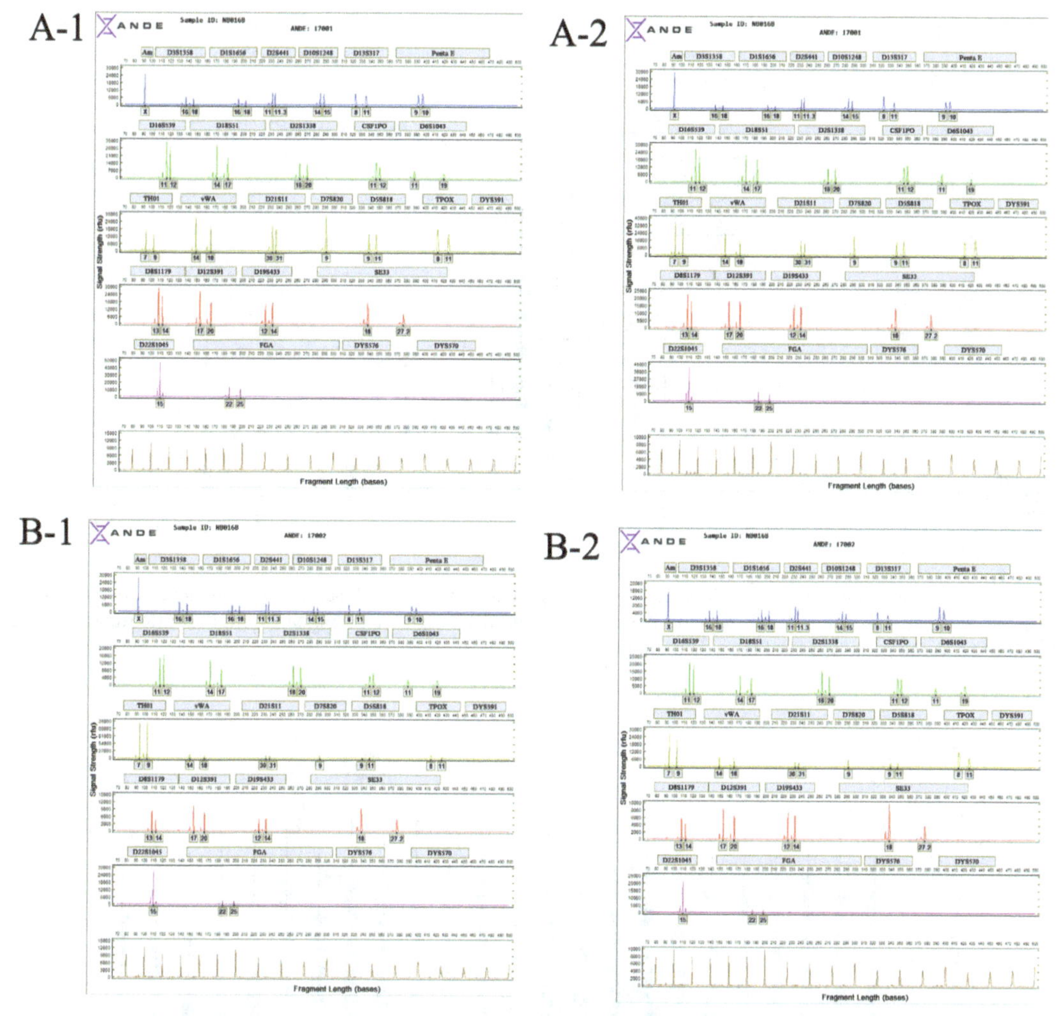

FIG. 13 Representative full DNA IDs generated from processing 5 mg hammered fresh femur (from the same donor shown in Fig. 12), after brief demineralization in 120 μL of ANDE Bone Solution. 15 μL of the solution was added onto each ANDE swab for Rapid DNA analysis using two different instruments. These instruments were different from the systems used to generate the data shown in Fig. 12. Samples 1 and 2 were 15 μL aliquots from different tubes of 5 mg femur samples.

Considered in concert, these data show that the ANDE 6C System has a wide dynamic range to successfully process fresh and aged bone samples.

Resolution and precision

The ability to resolve two adjacent fragments across the STR sizing range was calculated for all 100 independent bone samples. Resolution for the DNA IDs with passing ILS was calculated as described in Luckey et al. (1993) with R (resolution) values ≥ 0.2 bases indicating

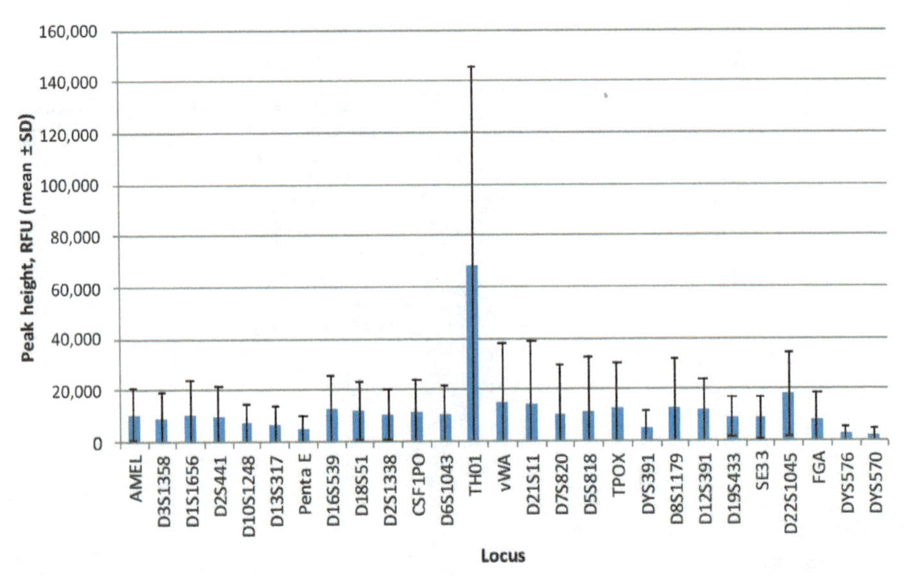

FIG. 14 Peak height analysis for 100 bone samples. Signal strength for autosomal loci was determined by summing the signal strengths for all called peaks in each locus and dividing by two. Signal strength for hemizygous loci was defined as the peak height of the called allele.

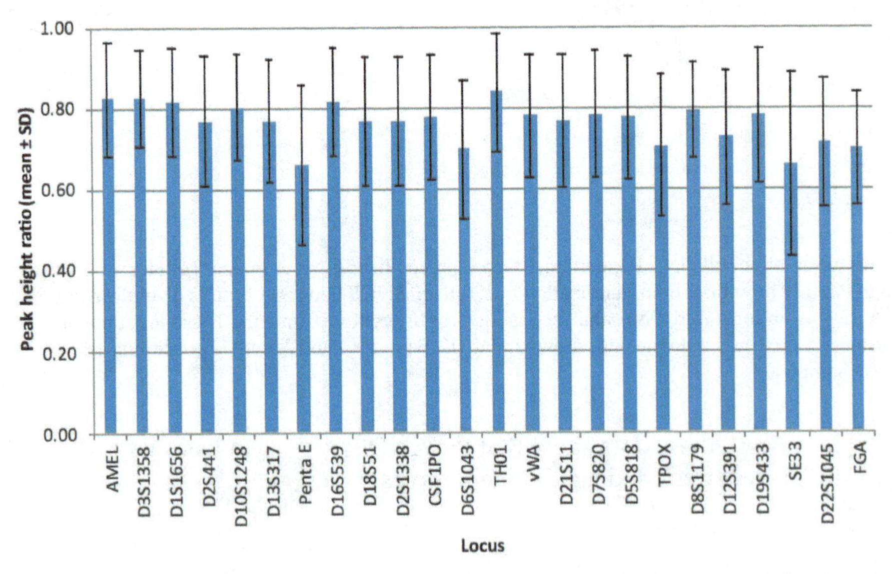

FIG. 15 Heterozygous peak height ratio (PHR) analysis for 100 bone samples. Average peak height ratio by locus is displayed.

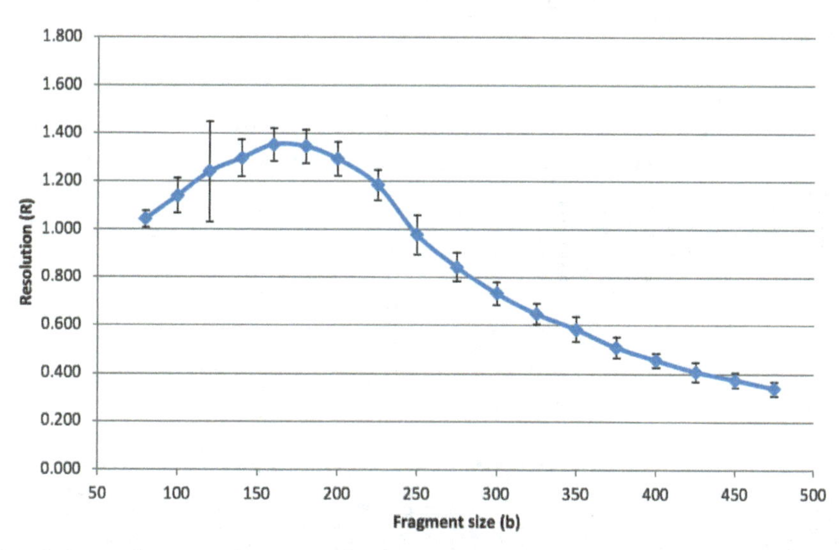

FIG. 16 Resolution analysis. Resolution for all 100 passing samples with R (resolution) values ≥ 0.2 indicating single base pair resolution.

single base resolution. Resolution calculated based on these samples shows that the system is capable of resolving fragments that are separated by 1 base across the separation range to greater than 500 bases (Fig. 16).

Interrun precision was calculated by determining the standard deviation of the fragment sizes (in bases) of each of the allelic ladder fragments for the 49 passing allelic ladders. The standard deviation in bases ranged from 0.008 bases at D7S820 to 0.064 bases at DYS570 (Fig. 17). The variation at three standard deviations ranges from 0.024 bases to 0.192 bases and is well within the 0.5 base limits for STR analysis. Taken together, the developmental validation data demonstrate that the ANDE 6C system meets the requirements for the generation of STR profiles from bones and teeth.

Conclusions

The impact of Rapid DNA on the ability to process bone and tooth samples in a broad array of applications cannot be overstated. The reasons for conventional laboratories having traditionally avoided bone processing are eliminated by the automated instrument, consumable chip, and analytic software of the ANDE Rapid DNA Identification System. The system's advantages can no longer be considered theoretical as it has been routinely demonstrated in evidentiary and DVI settings by nontechnical operators. These successes are unsurprising in light of previous developmental validation data in general and by the extensive developmental validation data presented herein. Rapid DNA Identification is ushering in a new era of dramatically enhanced utility for bone and tooth samples in day-to-day law enforcement, military, and DVI applications.

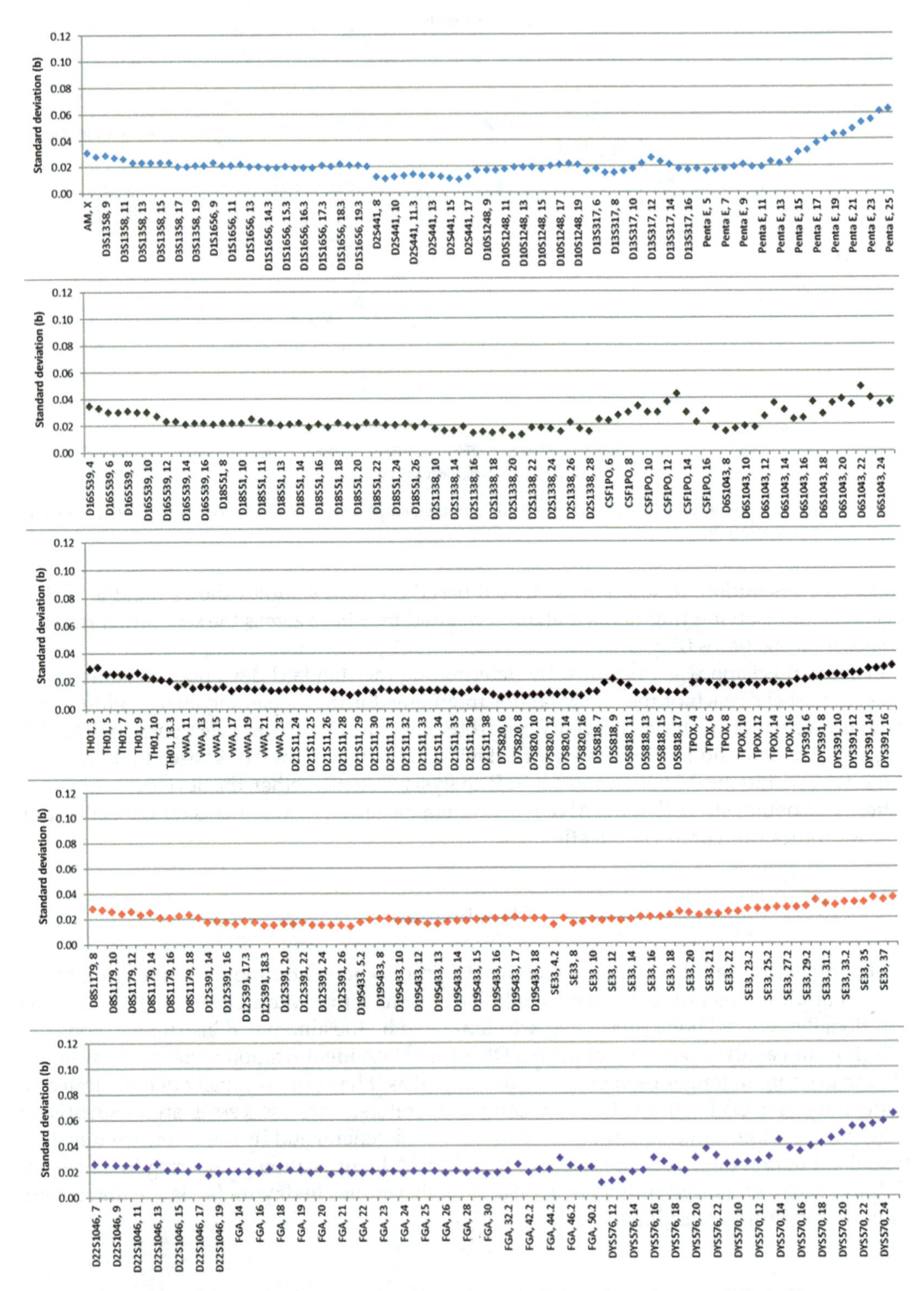

FIG. 17 Sizing variation at a single standard deviation for each allele with passing run allelic ladders.

References

Alonso, A., Andelinovic, S., Martín, P., Sutlovic, D., Erceg, I., Huffine, E., Fernández de Simón, L., Albarrán, C., Definis-Gojanovic, M., Fernández-Rodriguez, A., García, P., Drmic, I., Rezic, B., Kuret, S., Sancho, M., Primorac, D., 2001. DNA typing from skeletal remains: evaluation of multiplex and megaplex STR systems on DNA isolated from bone and teeth samples. Croat. Med. J. 42 (3), 260–266.

Ambers, A., Gill-King, H., Dirkmaat, D., Benjamin, R., King, J., Budowle, B., 2014. Autosomal and Y-STR analysis of degraded DNA from the 120-year-old skeletal remains of Ezekiel Harper. Forensic Sci. Int. Genet. 9, 33–41.

Ambers, A., Churchill, J.D., King, J.L., Stoljarova, M., Gill-King, H., Assidi, M., Abu-Elmagd, M., Buhmeida, A., Budowle, B., 2016. More comprehensive forensic genetic marker analyses for accurate human remains identification using massively parallel sequencing (MPS). BMC Genomics 17 (Suppl 9), 750.

Ambers, A., Vanek, D., Votrubova, J., Sajantila, A., Budowle, B., 2018. Improved Y-STR typing for disaster victim identification, missing persons investigations, and historical human skeletal remains. Int. J. Leg. Med. 132 (6), 1545–1553.

Ambers, A., Bus, M.M., King, J.L., Jones, B., Durst, J., Bruseth, J.E., Gill-King, H., Budowle, B., 2019a. Forensic genetic investigation of human skeletal remains recovered from the *La Belle* shipwreck. Forensic Sci. Int. 306, 110050.

Ambers, A., Votrubova, J., Zeng, X., Vanek, D., 2019b. Enhanced interrogation of degraded DNA from human skeletal remains: increased genetic data recovery using the expanded CODIS loci, multiple sex determination markers, and consensus testing. Anthropol. Anz. 76 (4), 333–351.

Amory, S., Huel, R., Bilic, A., Loreille, O., Parsons, T.J., 2012. Automatable full demineralization DNA extraction procedure from degraded skeletal remains. Forensic Sci. Int. Genet. 6 (3), 398–406.

Budowle, B., 1997. Studies for selecting core STR loci for CODIS. In: Proceedings of Cambridge Health Tech Institute's Second Annual Conference on DNA Forensics: Science, Evidence, and Future Prospects. McLean, VA.

Carney, C., Whitney, S., Vaidyanathan, J., Persick, R., Noel, F., Vallone, P.M., Romsos, E.L., Tan, E., Grover, R., Turingan, R.S., French, J.L., Selden, R.F., 2019. Developmental validation of the ANDE™ rapid DNA system with FlexPlex™ assay for arrestee and reference buccal swab processing and database searching. Forensic Sci. Int. Genet. 40, 120–130.

Crainic, K., Paraire, F., Lettereux, M., Durigon, M., de Mazancourt, P., 2002. Skeletal remains presumed submerged in water for three years identified using PCR-STR analysis. J. Forensic Sci. 47 (5), 1–3.

D'Errico, G., Zanon, C., Lago, G., Vecchio, C., Garofano, L., 1998. DNA analysis on skeletal remains after months of permanence in sea-water. In: Progress in Forensic Genetics No. 7. Elsevier, Amsterdam, pp. 106–108.

Della Manna, A., Nye, J.V., Carney, C., Hammons, J.S., Mann, M., Al Shamali, F., Vallone, P.M., Romsos, E.L., Marne, B.A., Tan, E., Turingan, R.S., Hogan, C., Selden, R.F., French, J.L., 2016. Developmental validation of the DNAscan™ rapid DNA analysis™ instrument and expert system for reference sample processing. Forensic Sci. Int. Genet. 25, 145–156.

Deng, Y.-J., Li, Y.-Z., Yu, X.G., Li, L., Wu, D.-Y., Zhou, J., Man, T.-Y., Yang, G., Yan, J.-W., Cai, D.-Q., Wang, J., Yang, H.-M., Li, S.B., Yu, J., 2005. Preliminary DNA identification for the tsunami victims in Thailand. Genomics Proteomics Bioinformatics 3 (3), 143–157.

Donkervoort, S., Dolan, S.M., Beckwith, M., Pruet Nothrup, T., Sozer, A., 2008. Enhancing accurate data collection in mass fatality kinship identifications: lessons learned from hurricane Katrina. Forensic Sci. Int. Genet. 2 (4), 354–362.

Edson, S.M., Roberts, M., 2019. Determination of materials present in skeletonized human remains and the associated DNA: development of a GC/MS protocol. Forensic Sci. Int. Synergy 1, 170–184.

Edson, S.M., Ross, J.P., Coble, M.D., Parsons, T.J., Barritt, S.M., 2004. Naming the dead—confronting the realities of rapid identification of degraded skeletal remains. Forensic Sci. Rev. 16 (1), 63–90.

Edwards, A., Civitello, A., Hammond, H.A., Caskey, C.T., 1991. DNA typing and genetic mapping with trimeric and tetrameric tandem repeats. Am. J. Hum. Genet. 49 (4), 746–756.

Ellingham, S., 2018. Fire Destruction of the Body and Biochemical and Structural Changes to the Bone Matrix. American Academy of Forensic Sciences (AAFS), Seattle, WA.

Gielda, L., Rigg, S., 2017. Extraction of amplifiable DNA from embalmed human cadaver tissue. BMC. Res. Notes 10 (1), 1–5.

Gin, K., Tovar, J., Bartelink, E.J., Kendell, A., Milligan, C., Willey, P., Wood, J., Tan, E., Turingan, R.S., Selden, R.F., 2020. The 2018 California wildfires: integration of rapid DNA to dramatically accelerate victim identification. J. Forensic Sci. 65 (3), 791–799.

Grover, R., Jiang, H., Turingan, R.S., French, J.L., Tan, E., Selden, R.F., 2017. FlexPlex27—highly multiplexed rapid DNA identification for law enforcement, kinship, and military applications. Int. J. Leg. Med. 131 (6), 1–13.

Harbeck, M., Schleuder, R., Schneider, J., Wiechmann, I., Schmahl, W.W., Grupe, G., 2011. Research potential and limitations of trace analyses of cremated remains. Forensic Sci. Int. 204 (1–3), 191–200.

Hartman, D., Drummer, O.H., Eckhoff, C., Scheffer, J.W., 2011. The contribution of DNA to the disaster victim identification (DVI) effort. Forensic Sci. Int. 205 (1–3), 52–58.

Hayatsu, H., Pan, S., Ukita, T., 1971. Reaction of sodium hypochlorite with nucleic acids and their constituents. Chem. Pharm. Bull. 19 (10), 2189–2192.

Hochmeister, M.N., Budowle, B., Borer, U.V., Eggmann, U., Comey, C.T., Dirnhofer, R., 1991. Typing of deoxyribonucleic acid (DNA) extracted from compact bone from human remains. J. Forensic Sci. 36 (6), 1649–1661.

Holland, M.M., Cave, C.A., Holland, C.A., Bille, T.W., 2003. Development of a quality, high throughput DNA analysis procedure for skeletal samples to assist with the identification of victims from the World Trade Center attacks. Croat. Med. J. 44 (3), 264–272.

Hsu, C.M., Huang, N.E., Tsai, L.C., Kao, L.G., Chao, C.H., Linacre, A., Lee, J.C., 1999. Identification of victims of the 1998 Taoyuan airbus crash accident using DNA analysis. Int. J. Leg. Med. 113 (1), 43–46.

Jeffreys, A.J., Wilson, V., Thein, S.L., 1985. Hypervariable 'minisatellite' regions in human DNA. Nature 314 (6006), 67–73.

Kemp, B.M., Smith, D.G., 2005. Use of bleach to eliminate contaminating DNA from the surface of bones and teeth. Forensic Sci. Int. 154 (1), 53–61.

Krings, M., Stone, A., Schmitz, R.W., Krainitzki, H., Stoneking, M., Paabo, S., 1997. Neandertal DNA sequences and the origin of modern humans. Cell 90 (1), 19–30.

Leclair, B., Fregeau, C.J., Bowen, K.L., Fourney, R.M., 2004. Enhanced kinship analysis and STR-based DNA typing for human identification in mass fatality incidents: the Swissair flight 111 disaster. J. Forensic Sci. 49 (5), 939–953.

Lee, E.J., Leudtke, J.G., Allison, J.L., Arber, C.E., Merriwether, D.A., Wolfe Steadman, D., 2010a. The effects of different maceration techniques on nuclear DNA amplification using human bone. J. Forensic Sci. 55 (4), 1032–1038.

Lee, H.Y., Park, M.J., Kim, N.Y., Sim, J.E., Yang, W.I., Shin, K.-J., 2010b. Simple and highly effective DNA extraction methods from old skeletal remains using silica columns. Forensic Sci. Int. Genet. 4 (5), 275–280.

Litt, M., Luty, J.A., 1989. A hypervariable microsatellite revealed by in vitro amplification of a dinucleotide repeat within the cardiac muscle actin gene. Am. J. Hum. Genet. 44 (3), 397.

Loreille, O.M., Diegoli, T.M., Irwin, J.A., Coble, M.D., Parsons, T.J., 2007. High efficiency DNA extraction from bone by total demineralization. Forensic Sci. Int. Genet. 1 (2), 191–195.

Luckey, J.A., Norris, T.B., Smith, L.M., 1993. Analysis of resolution in DNA sequencing by capillary gel electrophoresis. J. Phys. Chem. 97 (12), 3067–3075.

McEntire, D.A., Sadiq, A.A., Gupta, K., 2012. Unidentified bodies and mass-fatality management in Haiti: a case study of the January 2010 earthquake with a cross-cultural comparison. Int. J. Mass Emerg. Disasters 30 (1), 301–327.

Miloš, A., Selmanovic, A., Smajlovic, L., Huel, R.L.M., Katzmarzyk, C., Rizvic, A., Parsons, T.J., 2007. Success rates of nuclear short tandem repeat typing from different skeletal elements. Croat. Med. J. 48 (4), 486.

Moreno, L.I., Brown, A.L., Callaghan, T.F., 2017. Internal validation of the DNAscan/ANDE™ rapid DNA analysis™ platform and its associated PowerPlex® 16 high content DNA biochip cassette for use as an expert system with reference buccal swabs. Forensic Sci. Int. Genet. 29, 100–108.

Moretti, T., Budowle, B., 1998. The CODIS STR project: evaluation of fluorescent multiplex STR systems. In: Proceedings of the 50th Annual Meeting of the American Academy of Forensic Sciences. San Francisco, CA.

Mundorff, A., Davoren, J.M., 2014. Examination of DNA yield rates for different skeletal elements at increasing post mortem intervals. Forensic Sci. Int. Genet. 8 (1), 55–63.

Mundorff, A.Z., Bartelink, E.J., Mar-Cash, E., 2009. DNA preservation in skeletal elements from the World Trade Center disaster: recommendations for mass fatality management. J. Forensic Sci. 54 (4), 739–745.

Mundorff, A.Z., Davoren, J., Weitz, S., 2013. Developing an Empirically Based Ranking Order for Bone Sampling: Examining the Differential DNA Yield Rates between Human Skeletal Elements over Increasing Post Mortem Intervals. Department of Justice, Washington.

Pajnič, I.Z., 2016. Extraction of DNA from human skeletal material. In: Forensic DNA Typing Protocols, pp. 89–108.

Palmbach, T., Bloom, J., Hoynes, E., Primorac, D., Gaboury, M.T., 2014. Utilizing DNA analysis to combat the world-wide plague of present day slavery—trafficking in persons. Croat. Med. J. 55 (1), 3–9.

Parsons, T.J., Weedn, V.W., 1996. Preservation and recovery of DNA in postmortem specimens and trace samples. In: Advances in Forensic Taphonomy: The Fate of Human Remains. CPR Press, New York, pp. 109–138.

Rassmann, K., Schlotterer, C., Tautz, D., 1991. Isolation of simple-sequence loci for use in polymerase chain reaction-based DNA fingerprinting. Electrophoresis 12 (2–3), 113–118.

Sarnecka, A.K., Nawrat, D., Piwowar, M., Ligeza, J., Swadsba, J., Wajcik, P., 2019. DNA extraction from FFPE tissue samples—a comparison of three procedures. Contemp. Oncol. 23 (1), 52.

Sudoyo, H., Widodo, P.T., Suryadi, H., Lie, Y.S., Safari, D., Widjajanto, A., Aji Kadarmo, D., Hidayat, S., Marzuki, S., 2008. DNA analysis in perpetrator identification of terrorism-related disaster: suicide bombing of the Australian Embassy in Jakarta 2004. Forensic Sci. Int. Genet. 2 (3), 231–237.

Tan, E., Turingan, R.S., Hogan, C., Vasantgadkar, S., Palombo, L., Schumm, J.W., Selden, R.F., 2013. Fully integrated, fully automated generation of short tandem repeat profiles. Investigative Genet. 4 (1), 1–15.

Turingan, R.S., Vasantgadkar, S., Palombo, L., Hogan, C., Jiang, H., Tan, E., Selden, R.F., 2016. Rapid DNA analysis for automated processing and interpretation of low DNA content samples. Investigative Genet. 7 (1), 1.

Turingan, R.S., Brown, J., Kaplun, L., Smith, J., Watson, J., Boyd, D.A., Wolfe Steadman, D., Selden, R.F., 2019. Identification of human remains using rapid DNA analysis. Int. J. Leg. Med. 134, 863–872.

Turingan, R.S., Tan, E., Jiang, H., Brown, J., Estari, Y., Krautz-Peterson, G., Selden, R.F., 2020. Developmental validation of the ANDE 6C system for rapid DNA analysis of forensic casework and DVI samples. J. Forensic Sci. 65 (4), 1056–1071.

Uzair, A., Rasool, N., Wasim, M., 2017. Evaluation of different methods for DNA extraction from human burnt bones and the generation of genetic profiles for identification. Med. Sci. Law 57 (4), 159–166.

Vacchiano, G., Vyshka, G., 2015. Skeletal remains submerged in the Mediterranean Sea for eight years: histological observations. Anthropology 2 (137). https://doi.org/10.4172/2332-0915.1000137.

von Wurmb-Schwark, N., Simeoni, E., Ringleb, A., Oehmichen, M., 2004. Genetic investigation of modern burned corpses. Int. Congr. Ser. 1261, 50–52.

Weber, J.L., May, P.E., 1989. Abundant class of human DNA polymorphisms which can be typed using the polymerase chain reaction. Am. J. Hum. Genet. 44 (3), 388.

Wheeler, A., Czado, N., Gangitano, D., Turnbough, M., Hughes-Stamm, S., 2017. Comparison of DNA yield and STR success rates from different tissues in embalmed bodies. Int. J. Leg. Med. 131 (1), 61–66.

Whiteman, M., Hong, H.S., Jenner, A., Halliwell, B., 2002. Loss of oxidized and chlorinated bases in DNA treated with reactive oxygen species: implications for assessment of oxidative damage *in vivo*. Biochem. Biophys. Res. Commun. 296 (4), 883–889.

Yokoi, T., Aoki, Y., Sagisaka, K., 1989. Human identification and sex determination of dental pulp, bone marrow, and blood stains with a recombinant DNA probe. Z. Rechtsmed. 102 (5), 323–330.

CHAPTER

16

Emerging technologies for DNA analysis of challenged samples

Nicole Novroski PhD

Department of Anthropology, Forensic Science Program, University of Toronto Mississauga, Toronto, ON, Canada

Introduction

The forensic genetics community commonly relies on the polymerase chain reaction (PCR) coupled with capillary electrophoresis (CE) for forensic human identity testing of skeletal remains (Moretti et al., 2001). A common issue with highly decomposed or skeletonized human remains is the possibility for little-to-no DNA recovery following extraction procedures (Lorente et al., 2000). Current methodologies are limited in their application to (and success with) highly compromised samples. Recent technological and chemistry advances in the field of forensic genetics have drastically improved the sampling, preservation, and processing of samples for DNA profiling. Such improvements now allow for typing and characterization of a vast array of new genetic markers, coupled with improved genotyping and sequencing techniques. The result is a notable increase in the amount of genetic information that can be extracted from a single sample, which has led to improved accuracy in human identification methodologies, even when the starting material is degraded, otherwise environmentally insulted, or limited in quantity (Ambers et al., 2016, 2020).

Massively parallel sequencing

High-throughput DNA sequencing technologies have rapidly evolved over the past decade. New multiplexes and instrumentation are capable of producing large volumes of data at relatively low costs (compared to traditional DNA typing methodologies and platforms). Next-generation sequencing (NGS), second-generation sequencing (SGS), and massively parallel sequencing (MPS) all refer to the current generation of post-Sanger sequencing technologies being broadly adapted to various applications in evolutionary biology, molecular anthropology, metagenomics, medical genetics, epidemiology, and forensic genetics (Bentley, 2006; Mardis,

Forensic Genetic Approaches for Identification of Human Skeletal Remains
https://doi.org/10.1016/B978-0-12-815766-4.00016-9

2008; Voelkerding et al., 2009; Metzker, 2010; Glenn, 2011; Fordyce et al., 2011; Irwin et al., 2011; Quail et al., 2012; Van Neste et al., 2012; Tan et al., 2013; van Dijk et al., 2014; Ambers et al., 2016, 2020). MPS technologies utilize several high-throughput approaches to sequence DNA using a massively parallel processing concept which allows for both de novo characterization and re-sequencing of genomic targets for a variety of species (i.e., sequencing one or more segments of an individual's genome in order to investigate sequence differences between the individual and a reference or standard genome of the given species) (Kircher and Kelso, 2010; Goodwin et al., 2016; Bruijns et al., 2018). Since MPS offers a compendium of uses in clinical and applied genetics, there is obvious value in exploiting MPS chemistries and technologies for missing persons, disaster victim identification, and unidentified human remains purposes.

Current MPS technologies employ three main steps: (1) DNA sample preparation (commonly referred to as *library preparation*), using either a commercially available or custom multiplex/panel; (2) immobilization of DNA libraries; and (3) simultaneous sequencing of millions of DNA libraries in a massively parallel fashion (Liu et al., 2012; Goodwin et al., 2016; Bruijns et al., 2018). Library preparation describes the preparation of the DNA extract for sequencing. A vital step when preparing libraries involves the addition of known, defined sequences (commonly referred to as *adapters*) to the ends of the newly fragmented DNA (i.e., either random enzymatically fragmented via a process known as *tagmentation* or using specific primers to amplify only genomic fragments of interest). Once the adapters have successfully been added to each of the fragments from a single sample from an individual, the totality of fragments in each sample becomes known as a *library*. Addition of the adapters is necessary to ensure that the new DNA fragments, or libraries, can be anchored to a solid surface that ultimately defines the site in which the sequencing reactions will begin (i.e., *immobilization*). Prior to the sequencing reaction, most high-throughput sequencing systems require amplification of the sequencing library (i.e., DNA fragments) in order to create spatially distinct and detectable clusters of clonal DNA copies (i.e., identical DNA strands attached to the sequencing surface in bundles analogous to patches of land dedicated to growing specific crops). Sequencing then is performed for millions of DNA templates (as opposed to only 96 samples using CE) using one of many sequencing platforms that utilize either pyrosequencing, sequencing-by-synthesis, or ion semiconductor sequencing technologies (Ronaghi, 2001; Goodwin et al., 2016; Bruijns et al., 2018).

MPS applications provide a number of advantages over current DNA typing methods. Traditional PCR-CE for human identity typing is typically restricted in the following ways: (1) the total number of targets captured in a single assay is limited due to dye-channel and amplicon size restrictions; (2) lack of ability to type multiple targets in a single reaction (i.e., STRs and SNPs cannot be genotyped simultaneously); and (3) lack of ability to use a single platform for all genomic targets. In regard to the latter limitation, PCR-CE is currently used for STRs, single nucleotide polymorphisms (SNPs), and insertions/deletions (indels), while Sanger sequencing is currently the standard approach for characterizing the control region (specifically hypervariable regions I and II) of the mitochondrial genome (Kircher and Kelso, 2010; Børsting et al., 2014; Churchill et al., 2015; Larmuseau et al., 2015). An additional hindrance is the amount of potential biological material, sample, and/or resultant DNA extract available for testing, considering that many human skeletal remains cases present with limited DNA quantities or degraded templates (Ambers et al., 2016; Kulstein et al., 2017). In contrast to current CE-based approaches, MPS provides a substantial increase in throughput with regard to multiplex design, by allowing both a greater number of targets (i.e., well

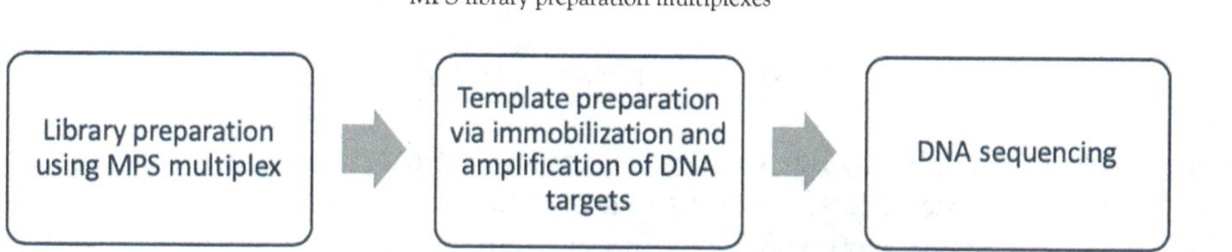

FIG. 1 Overview of the general workflow in massively parallel sequencing (MPS).

beyond the current constraint of 30–40 targets per CE multiplex) and a diverse combination of marker types (i.e., STRs, SNPs, Y- and X-chromosome markers) to be processed in the same reaction, while generating a significant depth of coverage for a relatively low cost-per-sample (Churchill et al., 2015, 2017a; Oldoni et al., 2018). Furthermore, MPS now affords the ability to sequence the entirety of the mitochondrial DNA (mtDNA) genome in a single assay, thereby reducing overall consumption of the already potentially limited DNA sample (King et al., 2014; Hickman et al., 2018; Strobl et al., 2019; Ambers et al., 2020).

While MPS technologies emerged as early as 1994–98 (Berglund et al., 2011), commercial availability of some of the currently used systems began around 2005 (Berglund et al., 2011; Goodwin et al., 2016; Bruijns et al., 2018). While each MPS instrument may differ in the chemistry and overall technological approach, all platforms share a common goal of sequencing (in parallel) a substantial number of DNA copies and generating millions-to-billions of sequencing reads per instrument run. Fig. 1 summarizes the basic workflow in massively parallel sequencing.

MPS library preparation multiplexes

Forensic MPS multiplexes allow for capture of hundreds-to-thousands of unique genomic targets simultaneously in a single sequencing run. Many multiplexes utilize a PCR-based approach for amplifying the genomic targets of interest. However, some genomic targets will exceed instrument sequencing chemistry and/or technological capabilities. Therefore, depending on the sequencing platform being used, genomic amplicons of interest may (1) be captured in a single sequencing read (~ 300 nucleotides or less), or (2) larger DNA fragments will likely undergo tagmentation (enzymatic fragmentation) and bioinformatic reassembly or targeted via the use of multiple primer sets (as in the case of the mtDNA genome) (Goodwin et al., 2016; Bruijns et al., 2018).

During the library preparation workflow, DNA fragments of interest undergo a series of molecular biology manipulations to prepare each fragment of template DNA for sequencing. Indexes are often added to each fragment, which serve as unique barcodes for each piece of DNA (where all fragments from the same sample will have the same barcode) (Smith et al., 2010). In other words, a single well of a 96-well plate contains the fragmented and indexed DNA from a single individual. Further, adapters are ligated to the ends of each fragment, which allows for the DNA fragments to become fixed onto the DNA sequencing surface (i.e., during the immobilization step; typically a bead for emulsion PCR or to the flow cell during bridge amplification, discussed later in this chapter) (Syed et al., 2009; Head et al., 2014). There are additional checkpoints to ensure that each sample is purified and normalized, so that all fragments for all samples are (in theory) equally represented in the DNA sequencing reaction.

STR multiplexes for MPS

STR genotyping using MPS has drastically improved the usefulness and discrimination power of human microsatellites. MPS allows for the characterization of both length- and sequenced-based alleles at each STR locus in each sample; as such, a majority of STRs have revealed extensive sequenced-based SNP variation both within and outside of the repeat region that was not previously observable using length-based (CE) approaches. Thus, STR genotyping using MPS is generally considered more informative for most loci over current PCR-CE methods. In 2015, Gettings et al. reported one of the first comprehensive studies of sequence-based allele variation using MPS in forensic genetics (Gettings et al., 2015a). The study documented what was known about STR allelic sequence variation in (in the repeat region) and around (in the flanking region) 24 of the most commonly used loci in forensic genetic multiplexes. The STR loci included D1S1656, TPOX, D2S441, D2S1338, D3S1358, FGA, CSF1PO, D5S818, SE33, D6S1043, D7S820, D8S1179, D10S1248, TH01, vWA, D12S391, D13S317, Penta E, D16S539, D18S51, D19S433, D21S11, Penta D, and D22S1045; a subset of reported variant alleles were compiled by incorporating data from public databases such as GenBank, dbSNP, and the 1000 Genomes Project (Gettings et al., 2015a). In a subsequent study in 2016, Novroski et al. completed an extensive length- and sequenced-based characterization of the 59 STRs contained within the ForenSeq DNA Signature Prep Kit (Novroski et al., 2016). These findings demonstrated a remarkable increase in discrimination power for autosomal STRs when using alleles characterized by MPS. Further, this study highlighted that STRs tend to cluster within four main categories, in which sequence variation exists: (1) either in both the repeat region and the flanking region; (2) in the repeat region only; (3) in the flanking region only; or (4) not at all in the amplicon. These results identified that MPS-based approaches for all currently employed loci may not be valuable, and that additional non-CODIS markers may be necessary for increasing human identification efforts using MPS-based typing methods (Novroski et al., 2016). Novroski et al. subsequently created a custom PCR-based multiplex using the AmpliSeq for Illumina chemistry which allowed for the capture of 73 novel STRs with the increased diversity that may be suitable for future forensic assays where individualization between closely related individuals or commingled human remains is necessary (Novroski et al., 2018a,b).

Current commercial STR multiplexes include the ForenSeq DNA Signature Library Prep Kit (formerly Illumina, now manufactured by Verogen Inc., San Diego, California, United States), which encompasses a total of 230 genomic targets (58 of which are STRs, i.e., 27 autosomal STRs, 7 X-chromosome STRs, 24 Y-chromosome STRs) (Churchill et al., 2015, 2017a; Gettings et al., 2015a; Hussing et al., 2015; Novroski et al., 2016). An extensive compendium of literature exists for the ForenSeq Kit, where both a wide variety of panel applications and sample types have been evaluated using in-house software systems in addition to the commercial Universal Analysis Software (UAS) supplied with the MiSeq FGx Forensic Genomics System (Verogen Inc., San Diego, California, United States). The ForenSeq Kit, coupled with the MiSeq FGx Forensic Genomics System, is the first fully validated next-generation DNA system designed specifically for forensic genetic use and is currently the only MPS-based system approved for use in the National DNA Index System (NDIS) CODIS database in the United States (England and Harbison, 2019; Verogen, 2019). The MiSeq FGx Forensic Genomics System also has been approved for use in the national DNA databases of France and the Netherlands.

Thermo Fisher Scientific developed the Precision ID GlobalFiler NGS STR Panel v2, which includes 35 genomic targets comprised of 21 CODIS STR markers, 9 additional multiallelic STR markers, and four sex determination markers (Fordyce et al., 2015; Barrio et al., 2019; Tao, 2019). The multiplex was designed for optimal performance using the Ion Chef System and the Converge NGS Analysis Module (Applied Biosystems, Thermo Fisher Scientific, Foster City, California, United States). Promega's PowerSeq 46GY System contains 23 autosomal STRs, 22 Y-STRs, and amelogenin, which is currently the largest combination of autosomal and Y-STR loci in a single commercial kit (Promega Corporation, Madison, Wisconsin, United States) (Montano et al., 2018). Since Promega does not currently offer a proprietary sequencing instrument, the PowerSeq 46GY multiplex is meant to be coupled with the library preparation chemistries of one of the leading manufacturers for sequencing instrumentation in forensic genetics (Verogen or Thermo Fisher Scientific).

SNP multiplexes

Similar to STRs, SNPs lend well to MPS multiplexing due to the increased throughput capabilities of a single multiplex. Whereas the 41-target HIrisPlex-S System required two SNaPshot multiplex assays using CE, MPS panels now offer the capability of single large-scale multiplexing in a single reaction and higher throughput during a single sequencing run (Chaitanya et al., 2018; Breslin et al., 2019). Furthermore, the ability to combine SNPs (identity-informative [ii]; ancestry-informative [ai]; phenotype-informative [pi]; lineage-informative [li]; etc.) with other genomic targets (e.g., STRs, indels) into a single multiplex allows for minimal consumption of the original DNA sample, while simultaneously increasing the totality of information generated on a per sample and per run basis (Churchill et al., 2015; Gettings et al., 2015b; Ambers et al., 2016; Bruijns et al., 2018). In human skeletal remains cases where it is common to obtain samples in limited quantity or quality, the use of large MPS multiplexes may offer more potential over traditional PCR-CE or Sanger sequencing workflows.

There are a variety of commercially available SNP multiplexes designed specifically for the forensic genetics community. The ForenSeq DNA Signature Library Prep Kit (Verogen Inc.) contains two primer sets that allow for flexibility in which SNP targets are amplified and captured during the sequencing run (Churchill et al., 2015). DNA Primer Mix A contains primer pairs for 58 STRs and 94 iiSNPs only. Alternatively, DNA Primer Mix B contains all markers included in DNA Primer Mix A and an additional set of primer pairs to capture 56 biogeographical aiSNPs and 22 piSNPs (where 2 of these aiSNPs are also used for phenotype estimation). Thermo Fisher Scientific manufactures the Applied Biosystems Precision ID Identity Panel and the Precision ID Ancestry Panel (Meiklejohn and Robertson, 2017; Al-Asfi et al., 2018). The Precision ID Identity Panel contains 34 Y-chromosomal SNPs and 90 autosomal SNPs that have been demonstrated through empirical testing by Kenneth Kidd from Yale University and the SNPForID Consortium to have both high heterozygosity and a low fixation index (Fst) (Phillips et al., 2007; Karafet et al., 2008; Pakstis et al., 2010). The Precision ID Ancestry Panel encompasses 165 autosomal aiSNPs that infer biogeographic ancestry information for typed samples (described in more detail in Chapter 12) (Kosoy et al., 2009; Kidd et al., 2012). Finally, QIAGEN (Hilden, Germany) allows for custom panel design. de la Puente et al. (2017) created the Qiagen SNP-ID kit, a 140-SNP forensic identification MPS multiplex initially designed for the Thermo Fisher MPS workflow. The International Commission on Missing Persons (ICMP), in conjunction with QIAGEN and numerous leading forensic geneticists in the field, has developed a > 1000 target autosomal SNP panel designed specifically for Missing Persons investigations (Peck et al., 2018). This panel exploits the capabilities of single nucleotide characterization offered by MPS technology and targets tri- and tetranucleotide SNPs, with the end goal of providing unparalleled discrimination power in a single SNP assay.

Indel and microhaplotype multiplexes

Characterizing insertions and deletions (indels) can also be accomplished using multiplexes designed for MPS platforms. Current methods for indel genotyping exploit the low mutation frequency of insertions and deletions, making these targets suitable for forensic and parentage testing with traditional PCR-CE approaches (Hwa et al., 2018). Utilizing MPS multiplexes on the same targets, indels can be better defined and this provides the potential to identify proximal SNPs that can in turn increase the power of discrimination of each respective indel.

Microhaplotypes, a cluster of 2 or more SNPs in a region of 200 nucleotides or less, are a common forensic target due to their increased heterozygosity relative to their biallelic counterparts, coupled with minimal stutter and stochastic effects during amplification, library preparation, and sequencing (Oldoni et al., 2018; Bennett et al., 2019). MPS, therefore, makes it possible to genotype microhaplotypes, either as custom microhaplotype multiplexes or coupled with other genomic targets, resulting in improved genotype/profile determinations in order to increase identification capabilities and discrimination power of the panels, especially when the DNA samples are compromised or potentially of poor quality. It should be noted that these combined loci (i.e., SNP + SNP, SNP + indel, SNP + STR, STR + indel, indel + indel) are extremely powerful for improving the per-locus discrimination power of many common marker types and are currently of high interest in the forensic science community.

Mitochondrial genome multiplexes

Traditional sequencing of the mitochondrial DNA (mtDNA) genome involves Sanger sequencing of the control region, including specifically hypervariable regions I/II and comparison to a human reference genome (Davis et al., 2015). MPS allows for remarkable increases in mtDNA genome sequencing, allowing for the capture of the entirety of the mitochondrial genome (i.e., 16,569 base pairs (bp) of information) (King et al., 2014; Churchill et al., 2017b; Ambers et al., 2020). The increased genetic information can thus be used to provide the highest level of maternal lineage discrimination (Stoljarova et al., 2016; Strobl et al., 2019). Capturing the entire mtDNA genome can be achieved using long-range PCR with large DNA fragments to generate large amplicons, although this method is best suited for high-quality DNA (King et al., 2014). Keeping in mind that mtDNA analysis is frequently employed when samples are compromised, degraded, or available only in small quantities, commercial manufacturers have created a tiling process of small overlapping amplicons which employs a PCR-based approach using a set of overlapping primer pairs to capture the entirety of the 16,569bp mtDNA genome. Thermo Fisher Scientific has developed two unique mtDNA multiplexes for MPS. The

Precision ID mtDNA Whole Genome Panel is currently the only approved mtDNA MPS panel for inclusion in the NDIS CODIS database (Churchill et al., 2017b; Strobl et al., 2018; Woerner Ambers et al., 2018; Ambers et al., 2018). Thermo Fisher Scientific also manufactures a Precision ID mtDNA Control Region Panel, a two-pool tiled multiplex assay that targets the entirety of the 1.2-kb control region of the human mitochondrial genome (encompassing hypervariable regions I, II, and III). This panel, consistent with the Precision ID mtDNA Whole Genome Panel, was designed using a tiling approach to obtain maximum coverage (i.e., high numbers of sequencing reads per target region) from highly compromised, degraded samples common in human skeletal remains cases. Verogen Inc. has also designed both the ForenSeq mtDNA Control Region Kit and the ForenSeq mtDNA Whole Genome Kit, as alternative choices to the multiplexes designed by Thermo Fisher Scientific (Holt et al., 2019). The ForenSeq mtDNA Control Region Kit utilizes approximately 120 primer pairs (designed according to the most recent and well-curated mtDNA variant and frequency data) to generate 18 amplicons that span the complete mtDNA control region; all amplicons are less than 150 nucleotides in length, in order to maximize sequencing success with degraded samples. Finally, Promega offers the PowerSeq CRM Nested System, which captures the control region of the mitochondrial genome using a nested amplification approach (Promega Corporation) (Holland et al., 2019; Huszar et al., 2019).

The development and implementation of MPS multiplexes for the forensic sciences is constantly evolving with respect to methodology and legal policy. With constant improvement in genomic chemistry and technological advancement, the design of improved custom panels containing a variety of genomic targets in different combinations is consistently possible, and the potential forensic applications of custom multiplexes are endless. As researchers uncover new genomic markers, the ability to multiplex current markers with future candidate markers will continue to expand (Wendt et al., 2016; Novroski et al., 2018a,b; Gettings et al., 2019).

MPS platforms

Over the past decade, a diversity of high-throughput sequencing platforms have been developed which integrate a variety of fluidic and optic technologies to perform and monitor the molecular sequencing reactions known as massively parallel sequencing. However, despite the variability in engineering configuration and/or sequencing chemistry, the technical paradigm of MPS is to perform sequencing in parallel for spatially separated, clonally amplified DNA templates and/or single DNA molecules in a flow cell or on a bead chip. MPS differs greatly from Sanger sequencing (often referred to as first-generation sequencing), which relies on electrophoretic separation of chain-termination products produced in individual sequencing reactions, which in practice represent a single read of each DNA molecule (Sanger et al., 1977).

Template preparation methods for MPS

Emulsion PCR

Emulsion PCR (emPCR or ePCR) is a common template preparation method used to clonally amplify DNA targets prior to MPS (Shao et al., 2011; Kanagal-Shamanna, 2016). Emulsion PCR is the current methodology used to prepare DNA templates in Thermo Fisher Scientific

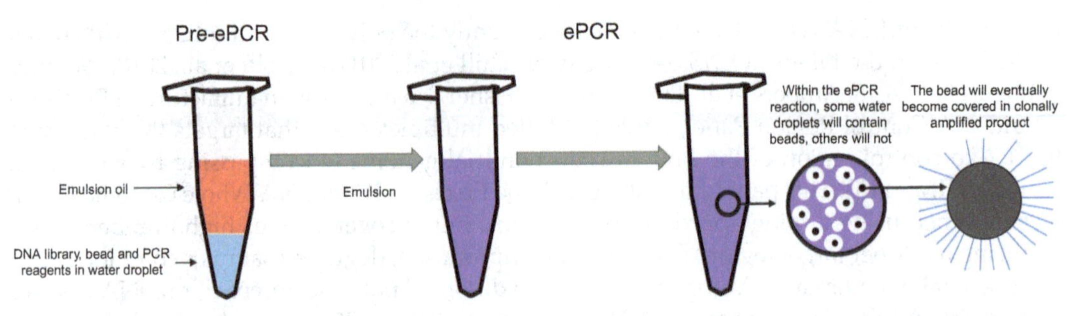

FIG. 2 Schematic of the emulsion PCR (ePCR) reaction setup. From left to right, the ePCR reagents are added into a solution and, once the emulsion is created, the amplification reaction can be carried out on beads containing a template DNA fragment. *Modified from Vierstraete, A., 2012. Next Generation Sequencing. Personal Website. Available from: https://users.ugent.be/~avierstr/nextgen/nextgen.html.*

sequencing workflows (i.e., the Precision ID MPS workflow) (Bruijns et al., 2018). The goal of ePCR is to dilute and compartmentalize each template DNA molecule into its own water droplet that will eventually become a water-in-oil emulsion (Fig. 2).

To prepare samples for emulsion PCR, DNA libraries are fragmented during the library preparation protocol, either by sonication (high sound energy) or via nebulization (forcing DNA through a small hole), to produce amplicons ranging from 300 to 800bp in length (Knierim et al., 2011). Adapters then are ligated onto each end of the DNA fragments. Adapters allow the DNA strands to bind to the emulsion beads, a step necessary for successful amplification prior to sequencing. Next, double-stranded DNA (dsDNA) molecules with bound adapters are denatured into single-stranded DNA (ssDNA) using heat (95°C). The formation of clonal bead populations is achieved using millions of individual beads; each bead begins with a single ssDNA fragment bound via the adapter. The beads are coated with streptavidin, a protein resistant to detergents, organic solvents, denaturants, proteolytic enzymes, and extremes in temperature or pH. Individual beads then become emulsified in water-in-oil droplets that contain all of the necessary reagents to carry out the emulsion PCR reaction (i.e., the DNA polymerase, primers, buffers, dNTPs) (Shao et al., 2011; Kanagal-Shamanna, 2016).

During emulsion PCR (ePCR), DNA strands are amplified on the bead surface within the water-in-oil droplet, which measures approximately 1 μm in diameter. Similar to traditional PCR, the ePCR amplification process involves the denaturation, annealing, and extension steps, resulting in > 1 million copies of the target DNA template amplified on the surface of each bead (Bruijns et al., 2018). Each bead will have a unique template, resulting in millions of clonally amplified copies of the respective template DNA fragments. After amplification is complete, the emulsion is broken using detergents, the emulsion product is vortexed and centrifuged, and enrichment for only the beads that were successfully clonally amplified is achieved using magnetic separation (Shao et al., 2011; Kanagal-Shamanna, 2016). Finally, beads containing DNA template undergo a 3′-end modification to prevent nonspecific ligation to template fragments.

Rolling circle amplification

In instances when highly fragmented and/or low copy number DNA are the only available substrates for human identification, rolling circle amplification (RCA) offers promise for the development of a DNA profile when all other methods fail (Wang et al., 2004; Kieser and Budowle, 2019). RCA involves an enzymatic reaction using circular DNA molecules as templates for exponential amplification of small genomic targets-of-interest. RCA utilizes a highly processive DNA polymerase to initiate DNA amplification by incorporating random primers to the circularized DNA template. A cascade of strand displacement events ensues, and the resultant product contains long tandem copies of the original template sequence. The benefit of RCA allows the circular molecule to serve as an infinite linear template. Although autosomal human DNA does not naturally exist in a circular form, ligation must be introduced in order to generate the circularized template necessary for the RCA reaction. RCA has demonstrated promising results with respect to individualization and source attribution, mixture and contamination detection, and in ancient, archeological, skeletonized human remains work (Kieser and Budowle, 2019).

DNA colony generation using bridge amplification

During library preparation of DNA targets for Illumina and Verogen instrumentation, each fragment undergoes reduced cycle amplification. It is during this amplification reaction that sequences used for primer binding to the acrylamide-coated glass flow cell (known as indices, or indexes) and terminal sequences are added to the 5′- and 3′ ends of each fragment bead (Schroth et al., 2007; Bentley et al., 2008; Metzker, 2010; Bruijns et al., 2018). Two indices are used for each sample, where each index is 8bp in length, and the unique combination of paired indices allows for sample identification during bioinformatic analysis (i.e., the analysis software will group all reads with the same indices together, to be interpreted as the same sample for all alleles at all loci). Using this chemistry, a total of 96 different samples can be run in a single sequencing reaction (which is comparable to CE). The terminal sequences attached during the reduced amplification reaction are used for attaching each DNA fragment to the flow cell, so that sequencing-by-synthesis (Illumina, Verogen) can be carried out (Fedurco et al., 2006; Liu et al., 2007). The bottom of each flow cell is coated with oligonucleotides (short nucleotide sequences that are complementary to the terminal sequences attached during amplification) that serve to hold the DNA fragments in place during sequencing. When DNA fragments enter the flow cell prior to bridge amplification, one of the adapters hybridizes to its complementary oligonucleotide. Hybridized DNA libraries are then covalently attached to the flow cell during the first round of amplification. After denaturing the new dsDNA, a single DNA molecule is covalently bound to the flow cell and serves as template DNA for clonal amplification (Fedurco et al., 2006; Liu et al., 2007). Cluster generation via bridge amplification then commences.

The goal of bridge amplification (similar to the goal of ePCR) is to create hundreds to thousands of identical strands of DNA known as "clusters" on the flow cell. Some clusters will be the forward strand of a DNA fragment, whereas the rest of the DNA fragments will be clusters of the reverse DNA strand. Cluster generation is accomplished using polymerases that move along a strand of DNA, creating a complementary strand (in the same way that traditional amplification typically occurs). The adapter sequence at the top of the DNA strand bends (like a bridge) and attaches to an oligonucleotide on the bottom of the flow cell. The polymerase attaches to the reverse strand and then (again) replication of the sequence occurs, resulting in a copy of the original template strand in the form of dsDNA. This dsDNA

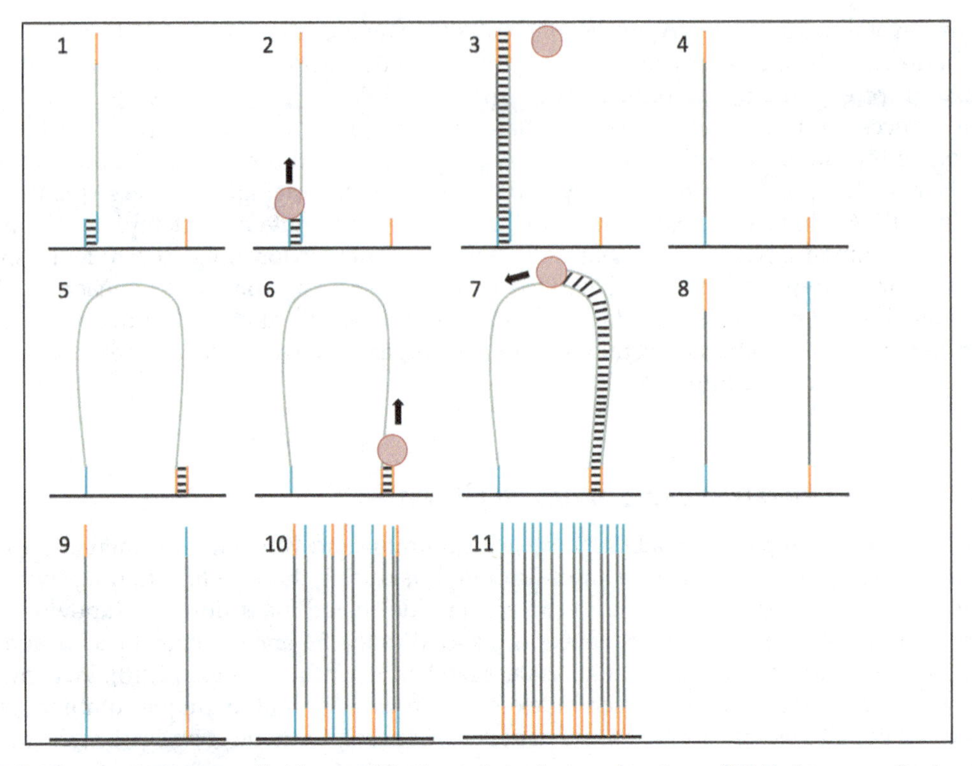

FIG. 3 Step-by-step process of bridge amplification to form clusters of a colony of DNA clones on a fixed surface, such as a flow cell.

fragment is denatured, and each strand then bends to become reanchored to the flow cell at a different oligonucleotide in the newly forming cluster (Fedurco et al., 2006; Liu et al., 2007) (Fig. 3).

Clusters of each unique DNA fragment will be formed via bridge amplification over the entirety of the flow cell all simultaneously. After sufficient cycles of clonal amplification, each cluster will be a combination of forward and reverse strands of DNA in the same region. To produce concordant strand clusters that can fluoresce uniformly during sequencing, one strand is washed off the flow cell (Fedurco et al., 2006; Liu et al., 2007; Schroth et al., 2007; Bentley et al., 2008).

Single molecule templates

Single molecule templates can be prepared for Single Molecule, Real-Time (SMRT) Sequencing, which employs a long-read sequencing chemistry (i.e., feasibly much longer than 300 bp) (Glenn, 2011). The DNA template is treated to remove any potential DNA damage, and the fragment ends are repaired and circularized using SMRTbell adapters to create circularized SMRTbell libraries (PACBIO [Pacific Biosciences of California Inc.], Menlo Park, California, United States). SMRTbell templates result in a segment of single-stranded DNA to which a sequencing primer (and associated barcoded adapter) can be annealed. Finally, a polymerase is bound to the primer-annealed template and the resultant complex can be loaded onto SMRT Cells (where the sequencing reaction takes place on a PACBIO sequencing instrument) (Glenn, 2011). This method is ideal for long-read sequencing, which is not currently available using many of the mainstream sequencing instrumentations (Glenn, 2011; Goodwin et al., 2016).

Sequencing approaches

Pyrosequencing

DNA sequencing via pyrosequencing is a method to determine the order of nucleotides in a DNA sequence based on a sequencing-by-synthesis method. The pyrosequencing process detects light emitted with sequential addition of nucleotides during the synthesis of a complementary strand of DNA. The value of pyrosequencing in forensic genetics is multifold. Pyrosequencing allows for the detection of SNPs, InDels, STRs, and other sequence variation, as well as provides the capability to quantify DNA methylation and allele frequency. The concept of pyrosequencing was first proposed as a combined solid-phase sequencing method using streptavidin-coated magnetic beads with a recombinant DNA polymerase lacking 3′-to-5′proofreading activity, coupled with firefly luciferase enzyme detection (Ronaghi et al., 1996). In 1998, Ronaghi, Uhlen, and Nyren proposed a second solution-based pyrosequencing method which incorporated an additional enzyme (apyrase) in order to remove unincorporated nucleotides. The addition of apyrase improved the simplicity of the reaction preparation, and it was amenable to automation (Ronaghi et al., 1998a,b). Finally, in 2005, a third pyrosequencing technology using microfluidics was proposed (Margulies et al., 2005; Rothberg and Leamon, 2008). In this third iteration, the template DNA was fixed to a solid support to allow for parallel sequencing of multiple DNA templates using a microfabricated microarray. This drastically increased throughput capabilities for pyrosequencing and was the first MPS instrument introduced to the commercial market. A consequence of this innovation was a new era in genomics research, competitive pricing for DNA sequencing, and the capability to perform whole genome sequencing on countless species in nominal time frames with minimal cost/expense. Although pyrosequencing is still a viable sequencing method, the chemistry discussed herein was discontinued by the manufacturer in 2013 when the technology became noncompetitive.

The concept of pyrosequencing involves six major steps (Ronaghi, 2001; Rothberg and Leamon, 2008). First, the DNA template is fragmented into 100-bp amplicons of ssDNA. PCR amplification then is performed to create millions of copies of each fragment, which are split across thousands of wells in the sequencing instrument, with only one type of DNA fragment present in each well. Incubation of the DNA fragments takes place with a concoction of reagents including DNA polymerase, apyrase, ATP sulfurylase, adenosine 5′ phosphosulfate, and luciferin.

One nucleotide (dATP, dTTP, dCTP, or dGTP) is added to each of the wells containing the template DNA strand and, when the nucleotide is complementary to the template, becomes incorporated onto the complementary strand via a DNA polymerase at the 3′ end. The dNTP incorporation causes the release of pyrophosphate (PPi), which, in the presence of adenosine 5′ phosphosulfate (APS), is converted to adenosine triphosphate (ATP) via the

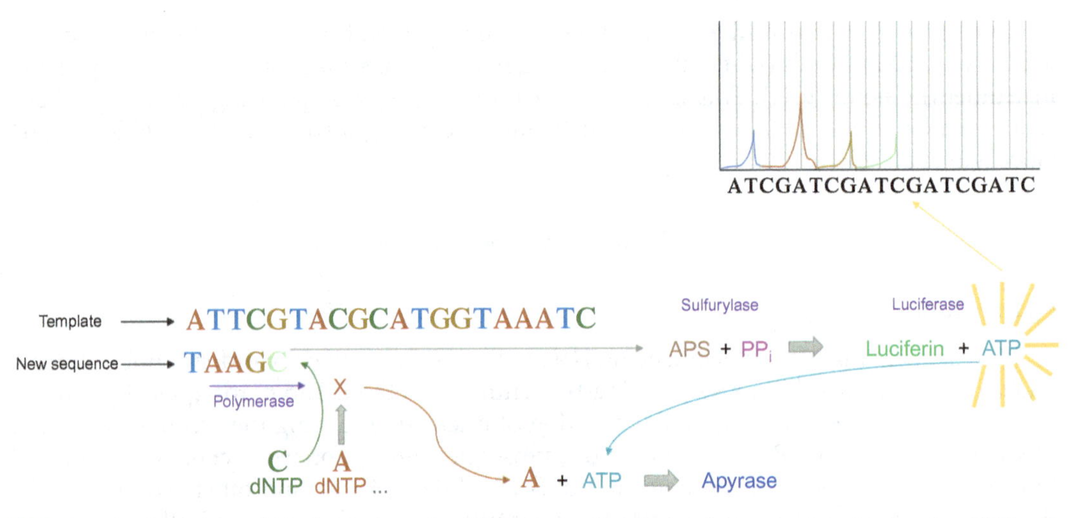

FIG. 4 Visual representation of a single nucleotide incorporation during the pyrosequencing reaction (*APS*, adenosine 5′ phosphosulfate; *ATP*, adenosine triphosphate; *dNTP*, deoxynucleotide triphosphate; *PPi*, pyrophosphate).

ATP sulfurylase enzyme. The newly created ATP becomes the substrate for the luciferase-mediated conversion of luciferin into oxyluciferin. Visible light is emitted from oxyluciferin that is proportional to the number of nucleotides incorporated into the sequence, which is captured by the pyrosequencing instrument. Finally, any unincorporated nucleotides and ATP get degraded by the apyrase reaction, where the reaction is complete and a new dNTP can be incorporated (Fig. 4). The sequencing reaction continues via this mechanism until the template strand synthesis is complete, which is limited to 300–500 bp (Shendure and Ji, 2008). Finally, a detector within the instrument identifies the intensity of light emitted during the reaction and translates this information into the number and type of nucleotides added. Wherever repeating nucleotides in the sequence occur, an increase in signal corresponding to the number of repeating nucleotides will be detected (Ronaghi, 2001; Rothberg and Leamon, 2008). For example, a well incorporating multiple dATP molecules into the complementary DNA fragment will be brighter than a well incorporating a single dATP molecule.

Sequencing-by-synthesis

Sequencing-by-synthesis (SBS) was developed by Balasubramanian and Klenerman using the reversible termination chemistry concept originally conceived by Canard and Sarfati at the Pasteur Institute in Paris (Canard and Sarfati, 1994; Bentley et al., 2008). Solexa (now a subsidiary of Illumina) was the first manufacturer to develop SBS for genomic sequencing applications, which relies on reversible dye-terminators that enable the identification of single bases as they are incorporated into DNA strands (Balasubramanian, 2015). SBS has applications in whole genome as well as targeted re-sequencing, analysis of transcriptomes and methylation, metagenomics, small RNA discovery, and genome-wide protein-nucleic acid interaction analyses.

Following one of many possible library preparations that takes purified DNA extracts and results in normalized pooled DNA libraries ready for sequencing, each fragment is anchored to a flow cell and then cluster generation via bridge amplification is conducted. Prior to commencement of the sequencing reaction, all of the "reverse" strands are washed off the flow cell, leaving behind only "forward" DNA strands (Fedurco et al., 2006; Liu et al., 2007). Primers then attach to the forward strands and a polymerase adds fluorescently tagged terminators (modified nucleotides) to the DNA strand. The instrument flows solutions of all four nucleotides (dATP, dTTP, dCTP, dGTP), one at a time for each cycle, to allow all DNA template strands to incorporate the appropriate nucleotide into the growing complementary sequence. A reversible terminator is on every nucleotide to prevent multiple additions in one round. Using four-color chemistry, each of the four terminators has a unique emission; hence, after each round, the instrument charge-coupled device (CCD) camera records the specific nucleotide added. Once the DNA strand has been "read" by the instrument, the strand that was just incorporated (sequenced) is washed off the flow cell. Next, a primer complementary to the first index binds and polymerizes the first index sequence before being washed away. The template strand then forms a new bridge, where the 3′ end of the DNA strand attaches to a different oligo on the flow cell. A primer complementary to the second index binds and again polymerizes the sequence, before being washed off the flow cell. The polymerase incorporates a complementary strand to the arched template strand. The two strands then separate, and the 3′ end of each strand is blocked to prevent any further synthesis. The forward strand is washed off the flow cell, and the SBS process restarts to copy the reverse strand (Canard and Sarfati, 1994; Bentley et al., 2008; Meyer and Kircher, 2010). The SBS reaction occurs for millions of clusters simultaneously, and each cluster has approximately 1000 identical copies of the template DNA (i.e., a library). At the end of the SBS reaction, the indices are used to associate individual sequence reads back to a particular sample for bioinformatic alignment and processing (Fig. 5).

Ion semiconductor sequencing

Ion semiconductor sequencing is an alternate approach to MPS and relies on the detection of hydrogen ions released during polymerization, or during synthesis of new DNA strands (Rothberg et al., 2011; Merriman et al., 2012). This method also involves a complementary strand that is built using the sequence of a template strand. Following ePCR, beads with clonally amplified template DNA are washed over a chip containing millions of microwells. Only one bead can settle into each microwell. Microwells containing template DNA are flooded with a single type of deoxyribonucleotide triphosphate (dNTP; dATP, dTTP, dCTP, or dGTP). If the dNTP present in the microwell is complementary to the leading nucleotide on the template strand, the dNTP will become incorporated into the growing complementary strand. When a dNTP is successfully incorporated, there is a release of a hydrogen ion, which creates a pH change that triggers an ion-sensitive field-effect transistor (ISFET) sensor in the instrument. The detection of the ion release and consequent pH change is the signal that a reaction has occurred. The complementary metal-oxide-semiconductor (CMOS) sequencing chip is comprised of many layers; under the layer of microwells is an ion-sensitive layer, below which is the ISFET ion sensor to detect individual reactions. If a template sequence presents with a homopolymer stretch (i.e., multiple repeating nucleotides), multiple dNTP

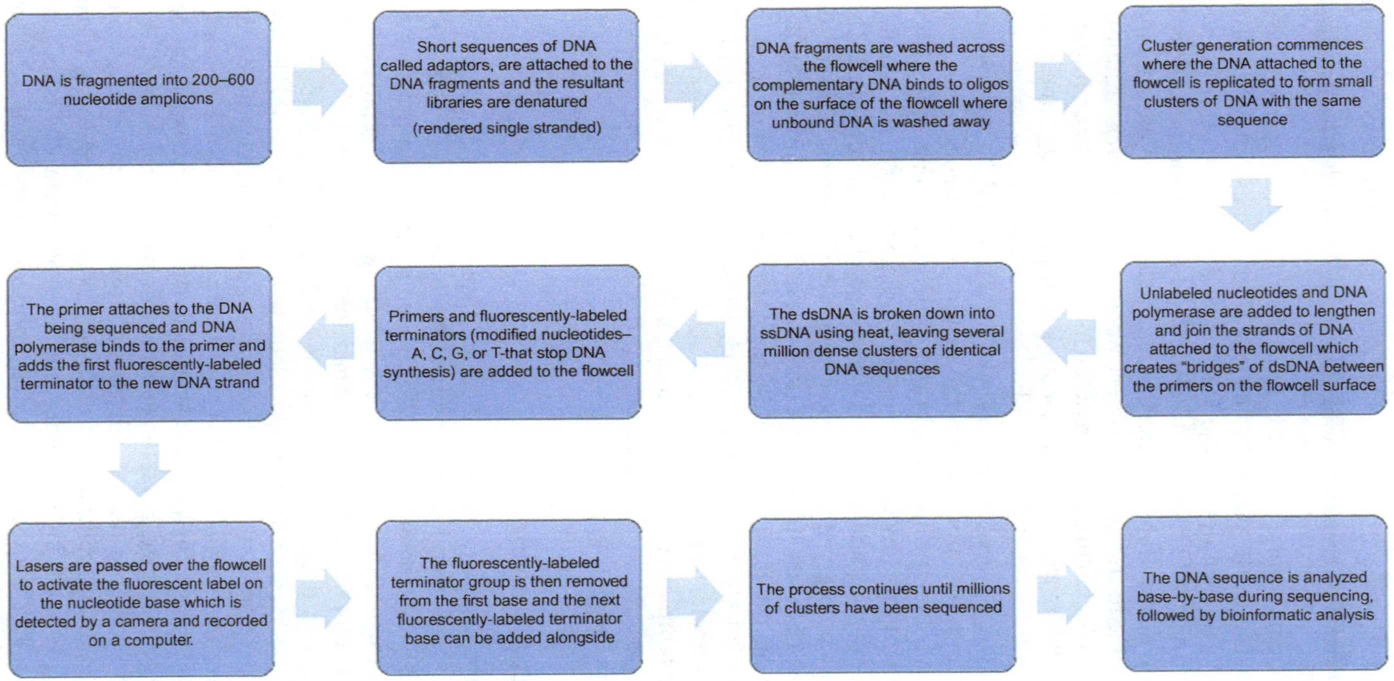

FIG. 5 General overview of the sequencing-by-synthesis (SBS) workflow.

molecules (consistent with the number of repeats in the homopolymer stretch) will be incorporated during a single cycle. Consequently, a corresponding number of hydrogen ions will be released, which will result in a proportional electronic signal that is detected by the instrument. The unattached dNTP molecules are washed away prior to the next cycle (i.e., when a different dNTP type is introduced). This process is repeated until the entire DNA template has been sequenced. Ion semiconductor sequencing technology differs from other SBS technologies in that no modified nucleotides (i.e., reversible terminator sequences) or optics (fluorescence) are used (Rothberg et al., 2011; Merriman et al., 2012). Ion semiconductor sequencing is commonly referred to as Ion Torrent sequencing, pH-mediated sequencing, silicon sequencing, or semiconductor sequencing, and is currently manufactured by Thermo Fisher Scientific.

The major benefits of ion semiconductor sequencing are rapid sequencing speed (compared to fluorescence-based SBS systems) and low upfront and operating costs. Also, this technology records natural polymerase-mediated nucleotide incorporation events, allowing for sequencing to occur and be detected in real-time (Rothberg et al., 2011). Ion semiconductor sequencing rates are limited primarily by the cycling of substrate nucleotides (A, T, C, and G) through the system. Manufacturers of ion semiconductor sequencing technologies have alleged that each incorporation event takes 4 seconds, whereas each sequencing run can be completed in as little as 1 hour, during which 100–200 nucleotides are sequenced. As the semiconductor chips and overall SBS chemistry are improved, the number of reads per chip (and therefore per run) will likely continue to increase.

This sequencing approach does present some well-documented limitations (Rothberg et al., 2011; Seo et al., 2013; Feng et al., 2016). If homopolymer repeats of the same nucleotide (e.g., AAAAAAAA) are present on a template strand, the resultant complementary strand will in turn incorporate multiple identical (complementary) nucleotides, and a proportional number of hydrogen ions are released during a single cycle. This will result in a greater pH change and a proportionally greater electronic signal. However, it can become difficult to enumerate long repeats where a percentage of the signal may be lost with each sequential repeating nucleotide incorporation. In other words, the detected change in pH is not robustly converted in a 1:1 ratio to the number of hydrogen ions released per reaction. Although not exclusive to semiconductor sequencing (i.e., this phenomenon can also be observed in pyrosequencing), signals generated from a homopolymer stretches (i.e., nucleotides that repeat in a high number) are difficult to differentiate from repeats of a similar but different number (e.g., a homopolymer repeat of seven nucleotides is difficult to differentiate from a homopolymer repeat of eight nucleotides) (Rothberg et al., 2011; Seo et al., 2013; Feng et al., 2016). Another limitation of ion semiconductor sequencing technology is the short read length (~ 400 nucleotides per read) compared to other sequencing methods (e.g., Sanger sequencing or pyrosequencing).

Multiplex and platform compatibility

Current manufacturers have designed comprehensive and user-friendly software that provides a seamless workflow for manufacturer-specific multiplexes coupled with their instrumentation. However, several multiplexes can be optimized and/or modified for use on other platforms. For example, Woerner Ambers et al. (2018) and Ambers et al. (2018) demonstrated the flexibility of the Precision ID mtDNA whole genome panel to be utilized on both the Ion S5 (Thermo Fisher Scientific) and the MiSeq (Illumina) systems. However, an important note must be made, i.e., when deviating outside of manufacturer-combined multiplex-instrument combinations, the user and/or laboratory will have to be bioinformatically competent or rely on external tools in order to interpret the resultant sequence data.

Direct PCR

In recent years, a shift to Direct PCR approaches has been initiated with the goal of increasing throughput of evidentiary samples. Not only has instrumentation been developed that focuses on generating an on-the-spot DNA profile in as little as 90 minutes, but new approaches to sample collection have also been refined (Hall and Roy, 2014; Holland and Wendt, 2015; Turingan et al., 2016, 2019; Ambers et al., 2018; Sherier et al., 2019). Direct PCR preparations typically involve a protocol that moves a sample likely containing forensically relevant biological material directly into a PCR tube for amplification, bypassing both the extraction and quantification stages (Ambers et al., 2018; Sherier et al., 2019). Successful direct-PCR DNA profiles have been demonstrated for blood, saliva, semen, fingernails, and hair, including low-template samples such as single hair follicles (Mercier et al., 1990; Gray et al., 2014; Ottens et al., 2013a,b; Ambers et al., 2018; Sherier et al., 2019). Further, Habib et al. (2017) demonstrated successful Direct PCR findings with both reference and/or postmortem samples, including challenging substrates such as toothbrush bristles, muscle tissue, and bone shavings, and in which only slight modifications were made to traditional Direct PCR protocols to obtain success. Furthermore, Direct PCR methodologies have been employed using many available and forensically relevant PCR-CE multiplexes, including the PowerPlex 21 System (Promega Corporation, Madison, Wisconsin, United States), the AmpFLSTR Identifiler Plus PCR Amplification Kit (Applied Biosystems [Thermo Fisher Scientific], Foster City, California, United States), and the GlobalFiler PCR Amplification Kit (Applied Biosystems) (Berry, 2014; Kingston, 2014; Altshuler and Roy, 2015; Sorensen et al., 2016). Direct PCR preparations have also been successfully incorporated into Y-chromosome marker workflows, including both the PowerPlex Y23 System (Promega) and the AmpFLSTR Yfiler PCR Amplification Kit (Applied Biosystems) (Altshuler and Roy, 2015).

Direct PCR preparations typically generate significantly increased peak heights in an electropherogram when using traditional CE approaches. This is a consequence of a nonspecific DNA input amount, which may exceed the ideal quantity of DNA in the reaction. Studies have revealed that only 17 cells are necessary to yield a viable full-STR profile using a Direct PCR approach compared to an average of 250 cells typically required for standard extraction methodologies in the forensic biology pipeline, where some loss of DNA is expected during the extraction process (e.g., when the extraction solution is transferred between multiple tubes or columns) (Ottens et al., 2013a).

Despite the immediate (and obvious) advantages of Direct PCR approaches, a major limitation is input DNA. The success and/or quality of the resultant DNA profile generated from a sample processed using Direct PCR will be dependent on the nature and quantity of the biological material sampled (Templeton et al., 2015). This is especially true for skeletonized human remains, which are typically abundant in PCR inhibitors (e.g., humic acids from soil). Since inhibitors and other debris are traditionally removed during a standard DNA extraction, the potential for interference in the generation and interpretation of a resultant DNA profile using a Direct PCR method is probable (Templeton et al., 2015). However, as Direct PCR methodologies continue to improve, optimization of the DNA extract, amplification reaction, and/or the PCR product prior to analysis may improve the completeness of the resultant profile, the ease of profile interpretation, and the statistical weight of the analyzed evidence (Ottens et al., 2013a).

Rapid DNA testing

The current workflow for forensic DNA typing is not a quick process. In general, the entire DNA analysis workflow from sample processing to interpretation can take numerous hours of labor (sometimes even days or weeks), depending on the reagents, equipment, and personnel available to carry out each protocol (Roeder et al., 2009; Latham and Miller, 2019). In contrast, the introduction of Rapid DNA analysis systems now allows for the quick (typically under 90 minutes) generation of a DNA profile from a variety of common forensic substrates in a manner amenable to expert systems (i.e., those DNA typing strategies with little-to-no human decision making) (Holland and Wendt, 2015; Turingan et al., 2016). Rapid DNA systems can be considered ideal for mass disaster events since the platform is suitable for mobile use, i.e., Rapid DNA instrumentation can be implemented and used in a field setting (Turingan et al., 2019). Furthermore, Rapid DNA analysis systems are simple enough that nonforensic genetic experts can operate the instrumentation with relative ease after minimal training.

Among the Rapid DNA analysis platforms currently available are: (1) the RapidHIT ID System (originally manufactured by IntegenX, Inc., now Thermo Fisher Scientific); (2) the ANDE 6C Instrument (ANDE, Longmont, Colorado, United States); (3) the ParaDNA Intelligence System (LGC Forensics, Abingdon, United Kingdom); and (4) the DNAscan Rapid DNA Analysis instrument (NetBio, Waltham, Massachusetts, United States) (Holland and Wendt, 2015; Della Manna et al., 2016; Turingan et al., 2016; Grover et al., 2017; Moreno et al., 2017; Federal Bureau of Investigation, 2018; Carney et al., 2019). Although initial validations focused on utilizing Rapid DNA systems primarily for single source buccal swabs (Holland and Wendt, 2015), additional studies have explored how Rapid DNA instrumentation can process saliva, semen, skin, and hair samples (Moreno et al., 2017; Turingan et al., 2019). Currently, many research institutions are internally validating the sensitivity and profile generation capabilities of a variety of challenging forensic samples on these platforms, including materials that contain degraded and low quantity DNA, as well as samples subject to contamination and inhibition.

Effective June 1, 2018, the Federal Bureau of Investigation (FBI) approved the ANDE 6C Instrument as the only Rapid DNA system approved for use at the National DNA Index System by accredited forensic DNA laboratories (Moreno et al., 2017; Federal Bureau of Investigation, 2018). Consequently, the first major use of Rapid DNA technology for the identification of human remains was in November of 2018 in Paradise, California, United States, when deadly destructive wildfires claimed the lives of 85 individuals in a matter of hours. The Sherriff's Office of Butte County, California, with the assistance of ANDE Corporation, analyzed many of the DNA samples recovered from the debris (Department of

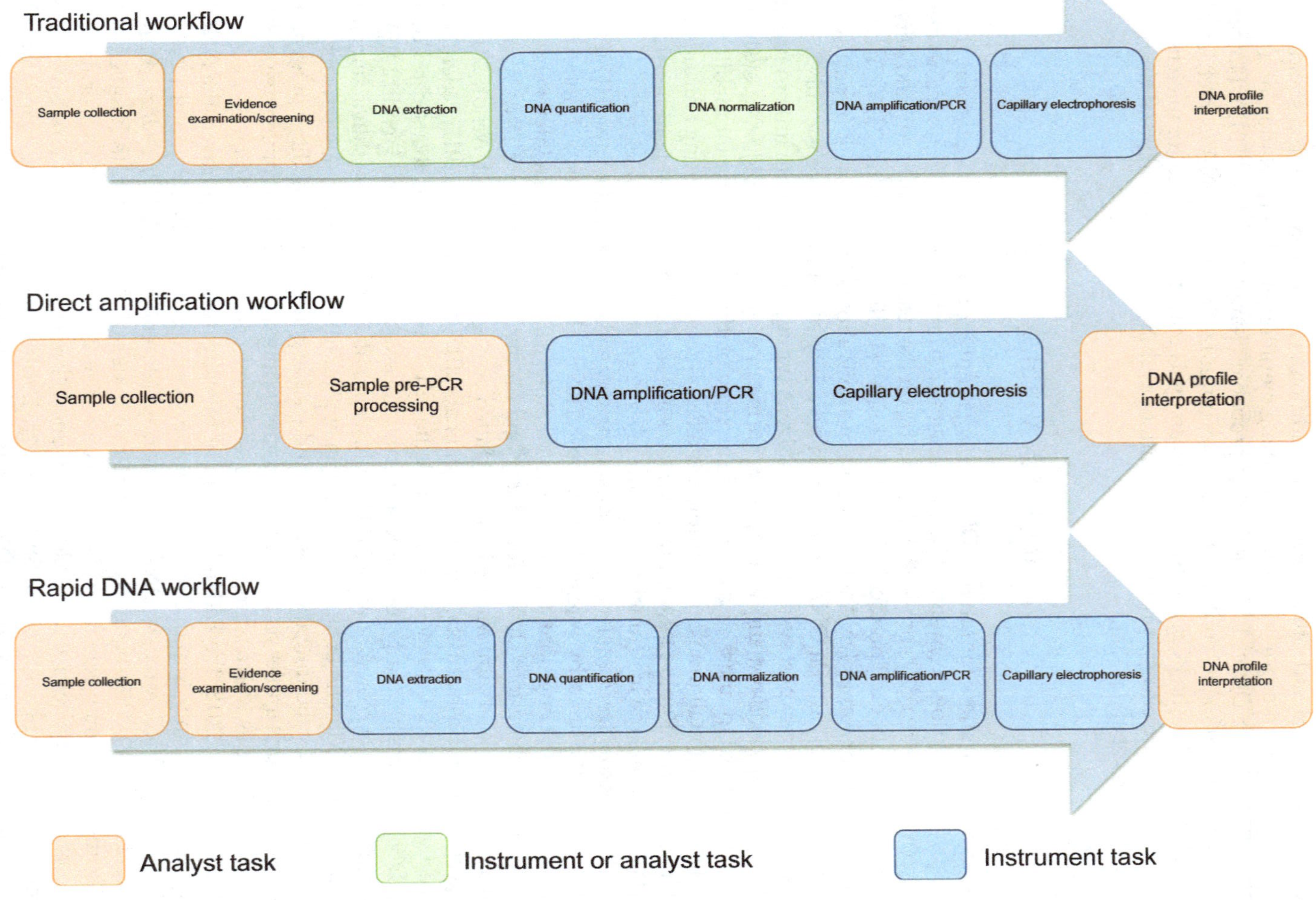

FIG. 6 Comparison of traditional, Direct PCR, and Rapid DNA analysis workflows.

Homeland Security, 2019). In many instances, the remains were so badly damaged that dental records and fingerprints were not viable methods of identification. Using the resultant DNA profiles, comparisons were made to reference samples collected from family members of the missing. Over 300 family reference profiles were submitted, and many of those samples were also processed using the ANDE system.

The greatest limitations associated with Rapid DNA testing instruments are that: (1) the sample is consumed during generation of a DNA profile and, hence, no sample is remaining for additional testing or other applications; (2) Rapid DNA devices are not as sensitive as current forensic laboratory analyses; and (3) Rapid DNA software is not amenable to analysis and interpretation of DNA mixtures (Thong et al., 2015; Scientific Working Group on DNA Analysis Methods (SWGDAM), 2017). Fig. 6 provides an overview and comparison of traditional, Direct PCR, and Rapid DNA workflows in forensic casework.

Conclusion

The diversity and capability of emerging technologies will continue to improve in the coming years. In the last decade, the field of forensic genetics has expanded to utilize massively parallel sequencing (MPS), Direct PCR, and Rapid DNA instrumentation in a variety of forensic genetic applications. As the instrumentation and chemistry associated with each method continue to evolve, greater capabilities and applications for the identification of highly decomposed or skeletonized remains will emerge.

References

Al-Asfi, M., McNevin, D., Mehta, B., Power, D., Gahan, M.E., Daniel, R., 2018. Assessment of the precision ID ancestry panel. Int. J. Legal Med. 132 (6), 1581–1594.

Altshuler, H., Roy, R., 2015. Evaluation of direct PCR amplification using various swabs and washing reagents. J. Forensic Sci. 60 (6), 1542–1552. https://doi.org/10.1111/1556-4029.12865.

Ambers, A., Churchill, J.D., King, J.L., Stoljarova, M., Gill-King, H., Assidi, M., Abu-Elmagd, M., Buhmeida, A., Budowle, B., 2016. More comprehensive forensic genetic marker analyses for accurate human remains identification using massively parallel DNA sequencing. BMC Genomics 17 (9), 21–30.

Ambers, A., Wiley, R., Novroski, N., Budowle, B., 2018. Direct PCR amplification of DNA from human bloodstains, saliva, and touch samples collected with microFLOQ® swabs. Forensic Sci. Int. Genet. 32, 80–87. https://doi.org/10.1016/j.fsigen.2017.10.010.

Ambers, A., Bus, M.M., King, J.L., Jones, B., Durst, J., Bruseth, J.E., Gill-King, H., Budowle, B., 2020. Forensic genetic investigation of human skeletal remains recovered from the *La Belle* shipwreck. Forensic Sci. Int. 306, 110050.

Balasubramanian, S., 2015. Solexa sequencing: decoding genomes on a population scale. Clin. Chem. 61 (1), 21–24.

Barrio, P.A., Martín, P., Alonso, A., Müller, P., Bodner, M., Berger, B., Parson, W., Budowle, B., DNASEQEX Consortium, 2019. Massively parallel sequence data of 31 autosomal STR loci from 496 Spanish individuals revealed concordance with CE-STR technology and enhanced discrimination power. Forensic Sci. Int. Genet. 42, 49–55.

Bennett, L., Oldoni, F., Long, K., Cisana, S., Maddela, K., Madella, K., Wootton, S., Chang, J., Hasegawa, R., Lagacé, R., Kidd, K.K., Podini, D., 2019. Mixture deconvolution by massively parallel sequencing of microhaplotypes. Int. J. Legal Med. 133 (3), 719–729.

Bentley, D.R., 2006. Whole genome re-sequencing. Curr. Opin. Genet. Divers. 16 (6), 545–552.

Bentley, D.R., Balasubramanian, S., Swerdlow, H.P., Smith, G.P., Milton, J., Brown, C.G., Hall, K.P., Evers, D.J., Barnes, C.L., Bignell, H.R., Boutell, J.M., Bryant, J., Carter, R.J., Cheetham, R.K., Cox, A.J., Ellis, D.J., Flatbush, M.R., Gormley, N.A., Humphray, S.J., Irving, L.J., Karbelashvili, M.S., Kirk, S.M., Li, H., Liu, X., Maisinger, K.S., Murray, L.J., Obradovic, B., Ost, T., Parkinson, M.L., Pratt, M.R., Rasolonjatovo, I.M.J., Reed, M.T., Rigatti, R., Rodighiero,

C., Ross, M.T., Sabot, A., Sankar, S.V., Scally, A., Schroth, G.P., Smith, M.E., Smith, V.P., Spiridou, A., Torrance, P.E., Tzonev, S.S., Vermaas, E.H., Walter, K., Wu, X., Zhang, L., Alam, M.D., Anastasi, C., Aniebo, I.C., Bailey, D.M.D., Bancarz, I.R., Banerjee, S., Barbour, S.G., Baybayan, P.A., Benoit, V.A., Benson, K.F., Bevis, C., Black, P.J., Boodhun, A., Brennan, J.S., Bridgham, J.A., Brown, R.C., Brown, A.A., Buermann, D.H., Bundu, A.A., Burrows, J.C., Carter, N.P., Castillo, N., Catenazzi, M.C.E., Chang, S., Cooley, R.N., Crake, N.R., Dada, O.O., Diakoumakos, K.D., Dominguez-Fernandez, B., Earnshaw, D.J., Egbujor, U.C., Elmore, D.W., Etchin, S.S., Ewan, M.R., Fedurco, M., Fraser, L.J., Fuentes Fajardo, K.V., Furey, W.S., George, D., Gietzen, K.J., Goddard, C.P., Golda, G.S., Granieri, P.A., Green, D.E., Gustafson, D.L., Hansen, N.F., Harnish, K., Haudenschild, C.D., Heyer, N.I., Hims, M.M., Ho, J.T., Horgan, A.M., Hoschler, K., Hurwitz, S., Ivanov, D.V., Johnson, M.Q., James, T., Huw Jones, T.A., Kang, G.-D., Kerelska, T.H., Kersey, A.D., Khrebtukova, I., Kindwall, A.P., Kingsbury, Z., Kokko-Gonzales, P.I., Kumar, A., Laurent, M.A., Lawley, C.T., Lee, S.E., Lee, X., Liao, A.K., Loch, J.A., Lok, M., Luo, S., Mammen, R.M., Martin, J.W., McCauley, P.G., McNitt, P., Mehta, P., Moon, K.W., Mullens, J.W., Newington, T., Ning, Z., Ng, B.L., Novo, S.M., O'Neill, M.J., Osborne, M.A., Osnowski, A., Ostadan, O., Paraschos, L.L., Pickering, L., Pike, A.C., Pike, A.C., Pinkard, D.C., Pliskin, D.P., Podhasky, J., Quijano, V.J., Raczy, C., Rae, V.H., Rawlings, S.R., Rodriguez, A.C., Roe, P.M., Rogers, J., Rogert Bacigalupo, M.C., Romanov, N., Romieu, A., Roth, R.K., Rourke, N.J., Ruediger, S.T., Rusman, E., Sanches-Kuiper, R.M., Schenker, M.R., Seoane, J.M., Shaw, R.J., Shiver, M.K., Short, S.W., Sizto, N.L., Sluis, J.P., Smith, M.A., Sohna Sohna, J.E., Spence, E.J., Stevens, K., Sutton, N., Szajkowski, L., Tregidgo, C.L., Turcatti, G., vandeVondele, S., Verhovsky, Y., Virk, S.M., Wakelin, S., Walcott, G.C., Wang, J., Worsley, G.J., Yan, J., Yau, L., Zuerlein, M., Rogers, J., Mullikin, J.C., Hurles, M.E., McCooke, N.J., West, J.S., Oaks, F.L., Lundberg, P.L., Klenerman, D., Durbin, R., Smith, A.J., 2008. Accurate whole human genome sequencing using reversible terminator chemistry. Nature 456 (7218), 53–59.

Berglund, E.C., Kiialainen, A., Syvänen, A.-C., 2011. Next-generation sequencing technologies and applications for human genetic history and forensics. Investig. Genet. 2, 23. https://doi.org/10.1038/nature09807.

Berry, C., 2014. The Use of a Direct to PCR Analysis Method for Disaster Victim Identification. National Centre for Forensic Science. University of Canberra, Bruce.

Børsting, C., Fordyce, S.L., Olofsson, J., Mogensen, H.S., Morling, N., 2014. Evaluation of the Ion Torrent™ HID SNP 169-plex: a SNP typing assay developed for human identification by second generation sequencing. Forensic Sci. Int. Genet. 12, 144–154.

Breslin, K., Wills, B., Ralf, A., Ventayol Garcia, M., Kukla-Bartoszek, M., Pospiech, E., Freire-Aradas, A., Xavier, C., Ingold, S., de La Puente, M., van der Gaag, K.J., Herrick, N., Haas, C., Parson, W., Phillips, C., Sijen, T., Branicki, W., Walsh, S., Kayser, M., 2019. HIrisPlex-S system for eye, hair, and skin color prediction from DNA: massively parallel sequencing solutions for two common forensically used platforms. Forensic Sci. Int. Genet. 43, 102152. https://doi.org/10.1016/j.fsigen.2019.102152.

Bruijns, B., Tiggelaar, R., Gardeniers, H., 2018. Massively parallel sequencing techniques for forensics: a review. Electrophoresis 39, 2642–2654.

Canard, B., Sarfati, R.S., 1994. DNA polymerase fluorescent substrates with reversible 3′-tags. Gene 148 (1), 1–6.

Carney, C., Whitney, S., Vaidyanathan, J., Persick, R., Noel, F., Vallone, P.M., Romsos, E.L., Tan, E., Grover, R., Turingan, R.S., French, J.L., Selden, R.F., 2019. Developmental validation of the ANDE™ rapid DNA system with FlexPlex™ assay for arrestee and reference buccal swab processing and database searching. Forensic Sci. Int. Genet. 40, 120–130.

Chaitanya, L., Breslin, K., Zuñiga, S., Wirken, L., Pośpiech, E., Kukla-Bartoszek, M., Sijen, T., Knijff, P., Liu, F., Branicki, W., Kayser, M., Walsh, S., 2018. The HIrisPlex-S system for eye, hair and skin colour prediction from DNA: introduction and forensic developmental validation. Forensic Sci. Int. Genet. 35, 123–135.

Churchill, J.D., Schmedes, S.E., King, J.L., Budowle, B., 2015. Evaluation of the Illumina® Beta Version ForenSeq™ DNA Signature Prep Kit for use in genetic profiling. Forensic Sci. Int. Genet. 20, 20–29.

Churchill, J.D., Novroski, N.M.M., King, J.L., Seah, L.H., Budowle, B., 2017a. Population and performance analyses of four major populations with Illumina's FGx Forensic Genomics System. Forensic Sci. Int. Genet. 30, 81–92.

Churchill, J., Peters, D., Capt, C., Strobl, C., Parson, W., Budowle, B., 2017b. Working towards implementation of whole genome mitochondrial DNA sequencing into routine casework. Forensic Sci. Int. Genet. Suppl. Ser. 6, e388–e389.

Davis, C., Peters, D., Warshauer, D., King, J., Budowle, B., 2015. Sequencing the hypervariable regions of human mitochondrial DNA using massively parallel sequencing: enhanced data acquisition for DNA samples encountered in forensic testing. Legal Med. (Tokyo) 17, 123–127.

de la Puente, M., Phillips, C., Santos, M., Fondevila, M., Carracedo, Á., Lareu, M.V., 2017. Evaluation of the Qiagen 140-SNP forensic identification multiplex for massively parallel sequencing. Forensic Sci. Int. Genet. 28, 35–43.

Della Manna, A., Nye, J.V., Carney, C., Hammons, J.S., Mann, M., Al Shamali, F., Vallone, P.M., Romsos, E.L., Marne, B.A., Tan, E., Turingan, R.S., Hogan, C., Selden, R.F., French, J.L., 2016. Developmental validation of the DNAscan rapid DNA analysis instrument and expert system for reference sample processing. Forensic Sci. Int. Genet. 25, 145–156.

Department of Homeland Security, 2019. Available from: https://www.dhs.gov/science-and-technology/news/2019/04/23/snapshot-st-rapid-dna-technology-identified-victims.

England, R., Harbison, S., 2019. A review of the method and validation of the MiSeq FGx™ Forensic Genomics Solution. WIREs Forensic Sci. 2 (1). https://doi.org/10.1002/wfs2.1351. Available from:.

Federal Bureau of Investigation, 2018. Rapid DNA. Available from: https://www.fbi.gov/services/laboratory/biometric-analysis/codis/rapid-dna.

Fedurco, M., Romieu, A., Williams, S., Lawrence, I., Turcatti, G., 2006. BTA, a novel reagent for DNA attachment on glass and efficient generation of solid-phase amplified DNA colonies. Nucleic Acids Res. 34 (3), e22.

Feng, W., Zhao, S., Xue, D., Song, F., Li, Z., Chen, D., He, B., Hao, Y., Wang, Y., Liu, Y., 2016. Improving alignment accuracy on homopolymer regions for semiconductor-based sequencing technologies. BMC Genomics 17 (Suppl. 7), 521. https://doi.org/10.1186/s12864-016-2894-9.

Fordyce, S.L., Avila-Arcos, M.C., Rockenbauer, E., Børsting, C., Frank-Hansen, R., Petersen, F.T., Willerslev, E., Hansen, A.J., Morling, N., Gilbert, M.T., 2011. High-throughput sequencing of core STR loci for forensic genetic investigations using the Roche Genome Sequencer FLX platform. BioTechniques 51, 127–133.

Fordyce, S.L., Mogensen, H.S., Børsting, C., Lagace, R.E., Chang, C.W., Rajagopalan, N., Morling, N., 2015. Second-generation sequencing of forensic STRs using the ion torrent HID STR, 10-plex and the ion PGM. Forensic Sci. Int. Genet. 14, 132–140.

Gettings, K.B., Aponte, R.A., Vallone, P.M., Butler, J.M., 2015a. STR allele sequence variation: current knowledge and future issues. Forensic Sci. Int. Genet. 18, 118–130.

Gettings, K.B., Kiesler, K.M., Vallone, P.M., 2015b. Performance of a next-generation sequencing SNP assay on degraded DNA. Forensic Sci. Int. Genet. 19, 1–9.

Gettings, K.B., Borsuk, L.A., Zook, J., Vallone, P.M., 2019. Unleashing novel STRs via characterization of genome in a bottle reference samples. Forensic Sci. Int. Genet. Suppl. Ser. 7 (1), 218–220. https://doi.org/10.1016/j.fsigss.2019.09.084.

Glenn, G.C., 2011. Field guide to next-generation DNA sequencers. Mol. Ecol. Resour. 11, 759–769.

Goodwin, S., McPherson, J.D., McCombie, W.R., 2016. Coming of age: ten years of next-generation sequencing technologies. Nat. Rev. Genet. 17, 333–351.

Gray, K., Crowle, D., Scott, P., 2014. Direct amplification of casework bloodstains using the Promega PowerPlex(®) 21 PCR amplification system. Forensic Sci. Int. Genet. 12, 86–92.

Grover, R., Jiang, H., Turingan, R.S., French, J.L., Tan, E., Selden, R.F., 2017. Flexplex27—highly multiplexed rapid DNA identification for law enforcement, kinship, and military applications. Int. J. Legal Med. 131 (6), 1489–1501.

Habib, M., Pierre-Noel, A., Fogt, F., Budimlija, Z., Prinz, M., 2017. Direct amplification of biological evidence and DVI samples using the Qiagen Investigator 24plex GO! Kit. Forensic Sci. Int. Genet. Suppl. Ser. 6, e208–e210.

Hall, D.E., Roy, R., 2014. An evaluation of direct PCR amplification. Croat. Med. J. 55 (6), 655–661. https://doi.org/10.3325/cmj.2014.55.655.

Head, S.R., Komori, H.K., LaMere, S.A., Whisenant, T., Van Nieuwerburgh, F., Salomon, D.R., Ordoukhanian, P., 2014. Library construction for next-generation sequencing: overviews and challenges. Biotechniques 56 (2). https://doi.org/10.2144/000114133. 61–passim.

Hickman, M.P., Grisedale, K.S., Bintz, B.J., Burnside, E.S., Hanson, E.K., Ballantyne, J., Wilson, M.R., 2018. Recovery of whole mitochondrial genome from compromised samples via multiplex PCR and massively parallel sequencing. Futur. Sci. OA 4 (9). https://doi.org/10.4155/fsoa-2018-0059.

Holland, M., Wendt, F., 2015. Evaluation of the RapidHIT™ 200, an automated human identification system for STR analysis of single source samples. Forensic Sci. Int. Genet. 14, 76–85.

Holland, M.M., Bonds, R.M., Holland, C.A., McElhoe, J.A., 2019. Recovery of mtDNA from unfired metallic ammunition components with an assessment of sequence profile quality and DNA damage through MPS analysis. Forensic Sci. Int. Genet. 39, 86–96.

Holt, C., Walichiewicz, P., Eagles, J., Daulo, A., Didier, M., Edwards, C., Fleming, K., Han, Y., Hill, T., Li, S., Rensfield, A., Sa, D., Stephens, K., 2019. Mitochondrial DNA data analysis strategies that inform MPS-based forensic casework implementation. Forensic Sci. Int. Genet. Suppl. Ser. 7 (1), 389–391.

Hussing, C., Børsting, C., Mogensen, H.S., Morling, N., 2015. Testing of the Illumina® ForenSeq™ kit. Forensic Sci. Int. Genet. Suppl. Ser. 5, e449–e450.

Huszar, T.I., Wetton, J.H., Jobling, M.A., 2019. Mitigating the effects of reference sequence bias in single-multiplex massively parallel sequencing of the mitochondrial DNA control region. Forensic Sci. Int. Genet. 40, 9–17.

Hwa, H.L., Chung, W.C., Chen, P.L., Lin, C.P., Li, H.Y., Yin, H.I., Lee, J.C., 2018. A 1204-single nucleotide polymorphism and insertion-deletion polymorphism panel for massively parallel sequencing analysis of DNA mixtures. Forensic Sci. Int. Genet. 32, 94–101.

Irwin, J., Just, R., Scheible, M., Loreille, O., 2011. Assessing the potential of next-generation sequencing technologies for missing persons identification efforts. Forensic Sci. Int. Genet. Suppl. Ser. 3 (1), E447–E448.

Kanagal-Shamanna, R., 2016. Emulsion PCR: techniques and applications. Methods Mol. Biol. 1392, 33–42.

Karafet, T.M., Mendez, F.L., Meilerman, M.B., Underhill, P.A., Zegura, S.L., Hammer, M.F., 2008. New binary polymorphisms reshape and increase resolution of the human Y chromosomal haplogroup tree. Genome Res. 18, 830–838.

Kidd, K.K., Kidd, J.R., Pakstis, A.J., Speed, W.C., 2012. Poster: Better SNPs for Better Forensics: Ancestry, Phenotype, and Family Identification. Shown at National Institute of Justice annual meeting, Arlington VA. Available from: https://medicine.yale.edu/lab/kidd/publications/NIJposter2012_Minihaps_237328_284_29491_v1.pdf.

Kieser, R., Budowle, B., 2019. Rolling circle amplification: a (random) primer on the enrichment of an infinite linear DNA template. WIREs Forensic Science 2 (1), e1359. Available from: https://doi.org/10.1002/wfs2.1359.

King, J.L., LaRue, B.L., Novroski, N.M., Stoljarova, M., Seo, S.B., Zeng, X., Warshauer, D.H., Davis, C.P., Parson, W., Sajantila, A., Budowle, B., 2014. High-quality and high-throughput massively parallel sequencing of the human mitochondrial genome using the Illumina MiSeq. Forensic Sci. Int. Genet. 12, 128–135.

Kingston, J., 2014. DNA Recovery from Aged Blood Stains. National Centre for Forensic Science. University of Canberra, Bruce.

Kircher, M., Kelso, J., 2010. High-throughput DNA sequencing—concepts and limitations. BioEssays Methods Models Tech. 32 (6), 524–536.

Knierim, E., Lucke, B., Schwarz, J.M., Schuelke, M., Seelow, D., 2011. Systematic comparison of three methods for fragmentation of long-range PCR products for next generation sequencing. PLoS One 6 (11), e28240. https://doi.org/10.1371/journal.pone.0028240.

Kosoy, R., Nassir, R., Tian, C., White, P.A., Butler, L.M., Silva, G., Kittles, R., Alarcon-Riquelme, M.E., Gregersen, P.K., Belmont, J.W., De La Vega, F.M., Seldin, M.F., 2009. Ancestry informative marker sets for determining continental origin and admixture proportions in common populations in America. Hum. Mutat. 30 (1), 69–78.

Kulstein, G., Hadrys, T., Wiegand, P., 2017. As solid as a rock—comparison of CE- and MPS-based analyses of the petrosal bone as a source of DNA for forensic identification of challenging cranial bones. Int. J. Legal Med. 132 (1), 13–24.

Larmuseau, M.H.D., Van Geystelen, A., Kayser, M., Van Oven, M., Decorte, R., 2015. Towards a consensus Y-chromosomal phylogeny and Y-SNP set in forensics in the next-generation sequencing era. Forensic Sci. Int. Genet. 15, 39–42.

Latham, K.E., Miller, J.J., 2019. DNA recovery and analysis from skeletal material in modern forensic contexts. Forensic Sci. Res. 4 (1), 51–59.

Liu, X., Milton, J., Smith, G.P., Barnes, C., Rasolonjatovo, I.M.J., Rigatti, R., Wu, X., Ost, T.W.B., Worsley, G.J., Earnshaw, D.J., Turcatti, G., Romieu, A., 2007. Preparation of single-stranded templates for nucleic acid sequencing. PCT Int. Appl., WO2006–GB2687.

Liu, L., Li, Y., Li, S., Hu, N., He, Y., Lin, D., Lu, L., Law, M., 2012. Comparison of next-generation sequencing systems. J. Biomed. Biotechnol. https://doi.org/10.1155/2012/251364, 251364.

Lorente, J.A., Entrala, C., Alvarez, J.C., Lorente, M., Villanueva, E., Carrasco, F., Budowle, B., 2000. Missing persons identification: genetics at work for society. Science 290, 2257–2258.

Mardis, E.R., 2008. Next-generation DNA sequencing methods. Annu. Rev. Genomics Hum. Genet. 9, 387–402.

Margulies, M., Egholm, M., Altman, W.E., Attiya, S., Bader, J.S., Bemben, L.A., Berka, J., Braverman, M.S., Chen, Y.J., Chen, Z., Dewell, S.B., Du, L., Fierro, J.M., Gomes, X.V., Godwin, B.C., He, W., Helgesen, S., Ho, C.H., Irzyk, G.P., Jando, S.C., Alenquer, M.L., Jarvie, T.P., Jirage, K.B., Kim, J.B., Knight, J.R., Lanza, J.R., Leamon, J.H., Lefkowitz, S.M., Lei, M., Li, J., Lohman, K.L., Lu, H., Makhijani, V.B., McDade, K.E., McKenna, M.P., Myers, E.W., Nickerson, E., Nobile, J.R., Plant, R., Puc, B.P., Ronan, M.T., Roth, G.T., Sarkis, G.J., Simons, J.F., Simpson, J.W., Srinivasan, M., Tartaro, K.R., Tomasz, A., Vogt, K.A., Volkmer, G.A., Wang, S.H., Wang, Y., Weiner, M.P., Yu, P., Begley, R.F., Rothberg, J.M., 2005. Genome sequencing in microfabricated high-density picolitre reactors. Nature 437, 376–380.

Meiklejohn, K.A., Robertson, J.M., 2017. Evaluation of the Precision ID Identity Panel for the Ion Torrent™ PGM™ sequencer. Forensic Sci. Int. Genet. 31, 48–56.

Mercier, B., Gaucher, C., Feugeas, O., Mazurier, C., 1990. Direct PCR from whole blood, without DNA extraction. Nucleic Acids Res. 18, 5908. https://doi.org/10.1093/nar/18.19.5908.

Merriman, B., Ion Torrent R&D Team, Rothbery, J.M., 2012. Progress in Ion Torrent semiconductor chip based sequencing. Electrophoresis 33 (23), 3397–3417.

Metzker, M.I., 2010. Sequencing technologies—the next generation. Nat. Rev. Genet. 11, 31–46.

Meyer, M., Kircher, M., 2010. Illumina sequencing library preparation for highly multiplexed target capture and sequencing. Cold Spring Harb. Protoc. https://doi.org/10.1101/pdb.prot5448.

Montano, E.A., Bush, J.M., Garver, A.M., Larijani, M.M., Wiechman, S.M., Baker, C.H., Wilson, M.R., Guerrieri, R.A., Benzinger, E.A., Gehres, D.N., Dickens, M.L., 2018. Optimization of the Promega PowerSeq™ Auto/Y system for efficient integration within a forensic DNA laboratory. Forensic Sci. Int. Genet. 32, 26–32.

Moreno, L.I., Brown, A.L., Callaghan, T.F., 2017. Internal validation of the DNAscan/ANDE rapid DNA analysis platform and its associated PowerPlex 16 high content DNA biochip cassette for use as an expert system with reference buccal swabs. Forensic Sci. Int. Genet. 29, 100–108.

Moretti, T., Baumstark, A., Defenbaugh, D., Keys, K., Smerick, J., Budowle, B., 2001. Validation of short tandem repeats (STRs) for forensic usage: performance testing of fluorescent multiplex STR systems and analysis of authentic and simulated forensic samples. J. Forensic Sci. 46 (3), 647–660.

Novroski, N.M.M., King, J.L., Churchill, J.D., Seah, L.H., Budowle, B., 2016. Characterization of genetic sequence variation of 58 STR loci in four major population groups. Forensic Sci. Int. Genet. 25, 214–226.

Novroski, N.M.M., Woerner, A.E., Budowle, B., 2018a. Potential highly polymorphic short tandem repeat markers for enhanced identity testing. Forensic Sci. Int. Genet. 37, 162–171.

Novroski, N.M.M., Wendt, F.R., Woerner, A.E., Bus, M.M., Coble, M., Budowle, B., 2018b. Expanding beyond the current core STR loci: an exploration of 73 STR markers with increased diversity for enhanced DNA mixture deconvolution. Forensic Sci. Int. Genet. 38, 121–129.

Oldoni, F., Kidd, K.K., Podini, D., 2018. Microhaplotypes in forensic genetics. Forensic Sci. Int. Genet. 38, 54–69.

Ottens, R., Taylor, D., Abarno, D., Linacre, A., 2013a. Optimising direct PCR from anagen hair samples. Forensic Sci. Int. Genet. Suppl. Ser. 4, e109–e110.

Ottens, R., Taylor, D., Abarno, D., Linacre, A., 2013b. Successful direct amplification of nuclear markers from a single hair follicle. Forensic Sci. Med. Pathol. 9 (2), 238–243.

Pakstis, A.J., Speed, W.C., Fang, R., Hyland, F.C., Furtado, M.R., Kidd, J.R., Kidd, K.K., 2010. SNPs for a universal individual identification panel. Hum. Genet. 127, 315–324.

Peck M., Idrizbegovic S., Bittner F., Parsons T., 2018. Optimization and performance of a very large MPS SNP panel for missing persons, in: Presented at 7th QIAGEN Investigator Forum, San Antonio, TX. Available from: https://www.slideshare.net/QIAGENscience/icmp-mps-snp-panel-for-missing-persons-michelle-peck-et-al.

Phillips, C., Fang, R., Ballard, D., Fondevila, M., Harrison, C., Hyland, F., Musgrave-Brown, E., Proff, C., Ramos-Luis, E., Sobrino, B., Carracedo, A., Furtado, M.R., Syndercombe Court, D., Schneider, P.M., SNPforID Consortium, 2007. Evaluation of the GenPlex SNP typing system and a 49plex forensic marker panel. Forensic Sci. Int. Genet. 1, 180–185.

Quail, M.A., Smith, M., Coupland, P., Otto, T.D., Harris, S.R., Connor, T.R., Bertoni, A., Swerdlow, H.P., Gu, Y., 2012. A tale of three next generation sequencing platforms: comparison of Ion Torrent, Pacific Biosciences and Illumina MiSeq sequencers. BMC Genomics 13, 341.

Roeder, A.D., Elsmore, P., Greenhalgh, M., McDonald, A., 2009. Maximizing DNA profiling success from sub-optimal quantities of DNA: a staged approach. Forensic Sci. Int. Genet. 3, 128–137.

Ronaghi, M., 2001. Pyrosequencing sheds light on DNA sequencing. Genome Res. 11, 3–11.

Ronaghi, M., Karamohamed, S., Pettersson, B., Uhlén, M., Nyrén, P., 1996. Real-time DNA sequencing using detection of pyrophosphate release. Anal. Biochem. 1 (1), 84–89.

Ronaghi, M., Pettersson, B., Uhlen, M., Nyren, P., 1998a. PCR-introduced loop structure as primer in DNA sequencing. BioTechniques 25, 876–884.

Ronaghi, M., Uhlen, M., Nyren, P., 1998b. A sequencing method based on real-time pyrophosphate. Science 281, 363–365.

Rothberg, J.M., Leamon, J.H., 2008. The development and impact of 454 sequencing. Nat. Biotechnol. 26, 1117–1124.

Rothberg, J.M., Hinz, W., Rearick, T.M., Schultz, J., Mileski, W., Davey, M., Leamon, J.H., Johnson, K., Milgrew, M.J., Edwards, M., Hoon, J., Simons, J.F., Marran, D., Myers, J.W., Davidson, J.F., Branting, A., Nobile, J.R., Puc, B.P., Light, D., Clark, T.A., Huber, M., Branciforte, J.T., Stoner, I.B., Cawley, S.E., Lyons, M., Fu, Y., Homer, N., Sedova, M., Miao, X., Reed, B., Sabina, J., Feierstein, E., Schorn, M., Alanjary, M., Dimalanta, E., Dressman, D., Kasinskas,

R., Sokolsky, T., Fidanza, J.A., Namsaraev, E., McKernan, K.J., Williams, A., Roth, G.T., Bustillo, J., 2011. An integrated semiconductor device enabling non-optical genome sequencing. Nature 475, 348–352.

Sanger, F., Nicklen, S., Coulson, A.R., 1977. DNA sequencing with chain-terminating inhibitors. Proc. Natl. Acad. Sci. 74 (12), 5463–5467.

Schroth, G.P., Lloyd, D., Zhang, L., 2007. Isothermal solid-phase amplification methods for creating clonal single nucleic acid molecule arrays on a planar surface using universal primers. PCT Int. Appl. WO2007107710A1.

Scientific Working Group on DNA Analysis Methods (SWGDAM), 2017. Scientific Working Group on DNA Analysis Methods Position Statement on Rapid DNA Analysis. Available from: https://docs.wixstatic.com/ugd/4344b0_f84df0465a2243218757fac1a1ccffea.pdf.

Seo, S.B., King, J.L., Warshauer, D.W., Davis, C.P., Ge, J., Budowle, B., 2013. Single nucleotide polymorphism typing with massively parallel sequencing for human identification. Int. J. Legal Med. 127 (6), 1079–1086.

Shao, K., Ding, W., Wang, F., Li, H., Ma, D., Wang, H., 2011. Emulsion PCR: a high efficient way of PCR amplification of random DNA libraries in aptamer selection. PLoS One 6 (9), e24910. https://doi.org/10.1371/journal.pone.0024910.

Shendure, J., Ji, H., 2008. Next-generation DNA sequencing. Nat. Biotechnol. 26, 1135–1145.

Sherier, A.J., Kieser, R.E., Novroski, N.M.M., Wendt, F.R., King, J.L., Woerner, A.E., Ambers, A., Garofano, P., Budowle, B., 2019. Copan microFLOQ® Direct Swab collection of bloodstains, saliva, and semen on cotton cloth. Int. J. Legal Med. 134 (1), 45–54. https://doi.org/10.1007/s00414-019-02081-6.

Smith, A.M., Heisler, L.E., St Onge, R.P., Farias-Hesson, E., Wallace, I.M., Bodeau, J., Harris, A.N., Perry, K.M., Giaever, G., Pourmand, N., Nislow, C., 2010. Highly-multiplexed barcode sequencing: an efficient model for parallel analysis of pooled samples. Nucleic Acids Res. 38 (13), e142.

Sorensen, A., Berry, C., Bruce, D., Gahan, M.E., Hughes-Stamm, S., McNevin, D., 2016. Direct-to-PCR tissue preservation for DNA profiling. Int. J. Legal Med. 130, 607–613.

Stoljarova, M., King, J.L., Takahashi, M., Aaspollu, A., Budowle, B., 2016. Whole mitochondrial genome genetic diversity in an Estonian population sample. Int. J. Legal Med. 130 (1), 67–71.

Strobl, C., Eduardoff, M., Bus, M.M., Allen, M., Parson, W., 2018. Evaluation of the precision ID whole MtDNA genome panel for forensic analyses. Forensic Sci. Int. Genet. 35, 21–25.

Strobl, C., Cihlar, J.C., Lagace, R., Wootton, S., Roth, C., Huber, N., Schnaller, L., Zimmermann, B., Huber, G., Hong, S.L., Moura-Neto, R., Silva, R., Alshamali, F., Souto, L., Anslinger, K., Egyed, B., Jankova-Ajanovska, R., Casas-Vargas, A., Usaquén, W., Silva, D., Barletta-Carrillo, C., Tineo, D.H., Vullo, C., Würzner, R., Xavier, C., Gusmão, L., Niederstätter, H., Bodner, M., Budowle, B., Parson, W., 2019. Evaluation of mitogenome sequence concordance, heteroplasmy detection, and haplogrouping in a worldwide lineage study using the Precision ID mtDNA Whole Genome Panel. Forensic Sci. Int. Genet. 42, 244–251.

Syed, F., Grunenwald, H., Caruccio, N., 2009. Next-generation sequencing library preparation: simultaneous fragmentation and tagging using in vitro transposition. Nat. Methods 6, i–ii. https://doi.org/10.1038/nmeth.f.272.

Tan, E., Turingan, R.S., Hogan, C., Vasantgadkar, S., Palombo, L., Schumm, J.W., Selden, R.F., 2013. Fully integrated, fully automated generation of short tandem repeat profiles. Investig. Genet. 4 (1), 1–15.

Tao, R., 2019. Pilot study for forensic evaluations of the Precision ID GlobalFiler™ NGS STR Panel v2 with the Ion S5™ system. Forensic Sci. Int. Genet. 43, 102147.

Templeton, J.E., Taylor, D., Handt, O., Skuza, P., Linacre, A., 2015. Direct PCR improves the recovery of DNA from various substrates. J. Forensic Sci. 60, 1558–1562.

Thong, Z., Phua, Y.H., Loo, E.S., Shue, B.H., Syn, C.K.C., 2015. Investigative leads from DNA: casework experience from the IntegenX RapidHIT™ 200 System. Forensic Sci. Int. Genet. Suppl. Ser. 5, E69–E70.

Turingan, R.S., Vasantgadkar, S., Palombo, L., Hogan, C., Jiang, H., Tan, E., Selden, R.F., 2016. Rapid DNA analysis for automated processing and interpretation of low DNA content samples. Investig. Genet. 7, 2.

Turingan, R.S., Brown, J., Kaplun, L., Smith, J., Watson, J., Boyd, D.A., Steadman, D.W., Selden, R.F., 2019. Identification of human remains using rapid DNA analysis. Int. J. Legal Med. https://doi.org/10.1007/s00414-019-02186-y. Available from:.

van Dijk, E.L., Auger, H., Jaszczyszyn, Y., Thermes, C., 2014. Ten years of next-generation sequencing technology. Trends Genet. 30, 418–426.

Van Neste, C., Van Nieuwerburgh, F., Van Hoofsat, D., Deforce, D., 2012. Forensic STR analysis using massive parallel sequencing. Forensic Sci. Int. Genet. 6, 810–818.

Verogen, 2019. FBI Approves Verogen's Next-Gen Forensic DNA Technology for National DNA Index System (NDIS). Available from: https://verogen.com/ndis-approval-of-miseq-fgx/.

Voelkerding, K.V., Dames, S.A., Durtschi, J.D., 2009. Next-generation sequencing: from basic research to diagnostics. Clin. Chem. 55, 641–658.

Wang, G., Maher, E., Brennan, C., Chin, L., Leo, C., Kaur, M., Zhu, P., Rook, M., Wolfe, J.L., Makrigiorgos, G.M., 2004. DNA amplification method tolerant to sample degradation. Genome Res. 14 (11), 2357–2366.

Wendt, F.R., Zeng, X., Churchill, J.D., King, J.L., Budowle, B., 2016. Analysis of short tandem repeat and single nucleotide polymorphism loci from single-source samples using a custom HaloPlex target enrichment system panel. Am J Forensic Med Pathol 37 (2), 99–107.

Woerner, A.E., Ambers, A., Wendt, F.R., King, J.L., Moura-Neto, R.S., Silva, R., Budowle, B., 2018. Evaluation of the precision ID mtDNA whole genome panel on two massively parallel sequencing systems. Forensic Sci. Int. Genet. 36, 213–224.

Further reading

Vierstraete, A., 2012. *Next Generation Sequencing*. Personal Website. Available from: https://users.ugent.be/~avierstr/nextgen/nextgen.html.

SECTION IV

Analysis of genetic data recovered from skeletonized human remains

CHAPTER

17

Best practices in the development and effective use of a forensic DNA database for identification of missing persons and unidentified human remains

Michael Hennessey BGS, MBA

Human Identification Projects, Gene Codes Forensics, Inc., Ann Arbor, MI, United States

Introduction

When a genetic analyzer produces an output file with the deoxyribonucleic acid (DNA) profiles from a set of unknown samples, the next step is to import these results into a database to see if they match any other profiles. But where did these samples come from? How were they collected, labeled, and processed? And how might that affect the ability of the database to find a match?

If there is a match, how will it be reported to the appropriate department or agency for confirmation? How will the DNA results be integrated with the other methods of identification, such as anthropology and odontology? Can the DNA database provide the necessary information to support this next step in the process? In the next two sections, we will review key elements of these upstream and downstream activities to better understand the burdens they place on a forensic DNA database.

Upstream context—Sample collection

There are two parallel sample collection tracks in every human identification (HID) project: (1) antemortem (AM) sample collection and (2) postmortem (PM) sample collection. While similar in nature, there are a few key differences that will confront the database manager. We will examine each in turn, beginning with the AM sample collection process.

https://doi.org/10.1016/B978-0-12-815766-4.00018-2

Upstream context—Antemortem sample collection

In a typical disaster victim identification (DVI) project or missing persons investigation, an informant (usually a family member) will report someone missing and provide the responsible agency with information regarding the missing person, along with reference material that can be used in the identification effort (e.g., objects that can be tested for DNA, fingerprint records, dental X-rays, etc.). This intake process generates four types of data:

1. Information about the missing person that can be used *to facilitate an identification*: (a) physical description (e.g., height, biological sex, ethnicity, hair color, eye color), (b) body modifications (e.g., tattoos, body piercings), (c) associated property, and (d) signifying behaviors (e.g., religion, medical conditions, military service, etc.).
2. Information about the missing person that can be used *to manage the case file*: (a) name, (b) national identification number, (c) contact information, (d) date of birth (DOB), (e) employer, (f) marital status, etc.
3. Information about the *informant*: (a) name, (b) relation to the missing person, (c) contact information, etc.
4. Information *about the interview itself*: (a) location, (b) time and date, (c) name of interviewer, etc.

In computer science, the term "metadata" means "data about the data". For example, the timestamp when data is entered into a database is metadata. Based on this definition, only the information *about* the interview qualifies as metadata (point #4).

However, in the field of HID, the term "metadata" is widely used to mean *any* data collected during the AM interview. Thus, all four data types listed above are often referred to as "metadata" by practitioners in the field. You will need to keep this in mind when discussing the operation of your database, as IT professionals will not consider the missing person's name or age as being part of the metadata, but an analyst will. For the purposes of this chapter, we will use the term "metadata" in the broader sense, as commonly understood by the human identification community.

AM information is often recorded by hand and transferred to a computer later so that it exists in both analog and digital format. For some projects, intakes are recorded directly on a computer; in others, the information exists only on paper. The AM intake will also generate a critical component of the identification process: tracking numbers. Each missing person case is assigned a unique identifier (ID). In this chapter, we will refer to this as the Reported Missing (RM) ID. In addition, each family is often assigned an ID since there may be more than one RM ID per family. And each item collected for DNA testing is assigned a tracking ID, often (but not always) a barcode number.

Once the AM interview is completed, the items collected for DNA testing will be forwarded to the forensic biology lab, while the other reference items will proceed to their respective departments, i.e., odontology, fingerprints, etc. Meanwhile, the information collected during the intake is transferred to a central records section. Typically, only a subset of the interview data accompanies the items distributed to the different sections of the medical-legal investigative process. As such, a buccal swab accessioned by forensic biology may only have a minimal amount of metadata present.

A critical dynamic of the AM intake process is that multiple informants often provide reports for the same victim, usually in separate interviews. And quite often, they provide conflicting answers. Some of these variations are due to misunderstandings in the interview process, while at other times a given informant may not possess the most current information about the victim. For example, many parents do not know that their adult child has tattoos. In addition, there can be more than one correct answer to a question, and different informants will provide conflicting (yet accurate) answers. For example, expatriates will often go by a nickname in the local language. A missing person report filed by an employer might provide the nickname, while the spouse will give the full legal (birth) name.

Furthermore, the intake process requests a lot of information that an informant may not have readily available. And these interviews can be quite time-consuming and emotionally exhausting. As a result, many informants are unable to complete the interview in a single session and thus participate in multiple sessions spread out over time.

Even the most robust file management system will experience some errors when trying to consolidate all of these separate intakes from all of the different interviews. This can affect the data in two ways: (1) Reference samples and interview information may be filed in the wrong case folder. As a result, a razor collected for Victim A may be filed in the folder for Victim B. The subsequent DNA match of this razor to an unidentified human remain (UHR) will be linked to Victim B; and although this match may be accurate from a biological perspective, it will result in a misidentification; and (2) Duplicate case files for the same missing person may be created. When this happens, the reference material and personal information for a given person will be disbursed across separate case folders, with none of them having sufficient data to produce an identification.

These two issues affect the DNA database in different ways. In the first instance, the analyst cannot assume that a match in the DNA database is an identification. The chain of custody for each sample will need to be reviewed to ensure its integrity. In addition, a review of the entire case file can allow the analyst to recognize situations where data and/or material may have been mishandled. For example, if the intake data for Victim B mentioned that he did not shave due to religious beliefs, that should cast doubt on the reliability of the razor in his respective file. In the second instance, the DNA database manager needs to periodically audit AM data for cases that lack sufficient information to make an identification. This could indicate that information or reference samples are spread out over multiple RM IDs. Reviewing the intakes for missing persons with similar names and RM IDs can reveal duplicate cases, and thus can allow for consolidation within the system.

Upstream context—Postmortem sample collection

PM data is similar to AM data in several ways:

1. Recovery and analysis of UHRs will produce data that can be used to make an identification and manage a case file.
2. PM data may be recorded manually and/or electronically.
3. Tracking numbers should be generated for the UHR and the specimens collected for testing.

4. Specimens collected for testing are forwarded to the appropriate sections for further analyses (e.g., DNA, odontology, etc.), and should be accompanied by a subset of the PM data.
5. All PM case data should be filed in central records (e.g., in a database, on paper, or both).

A fundamental difference between AM and PM sample collections is that PM data collections do not have to contend with different family members participating in numerous interviews that provide conflicting information. However, the PM intake process presents its own data management challenges, not the least of which involves the fact that UHRs can be subject to repeated examinations and multiple sample collection events. These additional instances of UHR analysis can produce conflicting data, some of which can be attributed to variations in analytic methods or the manner in which data is recorded (e.g., recording height and weight using Imperial vs metric measurements, or estimating the age of a body as "10–14 years of age" vs "juvenile"). In other cases, discrepancies between different postmortem reports might indicate that one or more of the PM intakes contains inaccurate data. For this reason, all PM intakes for a given UHR need to be reconciled against each other to account for any discrepancies.

In addition, pulling UHRs from mortuary storage for resampling and reexaminations increases the likelihood of material handling errors, such as PM specimens being tagged with barcodes from the wrong body bag or returning UHRs to the wrong storage locker. For example, a body recovered from a mass grave might be given a body bag tracking ID of UHR-101 (i.e., because it was the one hundred and first body recovered). After the initial autopsy, it is stored in bin 101 in the morgue. It may be pulled for re-analysis a few times, and on one occasion a sample for DNA is collected but accidentally given the label UHR 110 before being sent to the lab for analysis. The subsequent DNA match of sample UHR 110 to reference DNA from "Family X" will prompt the agency to remove the UHR from mortuary storage bin 110 and deliver it to "Family X." Although the DNA match is scientifically accurate, it is also a misidentification.

Of particular importance are human ID projects where commingling is present due to high impact trauma (e.g., airplane or train crashes, explosions, building collapses, etc.). In such cases, the tissue of more than one person may become fused together in a manner not apparent to a pathologist. Two samples taken from such a set of remains may yield different DNA profiles for the simple reason that the tissues of two different people are present.

A human identification project has an increased vulnerability to these PM data and material handling errors when multiple agencies or organizations are involved in the recovery and processing of remains. Hence, it probably goes without saying that this potential is amplified in a multinational effort. Simply put, there are multiple ways of performing the same type of analysis and a myriad number of ways to record similar results. The acceptable and allowable variations within human identification analysis almost guarantee that no two agencies will process a body in exactly the same way. Even DNA is vulnerable, as different agencies may have different strategies regarding which types of biological material to sample for testing, how that sample is excised from the body, how it is packaged for delivery to the lab, and so on.

Ultimately, when a reference item or UHR is received by the forensic biology laboratory, it will be accessioned in the lab's evidence intake system along with any accompanying metadata. This accessioning data will be stored in a repository: a database, a manual filing system, or a combination of the two. After accession, the item will go through an evidence exam, extraction, quantification, amplification, and analysis. If the resulting DNA profile meets the validation standards and technical review requirements of the lab, then it is eligible for upload into the forensic genetic database for matching.

Downstream context—Match confirmation

A match in the DNA database is not an identification; it is only an investigational lead. As the preceding sections have made clear, any number of material handling or data handling errors can result in a DNA match that is incorrect. The process of confirming that a match is a valid identification is the *downstream* context in which the forensic DNA database exists. This process can go by many different names (e.g., match validation, administrative review, case audit, match confirmation, reconciliation, etc.). Regardless of the name, it consists of four steps, starting with the narrowest scope of data (i.e., the genetic profiles), and broadening out with each additional step to eventually encompass the results of other identification modalities (e.g., fingerprints, odontology, etc.).

The first step is the *DNA sample concordance check*. In this step, all of the DNA results for a case (both AM and PM) must be reviewed and any discrepancies in the profiles accounted for. For example, two personal effects are tested for RM 200 but only one matches UHR 101. This discrepancy is harmless if it can be shown that the family accidentally donated personal effects from other (living) members of the family in addition to the missing person. However, sometimes the discrepancy can be more serious (i.e., the second personal effect may belong to a different missing person and was mistakenly filed in the case folder for RM 200).

On the PM side, another consideration is whether every sample taken from the UHR yields concordant results. For example, three different samples may be taken from a body for DNA testing: muscle tissue, a bone, and a tooth. If the family references only match to the bone sample, then the results of the other two samples must be located and reviewed. It may be that the other samples yielded partial results and are below the match threshold, but are otherwise concordant with the bone sample. However, if the three samples possess conflicting alleles, it might be a case of commingled remains or, alternatively, could mean that an error has occurred during the testing process.

The second step *confirms the chain of custody* for all AM and PM samples in the case. Each item in the lab for a given case must be traced back to its original collection event. Then an audit must be conducted in the other direction, i.e., all reference items collected from the family and all PM samples taken at autopsy must be accounted for in the DNA lab's accessioning system (and DNA database if they were tested). It is not enough to count the samples; one must check the tracking ID numbers, descriptions, and even their time stamps. For example, a bone sample accessioned by the lab on a date *preceding* the date the body was autopsied would indicate an error has taken place.

The third step *checks for consistency among the AM and PM intakes* to see if there are any issues that could cast doubt on the reliability of the information. If the first autopsy report describes a torso with no limbs and a second autopsy notes that the body was examined for fingerprints, then there is a problem. In an example described previously, a direct match between a UHR to a razor for a man who does not shave due to religious reasons should be a cause for concern. As noted previously, only a small subset of the metadata from the collection process will be transferred to the DNA lab. From there, only a subset of that data will be imported into the forensic DNA database. This means that in order to perform step three, the analyst in forensic biology will need access to data not typically available in the laboratory's information system. The alternative is for the lab to delegate this step to the section/part of the agency where this data is stored. There is no particular advantage to either method. In larger agencies, this step is generally performed by a dedicated department that may not actually be part of the DNA testing lab; in smaller agencies, the forensic biologist examining the case may execute this task. Either way, it is not possible to perform this step if the DNA results cannot be traced back to the original intake files.

The fourth step is the *interdisciplinary review* to see if the other identification methods yielded results concordant with DNA. If there is a DNA match between UHR 456 and RM 777, but odontology links the same remains to RM 988, then the discrepancy will need to be resolved. However, this step might have limited applicability depending on the availability of reference materials or the conditions of the UHR (e.g., if the remains are skeletonized, then fingerprint analysis is not possible; if a mandible or maxilla is not present, then odontology is not relevant; and so on).

One modality that is almost always relevant is forensic anthropology. Suppose DNA matches UHR 456 to RM 777 and according to the AM intake record, RM 777 is an adult Caucasian male. If the anthropology examination of the case concludes that UHR 456 is a juvenile Asian female, then the identification should be placed on hold until the matter can be resolved. Since there is no way to assess ahead of time which cases will benefit from such analysis, it is strongly recommended that an anthropological review of all DNA matches be included as part of the standard operating process. This is especially critical in cases of skeletonized and fragmented remains where other identification methods are usually not possible. In short, forensic anthropology might be the only method available to cross-check the DNA results.

Like step three (*checking for intake consistency*), the fourth step is often delegated to a department outside of the DNA lab, especially in larger agencies. However, this does not mean that the DNA lab will not participate in this process in some manner, especially if this step reveals conflicting results between the various identification disciplines used in the case.

Up to this point, we have looked at the process from the point of view of the forensic genetics lab. That is, the starting point for this discussion is a DNA match. But all identifications in any agency should trigger an *interdisciplinary review*, which means that an identification originating in another part of the agency should eventually be brought to the attention of the DNA lab, where (at the very least) the database manager will need to perform the *DNA sample concordance check* (and quite possibly the *chain of custody check* as well).

What each of these four steps in the confirmation process has in common is that they rely on the ability of an analyst to reliably (and efficiently) locate all of the samples and records for a case. For example, in the *sample concordance check*, an analyst needs to locate all of the reference samples for a given RM ID in the DNA database. There are two common methods to enable this. In the first method, the unique ID number for each RM is inserted in the sample name of each profile for that case. Then the user can query the sample name field of the database for all samples that contain that RM ID, usually in combination with wildcard characters.

However, embedding the RM ID in the sample name may not be an option. In some projects, the tracking number generated for the reference item at the collection step may not contain the RM ID. Replacing the collection ID with a new label that includes the RM ID will make it difficult to trace backward from the lab to the collection step when performing the chain of custody check. In addition, some laboratories assign each sample a unique tracking number at accessioning and *that* number becomes the sample name in the DNA database.

Appending the collection ID (or lab accession ID) with the RM ID is an option, although not ideal because this can produce overly long sample names that are cumbersome to work with. An alternate solution is to create a separate field in the forensic DNA database to store the RM ID for each sample. Then one must ensure that the RM ID is entered into that field for each sample when the profile is loaded in the database. With each sample linked to its RM ID, the analyst only needs to search the database's RM ID field to find all of the reference samples for a specific missing person.

While including the RM ID in the sample name is functional, the author suggests including a dedicated RM ID field in the DNA database rather than relying on the RM ID being in the sample name. For one thing, embedding the RM ID in the sample name means the analyst will almost always have to use some form of a wild card in their sample name search. For example, assume three reference samples are collected from RM 5256, and they are given the sample names AM-5256-1, AM-5256-2, and AM-5256-3. The user would then search by "AM 5256-*" to find all three. While this is not overly burdensome, it is also not overly efficient and can fail to locate relevant samples if there are any inconsistencies in the execution of the sample naming system.

A more serious concern is that, as noted, mistakes can occur in the handling of reference samples, with the result that a sample can have the wrong case attribution. If it turns out that AM-5256-3 actually belongs to RM 5255, then there is a significant problem. By the time the mistake is discovered, the errant sample name will already exist in different databases in the identification process (including, possibly, the DNA database itself), as well as in myriad hard copy reports. Changing the sample's name to AM-5255-1 will make it very difficult for anyone working on the case to trace/track its chain of custody.

Furthermore, even when not performing a dedicated chain of custody check, an analyst working with AM-5255-1 will be puzzled at the lack of "history" for the sample (i.e., no collection records, no accessioning record, etc.). The sample will seem to have appeared in the database out of thin air. Conversely, an analyst working on the case that starts with AM-5256-3 from either the collection records or laboratory accessioning database will be dismayed to discover that its profile cannot be found in the forensic DNA database. It will seem to have disappeared.

However, if there is a dedicated RM ID field, then this value can be updated and the sample name can remain consistent throughout the identification process. Although it is generally not a good idea to change tracking numbers for an object in the middle of a project, updating the RM ID for a reference sample will likely cause less confusion than editing the sample name itself. And as might be expected, a format that embraces both methods (RM ID as a dedicated field and embedding the RM ID in the sample name) offers the most robust solution.

If the AM samples for a given missing person are not linked to their RM ID in the database by either method previously described, then the analyst will have to look up the list of samples collected for that missing person in a different database or manually in the hard copies of the collection intakes. They will then have to search the DNA database one sample at a time. Such an approach is not efficient or reliable. The same logic applies to the PM results.

If samples in the DNA database are not linked to the body that they were taken from, the analyst will have to consult other resources to determine the list of profiles that need to be reviewed and then manually query the DNA database to resolve each of the PM samples that were not included in the original match. Again, this would be error-prone and slow.

A similar concern exists at the *chain of custody* (second) step. Whether an individual analyst queries the AM and PM collection records to verify the chain of custody or delegates this task to another part of the agency, the process will still rely on an efficient method of connecting the sample in the DNA database to the original collection event. If a UHR is given the body bag number ABC-123 and the profile from that body's sample is in the DNA database as 19-65-WXWZ, then executing the chain of custody review will be problematic at best.

We will omit here a detailed discussion of the remaining steps in the match confirmation process. This is not only for the sake of brevity but also in recognition of the dynamic that the later steps are more removed from the actual DNA lab setting and are probably handled by a different department within the agency. For our purposes, it is sufficient for the DNA analyst to be aware that their work will need to be connected to a general case audit and an interdisciplinary review. The reader can probably discern for themselves the possible pitfalls that could paralyze the process at these steps if the match results in the DNA database cannot be efficiently linked to the overall case data via common tracking ID codes.

Context—Conclusion

A forensic genetics database sits in the middle of a human identification process and, in order to maximize its utility, it is necessary to account for how the samples were handled in the preceding steps by other sections of the lab, other departments in the agency, and other agencies in the identification effort (not to mention supporting NGOs and possibly even foreign governments). The reality is that management of the data takes place well before the profiles are loaded into the database. At the other end of the process, the results of a DNA match will have to be reconciled against the other genetic profiles associated with the case as well as with the metadata. This is the context in which a forensic genetics database operates. The rest of this chapter will discuss a few of the more common challenges this context creates and how a database manager can overcome them, in order to not only improve the functionality of the database but to enhance the effectiveness of the overall identification effort. But first, we need to develop an understanding of how a database actually operates.

Relational databases

There are many moving parts in the identification process (i.e., the person reported missing, the remains, the DNA profile obtained, etc.). A basic concept in data management is quantifying the relationships between different parts of a process in terms of:

- one-to-one (1:1)
- one-to-many (1:M)
- many-to-many (M:M)
- many-to-one (M:1)

For example, there may be numerous family members who provide reference samples for each missing person. Thus, the "reference sample to missing person" relationship is M:1. On the other hand, a mass grave with multiple skeletal remains would have a relationship of 1:M between the recovery site and the UHRs. Meanwhile, each person should have one national identity number and a national ID should correspond to just one person, making this a 1:1 relationship. When discussing data management challenges with IT staff and consultants, not only will you need to list the parts of the process, but it will help to be able to quantify the relationships between them.

In data management terms, the parts of a process are often referred to "objects." These "objects" can be people, things, activities, events, or even abstract concepts. The list of objects can be articulated at both at a high level and at a granular level. It is not an exaggeration to say that if something can be referred to with a label, then it is an "object" in a database. For example, the assay kit used to create a genetic profile would be the "profile assay kit object," while the family interview could be called the "AM intake object."

A common method of recording data is to create a table with columns for each characteristic that the user wishes to record and a row for each instance of that object. This is the format used by popular spreadsheet programs such as Excel, and most readers are probably familiar with this basic concept. For example, the table below contains some basic information regarding the DNA testing of a sample (where "PPFUSION" refers to the PowerPlex Fusion kit by Promega Corporation, and "3100" refers to the Applied Biosystems 3100 Genetic Analyzer).

Sample name	Analyst	Date	Kit	Instrument
DNA-1234	CJR	5/22/20	PPFUSION	3100-A
DNA-1235	KB	5/22/20	PPFUSION	3100-A
DNA-1236	CJR	5/22/20	PPFUSION	3100-B

This type of table efficiently connects a specific sample to the analyst who performed the analysis, the date the sample was run, the kit that was used, and the instrument on which the analysis took place. Because the data in a table is organized along two dimensions (i.e., horizontal for the characteristics, vertical for each instance of the object), spreadsheets are sometimes referred to as 2D databases. While easy to create and understand, 2D databases suffer from a basic limitation.

Suppose sample DNA-1234 only yielded a partial DNA profile and had to be re-analyzed. There are two ways to add information about the second test in a 2D database. One way would be to add columns to the table to record the details of the second test:

Sample name	Analyst 1	Date 1	Kit 1	Instrument 1	Analyst 2	Date 2	Kit 2
DNA-1234	CJR	5/22/20	PPFUSION	3100-A	MJH	5/25/20	GFiler
DNA-1235	KB	5/22/20	PPFUSION	3100-A			
DNA-1236	CJR	5/22/20	PPFUSION	3100-B			

A second method would be to add another row in the table for the same sample:

Sample name	Analyst	Date	Kit	Instrument
DNA-1234	CJR	5/22/20	PPFUSION	3100-A
DNA-1235	KB	5/22/20	PPFUSION	3100-A
DNA-1236	CJR	5/22/20	PPFUSION	3100-B
DNA-1234	MJH	5/25/20	GFiler	3100-A

The first method has the advantage of allowing the user to see all of the related activities for a given sample in one row. But if the sample is tested multiple times (as is frequently the situation when working with skeletonized remains), then the table will need the complete set of characteristics in the header row to be repeated for each test result. If there are a dozen parameters and a set of bones is tested 5 times, then the table will have 60 columns. Such a table would be cumbersome to use, as once you start scrolling to the right to see the data associated with the results, it is easy to lose your place.

The second method has the advantage of keeping the table in manageable proportions. And since you can sort it by sample name, it is possible to cluster the data in such a way as to see all relevant records in one place. For this reason, adding rows to a table to account for multiple versions of the same object is the solution most often adopted. However, this method also has some limitations, not the least of which is that any variation in the way sample names are recorded can make it difficult to reliably find all of the related records when sorting the table by sample name.

The main point here is that a 2D database is best suited for data where there is a 1:1 relationship between the objects. That is, as long as there is only one test attempt per sample, then a 2D database is probably going to be sufficient for a project with a limited number of samples. However, if the relationship is 1:M, M:M, or M:1, then the tabular structure of a spreadsheet becomes a serious limitation. While there are many tricks a user can employ to manage such data, it can quickly become a complicated file that is inflexible and difficult to maintain. A better tool is a *relational database.*

Rather than using a single table that includes all relevant characteristics for an object, a relational database relies on numerous tables, each with a dedicated purpose. For example, there could be a table called "Sample Runs" which lists all of the runs a sample was tested in, as follows:

Sample name	Run 1	Run 2	Run 3	Run 4	Run 5
DNA-1234	67	68	71	83	
DNA-1235	67				
DNA-1236	67	69	83		

Next, there could be a separate table called "Instrument Runs" which lists the details of each individual run:

Run	Instrument	Date	Sample name 1	Sample name 2	Sample name 3
Run 66	3100B	6/7/20	DNA-0011	DNA0-0012	DNA-0013
Run 67	3100A	6/9/20	DNA-1234	DNA-1235	DNA-1236
Run 68	310	6/10/20	DNA-1234		

One could query the first table to see all of the runs in which sample DNA-1234 was tested. Or you could query the second table to see all of the samples tested in run number 67. These queries use Structured Query Language (SQL) and thus it is common to hear IT professionals refer to "using SQL" to retrieve data from a database.

The real power of relational databases is the ability to combine data from different tables. One example is to look up a piece of information in one table and use that result to query a different table to answer a second question. An example of a two-part query would be where you want to know the date that the second profile for DNA-1234 was produced. The first table could be queried for that sample's "Run 2" number (which is 68). Then the second table is queried for the run 68's "Date" to retrieve the answer of "6/10/20." Another example of structuring data from different tables would be to combine the sample name with the run number to produce a ID for each sample result, and then use that value to query a third table that has the allele values at each locus for each test result in the database. Relational database software can query the tables in any combination imaginable to retrieve discrete pieces of information.

The software in a relational database connects (or "relates") the relevant details of a specific object from different tables and displays them in a user-friendly format onscreen. This creates the illusion that all of the data exists in situ like a spreadsheet, and that it is just a matter of "opening" the page to see the information. In reality, the data in a relational database is fragmented into many small tables and must be assembled "on the fly" when queried by a command that the user invokes.

While finding and displaying specific pieces of data in an orderly fashion is certainly an important function of a database, even more critical is the ability of a database to perform operations on disparate pieces of data to create new data. Consider a table that records the DNA donors for each RM and a separate table that records the relationship of each family reference to the missing person. For example, Donor 1 and Donor 2 both gave reference DNA in the search for RM 1. Donor 1 is listed as the mother, and Donor 2 is listed as the father. Using this information, the software can create a pedigree with both donors in the correct familial roles.

Next, a query can look up DNA profiles of the donor samples after they have been tested. Based on the rules of genetic inheritance that a child must inherit an allele from each parent at each locus (barring mutations), the database software can calculate all of the possible allele combinations that a child of Donor 1 and Donor 2 could (and must) have at each locus. It can apply these results to the screening of DNA profiles from UHRs to exclude those samples that are not consistent with the two donors in the role of "parents."

The key point here is that the user does not create the pedigree or the screening criteria. The software creates this data after combining information from three different tables: (1) which

RM the donors are linked to, (2) their relationship to the missing person, and (3) the DNA profiles linked to their samples. However, if the database lacks the necessary information to relate these tables to each other, then there is no way for the software to create this data.

A frequent discussion point when considering the workflow in the lab is the desirability of automatically downloading results from the genetic analyzer directly into the DNA database. While such a seamless and automated flow of data certainly sounds appealing, it misunderstands the fundamental nature of how a database functions. The data file from the genetic analyzer usually does not include any metadata provided by the family and instead contains only the sample name, markers tested, and alleles recovered. In order to relate the data tables to each other in a manner that the user finds valuable and then perform operations on the data (e.g., creating family pedigrees), the database needs to know who the donor samples are related to and in what manner. This is why it is essential to have a step in the process before the profiles are loaded into the genetic database in which some of the metadata is integrated with the DNA profile. In other words, importing the profiles directly from the genetic analyzer into a database without any metadata will result in a list of samples in a computer, not a relational database.

For software to be able to query individual tables for discrete pieces of data, it is necessary for the first column in each table to contain a unique value in each row (or "record"). In database terms, this unique value is called a "key." This is why the tracking numbers generated in the upstream sample collection and testing process are so important (i.e., RM IDs, informant IDs, references sample IDs, body bag IDs, etc.), as they all can serve as keys in the various tables of a relational database (but only if they are unique).

Consider a tracking system that assigns each bone fragment a consecutive counting number as they are recovered from a mass grave, e.g., 1, 2, 3, etc. If there are multiple mass graves and this same approach is used for each, then there will be several bones with the tracking ID of "1." For this reason, for a project in which objects can enter the system from more than one source, there should be a prefix code to denote the source of the object. For example, in Spanish speaking countries, the recovery of remains might use the prefix "F" for "Fosa" (grave) followed by an ID for the grave (such as a counting number or village name). Hence, the 200th bone fragment from grave 16 would be F16-200.

A similar logic applies to the creation of RM IDs and DNA reference collection tracking numbers. If there are multiple intake centers and each center uses the same method of issuing incremental counting numbers, then there will multiple RM 1's, multiple donor 1's, etc. Including a prefix code to represent each intake center is a simple and reliable method to ensure that each family interview will produce unique tracking numbers across the entire project, and thus allows the creation of keys in relational database tables.

While creating unique tracking numbers for the main objects in the system is critical, a balance needs to be struck between the needs of the database and the needs of the human operator. It is tempting to create 16-digit bar codes for every missing person, reference sample, DNA profile, and bone fragment, as this would guarantee unique keys for every table in the database. However, a tracking ID of "1234-5678-9876-5432" is not very user-friendly because it not does not convey enough information to allow a person to understand what they are looking at. This is an important point to consider for human identification projects, as many IT professionals will default to implementing such bar codes across all phases of the project unless directed otherwise.

At a bare minimum, the format of the tracking codes for different objects in a human identification project should be project-specific, so that they cannot be confused with samples and cases from different projects handled by the same agency. The codes for DNA samples should indicate if they are PM or AM profiles and which case they belong to, as well as a suffix to indicate the test attempt on the sample. This last point is important because skeletonized remains often yield partial DNA profiles so they may be tested several times. In such cases, it will be quite helpful for the user to be able to see the case number in the sample name of each result, along with a unique suffix to be able to tell the results apart.

While inclusion of some data in the ID codes to allow a human operator to understand what they are looking at is beneficial, there is a limit to what is practical. Sample names that include the biological relationship of the donor to the missing should be avoided, as this information is self-reported by the donor and can be wrong. Additionally, there will be situations in which the donor is related to more than one missing person, making the "single relationship" code in the sample name misleading.

The author has also observed sample naming systems which attempt to include all iterations of the different steps in the laboratory process (i.e., extraction, amplification, etc.), so that the sample name becomes a faux laboratory information management system (LIMS). Attempting to include such information, as well as the name of the analyst, the kit used, the instrument name, etc. will produce a cumbersome tracking code that makes it more difficult for the human operator, not easier.

Data mapping

A useful way to think about data mapping is to consider the analogy of moving from one house to another. For example, assume a house in New York City has a single bedroom, bathroom, kitchen, living room, and basement. Suppose the resident of this house in New York City decides to move to Guatemala and finds a two-bedroom house with a kitchen, bathroom, and living room. The house in Guatemala does not have a basement, but it does have a patio in back. When it comes time to move, the resident packs his/her possessions in boxes and ships them to Guatemala where a moving company will deliver them to the new house.

How will the moving company know which box goes in which room of the new place? The answer seems easy, i.e., each box can be labeled by room (i.e., "bedroom," "bathroom," "living room," etc.). However, it turns out this solution is not so easy after all because the movers cannot read English (i.e., they only understand Spanish, so the box labels will not help). If the resident wants the move to be executed correctly, he/she will need to translate the box labels into Spanish. Assume the resident looks up the word "bedroom" using an online translator, and it generates the answer "habitación" (and so this label is added to the "bedroom" boxes).

While "habitación" might commonly be used in the Spanish-speaking world to refer to a bedroom, it also has the generic meaning of "room" and thus does not really help the moving crew. Instead, upon further consultation with native Spanish speakers, more accurate, and specific labels can be added to the boxes (e.g., "dormitorio" for the "bedroom" boxes, "cocina" for the "kitchen" boxes, "baño" for the "bathroom" boxes, etc.).

Now that the moving team can accurately read the box labels, new questions may arise. For example, recall that the home in New York City has one bedroom but the new place in Guatemala has *two* bedrooms. Which bedroom should the moving company use to store the boxes labeled "dormitorio"? If the resident wants some of the boxes delivered to the first bedroom and others to the second bedroom, then this should be specified on the label. Another question from the moving crew might be how to handle boxes labeled "sótano," which means "basement." Since the new place in Guatemala does not have a basement, it is not clear where these boxes should go. Until the instructions are clarified, these boxes will remain on the truck. Additionally, the moving crew might be concerned that not a single box is destined for the patio (and might conclude that the boxes marked "patio" were lost in transit, or that a mistake was made). The moving crew might perhaps assume the boxes from the basement should go on the patio.

Moving data from one place to another is like moving your household possessions from New York City to Guatemala, i.e., the rooms in your house are the tables in your database, and the boxes are the data. And, just like boxes of dishes and bowls need to be correctly labeled in order for them to be delivered/stored in the new kitchen, project data needs proper labels in order for it to be stored in the correct tables of a new database. *Data mapping* is the process of "mapping" the data from one source to another by identifying which tables in the two systems store equivalent data.

A basic example of *data mapping* is the transfer of results from a genetic analyzer (instrument) to the genetic database. The output file from the instrument may store the sample name in a field labeled "Sample ID," while the database might use a field labeled "Sample Info" for the same purpose. While such differences may seem slight to a person, they are significantly different for a computer. For this reason, the analyst will need to populate an import template for the database with information from the analyzer (instrument) by mapping the entries in "Sample ID" to "Sample Info."

Much like moving from one house to another, you will need to account for situations in which the genetic database does not have a "room" that corresponds to a "room" in the instrument's data. While problematic in some instances, such gaps might not be an issue, as it may not be necessary to have direct fidelity between different databases. In other words, it may not be necessary to migrate all data from one system to another. Another challenge involves a situation in which one system uses a single field to store data that another system uses multiple fields to store. For example, the source data might have a single column for both allele values at a DNA marker, while the destination database has a dedicated (separate) column for each allele at each DNA marker.

Finally, generic terms like "case," "sample," or "evidence" have to be treated with caution as they can be used to mean different things, even within the same agency. If the department that collects reference samples from families uses the term "case" to refer to each missing person, and mortuary operations use the term "case" to refer to each body recovered, then the genetic database will need a method of differentiating the two different usages of the same term. Otherwise, when a match is made in the DNA database and the confirmation process begins, the analyst might mistakenly use the *family* case number in place of the *mortuary* case number, or vice versa. One suggestion is to consider using prefixes such as "AM" or "PM" to denote different meanings of such common terms.

Database performance

In discussing database performance, users often focus on the hardware, software, and size of the database. While each of these attributes can certainly impact the speed with which a database executes a task, there are two other factors that are often overlooked which can also have a significant impact on performance: (1) family structure, and (2) match criteria. Family structure mainly impacts pedigree screening, while match criteria can affect both kinship searches and direct matching.

In pedigree screening, the software creates a family tree based on the familial relationships reported by the donors and then screens UHRs to see which ones are not excluded as the missing family member. It performs this step by applying the rules of genetic inheritance to the alleles at each marker where the family reference sample and the UHR both have results. Consider, for example, a family trio in which two parents are searching for a missing child. A child inherits one allele from each parent at each marker. At the first genetic marker examined, if the mother has allele values A and B and the father has allele values C and D, then all of their children must have one of the following four allele combinations at this marker: AC, AD, BC, or BD.

When the software screens UHR samples, it will examine the alleles at the first marker. If that sample has any one of these four combinations, then that UHR is not excluded from the family tree at that particular genetic marker. In this case, the software will then repeat the analysis at the second DNA marker, then the third DNA marker, and so on. If the analysis at each marker does not produce an exclusion (i.e., a difference), then the UHR is not excluded as a match candidate to that family pedigree. However, if the unidentified sample has an allele combination other than AC, AD, BC, or BD at the first genetic marker, then it is not possible for that sample to be a child of those parents. The sample would be excluded and there would be no need for the software to repeat this analysis at the rest of the DNA markers for that profile. When working with a pedigree that includes the parents, it is usually not necessary for the software to examine the alleles at each marker for each unidentified sample in the database, because most such samples will be quickly excluded after only a few markers (note that this does not take into account mutations or the possibility that the relationships as reported by the family are incorrect).

The rules of genetic inheritance do not require that siblings share any alleles at a given DNA marker. Therefore, in a pedigree in which the sole family member (donor) is a sibling, it is not possible to exclude UHR from consideration after analyzing the first few DNA markers. In this scenario, it is necessary to examine all of the alleles at each genetic marker for each UHR in the database. The same logic also extends to pedigrees consisting solely of distant relatives, such as cousins, aunt/uncles, nieces/nephews, grandparents, and grandchildren.

As a result, in two databases with the same number of profiles, using the same software, and running on the same hardware, a database consisting of family trios will complete a pedigree screen much faster than a database made up of single sibling pedigrees or families that only have distant relations. This difference becomes magnified as kits with more markers are analyzed. That is, a database in which the samples are only tested at the original core 13 CODIS markers will operate faster than one in which the profiles were tested at 23 or 27 loci (the latter of which would incorporate the newly expanded 20 CODIS core set of loci).

Match criteria can also have a significant impact on performance. Consider a pedigree screen for a family in which, at the first genetic marker tested, the mother's genotype is 10,14 and the father's genotype is 16,20. The screening step only needs to consider four possible allele combinations among the UHRs for this DNA marker:

10,16
10,20
14,16
14,20

However, if the system allows for the possibility of a single step mutation for any of the parental alleles, then the possible alleles that could be inherited from the mother are 9, 10, 11, 13, 14, and 15. From the father, the possible alleles would be 15, 16, 17, 19, 20, and 21. Now the possible allele combinations that any of their children could have expands to include:

9,15
9,16
9,17
9,19
9,20
9,21
etc.

Allowing for mutations can dramatically increase the time necessary to conduct a pedigree screen because the number of UHR samples that can be excluded at each marker will be significantly reduced. In direct matching, lowering the number of markers needed to make a match will have a similar impact, especially in projects with skeletonized remains where one would expect a high percentage of samples to produce partial profiles. Consider a database with 10,000 UHRs in which half of the samples tested yielded results at fewer than 10 markers. Dropping the match criteria from 11 to 9 will effectively double the number of comparisons that need to take place.

This is not an online search

It is easy to become frustrated with the length of time that a genetic database needs to search for matches between profiles, especially when one has become conditioned to expect nearly instant results using Online Search Engines (OSE) such as Google or Bing to find the latest sports scores, the weather forecast, or to look up historical information. However, comparing the performance of these two types of searches is a bit misleading because they operate in fundamentally different ways.

First, an OSE does not have to restrict itself to providing precise results. A search for tomorrow's weather forecast for Moscow may return results for both the Russian capital and the small U.S. city of the same name in the state of Idaho. In addition, search results will likely include a mix of historical weather reports and current conditions, in addition to the requested future prediction. In short, an OSE can quickly display a list of "matches" because it does not have the burden of needing to filter out material that is only marginally related. Imagine searching for a match to a genetic profile that was generated using the PowerPlex 16 kit, in which the matching algorithm decided to return results for *any* profile that contained results at the pentanucleotide (Penta) markers based on the fact that your query included Pentas, even if those markers do not actually match your query. The precision required for a DNA match imposes extra work that an OSE can ignore.

Second, an OSE does not need to list every website that matches your query, i.e., it just needs to provide enough links for you to be able to find the information of interest. Obviously, this would not be an acceptable performance standard for a forensic DNA database where the user needs a complete list of all samples that match the query. This is especially true when working with skeletonized remains in which there is a high level of fragmentation and thus dozens of UHR samples may match the reference samples. In that setting, it would not be acceptable to display just the "top ten" matching remains. The need for comprehensive match results requires that a forensic genetic database perform an exhaustive evaluation of all possible data, not just a representative sample of the most likely matches.

Third, an OSE can deploy a host of methods to accelerate the delivery of match results that would be inappropriate in a forensic search. For example, many OSEs use algorithms to identify the most popular answer to a query based on the behavior of other users who asked the same or related questions. Obviously, a forensic database should not filter its match results based on what was "most popular" with other analysts.

In addition, the operators of an OSE can take into account the search context to anticipate queries and cache answers ahead of time to questions it anticipates a user will ask. For example, the World Cup is a very popular global sports competition, but it only takes place once every four years. Thus, the volume of queries related to the World Cup can vary significantly over time, with interest being correlated to the countries that qualify for the tournament. Because of this, service providers can reasonably forecast the volume of queries ahead of time, where the queries will originate from, and what topics will be searched. The software in a forensic genetic database cannot forecast ahead of time which bone samples will receive the most attention, and the pattern of queries does not tend to be seasonal. Therefore, many of the methods used by an OSE to optimize performance are not available to a forensic DNA database.

Expunging records

In response to privacy concerns, one functionality considered for many forensic genetic databases is the ability to expunge a record. That is, it should be possible to remove all of the entries in a database related to a specific person so that they cannot be searched or used in a search. While the intent of such functionality is laudable, expunging a person may not be a practical solution. Even worse, it may not provide the privacy one expects.

As previously discussed, a relational database consists of numerous tables, each dedicated to a discrete "object," such as DNA test results or a list of the reported missing. These tables are connected to each other by keys (unique entries in the first column of the table), and queries pull discrete pieces of data from separate tables to answer specific questions. This creates a web of interdependencies between tables, such that removing the record of one person from one table can inadvertently disrupt the behavior of the database by deleting a connection that another query relies upon for proper functioning. For example, removing a reference donor will disrupt any pedigree that they have previously been a member of, and this in turn will alter subsequent match results associated with that family. If such results had previously produced an identification, it will not be possible in the future to recreate that identification.

In addition, if "Person A" provided "Reference Sample 1" and the decision is made to remove "Person A" from the database, does this mean that "Reference Sample 1" should also be removed? Without the name of "Person A" in the database, it is easy to argue that their privacy will not be compromised if the anonymous sample persists in the database. On the other hand, if "Reference Sample 1" remains in the database and is included in a pedigree in which the other members are known, it might be possible to discern the identity of the donor of "Reference Sample 1" through extrapolation.

If "Reference Sample 1" is removed from the database, how will this impact the quality assurance tables that track the results of instrument runs? Will the sudden absence of that sample from this table be treated as failed result? Will there now be an unexplained gap in the sample upload history associated with the analyst who originally imported the profile into the database? Such factors are important considerations of "simply removing a person" from the database. Even if one successfully expunges a person from the database (along with all of their attendant metadata, connections, and queries), that person still exists in every database backup executed since they were imported into the database. For completeness, every backup file would have to be scrubbed of all records related to the person in question. Otherwise, a database restore operation using one of those backup files will inadvertently "undo" the expunge. A similar concern exists regarding the original import files. After expunging someone from the database, an analyst could mistakenly (or maliciously) reintroduce them into the database by reexecuting that import file. This is important because, in some labs, new samples are appended to existing files and then those files are imported again.

For these reasons, it is worth considering an alternative to expunging records to protect the privacy of individuals in a database. This could involve the implementation of a "flag" in the code which prevents that genetic profile from being used in any searches and prevents any personal data associated with that person to be displayed onscreen or in any reports. This would protect the integrity of the database, could not be accidentally undone, and would provide a reasonable level of privacy/protection to the person in question.

Specifying capacity

It is hardly an insight to point out that computing power advances with every new generation of hardware. One needs to look no further than smartphones to observe continuous improvements in memory and processing power with each new release. Unfortunately, many agencies do not take this reality into account when specifying the requirements for their genetic database(s). It is not uncommon to see an agency release a Request for Proposal (RFP), specifying a system that must be able to handle a large arbitrary number of profiles (e.g., one million). However, the agency's legacy database may only be a fraction of that size, and their lab's monthly throughput is such that it would literally take decades to reach the capacity requested in the RFP. This is wasteful on two counts.

Consider a simple piece of technology—the USB memory stick. It is no surprise that, at any point in time, the most advanced models on the market are the most expensive as well. When a more advanced successor is released, the legacy model drops in price. Subsequently, the model from two generations back can drop to a fraction of the cost of the current, most advanced model. If a 1 gigabyte (GB) thumb drive costs $10 and has sufficient memory to back up files on an agency's computer, but the agency pays $100 for a 1 terabyte (TB) memory stick, then the agency has overpaid by $90. Moreover, by the time the agency eventually reaches the stage of actually needing a full TB of portable data, the original memory stick purchased may be obsolete. Like any other piece of equipment, computing systems wear out over time and eventually fail. Even if the hardware is reliable for many years into the future, they rely on operating system software to make them work. In the past 20 years alone, Microsoft has discontinued updating several operating systems (e.g., Windows 7, Windows XP, Windows 2000, etc.). Despite these factors, agencies are reluctant to procure systems with just enough capacity to satisfy their near-term needs. This is understandable on some level, as one does not want to have to buy new equipment every 3–5 years if it is not necessary. However, the reality of technology "life cycles" is that very few commercially available systems on the market today will be functional 10–20 years from now, and will have to be replaced at that time anyway.

Conclusion

Database management is a discipline unto itself, and the analyst called upon to manage a forensic genetic database will have to devote substantial time and energy to understanding its behavior in order to maximize its effectiveness. Hopefully, this chapter has been an effective introduction to the subject, and the analyst now has a better appreciation for the challenge at hand. The principal lesson set forth is that a genetic DNA database operates in a larger context, and this reality informs how the database will (and should) be used. Of particular importance is the need to connect results of DNA match to the overall investigation, placing a premium on the ability to look up the corresponding sample names and case IDs throughout an agency or agency's records.

CHAPTER

18

Software and database functionality for direct identification and kinship analysis: The Mass Fatality Identification System (M-FISys)

Howard D. Cash CEO

Gene Codes Forensics, Inc., Ann Arbor, MI, United States

Introduction

This chapter discusses how software and databases help analysts use reference deoxyribonucleic acid (DNA) to assign an identity to unidentified human remains (UHRs), through Direct or Kinship matching. It also points out that DNA does not stand on its own, and illustrates why interdisciplinary work between different forensic specialties is such an important principle in human identification.

Forensics is all about comparing knowns to unknowns. A known dental record with a patient's name can be compared to a postmortem dental chart for an unidentified victim. A fingerprint record from a past arrest can be compared to latent prints from a crime scene. Genetic information is different than most forensic specialties; a direct comparison can be made between DNA from UHRs and a known source (e.g., a previously collected postconviction profile, a retained surgical sample, or a personal effect such as a toothbrush). However, if a direct reference sample is unavailable, a comparison to familial exemplars can be performed. Since we inherit half of our DNA from each of our two parents, it is possible to calculate the statistical significance of the DNA shared between an unknown source and various people who are merely genetic relatives of an individual. An allele at any locus in a child should match an allele at the same locus in one or both parents (ignoring mutations for now). This second method, using family members as reference material, is sometimes referred to as "indirect matching" or a part of the field of "kinship analysis." A simple form of indirect matching is used in common paternity tests. A newer approach, referred to as investigative genealogy, does not test individual loci but rather measures the total amount of shared DNA

https://doi.org/10.1016/B978-0-12-815766-4.00019-4

(in centiMorgans, cM) between an unknown sample and a reference sample. Investigative Genetic Genealogy is discussed in Chapter 20 of this book.

There are operational and tactical differences when identifying what appears to be the skeleton of one individual, and identifying potentially commingled bones in a common location like a mass grave or accident site. The majority of this chapter focuses on multivictim scenarios; however, simplifications that can be used for an individual (single) set of remains are also highlighted.

Several databases and software tools can be used, either alone or in combination, to identify skeletonized remains. The content of this chapter will illustrate some of the approaches to direct and indirect profile matching using a database and software system originally designed to help identify the badly decomposed and commingled remains at the site of the 9-11 attack on New York City's World Trade Center. This tool is called "M-FISys" (pronounced like *emphasis*) and is an acronym for the Mass-Fatality Identification System. Some features in M-FISys are particularly relevant to the study of skeletal remains because it was designed as a resource to help identify highly fragmented, highly compromised human remains. This origin notwithstanding, M-FISys has expanded over nearly two decades of continuous development and is now used for a wide range of applications (e.g., criminalistics, disaster victim identification [DVI], paternity testing, missing person investigations, child trafficking) at various state and national government laboratories around the world. This chapter will not cover all aspects of the system but will concentrate on those tools most relevant to skeletonized, commingled, and degraded remains, with a few references to project tracking that directly impact such projects.

Part of the design philosophy of M-FISys is worth explaining at the outset. The World Trade Center attack was a single, large mass fatality incident in which all victims' remains were present in a relatively contained area. In this event, 2749 people were killed and recovered as > 20,000 individual remains. The statutory responsibility for identifying the victims fell on a single agency, the New York City Office of the Chief Medical Examiner (OCME); this agency had the authority to recruit and assign collaborator labs and to require that those collaborators adopt common procedures. The OCME, as a crime laboratory, was staffed with scientists trained to follow investigations from evidence to a conclusion on a system of rotating assignments. An M-FISys design goal was to minimize the time for a new investigator to be able to begin work using a new set of tools when they rotated onto the World Trade Center DNA Identification Unit. To meet that goal, the program is broken up into functional areas so that a new team member might be trained initially on only a subsystem, such as profile uploading, data quality assurance (QA), direct matching, or kinship matching. Access to different subsystems is through the Control Panel, shown in Fig. 1.

This generally allowed new staff to begin providing productive input by the afternoon of their first day, even if they had been introduced to only a small part of the toolset. This functional organization varies from other forensic data management systems that break access down by areas of authority, such as ante-mortem processing, postmortem processing, and matching/reconciliation. One approach is not inherently superior to the others, and different levels of access can be granted to an individual user of M-FISys by the system administrator (if deemed a priority).

The previous chapter (Chapter 17, by M.J. Hennessey, Director of Human Identification Projects, Gene Codes Forensics, Inc.) discusses the process of adding metadata ("data about

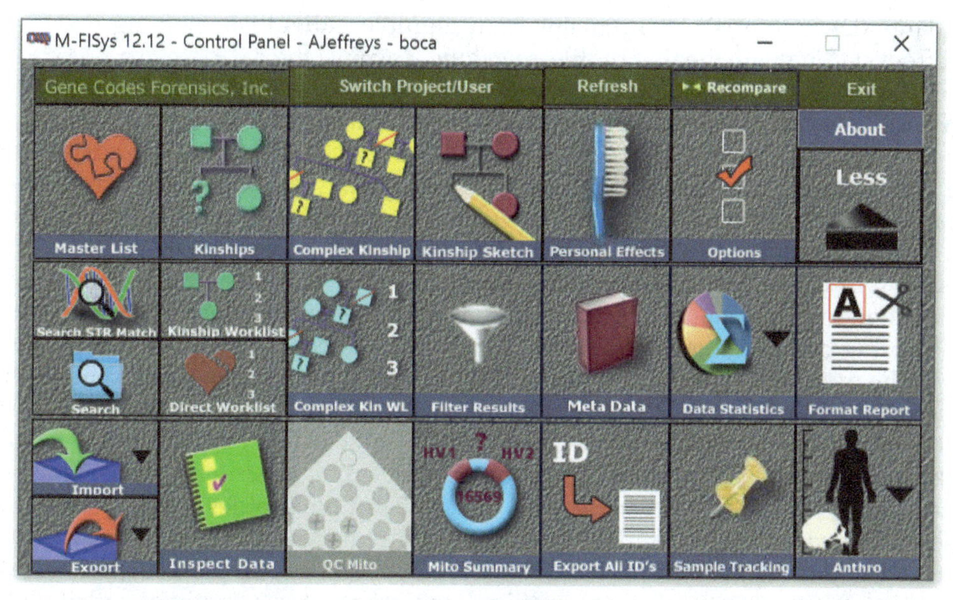

FIG. 1 Expanded control panel of the Mass Fatality Identification System (M-FISys): Direct match master list *(upper left)* and Anthro tools *(lower right)*.

the data") to genetic profiles that are developed in a laboratory. This metadata can include the multiplex reagent kit used to generate the profile, names and genetic relationships of familial sample donors, and descriptions of personal effects of missing people that might be used to generate reference profiles. In addition to descriptive data, the relational database at the core of M-FISys allows a laboratory to associate documents (e.g., digitized photographs, PDFs) with individual profiles.

Representing and directly comparing profiles

In situations where there is no presumptive ID based on non-DNA factors, there are two common ways to search a database of DNA profiles for a direct match: (1) a One-to-Many (1-M) comparison, and (2) Many-to-Many (M-M) comparisons. The One-to-Many approach (1-M) compares a profile from an unknown source against a database. This is typical in a crime scene investigation when a profile might be developed from an item of recovered evidence. When searched against a criminalistics database, one might be expecting to get one of only three categories of answers:

1. This is a *novel profile*. It is not in the current database.
2. This is a profile from an unknown person, but the same profile has been previously reported from, for example, four other crime scenes and uploaded to the database. This is an *investigational lead*. Perhaps investigators can identify an individual who has been at all five locations and might be the source of the sample.
3. This is a *known* person in the database. For example, it could be a previously convicted criminal. This is a person who may need to be located.

In a large database of skeletal remains from multiple individuals, nothing prevents an analyst from starting with the first profile, searching it against the database in a 1-M strategy, and then moving to the next profile; however, this would be grossly inefficient. One reason to take a Many-to-Many (M-M) approach is that it can reduce the search space. Two, twenty, or two-hundred remains from the same individual can be collapsed into a single aggregate profile if the samples are found to have the same genotype. A database with many duplicate profiles, as might be observed in scenarios with disarticulated and commingled skeletal remains, can thus be reduced to the unique profiles.

As shown in Fig. 2, each row displays a unique profile in the database, with the right half of the window showing allele values. In this case, the first column on the left contains an RM number based on the Reference Material from the person who has been Reported Missing. The number in parentheses (in the first column) shows how many samples have been tested and found to have congruent or at least consistent profiles (either ante-mortem or postmortem). For example, the first row labeled "RM 022 (15)" indicates that a total of 15 tested items share the same profile (within the selected match criteria), as the Reference Material for Reported Missing person #022. These 15 sample profiles collapsed into a single line are called an Aggregate Profile. As skeletal remains are being tested, they may be matched to each other even though there is no reference sample by which to identify the source. If there is a reference profile in the group, it is sometimes called an "RM Aggregate." If there is no reference sample, it is called an "Anonymous Set" or merely an "Aggregate Set."

To maintain scientific validity and data integrity, the User Interface (UI) allows the operator to drill down to the individual components of an aggregate. This was a fairly novel approach to data presentation when it was introduced for the World Trade Center identification effort. User Experience (UX) engineers include this in the category of "progressive disclosure," providing enough information in the UI to recognize the significance of the match while making it easy to see the supporting data. Clicking on the turnstile (⊞) at the leftmost edge of the profile exposes the substituent profiles, as shown in Fig. 3.

For training purposes, the names of postmortem samples are often prefaced with a "V-" for Victim, and ante-mortem references are prefaced with "PE-" for Personal Effects or "FR-"

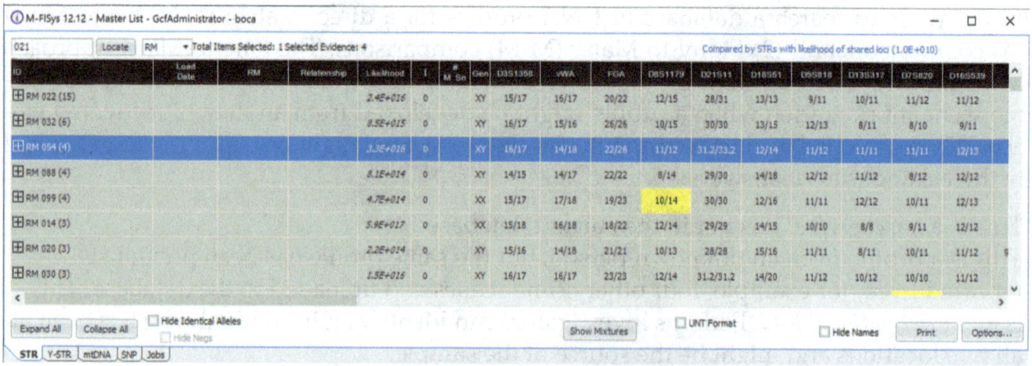

M-FISys 12.12 - Master List - GcfAdministrator - boca

021 Locate RM Total Items Selected: 1 Selected Evidence: 4

Compared by STRs with likelihood of shared loci (1.0E+010)

ID	Load Date	RM	Relationship	Likelihood	I	# M Sn	Gen	D3S1358	vWA	FGA	D8S1179	D21S11	D18S51	D5S818	D13S317	D7S820	D16S539
⊞ RM 022 (15)				2.4E+016	0		XY	15/17	16/17	20/22	12/15	28/31	13/13	9/11	10/11	11/12	11/12
⊞ RM 032 (6)				8.5E+015	0		XY	16/17	15/16	26/26	10/15	30/30	13/15	12/13	8/11	8/10	9/11
⊞ RM 054 (4)				3.3E+016	0		XY	16/17	14/18	22/26	11/12	31.2/33.2	12/14	11/12	11/11	11/11	12/13
⊞ RM 088 (4)				8.1E+014	0		XY	14/15	14/17	22/22	8/14	29/30	14/18	12/12	11/12	8/12	12/12
⊞ RM 099 (4)				4.7E+014	0		XX	15/17	17/18	19/23	10/14	30/30	12/16	11/11	12/12	10/11	12/13
⊞ RM 014 (3)				5.9E+017	0		XX	15/18	16/18	18/22	12/14	29/29	14/15	10/10	8/8	9/11	12/12
⊞ RM 020 (3)				2.2E+014	0		XY	15/16	14/18	21/21	10/13	28/28	15/16	11/11	8/11	10/11	11/12
⊞ RM 030 (3)				1.5E+016	0		XY	16/17	16/17	23/23	12/14	31.2/31.2	14/20	11/12	10/12	10/10	11/12

Expand All | Collapse All | Hide Identical Alleles | Hide Negs | Show Mixtures | UNT Format | Hide Names | Print | Options...

STR | Y-STR | mtDNA | SNP | Jobs

FIG. 2 Fully collapsed Master List of matched aggregate profiles in M-FISys.

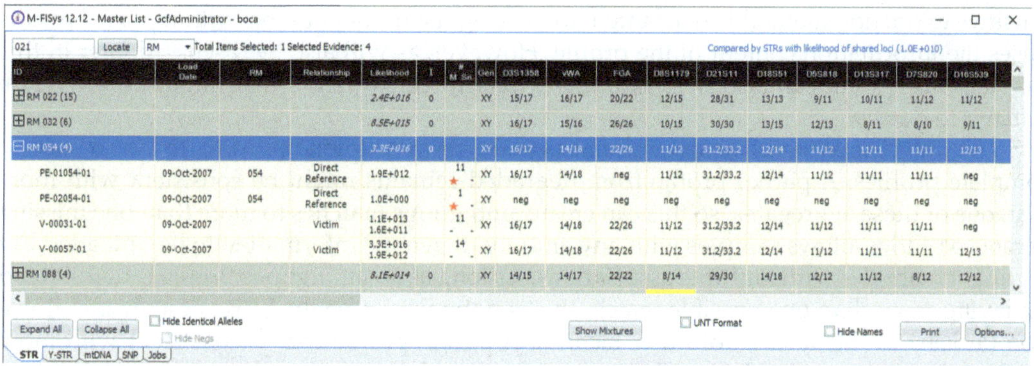

M-FISys 12.12 - Master List - GcfAdministrator - boca

021 Locate RM Total Items Selected: 1 Selected Evidence: 4

Compared by STRs with likelihood of shared loci (1.0E+010)

ID	Load Date	RM	Relationship	Likelihood	I	# M Sn	Gen	D3S1358	vWA	FGA	D8S1179	D21S11	D18S51	D5S818	D13S317	D7S820	D16S539
RM 022 (15)				2.4E+016	0		XY	15/17	16/17	20/22	12/15	28/31	13/13	9/11	10/11	11/12	11/12
RM 032 (6)				8.5E+015	0		XY	16/17	15/16	26/26	10/15	30/30	13/15	12/13	8/11	8/10	9/11
RM 054 (4)				3.3E+016	0		XY	16/17	14/18	22/26	11/12	31.2/33.2	12/14	11/12	11/11	11/11	12/13
PE-01054-01	09-Oct-2007	054	Direct Reference	1.9E+012		11 ★ -	XY	16/17	14/18	neg	11/12	31.2/33.2	12/14	11/12	11/11	11/11	neg
PE-02054-01	09-Oct-2007	054	Direct Reference	1.0E+000		1 ★ -	XY	neg	neg	neg	neg	neg	neg	neg	neg	neg	neg
V-00031-01	09-Oct-2007		Victim	1.1E+013 1.6E+011		11 - -	XY	16/17	14/18	22/26	11/12	31.2/33.2	12/14	11/12	11/11	11/11	neg
V-00057-01	09-Oct-2007		Victim	3.3E+016 1.9E+012		14 - -	XY	16/17	14/18	22/26	11/12	31.2/33.2	12/14	11/12	11/11	11/11	12/13
RM 088 (4)				8.1E+014	0		XY	14/15	14/17	22/22	8/14	29/30	14/18	12/12	11/12	8/12	12/12

Expand All · Collapse All · Hide Identical Alleles · Hide Negs · Show Mixtures · UNT Format · Hide Names · Print · Options...

STR · Y-STR · mtDNA · SNP · Jobs

FIG. 3 Master List with one RM Aggregate expanded.

for Family References. The M-FISys program ignores those characters and we only adopt these conventions to facilitate classroom learning. In the example shown in Fig. 3, an aggregate of matching profiles has been detected for two postmortem samples from a victim (i.e., V-00031-01 and V-00057-01) and two personal effects from a missing person (i.e., PE-01054-01 and PE-02054-01). The header row (highlighted in blue in Fig. 3) has a "(4)" notation after the name, so without even disclosing the contents to the aggregate, the operator knows how many consistent profiles have been found. This also shows an example where one of the two Direct References gave almost no results; i.e., in this case, only the gender locus yielded a result and all other loci are negative or below the laboratory's established reporting threshold (displayed here as "neg"). Given that an alternative Direct Reference yielded a full profile, it would be a policy decision whether or not to make a second attempt to generate a profile from the material that failed. PE-02054-01 is included in this aggregate because non-DNA information is being used in the absence of a profile. Laboratory analysis of sample "PE-02054-01" yielded no useful results, but the person who contributed the evidence declared that is was a "known" reference sample for this particular missing person, and it is therefore associated with the other reference sample. If these two items were to produce data that was in conflict, Quality Control (QC) tools (discussed later in this chapter) would alert the analyst.

When M-FISys is first launched, the default behavior is to compare every profile in the database to every other profile (i.e., a Many-to-Many, or M-M, comparison). This has the effect of re-associating profiles from all tested skeletal elements from the same individual in a single grouping. It is possible that not every sample will contain data at every locus and, if different laboratories contribute data used in a singular project, some items might not even be tested with the same multiplex kits. M-FISys distinguishes between loci that have no data because a locus was never tested (indicated by a dash, "–") and when a locus was tested but the experiment produced no data or a "negative" result that was below the laboratory's RFU threshold for reporting (indicated by the string "neg").

When an item of evidence is retested, the same principle of progressive disclosure can be applied. Imagine that a single piece of evidence (e.g., an ante-mortem reference or a postmortem sample) is tested a second or third time, perhaps because it is compromised and a more sensitive extraction method is warranted. Increasing the number of loci with reported results raises the information content of the profile. However, as will also be discussed later in this chapter, this testing can also help illuminate data integrity errors that might otherwise be difficult to detect.

Some database systems or operational policies restrict uploading data to full or nearly complete profiles. A partial profile from degraded remains might be consistent with more than one of these aggregates, so this can create ambiguous matches to more than one missing person. M-FISys allows samples with any amount of genetic information to be uploaded but highlights samples with multiple associations. As long as an ambiguous profile is highlighted as having more than one possible match, this is not going to be the cause of a misidentification (at least in the absence of other factors). A partial profile that is consistent with more than one aggregate is marked with a yellow flag at the beginning of the row, and tools are available to use other evidence to include or exclude that ambiguous match from one or more identifications.

Finally, the criteria for a match can be set individually or globally with a high degree of flexibility. Assignment to aggregate can be limited to only those matches that have a certain number of matching loci (with a specified number of mismatches allowed), or with a certain floor for how low the likelihood ratio (LR) is that this match would be found (Fig. 4).

The likelihood approach is preferred. If some allele values are rare, the statistical strength of a match can be high, even if the laboratory was able to determine allele values for fewer loci than another sample with more common alleles. The statistical value of a given allele at any locus is heavily influenced by the ethnic group studied to find population statistics. M-FISys can compute the likelihood of finding a particular profile in many ethnic groups simultaneously and, by default, displays the most conservative statistic found in any of the selected population studies.

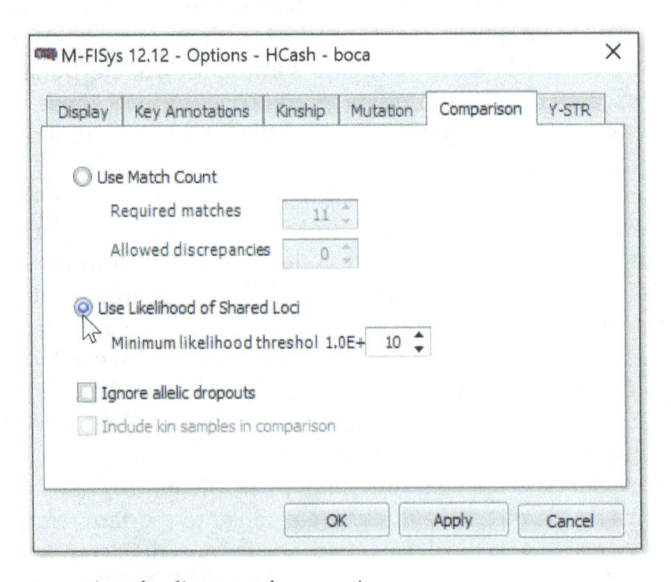

FIG. 4 Match parameter settings for direct match comparisons.

Direct matching

Once an M-M search has been performed, a Master List shows all the data in the database, with all aggregates collapsed. Search tools permit an operator to quickly look for a particular sample of interest by RM number, by the ID assigned to the sample in the lab, and even by the name of the missing person or family-reference donor. By the time the program first launches, all of the direct matching has already been done using the match parameters set by the laboratory or individual analyst. It is important to note that a "match" should not be used as a synonym for "an identification." Software can find a match; however, only an authorized analyst can declare an identification, and this is done after a review of all of the supporting data. When identification is declared and recorded in M-FISys, it is documented in the database with a time-and-date stamp along with the name of the analyst who confirmed the declaration.

In some organizations, individual cases are assigned to individual analysts, and they may choose to review their cases whenever new data is uploaded. Other organizations take advantage of a Direct Match Worklist, which shows all new matches which have not yet been declared as identified, starting with the matches with the greatest statistical strength. Within the worklist, cases can be assigned to individual lab members, preventing multiple staff members from inadvertently duplicating the work of others. If a staff member is not available on a certain day, or if he/she leaves the project, a supervisor can reassign those cases. In Fig. 5, a direct match worklist is shown with assignments listed in the column on the far right. The match with the highest likelihood (i.e., 2.8E + 021, 2.8×10^{21}) is the first item in the list, and other matches are listed beneath it, sorted by decreasing statistical strength.

Profiles are generally uploaded into M-FISys in a tab-delimited format, often organized using a spreadsheet program. The upload file includes the profiles and certain meta-data, such as the multiplex reagent kit used, the genetic relationship of a family member for a kinship reference, and/or an optional description of the item being tested. An external software tool

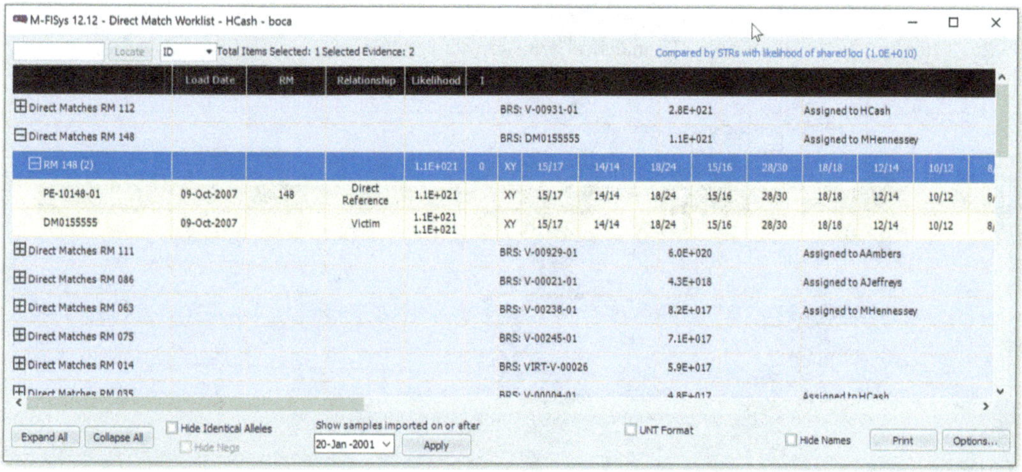

FIG. 5 Direct match worklist with case assignments.

called the "M-FISys Data Validator" can be run on the data before upload, in order to detect formatting errors and certain common mistakes such as data listed for alleles that are not used in the specified kit. Samples can also be manually typed or copied into a profile form in a process similar to what the FBI's CODIS software calls a "keyboard search," thereby checking a profile against the database without actually making that profile available in the database.

There are times when laboratory personnel want to look for matches that might be below the match threshold set for identifications. For example, if a high priority individual is being searched for among the remains of multiple victims, or if a family member of a victim is scheduled to visit the laboratory, it might be appropriate to see if there are any partial matches that are worthy of further investigation or analysis. Rather than re-compare the entire database, M-FISys has a facility to select individual cases and "Search for all possible matches with adjustable settings" (Fig. 6).

In another example (Fig. 7), selecting "Ignore Allelic Dropout" at the bottom right corner of the page allows the Direct Reference to match to Victim sample V-00224-01, despite the false homozygotes at FGA and D18.

Non-STR direct match assays

This section has primarily focused on autosomal STR profiles since STRs are the most commonly used data; however, M-FISys can also collect and search for matches and conflicts with Y-chromosome STR (Y-STR) data, mitochondrial DNA (mtDNA) profiles, and panels of single nucleotide polymorphisms (SNPs).

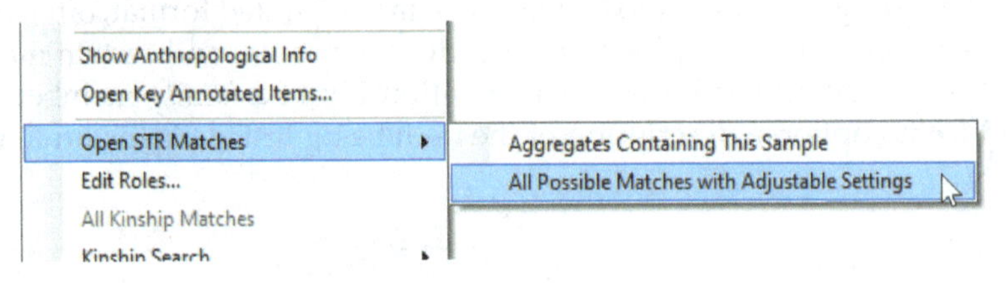

FIG. 6 Choosing "open all possible STR matches" with adjustable settings.

M-FISys 12.12 - M-FISys Matching STR Samples for PE-01034-01 - HCash - boca

Locate ID

Compared by STRs with likelihood of shared loci (1.0E+010)

ID	Load Date	RM	Relationship	Likelihood	I	# M Sn	Gen	D3S1358	vWA	FGA	D8S1179	D21S11	D18S51	D5S818
RM 034 (2)				3.4E+017	0		XY	16/17	16/16	20/24	10/14	32.2/33.2	12/14	12/12
PE-01034-01	09-Oct-2007	034	Direct Reference	3.4E+017		14 ★ -	XY	16/17	16/16	20/24	10/14	32.2/33.2	12/14	12/12
V-00224-01	09-Oct-2007		Victim	1.9E+018 3.0E+014	U	14 - -	XY	16/17	16/16	20/20	10/14	32.2/33.2	12/12	12/12

Expand All Collapse All Hide Identical Alleles Hide Negs UNT Format Hide Names Print Options...

STR Y-STR mtDNA SNP Jobs

Use Match Count Use Likelihood of Shared Loci Ignore allelic dropouts

Required matches 014 Minimum likelihood threshol 1.0E+ 10

Allowed discrepancies 1

FIG. 7 Example of adjusting match settings in a single case, to ignore allelic dropout.

Although the data is presented differently, the principles behind STR and SNP matching are essentially the same. The likelihood of a match is driven by the prevalence of each SNP type in a given ethnic group. If the loci are genetically unlinked, they can be treated as independent statistical events and their respective likelihoods can be multiplied together for an overall match likelihood. Allowing for experimental errors and allelic dropout, a large enough SNP array should be discriminating enough to present data that would be found to match a direct reference from a given person (and at the exclusion of all others). Allelic dropout, in the case of a given SNP, is more corrupting since the dynamic range of the locus is so much smaller than that of an STR. There may be dozens of measured alleles at a given STR locus, but each SNP locus can only be of type AA, BB, or the heterozygous AB. M-FISys has a unique capability to detect samples that are likely to be degraded and to optionally treat every homozygous locus as a suspect, not including it in the matching algorithm or likelihood calculation. By considering only heterozygous loci, the total likelihood statistic for the sample is reduced; however, even if only 60 or 70 SNP loci are tested, a sufficient number of heterozygous markers can be expected to be found to give a high likelihood of indicating a genetic match that is unique.

mtDNA and Y-STRs are not subject to sexual recombination, being inherited only down the maternal and paternal lines, respectively. We each have great, great, great, great-grandmothers, and grandfathers, so these haplotypes are shared among many people. These haplotypes are not differentiating enough to determine a unique identification on their own, in the absence of other data. However, they can be used to exclude other possibilities and (more importantly) to confirm or contradict matches found using other DNA tests or different forensic modalities.

A contradiction between different DNA assays (i.e., STR, Y-STR, mtDNA, SNP), warrants investigation, just like a discrepancy between DNA findings and other forensic disciplines (e.g., odontology, anthropology) would necessitate further investigation. If a DNA match and a dental match are in conflict, or an STR profile and a mitotype are discordant, these discrepancies should be resolved before an identification is declared. A broad range of data "consistency checks" in M-FISys help illuminate profile contradictions.

The Master List page (displayed in all preceding figures) shows an STR-*centric* view of the data. Y-STR, mtDNA, and SNP tabs at the bottom of the Master List bring the analyst to views that center on these other technologies. mtDNA is particularly applicable to skeletonized remains in an advanced stage of decomposition. There are several reasons that mtDNA is easier to recover than nuclear DNA. Only two copies of the autosomal (nuclear) genome are present in a human cell, and its size (~ 3.2 billion base pairs) makes it prone to fragmentation. By comparison, a single human cell may contain 1000–2000 mitochondria, each of which contains a relatively resilient, circular genome that is less than 17 kb long.

To make it easier to confirm the results of one assay with another, M-FISys sorts all data together, based on whatever criteria are chosen. It is conceptually similar to the way a spreadsheet works; however, with M-FISys, the sorted data can be on multiple tabs. If data is included in many columns in a Microsoft Excel spreadsheet, a user can sort by any column and all data in the same row stay together. By selecting a row in one tab (for example, the STR view), that same sample will be selected in all other technology windows where the data from that assay is available. This allows the operator to quickly move through all the tests that have been performed on an item of evidence, without having to search each tab for the same sample by name.

Fig. 8 shows an example displaying STR data in which a Direct Reference for RM 1791, Wolfgang Balker, matches six victim sample profiles. One of these samples (V-1766) was tested twice. The header row at the top of the window has a tri-glyph column indicating that in that column, the number ("#") of STR loci that have allele values is displayed above indicators that mtDNA and SNP data are available for the same sample (i.e., a red star if data is available, and a dash if there is none). In this example, the Direct Reference in the top row yielded results at 14 STR loci and there is available mtDNA data (indicated by the star symbol); however, there is no SNP data available (indicated by a dash symbol). Hence, of the six samples that match the Direct Reference by STR's, the Direct Reference and four of the matching postmortem samples (V-1766, V-1756, V-1771, V-1789) have available mtDNA data for comparison.

If all of the samples in this RM aggregate are selected in the STR tab, and then the mtDNA tab at the bottom of the window is clicked, any of the items that have mitochondrial profiles in the database will be selected in the mtDNA view. By starting with STR-centric matching, the mtDNA will be assembled in the same aggregates, whether the mitotypes match or not. If it seems that this could cause conflicting profiles in mtDNA, Y-STR, or SNP tabs to end up in the same aggregate, it is because that is the entire purpose of this organization, i.e., it is meant to uncover contradictions between different assays. The tab is selected at the bottom of the page to expose the mtDNA-centric view of the data (Fig. 9).

In this figure, a graphic cartoon and a tooltip show how much of the valid range of mtDNA has been successfully sequenced. Refer to the bottom row in this grouping. It reports that positions 16013-16378 of Hypervariable Region 1 [HV1] have been successfully sequenced for sample V-1789-01-Mito, as well as positions 63-343 for Hypervariable Region 2 [HV2]. Following standard nomenclature, the mitochondrial sequence is represented not as a full-length DNA string, but as a list of differences from the Revised Cambridge Reference Sequence (rCRS). In this example, 4-out-of-5 items are in perfect agreement on the basis of the mitochondrial profile, but one of the postmortem samples (V-1771-01) has a conflicting mitotype that must be investigated and resolved.

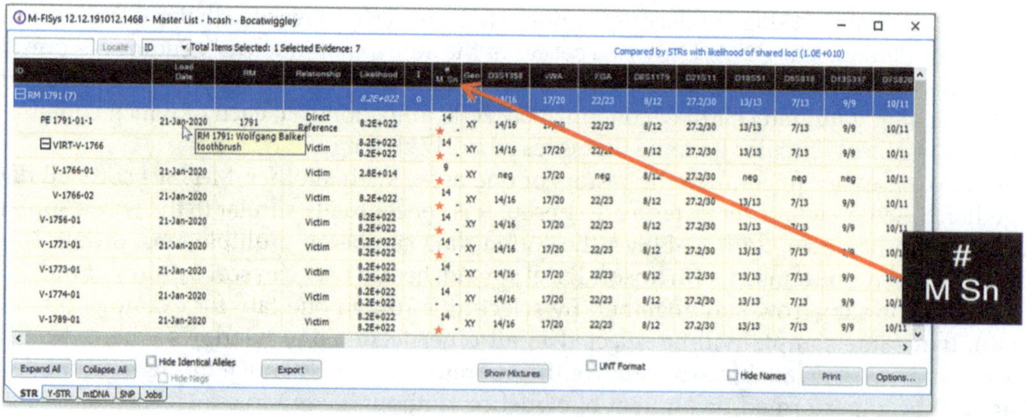

FIG. 8 Tri-glyph column showing data in other technology tabs (e.g., Y-STR, mtDNA, SNP).

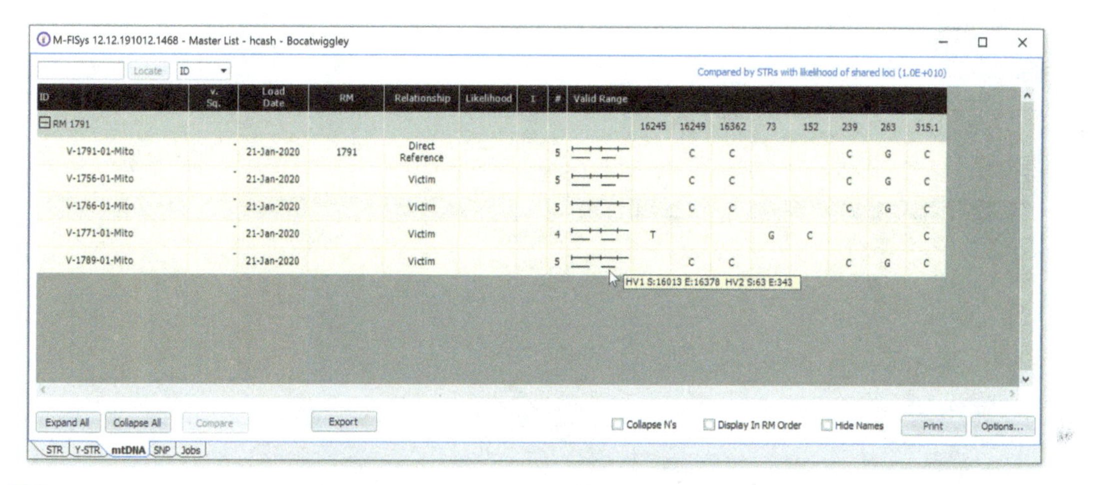

FIG. 9 mtDNA view of Wolfgang Balker's reference data and matching postmortem (PM) profiles.

If a full STR profile containing 14, 16, or 23 loci from a postmortem sample matches an antemortem reference sample, the evidence is very strong that they came from the same individual. If that assumption proves accurate, one would expect the profiles of other marker types (e.g., mtDNA, Y-STRs) to also match, which is why the example shown in Fig. 9 would merit special investigation. However, this strategy is not necessarily true in the reverse. An mtDNA match is not as discriminating as a full STR profile. The most common mtDNA type across HV1 and HV2 has been found to be shared by ~7% of the Caucasian population. Because mtDNA is not as specific (i.e., all children of the same mother are generally expected to inherit the same mtDNA profile), grouping all mtDNA matches does not have the same utility for re-associating disarticulated skeletal elements as some of the more individualizing tests, such as autosomal STRs. A full mtDNA match does not necessarily indicate that STRs or SNPs should also match; however, if they do, it is strong evidence to support identification. Considered another way, if only a partial STR profile can be easily developed from a bone in a more advanced state of decomposition, the mitotype is not enough to justify an identification on its own, but it adds statistical strength to any match that might be made using other modalities.

Quality control (QC) issues to consider in direct match data

Other QA tests for direct matching are built into M-FISys. Following are two examples of the types of conflicts that should be looked for, regardless of the database system being used.

In the QC STR report, warnings are presented any time a particular sample is tested more than once and yields conflicting results. In Fig. 10, conflicts between the two cases are highlighted in yellow. Both are based on actual cases; however, for confidentiality reasons, the sample profiles are computer-generated.

In the first grouping, victim sample V-00015 has been tested five times. The first attempt (V-00015-01) yielded results at only three loci: amelogenin (sex determination), D5S818, and TH01. In the second attempt (V-00015-02), results were obtained for eight loci (seven of which are visible in the figure). The only conflicts involve homozygous results (i.e., 30/30 at D21S11

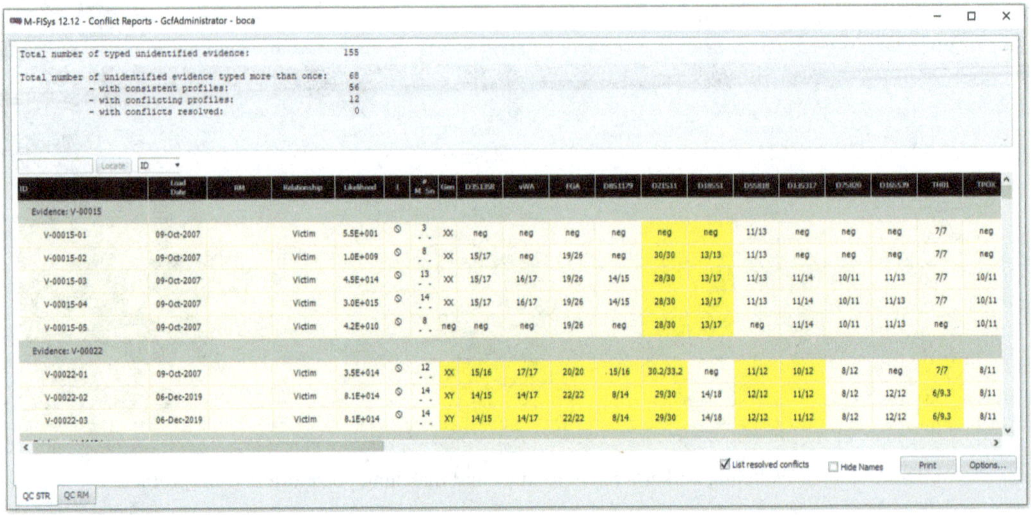

M-FISys 12.12 - Conflict Reports - GcfAdministrator - boca

```
Total number of typed unidentified evidence:                155

Total number of unidentified evidence typed more than once:  68
         - with consistent profiles:                          56
         - with conflicting profiles:                         12
         - with conflicts resolved:                            0
```

Locate ID

ID	Load Date	RM	Relationship	Likelihood	I	# M Sn	Gen	D3S1358	vWA	FGA	D8S1179	D21S11	D18S51	D5S818	D13S317	D7S820	D16S539	TH01	TPOX
Evidence: V-00015																			
V-00015-01	09-Oct-2007		Victim	5.5E+001	⊘	3	XX	neg	neg	neg	neg	neg	neg	11/13	neg	neg	neg	7/7	neg
V-00015-02	09-Oct-2007		Victim	1.0E+009	⊘	8	XX	15/17	neg	19/26	neg	30/30	13/13	11/13	neg	neg	neg	7/7	neg
V-00015-03	09-Oct-2007		Victim	4.5E+014	⊘	13	XX	15/17	16/17	19/26	14/15	28/30	13/17	11/13	11/14	10/11	11/13	7/7	10/11
V-00015-04	09-Oct-2007		Victim	3.0E+015	⊘	14	XX	15/17	16/17	19/26	14/15	28/30	13/17	11/13	11/14	10/11	11/13	7/7	10/11
V-00015-05	09-Oct-2007		Victim	4.2E+010	⊘	8	neg	neg	neg	19/26	neg	28/30	13/17	neg	11/14	10/11	11/13	neg	10/11
Evidence: V-00022																			
V-00022-01	09-Oct-2007		Victim	3.5E+014	⊘	12	XX	15/16	17/17	20/20	15/16	30.2/33.2	neg	11/12	10/12	8/12	neg	7/7	8/11
V-00022-02	06-Dec-2019		Victim	8.1E+014	⊘	14	XY	14/15	14/17	22/22	8/14	29/30	14/18	12/12	11/12	8/12	12/12	6/9.3	8/11
V-00022-03	06-Dec-2019		Victim	8.1E+014	⊘	14	XY	14/15	14/17	22/22	8/14	29/30	14/18	12/12	11/12	8/12	12/12	6/9.3	8/11

☑ List resolved conflicts ☐ Hide Names Print Options...

QC STR QC RM

FIG. 10 Quality control (QC) function noting conflicting STR alleles (highlighted in *yellow*) for a single tested bone sample.

and 13/13 at D18S51) that are in conflict with the subsequent tests. This is most likely a case of allelic dropout in degraded biological material. A review of the electropherograms might support this explanation. In M-FISys, when an analyst documents the homozygous loci as false homozygotes, a signed and dated entry will automatically be entered into the audit log and, as shown in Fig. 11, a green "check mark" on the profile will indicate that the discrepancy has been resolved by manual review.

The three tests of victim sample V-00022 (shown in the second group in Fig. 10) present a different type of conflict. Three tests have been performed, and profile V-00022-02 and V-00022-03 are concordant; however, sample V-00022-01 shows a distinctly different profile. Among the possible sources of this conflict are: (1) data management errors (e.g., the same ID was assigned to two different samples); (2) material handling errors (e.g., samples were mistakenly switched in a laboratory); (3) results were imported incorrectly into the database; and/or (4) commingling of remains prior to testing.

In the case that this "training" example is based on, the conflict was found to have arisen from the last of these possible sources of error. Bone and apparently attached muscle tissue were both tested, but in fact they came from two different victims. Different software systems have different procedures for documenting this class of error. In M-FISys, it would typically be handled by either: (1) invalidating the sample and retesting the remains with all new ID numbers, or (2) using the "Rename" function to replace the ID of either the bone or tissue sample with a new ID. These changes will be logged and signed in the M-FISys audit trail. Regardless of how the change is managed by the DNA laboratory, it is important to communicate these corrections to other departments so that samples, which were submitted to forensic biology for testing do not seem to have simply disappeared.

Another example detected a case in which different profiles were developed from different tests on the same sample. In this case, it was called an "*STR conflict*." In M-FISys, the term "*RM*

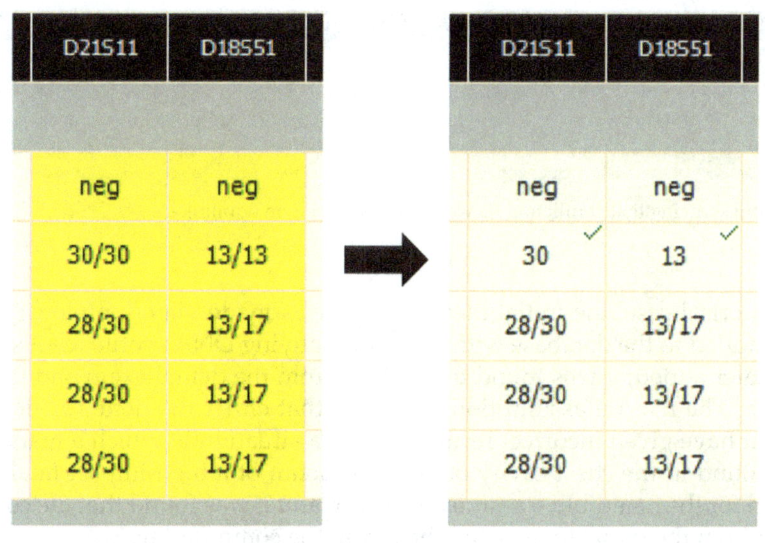

FIG. 11 Screenshot documenting the presence of allelic dropout: (1) before manual review *(left)* and (2) after resolution via manual review *(right)*.

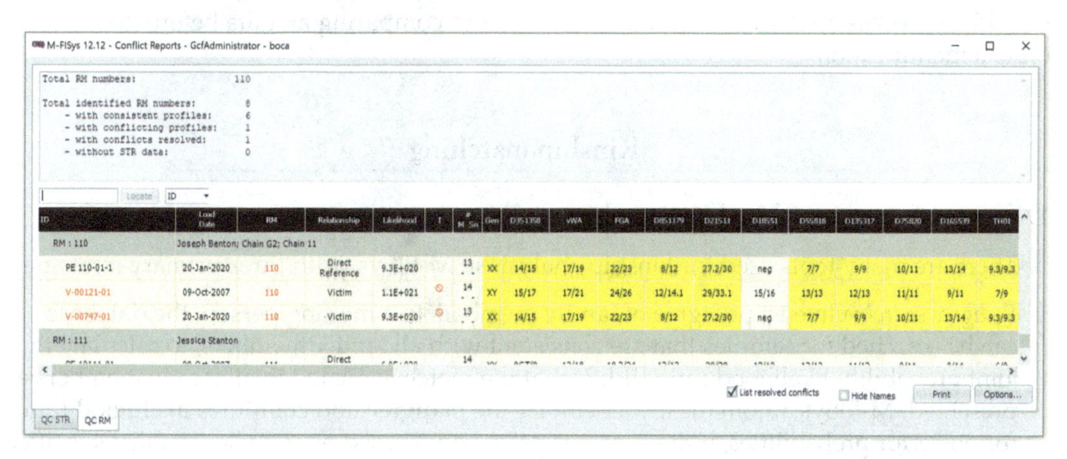

FIG. 12 Example of an RM conflict with items assigned the same ID despite conflicting STR profiles.

conflict" is used to represent the contradiction when two different profiles developed from different samples are identified as the same individual.

Consider the example in Fig. 12. Two victim profiles (V-00121-01 and V-00747-01) have clearly conflicting profiles. V-00747-01 is consistent with the personal effect exemplar (PE-110-01-1) used as a direct reference. A reviewer would want to know how postmortem sample V-00121-01 could have been assigned the identification of RM 110, Joseph Benton.

A review of the audit log (or simply hovering the mouse over the profile to expose a Tooltip) shows that sample V-00121-01 was assigned the ID before the data was uploaded to the database. The tooltip shown in Fig. 13 reports the identification method as "Identified through Dental X-Ray."

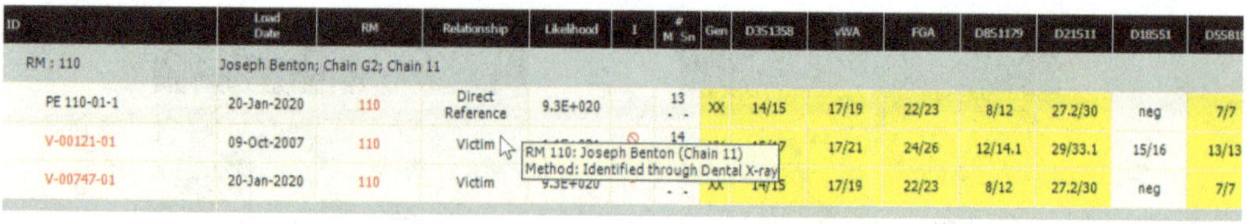

ID	Load Date	RM	Relationship	Likelihood	I	# M Sn	Gen	D3S1358	vWA	FGA	D8S1179	D21S11	D18S51	D5S81
RM : 110	Joseph Benton; Chain G2; Chain 11													
PE 110-01-1	20-Jan-2020	110	Direct Reference	9.3E+020		13 . .	XX	14/15	17/19	22/23	8/12	27.2/30	neg	7/7
V-00121-01	09-Oct-2007	110	Victim	[illegible]		14	[illegible]	[illegible]	17/21	24/26	12/14.1	29/33.1	15/16	13/13
V-00747-01	20-Jan-2020	110	Victim	9.3E+020		. .	XX	14/15	17/19	22/23	8/12	27.2/30	neg	7/7

RM 110: Joseph Benton (Chain 11)
Method: Identified through Dental X-ray

FIG. 13 Tooltip showing that conflicting ID was assigned based on odontology.

In this historical case, the initial identification was made via odontology, and this information was added to the database with no accompanying DNA profile results. When a DNA profile was later added, it was found that results from the dental exam and forensic biology did not agree. The natural assumption would be that either the dental review or the DNA analysis must have given incorrect results. This is a situation in which a mandible from one victim was found in the chest cavity of another victim of a high-impact fatality. The dental identification for the mandible was indeed correct, and it was found that the conflicting DNA profile came from the tissue that coated the jaw in the commingling.

These and other examples are meant to illustrate the value of not treating a DNA match as the end of an investigation or the sole source of data for identification. Rather, a DNA match should be used as a starting point for reviewing and comparing all data before declaring an official identification.

Kinship matching

Different formats of kinship searching

There are many tools used for kinship analysis in M-FISys, with three primary strategies:

- *Pedigree search*: Given a pedigree of family members to a missing person, the database can be searched for samples that are consistent with all family members, to calculate the joint probability likelihood ratio (JPLR) that the match is related to all family members as presented. M-FISys automatically assembles the pedigree and computes likelihood ratios (or posterior probabilities, if the user sets prior odds) based on the relationships specified in the input files.
- *Hypothesis search*: The user can draw a pedigree using graphic tools in M-FISys and fill in any family node with any profile contained in the database. This can be a new pedigree, not contemplated at the time of import. It is also possible to manually enter or copy-and-paste profiles that are not already in the database into the nodes of the proposed pedigree.
- *Familial search*: Rather than searching for a profile that is itself consistent with all of the family profiles in a pedigree, search for profiles that might not match on their own, but could represent a sibling or parent/child of a person for whom reference samples are available. This is more commonly used to develop leads for criminal investigation than for reviewing skeletal remains. Familial searching is a valuable criminalistics tool, but it is not detailed in this chapter.

Pedigree search

The Pedigree search is the most often used. An example is shown in Fig. 14.[a] The RM list in the upper left of the screenshot (marked "A") shows that the pedigree for RM 047 is being used as the criteria to search the database. This is for a missing person named *April Mullins*, a name shown near the top of the window and assigned when the case was created.

Below the RM list is a list of candidate matches to this pedigree (marked "B"). In this case, the single good candidate match found was victim sample V-10056-01. In the lower-left corner is a graphic presentation of the pedigree (marked "C"), automatically assembled based on the relationships that were specified when the profiles were uploaded.

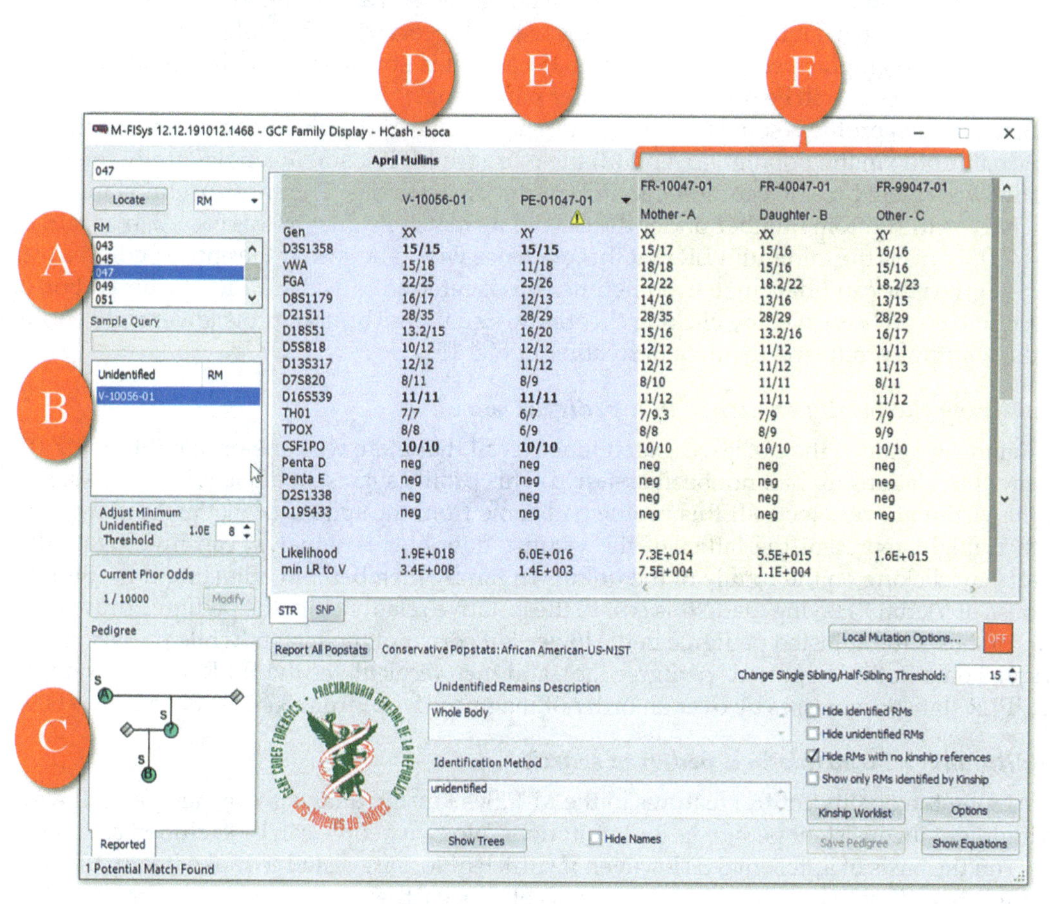

FIG. 14 Elements of the Pedigree Search window in M-FISys.

[a] There are two versions of the Pedigree Search tool in M-FISys, one of which is shown here and is only applicable to simpler pedigrees with close relatives and a single missing person. The complex kinship search window has the same operations, but includes additional tools for specifying arbitrarily distant or complex relationships, and supports multiple nodes for unknown persons if more than one individual in a family are among the missing.

In the middle section (blue area) the profile in the column marked "D" is the profile from the candidate postmortem sample, V-10056-01. All profiles to the right of this are reference samples. QC tools are spread throughout the program and, even though this is a kinship review, the system recognizes that RM 047 has not only kinship references, but also a Personal Effect (a direct reference) with the sample name PE-0104701, as shown in the column marked "E." The yellow caution triangle next to the direct reference warns that the candidate match is inconsistent with the direct reference. Since April is known to be female, and the direct reference sample typed as a male (XY) at the sex-determining locus, one might guess that the Direct Reference is wrong. This would be worth confirming.

To the right of the window is a light-yellow section ("F") containing the profiles of family references: the mother of the missing person, a daughter, and a third reference where the relationship is unknown, ambiguous, or perhaps entered incorrectly. The unknown profile is not included in the graphic pedigree illustration because it is not clear where it should be placed correctly. This issue will be revisited in a later paragraph.

Beneath each profile is some useful statistics, including (1) the likelihood of finding that individual profile in the population, and (2) the pairwise LR if just that single-family reference profile was related to the candidate match profile as reported. The most useful statistic on this page is the bottom number under the candidate match profile (in this case, 3.4E + 008, or 3.4×10^8). This is the minimum LR that this evidence would be seen if the entire pedigree was correct given the candidate match. Sometimes this is referred to as the JPLR. The derivation of this number can be viewed by clicking the "Show Equations" button in the lower right corner of the window, producing the data derivation in Fig. 15.

Modifying the family structure in a pedigree search

Referring again at the pedigree search tools, recall that there was one profile (FR-99047-01) where the relationship had not been assigned. This profile shares an allele with the daughter of the victim at every locus. If this profile had come from the spouse of victim April Mullins, then it might represent the father of the victim's daughter. An analyst can investigate that theory by clicking on the profile of the unknown family member and adjusting it to "Become Spouse of Victim." Having made this edit to the putative relationship, the pedigree drawing in Fig. 16 shows the adjusted pedigree in the lower-left corner of the screen (while preserving the original data in the "Reported" pedigree tab), and then recalculates the JPLR. In this example, the JPLR statistic increases by over an order of magnitude, i.e., from 3.4E + 008 to 7.0E + 009.

Additional QC controls in a pedigree search

Among the quality control features in the M-FISys kinship analysis suite are various ways to highlight inconsistent pedigrees and mutations. No two people can be excluded as full siblings on the basis of autosomal STRs, even if sufficient loci are tested to make it very unlikely that it is true. If both parents are heterozygous at a locus, there is no genetic requirement that their two children share any alleles. However, an arbitrary number of children can still only have a maximum of four different alleles among them at any locus, drawn from the two contributed by each parent. Some systems allow any random four people to be entered as full siblings of a missing person, even if they have eight different alleles between them at a locus. This of course is genetically impossible, and such inconsistencies are easy for a database to point out. In Fig. 17, M-FISys searches the database for a profile consistent with the profile of a victim's father and two reported full siblings. The victim sample (PE-01067-01) is not excluded

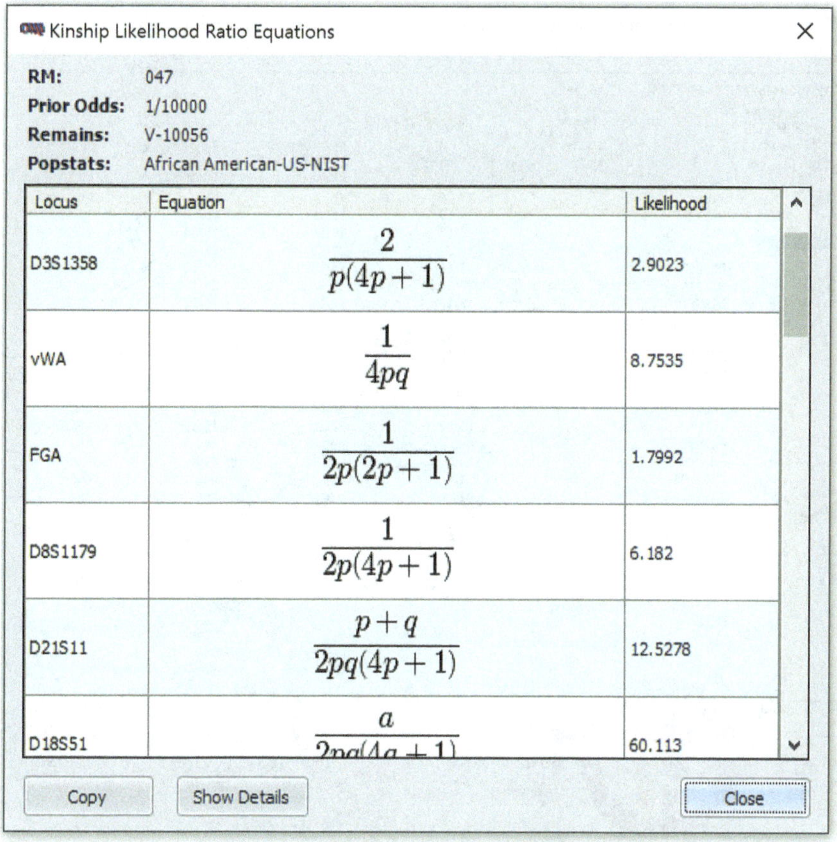

Kinship Likelihood Ratio Equations

RM: 047
Prior Odds: 1/10000
Remains: V-10056
Popstats: African American-US-NIST

Locus	Equation	Likelihood
D3S1358	$\frac{2}{p(4p+1)}$	2.9023
vWA	$\frac{1}{4pq}$	8.7535
FGA	$\frac{1}{2p(2p+1)}$	1.7992
D8S1179	$\frac{1}{2p(4p+1)}$	6.182
D21S11	$\frac{p+q}{2pq(4p+1)}$	12.5278
D18S51	$\frac{q}{2pq(4q+1)}$	60.113

Copy | Show Details | Close

FIG. 15 Statistical derivation of the Joint Probability Likelihood Ratio (JPLR) for RM 047 before adjusting the pedigree.

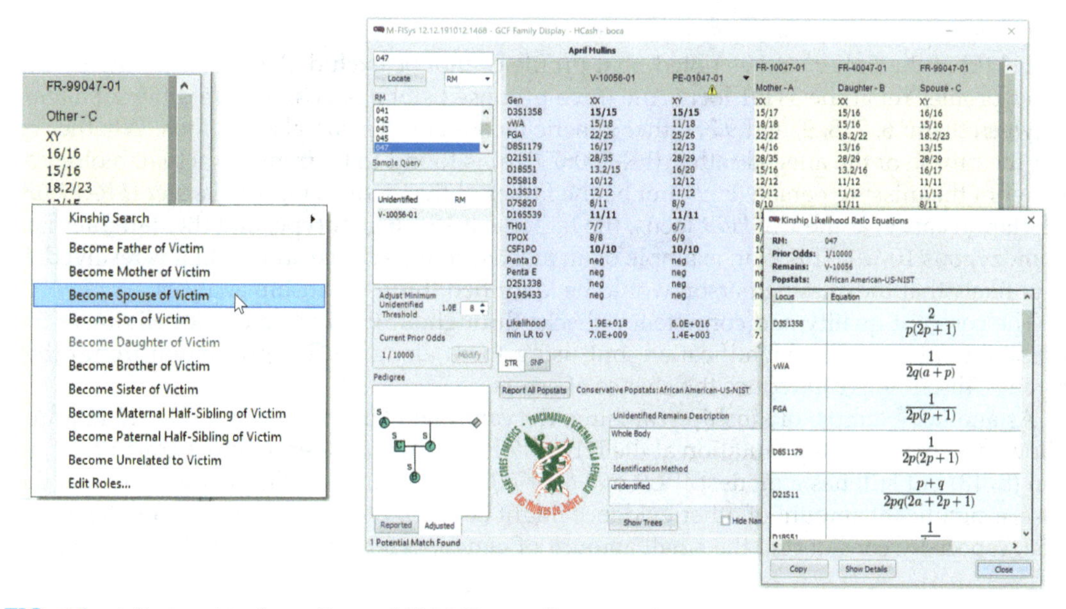

FIG. 16 Adjustment to the pedigree of RM 047, as well as re-configuration of equations and re-calculation of the JPLR.

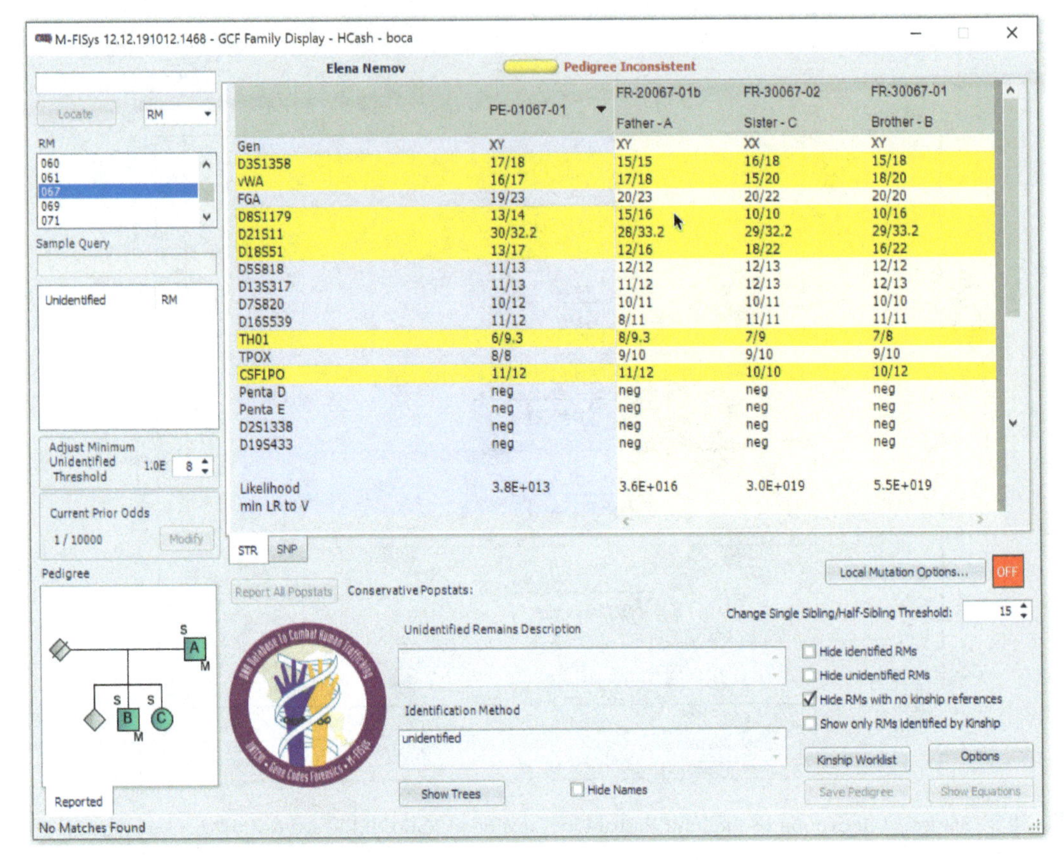

FIG. 17 Kinship data quality checks highlighting genetic inconsistencies.

as a child of the father, and as noted, that profile cannot be excluded as a full sibling of any other profile. Yet at the TH01 locus, the three proposed siblings possess five different alleles between them: 6, 7, 8, 9, and 9.3. Other genetic inconsistencies are also obvious. Whether or not the profile of the alleged father (FR-20067-01b) is found to be from the actual, biological father of the missing person, it cannot be the father of the victim's reported sister (FR-30067-02). For instance, at the D8S1179 locus, the father is a 15/16 genotype, and the daughter is a homozygous 10/10. This is an example of an inconsistent pedigree and, until it is resolved, it is unlikely that the missing person would be identified through kinship analysis.

The constant quality and consistency checks throughout the system cannot eliminate the possibility of mistaken identification, but using them along with confirmation from other forensic disciplines can reduce the chance of errors.

As another example of kinship searching, the candidate kinship match in Fig. 18 requires allowance for a one-step mutation at the D13S317 locus from mother (homozygous 12/12) to son (8/13) but still has a modest JPLR of 4.8E + 003. Depending on the circumstances of the case, a significant amount of other evidence might be warranted to declare this an identification, especially considering the small amount of genetic reference material, the requirement

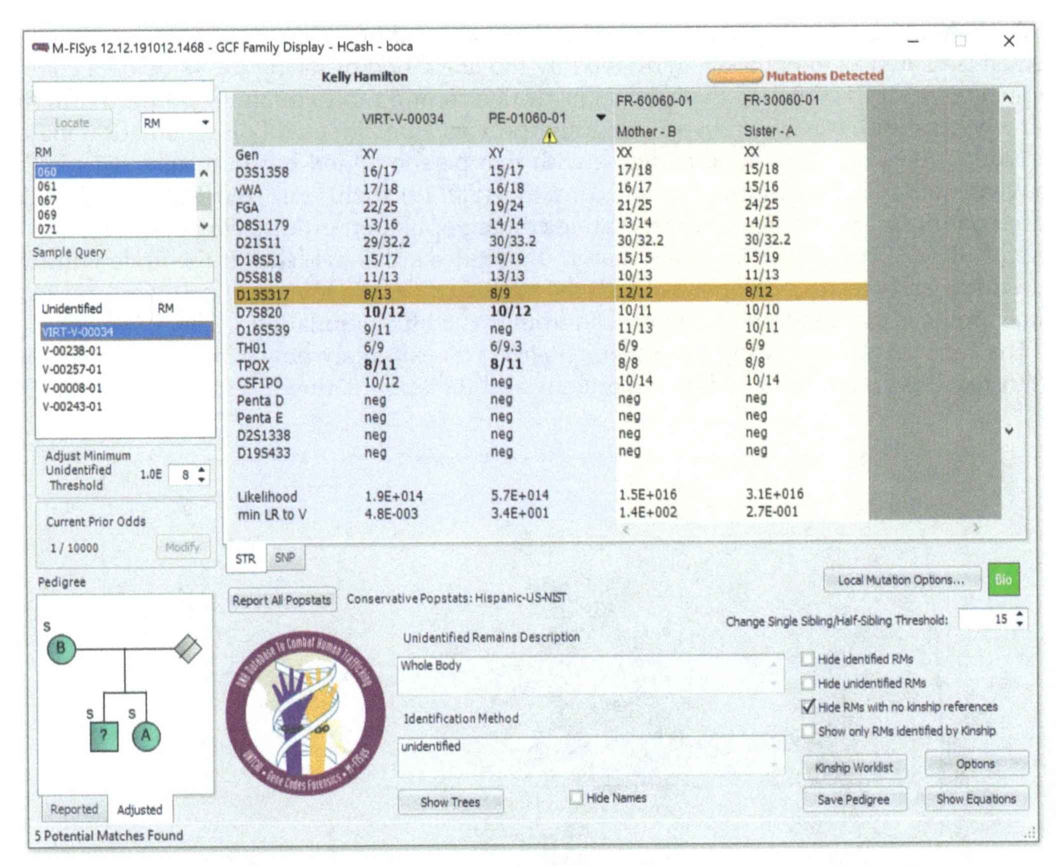

FIG. 18 Possible adventitious "false match" with a mutation.

for a mutation and this likelihood ratio, and the fact that the warning triangle indicates that the candidate match is in conflict with a personal effect that serves as a direct reference. In fact, the direct reference associated with the son (PE-01060-01) also seems to be excluded as having the profile of a person with this reported mother. Cases with multiple evidentiary conflicts plague many, large-scale identification efforts. This case would be worthy of intensive review, all the way back to the original collection of information and reference samples.

The *Kinship Pedigree Search* tool makes it simple to look at an organized presentation of the profiles of family members who are offered as genetic references, search the database for matches consistent with those relatives, and make modifications as necessary to account for unknown or mistaken relationships.

Kinship hypothesis

The *Kinship Hypothesis* approach in M-FISys is the opposite of the pedigree search, in that it assumes the analyst does not start out with a well-defined genetic pedigree already in the database. A tool called *Kinship Sketch* can be used to draw a pedigree of hypothetical relationships, and then populate the individual nodes with profiles from the database.

A LR, as used in kinship identification, is the likelihood of seeing the observed data if the person is related as hypothesized, divided by the likelihood of seeing the same data coincidentally (e.g., if the data for that person just came from a random, unrelated person in the same population). It is often displayed graphically as a pedigree with a hypothetical match included, divided by the same pedigree with that person added but unrelated. This is the convention used in the M-FISys *Kinship Sketch* tool (Fig. 19). In this case, the JPLR is 1.0E + 006. Based on the most statistically conservative ethnic population group being considered, it is roughly 1 million times more likely that you would see this evidence if the node with the question mark was from the person with the profile in V-00061 than that you would see the same profile in any random person chosen from that ethnic population.

The *Hypothesis Search* using the *Kinship Sketch* is an extremely powerful and flexible tool, but it requires setting up a pedigree manually and then populating it with profiles from the

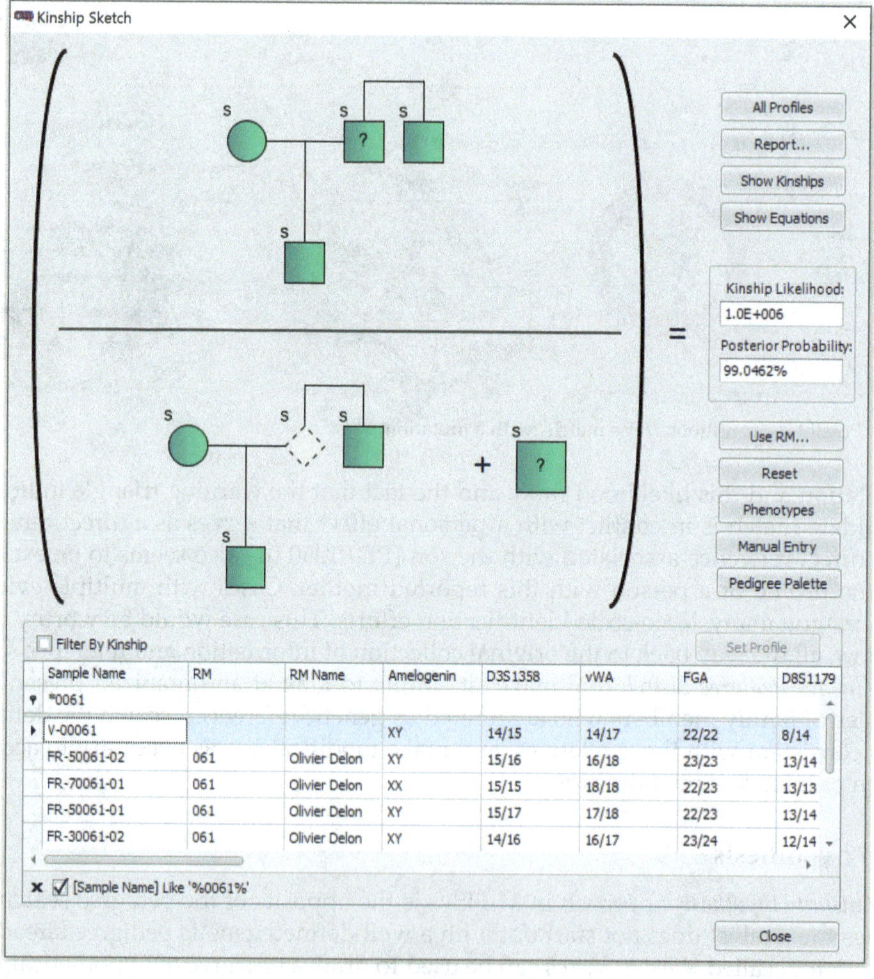

FIG. 19 Pedigree Search uses *Kinship Sketch* tool to define a likelihood ratio (LR).

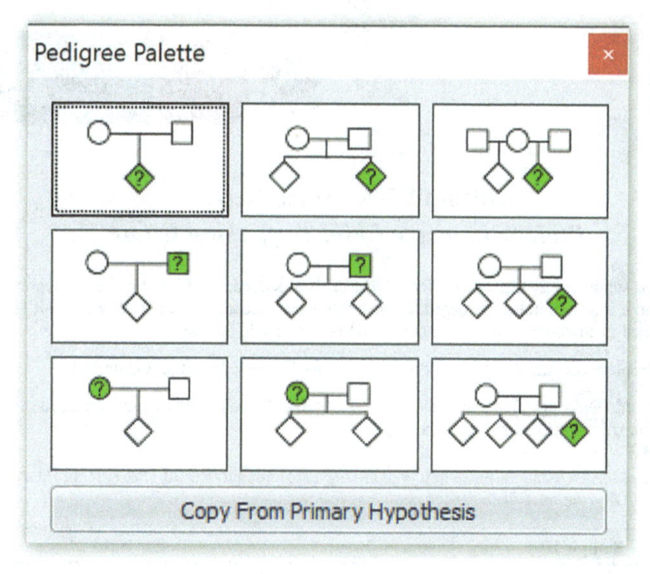

FIG. 20 Predesigned pedigree palette (i.e., templates) in M-FISys.

database (or typing them in by hand). It is more time-consuming than doing a pedigree search in which the relationships and profiles were all defined at the time the data was imported. To reduce the time needed to draw a pedigree, some common structures are available as prepared starting points for extending the family tree, as shown in Fig. 20.

Reports

Reports are not only important for documenting work in an identification effort, but can be an important mechanism for communicating with other stakeholders, such as laboratory and municipal administrators, collaborating research groups, families of decedents, and providers of funding and other resources. A first-order report might be one that confirms or excludes an ID. A kinship match report, while fully editable, might look like Fig. 21.

This report documents the resolution of a specific case. However, reports which track the progress of the overall effort are often valuable in any project that is larger than a single case. Work on decomposed and/or disarticulated human remains can often be characterized by multiple tests and multiple sources of reference material. Examples of some reports include:

1. *Evidence By Class* (Fig. 22)
   What fraction of the available DNA data represents postmortem, direct reference, and kinship reference data?
2. *Samples By Technology* (Fig. 23)
   How many profiles have been added to the database using STR, Y-STR, and mtDNA assays?

Kinship DNA Match Report

Project: Example Child Trafficking Effort

This report is submitted for the information of RM Id 102: Kimchi Daorung. I make this report knowing that if it is used in legal proceedings and entered as evidence then I shall be liable if I have willfully stated in anything I know to be false or do not believe to be true.

I Richard Peppleton of Ann Arbor State Police Center for Forensic Investigation declare:

I am a qualified DNA specialist and authorized by ASCL-LAB to provide expert evidence in relation to human identification based on DNA.

I have reviewed DNA data linking sample V-00900-01, attributed to RM Id 102: Kimchi Daorung to the following DNA samples from related people as follows:

Reference Sample ID	Name of Donor	Donor is the ...
FR-70003-01	Min Daorung	Mother
FR-40003-01	Simchi Daorung	Sister
FR-40003-02	Yaoxing Daorung	Sister

The likelihood ratio that the profile in V-00900-01 is related to the reference samples as reported is:

Popstats	Statistics
Hispanic-US-NIST	5.1E+015
Caucasian-US-NIST	1.0E+015
Asian-US-NIST	5.3E+016
African American-US-NIST	4.0E+013

In my opinion, this evidence strongly supports the identification of the source of V-00900-01 as RM Id 102: Kimchi Daorung.

Declared by: Richard Peppleton

Signature:_______________________ Date:___________

Witnessed by: Howard Cash

Signature:_______________________ Date:___________

Reviewed by: Mike Hennessey

Signature:_______________________ Date:___________

Database : boca

Tuesday, 21 January 2020 04:38 PM   Page 1/1

FIG. 21 Kinship Match Report for RM 102.

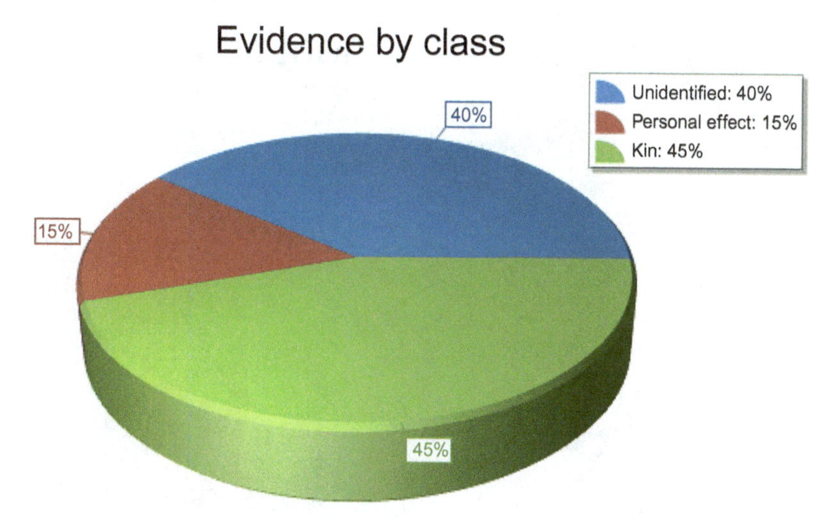

FIG. 22 The relative quantity of profiles from different source classes.

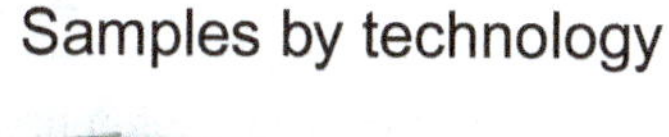

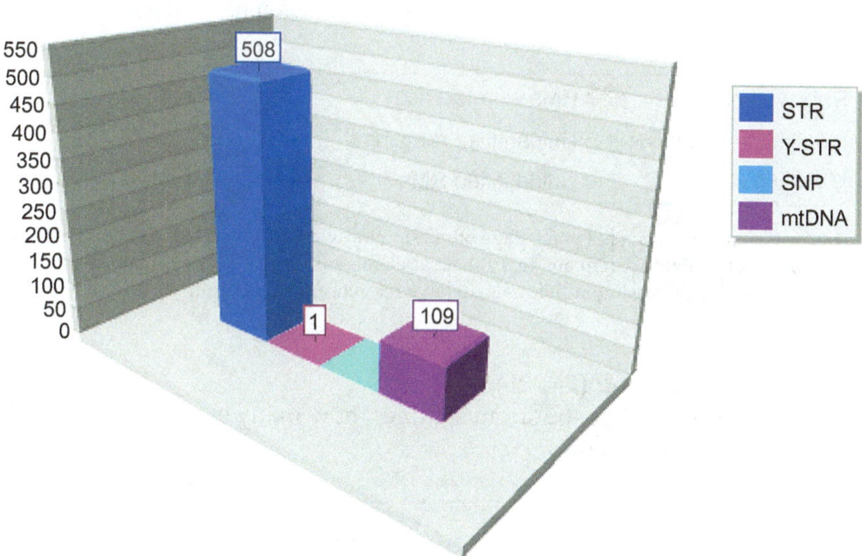

FIG. 23 In this case, how many STR, Y-STR, SNP, and mtDNA profiles are in the database

3. *DVI Statistics* (Fig. 24)
   a. How many missing individuals are assumed and how many have been identified?
   b. How many remains have been collected and how many have been identified?
   c. How many of the missing do and do not have reference samples available for matching?

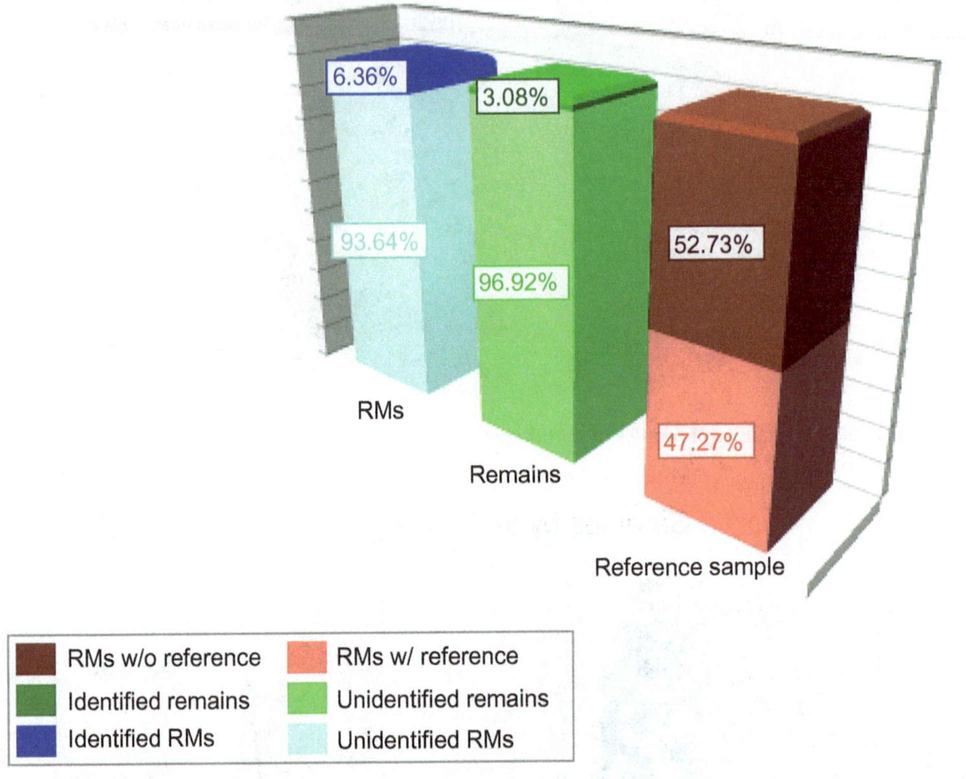

FIG. 24 The first column shows what percentage of missing individuals have been identified; the second column shows the proportion of individual remains (e.g., skeletal elements) that have been identified; and the third column keeps track of how many missing individuals do have reference material with which to attempt a DNA-based identification.

4. *STR Loci Count By Test Result* (Fig. 25)
   For both ante-mortem and postmortem samples, how many samples gave full results or different levels of partial profiles?

Advanced topics

Complexity of pedigrees

The efficiency of kinship searching is impacted much more by the complexity of pedigrees than by the size of the database being searched. Searching with only a single sibling as a reference sample can be time-consuming with a large database because, as already noted, siblings do not need to share any alleles at any given locus. It is not possible to screen out any candidate matches.

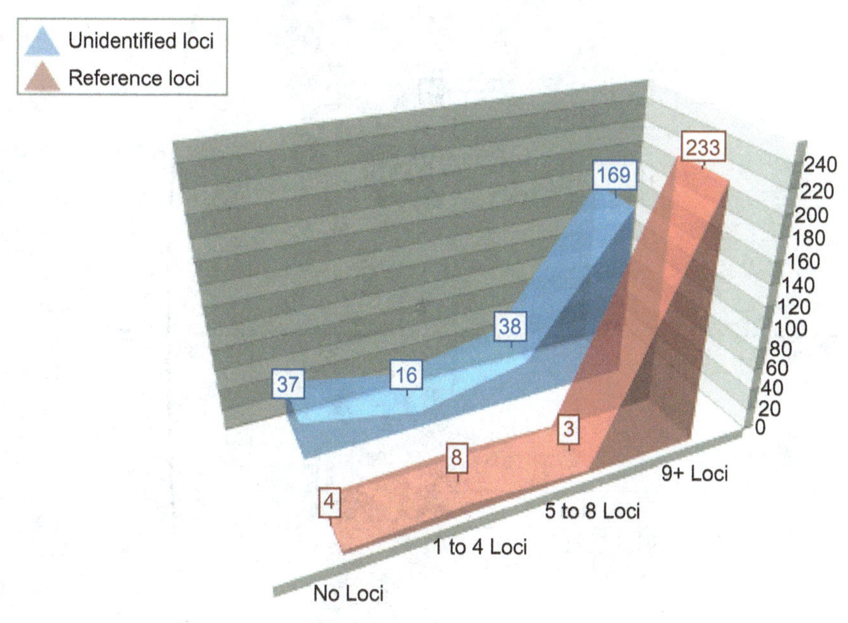

FIG. 25 This graph shows the number of STR loci successfully typed from postmortem samples (in *blue*) and reference samples (in *red*).

It may seem counterintuitive, but larger numbers of family references can make the search either more processor-intensive or less so. The genetic complexity of a pedigree is driven by how far back in the family tree the algorithm has to go to find a common ancestor who may have contributed alleles to both, and then determine the odds that the allele was transmitted down through the generations. There can be a great deal of genetic complexity in a pedigree with very few donors, while a pedigree with many genetic donors can be quite simple. Consider the two examples shown in Fig. 26 and Fig. 27.

In the pedigree shown in Fig. 26, the hypothesis is that the UHR (outlined in red) is from a pedigree that includes four siblings of a missing person, two half-siblings, both parents, and all four grandparents. When calculating the JPLR that the profile from the UHR has all the relatives as put forth in the hypothesis, the equations are very simple. This is because if profiles are available for both parents of a missing person, no number of siblings and earlier ancestors will contribute any additional genetic information whatsoever about the unknown case (assuming that full profiles are available for both parents). The equations in Fig. 27 exemplify what might be seen in an ordinary paternity testing case.

A properly designed database will confirm that the pedigree is internally consistent, but then will ignore all redundant nodes in the tree and, in this case, reduce the problem to only a parent-child trio. The record-keeping for so many donors is cumbersome, but the genetic analysis is not difficult.

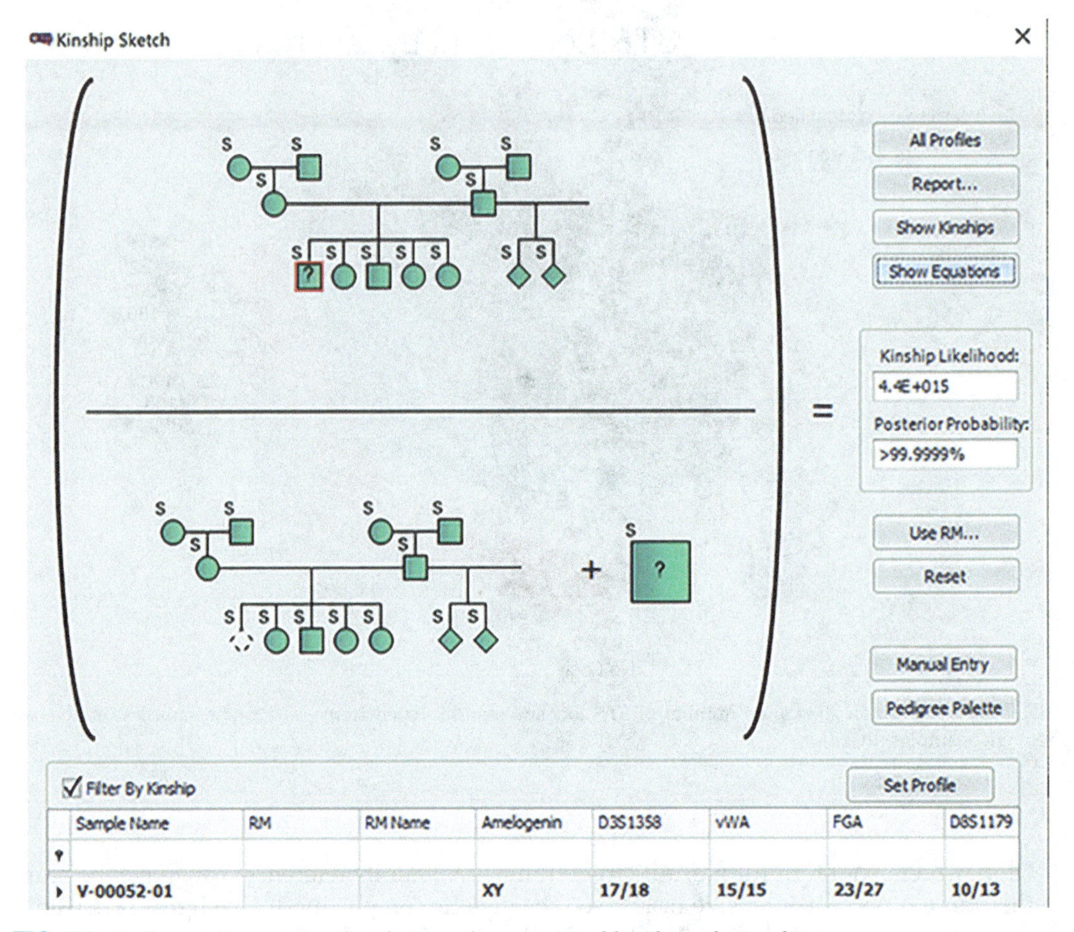

FIG. 26 Pedigree with many family references that may not add to the probative data.

Now consider the pedigree shown in Fig. 28. This case seems simple at a glance. It contains only first-order relatives: a parent, a sibling, and a child of a missing person. The paperwork for keeping track of the donors is easy, but the algebra in the calculations for a JPLR is more complex, as shown in Fig. 29.

A diligent undergraduate can learn the tedious steps to derive these equations. In fact, Gene Codes Forensics, Inc. has trained high school students to solve these problems consistently and correctly. A database and analysis system like M-FISys, by presenting the equations automatically and in an easily readable form, saves time.

Pedigrees with multiple hypotheses

As previously stated, a LR is the likelihood of one hypothesis (e.g., "this skeletal element comes from the father of three children who donated reference samples") divided by the likelihood of an exclusion hypothesis (e.g., "this bone comes from another unrelated person in the study's population who shares certain alleles by random chance with the family references").

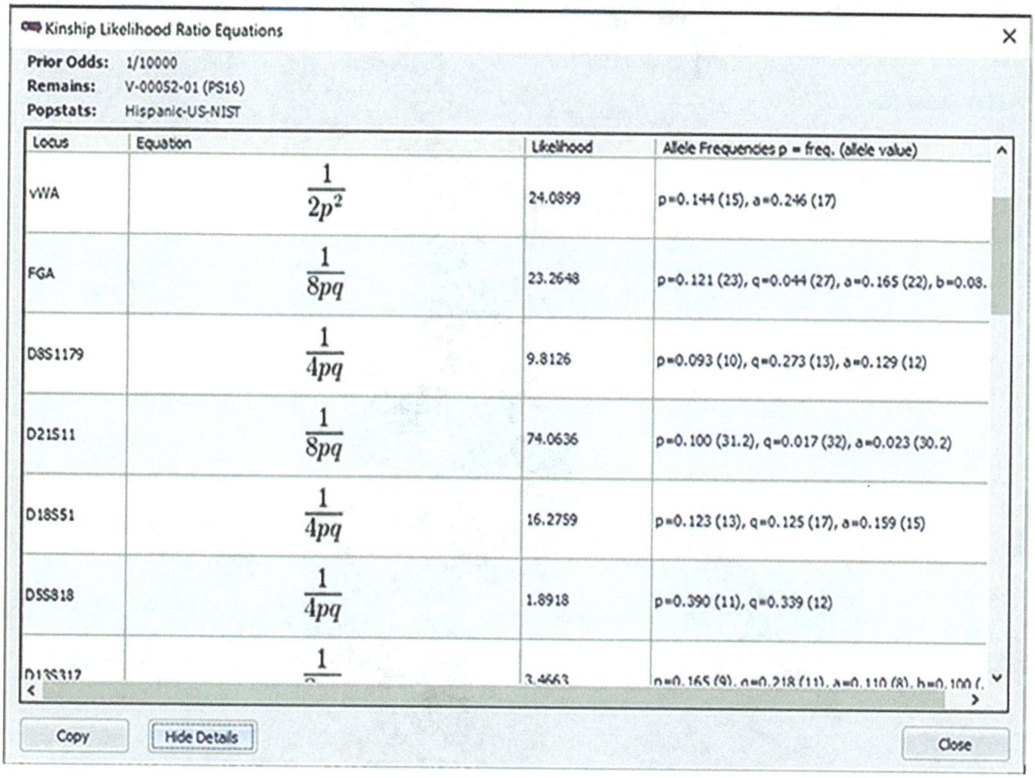

Kinship Likelihood Ratio Equations

Prior Odds: 1/10000
Remains: V-00052-01 (PS16)
Popstats: Hispanic-US-NIST

Locus	Equation	Likelihood	Allele Frequencies p = freq. (allele value)
vWA	$\frac{1}{2p^2}$	24.0899	p=0.144 (15), a=0.246 (17)
FGA	$\frac{1}{8pq}$	23.2648	p=0.121 (23), q=0.044 (27), a=0.165 (22), b=0.08.
D8S1179	$\frac{1}{4pq}$	9.8126	p=0.093 (10), q=0.273 (13), a=0.129 (12)
D21S11	$\frac{1}{8pq}$	74.0636	p=0.100 (31.2), q=0.017 (32), a=0.023 (30.2)
D18S51	$\frac{1}{4pq}$	16.2759	p=0.123 (13), q=0.125 (17), a=0.159 (15)
D5S818	$\frac{1}{4pq}$	1.8918	p=0.390 (11), q=0.339 (12)
D13S317	1 / [illegible]	3.4663	p=0.165 (9), q=0.218 (11), a=0.110 (8), b=0.100 (.

Copy | Hide Details | Close

FIG. 27 Simple equations for a pedigree with many members, but including both parents. This can effectively be reduced to a standard paternity trio.

This is a good measure of the statistical strength of a match when the distribution of alleles is in line with what has been measured in the population. However, if two or more candidate matches may come from people who are related to each other, this model is no longer sufficient.

In many mass fatality situations, more than one member of the same family can be among the decedents. In a case worked on by Gene Codes Forensics (using M-FISys), a mother was killed in an aircraft accident along with her 8-year-old daughter. Remains of the mother were identified first, using dental records. The DNA profile from the mother's remains was used in a pedigree to identify the remains of her daughter. If an administrative review were to consider only the mathematical model, it might seem to a simple-minded analyst that this identification was flawed because it failed to take into account the possibility that the mother and daughter profiles could be inverted. It is important to remember that even in a forensic biology laboratory, non-DNA evidence can exclude an identification. Even if there were no genetic information to differentiate between the two victims, unimpeachable physical evidence that the dental records from the adult could not be confused with the 8-year-old child is as much of exclusion as any genetic data that can be applied to this case.

In a kinship relationship where there is more than one possible postmortem match, the first step to simplifying the problem would be to test the samples for more loci to create an exclusion for all but one possibility. If additional testing is not practical, would not be probative,

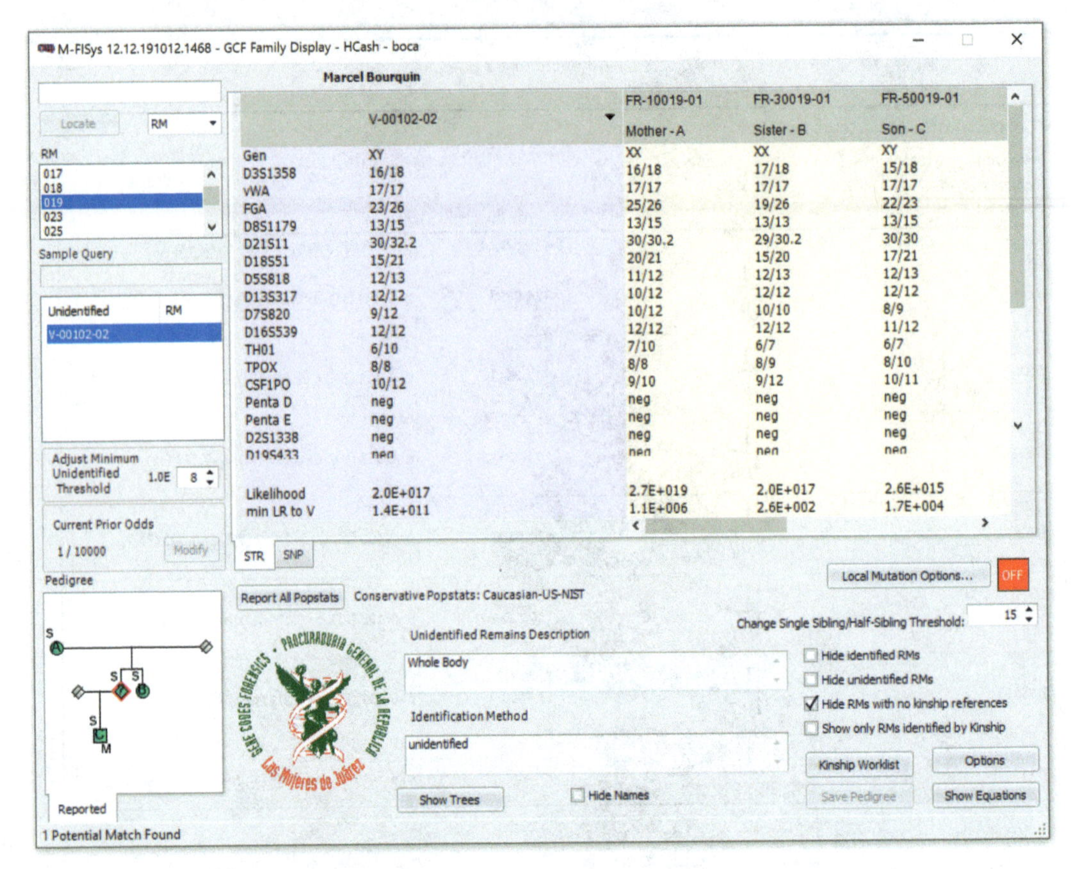

FIG. 28 A deceptively complex genetic pedigree for calculation purposes.

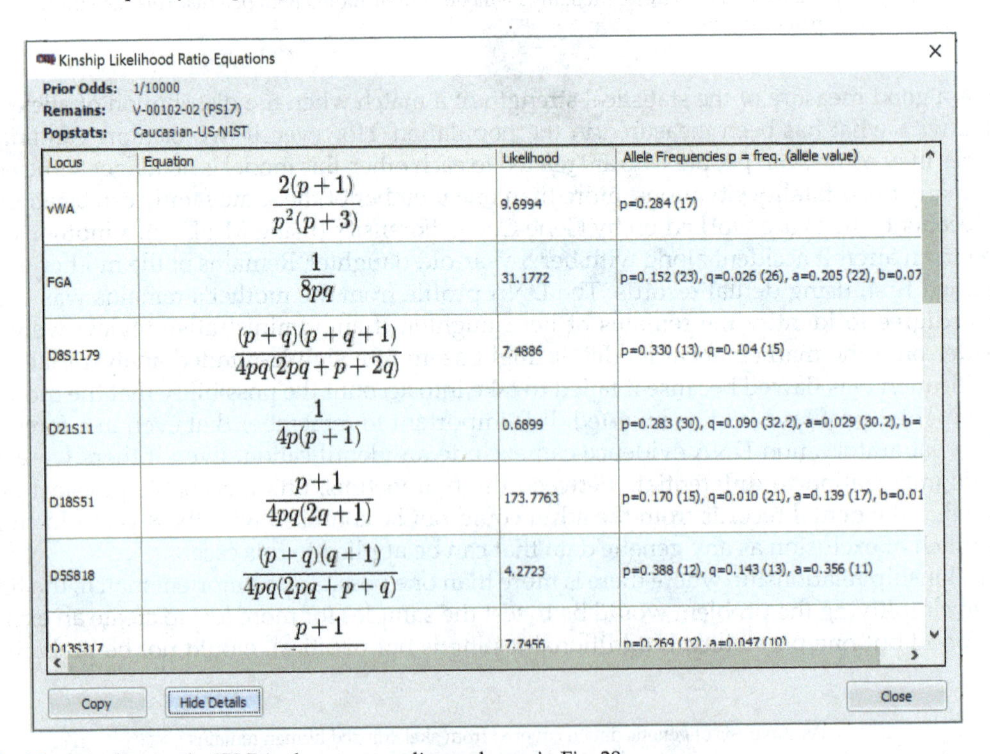

Kinship Likelihood Ratio Equations

Prior Odds: 1/10000
Remains: V-00102-02 (PS17)
Popstats: Caucasian-US-NIST

Locus	Equation	Likelihood	Allele Frequencies p = freq. (allele value)
vWA	$\frac{2(p+1)}{p^2(p+3)}$	9.6994	p=0.284 (17)
FGA	$\frac{1}{8pq}$	31.1772	p=0.152 (23), q=0.026 (26), a=0.205 (22), b=0.07
D8S1179	$\frac{(p+q)(p+q+1)}{4pq(2pq+p+2q)}$	7.4886	p=0.330 (13), q=0.104 (15)
D21S11	$\frac{1}{4p(p+1)}$	0.6899	p=0.283 (30), q=0.090 (32.2), a=0.029 (30.2), b=
D18S51	$\frac{p+1}{4pq(2q+1)}$	173.7763	p=0.170 (15), q=0.010 (21), a=0.139 (17), b=0.01
D5S818	$\frac{(p+q)(q+1)}{4pq(2pq+p+q)}$	4.2723	p=0.388 (12), q=0.143 (13), a=0.356 (11)
D13S317	$p+1$	7.7456	p=0.269 (12), a=0.047 (10)

FIG. 29 Algebra for the JPLR in the sparse pedigree shown in Fig. 28.

or if there are not other family references to help triangulate a genetic difference between the two individuals, M-FISys can calculate statistics for multiple hypotheses.

Consider a simplistic case in which David is undergoing a paternity test. Assume that David is not excluded as the father, but neither is his brother Richard and that for some reason, additional testing is not possible. Compared only to an unrelated person in the population, the posterior probability that David is the father might be calculated as 99.2%, while the posterior odds that his brother Richard is the father is 99.5%. Using these percentages as a layperson might, it makes no sense that two different people have a greater than 99% chance of "being the father" of the same child. It might be useful to know the posterior probability, paternity index, or likelihood ratio that either man is the father compared to an *unrelated* person but, given the two plausible hypotheses, it is also necessary to consider the overall posterior probability given that another, *related* person is the possible parent. In this case, M-FISys can calculate statistics for two or more scenarios in parallel, illustrated as Hypothesis 1 (H1) (Fig. 30) and Hypothesis 2 (H2) (Fig. 31). While this is mathematically more correct than treating the two cases separately, it is not enough information to make a determination of which man is the father. One has a stronger statistical metric (60.6% vs. 39.4%), but that only means that one brother shares alleles that are more common with the child than his brother does. It does not make one more likely to be the father than the other.

This is why interpreting analysts who must give courtroom testimony are counseled against using formulations such as "*x* is more than 45,000 times more likely to be the contributor of the DNA than somebody else." A better way to communicate this information would be to say: "The evidence shown would be 45,000 times more likely to be seen *if x* was the contributor than if some other, unrelated person had the same alleles by chance." Using a

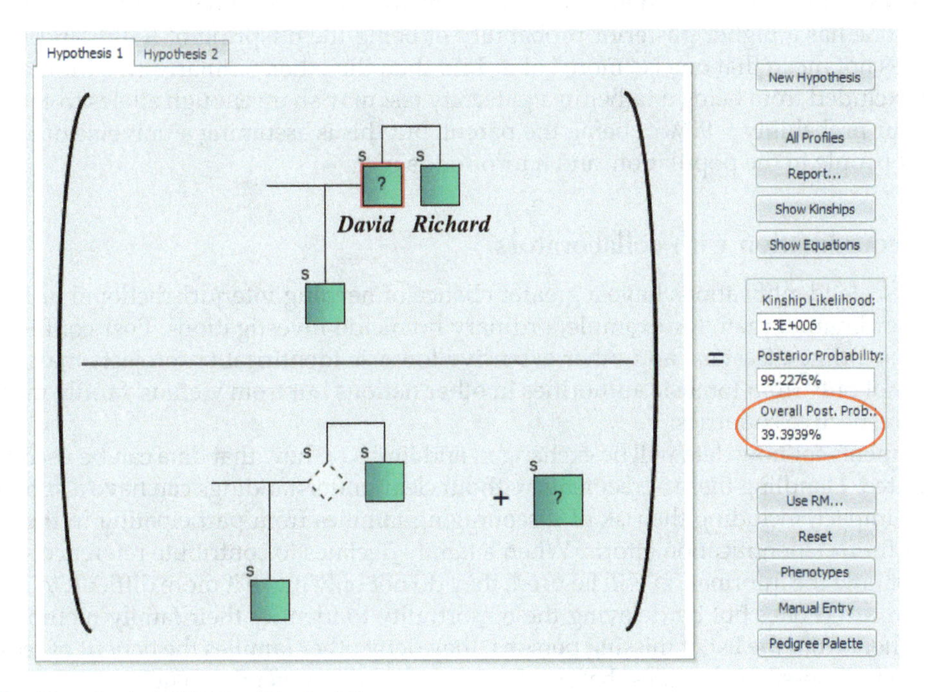

FIG. 30 Hypothesis 1 (H1): David is the father.

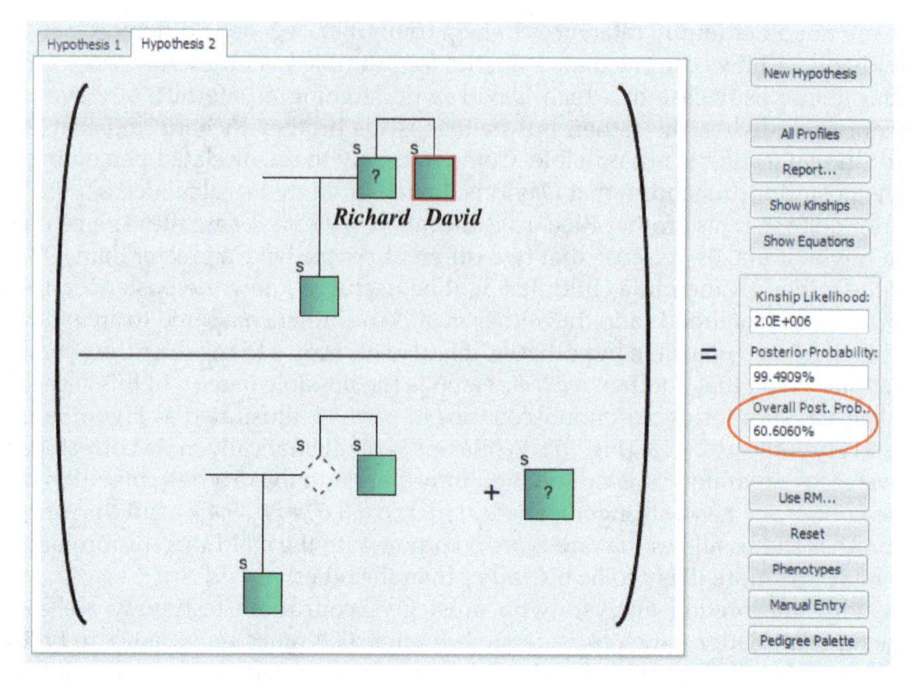

FIG. 31 Hypothesis 2 (H2): Richard is the father.

personal example, even though my sister and I are both true children of our two parents, the fact that one has a higher posterior probability of being the offspring of Ralph and Barbara Cash does not mean that one is "more" of a child than the other. Similarly, two brothers who are not excluded from being a father in a paternity test may share enough alleles to each have a posterior probability > 90% of being the parent, but this is assuming a universe of random, *unrelated* people in the population, and a prior odds of ½.

Secure coordination with collaborators

Large scale identifications have a greater chance of needing interjurisdictional or international coordination than, for example, ordinary homicide investigations. Post conflict identifications, airline disasters, and other extensive forensic identification projects may require support not only from forensic authorities in other nations but from victims' family members in those respective countries.

Agreements for how data will be exchanged and limits on how that data can be used can be complicated. Handling these agreements without clear understandings can have a substantial negative impact, including the risk of discouraging families from participating in the immediate (or future) identification efforts. When a family declines to contribute reference samples for fear of how the information will be used, they do not only make it more difficult to identify their own loved ones but by delaying the opportunity to identify their family members and remove them from the list of missing persons, they deny other families the benefit of reducing the number of possible identities that need to be searched as potential matches.

M-FISys has a unique capability to do a confidential search on remote databases without unnecessarily sharing/releasing the profiles of family members. The following is an example of how it can work, based on current events. Imagine that in the year 2020, skeletal remains were to be found in the desert near the Texas border with Mexico, along with clothing, property, and other circumstantial evidence suggesting that the decedent might be a person who had crossed the border into the United States illegally. Texas authorities might begin by reviewing local cases of missing persons. If this failed, they could use a feature in M-FISys to solicit information from authorities in foreign countries to the South. The laboratory in Texas could package one or more UHR profiles into a secure file (encoded with AES 256-bit encryption and with all identifiers other than the sample name stripped away) and send it to collaborators in Mexico, Guatemala, El Salvador, Honduras, Costa Rica, or other counterpart agencies in jurisdictions that might have profiles representing reference material for a missing person. Each receiving laboratory could access the data in the file only to compare it to profiles in their own database or databases. If no match were to be discovered, the contributed profiles would remain locked and inaccessible, in an encrypted file. However, if a potential match were to be found, the receiving lab would only be able to view a report that the file they received contained a profile with a particular name and that this profile was consistent with ante-mortem data from a particular case of their own. At this point, additional "information exchange" could be negotiated based on plausible evidence that a specific person might be identified through a collaborative review of the material.

This procedure, supported by the confidential collaboration tool in M-FISys, highlights a difference between the goals that a humanitarian or nongovernmental organization (NGO) might have, compared to what an investigative intelligence or law enforcement agency might have. A law enforcement organization might want to maximize the amount of data that can be collected from a remote collaborator because the data could later be found to be useful in the current or in a future investigation. By contrast, a health department or humanitarian NGO might try to exchange only enough information to minimize or eliminate any chance that identification would be missed due to insufficient knowledge sharing. One approach is not inherently more efficient or more ethical than the other. The goals and responsibilities of the two types of organizations are simply different. M-FISys allows partner laboratories to collaborate while sharing the minimal amount of useful data, but it does not in any way prevent a laboratory from sharing more if that is found to be appropriate.

Anthropological information

It was mentioned earlier in this chapter that M-FISys includes the capability of associating documents such as photographs and PDFs with individual profiles. This does not make it a document management system but does allow analysts to keep documentation relevant to the ID process close at hand.

One area where the support is a bit more extensive is basic anthropological and demographic information. Ante-mortem information regarding height, age, gender, and date of disappearance can be compared to estimated height, age, gender, date of recovery, and post-mortem interval of a recovered skeletal case. It can also be used to search the database for other cases that have consistent descriptions along selected parameters (Fig. 32).

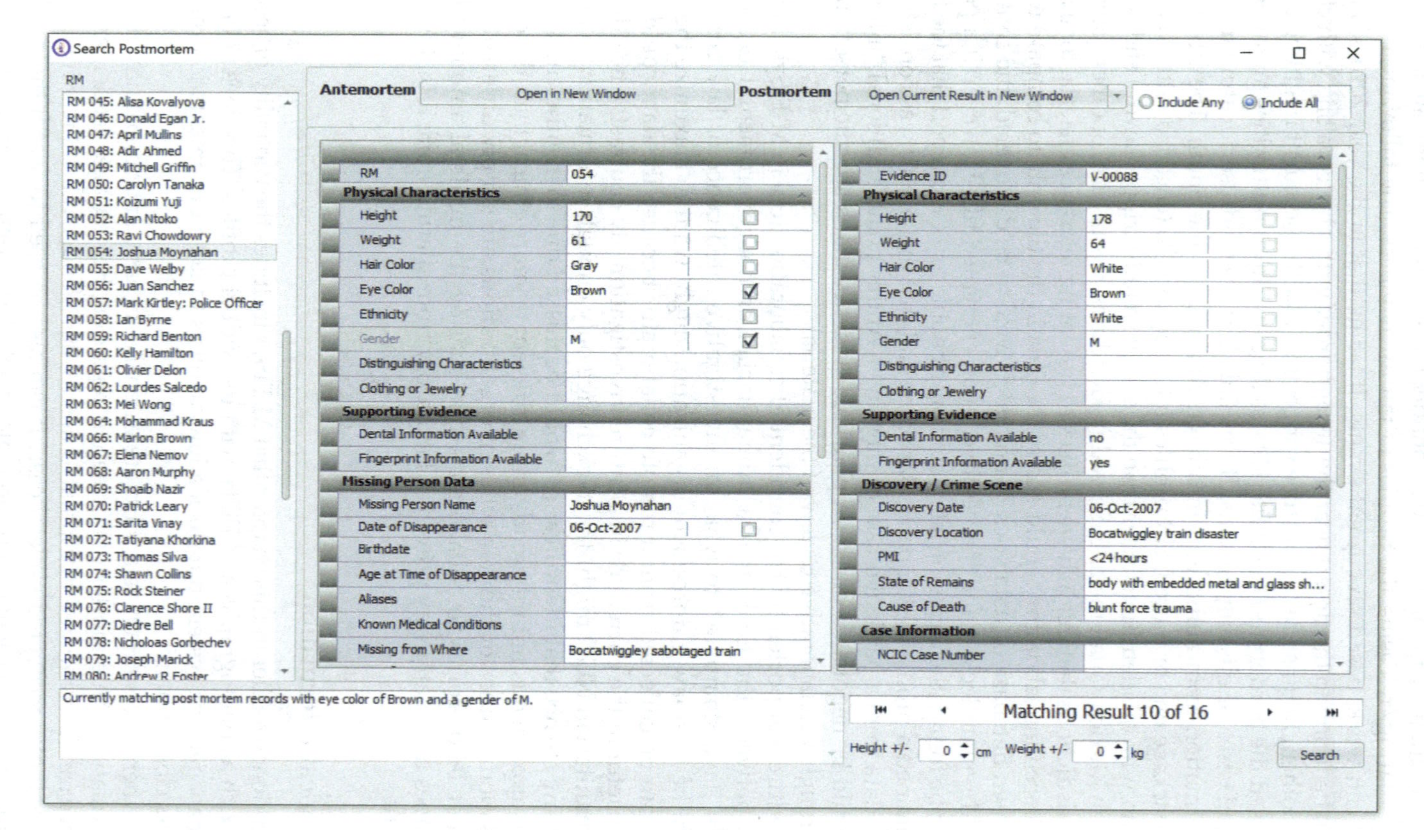

FIG. 32 Anthropological and demographic documentation and search tools in M-FISys.

The anthropology tools in M-FISys are intended as a support for direct- and kinship DNA analysis, and at this time are not being developed as a full anthropological information database. Search fields have been added and modified over time to support the work of particular client investigators.

Conclusion

DNA is an extremely powerful and facile tool for identifying UHRs and can be used in a variety of cases where other forensic specialties are not applicable. However, experience has shown that the most reliable identification projects are characterized by professional collaboration between DNA analysts, anthropologists, dentists/odontologists, fingerprint experts, medico-legal investigators, and others whose expertise can be fruitfully applied to the circumstance at hand. In cases involving skeletal remains, DNA, odontology, and anthropology experts should collaborate closely, and a final anthropological review of remains before they are released is a valuable safeguard against erroneous identifications.

Purpose-designed software and databases have a primary task of comparing the DNA profiles of remains and of reference samples, and subsequently presenting matches that can be confirmed by examiners as identifications. Ideally, the software will do more than perform direct matching and kinship matching. Reassociating skeletal remains into aggregate profiles and condensing multiple tests on the same elements into virtual profiles substantially simplifies data handling and presentation. A system can undergo continuous improvement and enhancement in order to highlight potential data conflicts and errors based on experiences in new projects and on new technologies developed in the scientific community. Finally, a well-structured system should not only *allow* collaboration between different forensic disciplines and between forensic laboratories in distant locations but should *promote* and *facilitate* these interactions. M-FISys has benefited from the experience and recommendations of a large number of collaborating scientists over decades of work and in countries with a wide variety of investigative traditions (for which the author is very grateful).

CHAPTER

19

Bioinformatic tools for interrogating DNA recovered from human skeletal remains

Frank R. Wendt PhD

Department of Psychiatry, Yale School of Medicine, U.S. Department of Veteran Affairs Connecticut Healthcare System, New Haven, CT, United States

Introduction

Genetic data obtained from human skeletal remains are often incomplete or missing relative to the pristine, single-source DNA obtained from reference materials (Ambers et al., 2014a, 2016, 2017, 2019, 2020; Ambers, 2019; Harrel and Hughes-Stamm, 2020; Holmes et al., 2018; Kieser et al., 2019; Olivieri et al., 2018). One of the penultimate goals of human remains work is to answer the question "who was this person?" Answering this question becomes increasingly complicated as human remains are exposed to environmental insults that break down endogenous DNA over time. Fortunately, different types of DNA and loci within the DNA molecule are sensitive or robust to these environmental insults (Alaeddini et al., 2010; Ambers et al., 2014b). These sensitivities and marker types are described in detail elsewhere in the text, but will be briefly touched upon in the context of the analytic pipelines discussed herein. In this chapter, computational and bioinformatic tools are discussed which facilitate and maximize the information obtained from the DNA of human skeletal remains.

This chapter contains three elements relevant for analysis and interpretation of genetic data. First, the relevant marker types discussed in this chapter will be reintroduced to complement their individual chapters throughout the text focusing on information relevant for understanding the tools introduced. Second, file formats commonly encountered in genetic data analysis will be described such that an operational bioinformatic lexicon may be developed. Lastly, bioinformatic and computational tools for extracting and utilizing genetic data will be introduced and described in a manner designed to facilitate informed decision-making during data analysis phases of historical, forensic, and anthropologic work.

Forensic Genetic Approaches for Identification of Human Skeletal Remains
https://doi.org/10.1016/B978-0-12-815766-4.00021-2

Relevant loci

Short tandem repeats

Short tandem repeats (STRs) contain a sequence of nucleotides recurring adjacent to one another at a locus (Gymrek, 2017). The STR core repeat unit may range from 1 to 6 base pairs (bp) in length. STRs are found across the human autosomes and sex chromosomes. The specific attributes of analyzing autosomal STRs versus X-chromosome STRs (X-STRs) and Y-chromosome STRs (Y-STRs) are described here.

Autosomal STRs

Autosomal STRs are the primary locus utilized for source attribution of biological material to an individual (Butler and Hill, 2012). Individualization relies on (1) the highly heterozygous nature of select autosomal STRs, (2) the independent assortment of information at those select loci, and (3) the population frequency differences of autosomal STRs alleles (Avila et al., 2019; Barrio et al., 2019; Borsuk et al., 2018; Gettings et al., 2018; Ng et al., 2018; Novroski et al., 2019; Oldt and Kanthaswamy, 2019; Wendt et al., 2016). These features of autosomal STRs in human skeletal remains work enable the use of the statistical principle called the product rule. The product rule states that the frequency of independent genetic events may be multiplied together to produce the relative frequency of an autosomal STR profile in a given population. Much human skeletal remains work relies on length-based changes in autosomal STRs to make such inferences about human identity.

Massively parallel sequencing (MPS) has revealed a unique feature of STRs which makes the autosomal variety of DNA marker particularly informative. MPS produces sequence-level information for each locus of interest such that the same length-based allele (e.g., allele 11 at the TPOX locus) is different on the sequence level (e.g., TPOX-11 version 1 and TPOX-11 version 2). This improved resolution of length-based alleles using sequence-level information improves the population-level heterozygosity of autosomal STRs, resulting in a general decrease in the likelihood of another person from the general population having the same DNA profile (Barrio et al., 2019; Gettings et al., 2018; Novroski et al., 2019; Wendt et al., 2016). In other words, sequence-level data makes a given DNA profile more rare.

Y-chromosome STRs

Unlike autosomal STRs, Y-chromosome STRs (Y-STRs) cannot be used for individualization of a substrate to a specific individual (Kayser, 2017). The Y chromosome is inherited through paternal lineages and lacks recombination events (Page et al., 1982). This means that changes to Y-chromosomal information in a paternal lineage may only arise through relatively rare mutational events (Domingues et al., 2007; Fernandes et al., 2008; Gusmao et al., 2002; Lee et al., 2007). STRs assayed from the Y-chromosome are not independent pieces of genetic information, and therefore, the product rule of autosomal STR analysis cannot be applied here. The Y-STR profile is most appropriately analyzed as a haplotype (i.e., a constellation of DNA changes in a given region). For statistical interpretation of Y-STR information, the relative rarity of a haplotype is reported rather than the product of STR genotype relative rarities (Ballantyne and Kayser, 2012; Davis et al., 2011; Kayser et al., 1997; Roewer et al., 1992). Y-STR typing is commonly employed to answer questions related to human identification, kinship and paternity and fertility, sexual assault legal cases, and human migratory events over contemporary and evolutionary history.

X-chromosome STRs

The X-chromosome is unique in that males carry a single X-chromosome while females carry two X-chromosomes. This differential dosage of X-chromosomal STRs requires more complex statistical interpretation than that of autosomal or Y-chromosomal loci and results in different X-STR power of discrimination between sexes. In particular, the X-chromosome is quite large relative to the Y-chromosome (Goodfellow et al., 1985; Wilson Sayres, 2018). The size and distribution of the X-chromosome and STRs on the chromosome mean that some STRs may be linked and some may be independent. In instances of independence, the product rule may be applied to X-STRs (Diegoli, 2015; Fukuta et al., 2019; Lang et al., 2019; Roberts et al., 1987; Salvador et al., 2018). For matching male biological material to a donor, X-STRs are generally less powerful than autosomal STR counterparts. However, X-STRs may be highly informative with respect to missing person events (see also "Mitochondrial DNA" section) (Chen et al., 2014; Prieto-Fernandez et al., 2016; Zidkova et al., 2014).

Single nucleotide polymorphisms

Single nucleotide polymorphisms (SNPs) are individual DNA base changes occurring frequently across the human genome with established frequencies in the general population (i.e., unlike mutational events, SNPs are regularly observed in the general population). Unlike STRs, SNPs generally have only two alleles, meaning the per-locus population-level heterozygosity is much lower than STR loci. SNPs have added benefit, however, in their relatively small amplicon size and amenability to inferring human identity, biogeographic ancestry, and even physical appearance (Budowle and Van Daal, 2008).

Identity-informative SNPs

Identity-informative SNPs (iiSNPs) are autosomal loci that can attribute a source of biological material to a single individual (Churchill et al., 2016, 2017; Tasker et al., 2017). Because of the decreased allele distribution relative to autosomal STRs (Li et al., 2017), many more iiSNPs are required to achieve comparable random match probability (RMP) statistics. In contrast, their small amplicon size makes them an ideal candidate marker for comparing reference material to DNA obtained from highly degraded skeletal remains.

Ancestry-informative SNPs

Certain SNPs across the genome exhibit vastly different frequencies in one population relative to another population. Collecting genotype information about enough of these ancestry-informative SNPs (aiSNPs) makes it possible to attribute a biological sample to some global geographic region (Bulbul and Filoglu, 2018; Bulbul et al., 2018a; Pakstis et al., 2017). This type of inference is quite accurate on a large scale, but research efforts are actively identifying ways to make more nuanced predictions about the origin of skeletonized remains such that historical information may be returned to appropriate countries, regions, and/or minority racial and ethnic factions (Ambers et al., 2016, 2017; Esposito et al., 2018; Huckins et al., 2014). Analysis of aiSNPs typically relies on principal component analysis

(PCA) whereby an unknown sample is plotted against known reference population samples in a feature space that captures the variance in aiSNP genotypes across populations.

Phenotype-informative SNPs

In the context of using DNA to make conclusions about historical, anthropological, and missing persons skeletal remains, phenotype-informative SNPs (piSNPs) have garnered substantial attention for their ability to predict the outwardly visible physical attributes of an individual (Breslin et al., 2019; Kayser, 2015; Liu et al., 2019; Xiong et al., 2019). It should be noted that piSNPs may provide putative information about the physical appearance of an individual as indirect measures of ancestry based on ancestry-phenotype relationships. The group of SNPs classified as piSNPs demonstrates direct relationships and predictive potential for specific features of human physical appearance including eye color, hair color, skin color, and height (Breslin et al., 2019; Liu et al., 2019). Many analytic methods have been employed to make phenotype predictions using DNA, including machine learning and polygenic risk scoring.

Mitochondrial DNA

The mitochondria are cellular organelles independent of the nucleus (i.e., the origin of autosomal, Y-chromosomal, and X-chromosomal information) found in much greater copy numbers than the nuclear genome. A single cell may contain hundreds to thousands of mitochondria (Robin and Wong, 1988). Mitochondrial DNA (mtDNA) is transmitted through maternal lineages and, given its lack of recombination, must be statistically interpreted as a haplotype (Andrews et al., 1999; Holland et al., 2018; Ivanov et al., 1996). In the same manner as Y-STRs, the primary reporting statistic of mtDNA is haplotype rarity.

Relevant data file formats

To efficiently digest the breadth of bioinformatic tools available, it becomes important to have an operational vocabulary of relevant data and file types commonly encountered in genetic data analysis. Some of these file types are described below and various DNA-related file formatting guides are available. See https://www.ncbi.nlm.nih.gov/sra/docs/submitformats/ for one example.

Common file formats

General data file formats relevant for the bioinformatic methods presented here include comma-separated value files (.csv) and text files (.txt). The .csv file can store information in tabular format and can be opened with standard spreadsheet programs such as Microsoft Excel and Google Spreadsheets. Unlike .csv files, .txt files contain unformatted text which can be reorganized using text editing and word processing programs.

The current standard forensically relevant file format relates to size-separated PCR amplicons from a genetic analyzer (e.g., 3130xl and 3500xl). The fragment analysis file (FSA; .fsa) contains electropherogram records of fragment separation and relevant instrumentation characteristics. Fragment analysis files may be read and analyzed by commercially available software platforms and/or open-source platforms such as R (Covarrubias-Pazaran et al., 2016).

Sequence-specific file formats

With the application of MPS to questions of anthropological, forensic, and historical relevance came a new collection of file types and bioinformatic approaches to analyze and store information. Because MPS relies on sequencing DNA in parallel, the files types discussed in this section are generally very large and their use requires familiarity with command line interfaces.

The .fastq file format (FASTQ) stores DNA sequence information and per-base quality information for each read generated in an MPS experiment (Fig. 1) (Cock et al., 2010). Each FASTQ file entry contains four lines of relevant information per read. The first line of a FASTQ file entry always begins with an "@" (read "at") character and an identifier. This identifier may contain information about the instrument used to sequence the DNA, which lane of the flow cell was used for that read, the x- and y-coordinates for the relative position of the clonal DNA cluster on a flow cell, index numbers for sequencing experiments containing many multiplexed samples, and a numeric indicator of whether the sequence in question is part of a paired-end read. The second line in a FASTQ file entry contains the string sequence of base calls at each position in a read. The third line always begins with a "+" (read "plus") character followed by either (1) no information or (2) the same header information found in the first line. The fourth line of a FASTQ file entry contains a string of characters representing the quality values for each base listed in line two of the same entry. Characters used to represent base quality are based on the American Standard Code for Information Interchange (ASCII) character coding used to represent text in computers and telecommunications.

Sequence alignment/map format files (SAM; .sam) are tab-delimited text files used for storing information about FASTQ base alignments to some reference sequence (Fig. 1) (Li et al., 2009). Header sections of SAM files always begin with the "@" character followed by

FASTQ

```
@M01451: 159:000000000-AW41W:1:1109:20256:4529 1:N:0:ATATCAGG+ACTGCATA
GTTCTTGCCCCTGCTTCCTGCTCCAGGCCCTTACCAAGGGTAGGCCGGTG
+
CBCCCAFF<FGGGGCACFG<@<E<EF8C..89@CAA,;CFBCF;C<@FEF
```

SAM

```
@PG ID:bwa PN:bwa VN:0.7.15-r1140 CL:bwa mem –t 100 –A 1 –B 12 –O 6 –E 1 –M hg38.fa S1_Read1.fastq S1_Read2.fastq
S1_Aligned
M01451:159:000000000-AW41W:1:1104:6399:18125 369 chr1 83798 0 110H39M152H chr21 8464885 0 AGAGAGAG >,E9FCF=
M01451:159:000000000-AW41W:1:2116:10380:25023 385 chr1 99040 0 99H32M170H chr22 42139894 0 TTTTTCCT ,,,++,4,
```

VCF

```
##fileformat=VCFv4.2
##source=PLINKv1.90
##FORMAT=<ID=AD,Number=R,Type=Integer,Description="Allelic depths for the ref and alt alleles in the order listed">
##FORMAT=<ID=GT,Number=1,Type=String,Description="Genotype">
##INFO=<ID=AC,Number=A,Type=Integer,Description="Allele count in genotypes, for each ALT allele">
##INFO=<ID=AF,Number=A,Type=Float,Description="Allele frequency, for each ALT allele">
#CHROM  POS       ID         REF  ALT  QUAL      FILTER  INFO             FORMAT  SAMPLE1      SAMPLE2
10      92314597  rs7094433  A    G    13451.72  .       AC=5;AF=0.086    GT:AD   0/0:53,38    0/1:204,200
12      23517084  Rs4604980  C    T    2203.85   .       AC=2;AF=0.034    GT:AD   0/1:273,212  1/1:100
```

FIG. 1 Visual representation of human readable file types relevant for sequence-based genetic data analysis.

at least one two-character string coding the format and/or content of the section. Alignment sections of the SAM file format represent a linear alignment of segments in a tab-separated format. These files may also contain, where relevant, information about insertions, deletions, chimeric alignments (i.e., a read that cannot be represented as aligned to a single reference), and multiple mapping (i.e., ambiguous placement of a single read with respect to a reference). The SAM file format is human readable, but does not compress well for efficient data storage. To compact sequence alignments from the SAM file, the binary representation of SAM (BAM; .bam files) files may be used. The BAM file format contains identical information as SAM files, but does so in a binary format that is not human readable, is easily indexable, and is easy to navigate without demanding large computational memory (Barnett et al., 2011; Carver et al., 2010). Accompanying each BAM file may be an index (.bai) file. Index files allow programs to quickly navigate parts of the BAM file without individually reading all sequences.

Variant call files (VCF; .vcf) store information regarding variable positions in a sequence relative to some reference (Fig. 1) (Danecek et al., 2011). The header of a VCF file contains several lines of metainformation describing the data fields included in the variant calling section of the file. Directly below the header of a VCF file is a series of lines, each describing a single position in the genome. The first few columns associated with each row describe the chromosome, position, reference SNP cluster identifier (rs) number (if known), reference allele, alternate allele, quality information, allele frequency information, and other optional column flags. The remaining columns are labeled with an individual unique sample identifier. The intersecting cells between genomic locus row and sample identifying column contain genotype information, quality information, and read depth information for each locus, among additional optional content flags.

Extracting genotype information

The bioinformatic tools used to analyze all marker types described in this chapter rely on accurate extraction of genotype information from size-separated (capillary electrophoresis (CE)) or massively parallel sequenced samples. Several approaches for obtaining these data are described here.

Capillary electrophoresis platforms

CE procedures are currently the gold standard platform for forensic DNA analysis regardless of marker type. CE instrumentation can be used for size separating STR and SNP alleles, and for sequencing mitochondrial DNA via the Sanger sequencing method.

STR allele calling

Two major commercially available STR analysis platforms are GeneMarker HID (Holland and Parson, 2011) and GeneMapper ID-X (Hansson and Gill, 2011). These computational tools analyze size-separated STR amplicons from the autosome, Y-chromosome, and X-chromosome. GeneMapper and GeneMarker read FSA files from the CE instrument and interpret the fluorescent dye information associated with each PCR amplicon to produce an electropherogram. Electropherograms are visual representations of size- and fluorescence-separated PCR amplicons from multiplexed STR loci. The number of dyes used to tag multiplexed loci dictates the number of rows in an electropherogram. Historically, four-dye chemistries have been used to produce electropherograms with four distinct rows of STR data (typically shown in red, blue, black, green) plus one row for an internal lane standard for quality control evaluation. Note that five-dye and six-dye chemistries have been and are currently being developed to increase the number of loci analyzed per sample, and therefore, the relative informativeness of a DNA profile (Ludeman et al., 2018). On the top of each row is a locus identifier using either DNS nomenclature ("D" for DNA, chromosome N, "S" for single copy) or a conventional nomenclature affiliated with the initial discovery of the locus (e.g., vWA is named due to its physical proximity to the region of the genome encoding the von Willebrand factor antigen blood biomarker). The orientation of STR loci in each row of an electropherogram is unique to a single commercially available STR kit. Each row of an electropherogram shows size-separated alleles along the x-axis and the relative fluorescence units (RFUs) of each allele along the y-axis. Both GeneMapper and GeneMarker allow the user to identify analytic, stochastic, and per-locus stutter thresholds to guide interpretation of DNA profiles. Each software has predetermined suggested thresholds; however, it is considered best practice to adjust these thresholds according to your laboratory's internally validated interpretation and analytical thresholds.

There are several features of an electropherogram common to DNA profiles from human remains. The first of these is a degradation pattern also referred to as a ski-slope pattern (Hansson et al., 2017; Shved et al., 2014). This pattern visually manifests as the relative decrease in RFUs with increasing locus size. In other words, as the profile is read from left to right, the intensity of fluorescent signal from each locus decreases. The cause of this pattern is the more rapid degradation of longer DNA when exposed to environmental insults. Degraded DNA profiles may (1) be difficult to interpret and/or (2) provide little source attributive information if allele RFUs approach the laboratory-validated analytic threshold for distinguishing allelic signal from instrument noise. The second feature of DNA profiles from human remains is heterozygote imbalance (Hansson et al., 2017; Kelly et al., 2012). In instances where an individual has inherited two different size alleles from each parent, commercially available STR kits amplify the loci from high-quality samples in an unbiased manner such that the balance between allele RFUs is relatively close to 1 (i.e., the ratio of RFU for allele 1 to the RFU of allele 2 equals 1). In DNA profiles from human remains, it is common to observe ratios much less than 1 due to, for example, preferential amplification of the smaller allele. This can be problematic for several reasons (Green et al., 2013; Walsh et al., 1992). First, the minor allele may not meet the analytical threshold, and therefore, may be considered noise instead of an allele. Second, if several imbalanced loci are detected in a profile, the combination of analytical and stochastic thresholds may lead an analyst to interpret the profile as a mixture. For these reasons, generating and interpreting STR profiles from human skeletal remains are typically performed by analysts trained to recognize and interpret these CE results using well-validated decision-making protocols.

Sequencing with CE

CE-based fragment separation also may be used for genotyping SNP loci and assessing sequencing variation in the mitochondrial genome. The same companies providing analytic support for STR data (e.g., SoftGenetics LLC and Applied Biosystems) have actively contributed software solutions to the analysis and interpretation of SNPs and mtDNA variants. GeneMapper and GeneMarker software suites support SNP-based genotype analysis in a similar fashion as described with STR loci (Chaitanya et al., 2018; LaRue et al., 2012a,b, 2014; Martins et al., 2019; Moura-Neto et al., 2018; Ristow et al., 2017).

Calling variants in mtDNA is a powerful approach in analyzing human remains due to detection of maternal lineage information. The chemistry used to assay mtDNA is called Sanger sequencing. This method produces strings of DNA of varying length, which must be aligned to a reference sequence. The Mutation Surveyer (Minton et al., 2011) software contains a graphical user interface designed similarly to GeneMarker and is a commonly employed tool for variant calling from mtDNA sequence data with reference to the revised Cambridge Reference Sequence (rCRS) (Andrews et al., 1999). While whole-mtDNA sequencing via the Sanger method is feasible, typical forensic application of this method focuses on hypervariable regions 1 and 2 (HV1 and HV2, respectively).

CE output

The genotype results from CE analyses are often reported in tabular format for side-by-side comparison of evidentiary, suspect, and reference profile information. When deemed appropriate (e.g., when an evidentiary profile is consistent with a suspect profile), the genotype data may be used to calculate a random match probability (RMP) using population-level allele frequencies from a laboratory database or national STR or SNP allele frequency database.

CE-based mtDNA analysis often reports two pieces of information. First, mtDNA positions of the questioned or evidentiary samples, which vary with respect to the revised Cambridge Reference Sequence (rCRS), are typically reported in tabular format for upload to various programs that determine mtDNA haplotype/haplogroups. Second, an electropherogram showing the peak intensities of these variable regions may be reported to facilitate future technical and/or case review.

Massively parallel sequencing platforms

The rising popularity and utility of MPS platforms for analyzing forensically relevant sample types, including contemporary and skeletonized human remains, has fostered the development of a deep catalog of resources for extracting useful information from forensically relevant marker types (Ambers et al., 2016, 2017, 2019, 2020; Elwick et al., 2018, 2019; Gaudio et al., 2019; Parson et al., 2018; Zavala et al., 2019; Zeng et al., 2019). Several of these are discussed in this section, but interested parties are encouraged to perform regular literature searches through the PubMed resource to remain abreast with this rapidly growing discipline within the forensic genetics community.

STRait Razor

One bioinformatic platform for extracting length and sequence-based information from MPS data is the short tandem repeat allele identification tool (STRait Razor) (King et al., 2017; Warshauer et al., 2013, 2015; Woerner et al., 2017). STRait Razor was initially designed to extract STR allele calls from MPS data, but has since been improved to also call SNP and insertion/deletion polymorphisms (indels). The STRait Razor suite of tools also houses an internal sequence database from which previously observed alleles may be called and to which new high confidence alleles may be added.

STRait Razor reads MPS output files in the FASTQ format to find an anchor sequence near the target locus motif. Anchor sequences are typically about 30 nucleotides in length. The combination of correct anchor sequence identification and a locus motif (e.g., STR repeat motif:

GATA; SNP motif: A/T; indel motif: AT_TA/ATTTA) is used to assign an appropriate locus name to each string sequence entered into STRait Razor. The name assigned to each string sequence may be reduced to a length-based allele call for genotype comparisons with CE-generated genotypes (see "Capillary Electrophoresis Platforms: STR Allele Calling" section).

An interesting component of STRait Razor is the user-friendly customizability of locus motifs and identity of anchor sequences. Though locus identities are considered stable in the software's current form, novel alleles are likely to arise as MPS is applied to previously under-represented populations in human genetics (e.g., Native American and continental African populations). STRait Razor facilitates easy identification and inclusion of these alleles into the existing infrastructure of the tool. Furthermore, for those interested in research applications of MPS data, the configuration of motif and anchor sequences may be customized for novel STRs, SNPs, and/or indels of interest to a laboratory.

STRinNGS

The STRinNGS (STRs in next-generation sequencing) is a python script which also relies on the detection of anchor sequences in an MPS read (from BAM or FASTQ file formats) (Friis et al., 2016). STRinNGS aligns input files to a human reference genome to identify anchor sequences, call STRs, and interrogate flanking region SNPs/indels. Like STRait Razor, STRinNGS uses a customizable configuration file containing STR motifs and anchors of interest. Again, considered generally stable, these details may be modified by the user should genotype data suggest modifications need to be made (e.g., novel alleles are uncovered).

lobSTR

Similar to STRait Razor and STRinNGS, the lobSTR tool serves as a valuable resource to extracting STR information from MPS data (Gymrek et al., 2012). A key difference is that lobSTR evaluates STRs across the genome, while STRait Razor and STRinNGS are specifically tailored for forensically relevant STR, SNP, and indel classes. Fortunately, STRait Razor and STRinNGS may be expanded with inclusion of additional loci as the user sees fit. The methods underlying STRait Razor, STRinNGS, and lobSTR are similar. The lobSTR tool employs a three-step approach to call STRs from FASTQ or BAM file formats. First in the process is sensing: identify reads that fully encompass an STR locus. This step is critical as only those reads spanning the entire length of the repeat region are informative for allele and genotype calling. Those reads that lack full STR repeat region coverage are only capable of providing an underestimate (i.e., lower-bound) of the estimate repeat length. Second, lobSTR aligns flanking regions (i.e., analogous to anchor sequences in STRait Razor and STRinNGS literature) to a reference such that its position and length relative to that reference may be reported. Importantly, the alignment methods used in lobSTR (1) tolerate gap alignment concerns and (2) are insensitive to the magnitude of STR variation complexity. Because STRait Razor and STRinNGS have been designed to maximize information from a given read for a forensically relevant locus, the flanking regions adjacent to each STR, SNP, or indel may also be evaluated, while lobSTR solely reports genome-wide STR alleles.

Variant calling

STR loci require specific alignment strategies to overcome their complexity. The above tools were developed to overcome these limitations; however, robust SNP calling platforms are available and standardized in the human genetics community and may be used for forensic purposes as well. This can be done in three steps: sequence, alignment, and variant calling (Hansen, 2016; Kishikawa et al., 2019). For more detailed information, please refer to the in-depth pipeline described by Causey et al., 2018.

First, FASTQ files are generated by the MPS instrumentation. FASTQ files may be aligned to a human reference genome using the Burrows-Wheeler Aligner (BWA) (Li and Durbin, 2009). BWA is a software package for mapping sequences to a large reference genome (referred to as an index in BWA literature). BWA has three algorithms designed to tackle specific alignment-related questions. First, BWA-backtrack aligns relative small reads (e.g., less than 100 bp in length). BWA-SW and BWA-MEM are capable of aligning longer reads up to 1 Mbp. The primary difference between BWA-SW and BWA-MEM is speed and accuracy, with BWA-MEM typically recommended for its superior performance over BWAS-SW. The output from BWA is a SAM file.

FASTQ files may also be aligned to a reference genome using Bowtie. Bowtie is a short-read aligner that relies on similar principles as BWA. Bowtie is most efficient when reads have a single high-quality alignment in the genome. This makes Bowtie run times extremely fast.

For more efficient long-term storage and utility, SAM files may be converted to BAM and sorted BAM files using SAMtools. First, the SAMtools `view` command is used to convert SAM to BAM. Second, the SAMtools `sort` command is used to arrange BAM files for faster variant discovery and calling (Li et al., 2009).

Variant calling relative to a reference genome may be performed in SAMtools with the `mpileup` command. Furthermore, variant calling may be performed with the Genome Analysis Toolkit (GATK) (Van Der Auwera et al., 2013). To perform quality control (e.g., allele frequency filtering, heterozygosity filtering, etc.) and analyses on the VCF variant calling output file, VCFtools is a user-friendly and well-documented toolset (Danecek et al., 2011).

For mtDNA analysis, the Microsoft Excel-based tool mitoSAVE is one option to quickly convert the contents of mtDNA variants into mtDNA haplotype assignments using field-specific variant calling rules (King et al., 2014). The mitoSAVE tool requires certain information from the VCF file format and care must be taken to ensure the appropriate information is provided during variant calling pipelines. These required items include genotype, allelic depth, read depth, and genotype quality. To call mtDNA haplotypes, the contents of a VCF file may be pasted into mitoSAVE. The tool then calls haplotypes using the least number of differences between the input and the rCRS. MitoSAVE contains a watchlist for commonly encountered alignment flags (e.g., homopolymer regions) which may contain variable alignments relative to the reference. A haplotype is then generated and can be evaluated by the user considering user-defined quality, heteroplasmy, and depth of coverage thresholds.

Bioinformatic tools

Now that genotype data have been extracted from results of bench experiments commonly performed on forensically relevant substrates, there is quite a bit of information that can be generated from these genotype calls. The breadth of bioinformatic tools available to analyze genotype data is quite large. This section describes many commonly used tools for each marker type. Note that many of these tools are under regular development and their utilities will likely improve over time as research continues to tackle questions related to how forensic DNA data may be used to maximize evidentiary value and protect the privacy rights of victims and the accused.

Population genetic parameters

When testing a set of loci on a new population group, it becomes relevant to establish that the population genetic assumptions underlying our locus interpretation statistics are upheld. These include linkage disequilibrium and Hardy-Weinberg equilibrium. Two commonly used platforms for these types of analyses are Genetic Data Analysis (GDA; Weir, 1996) and Arlequin (Excoffier et al., 2007). Both tools exist as graphical user interfaces, making them user-friendly ways to extract large amounts of population-relevant information from a marker set.

Individualization and source attribution

A random match probability (RMP) describes the probability that the DNA in a random sample from the population has the same profile as the DNA in the evidentiary sample. RMPs are the primary statistic reported for single-source samples and rely on autosomal STR and/or SNP genotype data. These values represent the relative rarity of a DNA profile given the frequency of the observed alleles in a reference population. Individual laboratories often curate their own catalog of allele frequency statistics; however, it has been demonstrated that more centralized quality control and data curation are essential for minimizing error in frequency databases. One relatively large and well-curated allele frequency database is maintained as part of the STRs for identity ENFSI (European Network of Forensic Science Institutes) Reference database (STRidER) tool (Bodner et al., 2016). STRidER is a curated and freely accessible online tool for browsing and utilizing autosomal STR allele frequency information. From STRidER's query page, users may input genotype calls for autosomal STR genotyped by 15 commercially available STR kits. Note that STRidER has been extensively curated with European STR allele frequencies, but other tools enable RMP calculation with allele frequencies in North America. The University of North Texas Health Science Center maintains a random match probability (RMP) calculator with curated allele frequencies from African-Americans, European-Americans, and Asian-Americans (see https://www.unthsc.edu/graduate-school-of-biomedical-sciences/laboratory-faculty-and-staff/excel-workbooks/). Note that, due to regional differences in allele frequencies, it may be appropriate for laboratories to generate their own databases (including employee-reference samples) to better represent the demography of their region. These two resources may serve as suitable guides for design, quality procedures, and implementation of locally curated frequency databases.

Ancestry inference

Several marker types have utility for ancestry inference in forensic DNA casework and, when available, provide powerful information about the global and/or regional origins of human remains. These analyses become particularly useful with respect to human trafficking and human biological material recovered from anthropological, historical, and archeological sites.

Global ancestry estimation

Ancestry of an individual may be inferred using unlinked (i.e., loci in linkage equilibrium) SNPs and/or STR loci. Most methods for inferring ancestry rely on statistical clustering of samples into groups of similar genetic architecture. One such method is with the free software package STRUCTURE (Porras-Hurtado et al., 2013; Rosenberg et al., 2002). STRUCTURE uses a model-based approach assuming a cluster (also referred to as a population) is modeled by a characteristic set of allele frequencies at a set of loci. As the name suggests, a powerful feature of STRUCTURE lies in its ability to model admixture within the populations being analyzed. In doing so, STRUCTURE probabilistically assigns a single individual from the analysis to one or more of K populations (K may be statistically inferred within STRUCTURE or provided by the user). Using a similar model-based clustering approach, the ADMIXTURE software assigns a probability of ancestry or ancestry proportion using population allele frequencies. A key difference between the two is that ADMIXTURE employs a maximum likelihood high-dimensional optimization approach which is computationally faster than the Markov chain Monte Carlo algorithm underlying STRUCTURE (Alexander et al., 2009).

Using aiSNP genotype data, an individual profile may be probabilistically assigned to one of over 100 global populations housed in The ALlele FREquency Database (ALFRED; a resource of gene frequency data on human populations) using the Forensic Resource Reference On Genetics knowledge base (FROG-kb; Fig. 2) (Rajeevan et al., 2003, 2007, 2012a,b). To perform an ancestry search in FROG-kb, a user may search aiSNP profiles from over 10 different aiSNP sets using either genotype file upload or radio-button selections (Bulbul et al., 2018b; Eduardoff et al., 2016; Gettings et al., 2014; Kidd et al., 2011; Kosoy et al., 2009; Lao et al., 2010;

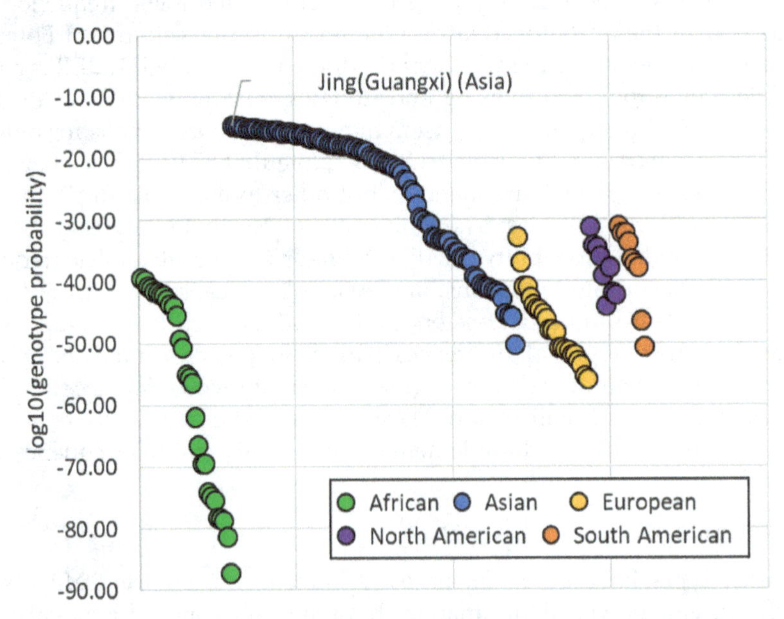

FIG. 2 Example ancestry assignment with FROG-kb using 55 ancestry-informative single nucleotide polymorphisms (SNPs) from the Kidd panel for a Korean sample. The highest probability assignment for an unknown sample is labeled.

Li et al., 2016; Nievergelt et al., 2013; Pakstis et al., 2015; Phillips et al., 2007, 2013; Santos et al., 2016). After selecting "compute," the user can view tabulated probabilities of population assignment as well as a graph showing probability of assignment to all reference populations.

Principal component analysis (PCA) is a popular option for ancestry inference using SNP data and is commonly reported in the forensic and general human genetics communities (Byun et al., 2017; Zeng et al., 2016a,b). To start explaining the utility of PCA, let us consider the problem that PCA aims to solve. If one considers that many loci are genotyped in many individuals, the matrix of loci × individuals would be quite large (Fig. 3). Now consider plotting the genotypes of locus 1 (x-axis) and locus 2 (y-axis) for all individuals with a genotype at both loci. This is a straightforward scatter plot, which may show some separation of the large number of genotyped individuals based on population distributions of alleles at loci 1 and 2. Now consider also plotting locus 3 in a third dimension (z-axis). Doing so may separate individuals even further based on allele distributions at locus 3. If more loci are added in additional dimensions, a scatter plot is generated which may be nearly uninterpretable to the human eye. One solution to this problem is called dimensional reduction, whereby highly dimensional data (e.g., 100 aiSNPs) are reduced to several new unobserved variables called principal components (PCs). PCs are representative to many aiSNPs and are designed to maximize the variance in the original data. When individuals from a population are plotted using their PCs instead of information directly from their aiSNPs, an easily interpretable 2-dimensional image (or 3-dimensional if desired) is created which represents a large portion of the variance in the original 100-aiSNP dimensional space (Fig. 3).

Maternal lineage estimation

Using output from mtDNA variant calling and haplotype assignments, the maternal lineage of an unknown sample may be evaluated. Several tools utilize either FASTQ and/or VCF files for performing haplotype and haplogroup assignments of the mtDNA genome.

HaploGrep (Kloss-Brandstatter et al., 2011) and HaploGrep 2 (Weissensteiner et al., 2016) are web-based applications based on Phylotree (a classification tree estimated from worldwide mtDNA data) (Van Oven and Kayser, 2009). Haplogroup assignment is performed automatically upon upload of variant calls or sequence reads. HaploGrep's tabular output shows information about sample identifier, the haplogroup assignment for that sample's haplotype, and the polymorphic regions of the sample's mitochondrial haplotype. Phylogenetic information also is presented visually with spatial representation of the sample's phylogeny relative to other members of the same haplogroup. This visual shows subhaplogroups and various labels for polymorphisms (back mutation events, SNPs associated with the haplotype assignment, SNPs not observed in the sample, and missing haplogroup-associated SNPs). Furthermore, HaploGrep assigns polymorphisms to several categories: (1) polymorphism not described in Phylotree, (2) polymorphism not associated with the assigned haplogroup, (3) polymorphism is a hotspot mutation described in Phylotree, and (4) polymorphism is out of the analysis range.

A third tool is called EMPOP (Huber et al., 2018; Parson et al., 2004; Parson and Dur, 2007; Rock et al., 2011; Scheible et al., 2011). The goal of EMPOP is to collect, curate, and maintain high-quality searchable mtDNA haplotypes. EMPOP search queries are based on the SAM 2 search engine, which curates full mtDNA sequences to perform unbiased and conservative alignments for forensic purposes (i.e., nomenclature is consistent with forensic variant calling

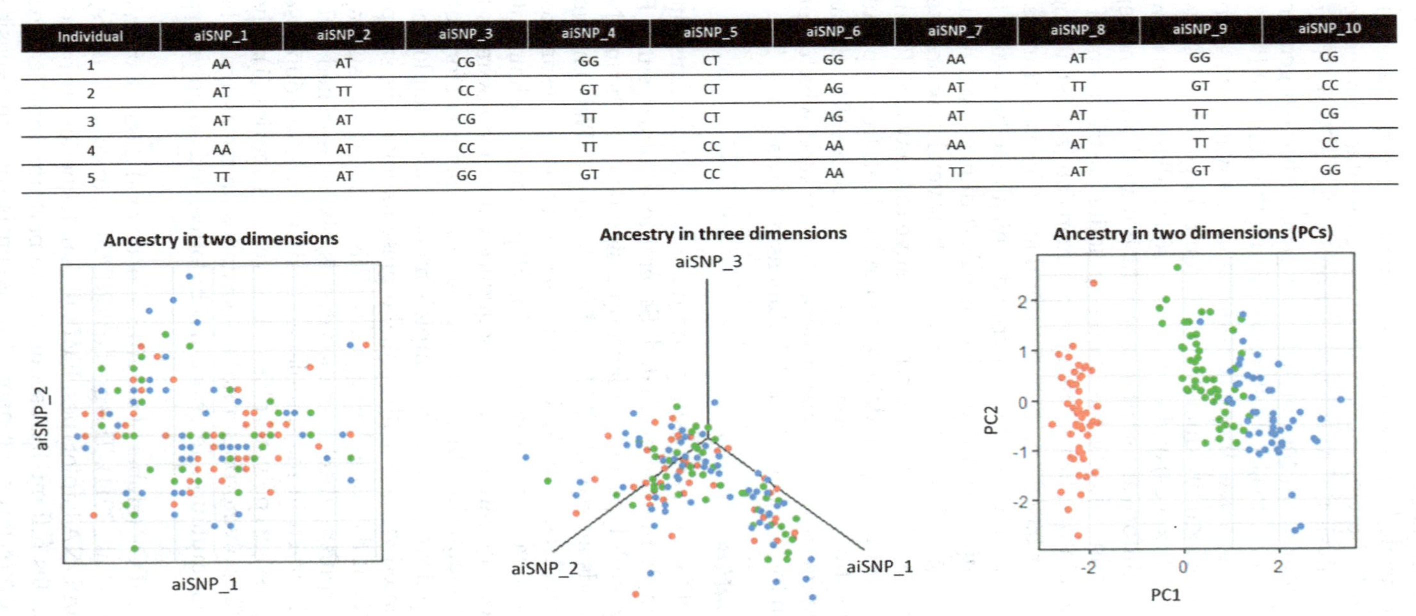

Individual	aiSNP_1	aiSNP_2	aiSNP_3	aiSNP_4	aiSNP_5	aiSNP_6	aiSNP_7	aiSNP_8	aiSNP_9	aiSNP_10
1	AA	AT	CG	GG	CT	GG	AA	AT	GG	CG
2	AT	TT	CC	GT	CT	AG	AT	TT	GT	CC
3	AT	AT	CG	TT	CT	AG	AT	AT	TT	CG
4	AA	AT	CC	TT	CC	AA	AA	AT	TT	CC
5	TT	AT	GG	GT	CC	AA	TT	AT	GT	GG

FIG. 3 Visual explanation of the utility of dimensional reduction for ancestry assignment using ancestry-informative single nucleotide polymorphisms (aiSNPs) in human identification. A high dimensional matrix of samples and genotypes is complicated to visualize in two- or three-dimensional space with meaningful resolution of populations *(red, green, and blue)*. After dimensional reduction and identification of principal components (PCs 1 and 2), the *red population* clearly separates from *blue and green*.

nomenclature to facilitate casework reporting). If desired, HV1, HV2, the control region, and/or individual SNPs may be searched instead of the full mt-genome. The nucleotide string sequence of interest is pasted into the Query page and differences relative to the rCRS are reported in tabular form. Due to stochastic sampling of mtDNA during tissue development, different tissues may have different levels of point (mixture of nucleotides at given positions) and length heteroplasmies (change in number of nucleotides in length variant regions). In practice, these differences (particularly length heteroplasmies) may lead to different alignments of mtDNA from two different tissues of the same individual. To facilitate this, EMPOP offers a feature for neighbor searching such that decisions of "match" versus "no-match" are enhanced and more reliable when length-based difference between a reference sample and an evidentiary sample is observed. The following pieces of information are reported for a query in EMPOP: (1) the number of mtDNA haplotype matches in the entire database, (2) the sample's mtDNA haplogroup assignment, (3) the number of observed matches stratified by geographic region (e.g., Africa, America, Asia, Europe), and (4) the number of matches by metapopulation affiliation (e.g., Sub-Saharan African, West Eurasian, South Asian, East Asian, etc.).

Several additional tools exist within the EMPOP framework, including the haplogroup browser, EMPcheck, and NETWORK. The haplogroup browser performs mtDNA queries (as described above). EMPcheck is a tool for evaluating the likelihood of a data table coded with respect to the rCRS. NETWORK is a tool for calculating and visually representing networks among mtDNA haplotypes. These are especially useful with respect to heteroplasmic sites and potential sequencing artifacts.

Paternal lineage estimation

Genotype output from Y-chromosomal analyses can be useful for estimating the paternal lineage of a questioned evidentiary item and/or skeletonized human remains. The online platform Y-chromosomal Haplotype Reference Database (YHRD; www.yhrd.org) is a freely accessible tool for using Y-STR (and where appropriate and available, Y-SNP) data to estimate the Y-chromosomal haplotype of a sample (Willuweit and Roewer, 2015). YHRD is an open access collection of population samples for Y-chromosomal variants. The database supports most currently employed commercially available Y-STR haplotype formats from Promega Corporation and Applied Biosystems. Depending on the Y-STR haplotype of interest (i.e., which kit is being used to generate Y-STR information), YHRD contains between 15,000 (maximal haplotype of 27 STRs) and 285,000 haplotype observations (minimal haplotype of 8 STRs).

To use the main feature of YHRD, a user may enter allele information at each STR in their Y-STR kit and search against the YHRD database. The resulting profile description contains (1) an observed haplotype frequency (i.e., the number of times the submitted haplotype is observed in YHRD) and a 95% confidence interval, and (2) three measures of expected haplotype frequencies: (i) the discrete laplace (DL) method which accounts for information regarding haplotype frequency in a metapopulation, (ii) an augmented haplotype frequency ($n+1/N+1$) which adds the observed haplotype to both the database and the number of observations, and (iii) the kappa estimate of haplotype frequency using the relative proportion of singleton observations in the population or metapopulation.

Haplogroup Predictor (http://www.hprg.com/hapest5/?hapest5) is a similar tool that takes input haplotype data for up to 111 input Y-STR loci (i.e., a full profile from Family Tree DNA). The tool calculates haplogroup probability assignments and goodness-of-fit measures. Note that the goodness-of-fit score reported for a given haplotype in a haplogroup is independent of other goodness-of-fit scores reported for that haplotype in other haplogroups. Conversely, the reported haplogroup probabilities are dependent upon one another and must sum to 1. Probability of haplogroup assignment is interpreted as follows: in Fig. 4 an unknown sample has 88.7% probability of assignment to the Q-Z780 (Q1a2a1b) subclade, 5% probability of assignment to the Q-M3 subclade (Q1a2a1a1), 3% probability of assignment to the R1b subclade (ancestral Eurasians), 1.8% probability of assignment to the Q-L330 (Q1a2a1c), and 1.5% probability of assignment to the E1b1b subclade (northeast Africa). Subclades are branches of an individual haplogroup defined by the presence of SNPs relative to the parent clade. In the unknown sample from Fig. 4, there is a high probability of assignment to haplogroup Q1a (more specifically Q1a2), suggesting this individual's paternal lineage is that of Native American or East Asian origin (Huang et al., 2018; Malyarchuk et al., 2011).

Phenotype inference

In the forensic science community, phenotype inferences are typically restricted to using a small number of piSNPs to predict hair, eye, and skin color. One prominently used method is the HIris-Plex-S System for hair, eye, and skin color prediction, which is based on 41 SNPs in the human genome (Breslin et al., 2019; Chaitanya et al., 2018). These SNPs are primarily representative of the *MC1R* gene which encodes the melanocortin 1 receptor. This is a G-protein coupled receptor which binds the peptide hormone melanocortin at the surface of melanocytes. The receptor is a key component of the pathway responsible for development of the pigment melanin. The HIris-Plex-S webpage allows users to select the number of alleles (0: listed input allele is not present in the sample; 1: heterozygous; or 2: homozygous for the input allele) at a given piSNP. When the input table is completed, the user may display the phenotype predictions for a single sample (Fig. 5). Eye color prediction results assign a probability ranging from zero to one for three eye color categories: blue, intermediate (e.g., hazel, green), or brown. Hair color prediction results assign probabilities to two hair features: four categories of color (blond, brown, red, or black) and two categories of shade (light or dark coloring). Lastly, skin color prediction assigns probabilities to five categories: very pale skin, pale skin, intermediate skin, dark skin, and dark to black skin. Because the categorical outputs are mutually exclusive (i.e., an individual cannot have blond hair and black hair), the within-trait predictions add up to 1.

The Snipper tool allows for hair, eye, and skin color in a similar manner as HIris-Plex-S with the added functionality of using specific statistical classifier: Naïve Bayes, multinomial logistic regression, or genetic distance algorithms (Ruiz et al., 2013). Furthermore, Snipper enables age prediction using a set of seven methylated sites across the genome. Though not readily encountered in the forensic science community, methylation age has been validated for high-throughput epigenetic studies of human health and disease. Unlike the DNA changes in humans (e.g., SNPs are single base changes in DNA, indels are small insertions and deletions in DNA, STRs are larger repeat expansions in DNA), epigenetic modifications are changes to the human genome not occurring at the level of the DNA string sequence.

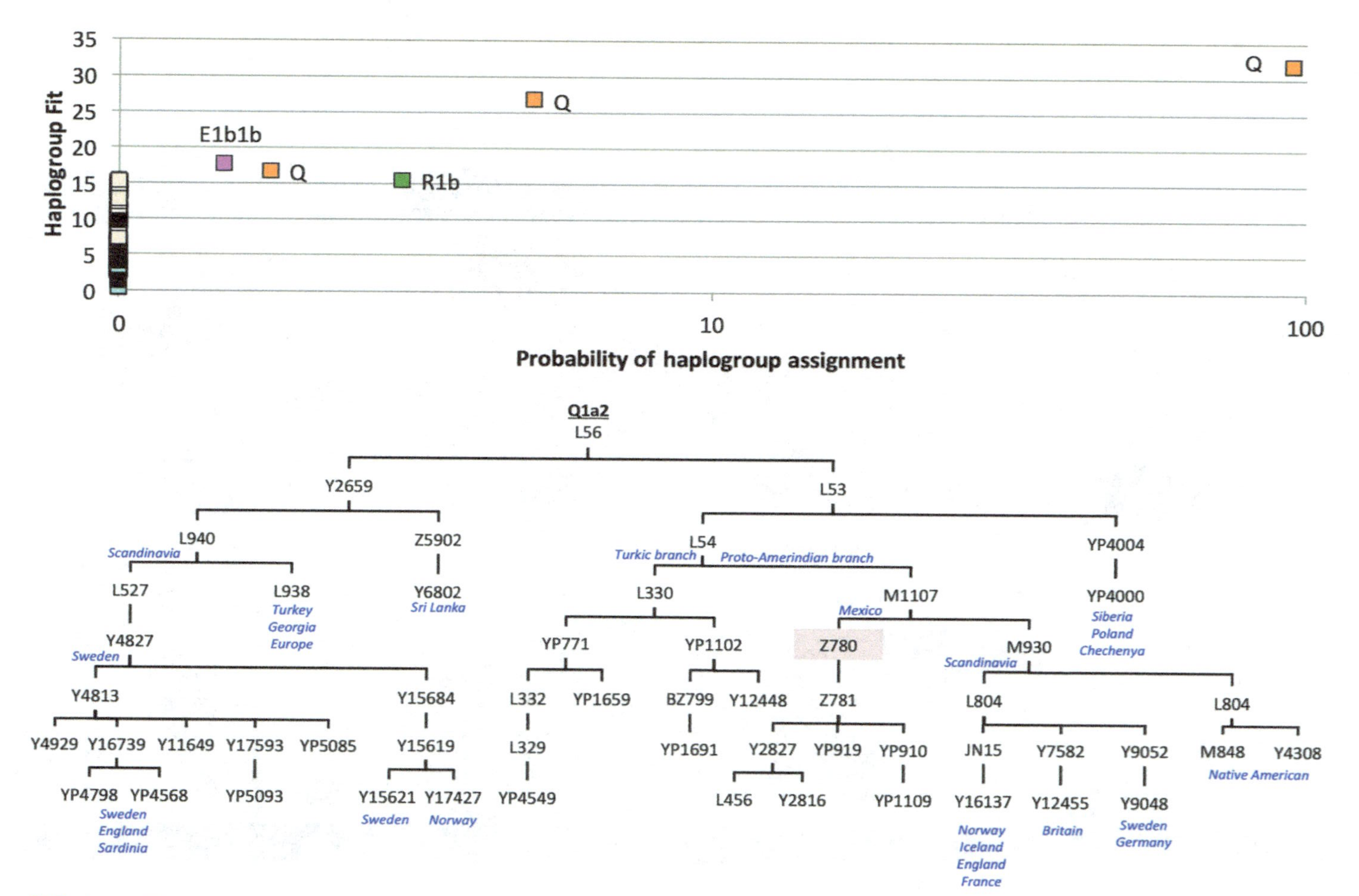

FIG. 4 Y-STR lineage estimate using the Haplogroup Predictor tool for an unknown sample. In the top panel, haplotypes belonging to the same haplogroup are shown in the *same color*. For this unknown sample, five haplotypes were assigned to three haplogroups (Q, E1b1b, and R1b). The *bottom panel* shows Y-chromosomal lineage and geographic information for the Q1a2 subclade of the Q haplogroup. The subclade with the highest probability assignment for this unknown sample is shaded in *orange*.

Gene	SNP ID	Input allele	Number of alleles	Gene	SNP ID	Input allele	Number of alleles	Gene	SNP ID	Input allele	Number of alleles	Gene	SNP ID	Input allele	Number of alleles
ANKRD11	rs3114908	T	2	*LOC105374875*	rs4959270	A	0	*OCA2*	rs1800407	A	0	*TUBB3*	rs1805009	C	2
ASIP	rs6119471	C	0	*MC1R*	rs312262906	A	1	*OCA2*	rs1800414	C	0	*TYR*	rs1042602	T	2
BNC2	rs10756819	G	0	*MC1R*	rs11547464	A	0	*OCA2*	rs1470608	A	0	*TYR*	rs1393350	T	2
DEF8	rs8051733	C	2	*MC1R*	rs885479	T	2	*OCA2*	rs1545397	T	1	*TYR*	rs1126809	A	0
HERC2	rs12913832	T	1	*MC1R*	rs1805008	T	2	*OCA2*	rs12441727	A	0	*TYRP1*	rs683	G	0
HERC2	rs2238289	C	1	*MC1R*	rs1805005	T	2	*PIGU*	rs2378249	C	0				
HERC2	rs6497292	C	0	*MC1R*	rs1805006	A	1	*RALY*	rs6059655	T	0				
HERC2	rs1129038	G	2	*MC1R*	rs1805007	T	0	*SLC24A4*	rs2402130	G	0				
HERC2	rs1667394	C	1	*MC1R*	rs201326893	A	0	*SLC24A4*	rs17128291	C	0				
IRF4	rs12203592	T	0	*MC1R*	rs2228479	A	2	*SLC24A5*	rs1426654	G	2				
KITLG	rs12821256	G	1	*MC1R*	rs1110400	C	0	*SLC45A2*	rs28777	C	0				
LOC105370627	rs12896399	T	0	*MC1R*	rs3212355	A	2	*SLC45A2*	rs16891982	C	2				

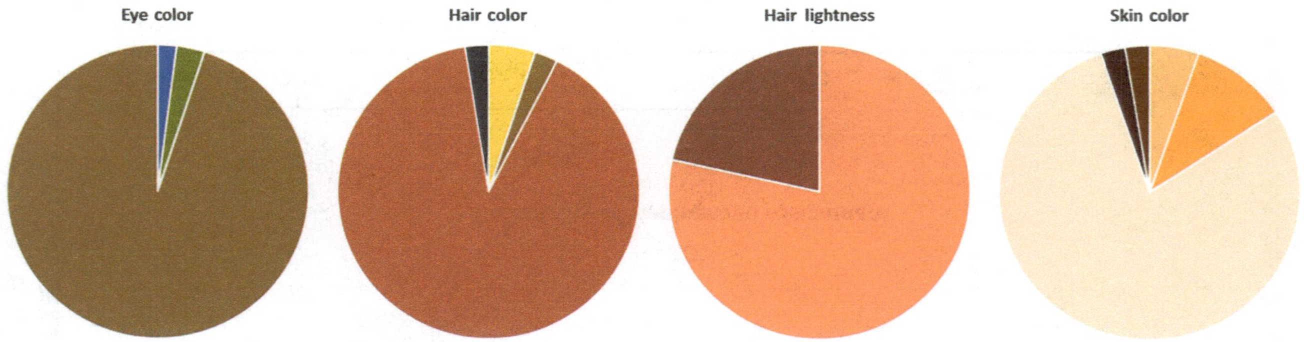

Eye color

Hair color

Hair lightness

Skin color

FIG. 5 Estimated eye, hair, and skin color probabilities (ranging from 0 to 1) for the given example genotype data using HIris-Plex-S. This individual likely has brown eyes, light red hair, and very pale skin.

Methylation age prediction relies on positions in the genome where a methyl group has been added to the DNA molecule. Methylation changes at these sites occur in a similar manner across individuals as we chronologically age such that chronological age may be calculated with relatively high accuracy using methylated genomic positions.

Emerging tools

As might be expected, the tools used by the forensic genetics community tend to be exceptionally well-developed for analyzing and interpreting the forensic context of a locus of interest; however, the larger human genetics community is simultaneously developing analogous tools for interrogating health and human disease. To explore how STR alignments are investigated in the larger human genetics community, consider reading about the tools RepeatSeq (Highnam et al., 2013), GangSTR (Mousavi et al., 2019), WebSTR (http://webstr.gymreklab.com/), and CookieMonSTR (https://github.com/gymreklab/strdenovotools). Those interested in the utility of SNP-based data are encouraged to familiarize themselves with genome-wide association studies and the tools plink, PRSice (Choi and O'Reilly, 2019), and Linkage Disequilibrium Score Regression (Bulik-Sullivan et al., 2015a,b). Though not specifically developed for forensic science practices, these tools are becoming increasingly appealing to the community to identify meaningful links between genetics and human identity, physical appearance, and biogeographic ancestry.

Summary

Human remains bring unique challenges to the laboratory with respect to degradation, inhibition, and limited sample availability. These features of contemporary and skeletonized human remains complicate the process of DNA profiling for source attribution of a sample. However, even with small amounts of genetic data, sophisticated and rigorously tested bioinformatic tools enable analysts to draw meaningful conclusions regarding (1) human identity, (2) general, maternal, and paternal ancestries, and (3) outwardly visible characteristics.

This chapter described many commonly utilized tools that aid in making decisions about the origin of unidentified human remains; however, this was only a sampling of the tools available and should not be considered an exhaustive list of forensically relevant tools. Bioinformatics and computer science are rapidly evolving, and novel methods, statistics, and tools are being developed daily to garner answers about the identity and/or biogeographic ancestry of an individual. Though offering much utility for the forensic sciences, these tools come with privacy and statistical limitations that should be considered. Readers of this chapter are encouraged to perform regular Internet searches for new tools and follow the progress of commonly used tools in the courtroom.

References

Alaeddini, R., Walsh, S.J., Abbas, A., 2010. Forensic implications of genetic analyses from degraded DNA—a review. Forensic Sci. Int. Genet. 4, 148–157.

Alexander, D.H., Novembre, J., Lange, K., 2009. Fast model-based estimation of ancestry in unrelated individuals. Genome Res. 19, 1655–1664.

Ambers, A., 2019. Forensic genetic investigation of human skeletal remains discovered in the Himalayas: Exploring the value of biogeographic ancestry in the absence of a DNA database. The ISHI Report: News from the World of DNA Forensics. https://promega.foleon.com/theishireport/april-2019-final/biogeographic-ancestry-in-absence-of-dna-database/.

Ambers, A., Gill-King, H., Dirkmaat, D., Benjamin, R., King, J., Budowle, B., 2014a. Autosomal and Y-STR analysis of degraded DNA from the 120-year-old skeletal remains of Ezekiel Harper. Forensic Sci. Int. Genet. 9, 33–41.

Ambers, A., Turnbough, M., Benjamin, R., King, J., Budowle, B., 2014b. Assessment of the role of DNA repair in damaged forensic samples. Int. J. Legal Med. 128, 913–921.

Ambers, A.D., Churchill, J.D., King, J.L., Stoljarova, M., Gill-King, H., Assidi, M., Abu-Elmagd, M., Buhmeida, A., Al-Qahtani, M., Budowle, B., 2016. More comprehensive forensic genetic marker analyses for accurate human remains identification using massively parallel DNA sequencing. BMC Genomics 17, 750.

Ambers, A.D., Churchill, J.D., King, J.L., Stoljarova, M., Gill-King, H., Assidi, M., Abu-Elmagd, M., Buhmeida, A., Al-Qahtani, M., Budowle, B., 2017. Erratum to: More comprehensive forensic genetic marker analyses for accurate human remains identification using massively parallel DNA sequencing. BMC Genomics 18, 312.

Ambers, A., Zeng, X., Votrubova, J., Vanek, D., 2019. Enhanced interrogation of degraded DNA from human skeletal remains: increased genetic data recovery using the expanded CODIS loci, multiple sex determination markers, and consensus testing. Anthropol. Anz. 76, 333–351.

Ambers, A., Bus, M.M., King, J.L., Jones, B., Durst, J., Bruseth, J.E., Gill-King, H., Budowle, B., 2020. Forensic genetic investigation of human skeletal remains recovered from the La Belle shipwreck. Forensic Sci. Int. 306, 110050.

Andrews, R.M., Kubacka, I., Chinnery, P.F., Lightowlers, R.N., Turnbull, D.M., Howell, N., 1999. Reanalysis and revision of the Cambridge reference sequence for human mitochondrial DNA. Nat. Genet. 23, 147.

Avila, E., Cavalheiro, C.P., Felkl, A.B., Graebin, P., Kahmann, A., Alho, C.S., 2019. Brazilian forensic casework analysis through MPS applications: statistical weight-of-evidence and biological nature of criminal samples as an influence factor in quality metrics. Forensic Sci. Int. 303, 109938.

Ballantyne, K.N., Kayser, M., 2012. Additional Y-STRs in forensics: why, which, and when. Forensic Sci. Rev. 24, 63–78.

Barnett, D.W., Garrison, E.K., Quinlan, A.R., Stromberg, M.P., Marth, G.T., 2011. BamTools: a C++ API and toolkit for analyzing and managing BAM files. Bioinformatics 27, 1691–1692.

Barrio, P.A., Martin, P., Alonso, A., Muller, P., Bodner, M., Berger, B., Parson, W., Budowle, B., 2019. Massively parallel sequence data of 31 autosomal STR loci from 496 Spanish individuals revealed concordance with CE-STR technology and enhanced discrimination power. Forensic Sci. Int. Genet. 42, 49–55.

Bodner, M., Bastisch, I., Butler, J.M., Fimmers, R., Gill, P., Gusmao, L., Morling, N., Phillips, C., Prinz, M., Schneider, P.M., Parson, W., 2016. Recommendations of the DNA Commission of the International Society for Forensic Genetics (ISFG) on quality control of autosomal Short Tandem Repeat allele frequency databasing (STRidER). Forensic Sci. Int. Genet. 24, 97–102.

Borsuk, L.A., Gettings, K.B., Steffen, C.R., Kiesler, K.M., Vallone, P.M., 2018. Sequence-based US population data for the SE33 locus. Electrophoresis 39, 2694–2701.

Breslin, K., Wills, B., Ralf, A., Ventayol Garcia, M., Kukla-Bartoszek, M., Pospiech, E., Freire-Aradas, A., Xavier, C., Ingold, S., De La Puente, M., Van Der Gaag, K.J., Herrick, N., Haas, C., Parson, W., Phillips, C., Sijen, T., Branicki, W., Walsh, S., Kayser, M., 2019. HIrisPlex-S system for eye, hair, and skin color prediction from DNA: massively parallel sequencing solutions for two common forensically used platforms. Forensic Sci. Int. Genet. 43, 102152.

Budowle, B., Van Daal, A., 2008. Forensically relevant SNP classes. BioTechniques 44. 603–608, 610.

Bulbul, O., Filoglu, G., 2018. Development of a SNP panel for predicting biogeographical ancestry and phenotype using massively parallel sequencing. Electrophoresis 39, 2743–2751.

Bulbul, O., Pakstis, A.J., Soundararajan, U., Gurkan, C., Brissenden, J.E., Roscoe, J.M., Evsanaa, B., Togtokh, A., Paschou, P., Grigorenko, E.L., Gurwitz, D., Wootton, S., Lagace, R., Chang, J., Speed, W.C., Kidd, K.K., 2018a. Ancestry inference of 96 population samples using microhaplotypes. Int. J. Legal Med. 132, 703–711.

Bulbul, O., Speed, W.C., Gurkan, C., Soundararajan, U., Rajeevan, H., Pakstis, A.J., Kidd, K.K., 2018b. Improving ancestry distinctions among Southwest Asian populations. Forensic Sci. Int. Genet. 35, 14–20.

Bulik-Sullivan, B., Finucane, H.K., Anttila, V., Gusev, A., Day, F.R., Loh, P.R., Duncan, L., Perry, J.R., Patterson, N., Robinson, E.B., Daly, M.J., Price, A.L., Neale, B.M., 2015a. An atlas of genetic correlations across human diseases and traits. Nat. Genet. 47, 1236–1241.

Bulik-Sullivan, B.K., Loh, P.R., Finucane, H.K., Ripke, S., Yang, J., Patterson, N., Daly, M.J., Price, A.L., Neale, B.M., 2015b. LD score regression distinguishes confounding from polygenicity in genome-wide association studies. Nat. Genet. 47, 291–295.

Butler, J.M., Hill, C.R., 2012. Biology and genetics of new autosomal STR loci useful for forensic DNA analysis. Forensic Sci. Rev. 24, 15–26.

Byun, J., Han, Y., Gorlov, I.P., Busam, J.A., Seldin, M.F., Amos, C.I., 2017. Ancestry inference using principal component analysis and spatial analysis: a distance-based analysis to account for population substructure. BMC Genomics 18, 789.

Carver, T., Bohme, U., Otto, T.D., Parkhill, J., Berriman, M., 2010. BamView: viewing mapped read alignment data in the context of the reference sequence. Bioinformatics 26, 676–677.

Causey, J.L., Ashby, C., Walker, K., Wang, Z.P., Yang, M., Guan, Y., Moore, J.H., Huang, X., 2018. DNAp: a pipeline for DNA-seq data analysis. Sci. Rep. 8, 6793.

Chaitanya, L., Breslin, K., Zuniga, S., Wirken, L., Pospiech, E., Kukla-Bartoszek, M., Sijen, T., Knijff, P., Liu, F., Branicki, W., Kayser, M., Walsh, S., 2018. The HIrisPlex-S system for eye, hair and skin colour prediction from DNA: introduction and forensic developmental validation. Forensic Sci. Int. Genet. 35, 123–135.

Chen, M.Y., Ho, C.W., Pu, C.E., Wu, F.C., 2014. Genetic polymorphisms of 12 X-chromosomal STR loci in Taiwanese individuals and likelihood ratio calculations applied to case studies of blood relationships. Electrophoresis 35, 1912–1920.

Choi, S.W., O'Reilly, P.F., 2019. PRSice-2: Polygenic Risk Score software for biobank-scale data. Gigascience 8 (7). https://doi.org/10.1093/gigascience/giz082.

Churchill, J.D., Schmedes, S.E., King, J.L., Budowle, B., 2016. Evaluation of the Illumina((R)) Beta Version ForenSeq DNA Signature Prep Kit for use in genetic profiling. Forensic Sci. Int. Genet. 20, 20–29.

Churchill, J.D., Novroski, N.M.M., King, J.L., Seah, L.H., Budowle, B., 2017. Population and performance analyses of four major populations with Illumina's FGx Forensic Genomics System. Forensic Sci. Int. Genet. 30, 81–92.

Cock, P.J., Fields, C.J., Goto, N., Heuer, M.L., Rice, P.M., 2010. The Sanger FASTQ file format for sequences with quality scores, and the Solexa/Illumina FASTQ variants. Nucleic Acids Res. 38, 1767–1771.

Covarrubias-Pazaran, G., Diaz-Garcia, L., Schlautman, B., Salazar, W., Zalapa, J., 2016. Fragman: an R package for fragment analysis. BMC Genet. 17, 62.

Danecek, P., Auton, A., Abecasis, G., Albers, C.A., Banks, E., Depristo, M.A., Handsaker, R.E., Lunter, G., Marth, G.T., Sherry, S.T., Mcvean, G., Durbin, R., 2011. The variant call format and VCFtools. Bioinformatics 27, 2156–2158.

Davis, C., Ge, J., Chidambaram, A., King, J., Turnbough, M., Collins, M., Dym, O., Chakraborty, R., Eisenberg, A.J., Budowle, B., 2011. Y-STR loci diversity in native Alaskan populations. Int. J. Legal Med. 125, 559–563.

Diegoli, T.M., 2015. Forensic typing of short tandem repeat markers on the X and Y chromosomes. Forensic Sci. Int. Genet. 18, 140–151.

Domingues, P.M., Gusmao, L., Da Silva, D.A., Amorim, A., Pereira, R.W., De Carvalho, E.F., 2007. Sub-Saharan Africa descendents in Rio de Janeiro (Brazil): population and mutational data for 12 Y-STR loci. Int. J. Legal Med. 121, 238–241.

Eduardoff, M., Gross, T.E., Santos, C., De La Puente, M., Ballard, D., Strobl, C., Borsting, C., Morling, N., Fusco, L., Hussing, C., Egyed, B., Souto, L., Uacyisrael, J., Syndercombe Court, D., Carracedo, A., Lareu, M.V., Schneider, P.M., Parson, W., Phillips, C., 2016. Inter-laboratory evaluation of the EUROFORGEN Global ancestry-informative SNP panel by massively parallel sequencing using the Ion PGM. Forensic Sci. Int. Genet. 23, 178–189.

Elwick, K., Zeng, X., King, J., Budowle, B., Hughes-Stamm, S., 2018. Comparative tolerance of two massively parallel sequencing systems to common PCR inhibitors. Int. J. Legal Med. 132, 983–995.

Elwick, K., Bus, M.M., King, J.L., Chang, J., Hughes-Stamm, S., Budowle, B., 2019. Utility of the Ion S5 and MiSeq FGx sequencing platforms to characterize challenging human remains. Legal Med. (Tokyo) 41, 101623.

Esposito, U., Das, R., Syed, S., Pirooznia, M., Elhaik, E., 2018. Ancient ancestry informative markers for identifying fine-scale ancient population structure in Eurasians. Genes (Basel) 9.

Excoffier, L., Laval, G., Schneider, S., 2007. Arlequin (version 3.0): an integrated software package for population genetics data analysis. Evol. Bioinform. 1, 47–50.

Fernandes, A.T., Rosa, A., Goncalves, R., Jesus, J., Brehm, A., 2008. The Y-chromosome short tandem repeats variation within haplogroup E3b: evidence of recurrent mutation in SNP. Am. J. Hum. Biol. 20, 185–190.

Friis, S.L., Buchard, A., Rockenbauer, E., Borsting, C., Morling, N., 2016. Introduction of the Python script STRinNGS for analysis of STR regions in FASTQ or BAM files and expansion of the Danish STR sequence database to 11 STRs. Forensic Sci. Int. Genet. 21, 68–75.

Fukuta, M., Gaballah, M., Takada, K., Miyazaki, H., Kato, H., Aoki, Y., Hamed, S.S., Elmorsi, D.A.A., Eldakroory, S.A., 2019. Genetic polymorphism of 27 X-chromosomal short tandem repeats in an Egyptian population. Legal Med. (Tokyo) 37, 64–66.

Gaudio, D., Fernandes, D.M., Schmidt, R., Cheronet, O., Mazzarelli, D., Mattia, M., O'Keeffe, T., Feeney, R.N.M., Cattaneo, C., Pinhasi, R., 2019. Genome-wide DNA from degraded petrous bones and the assessment of sex and probable geographic origins of forensic cases. Sci. Rep. 9, 8226.

Gettings, K.B., Lai, R., Johnson, J.L., Peck, M.A., Hart, J.A., Gordish-Dressman, H., Schanfield, M.S., Podini, D.S., 2014. A 50-SNP assay for biogeographic ancestry and phenotype prediction in the U.S. population. Forensic Sci. Int. Genet. 8, 101–108.

Gettings, K.B., Borsuk, L.A., Steffen, C.R., Kiesler, K.M., Vallone, P.M., 2018. Sequence-based U.S. population data for 27 autosomal STR loci. Forensic Sci. Int. Genet. 37, 106–115.

Goodfellow, P., Darling, S., Wolfe, J., 1985. The human Y chromosome. J. Med. Genet. 22, 329–344.

Green, R.L., Lagace, R.E., Oldroyd, N.J., Hennessy, L.K., Mulero, J.J., 2013. Developmental validation of the AmpFlSTR(R) NGM SElect PCR Amplification Kit: a next-generation STR multiplex with the SE33 locus. Forensic Sci. Int. Genet. 7, 41–51.

Gusmao, L., Alves, C., Costa, S., Amorim, A., Brion, M., Gonzalez-Neira, A., Sanchez-Diz, P., Carracedo, A., 2002. Point mutations in the flanking regions of the Y-chromosome specific STRs DYS391, DYS437 and DYS438. Int. J. Legal Med. 116, 322–326.

Gymrek, M., 2017. A genomic view of short tandem repeats. Curr. Opin. Genet. Dev. 44, 9–16.

Gymrek, M., Golan, D., Rosset, S., Erlich, Y., 2012. lobSTR: a short tandem repeat profiler for personal genomes. Genome Res. 22, 1154–1162.

Hansen, N.F., 2016. Variant calling from next generation sequence data. Methods Mol. Biol. 1418, 209–224.

Hansson, O., Gill, P., 2011. Evaluation of GeneMapper® ID-X mixture analysis tool. Forensic Sci. Int. Genet. Suppl. Ser. 3, 11–12.

Hansson, O., Egeland, T., Gill, P., 2017. Characterization of degradation and heterozygote balance by simulation of the forensic DNA analysis process. Int. J. Legal Med. 131, 303–317.

Harrel, M., Hughes-Stamm, S., 2020. A powder-free DNA extraction workflow for skeletal samples. J. Forensic Sci. 65 (2), 601–609. https://doi.org/10.1111/1556-4029.14197.

Highnam, G., Franck, C., Martin, A., Stephens, C., Puthige, A., Mittelman, D., 2013. Accurate human microsatellite genotypes from high-throughput resequencing data using informed error profiles. Nucleic Acids Res. 41, e32.

Holland, M.M., Parson, W., 2011. GeneMarker(R) HID: a reliable software tool for the analysis of forensic STR data. J. Forensic Sci. 56, 29–35.

Holland, M.M., Makova, K.D., McElhoe, J.A., 2018. Deep-coverage MPS analysis of heteroplasmic variants within the mtGenome allows for frequent differentiation of maternal relatives. Genes (Basel) 9 (3), 124. https://doi.org/10.3390/genes9030124.

Holmes, A.S., Roman, M.G., Hughes-Stamm, S., 2018. In-field collection and preservation of decomposing human tissues to facilitate rapid purification and STR typing. Forensic Sci. Int. Genet. 36, 124–129.

Huang, Y.Z., Pamjav, H., Flegontov, P., Stenzl, V., Wen, S.Q., Tong, X.Z., Wang, C.C., Wang, L.X., Wei, L.H., Gao, J.Y., Jin, L., Li, H., 2018. Dispersals of the Siberian Y-chromosome haplogroup Q in Eurasia. Mol. Gen. Genomics. 293, 107–117.

Huber, N., Parson, W., Dur, A., 2018. Next generation database search algorithm for forensic mitogenome analyses. Forensic Sci. Int. Genet. 37, 204–214.

Huckins, L.M., Boraska, V., Franklin, C.S., Floyd, J.A., Southam, L., Sullivan, P.F., Bulik, C.M., Collier, D.A., Tyler-Smith, C., Zeggini, E., Tachmazidou, I., 2014. Using ancestry-informative markers to identify fine structure across 15 populations of European origin. Eur. J. Hum. Genet. 22, 1190–1200.

Ivanov, P.L., Wadhams, M.J., Roby, R.K., Holland, M.M., Weedn, V.W., Parsons, T.J., 1996. Mitochondrial DNA sequence heteroplasmy in the Grand Duke of Russia Georgij Romanov establishes the authenticity of the remains of Tsar Nicholas II. Nat. Genet. 12, 417–420.

Kayser, M., 2015. Forensic DNA Phenotyping: predicting human appearance from crime scene material for investigative purposes. Forensic Sci. Int. Genet. 18, 33–48.

Kayser, M., 2017. Forensic use of Y-chromosome DNA: a general overview. Hum. Genet. 136, 621–635.

Kayser, M., Caglia, A., Corach, D., Fretwell, N., Gehrig, C., Graziosi, G., Heidorn, F., Herrmann, S., Herzog, B., Hidding, M., Honda, K., Jobling, M., Krawczak, M., Leim, K., Meuser, S., Meyer, E., Oesterreich, W., Pandya, A., Parson, W., Penacino, G., Perez-Lezaun, A., Piccinini, A., Prinz, M., Schmitt, C., Roewer, L., et al., 1997. Evaluation of Y-chromosomal STRs: a multicenter study. Int. J. Legal Med. 110. 125–133, 141–149.

Kelly, H., Bright, J.A., Curran, J.M., Buckleton, J., 2012. Modelling heterozygote balance in forensic DNA profiles. Forensic Sci. Int. Genet. 6, 729–734.

Kidd, J.R., Friedlaender, F.R., Speed, W.C., Pakstis, A.J., De La Vega, F.M., Kidd, K.K., 2011. Analyses of a set of 128 ancestry informative single-nucleotide polymorphisms in a global set of 119 population samples. Investig. Genet. 2, 1.

Kieser, R.E., Bus, M.M., King, J.L., Van Der Vliet, W., Theelen, J., Budowle, B., 2019. Reverse complement PCR: a novel one-step PCR system for typing highly degraded DNA for human identification. Forensic Sci. Int. Genet. 44, 102201.

King, J.L., Sajantila, A., Budowle, B., 2014. mitoSAVE: mitochondrial sequence analysis of variants in Excel. Forensic Sci. Int. Genet. 12, 122–125.

King, J.L., Wendt, F.R., Sun, J., Budowle, B., 2017. STRait Razor v2s: advancing sequence-based STR allele reporting and beyond to other marker systems. Forensic Sci. Int. Genet. 29, 21–28.

Kishikawa, T., Momozawa, Y., Ozeki, T., Mushiroda, T., Inohara, H., Kamatani, Y., Kubo, M., Okada, Y., 2019. Empirical evaluation of variant calling accuracy using ultra-deep whole-genome sequencing data. Sci. Rep. 9, 1784.

Kloss-Brandstatter, A., Pacher, D., Schonherr, S., Weissensteiner, H., Binna, R., Specht, G., Kronenberg, F., 2011. HaploGrep: a fast and reliable algorithm for automatic classification of mitochondrial DNA haplogroups. Hum. Mutat. 32, 25–32.

Kosoy, R., Nassir, R., Tian, C., White, P.A., Butler, L.M., Silva, G., Kittles, R., Alarcon-Riquelme, M.E., Gregersen, P.K., Belmont, J.W., De La Vega, F.M., Seldin, M.F., 2009. Ancestry informative marker sets for determining continental origin and admixture proportions in common populations in America. Hum. Mutat. 30, 69–78.

Lang, Y., Guo, F., Niu, Q., 2019. StatsX v2.0: the interactive graphical software for population statistics on X-STR. Int. J. Legal Med. 133, 39–44.

Lao, O., Vallone, P.M., Coble, M.D., Diegoli, T.M., Van Oven, M., Van Der Gaag, K.J., Pijpe, J., De Knijff, P., Kayser, M., 2010. Evaluating self-declared ancestry of U.S. Americans with autosomal, Y-chromosomal and mitochondrial DNA. Hum. Mutat. 31, E1875–E1893.

LaRue, B.L., Ge, J., King, J.L., Budowle, B., 2012a. A validation study of the Qiagen Investigator DIPplex(R) kit; an INDEL-based assay for human identification. Int. J. Legal Med. 126, 533–540.

LaRue, B.L., Sinha, S.K., Montgomery, A.H., Thompson, R., Klaskala, L., Ge, J., King, J., Turnbough, M., Budowle, B., 2012b. INNULs: a novel design amplification strategy for retrotransposable elements for studying population variation. Hum. Hered. 74, 27–35.

LaRue, B.L., Lagace, R., Chang, C.W., Holt, A., Hennessy, L., Ge, J., King, J.L., Chakraborty, R., Budowle, B., 2014. Characterization of 114 insertion/deletion (INDEL) polymorphisms, and selection for a global INDEL panel for human identification. Legal Med. (Tokyo) 16, 26–32.

Lee, H.Y., Park, M.J., Chung, U., Lee, H.Y., Yang, W.I., Cho, S.H., Shin, K.J., 2007. Haplotypes and mutation analysis of 22 Y-chromosomal STRs in Korean father-son pairs. Int. J. Legal Med. 121, 128–135.

Li, H., Durbin, R., 2009. Fast and accurate short read alignment with Burrows-Wheeler transform. Bioinformatics 25, 1754–1760.

Li, H., Handsaker, B., Wysoker, A., Fennell, T., Ruan, J., Homer, N., Marth, G., Abecasis, G., Durbin, R., 2009. The equence Alignment/Map format and SAMtools. Bioinformatics 25, 2078–2079.

Li, C.X., Pakstis, A.J., Jiang, L., Wei, Y.L., Sun, Q.F., Wu, H., Bulbul, O., Wang, P., Kang, L.L., Kidd, J.R., Kidd, K.K., 2016. A panel of 74 AISNPs: improved ancestry inference within Eastern Asia. Forensic Sci. Int. Genet. 23, 101–110.

Li, L., Wang, Y., Yang, S., Xia, M., Yang, Y., Wang, J., Lu, D., Pan, X., Ma, T., Jiang, P., Yu, G., Zhao, Z., Ping, Y., Zhou, H., Zhao, X., Sun, H., Liu, B., Jia, D., Li, C., Hu, R., Lu, H., Liu, X., Chen, W., Mi, Q., Xue, F., Su, Y., Jin, L., Li, S., 2017. Genome-wide screening for highly discriminative SNPs for personal identification and their assessment in world populations. Forensic Sci. Int. Genet. 28, 118–127.

Liu, F., Zhong, K., Jing, X., Uitterlinden, A.G., Hendriks, A.E.J., Drop, S.L.S., Kayser, M., 2019. Update on the predictability of tall stature from DNA markers in Europeans. Forensic Sci. Int. Genet. 42, 8–13.

Ludeman, M.J., Zhong, C., Mulero, J.J., Lagace, R.E., Hennessy, L.K., Short, M.L., Wang, D.Y., 2018. Developmental validation of GlobalFiler PCR amplification kit: a 6-dye multiplex assay designed for amplification of casework samples. Int. J. Legal Med. 132, 1555–1573.

Malyarchuk, B., Derenko, M., Denisova, G., Maksimov, A., Wozniak, M., Grzybowski, T., Dambueva, I., Zakharov, I., 2011. Ancient links between Siberians and Native Americans revealed by subtyping the Y chromosome haplogroup Q1a. J. Hum. Genet. 56, 583–588.

Martins, C., Ferreira, P.M., Carvalho, R., Costa, S.C., Farinha, C., Azevedo, L., Amorim, A., Oliviera, M., 2019. Evaluation of InnoQuant((R)) HY and InnoTyper((R)) 21 kits in the DNA analysis of rootless hair samples. Forensic Sci. Int. Genet. 39, 61–65.

Minton, J.A., Flanagan, S.E., Ellard, S., 2011. Mutation surveyor: software for DNA sequence analysis. Methods Mol. Biol. 688, 143–153.

Moura-Neto, R.S., Mello, I.C.T., Silva, R., Maette, A.P.C., Bottino, C.G., Woerner, A., King, J., Wendt, F., Budowle, B., 2018. Evaluation of InnoTyper(R) 21 in a sample of Rio de Janeiro population as an alternative forensic panel. Int. J. Legal Med. 132, 149–151.

Mousavi, N., Shleizer-Burko, S., Yanicky, R., Gymrek, M., 2019. Profiling the genome-wide landscape of tandem repeat expansions. Nucleic Acids Res. 47, e90.

Ng, J., Oldt, R.F., Kanthaswamy, S., 2018. Assessing the FBI's Native American STR database for random match probability calculations. Legal Med. (Tokyo) 30, 52–55.

Nievergelt, C.M., Maihofer, A.X., Shekhtman, T., Libiger, O., Wang, X., Kidd, K.K., Kidd, J.R., 2013. Inference of human continental origin and admixture proportions using a highly discriminative ancestry informative 41-SNP panel. Investig. Genet. 4, 13.

Novroski, N.M.M., Wendt, F.R., Woerner, A.E., Bus, M.M., Coble, M., Budowle, B., 2019. Expanding beyond the current core STR loci: an exploration of 73 STR markers with increased diversity for enhanced DNA mixture deconvolution. Forensic Sci. Int. Genet. 38, 121–129.

Oldt, R.F., Kanthaswamy, S., 2019. Expanded CODIS STR allele frequencies—evidence for the irrelevance of race-based DNA databases. Legal Med. (Tokyo) 42, 101642.

Olivieri, L., Mazzarelli, D., Bertoglio, B., De Angelis, D., Previdere, C., Grignani, P., Cappella, A., Presciuttini, S., Bertuglia, C., Di Simone, P., Polizzi, N., Iadicicco, A., Piscitelli, V., Cattaneo, C., 2018. Challenges in the identification of dead migrants in the Mediterranean: the case study of the Lampedusa shipwreck of October 3rd 2013. Forensic Sci. Int. 285, 121–128.

Page, D., De Martinville, B., Barker, D., Wyman, A., White, R., Francke, U., Botstein, D., 1982. Single-copy sequence hybridizes to polymorphic and homologous loci on human X and Y chromosomes. Proc. Natl. Acad. Sci. U. S. A. 79, 5352–5356.

Pakstis, A.J., Haigh, E., Cherni, L., Elgaaied, A.B.A., Barton, A., Evsanaa, B., Togtokh, A., Brissenden, J., Roscoe, J., Bulbul, O., Filoglu, G., Gurkan, C., Meiklejohn, K.A., Robertson, J.M., Li, C.X., Wei, Y.L., Li, H., Soundararajan, U., Rajeevan, H., Kidd, J.R., Kidd, K.K., 2015. 52 additional reference population samples for the 55 AISNP panel. Forensic Sci. Int. Genet. 19, 269–271.

Pakstis, A.J., Kang, L., Liu, L., Zhang, Z., Jin, T., Grigorenko, E.L., Wendt, F.R., Budowle, B., Hadi, S., Al Qahtani, M.S., Morling, N., Mogensen, H.S., Themudo, G.E., Soundararajan, U., Rajeevan, H., Kidd, J.R., Kidd, K.K., 2017. Increasing the reference populations for the 55 AISNP panel: the need and benefits. Int. J. Legal Med. 131, 913–917.

Parson, W., Dur, A., 2007. EMPOP—a forensic mtDNA database. Forensic Sci. Int. Genet. 1, 88–92.

Parson, W., Brandstatter, A., Alonso, A., Brandt, N., Brinkmann, B., Carracedo, A., Carach, D., Froment, O., Furac, I., Grzybowski, T., Hedberg, K., Keyser-Tracqui, C., Kupiec, T., Lutz-Bonengel, S., Mevag, B., Ploski, R., Schmitter, H., Schneider, P., Syndercombe-Court, D., Sorensen, E., Thew, H., Tully, G., Scheithauer, R., 2004. The EDNAP mitochondrial DNA population database (EMPOP) collaborative exercises: organisation, results and perspectives. Forensic Sci. Int. 139, 215–226.

Parson, W., Eduardoff, M., Xavier, C., Bertoglio, B., Teschler-Nicola, M., 2018. Resolving the matrilineal relationship of seven Late Bronze Age individuals from Stillfried, Austria. Forensic Sci. Int. Genet. 36, 148–151.

Phillips, C., Salas, A., Sanchez, J.J., Fondevila, M., Gomez-Tato, A., Alvarez-Dios, J., Calaza, M., De Cal, M.C., Ballard, D., Lareu, M.V., Carracedo, A., 2007. Inferring ancestral origin using a single multiplex assay of ancestry-informative marker SNPs. Forensic Sci. Int. Genet. 1, 273–280.

Phillips, C., Freire Aradas, A., Kriegel, A.K., Fondevila, M., Bulbul, O., Santos, C., Serrulla Rech, F., Perez Carceles, M.D., Carracedo, A., Schneider, P.M., Lareu, M.V., 2013. Eurasiaplex: a forensic SNP assay for differentiating European and South Asian ancestries. Forensic Sci. Int. Genet. 7, 359–366.

Porras-Hurtado, L., Ruiz, Y., Santos, C., Phillips, C., Carracedo, A., Lareu, M.V., 2013. An overview of STRUCTURE: applications, parameter settings, and supporting software. Front. Genet. 4, 98.

Prieto-Fernandez, E., Baeta, M., Nunez, C., Zarrabeitia, M.T., Herrera, R.J., Builes, J.J., De Pancorbo, M.M., 2016. Development of a new highly efficient 17 X-STR multiplex for forensic purposes. Electrophoresis 37, 1651–1658.

Rajeevan, H., Osier, M.V., Cheung, K.H., Deng, H., Druskin, N.L., Einzen, R., Kidd, J.R., Stein, S., Pakstis, A.J., Tosches, N.P., Yeh, C.C., Miller, P.L., Kidd, K.K., 2003. ALFRED: the ALelle FREquency database. Update. Nucleic Acids Res. 31, 270–271.

Rajeevan, H., Cheung, K.H., Gadagkar, R., Stein, S., Soundararajan, U., Kidd, J.R., Pakstis, A.J., Miller, P.L., Kidd, K.K., 2007. ALFRED: an allele frequency database for microevolutionary studies. Evol. Bioinformatics Online 1, 1–10.

Rajeevan, H., Soundararajan, U., Kidd, J.R., Pakstis, A.J., Kidd, K.K., 2012a. ALFRED: an allele frequency resource for research and teaching. Nucleic Acids Res. 40, D1010–D1015.

Rajeevan, H., Soundararajan, U., Pakstis, A.J., Kidd, K.K., 2012b. Introducing the Forensic Research/Reference on Genetics knowledge base, FROG-kb. Investig. Genet. 3, 18.

Ristow, P.G., Barnes, N., Murphy, G.P., Brown, H., Cloete, K.W., D'Amato, M.E., 2017. Evaluation of the InnoTyper((R)) 21 genotyping kit in multi-ethnic populations. Forensic Sci. Int. Genet. 30, 43–50.

Roberts, D.F., Papiha, S.S., Bhattacharya, S.S., 1987. A case of disputed maternity. Lancet 2, 478–480.

Robin, E.D., Wong, R., 1988. Mitochondrial DNA molecules and virtual number of mitochondria per cell in mammalian cells. J. Cell. Physiol. 136, 507–513.

Rock, A., Irwin, J., Dur, A., Parsons, T., Parson, W., 2011. SAM: string-based sequence search algorithm for mitochondrial DNA database queries. Forensic Sci. Int. Genet. 5, 126–132.

Roewer, L., Arnemann, J., Spurr, N.K., Grzeschik, K.H., Epplen, J.T., 1992. Simple repeat sequences on the human Y chromosome are equally polymorphic as their autosomal counterparts. Hum. Genet. 89, 389–394.

Rosenberg, N.A., Pritchard, J.K., Weber, J.L., Cann, H.M., Kidd, K.K., Zhivotovsky, L.A., Feldman, M.W., 2002. Genetic structure of human populations. Science 298, 2381–2385.

Ruiz, Y., Phillips, C., Gomez-Tato, A., Alvarez-Dios, J., Casares De Cal, M., Cruz, R., Maronas, O., Sochtig, J., Fondevila, M., Rodriguez-Cid, M.J., Carracedo, A., Lareu, M.V., 2013. Further development of forensic eye color predictive tests. Forensic Sci. Int. Genet. 7, 28–40.

Salvador, J.M., Apaga, D.L.T., Delfin, F.C., Calacal, G.C., Dennis, S.E., De Uungria, M.C.A., 2018. Filipino DNA variation at 12 X-chromosome short tandem repeat markers. Forensic Sci. Int. Genet. 36, e8–e12.

Santos, C., Phillips, C., Fondevila, M., Daniel, R., Van Oorschot, R.A.H., Burchard, E.G., Schanfield, M.S., Souto, L., Uacyisrael, J., Via, M., Carracedo, A., Lareu, M.V., 2016. Pacifiplex: an ancestry-informative SNP panel centred on Australia and the Pacific region. Forensic Sci. Int. Genet. 20, 71–80.

Scheible, M., Alenizi, M., Sturk-Andreaggi, K., Coble, M.D., Ismael, S., Irwin, J.A., 2011. Mitochondrial DNA control region variation in a Kuwaiti population sample. Forensic Sci. Int. Genet. 5, e112–e113.

Shved, N., Haas, C., Papageogopoulou, C., AkguelK, G., Paulsen, K., Bouwman, A., Warinner, C., Ruhli, F., 2014. Post mortem DNA degradation of human tissue experimentally mummified in salt. PLoS One 9, e110753.

Tasker, E., LaRue, B., Beherec, C., Gangitano, D., Hughes-Stamm, S., 2017. Analysis of DNA from post-blast pipe bomb fragments for identification and determination of ancestry. Forensic Sci. Int. Genet. 28, 195–202.

Van Der Auwera, G.A., Carneiro, M.O., Hartl, C., Poplin, R., Del Angel, G., Levy-Moonshine, A., Jordan, T., Shakir, K., Roazen, D., Thibault, J., Banks, E., Garimella, K.V., Altshuler, D., Gabriel, S., Depristo, M.A., 2013. From FastQ data to high confidence variant calls: the Genome Analysis Toolkit best practices pipeline. Curr. Protoc. Bioinformatics 43, 11.10.1–11.10.33.

Van Oven, M., Kayser, M., 2009. Updated comprehensive phylogenetic tree of global human mitochondrial DNA variation. Hum. Mutat. 30, E386–E394.

Walsh, P.S., Erlich, H.A., Higuchi, R., 1992. Preferential PCR amplification of alleles: mechanisms and solutions. PCR Methods Appl. 1, 241–250.

Warshauer, D.H., Lin, D., Hari, K., Jain, R., Davis, C., LaRue, B., King, J.L., Budowle, B., 2013. STRait Razor: a length-based forensic STR allele-calling tool for use with second generation sequencing data. Forensic Sci. Int. Genet. 7, 409–417.

Warshauer, D.H., King, J.L., Budowle, B., 2015. STRait Razor v2.0: the improved STR allele identification tool–razor. Forensic Sci. Int. Genet. 14, 182–186.

Weir, B.S., 1996. Genetic Data Analysis, second ed. Sinauer Associates, Sunderland, MA. 376 pp.

Weissensteiner, H., Pacher, D., Kloss-Brandstatter, A., Forer, L., Specht, G., Bandelt, H.J., Kronenberg, F., Salas, A., Schonherr, S., 2016. HaploGrep 2: mitochondrial haplogroup classification in the era of high-throughput sequencing. Nucleic Acids Res. 44, W58–W63.

Wendt, F.R., Churchill, J.D., Novroski, N.M.M., King, J.L., Ng, J., Oldt, R.F., McCulloh, K.L., Weise, J.A., Smith, D.G., Kanthaswamy, S., Budowle, B., 2016. Genetic analysis of the Yavapai Native Americans from West-Central Arizona using the Illumina MiSeq FGx forensic genomics system. Forensic Sci. Int. Genet. 24, 18–23.

Willuweit, S., Roewer, L., 2015. The new Y chromosome haplotype reference database. Forensic Sci. Int. Genet. 15, 43–48.

Wilson Sayres, M.A., 2018. Genetic diversity on the sex chromosomes. Genome Biol. Evol. 10, 1064–1078.

Woerner, A.E., King, J.L., Budowle, B., 2017. Fast STR allele identification with STRait Razor 3.0. Forensic Sci. Int. Genet. 30, 18–23.

Xiong, Z., Dankova, G., Howe, L.J., Lee, M.K., Hysi, P.G., De Jong, M.A., Zhu, G., Adhikari, K., Li, D., Li, Y., Pan, B., Feingold, E., Marazita, M.L., Shaffer, J.R., McAloney, K., Xu, S.H., Jin, L., Wang, S., De Vrij, F.M., Lendemeijer, B., Richmond, S., Zhurov, A., Lewis, S., Sharp, G.C., Paternoster, L., Thompson, H., Gonzalez-Jose, R., Bortolini, M.C., Canizales-Quinteros, S., Gallo, C., Poletti, G., Bedoya, G., Rothhammer, F., Uitterlinden, A.G., Ikram, M.A., Wolvius, E., Kusher, S.A., Nijsten, T.E., Palastra, R.T., Boehringer, S., Medland, S.E., Tang, K., Ruiz-Linares, A., Martin, N.G., Spector, T.D., Stergiakouli, E., Weinberg, S.M., Liu, F., Kayser, M., 2019. Novel genetic loci affecting facial shape variation in humans. Elife, 8. https://doi.org/10.7554/eLife.49898. e49898.

Zavala, E.I., Rajagopal, S., Perry, G.H., Kruzic, I., Basic, Z., Parsons, T.J., Holland, M.M., 2019. Impact of DNA degradation on massively parallel sequencing-based autosomal STR, iiSNP, and mitochondrial DNA typing systems. Int. J. Legal Med. 133, 1369–1380.

Zeng, X., Chakraborty, R., King, J.L., LaRue, B., Moura-Neto, R.S., Budowle, B., 2016a. Selection of highly informative SNP markers for population affiliation of major U.S. populations. Int. J. Legal Med. 130, 341–352.

Zeng, X., Warshauer, D.H., King, J.L., Churchill, J.D., Chakraborty, R., Budowle, B., 2016b. Empirical testing of a 23-AIMs panel of SNPs for ancestry evaluations in four major US populations. Int. J. Legal Med. 130, 891–896.

Zeng, X., Elwick, K., Mayes, C., Takahashi, M., King, J.L., Gangitano, D., Budowle, B., Hughes-Stamm, S., 2019. Assessment of impact of DNA extraction methods on analysis of human remain samples on massively parallel sequencing success. Int. J. Legal Med. 133, 51–58.

Zidkova, A., Capek, P., Horinek, A., Coufalova, P., 2014. Investigator(R) Argus X-12 study on the population of Czech Republic: comparison of linked and unlinked X-STRs for kinship analysis. Electrophoresis 35, 1989–1992.

CHAPTER

20

The emerging discipline of forensic genetic genealogy

Colleen Fitzpatrick PhD

Identifinders International LLC, Fountain Valley, CA, United States

Introduction

The advantages of applying genetic genealogical methods to forensic casework are numerous, as are the related controversies. Unlike CODIS, the successful use of a genealogical database does not require the DNA of an unidentified person, nor the DNA of his immediate family member, to have been previously entered into the system. The genealogical identification process can function using the DNA profiles of even distant relatives and has, therefore, proven to be an important new tool in forensic intelligence, as well as the reason for much of its success. At the same time, the possible involvement of extended family in forensic investigations has prompted much discussion among the genealogical, forensic, and legal communities about the proper role of genetic genealogy in forensic identification (Bettinger, 2019a; Guerrni et al., 2018; Lewis, 2019; Moltini, 2019).

The recent newsworthy identification of cold case assailants based on genetic genealogy (Anderson, 2019; Fuller, 2018; Hernandez, 2019a) belies the challenges of using the technology for a broader range of "cold case" work, including the identification of skeletal remains. Most cases that have appeared in the news are those that have been readily solved using semen or blood samples, and hence, are not necessarily representative of general casework that exhibits a broader range of difficulty. Many challenges are emerging as the more tractable and less time-consuming cases are cleared, bringing more difficult cases to the forefront. Some of these challenges reflect the limitations of databases used for profile comparison, while others relate to the quality and quantity of DNA required to generate genetic genealogy data.

The application of genetic genealogy to the identification of skeletal remains should be discussed within a framework that recognizes the differences between genealogical and forensic identification methods. Genetic genealogical research involves analyzing a saliva sample from a living person who usually has some information about his/her family history; in contrast, forensic identification of skeletal remains is based on an often compromised postmortem

https://doi.org/10.1016/B978-0-12-815766-4.00022-4

DNA sample, with little or no a priori knowledge of the identity of the deceased individual to whom the remains belong. Furthermore, genetic genealogical techniques were developed to solve cases of unknown parentage and rely on data produced by direct-to-consumer (DTC) DNA testing companies that are not accredited to do forensic casework, for comparison to data that appear in online repositories that are crowd-sourced and unregulated. Even so, genetic genealogy has proven useful to the forensic community as a valuable method of generating investigative leads for solving sometimes decades-old cases (Augenstein, 2018; CBS/AP, 2018; KTVL, 2019; Shapiro, 2019).

Genetic genealogy and Y-STR testing

Genetic genealogy was first introduced in 2000 by several DTC DNA testing companies as a means of complementing genealogical records with genetic data. While written documentation and oral family histories have always been staples of genealogical research, they are not always accurate and are never complete. DNA testing appealed to genealogists as an independent means of verifying presumed family relationships and of finding otherwise undocumented family connections where recorded history was unavailable.

The earliest products marketed to the genealogy community by DTC DNA testing companies were Y-chromosome short tandem repeat (Y-STR) test panels. The popularity of Y-STR analysis as a tool for genealogical research grew rapidly because of the patrilineal co-inheritance of the Y chromosome with the family surname. Product lines were eventually expanded to include panels of up to 111 Y-STR loci. Although testing of the mitochondrial DNA (mtDNA) hypervariable regions I and II (HVI, HVII) was also made available by DTC companies in the early days of genetic genealogy, it was not as widely adopted because of the slow mutation rate of mtDNA and the fact that it is maternally inherited and hence not commonly associated with transmission of the family surname.

The first Y-STR panel offered to consumers was a 12-locus panel that included: (1) the 9-locus European Minimal Haplotype (Kayser et al., 1997; Roewer et al., 2000); (2) the DYS438 locus, one of two loci recommended in 2003 for inclusion in forensic Y-STR test panels by the Scientific Working Group on DNA Analysis Methods (SWGDAM); and (3) two additional slowly mutating loci (DYS388, DYS426). Driven by a need to satisfy the increasing personal curiosity of genealogists via more discriminating tests, and without the validation and standardization requirements necessary for forensic casework, DTC Y-STR product lines expanded rapidly to include 25-, 37-, 67-, and 111-locus panels. Both slow- and rapidly mutating markers were included, many of which were also included in standardized Y-STR PCR amplification kits. The use of rapidly mutating (RM) loci offered the advantage of increasing the discriminatory power of paternal lines at different time depths. While loci with slower mutation rates are useful in probing deeply rooted pedigrees, loci with faster mutation rates allow differentiation among more recently related individuals. Table 1 shows the number of loci that overlap between various Y-STR amplification kits and DTC genetic genealogy Y-STR test panels. A list of loci included in each of the genealogy test panels can be found on the Family Tree DNA website (Family Tree DNA, 2019).

In order to better control the genetic genealogy customer base and as a means of organizing an unpaid sales force, DTC testing companies encouraged their clientele to form Y-STR

TABLE 1 Overlap in loci between standardized Y-STR amplification kits and direct-to-consumer (DTC) Y-STR test panels.

	Number of loci in genetic genealogy Y-STR test panels				
	12	25	37	67	111
PowerPlex Y (12)	10	11	12	12	12
YFiler (17)	10	13	16	17	17
PowerPlex Y23 (23)	10	13	19	20	23
YFiler Plus (27)	10	14	20	21	24

The number of loci amplified by each Y-STR amplification kit is shown in parentheses.

projects dedicated to common research interests that were managed by group administrators. Companies also hosted public websites to serve as repositories of project results. A haplotype is usually displayed in a genealogical Y-STR database with the name of the most distant ancestor of the donor, the ancestor's approximate year of birth, and other genealogical information; typically, donor names are not provided. For example, refer to the "Miller Surname Project" (Miller Surname Project, n.d., http://www.familytreedna.com/public/Miller?iframe=yresults).

As of 2019, an estimated total of 250,000–300,000 Y-STR profiles have been posted on thousands of public genetic genealogy web pages dedicated to groups researching specific surnames, nationalities, ethnicities, haplogroups, descendancy from historical figures, and other special interests; many have been tested on the now-standard 37-locus panel. The estimated number of publicly available genetic genealogy profiles compares favorably with the number of Y-haplotypes in the Y-chromosome Haplotype Reference Database (YHRD) for various Y-STR amplification kits. As of December 2019 the YHRD database includes over 307,000 (9-locus) minimal haplotypes, 267,000 (12-locus) PowerPlex Y haplotypes, 247,000 (17-locus) YFiler haplotypes, 72,000 (23-locus) PowerPlex Y23 haplotypes, and 73,000 (27-locus) YFiler Plus haplotypes, collected from a total of 33 metapopulation groups worldwide (Y-chromosome Users Group, 2019). The Applied Biosystems (AB) YFiler database is much smaller, containing 11,300 YFiler haplotypes from 13 different world population groups (Applied Biosystems, n.d.). Each type of database has its benefits. Because of the informal and somewhat disorganized means by which public genealogy Y-STR databases have grown over the years, a match can provide a possible surname for an unknown sample, but the data cannot be used to provide the statistical power of the match. On the other hand, comparison of a Y-STR profile to the YHRD and AB YFiler databases can provide the statistical power of a match, along with a possible nationality via comparison to various national Y-STR databases. However, unlike genetic genealogy Y-STR databases, the YHRD and AB YFiler databases are anonymous and cannot provide surname information.

Application of genetic genealogy Y-STR data to forensic casework

The public availability and increasing size of genetic genealogy Y-STR databases prompted their use by law enforcement as a source of forensic intelligence starting in 2011. Because of the overlap between loci included in the genetic genealogy Y-STR test panels and those included in Thermo Fisher Scientific's AmpFLSTR YFiler and Promega's PowerPlex Y23 PCR amplification kits, a forensic Y-STR profile could be compared to the genetic genealogy Y-STR databases without retesting the evidence. Genetic genealogy databases could be mined for matches to a Y-STR profile from a crime scene or from unidentified remains, to obtain a possible last name in the absence of a CODIS match, or when a CODIS profile was unavailable. The public genetic genealogy repositories of Y-STR results, YSearch (http://www.YSearch.org) and the Sorenson Molecular Genealogy Foundation database (http://www.smgf.org), could be searched for matches by simply typing in Y-profile alleles. If a match was found indicating a possible surname, a specific genealogical project could be identified that included that surname to allow for a more focused search for confirmatory information. Even if a match was not found, the nationality or ethnicity of near matches could be informative. Note that even though the YSearch and Sorenson Y-STR databases are no longer available, it is still possible to search larger projects associated with nationalities or haplogroups to find matching surnames for further analysis.

The first known forensic case where genetic genealogy was used to generate investigative leads was the 1991 murder of Sarah Yarborough, a high school student in Federal Way, Washington (Sullivan, 2012). Comparison of the 17-locus YFiler Y-STR profile obtained from DNA collected at the crime scene to online Y-STR genealogical databases produced the possible last name of Fuller for the killer. Matching profiles in the Fuller surname project study belonged to descendants of Robert Fuller, an early immigrant to Massachusetts in the late 1630s. Yarborough had a high school classmate named Elizabeth Fuller, who was one of five daughters of William Fuller. When William Fuller voluntarily submitted a DNA sample for CODIS testing, he was ruled out as Yarborough's killer and as the father of her killer. However, he was a Y-STR match for her killer and was determined genealogically to be a direct male descendant of the same Robert Fuller. Although Yarborough's killer was not identified at that time, his genealogy had been established back to the 1600s (Moran and Keneally, 2012).

The first known successful forensic identification using genetic genealogy was achieved by the Phoenix Police Department's 2014 investigation of the Phoenix Canal Murders. Comparison of the 17-locus YFiler Y-STR profile left at the 1992 and 1993 crime scenes with Y-STR genetic genealogy databases produced matches to six profiles associated with the surname Miller, narrowing the list of suspects from around 2000 to only five individuals with that name. This led to a match in the CODIS database between the DNA left at the scene and DNA obtained from Phoenix resident Brian Patrick Miller (Cassidy, 2016). Miller's arrest was the first made based on a lead generated by a match found in a genetic genealogy database.

It should be noted that if a genealogical match to a forensic Y-STR profile is found, only the surname of the match is significant; neither the identity of the genealogist who submitted the matching haplotype nor his family pedigree is usually relevant to an investigation. Since the nonrecombinant region of the Y-chromosome is passed intact along the paternal line for generations (except for occasional mutations), a genetic genealogy match found for a Y-STR profile could indicate a relationship that occurred centuries in the past, even before the adoption of surnames. This was found to be true for the 2018 identity theft case of a man living under the stolen identity of Joseph Newton Chandler III. A match between Chandler's Y-STR profile and a member of the Nicholas surname study indicated that as Chandler's possible original surname. When Chandler was finally identified as Robert Ivan Nichols, it was discovered that the connection between the matching Nicholas genealogist and Robert Nichols was at the latest in the late 1600s (Metzger, 2018).

It should also be understood that a match in a genetic genealogy database represents only a *possible* last name for an unknown. It is well-known in the genealogy community that a by-product of Y-STR testing is the potential discovery of misattributed paternity, where the genetic history of a family deviates from its historical version, resulting in a change in surname. Just as Y-STR testing has been useful for male adoptees to discover the possible surnames of their birth families, there have been cases where an individual who thought he knew his parentage was surprised to find out he was adopted or the product of an out-of-wedlock liaison.

One of the most prominent cases of misattributed paternity discovered through Y-STR analysis was the 2012 identification of remains found on the former grounds of Grey Friars friary in Leicester, England, as those of King Richard III. Although the pedigrees of the Plantagenet kings are among the most well-researched in the world, and the identification of the remains as those of Richard III was supported by a mitochondrial DNA (mtDNA) match to maternal-line descendants, archeological evidence, radiocarbon dating, isotopic analyses, and historical accounts of the fatal injuries sustained by the king, there was no Y-STR match found between the remains and five documented paternal-line relatives of the king. These five relatives descended from Richard's great grandfather Edward III (1312–77) through Edward's son John of Gaunt (1340–99) and then 13 generations later through Gaunt's patrilineal descendant Henry Somerset, 5th Duke of Beaufort (1744–1803) (see Fig. 1). In addition, one of the Duke's descendants did not match the other four. Therefore, Y-STR analyses indicated at least two misattributed paternities, one during the generations between the Duke and his mismatched descendant, and the other along the family line connecting Richard III with the Duke through their common patrilineal ancestor Edward III. Depending on the generation in the royal genealogy in which the earlier misattributed paternity occurred, a large part of English history could be rewritten, since the claim to the English throne by the entire Lancaster and Tudor dynasties, and even perhaps by Richard himself, could be called into question (King et al., 2014).

Massively parallel sequencing, Y-SNPs, and Y-STRs

In the 2000s, the decreasing cost of massively parallel sequencing (MPS) and the increase in bioinformatics computing resources eventually led to the viability of whole genome sequencing (WGS) as a commercial product. In 2007, the first whole genome sequencing service offered by Knome carried a price tag of $350,000 and could be afforded only by the wealthy few (Harmon, 2008); additionally, the service did not include analysis or interpretation. By 2013, however, the price had decreased to about $2000 (National Human Genome Research Institute, 2019), making it more affordable (although still pricey for the mass market) and only available for medical diagnostics through a doctor's order.

The development of MPS also facilitated the large-scale discovery of single nucleotide polymorphisms (SNPs). These include Ancestry Informative Markers (AIMs) used to establish biogeographical ancestry (Kidd et al., 2011, 2014; Nievergelt et al., 2013; Pakstis et al., 2012). The discovery of many new AIMs located on the Y-chromosome allowed the rearrangement and expansion of Y-chromosome phylogeny (Francalacci et al., 2013; Rocca et al., 2012;

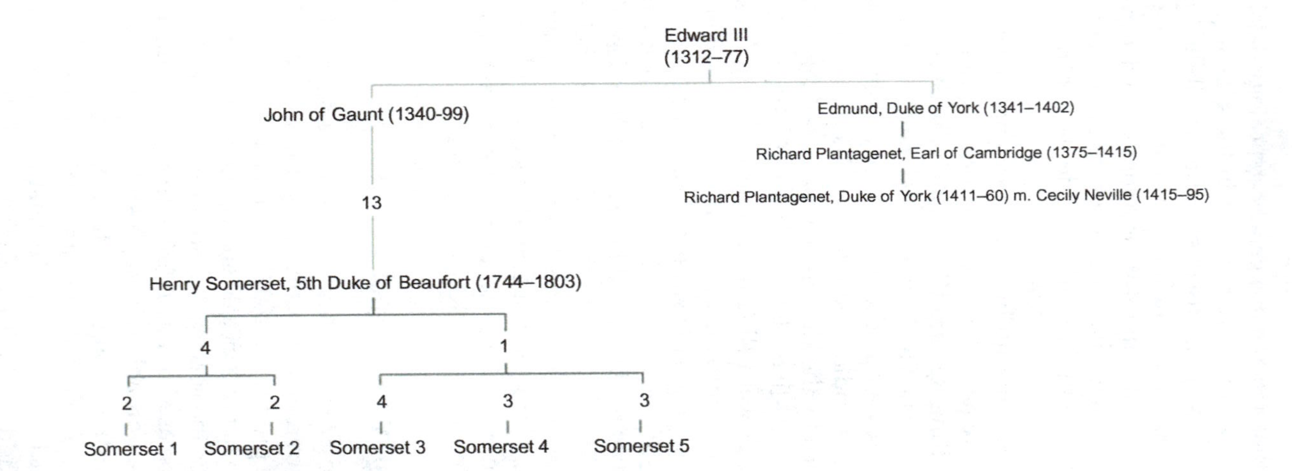

FIG. 1 Pedigree of King Richard III, indicating his relationship with the five Y-STR Somerset family references used for his identification. Numbers indicate generations between named individuals (King et al., 2014).

Van Geystelen et al., 2013). Tests on individual Y-chromosome SNPs (Y-SNPs) and panels of Y-SNPs based on traditional Sanger sequencing had been available to the genetic genealogy community as early as 2006 (International Society of Genetic Genealogy, 2019a); however, as sequencing costs continued to decrease, it was only a matter of time before targeted Y-chromosome sequencing became available to genealogists for SNP testing on a much larger scale. In January 2013, Full Genomes Corporation (FGC) became the first DTC company to offer targeted Y-chromosome sequencing to genetic genealogists. The price was $1499 and included an ancestry report. In November 2013, Family Tree DNA followed suit by discontinuing its Y-SNP testing based on Sanger sequencing and replacing it with its MPS-based Big Y product.

It has long been known that Y-STRs exhibit variability against the stable genetic background provided by Y-SNPs. One study has shown that 80% of Y-STR variation is found between haplogroups, with less than 4% of genetic variation attributed to differences within haplogroups (Bosch et al., 1999). As early as 2006, as Y-SNP testing was becoming more popular among genealogists, and as aficionados were testing both their Y-STRs and Y-SNPs, DTC testing companies began to offer Y-haplogroup predictions based on Y-STR test results. Two haplogroup prediction tools were created for genealogists to run their own haplogroup and subclade analysis: (1) Nevgen and (2) the Whit Athey haplogroup predictors. Both webtools use a Bayesian-allele-frequency approach to predict Y-haplogroup membership based on a goodness-of-fit score generated by comparison to modal haplotypes (Athey, 2005, 2006). Nevgen also includes a correlation algorithm to predict haplogroup based on frequency of allele pairs (Nevgen, 2019). In addition, services such as YFull were founded to assist with the interpretation of raw sequencing data and the discovery of new Y-SNPs. YFull offers an online Y-chromosome phylogenetic tree constructed from consumer data which is labeled with the nationality of individual contributors (YFull, n.d.).

The ability to predict Y-haplogroups based on Y-STRs increased the value of genetic genealogy databases for forensic casework, since information about an unknown's biogeographic ancestry along his direct male line could be obtained even in the absence of a name match. An example is the investigation of the 1984 Seminole County murder of Pamela Cahanes, which for over 30 years was based on the belief that her killer was Caucasian. The course of the investigation was changed in 2015 when genetic genealogy analysis of the Y-STR profile obtained from crime scene DNA indicated that the suspect was in fact African American along his patrilineal line. While Y-STR analysis cannot predict full biogeographic ancestry, nor phenotypic characteristics such as hair color and eye color, the prediction of patrilineal African American ancestry for Cahanes' killer broadened the focus of the investigation to individuals with at least mixed ancestry. An arrest was made in March 2019 of Thomas Lewis Garner, an African American who had been stationed at the Orlando Naval Training Station at the same time as Cahanes (Williams, 2019).

The value of using Y-STRs to predict both the biogeographic ancestry and surname of an unknown is illustrated by the 1999 murder of Marianne Vaatstra in The Netherlands. Although the case was solved without reference to genetic genealogy, the investigation followed a similar methodology. As Kayser describes the development of the case (Kayser, 2017), when no match was found in the Dutch national criminal database for DNA obtained from a semen stain at the crime scene, and after dozens of individuals potentially linked to the case were ruled out through standard autosomal STR testing, the killer's Y-STR profile was compared with the YHRD database. The results predicted that Vaatstra's killer was Northern European along his patrilineal line. This helped quiet anxiety among local residents who believed that the murder was associated with a nearby center for asylum seekers.

In 2012, when a familial search of the Dutch national criminal offender database failed to find close matches that might indicate a relative of the killer, a voluntary large-scale mass DNA screening effort of local residents within 5 miles of the crime scene was initiated as a last resort. As it would have been time-consuming and expensive to test all 6600 volunteers on autosomal STRs, a strategy was developed to test the samples only on Y-STRs to help narrow the list of candidates to only individuals who shared patrilineal descent with the killer. Two matches were discovered in the first round of testing with two different surnames. The investigation then focused on testing volunteers with those surnames on 61 additional Y-STR markers that included 13 rapidly mutating (RM) Y-STRs, increasing the chance of observing additional mutations that would rule out more distantly related individuals. Surprisingly, one of the remaining Y-STR matches was also an autosomal match to the crime scene DNA because the killer had voluntarily submitted his sample to the dragnet (Kayser, 2017).

In a sense, to solve the Vaatstra case, Dutch authorities developed a methodology parallel to that used by genetic genealogists to solve cases of misattributed paternity, which has recently been borrowed by the forensic community for cold case investigations. The YHRD database could be viewed as a type of peer-reviewed analogue of the Y-haplogroup predictors used by genealogists to predict biogeographic ancestry. DNA collected as a result of the large-scale voluntary mass screening on this case could be considered a type of ad hoc YSearch database that was used to discover a surname for the killer, after autosomal STR testing and familial searching had failed to produce investigative leads.

Y-STR analysis has been used in combination with mtDNA sequencing for the identification of human remains as a filter to rule out unlikely relationships in conjunction with autosomal STR analysis (Ambers, et al., 2018; Ge et al., 2011) for historical identifications (Ambers, et al., 2014; Bogdanowicz et al., 2009; Coble et al., 2009), for military identifications (Irwin et al., 2007; Lee et al., 2010), and for mass grave and disaster victim identifications (Biesecker, et al., 2005; Marjanović et al., 2015). In cases where autosomal STRs are unavailable or uninformative because of the number of generations separating an historical figure from his living descendants, Y-STRs and mtDNA analyses can often be the only means of identification.

Even so, Y-STR genetic genealogy databases and Y-haplogroup predictor tools have not realized their potential for the identification of skeletal remains, although they have become popular with law enforcement for investigating violent crimes. Analysis of degraded remains presents additional challenges because of environmental insults typically experienced through exposure to heat, direct sunlight, and water, which can randomly fragment and damage the chemical structure of DNA (Ambers et al., 2016). It is more difficult to obtain full autosomal and Y-STR profiles using conventional typing methods because of small template size and the low quantity of nuclear DNA (nDNA) typically available for analysis. Recently developed single nucleotide polymorphism (SNP) assays using MPS have demonstrated the potential of AIMs and phenotypic informative SNPs to accommodate degraded templates that benefit from smaller size amplicons (Ambers et al., 2016; Latham and Miller, 2019; Zavala et al., 2019). In cases where complete or nearly complete Y-STR haplotypes are not available, Y-STR genetic genealogy databases are not useful, nor are haplogroup predictors. The direct use of SNP testing has more potential for forensic identification.

Genetic genealogy autosomal SNP testing

More recently, DTC autosomal SNP testing has eclipsed genetic genealogy Y-STR testing in popularity. DTC autosomal SNP testing based on high-density SNP arrays was introduced in 2006 by 23andMe as the first commercial personal genome testing service, offering insight into genetic predisposition and ancestry (Top10DNATests, 2020). In 2010, Family Tree DNA introduced a similar product (International Society of Genetic Genealogy, 2019a), followed by Ancestry in May 2012 (Ancestry, 2012) and MyHeritage in 2016 (MyHeritage, 2016). Each company uses a customized Illumina platform to test approximately 600,000–700,000 SNPs (International Society of Genetic Genealogy, 2019b) to generate relationship estimates for pairs of individuals based on the detection of matching segments of DNA that are identical by descent (IBD) (i.e., genomic segments shared by the individuals that are indicative of shared family lines and common ancestry).

The primary value to genealogists of high-density autosomal SNP testing is that it is useful for discovering relationships along all branches of a family, both single- and mixed-gender lines. Because longer IBD segments are typically broken up by recombination, the size and number of IBD segments shared by individuals can indicate the number of generations that have elapsed since their common ancestor, to provide an estimate of their degree of relationship (Henn et al., 2012). IBD segments are usually greater than about 6 centiMorgans (cM) in length; those less than 6 cM are considered Identical by State (IBS) (i.e., shared by individuals because they are members of the same gene pool) (AncestryDNA, 2016).

Genealogists have found the pedigree-informative capabilities of DTC autosomal SNP testing to be a powerful tool to confirm known genealogies and to fill gaps in incomplete family pedigrees. In the case of a male or female with unknown or misattributed parentage (e.g., adoption and illegitimacy), a family pedigree can be reconstructed (and hopefully birth parents identified) based on (1) that person's estimated relationships with his autosomal matches, (2) the estimated relationships these matches have with each other, and (3) the overlap of the matches' family pedigrees. Information on Y-chromosome and mtDNA SNPs that define biogeographic ancestry along the patrilineal and matrilineal lines may also be useful.

Likely because of the complex analysis required, the popularity of DTC autosomal DNA testing grew slowly at first, along with the size of associated DTC databases. As of December 2012, the earliest date for which statistics are available (International Society of Genetic Genealogy, 2019c), the 23andMe database was the largest, with only about 180,000 members. The combined size of all DTC autosomal SNP databases was less than about 330,000 members. The chance of finding a close relative just through 23andMe was relatively low, estimated following the method presented by Henn et al. as 4.76% for a first cousin and 21.6% for a second cousin (Henn et al., 2012).

However, as genealogists came to understand the potential of autosomal SNP testing, the databases of the four main DTC DNA testing companies increased dramatically. In 2017,

Ancestry reported record fourth-quarter sales of 1.4 million kits, 390,000 more than sold in all of 2015 (Ancestry, 2017). As of April 2019, Ancestry reported that its database contained 15 million members (Ancestry, n.d.-a), and 23andMe reported its database contained 10 million members (23andme, n.d.). As DTC databases grew, the probability of finding close matches increased, so that DTC autosomal SNP testing became almost a necessity among male and female adoptees searching for their birth parents. As of 2019, a survey of 3410 adoptees who had taken a DTC DNA test found a high success rate: 92% of respondents reported that their closest matches were half second cousins or better, with 24% reporting that their closest matches were a parent or a sibling. A total of 77% reported they were able to find a parent and/or a sibling as a result of DTC autosomal DNA testing (Bettinger, 2019b).

Application of genetic genealogy autosomal testing to forensic casework

As autosomal SNP testing gained popularity among genetic genealogists, the possibility of applying the same techniques to the identification of unknown assailants and the unidentified dead attracted the attention of the forensic community. Unfortunately, none of the DTC companies allow law enforcement access to their testing services, nor use of their customer databases, citing privacy concerns (Ancestry, n.d.-b). To apply the same type of genetic genealogy analysis to forensic cases, two challenges had to be overcome: (1) a method had to be developed to produce DTC-equivalent autosomal SNP data through independent means, and (2) a database of DTC test results had to be available for comparison purposes.

DTC autosomal SNP testing was derived from high-density SNP genotyping methods developed not by the forensic community, but by the biomedical industry for molecular-level detection of genetic variants relating to genetic traits, disease etiology, and response to drug therapies. Forensically accredited laboratories are restricted to autosomal and Y-STR typing methods based on capillary electrophoresis (CE) which have been NDIS-approved and validated according to SWGDAM guidelines (SWGDAM, 2012). The infrastructure does not exist for forensically accredited laboratories to produce the type of autosomal SNP data that genetic genealogy analysis requires; hence, without the cooperation of DTC testing companies, the replication of DTC results from forensic samples must be performed by private laboratories or external service centers. Fortunately, by early 2017, when the first practical efforts were made towards applying genetic genealogy autosomal SNP testing to forensic casework, there were a variety of laboratories with the capability of replicating DTC SNP datasets (Glenn, 2011; MarketWatch, 2017).

Although all DTC DNA testing companies maintain proprietary databases, customers are permitted to download their raw data for personal use. As of 2010, the only public repository that could accommodate such data were GEDmatch (https://www.gedmatch.com); this provided genealogists an independent means of analyzing their DTC results and for cross-platform autosomal SNP data comparison. GEDmatch provides advanced analysis tools not offered elsewhere, such as a collection of ethnic admixture utilities, triangulation and phasing tools, and a utility called "are your parents related" to assess consanguinity, an important factor in estimating endogamous contributions to relationship estimates. GEDmatch accepts raw data from all DTC companies, but it is not affiliated with any DTC company; uploads are voluntary. As of January 2016, the earliest date for which statistics are available, GEDmatch contained about 225,000 members (Larkin, 2019a); as of

November 2019, it had grown to over 1.3 million (Rogers, 2019), increasing in size in parallel to the growth of DTC databases.

While genetic genealogy Y-STR results are publicly displayed in online databases controlled by group administrators, access to GEDmatch requires a user to set up a free account and to log in to the system. Once a raw dataset is uploaded, it is assigned a kit number and then tokenized. The original data are then deleted and is no longer available even to the owner. A kit may or may not be exposed to the entire database, depending on user-defined settings. A kit that is designated as "Private" will not be included in the database. A kit marked "Research" will be available for comparison to the database, but it will not be visible to other GEDmatch uses. ("Research" mode is required for law enforcement uploads). There are two "Public" settings implemented in May 2019, one that allows law enforcement access (the "opt-in" mode) and one that prohibits law enforcement access (the "opt-out" mode). As of November 2019, approximately 190,000 GEDmatch users have opted in (Rogers, 2019).

Unlike Y-STR data, in which a match on a common set of Y-STR alleles is indicative only of shared patrilineal ancestry, the analysis of autosomal SNP data depends on detection of shared IBD segments that could originate through one or more shared patrilineal, matrilineal, or mix-gender lines. Furthermore, while genetic genealogy Y-STR data are useful even without knowing the identity of profile donors, the use of genetic genealogy autosomal SNP data requires that contributors be identified, and that personal information be made available about their family pedigrees. Exposure of such information to law enforcement has raised many concerns over informed consent and the need to balance privacy concerns with public safety. A discussion of these issues is outside of the scope of the present work, but the reader is directed to Hill and Murphy (2019), Kolata and Murphy (2018), Larkin (2019b), and Saey (2019) for further information.

Feasibility of using GEDmatch for data obtained from degraded DNA

Even with the ability to create DTC-like autosomal SNP data through independent means, and the availability of GEDmatch as a database for comparison purposes, there were still technical challenges in demonstrating the feasibility of applying genetic genealogy to forensic casework. GEDmatch analysis tools were created to analyze DTC data generated from fresh saliva samples with high-confidence SNP call rates. While there are fewer technical issues with applying the tools to identify suspects based on DNA obtained from relatively fresh semen and blood samples, the application of genetic genealogy methods to compromised samples, especially those obtained from skeletal remains, raises important concerns.

Skeletal remains often experience fragmentation of DNA templates and structural modifications during the natural decomposition process (Ambers et al., 2016), in addition to degradation that can be experienced through exposure to a hostile environment. Both decomposition and environmental degradation can introduce errors in the detection of IBD segments, which can result in errors in relationship estimates. IBD segments shared by a pair of individuals are defined as segments greater than 6 cM in length that contain a matching series of identical alleles on one or both homologous chromosomes and that are bounded by opposite homozygotes (Henn et al., 2012). In general, the greater the number and length of segments shared by two individuals, the more closely they are related. Failure to detect mismatching homozygotes terminating an IBD segment in a degraded sample may cause the algorithm to incorrectly define the terminus of that segment as another pair of opposite homozygotes further downstream. The length of the segment will appear inflated.

Conversely, if the missing set of homozygotes is on the leading edge of a segment, and the algorithm fails to detect it, the otherwise terminating homozygote may be identified as the front end of a new IBD segment. If both sets of opposite homozygotes are absent, an IBD segment will not be detected. For highly fragmented DNA, only small segments may survive, and only reduced amounts of shared DNA will be detected, so that only distant cousins may be identified. Even if degraded samples meet initial quality control requirements in terms of average fragment length and the total amount of double-stranded DNA available, gaps in a degraded genome, low confidence reads, heterozygous bias, and contamination from human and nonhuman sources could lead to misinterpretation of IBD lengths. The resulting inaccuracy of relationship estimates and the generation of "ghost" matches, could disrupt the identification process.

The first step towards demonstrating the feasibility of using synthetic DTC data for forensic casework was to create data using a source other than a fresh saliva sample. This was accomplished in mid-2017 using biopsy tissue from a man who had recently died of cancer (Adams, 2017a). The tissue had been preserved as a formalin-fixed paraffin-embedded (FFPE) sample. It was uncertain whether cancer-related lesions and/or postmortem modifications to the subject's genome would influence SNP confidence levels generated by the sequencing process, and what affect, if any, this would have on the function of GEDmatch algorithms.

Because it was not possible to process the tissue through a DTC testing company, 23andMe and Ancestry-like SNP files were created based on 30X sequencing data obtained through an independent laboratory using an Illumina HiSeq x10 System. The sample contained 50 micrograms (μg) of high-quality DNA. GEDmatch datasets were then created from the resulting BAM files using the following two methods:

(1) A script was applied directly to the BAM file aligned to hg19 using a list of SNPs from a 23andMe reference file. BCFtools (part of Samtools) was then used to call SNPs from the raw reads at those locations to generate a pseudo-23andMe dataset (Loe, 2017; Krahn, 2017);

(2) A variant call file (.vcf) was generated from the BAM file consisting of 453k SNPs using proprietary software tools based on GATK from Broad Institute, after which 504,000 SNPs were added from the Ancestry reference file to generate a pseudo-Ancestry dataset (Press, 2017).

Both the pseudo-23andMe and pseudo-Ancestry files were successfully uploaded to GEDmatch and checked for the total number of usable SNPs and the presence of no-calls. The lists of autosomal matches for the two kits were compared and found to be substantially identical through third-cousin relationships.

Since several family members of the tissue donor (Fig. 2) had already been tested, the number of centiMorgans shared by the donor with each relative could be compared to the expected range predicted by statistics reported by the Shared cM Project (ScP) (see discussion of the ScP below.) (Adams, 2017b; Bettinger, 2017). In all cases, the observed amounts for each relationship fell comfortably within statistical predictions (Table 2).

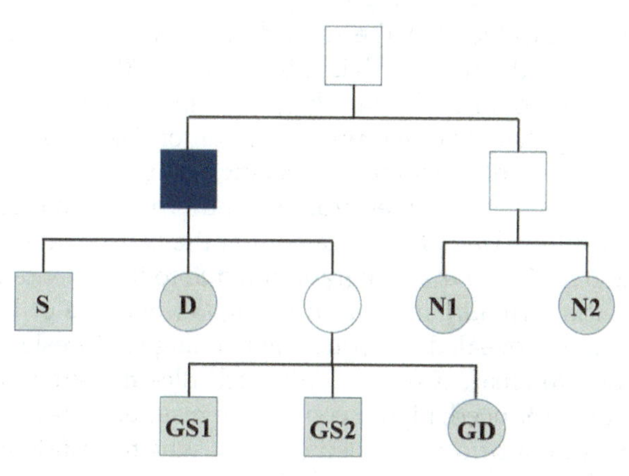

FIG. 2 The biopsy tissue donor is represented as a *dark square;* other family members who were tested are indicated in *gray;* family members who were not tested are indicated in *white.* (Refer to Table 2 for legend.)

TABLE 2 A number of cM shared by the tissue donor with various members of his family, compared to the observed average values and ranges reported by the Shared cM Project (ScP).

Relationship to donor	Designation in Fig. 1	Donor Ancestry kit (cM)	Donor 23andMe kit (cM)	ScP Average (cM)	ScP range (cM)
Son, test no. 1	S	3554	3527	3471	3266–3720
Son, test no. 2	S	3553	3504	3471	
Daughter	D	3516	3538	3471	
Niece no. 1	N1	1971	1940	1744	1301–2193
Niece no. 2	N2	1908	1883	1744	
Grandson no. 1	GS1	1599	1579	1765	1272–2365
Grandson no. 2	GS2	1888	1878	1765	
Granddaughter	GD	2208	2186	1765	

Although the success of the experiment made it clear that DTC-like data could be created from something other than a saliva sample, considering the large quantity of relatively fresh DNA that had been used, the question still remained whether GEDmatch algorithms could function on SNP datasets created from compromised DNA samples. This challenge was addressed by the DNA Doe Project with its first case, the identification of Joseph Newton Chandler III, a man who had committed suicide in Eastlake, Ohio in 2002. After Chandler's remains had been cremated, it was discovered that the real Joseph Newton Chandler III was a 9-year-old boy who had died in a car accident in 1945 outside of Dallas, Texas. Mr. X had stolen the child's identity in 1978 by claiming he was the real Chandler and obtaining a copy of the child's social security card in Rapid City, South Dakota (Caniglia, 2016; Cleveland.com, 2019).

By 2016, when attempts to identify Chandler by the U.S. Marshal's Office of Northern Ohio had failed, the deceased's Y-STR profile was compared to online Y-STR genetic genealogy databases. The results indicated that the man descended from a prominent 18th century Colonial Virginia family by the name of Nicholas (WYKC, 2016); however, attempts to identify him with the assistance of Nicholas family genealogists were unsuccessful.

Based on earlier results that indicated preserved tumor tissue could be used to generate a viable GEDmatch kit, 30X whole genome sequencing was performed on DNA obtained from cancer biopsy tissue that had been taken from Chandler in the late 1990s and preserved as FFPE. However, the resulting BAM file revealed that the DNA extracted from the tissue had experienced severe and highly uneven degradation due to the extended length of time it had been in contact with the formalin and paraffin (Fig. 3). Because of large gaps and uneven degradation in the genome revealed by sequencing, a range of thresholds were applied using bioinformatics software to create a set of GEDmatch files that traded off confidence levels with the amount of data produced; higher confidence resulted in few calls and a smaller data file. Even so, there was no a priori knowledge about which confidence level, if any, would generate useful and self-consistent GEDmatch data.

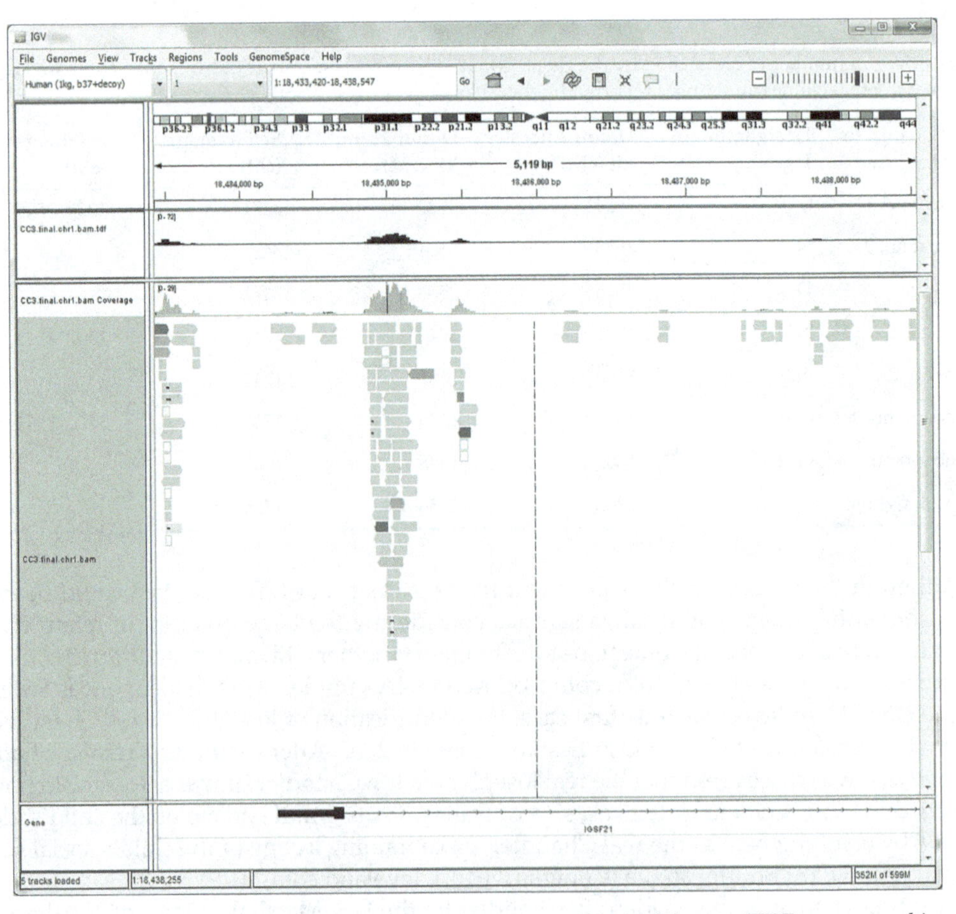

FIG. 3 BAM file showing uneven SNP coverage on Chromosome 1 derived from an FFPE premortem biopsy tissue sample obtained from the identity theft Joseph Newton Chandler III.

A normal GEDmatch kit will exhibit several thousand matches, but even Chandler's lowest confidence kit showed only 127, with no guarantee that the matches were genuine. According to GEDmatch kit diagnostics, the kit exhibited an 88% no-call rate; only 12% of his genome had survived the 14 years of exposure to formalin and paraffin.

To assess the function of GEDmatch algorithms on such highly degraded DNA, raw datasets obtained from two 23andMe customers were degraded in software to mimic the damage observed in Chandler's sample. If a SNP was missing from the Chandler sequence, it was removed from the authentic 23andMe datasets. The GEDmatch results derived from the two artificially degraded datasets were then compared to their corresponding original versions as a means of predicting how well Chandler's GEDmatch results based on his degraded sample might correspond to what could be obtained from his fresh DNA sample, if it were available.

Of primary concern was whether the matches to Chandler's degraded data were genuine, i.e., whether they would appear on the list of matches to his original, undegraded data. For the matches that did appear on both lists, an additional concern was reliability of the relationship estimates GEDmatch provided for them, based on the total number of shared cM and the lengths of their longest shared segments.

Fig. 4A shows a comparison between the order of the matches on the degraded 23andMe GEDmatch lists and their corresponding positions on the undegraded lists. Note that the reverse comparison was not of interest, since without a fresh sample of Chandler's DNA, it was not possible to know whether a match on Chandler's original list would appear on his degraded list. Of the

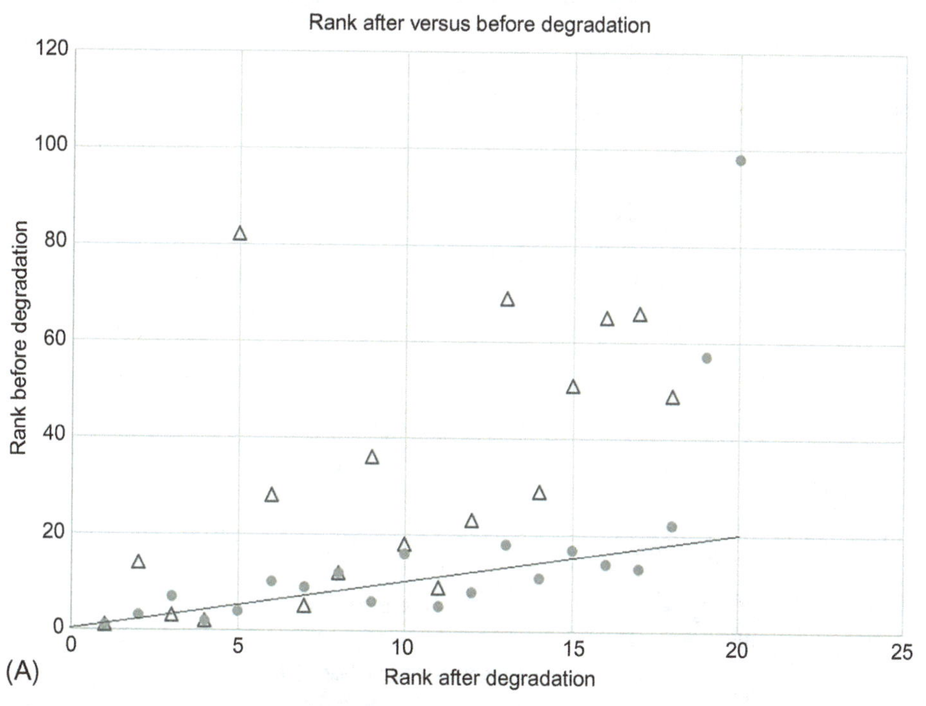

FIG. 4 Comparison of GEDmatch results after artificial degradation versus before degradation of two datasets D1 *(triangles)* and D2 *(circles)* obtained from two 23andMe customers: (A) rank of matches,

(Continued)

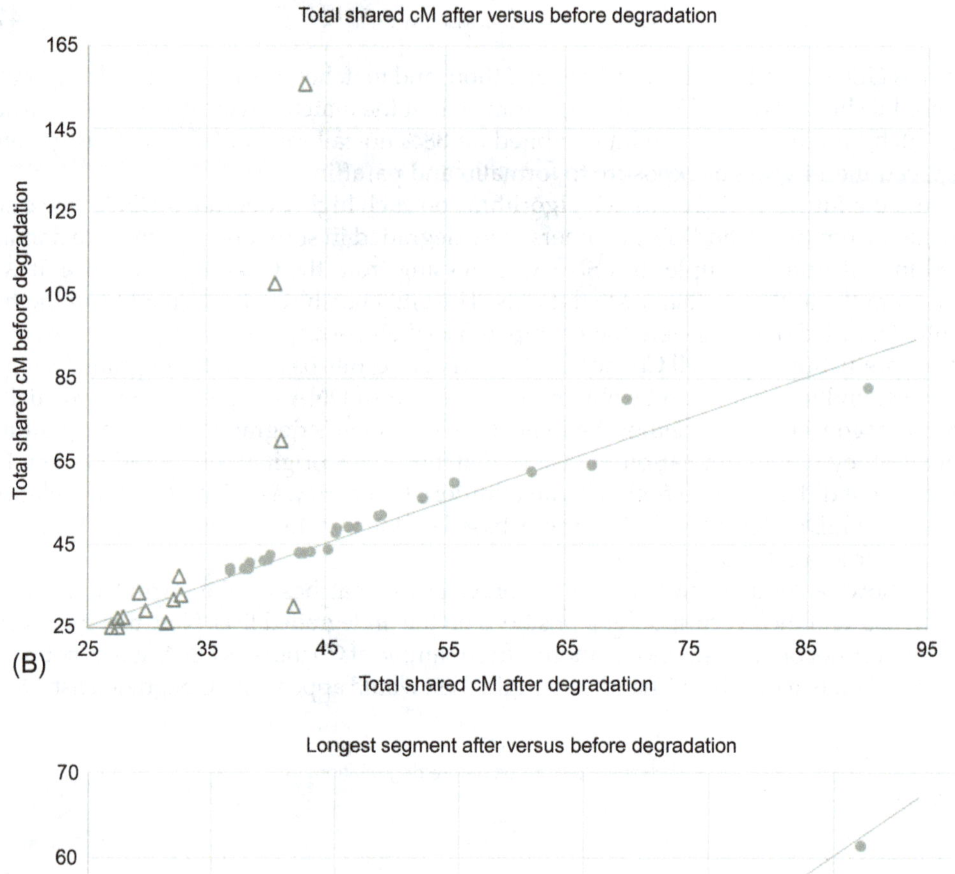

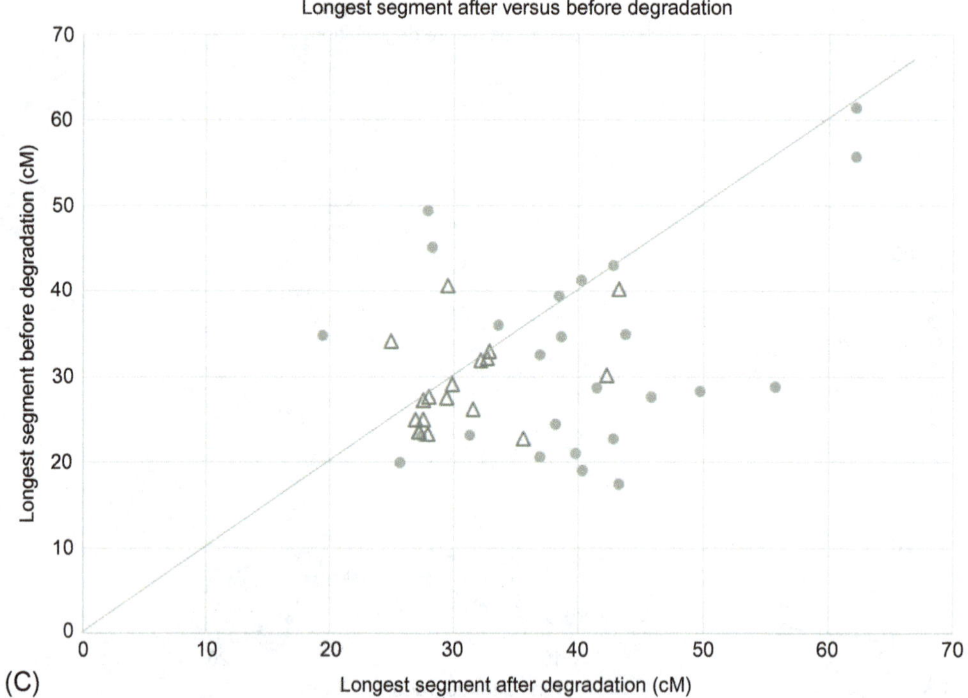

FIG. 4, CONT'D (B) total shared cM, and (C) longest shared segment (cM). *Lines* represent a slope of unity, representing no change after degradation.

top 50 matches to the degraded version of Dataset No. 1 (D1, triangles), only 35 of those appeared on the original list. Fifteen of the top matches on the degraded list were "ghosts" that were not on the original list. The first match on the degraded list that did not appear on the original list was match No. 19. A linear regression performed on the rank of the top 18 matches after degradation versus before degradation gave a slope of 3.20±0.46, indicating that many of the distant matches on the original list appear closer to the top of the degraded list, probably because the top matches on the original list were more affected by the degradation and moved down in rank.

Dataset No. 2 (D2, circles) was more well-behaved. Of the top 50 matches to the degraded version of D2, all 50 appeared on the original list. A linear regression performed on the rank of the first 25 datapoints after degradation versus before degradation gave a slope of 2.99±0.75, indicating, as in the case of D1, that many of the top 25 matches on the degraded list were distant matches on the original list. However, when only the top 18 matches were considered for comparison, a linear regression produced a slope of 1.01±0.08, indicating that on the average, the order of the top 18 matches after degradation was approximately the same as the order before degradation.

The behavior of the total number of shared cM and the length of the longest shared segment seem to support these observations. A linear regression on the total number of shared cM for D1 gave a slope of 1.37±0.22 based on the top 18 matches. When the top four original matches were omitted from the calculation (that is, matches with totals > 65 cM before degradation), the slope reduced to 0.92±0.04. The data suggest that matches closer to the top of the original list, sharing a higher number of cM, are more affected by degradation and appear in the degraded version as sharing a significantly lower number of cM. Therefore, the top matches on the original list appear on the degraded list among matches that originally shared much less DNA. (See Fig. 4B.) The order of matches for the degraded dataset therefore consists of a mix of matches that were originally of higher and lower rank, consistent with the data shown in Fig. 4A.

The effect of degradation on the total number of shared cM for D2 appears to be minimal and independent of the number of matches included in the analysis through the top 50. A linear regression on the total number of shared cM for D2 after degradation versus before degradation produced a slope of 1.01±0.03 based on the top 18 matches, consistent with the stability in their rank seen in Fig. 4A. Based on the top 25 matches, the slope was 0.98±0.03; based on the top 50 matches the slope was 0.96 ± 0.02.

For further insight into the degradation process, Fig. 4C shows a comparison between the longest shared segments after versus before degradation. A linear regression on the longest shared segments for D1 indicated a slope of 0.93±0.04. A linear regression on the longest shared segments for D2 based on the top 18 matches yielded a slope of 0.86±0.04; on the top 25 matches yielded a slope of 0.85±0.03; and on the top 50 matches yielded a slope of 0.86±0.02. There is scatter in the data that indicates both an increase in size of smaller segments and a decrease in size of longer segments observed for the matches in both datasets. However, the effect of degradation on the average on both datasets was to somewhat elongate the largest shared segments, although degradation did not have much effect on D2 in terms of the rank and the total amount of DNA shared with the top matches.

The data suggested that GEDmatch algorithms can function on autosomal SNP data created from degraded samples, although the order of the matches and the relationship estimates based on total shared cM may be affected. More detailed analysis of the underlying causes of this behavior was not necessary to show that Chandler's degraded GEDmatch dataset was probably usable. Future research on the effect on GEDmatch of varying levels and types of degradation is recommended.

The top matches to Chandler's GEDmatch kit were a mother M1 (b. 1953) and her daughter M2 (b. 1980) who both shared a single 50.1 cM segment with him, associated with a 30% chance of a 3C1R, ½ 3C, ½ 2C2R, or 2C3R relationship. (See discussion of the DNA Painter Shared cM Project tool below.) The fact that the mother and daughter shared the same single segment with Chandler and were estimated to have the same relationship with Chandler was not disturbing. It is not uncommon to have "sticky segments" along a family line, i.e., segments that remain intact and do not recombine as they are handed down in a family. Since Chandler was assumed to have been born in the mid-1930s, 3C1R was the most likely of the highest probability relationships for the mother.

After 6 months, the composite family tree created from the individual family trees of the 21 top matches included 16,000 individuals, but no one by the name of Nicholas. The decision to perform a second round of 30X sequencing was made, although to do so meant that the remaining library would be consumed and that the existing supply of DNA would be exhausted.

When the data generated by the second round of 30X sequencing produced virtually identical results, the two sets of read data were merged to create a 60X dataset. As hoped, read data from the second round of sequencing filled gaps in the read data from the first round of sequencing. Combining the two datasets increased the fraction of confident calls from 12% to 46%, resulting in a new match M3 who shared 49 cM with Chandler. Within a few hours, it was discovered that M3 was a second cousin once removed to a Mrs. Nichols. Her husband Mr. Nichols had already been placed in the tree as a third cousin once removed to the top match M1. The couple's four sons were, therefore, connected to all three top matches through their parents.

In March 2018, Chandler's case became the first forensic case solved using genetic genealogy autosomal SNP testing. One of the Nichols' sons Robert Ivan Nichols was identified as Joseph Newton Chandler III through a clue on a 1985 rental agreement. Nichols' childhood home was recorded on his birth certificate as 1823 Center St., New Albany, Indiana; on the rental agreement, Chandler referenced a fake sister living at 1823 Center St., Columbus, Ohio. Although Chandler's Y-STR profile had been an exact match to that of the descendants of the Colonial Virginia Nicholas family, his connection to the family was probably in the mid-to-late 1600s, prior to their arrival in America in about 1720. Apparently, the name Nicholas had been changed to the variant Nichols sometime in the intervening 350 years. Genealogical analysis showed that the predicted relationships of the three top matches to Chandler/Nichols were not far off from their true relationships (Table 3). More extensive kinship analysis of Chandler/Nichols matches is suggested as a topic for future research.

TABLE 3 Comparison of the top three matches to Joseph Newton Chandler/Robert Ivan Nichols.

Match	cM	Predicted relationship	Probability[a] (%)	Genealogical relationship
M1	50.1	3C1R	30.13	3C1R
M2	50.1	3C1R	16.23	3C
M3	49	3C1R	6.53	2C1R

[a] *As predicted by the ScP tool.*

The Chandler case is significant not only because it established the feasibility of applying genetic genealogy autosomal SNP testing to forensic casework. More importantly, it established the feasibility of applying genetic genealogy autosomal SNP testing to compromised DNA samples, with important implications for the identification of skeletal remains. It was only several weeks later that the Golden State Killer was identified using similar methods, based on a pristine DNA sample obtained from a 1981 rape kit (Arango et al., 2018).

Genetic genealogy autosomal SNP analysis tools

Because of the statistical nature of DNA inheritance, the amount of IBD DNA shared by a pair of individuals can be associated with multiple categories of relatedness and must be interpreted in the context of the relative probabilities of these categories. IBS segments are not usually included in relationship estimates.

Two references frequently cited for IBD relationship estimates are a simulation-based study conducted by Ancestry (AncestryDNA, 2016) and "The Shared cM Project" (ScP), a 2017 crowd-sourced survey based on the total number of shared cM reported for 16,000 pairs of respondents with known relationships (Bettinger, 2017). Fig. 5 shows (in table format) a summary of ScP data, including the maximum, minimum, and average number of cM shared by various relationship categories. The Shared cM Project tool is an online webtool based on the results of the Ancestry simulation that is available for free on the DNA Painter website (https://www.dnapainter.com). The tool allows a user to input a number of shared cM to obtain the relative probabilities of the possible relationship categories associated with that number.

A simple example

Genetic genealogy identification of an unidentified assailant or unidentified remains (a.k.a. a "John Doe") involves estimating his relationship to multiple individuals. Genealogies that include a hypothetical position for a John Doe in the known family pedigree of his so-called DNA-cousins must be evaluated by multiplying the probability of his relationship in that position for one of those DNA-cousins with the probabilities associated with his relationships in that position for all other DNA-cousins with the same common ancestor. If any one of those relationships is associated with a zero probability, the combined probability is zero and that position in the tree is prohibited. The position with the highest relative probability is the most likely based on the number of shared cM alone, but may be ruled out when other factors such as relative age or X-chromosome inheritance patterns are considered.

As a simple example, consider the case of a John Doe who shares 800 cM with Match A. Applying the ScP webtool reveals that the two could be connected in one of two ways: (1) with about 94% probability as first cousins, half uncle-half nephew, etc.; or (2) with about 6% probability as half first cousins, first cousins once removed, etc. (Fig. 6). Relationships in the first category are more likely, but not conclusive. Note that the sum of the probabilities in Fig. 6 is rounded to 100%, with the understanding that the likelihood of a relationship outside the observed ranges is small but nonzero.

Introducing a second DNA-cousin, Match B, who shares 250 cM with John Doe and who has a known relationship to Match A, creates a constraint that can be used to determine John's

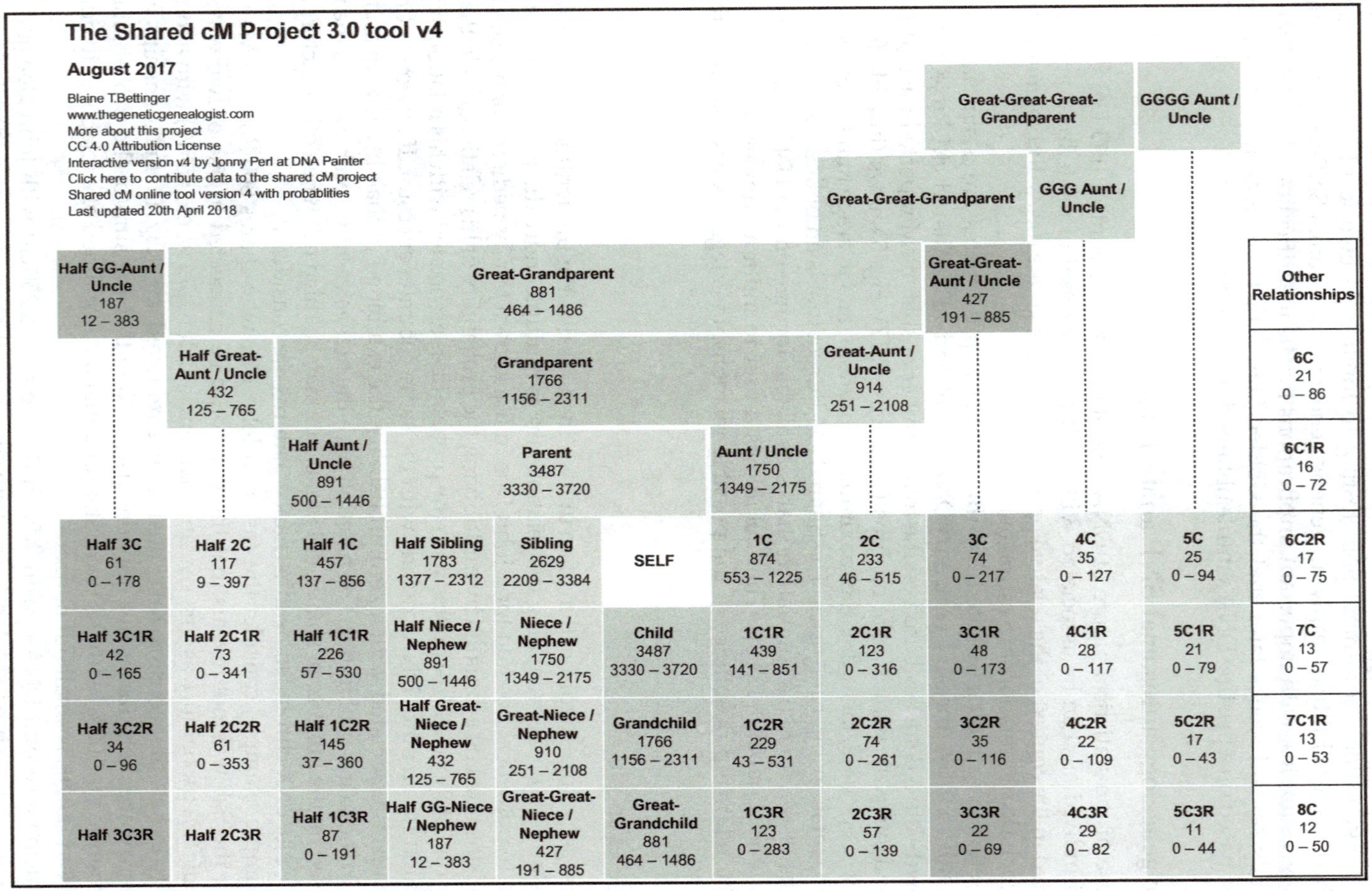

The Shared cM Project 3.0 tool v4

August 2017

Blaine T.Bettinger
www.thegeneticgenealogist.com
More about this project
CC 4.0 Attribution License
Interactive version v4 by Jonny Perl at DNA Painter
Click here to contribute data to the shared cM project
Shared cM online tool version 4 with probablities
Last updated 20th April 2018

Great-Great-Great-Grandparent

GGGG Aunt / Uncle

Great-Great-Grandparent

GGG Aunt / Uncle

Half GG-Aunt / Uncle 187 12 – 383

Great-Grandparent 881 464 – 1486

Great-Great-Aunt / Uncle 427 191 – 885

Half Great-Aunt / Uncle 432 125 – 765

Grandparent 1766 1156 – 2311

Great-Aunt / Uncle 914 251 – 2108

Half Aunt / Uncle 891 500 – 1446

Parent 3487 3330 – 3720

Aunt / Uncle 1750 1349 – 2175

Half 3C 61 0 – 178	**Half 2C** 117 9 – 397	**Half 1C** 457 137 – 856	**Half Sibling** 1783 1377 – 2312	**Sibling** 2629 2209 – 3384	**SELF**	**1C** 874 553 – 1225	**2C** 233 46 – 515	**3C** 74 0 – 217	**4C** 35 0 – 127	**5C** 25 0 – 94
Half 3C1R 42 0 – 165	**Half 2C1R** 73 0 – 341	**Half 1C1R** 226 57 – 530	**Half Niece / Nephew** 891 500 – 1446	**Niece / Nephew** 1750 1349 – 2175	**Child** 3487 3330 – 3720	**1C1R** 439 141 – 851	**2C1R** 123 0 – 316	**3C1R** 48 0 – 173	**4C1R** 28 0 – 117	**5C1R** 21 0 – 79
Half 3C2R 34 0 – 96	**Half 2C2R** 61 0 – 353	**Half 1C2R** 145 37 – 360	**Half Great-Niece / Nephew** 432 125 – 765	**Great-Niece / Nephew** 910 251 – 2108	**Grandchild** 1766 1156 – 2311	**1C2R** 229 43 – 531	**2C2R** 74 0 – 261	**3C2R** 35 0 – 116	**4C2R** 22 0 – 109	**5C2R** 17 0 – 43
Half 3C3R	**Half 2C3R**	**Half 1C3R** 87 0 – 191	**Half GG-Niece / Nephew** 187 12 – 383	**Great-Great-Niece / Nephew** 427 191 – 885	**Great-Grandchild** 881 464 – 1486	**1C3R** 123 0 – 283	**2C3R** 57 0 – 139	**3C3R** 22 0 – 69	**4C3R** 29 0 – 82	**5C3R** 11 0 – 44

Other Relationships
6C 21 0 – 86
6C1R 16 0 – 72
6C2R 17 0 – 75
7C 13 0 – 57
7C1R 13 0 – 53
8C 12 0 – 50

FIG. 5 Results of the shared cM project showing the number of cM shared as a function of degree of relationship (Bettinger, 2017).

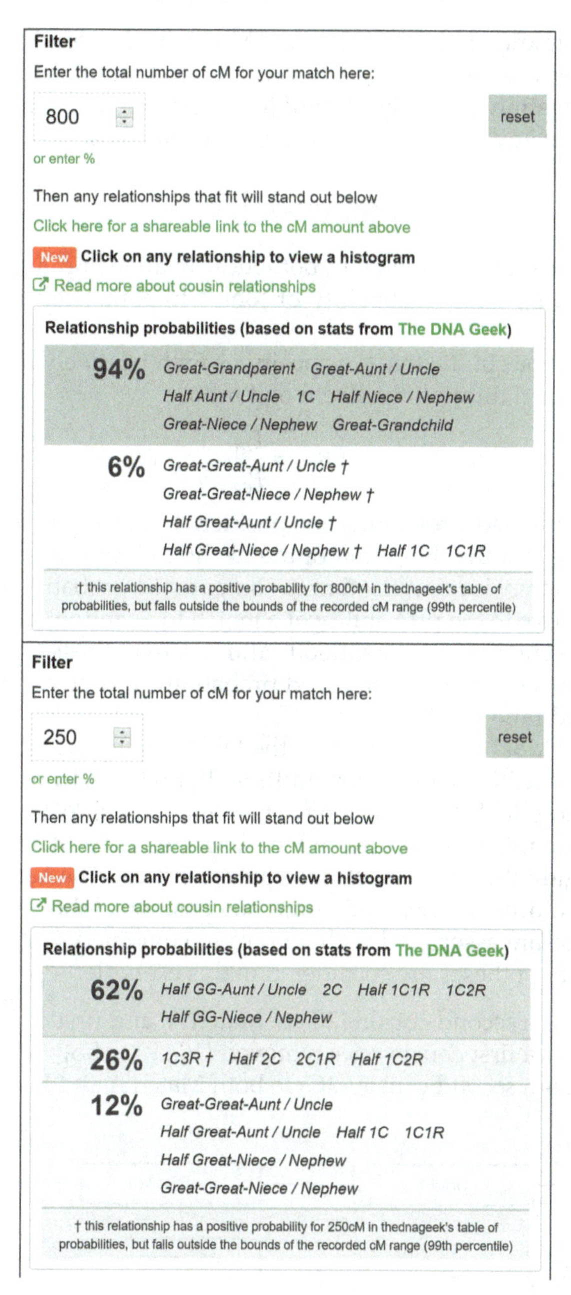

FIG. 6 Relationship probabilities based on 800 cM of shared DNA (top), and 250 cM of shared DNA (bottom).

more likely positions in their mutual family tree based on the product of the probabilities associated with his individual relationships with each of them. This assumes that the DNA shared by Match A, Match B, and John Doe has been randomly inherited so that the individual probabilities can be treated as independent.

Let's assume that it is known that Match A and Match B are second cousins, sharing a set of great grandparents. Because John shares DNA with both A and B, he must occupy a position in the overlap of their family trees. By placing John in various positions in that overlap, the combined probability of any given position can be calculated as:

$$P_C = P_A(M_A) \cdot P_B(M_B) \tag{1}$$

where P_A is the probability of John's hypothetical relationship to Match A based on M_A shared cM, and where P_B is the probability of John's hypothetical relationship to Match B based on M_B shared cM.

More favorable positions in the tree (i.e., positions with relatively higher combined probabilities) can be determined through the likelihood ratio:

$$LR_{1,2} = \frac{P_{C1}}{P_{C2}} \tag{2}$$

where P_{C1} is the combined probability for position 1, and P_{C2} is the combined probability for position 2. Note that the definition of the likelihood ratio used here is different from that normally used in forensic identification to compare the probability of a DNA match to a known individual relative to the probability of a match to a random individual in the general population. In the present case, the likelihood ratio of a hypothesis is defined relative to the hypothesis with the lowest nonzero combined probability. If only one nonzero hypothesis is available, its likelihood ratio will be unity.

The "What are the Odds" (WATO) tool on the DNA Painter website offers an automated means of calculating likelihood ratios for multiple hypothetical positions (hypotheses) for an unknown within an established genealogy (Perl and Larkin, 2018). The use of the WATO tool for the present example is illustrated in Fig. 7. Match A and Match B are second cousins who share a great grandfather as their common ancestor (CA). The number of cM John shares with each of them is indicated. There are numerous positions John could occupy in the tree, but only some of those are associated with a nonzero combined probability. For the sake of simplicity, only three hypotheses are considered in this example:

Hypothesis 1: John is a second cousin (2C) to Match A, and first cousin (1C) to Match B
Hypothesis 2: John is a first cousin once removed (1C1R) to both Match A and B
Hypothesis 3: John is a second cousin (2C) to both Match A and B

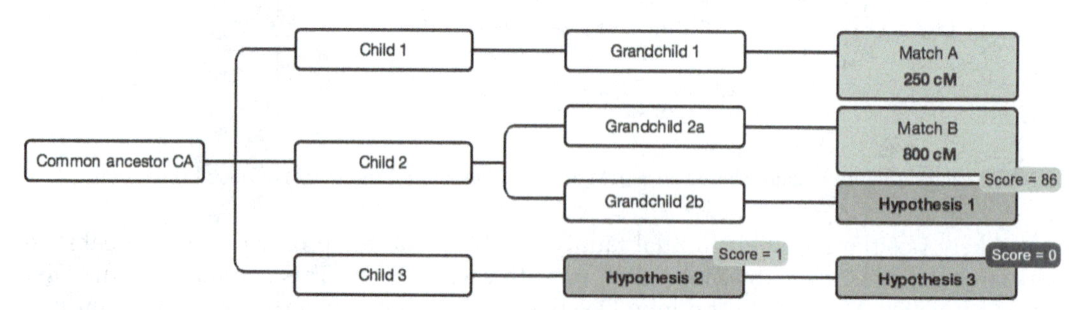

FIG. 7 WATO tree associated with the John Doe example described in the text.

TABLE 4 Combined probabilities associated with the John Doe example described in the text.

	Shared (cM)	Hypothesis 1	Hypothesis 2	Hypothesis 3
Match A	250	2C 0.63	1C 1x R 0.12	2C 0.63
Match B	800	1C 0.94	1C 1x R 0.056	2C 0.0
Combined probability		0.59	0.0067	0.0

The combined probabilities associated with these three hypotheses can be obtained from the ScP tool and are shown in Table 4. The probability of Hypothesis 3 is zero because a total of 800 cM does not support a 2C relationship to Match B. The likelihood ratio of Hypothesis 1 relative to Hypothesis 2 is:

$$L_{1,2} = \frac{0.59}{0.0067} = 86.7 \tag{3}$$

In other words, it is 86.7 times more likely that John is a second cousin to Match A and a first cousin to Match B than he is a first cousin once removed to both Match A and Match B.

This process can be generalized to the calculation of the combined probability of John's hypothetical position in the family pedigree relative to N matches, each of which shares M cM with him:

$$P_C = \prod_{i=1}^{N} P_i(M_i) \tag{4}$$

where P_i is the probability of John's hypothetical relationship with match i based on M_i shared cM. The likelihood ratio of any hypothesis relative to any other hypothesis can be calculated using Eq. (2), but the WATO tool only outputs likelihood ratios relative to the lowest nonzero combined probability.

Note that the hypothetical position selected for John need not be that of a known family member. The same procedure could be used to test whether John could be a family member who was given up for adoption or who was illegitimate. It can also happen that a hypothesis can be ruled out because it may not be physically possible. In the example, if grandchild 2b were male, and it had been determined that John shared X-DNA with Match B, Hypothesis 1 would not be feasible, as John could not have inherited X-DNA from his father. In this case, Hypothesis 1 would be discarded, and other hypothetical positions in the tree would be considered.

Existing challenges

There are still many challenges that have arisen as increasingly more difficult forensic identifications are attempted using genetic genealogy. Three of the most important challenges are manpower limitations, database size, and database composition.

The time scale related to genetic genealogy forensic casework can be much longer than conventional genetic genealogical research, requiring significantly more man hours of work, not including the initial investment of time needed to develop autosomal SNP data independent of established DTC testing company pipelines.

One reason for this is that, although the total membership in the DTC testing companies' databases has reached over 30 million, only two much smaller databases are available for forensic casework: (1) GEDmatch, and (2) Family Tree DNA (FTDNA). GEDmatch has about 1.3 million members as of December 2019. However, in May 2019, GEDmatch changed its terms of service to include an opt-out default setting to require users to voluntarily opt-in to law enforcement use of their data. The total number of kits accessible for forensic casework as of late 2019 is approximately 190,000, with 30,000 new opt-ins added each month (Rogers, 2019). In January 2019, it was revealed that FTDNA had been working with the FBI (Hernandez, 2019b). Shortly thereafter, FTDNA became the only DTC DNA testing company to accept data uploads from law enforcement agencies. FTDNA's terms of service include an opt-in *default setting*. The company has an estimated 1 million members in its database (International Society of Genetic Genealogy, 2019a). If the top matches include a close family member, the law enforcement agency is prohibited access to the match list and instead provided with a report.

Because of the relatively small sizes of the GEDmatch and FTDNA databases, top matches are usually more distant than those discovered through DTC databases, and therefore, more effort is required to build out and connect family pedigrees. Another limitation is that while working a forensic case, genealogists are not at liberty to contact matches to obtain family information or to request target testing, activities that could otherwise reduce the time and effort required to construct genealogical pedigrees.

Endogamy, or heavy intermarriage, also plays a role in the amount of time required to solve a case. In endogamous populations, relationship estimates based on the amount of shared DNA are typically inaccurate, as the DNA shared by two relatives is no longer inherited from a single common ancestor, but can be derived from multiple common ancestors connected through multiple family lines. It can be time-consuming to untangle endogamous family connections to determine individual contributions, if it is possible at all. The GEDmatch tool "Are Your Parents Related" (AYPR) can be useful in identifying cases where an individual's parents originate from an endogamous population. If the AYPR result is negative, endogamous matches (if present) will connect to only one side of the individual's family. Genealogical research can then focus on the remaining matches as a means of identifying the parent from the nonendogamous side, which may lead to the endogamous spouse. Their children will connect to both the endogamous and nonendogamous sides of the family, so that one of them may be identified as the Doe or the unknown assailant.

With little or no a priori information about an unknown, it is hard to predict in advance how long forensic genealogical analysis may take, or even whether a case may be solved at all. According to DNA Doe Project (DDP) statistics, the average time required to solve a Doe case is about 90 days (Press, 2019), including the two most difficult cases that could be considered statistical outliers: (1) the "Mill Creek Shed Man" case, that required 19 ½ months, and the "Belle in the Well" case, that required 16 months to solve. When these cases are omitted, the average time required to solve a case decreases to about 60 days.

DDP is a volunteer-staffed, nonprofit organization with unlimited manhours to devote to casework, so that these solve times reflect other limitations, such as the ethnic composition of the databases. Forensic genealogy companies that charge an hourly fee for research must confine their efforts to a cost-affordable number of labor hours, imposing an economic limit on the manpower they can devote to a case.

There have been discussions about accelerating forensic genealogy casework by automating the process of building and connecting family trees, but this is probably not practical. Most family trees are not well-documented and may contain misinformation based on family folklore and substandard research skills. There is also the possibility that a match or someone in a family line was adopted or born out of wedlock, in which case the documented family history will deviate from the genetic history. The misattributed paternity may be unrecorded by a published family pedigree, and therefore, will go undetected by an automated system.

The ethnic composition of the available databases can also present challenges. Based on the My Heritage database of about 1.28 million (approximately the same size as the currently available GEDmatch and FTNA databases combined), Erlich et al. (2018) demonstrated that 60% of all searches resulted in a match to a third cousin or closer, with approximately 15% probability of finding a match to a second cousin or closer. The search included all ethnicities.

However, when a search was focused on Caucasian Europeans, there was a 30% greater chance of finding a match than for a search based on individuals of African American ancestry. Ehrlich provides the only statistics publicly available on database composition, indicating that, as of 2018, the MyHeritage database was about 76% Northern European, 9.5% Southern European, 4.5% Sub-Saharan African, and 2.9% Native American, with smaller contributions from other ethnicities (Table 5). There are no statistics available that indicate the percentages of these ethnicities in the GEDmatch or the FTDNA databases, but it is reasonable to assume that they are similar since the companies draw on much the same customer base. Such an ethnicity breakout is consistent with the experience of many genetic genealogy researchers that most DTC databases are biased towards Caucasian European ancestry, and therefore, the solve rates for cases involving other ethnicities are much lower. For example, of the 12 Native American or Hispanic cases addressed by DDP, as of December 2019, 10 remain unsolved, some after more than a year.

TABLE 5 Percent ethnic composition of the MyHeritage database containing 1.28 million members, as of 2018 (Erlich et al., 2018).

Main DNA ethnicity	Percentage
North European	76.3%
South European	9.5%
Sub-Saharans	4.5%
Native American	2.9%
Ashkenazi Jewish	2.4%
South/West Asia	2.1%
East Asia	1.7%
North Africa	0.3%
Oceania	0.2%

The relatively small sizes of the GEDmatch and FTDNA databases, combined with the low contributions of underrepresented ethnicities, are limitations that will eventually be resolved as DTC DNA testing increases. However, the databases will always be biased towards Caucasian Europeans, considering this is the ethnicity of the majority of genealogists.

The future

There are still many questions that must be addressed about the limits to which genetic genealogy can be applied to compromised human remains. The most important of these relate to the lowest amount of DNA and the highest level of degradation that can be used to generate a usable GEDmatch kit. Samples will continue to be processed by independent laboratories to develop genetic genealogy data, but much work must be done to understand how to optimize the process, with the goal of doing more and more with less and less. Forensic laboratory pipelines that have been standardized to generate autosomal STR profiles for uploads to CODIS will inevitably develop the ability to generate DTC-like data for uploads to genetic genealogy autosomal databases. It is still to be determined what that process will involve and which data will be the best predictors of success. It will be exciting to see how genetic genealogy tools will evolve in parallel.

No matter how these questions are answered, it is clear that the use of forensic genetic genealogy will continue as one of the most significant developments in forensic identification since CODIS was introduced in the early 1990s.

References

23andme, (n.d.) About us. <https://mediacenter.23andme.com/company/about-us/> Accessed 10 December 2019.

Adams, J., 2017a. The Great Paternal DNA Quest. Accessed 14 June 2017.

Adams, J., 2017b. Private communication.

Ambers, A., Gill-King, H., Dirkmaat, D., et al., 2014. Autosomal and Y-STR analysis of degraded DNA from the 120-year-old skeletal remains of Ezekiel Harper. Forensic Sci. Int. Genet. 9, 33–41. https://doi.org/10.1016/j.fsigen.2013.10.014.

Ambers, A., Churchill, J., King, J., et al., 2016. More comprehensive forensic genetic marker analysis for accurate human remains identification using massively parallel sequencing. BMC Genomics 17 (9), 750. https://doi.org/10.1186/s12864-016-3087-2.

Ambers, A., Votrubova, J., Vanek, D., et al., 2018. Improved Y-Str typing for disaster victim identification, missing persons investigations, and historical human skeletal remains. Int. J. Legal Med. 132, 1545. https://doi.org/10.1007/s00414-018-1794-8.

Ancestry, (n.d.-a). We Unlock New Understanding and Meaningful Connections. <https://www.ancestry.com/corporate/about-ancestry/our-story> Accessed 10 December 2019.

Ancestry, (n.d.-b) Ancestry Guide for Law Enforcement. <https://www.ancestry.com/cs/legal/lawenforcement> Accessed 10 December 2019.

Ancestry, 2012. Ancestry.com Launches New AncestryDNA Service: The Next Generation of DNA Science Poised to Enrich Family History Research. https://www.ancestry.com/corporate/newsroom/press-releases/ancestry.com-dna-launches. (Accessed 10 December 2019).

Ancestry, 2017. Ancestry Sets AncestryDNA Sales Record Over Holiday Period and Fourth Quarter. https://www.ancestry.com/corporate/newsroom/press-releases/ancestry-sets-ancestrydna-sales-record-over-holiday-period-and-fourth>. (Accessed 10 December 2019).

AncestryDNA, 2016. AncestryDNA Matching White Paper, Discovering Genetic Matches Across a Massive, Expanding Database. https://www.ancestry.com/dna/resource/whitePaper/AncestryDNA-Matching-White-Paper.pdf>. (Accessed 10 December 2019).

Anderson, T., 2019. DNA leads to arrest in Warwick, R.I. murder case from 2013. Boston Globe. https://www.bostonglobe.com/metro/2019/02/06/dna-helps-crack-warwick-cold-case-grandfather-was-killed-with-hammer-local-park/BDJCUZ7ZPzold2SW36zIYK/story.html>. (Accessed 4 August 2019).

Applied Biosystems, (n.d.) YFiler Haplotype Database. <http://www6.appliedbiosystems.com/yfilerdatabase/index.jsp> Accessed 1 January 2020.

Arango, T., et al., 2018. To catch a killer: a fake profile on a DNA site and a pristine sample. New York Times. https://www.nytimes.com/2018/04/27/us/golden-state-killer-case-joseph-deangelo.html>. (Accessed 11 December 2019).

Athey, W., 2005. Haplogroup prediction from Y-STR values using an allele frequency approach. J. Genet. Genealogy 1, 1–7.

Athey, T.W., 2006. Haplogroup prediction from Y-STR values using a Bayesian-allele frequency approach. J. Genet. Genealogy 2, 34–39.

Augenstein, S., 2018. DNA Doe project names another, giving major piece in infamous Ohio mystery. Forensic Magazine. https://www.forensicmag.com/news/2018/06/dna-doe-project-names-another-giving-major-piece-infamous-ohio-mystery>. (Accessed 4 August 2019).

Bettinger, B., 2017. The Shared cM Project—Version 3.0 (August 2017). https://secureservercdn.net/184.168.47.225/35b.ded.myftpupload.com/wp-content/uploads/2017/08/Shared_cM_Project_2017.pdf.>. (Accessed 10 December 2019).

Bettinger, B., 2019a. Facing DNA Privacy Concerns Head-on With Informed Consent. *The Genetic Genealogist*, https://thegeneticgenealogist.com/2019/05/22/facing-dna-privacy-concerns-head-on. (Accessed 3 August 2019).

Bettinger, B., 2019b. Adoptee Success Using DNA Testing: Mid-2019 Analysis. https://www.dropbox.com/s/mwisyzx9eelc7ss/Adoptee%20Statistics%20Mid-2019.pdf?dl=0>. (Accessed 11 December 2019).

Biesecker, L., Baily-Wilson, J., Ballantyne, J., et al, 2005. DNA identifications after the 9/11 world trade center attack. Science, 310 (5751), 1122-1123. doi:https://doi.org/10.1126/science.1116608 Accessed 4 August 2019.

Bogdanowicz, W., Allen, M., Branicki, W., et al., 2009. Genetic identification of putative remains of the famous astronomer Nicolaus Copernicus. Proc. Natl. Acad. Sci. U.S.A. 106 (30), 12279–12282. doi:https://doi.org/10.1073/pnas.0901848106 Accessed 4 August 2019.

Bosch, E., Calafell, F., Santos, F.R., et al. (1999). Variation in short tandem repeats is deeply structured by genetic background on the human Y chromosome. Am. J. Hum. Genet., 65(6): 1623-1638. doi:https://doi.org/10.1086/302676 Accessed 4 August 2019.

Caniglia, J., 2016. Chasing a Ghost: U.S. Marshal Investigates Dead Eastlake Man Who Stole Boy's Identity; Believes he was a Violent Fugitive. https://www.cleveland.com/court-justice/2014/10/chasing_a_ghost_us_marshal_inv.html>. (Accessed 10 December 2019).

Cassidy, M., 2016. How forensic genealogy led an arrest in the Phoenix 'canal killer' case. The Republic. https://www.azcentral.com/story/news/local/phoenix/2016/11/30/how-forensic-genealogy-led-arrest-phoenix-canal-killer-case-bryan-patrick-miller-dna/94565410>. (Accessed 3 August 2019).

CBS/AP, 2018. "Buckskin Girl" Case: DNA Breakthrough Leads to ID of 1981 Murder Victim. https://www.cbsnews.com/news/buckskin-girl-case-groundbreaking-dna-tech-leads-to-id-of-1981-murder-victim/>. (Accessed 4 August 2019).

Cleveland.com, 2019. Chasing a Ghost: U.S. Marshal Investigates Dead Eastlake Man Who Stole Boy's Identity; Believes he was a Violent Fugitive. https://www.cleveland.com/court-justice/2014/10/chasing_a_ghost_us_marshal_inv.html>. (Accessed 4 January 2020).

Coble, M.D., Loreille, O.M., Wadhams, M.J., et al., 2009. Mystery solved: the identification of the two missing Romanov children using DNA analysis. PLoS One 4 (3), e4838. doi:https://doi.org/10.1371/journal.pone.0004838 Accessed 4 August 2019.

Erlich, Y., et al., 2018. Identity inference of genomic data using long-range familial searches. Science 362 (6415), 690-694. doi:https://doi.org/10.1126/science.aau4832 Accessed 11 December 2019.

Family Tree DNA, 2019. Family Tree DNA Learning Center. https://www.familytreedna.com/learn/y-dna-testing/y-str/y-dna-str-markers-family-tree-dna-test/>. (Accessed 4 August 2019).

Francalacci, P., Morelli, L., Angius, A., et al., 2013. Low-pass DNA sequencing of 1200 Sardinians reconstructs European Y-chromosome phylogeny. Science 341 (6145), 565-569. doi:https://doi.org/10.1126/science.1237947 Accessed 4 August 2019.

Fuller, T., 2018. How genealogy led to the front door of the Golden state killer. New York Times. https://www.nytimes.com/2018/04/26/us/golden-state-killer.html>. (Accessed 4 August 2019).

Ge, J., Budowle, B., Chakraborty, R., 2011. Choosing relatives for DNA identification of missing persons. J. Forensic Sci. 56 (1), 23-28. doi:https://doi.org/10.1111/j.1556-4029.2010.01631.x Accessed 4 August 2019.

Glenn, T.C., 2011. Field guide to next-generation DNA sequencers. Mol. Ecol. Resour. 11, 759–769.

Guerrni, C.J., Robinson, J.O., Petersen, D. et al., 2018. Should police have access to genetic genealogy databases? Capturing the Golden state killer and other criminals using a controversial new forensic technique. PLoS Biol. doi:https://doi.org/10.1371/journal.pbio.2006906 Accessed 3 August 2019.

Harmon, A., 2008. Gene map becomes a luxury item. New York Times. https://www.nytimes.com/2008/03/04/health/research/04geno.html>. (Accessed 3 August 2019).

Henn, B.M., et al., 2012. Cryptic distant relatives are common in both isolated and cosmopolitan genetic samples. PLoS One 7 (4). https://doi.org/10.1371/journal.pone.0034267. e34267.

Hernandez, L., 2019a. Suspect in 37-year-old unsolved Lake Tahoe homicide linked to Bay Area. San Francisco Chronicle. https://www.sf.chronicle.com/crime/article/Suspect-in-37-year-old-unsolved-Lake_Tahor-13827776.php. (Accessed 3 August 2019).

Hernandez, S., 2019b. One of the biggest at-home DNA testing companies is working with the FBI. BuzzFeed News. https://www.buzzfeednews.com/article/salvadorhernandez/family-tree-dna-fbi-investigative-genealogy-privacy>. (Accessed 11 December 2019).

Hill, K., Murphy, H., 2019. Your DNA profile is private? A Florida judge just said otherwise. New York Times. https://www.nytimes.com/2019/11/05/business/dna-database-search-warrant.html>. (Accessed 10 December 2019).

International Society of Genetic Genealogy, 2019a. Family Tree DNA. < https://isogg.org/wiki/Family_Tree_DNA#Early_testing_.282000_-_2006.29 > Accessed 4 January 2020.

International Society of Genetic Genealogy, 2019b. Autosomal DNA Testing Comparison Chart. https://isogg.org/wiki/Autosomal_DNA_testing_comparison_chart>. (Accessed 10 December 2019).

International Society of Genetic Genealogy, 2019c. Revision History of "Autosomal DNA Statistics". https://isogg.org/w/index.php?title=Autosomal_DNA_statistics&action=history>. (Accessed 10 December 2019).

Irwin, J., Edson, S., Loreille, O., et al., 2007. DNA identification of "Earthquake McGoon" 50 years postmortem. J. Forensic Sci. 52 (5). https://doi.org/10.1111/j.1556-4029.2007.00506.x.

Kayser, M., Caglià, A., Corach, D., et al., 1997. Evaluation of Y-chromosomal STRs: a multicenter study. Int. J. Legal Med. 110 (3), 125–133. https://doi.org/10.1007/s004140050051.

Kayser, M., 2017. Forensic use of Y-chromosome DNA: a general overview. Human Genet. 136 (5), 621–635. https://doi.org/10.1007/s00439-017-1776-9.

Kidd, K.K., Speed, W.C., Pakstis, A.J., et al., 2014. Progress towards an efficient panel of SNPs for ancestry reference. Forensic Sci. Int. Genet. 10, 23–32. https://doi.org/10.1016/j.fsigen.2014.01.002.

Kidd, J.R., Friedlaender, F.F., Speed, W.C., et al., 2011. Analyses of a set of 128 ancestry informative single-nucleotide polymorphisms in a global set of 119 population samples. Investig. Genet. 2, 1. https://doi.org/10.1186/2041-2223-2-1.

King, T., Fortes, G.G., Balaresque, P., et al., 2014. Identification of the remains of King Richard III. Nat. Commun. 5, 5631. https://doi.org/10.1038/ncomms6631.

Kolata, G., Murphy, H., 2018. Stores of DNA, for anybody to pore over, raise questions of privacy. Seattle Times. https://www.seattletimes.com/nation-world/stores-of-dna-for-anybody-to-pore-over-raise-questions-of-privacy/>.

Krahn, T., 2017. https://github.com/tkrahn/extract23.

KTVL, 2019. Jane "Annie' Doe" Identified as Anne Lehman in 47-Year-Old Cold Case. https://ktvl.com/news/local/jane-doe-identified-as-anne-lehman-in-47-year-old-josephine-county-cold-case>.

Larkin, L., 2019a. Private communication.

Larkin, L., 2019b. Informed consent: what it is, what it isn't, and why it's necessary. DNA Geek. https://thednageek.com/informed-consent-what-it-is-what-it-isnt-and-why-its-necessary/>. (Accessed 10 December 2019).

Latham, K., Miller, J., 2019. DNA recovery and analysis from skeletal material in modern forensic contexts. Forensic Sci. Res. 4 (1), 51–59. https://doi.org/10.1080/20961790.2018.1515594.

Lee, H., Kim, N., Park, M., et al., 2010. DNA typing for the identification of old skeletal remains from Korean War victims. J. Forensic Sci. 55 (6). https://doi.org/10.1111/j.1556-4029.2010.01411.x.

Lewis, R., 2019. DNA for the Greater Good: Should Police Have Access to Consumer DNA Databases? Genetic Literacy Project, https://geneticliteracyproject.org/2019/06/04/dna-for-the-greater-good-should-the-police-have-access-to-consumer-dna-databases. (Accessed 3 August 2019).

Loe, J., 2017. Private communication.

Marjanović, D., Metjahić, N., Čakar, J., et al., 2015. Identification of human remains from the Second World War mass graves uncovered in Bosnia and Herzegovina. Croat. Med. J. 56 (3), 257–262. https://doi.org/10.3325/cmj.2015.56.257.

MarketWatch, 2017. Global Next Generation Sequencing (NGS) Market—Analysis and Forecast (2017–2024). https://www.marketwatch.com/press-release/global-next-generation-sequencing-ngs-market-analysis-and-forecast-2017-2024-2017-12-06. (Accessed 10 December 2019).

Metzger, S., 2018. How did authorities solve the true identity of Joseph Newton Chander III? WKYC. https://www.wkyc.com/article/news/crime/how-did-authorities-solve-the-true-identity-of-joseph-newton-chandler-III/95-566345556>. (Accessed 3 August 2019).

Miller Surname Project—Y-DNA Classic Chart, (n.d.). <https://www.familytreedna.com/public/Miller?iframe=yresults> Accessed 3 August 2019.

Moltini, M., 2019. The US urgently needs new privacy Laws. Wired Science. https://www.wired.com/story/the-us-urgently-needs-new-genetic-privacy-laws/>. (Accessed 4 August 2019).

Moran, L., Keneally, M., 2012. The coldest case ever? Police trace DNA of 1991 killer back to 17th century family who came over on the mayflower. Daily Mail. https://www.dailymail.co.uk/news/article-2084692/Sarah-Yarborough-murder-Could-DNA-linked-17th-century-Mayflower-family-solve-1991-case.html>. (Accessed 3 August 2019).

MyHeritage.com, 2016. Introducing MyHeritageDNA. https://blog.myheritage.com/2016/11/introducing-myheritage-dna/>. (Accessed 10 December 2019).

National Human Genome Research Institute, 2019. The Cost of Sequencing a Human Genome. https://www.genome.gov/about-genomics/fact-sheets/Sequencing-Human-Genome-cost>. (Accessed 3 August 2019).

Nevgen, 2019. About Nevgen Haplogroup Predictor. https://www.nevgen.org/AboutNevGen.html>. (Accessed 3 August 2019).

Nievergelt, C.M., Maihofer, A.X., Shektman, T., et al., 2013. Inference of human continental origin and admixture proportions using a highly discriminative ancestry informative 41-SNP panel. Investig. Genet. 4, 13. https://doi.org/10.1186/2041-2223-4-13.

Pakstis, A.J., Fang, R., Furtado, M.R., et al., 2012. Mini-haplotypes as lineage informative SNPs and ancestry inference SNPs. Eur. J. Human Genet. 20 (11), 1148-1154. doi:https://doi.org/10.1038/ejhg.2012.69 Accessed 4 August 2019.

Perl, J., Larkin, L., 2018. What are the Odds? DNA Painter, https://dnapainter.com/tools/probability>. (Accessed 11 December 2019).

Press, M., 2019. Private communication.

Press, W., 2017. Private communication.

Rocca, R.A., Magoon, G., Reynolds, D.F., et al., 2012. Discovery of Western European R1b1a2 Y chromosome variants in 1000 Genomes Project data: An online community approach. PLoS One 7 (7), e41634. https://doi.org/10.1371/journal.pone.0041634.

Roewer, L., Kayser, M., de Knijff, P., et al., 2000. A new method for the evaluation of matches in non-recombining genomes: application to Y-chromosomal short tandem repeat (STR) haplotypes in European males. Forensic Sci. Int. 114 (1), 31–43. https://doi.org/10.1016/S0379-0738(00)00287-5.

Rogers, C., 2019. Private communication.

Saey, T.H., 2019. Genealogy companies could struggle to keep clients' data from police. Science News. https://www.sciencenews.org/article/forensic-genetic-genealogy-companies-police-privacy>. (Accessed 10 December 2019).

Shapiro, E., 2019. Jane Doe murder victim finally identified 3 decades later thanks to forensic genealogy: Sheriff. https://abcnews.go.com/US/jane-doe-murder-victim-finally-identified-decades-forensic/story?id=60447563. (Accessed 3 August 2019).

Sullivan, J., 2012. DNA may link teen's killer to a mayflower family. Seattle Times. https://www.seattletimes.com/seattle-news/dna-may-link-teens-killer-to-a-mayflower-family>. (Accessed 3 August 2019).

SWGDAM, 2012. SWGDAM Interpretation Guidelines for Autosomal STR Typing for Forensic DNA Testing Laboratories. Approved January 14, 2010, https://www.forensicdna.com/assets/swgdam_2010.pdf>. (Accessed 10 December 2019).

Top10DNATests, 2020. 23andMe Review. https://www.top10dnatests.com/reviews/23andme-review/. (Accessed 4 January 2020).

Van Geystelen, A., Decorte, T., Larmuseau, M.H.D., 2013. Updating the Y-chromosomal phylogenetic tree for forensic applications based on whole genome SNPs. Forensic Sci. Int. Genet. 7(6), 573-580. doi:https://doi.org/10.1016/j.fsigen.2013.03.010 Accessed 4 August 2019.

Williams, M., 2019. Pamela Cahanes cold case: DNA match leads to arrest in navy grad's killing—34 years later. Orlando Sentinel. https://www.orlandosentinel.com/news/breaking-news/os-ne-pamela-cahanes-cold-case-solved-20190314-story.html>. (Accessed 3 August 2019).

WYKC, 2016. U.S. Marshals have identified a possible last name for a John Doe who committed suicide in Eastlake back in 2002. WYKC Studios. https://www.wkyc.com/article/news/local/first-look-infamous-cleveland-cold-case-suspect-may-be-named/372013817. (Accessed 10 December 2019).

Y-chromosome Users Group, 2019. YHRD: Y-Chromosome STR Haplotype Reference Database. Release 62—2019/Dec/31. https://yhrd.org/>. (Accessed 1 January 2020).

YFull, (n.d.). Y-Chr Sequence Interpretation Service. <https://www.yfull.com/tree> Accessed 3 August 2019.

Zavala, E., Rajagopal, S., Perry, G., et al., 2019. Impact of DNA degradation on massively parallel sequencing-based autosomal STR, iiSNP, and mitochondrial DNA typing systems. Int. J. Legal Med. https://doi.org/10.1007/s00414-019-02110-4.

SECTION V

Complementary and multidisciplinary approaches to assist in identification of unidentified human skeletal remains

CHAPTER

21

Forensic anthropology in a DNA world: How anthropological methods complement DNA-based identification of human remains

Eric J. Bartelink PhD, D-ABFA

Department of Anthropology, Human Identification Laboratory, California State University, Chico, CA, United States

Introduction

The American Board of Forensic Anthropology (ABFA) defines forensic anthropology as "the application of the science of physical or biological anthropology to the legal process" (American Board of Forensic Anthropology, http://theabfa.org). Although this succinct definition broadly addresses what forensic anthropologists do, a more inclusive definition is needed to address the current breadth and scope of the field. Dirkmaat et al. (2008) define forensic anthropology more broadly as "the scientific discipline that focuses on the life, the death, and the postlife history of a specific individual, as reflected primarily in their skeletal remains and the physical and forensic context in which they are emplaced." This more promising definition closely matches the current practice of forensic anthropology and is inclusive of the role of practitioners in scene recovery context, analysis of the postmortem history of human remains, and the use of advanced methods in skeletal identification. Forensic anthropologists today work in a wide variety of settings, including universities, medical examiner (ME) offices, federal laboratories, state and federal agencies, museums, and humanitarian and human rights organizations.

Forensic anthropologists examine skeletal materials in order to: (1) determine if material is in fact skeletal in nature; (2) differentiate human from nonhuman remains; (3) identify remains as recent, historic, or prehistoric; (4) construct biological profiles of unidentified remains; (5) assess skeletal anomalies, antemortem trauma, pathological conditions, and evidence of surgical intervention that may be useful in personal identification; (6) assess postmortem changes to aid in documenting scene context, postmortem interval, and postmortem

https://doi.org/10.1016/B978-0-12-815766-4.00023-6

artifacts that can mimic skeletal trauma; and (7) describe and interpret evidence of perimortem trauma associated with the death event. Forensic anthropology casework is highly varied, with some practitioners working primarily on domestic cases and others working in international humanitarian contexts (e.g., on the recovery and identification of war dead, or on mass fatality incident identification efforts).

The aim of this chapter is to briefly review the scope of identification methods used in forensic anthropology in the United States, and more importantly, to highlight ways in which forensic anthropology assessments can guide procedures in DNA testing and identification. The forensic sciences do not operate in a vacuum. Today, personal identification of the deceased is both a multidisciplinary and interdisciplinary science. Despite the longstanding fear that the role of the forensic anthropologist would be minimized due to the expanding role of DNA technology, forensic anthropologists continue to make significant contributions in the identification of human remains. If anything, the collaboration between forensic anthropologists and DNA experts has only strengthened the role of forensic anthropology in the medicolegal system. This review emphasizes the recent literature on advances in identification methods from the skeleton.

History of forensic anthropology

Forensic anthropology emerged as a new discipline in the mid-to-late 1800s, and casework was typically conducted by anatomists and physical anthropologists. Most often, cases involved the assessment of remains to determine if they were human or nonhuman, or to determine the biological profile characteristics of unidentified individuals, including their sex, ancestry, age-at-death, and stature. A critical milestone was Wilton M. Krogman's *Guide to the Identification of Human Skeletal Material*, a pamphlet published in 1939 for the Federal Bureau of Investigation (FBI) (Krogman, 1939). The need to identify the dead from World War II and the Korean War further ushered in the need for trained skeletal anatomists. This represented a turning point and involved the development of new methods and reference standards for assessing biological profiles of unidentified remains. Key texts were published in the decades following, including *The Human Skeleton in Forensic Medicine* (Krogman, 1962) and *Essentials of Forensic Anthropology: Especially as Developed in the United States* (Stewart, 1979).

By the early 1970s, physical anthropologists began to come together at the annual meetings of the American Academy of Forensic Sciences (AAFS), forming the Physical Anthropology Section in 1972 (now referred to simply as the Anthropology Section). This was followed 5 years later by the formation of the American Board of Forensic Anthropology (ABFA) in 1977, the first board certification examination available in forensic anthropology. Today, board certification is considered the gold standard in forensic casework. As of 2019, the ABFA has certified 131 Diplomates, 91 of which are considered "active status." More recently, the Latin American Association of Forensic Anthropology (ALAF) and the Forensic Anthropology Society of Europe (FASE) have developed a board certification examination process, modeled in part on the framework of the ABFA. However, the ABFA is currently the only certification board that is accredited by an outside agency (i.e., the Forensic Specialties Accreditation Board).

In 2009, the National Research Council published the report *Strengthening Forensic Science in the United States: A Path Forward*, which provided a critical analysis of biases and limitations in the forensic sciences and ways to make substantive improvements (National Research Council, 2009). Although forensic anthropology was not a focus of the report, the community largely embraced these findings, as well as the charge to infuse more rigor to the analysis of human remains. With support of the FBI and the Department of Defense (DoD), the Scientific Working Group for Forensic Anthropology (SWGANTH) was formed in 2008 to develop best practices for the forensic anthropology community. By 2014, the SWGANTH effort was replaced by the federally funded Organization of Scientific Area Committees (OSAC), developed under the National Institute of Standards and Technology (NIST). The OSAC Anthropology subcommittee has drafted numerous standards and guidelines which are in varying stages of review. OSAC standards and guidelines must be approved by an external standards development organization, such as the AAFS's Academy Standards Board. Once approved, these documents will be published on an official registry and recognized as national standards or guidelines by the forensic community. These recent developments reflect forensic anthropology's strong academic emphasis on evidence-based research and the importance in quantifying certainty for various skeletal methods, including biological profile estimations and establishing personal identification.

Medicolegal significance

Not all requests received by forensic anthropologists are of medicolegal significance. For instance, forensic anthropologists are routinely asked to determine (1) whether possible evidence is actually skeletal material; (2) whether skeletal remains are human or nonhuman; and (3) whether human remains are recent, historic, prehistoric, or anatomical in origin. Forensic anthropologists have advanced training in human osteology, including the identification of highly fragmented and burned remains. Many practitioners also have advanced knowledge of nonhuman skeletal remains, gained through coursework and training in comparative vertebrate anatomy and zooarchaeology. A well-trained forensic anthropologist usually can readily identify material as skeletal remains and can differentiate human from nonhuman remains.

A more difficult task is determining whether remains are recent versus historic or prehistoric. Context of discovery is useful. For example, remains inadvertently discovered near an old pioneer cemetery or at a known archeological site are unlikely to be of medicolegal significance. Severe tooth wear, cranial morphology, the condition of the remains, and the presence of cultural material (e.g., artifacts) can help to establish that remains are historic or prehistoric. In addition, anatomical skulls or "trophy" skulls acquired from past wars are also unlikely to be of medicolegal significance (Pokines et al., 2017; Yucha et al., 2017). Another method, bomb pulse radiocarbon dating, can determine whether an individual was born prior to 1940 from analysis of bones or teeth, or instead can provide an estimated birth year based on early forming teeth. This method is based on comparing artificial radiocarbon concentrations in a bone or tooth sample to the bomb-curve, which is the global spike in artificial carbon caused by atomic bomb testing prior to the 1963 Nuclear Test Ban Treaty (Buchholz et al., 2018).

Nonmedicolegal cases still require analysis and may involve a written report to document the case findings. It is worth noting that some instances involving nonhuman remains may be related to animal poaching or animal abuse cases, which would be of medicolegal significance.

Similarly, historic or prehistoric remains illegally removed or looted from a known cemetery or archeological site may also constitute a crime, and therefore, fall under the purview of the medicolegal system.

Typically, a forensic anthropologist can easily make a determination of medicolegal significance through standard macroscopic analysis. However, in some cases this can be a challenge and may require the use of more advanced methods. Material that is small, nondiagnostic, or burned may be difficult to identify as skeletal remains. The use of microscopy, radiography, X-ray fluorescence spectrometry, and scanning electron microscope/energy dispersive X-ray spectroscopy are all useful for determining whether material is skeletal in origin. Once nondiagnostic or burned material is identified as skeletal remains, determining human versus nonhuman can be challenging. In the case of nondiagnostic skeletal fragments, the use of histology (Crowder et al., 2018), protein radioimmunoassay (Ubelaker, 2018a), and DNA testing (Edson et al., 2018) are all effective for differentiating human versus nonhuman remains. In addition, burned remains can still be analyzed histologically for determination of human versus nonhuman microstructure (Cattaneo et al., 1999).

Creating a biological profile for unidentified human remains

The ability to accurately and reliably estimate a decedent's biological sex, age-at-death, ancestry, and stature is important in the early stages of an investigation. The biological profile provides law enforcement with critical information for reviewing missing persons reports to identify possible matches to the decedent. Biological profiles that represent a smaller segment of the local demographic (e.g., individuals of very short or tall statures, the very young or very old, underrepresented ancestry groups) may be more useful than more common profiles that include a larger proportion of the missing population. Biological profile information on unidentified human remains can be entered into the National Missing and Unidentified Persons System (NamUs), where it can be searched against missing persons records (Murray et al., 2018). While constructing a biological profile provides a first step toward identification, this information is only presumptive. As discussed later in this chapter, personal identification should involve a scientific process in which antemortem records (e.g., fingerprints, dental records, medical records, DNA) are compared with postmortem data from the unidentified remains.

Sex determination

Forensic anthropology methods used for sex determination are predicated on the fact that humans are a sexually dimorphic species. The amount of dimorphism is relatively small and varies within the skeleton. Sex determination methods are focused on features that tend to show higher levels of dimorphism (e.g., the pelvis) and, to a lesser degree, measurements of the postcranial skeleton and characteristics of the skull. Most sexually dimorphic traits do not emerge until puberty, limiting the use of traditional methods of sex determination for subadult remains. However, recent osteometric sex determination methods on subadults have demonstrated moderate to high correct classification rates (Stull et al., 2017).

When a relatively complete skeleton is available, the greatest weight is placed on nonmetric characteristics of the pelvis (Phenice, 1969; Kenyhercz et al., 2017; Klales et al., 2012). Some features are usually expressed only in females (e.g., ventral arc, preauricular sulcus, dorsal pitting), whereas other features are broader (e.g., sciatic notch, subpubic angle, subpubic concavity) or are more sharply defined (e.g., ischiopubic ramus ridge) in females compared to males. Accuracy rates for the pelvis are greater than 95%. Postcranial osteometrics also provide high correct classification rates for sex, reflecting larger bone dimensions in males compared to females (Spradley and Jantz, 2011). In addition, sexual dimorphism in the skull shows relatively high correct classification rates for sex, although nonmetric skull traits are more difficult to reliably score (Langley et al., 2018; Lewis and Garvin, 2016; Walker, 2008). When skeletal remains are too fragmentary, incomplete, or ambiguous to make a reliable sex determination, molecular techniques can provide accurate results using the amelogenin gene in skeletal remains (Gibbon et al., 2009) or the expression of amelogenin proteins in tooth enamel (Parker et al., 2019). Although molecular methods are destructive, they can be valuable for sex determination when traditional anthropological methods are unsuccessful.

Ancestry estimation

Skeletal methods used for ancestry estimation are based on the relationship between measurements and traits of the skeleton and an individual's genetic heritage. Traditionally, ancestry estimations have focused on skull morphology, as there are clear and measureable ancestry differences in skull shape. By the 18th century, physical anthropologists and anatomists began to study "racial" differences between groups, often associating social and behavioral traits with different skull types (Little and Kennedy, 2010). Genetic studies conducted in the 1970s demonstrated that there is more variation *within* "racial" groups than there is *between* "racial" groups (Lewontin, 1972). This called into question the validity of race as a biological construct, as it fails to meet the taxonomic criteria for subspecies in biology. The dismantling of the racialized approach to human variation began in 1960s, and today, race is recognized primarily as a social construct by anthropologists. Most forensic anthropologists have moved away from a typological approach (e.g., the trait list method) and instead have developed statistical frameworks for assessing biogeographic ancestry, which often focus on how skull and dental morphology relates to a gene pool at a continental level (e.g., Europe, Africa, Asia, etc.) (Pilloud and Hefner, 2016). Current approaches that use craniometric, morphometric, and macromorphoscopic approaches have a strong statistical basis for estimating ancestry. Recent studies have also found a high level of concordance between genetic estimates and cranially based estimates of ancestry, further demonstrating that skull morphology can be used to infer genetic relationships (Carson, 2006; Harvati and Weaver, 2006; Smith, 2011). Several recent studies have also identified meaningful ancestry differences in postcranial remains, usually involving size or shape differences (Liebenberg et al., 2019; Meeusen et al., 2015; Spradley, 2014).

The skull and dentition of an unidentified forensic case can be assigned to a particular ancestry group using robust statistical software packages. Among these software tools include: (1) the craniometric program Fordisc 3.0 (Jantz and Ousley, 2005); (2) the cranial macromorphoscopic software applications hefne*R* (Hefner, 2009; Hefner and Ousley, 2014); (3) the dental morphology program rASUDAS (Scott et al., 2018); and (4) the nonmetric and metric mandible program (hu)MANid (Berg and Kenyhercz, 2017; Kenyhercz and Berg, 2018). These programs result in classification of an unknown skull, mandible, or dentition into the ancestry reference group to which it is most similar. Although based on different methods, these tools provide robust statistical information on the probability of group membership given the reference samples available for comparison. As with any estimation method, there are limitations and biases to metric, macromorphoscopic, and nonmetric approaches. For example, an individual of mixed ancestry will often be more difficult to accurately classify. Further, ancestry labels such as "Hispanic" are vague population descriptors and fail to accurately address the amount of underlying variation within the Spanish-speaking countries of Latin America. Ancestry estimation is complex and requires a nuanced understanding of reference samples, the biosocial diversity of missing persons in the local area, and the statistical bases and assumptions used in the analysis. Importantly, forensic anthropologists need to be able to translate ancestry verbiage into language that law enforcement can understand.

Age-at-death estimation

Estimation of age-at-death is a challenging aspect of the biological profile due to the imperfect relationship between biological age and chronological age. Age estimation of subadults is more accurate and precise, given that morphological changes in the skeleton progress in a more linear fashion than in adults. For subadults, age-at-death can be estimated from dental development, dental eruption, epiphyseal union, and using long bone length. Dental development—and to a lesser degree, dental eruption—is more canalized than bone growth. In other words, the development and eruption of teeth track chronological age more accurately than skeletal growth. There are a number of aging standards for teeth (AlQahtani et al., 2014; Ubelaker, 2018b), bone dimensions, and the sequence of epiphyseal union (Cunningham et al., 2016; Schaefer et al., 2018). Most often, the growth or developmental stage of an unidentified decedent is compared to a reference sample to produce a point estimate with an associated age interval (e.g., 5 years ± 2 years of age). Once a person is in their early twenties, the third molars are usually completely erupted and all epiphyses are completely fused, with the exception of the clavicle (which may fuse as late as 30 years of age). While most methods are developed around the analysis of dry bones, age estimation can be conducted using various radiological imaging techniques, including X-rays, computed tomography (CT), magnetic resonance imaging (MRI), and ultrasound. Further, radiological methods have also featured in age estimation of living individuals, especially in instances of asylum seekers, undocumented immigrants, sex trafficking cases, and other criminal cases involving possible juveniles (Black et al., 2010). Recently, there has been controversy over improper use of methods and reference standards for determining whether detained individuals are minors (i.e., younger than 18 years of age) as well as ethical concerns on exposing individuals to X-ray examinations for nonmedical purposes (Franklin et al., 2015).

For adults, most age-at-death estimation methods focus on signs of degenerative changes in the skeleton. These methods are macroscopic and involve assessing a joint surface (e.g., pubic symphysis, auricular surface of the sacroiliac joint, acetabulum) or the sternal rib ends for morphological and degenerative changes (Hartnett-McCann et al., 2018). Accuracy and precision are highest for younger individuals, but decrease with age. Due to the more erratic pattern of aging among older adults, age estimation intervals are notoriously wide (i.e., 20+ years).

Nutrition, activity levels, health, genetic factors, and use of drugs and alcohol all can influence the aging process. The pubic symphysis is the most common feature used for age estimation in forensic casework, and the symphyseal surface is assessed macroscopically and assigned to a specific phase that has a point estimate and an age interval (Brooks and Suchey, 1990; Hartnett, 2010a). The auricular surface is assessed in a similar fashion, although this method shows a greater degree of interobserver error than the pubic symphysis (Lovejoy et al., 1985; Osborne et al., 2004). The sternal rib ends also have a long history as an age indicator in forensic anthropology (Hartnett, 2010b; İşcan et al., 1984, 1985). The sternal rib ends undergo age-related metamorphosis as costal cartilage ossifies, and these degenerative changes can be compared against phases that have a point estimate and age interval. This method was originally based on the right fourth rib, but subsequent studies have found that it works reasonably well for other ribs that attach to costal cartilage (Hartnett-McCann et al., 2018). More advanced approaches for assessing these age indicators have been developed using three-dimensional (3D) scans and have shown promising results (Slice and Algee-Hewitt, 2015; Stoyanova et al., 2015).

All age estimation methods demonstrate inaccuracy and bias (e.g., underaging or overaging). It is, therefore, important to report age intervals for each method used in the analysis and to use aging standards that are most appropriate for a given case. However, combining various age estimates based on different methods into a single estimate is problematic. Recent age estimation methods, such as transition analysis, have attempted to circumvent this by using Bayesian statistics and by avoiding biases through the use of improper reference samples (Milner and Boldsen, 2012). These methods are still in development, and the most recent iteration involves scoring morphological changes throughout the whole skeleton (Getz, 2017). Transition analysis provides more accurate and precise age estimations for older individuals, which is promising.

Other skeletal aging methods based on degenerative changes are informative, but primarily serve as general indicators of advanced age, such as joint lipping in the vertebral column or in the appendicular skeleton. Obliteration of the cranial sutures also has been used for age estimation, but suture closure is highly variable and therefore unreliable for age estimation. More specialized age estimation methods such as histology focus on counting osteons in bone (Crowder et al., 2018) or cementum annulations in teeth (Naji et al., 2016; Wittwer-Backofen, 2012). These specialized histological methods are destructive and require advanced training in sample preparation, analysis, and interpretation.

Stature estimation

Stature refers to an individual's height during life and is most often estimated from the postcranial skeleton. Stature can be estimated either through the anatomical method or using linear regression. For the anatomical method, all bones that contribute to stature are measured, including the height of the cranium, vertebral body heights of C2 to L5, height of the first sacral segment, physiological length of the femur, tibial length, and articulated talocalcaneal height (Raxter and Ruff, 2018). A correction is also added to account for soft tissue, and equations are available that incorporate the effects of height loss with age. The anatomical method provides accurate stature estimates, but is only suitable for cases where all the necessary bones to measure are present and intact. The other stature method uses linear regression and is based on the relationship between the dimensions of postcranial bones and an individual's height. Skeletal elements that contribute to stature (e.g., femur, tibia) show the strongest correlation with height, although stature estimates can be made from other elements, such as arm bones (Ousley, 2012; Wilson et al., 2010). Regression equations are available for males and females from a number of different populations, reflecting the need for population-specific standards. Stature estimates should be reported with a prediction interval around the point estimate, using the equation with the smallest interval (Ousley, 2012). Fordisc 3.0 includes numerous regression formulae for males and females from different populations and also an "any" option to use when sex and ancestry are unknown. As a person ages, their stature will decline due to degenerative changes affecting the vertebral column. Some researchers have proposed equations to correct for age-related loss in stature (Galloway et al., 1990; Giles, 1991). However, a missing person's reported stature may reflect their maximum stature in life, not their current stature.

Personal identification using comparative radiography

In many U.S. jurisdictions, scientifically based identifications are required, which reduces the likelihood of a misidentification. The decedent identification process can proceed in a number of different directions depending on what information is available. In cases of decomposed, mummified, and burned remains, it is often possible to obtain fingerprints, as well as information on scars or tattoos. Investigators can submit fingerprint records to the Next Generation Identification (NGI) system, which now incorporates the Integrated Automated Fingerprint Identification System (IAFIS), or often can search missing persons records or databases for possible tattoo matches. Personal effects found in association with the remains, such as a driver's license or cellular phone, can provide tentative leads toward identification. However, if no information is available, the biological profile information can be searched against missing persons reports, the National Crime Information Center (NCIC), or NamUs to find possible matches (Murray et al., 2018; Osborn-Gustavson et al., 2018). Once a missing persons list is compiled, investigators can request antemortem medical or dental records on these individuals. Radiographs are especially useful for personal identification because they can be compared side-by-side with postmortem films or using radiographic superimposition methods. It is also routine for medical examiners or investigators to submit tissue samples (e.g., a bone, tooth, or nail sample) for DNA testing and to upload resultant profiles into the Combined DNA Index System (CODIS). These DNA results can be directly compared against family reference DNA samples or preexisting DNA exemplars (e.g., surgical tissue blocks or a toothbrush) belonging to the missing person. If all of these lines of evidence fail to yield results, other advanced methods can be used to provide additional investigative leads. This includes the use of stable isotopes as a gelocation tool for predicting a decedent's place of childhood residence (e.g., teeth), where they resided as an adult (e.g., bones), or their more recent travel history (e.g., hair, nails).

Comparative radiographic methods

The use of comparative radiography for personal identification has a long history dating back to the 1920s (Viner, 2018). Today, radiography is commonly used by forensic odontologists, medical examiners, forensic anthropologists, and forensic radiologists for identifying the dead. In cases where remains are decomposed, burned, fragmented, or skeletal, comparative radiography may be the fastest and least expensive option for identification. Although all applications involve comparing an antemortem radiograph to a postmortem radiograph, the specific approach used varies widely. For example, forensic odontologists conduct a direct comparison of antemortem and postmortem records and make a determination as to whether two sets of radiographs are from the same individual based on a number of individualizing characteristics (Clement, 2017). In their comparison, odontologists examine radiographs and dental records to identify virgin teeth, restored teeth, crowns, bridges, dentures, missing teeth, surgical hardware, disease of the jaws and teeth, as well as anatomical characteristics. An identification is typically made when there are no unexplainable discrepancies between the two sets of images. Although odontological comparisons rarely are quantified in the same way as DNA profiles or fingerprints, the method is believed to be highly accurate. However, methods such as the program Odontosearch allow a dental match to be quantified by comparing the pattern of missing, filled, and unrestored teeth for a given case against a large database of dental patterns (Adams, 2003). Anthropologists take a similar approach as odontologists, although they usually focus on other skeletal regions of the body, such as the head, neck, chest, abdomen, vertebral column, hand, and foot (Streetman and Fenton, 2018). The approach can involve comparison of normal anatomical features as well as pathological conditions, skeletal anomalies, and evidence of surgical intervention (e.g., sternotomy wires, orthopedic devices, etc.).

In recent years, forensic anthropologists have developed approaches for quantifying the strength of a radiographic comparison match, including (1) the frontal sinus (Christensen and Hatch, 2016, 2018); (2) the vertebral column (Derrick et al., 2015, 2018); and (3) the cervicothoracic region (Stephan et al., 2018), among others. Radiographic superimposition methods are also highly accurate tools, especially for comparing records of surgical treatment or healed fractures (Milligan et al., 2018; Stephan et al., 2011). Regardless of whether a strictly quantitative or qualitative approach is used, there are only three outcomes from a radiographic comparison: (1) an identification; (2) an exclusion; or (3) insufficient evidence (i.e., indeterminate) (Streetman and Fenton, 2018).

Stable isotope analysis as an investigative tool

The methods discussed previously in this chapter represent a diverse set of tools used by forensic anthropologists to (1) determine whether remains are of medicolegal significance, (2) to narrow down possible missing persons matches, or (3) to use comparative radiographic methods for personal identification. Despite the success of these approaches, each year in the United States there are about 4400 unidentified decedents, approximately 1000 of whom still remain unidentified 1 year later (www.namus.gov). For these cold cases, stable isotope analysis can provide new investigative leads.

Stable isotopes are atoms of the same element with the same number of protons and electrons, but a different number of neutrons in the atom's nucleus. Stable isotopes of the same element (e.g., ^{18}O and ^{16}O) have different atomic weights, and thus, travel at different rates in chemical reactions. For incomplete reactions such as photosynthesis, isotopic fractionation results in the differential incorporation of the heavy and light isotopes relative to the original substrate. Stable isotopes are incorporated into the body primarily through the consumption of food and beverages and do not undergo radioactive decay (Meier-Augenstein, 2018). Teeth, bones, hair, or nails are most commonly sampled for isotope analysis (Chesson et al., 2018a,b). Each tissue provides a different snapshot of an individual's life history. For example, isotope ratios in teeth provide a record of childhood diet and place of residence. Essentially, enamel locks in the isotopic signatures from food and beverage consumed during the period of tooth formation and does not change over time. In comparison, bone isotopes provide a bulk average of the last 10–20 years of diet and residence location, because bone is a continuously remodeling tissue. Finally, hair and nail keratin provide a recent snapshot of time, reflecting diet and travel history over the weeks and months prior to death (Meier-Augenstein, 2018).

Only a handful of elements are routinely used in provenancing (i.e., sourcing) studies. These include the bio-elements hydrogen (H), carbon (C), nitrogen (N), oxygen (O), and sulfur (S), and the geo-elements strontium (Sr) and lead (Pb) (Chesson et al., 2018b). Each isotope system provides different types of information. For example, C, N, and S isotopes provide information on dietary inputs, which may be geographically patterned due to ethnic and cultural differences in diet (Bartelink et al., 2014, 2018; Bartelink and Chesson, 2019; Hülsemann et al., 2015). Isotopes such as H and O track global meteoric precipitation patterns and reflect the source of drinking water (Bowen et al., 2007; Cerling et al., 2016). Sr isotopes originate from rocks and minerals and are transferred from soils into plants and then into animals (Bataille and Bowen, 2012). Pb is incorporated into the skeleton due to direct exposure from inhaling and ingesting small amounts of soil and dust that contain lead (Keller et al., 2016). Both Sr and Pb can substitute for calcium in the skeleton, and thus, serve as geographic tracers.

Stable isotope ratios are measured using isotope-ratio mass spectrometry and are compared against international standards of known isotope composition. Baseline oxygen, hydrogen, and strontium data are available for the continental United States and provide reference data from which to assign an unknown individual to a region-of-origin. The baseline data can be displayed graphically using geographic information system (GIS) maps, in which different color patterns reflect ranges of isotope values (Chesson et al., 2018a). These maps show isotopic landscapes and are referred to as *isoscapes*. For a human tissue sample (e.g., teeth), isotope values can be converted to isotope values for the source water and/or geology, and then a prediction region can be highlighted on a map. Using O and Sr together provides a more robust prediction tool than only using one isotope system. A brief case study is provided below to highlight how stable isotope data can provide additional investigative information beyond the biological profile.

Example case study

In 2006, human remains were accidentally discovered in the central Nevada desert along a defunct freeway off-ramp. The remains were buried in a shallow grave that had eroded, exposing the skull. The remains were transferred to the Human Identification Laboratory at California State University, Chico, to conduct a skeletal analysis. The remains consisted of a complete skeleton of an adult male, between 20 and 25 years of age, and with a stature of 5 ft, 6 in. ± 2 in. A craniometric assessment of the skull using Fordisc 3.0 indicated that the decedent was most likely Hispanic. The dentition consisted of all virgin teeth (i.e., no dental work), and the skeleton showed no evidence of antemortem skeletal conditions or surgeries. Despite an extensive search of missing persons records and submission of DNA samples to CODIS, the decedent still could not be identified. As all other identification leads were exhausted, the author requested permission to sample a bone and a first molar to predict possible regions of origin of the decedent using stable isotope analysis.

Investigators wanted to know whether isotopic analysis could determine whether the decedent was local or nonlocal to the area where his remains were recovered, or even if he could possibly be from Latin America. From the bone sample, collagen was extracted for C and N isotopes. Comparisons of these data with published reference samples (Chesson et al., 2018b) indicated that the decedent's diet was similar to other U.S. Americans and not to Latin American populations (Bartelink et al., 2018). Isotope data from the molar tooth was more useful. The *isoscape* prediction map generated from the oxygen and strontium isotope data show possible areas within the United States where the decedent may have acquired his drinking water and food (Fig. 1). The areas highlighted in red on the map represent locations

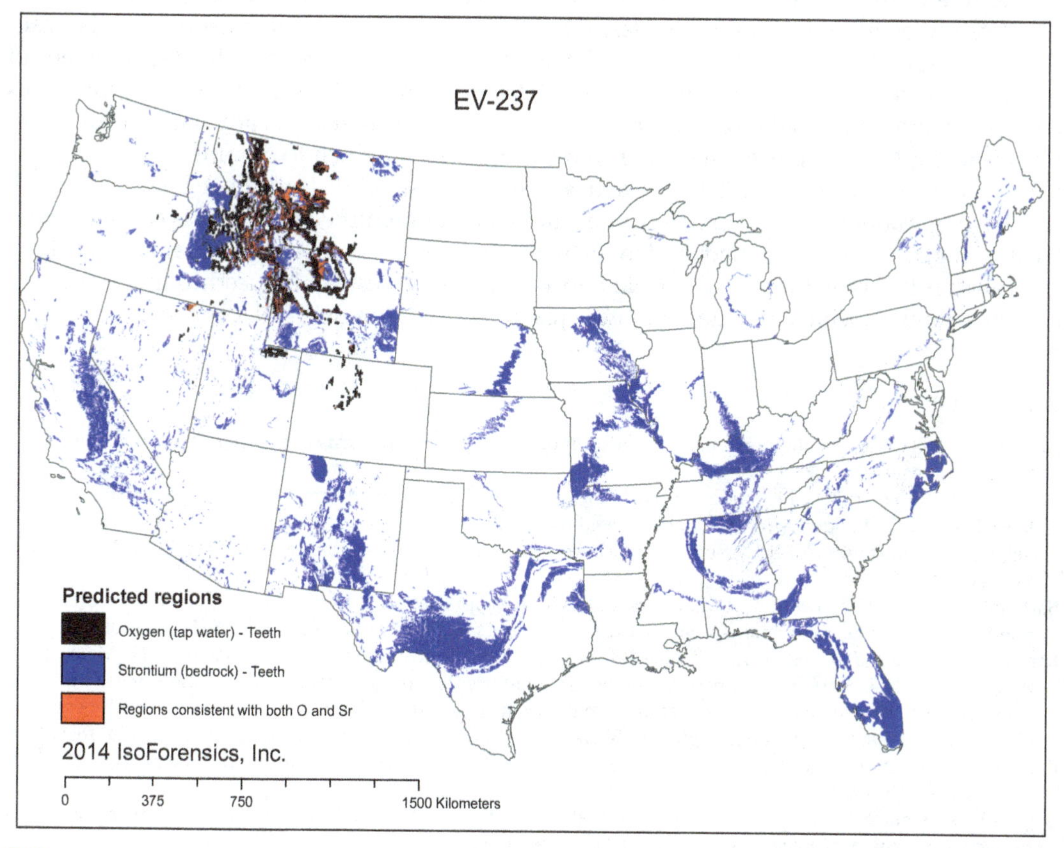

FIG. 1 Region-of-origin prediction map using the oxygen and strontium isotopic composition of a first molar tooth. [Water base layer data from Bowen et al. (2007) and geological strontium baseline layer from Bataille and Bowen (2012)]. *Areas highlighted in red* reflect locations that show overlap in both oxygen (O) and strontium (Sr) values and indicate where the decedent may have obtained his drinking water (based on O isotope ratios) and food (based on measured Sr isotope ratios). These possible regions include areas within the continental United States, including northern Nevada, Idaho, Montana, and Wyoming. *Map created by IsoForensics, Inc.*

where both O and Sr isotope values overlap, indicating the most likely possibilities for the childhood region-of-origin. This includes the northernmost region of Nevada, as well as areas within Idaho, Montana, and Wyoming. This information provided additional leads for investigators to pursue, including searching missing persons reports in these areas. Although stable isotope analysis cannot be used to identify an individual or even a specific location, it can help to narrow down a region of origin where a decedent was originally from or had lived previously. When used in conjunction with other identification methods, stable isotope analysis is an important tool for narrowing down search parameters for missing persons investigations.

Conclusion

The role of forensic anthropology in the identification of human remains has expanded significantly over the past four decades. Practitioners now have a wide array of tools available that allow detailed studies of even the smallest piece of evidence. The development of modern reference skeletal collections, large databases, and advanced statistical approaches has also greatly improved biological profile estimations. Furthermore, advances in comparative radiography have created a statistical framework from which to quantify the strength of a personal identification. For cold cases in which all investigative leads have been exhausted, stable isotope analysis can help determine whether an unidentified decedent is a local or non-local to the area where they died, and to further narrow down a more specific region-of-origin. Information from the biological profile and isotopic analyses can be used to guide missing persons searches, which may be ultimately resolved using forensic DNA analysis.

References

Adams, B.J., 2003. Establishing personal identification based on specific patterns of missing, filled, and unrestored teeth. J. Forensic Sci. 48 (3), 487–496.

AlQahtani, S.J., Hector, M.P., Liversidge, H.M., 2014. Accuracy of dental age estimation charts: Schour and Massler, Ubelaker, and the London Atlas. Am. J. Phys. Anthropol. 154 (1), 70–78.

Bartelink, E.J., Chesson, L.A., 2019. Recent applications of isotope analysis to forensic anthropology. Forensic Sci. Res. 4 (1), 29–44.

Bartelink, E.J., Berg, G.E., Beasley, M.M., Chesson, L.A., 2014. Application of stable isotope forensics for predicting region of origin of human remains from past wars and conflicts. Ann. Anthropol. Pract. 38 (1), 124–136.

Bartelink, E.J., Berg, G.E., Chesson, L.A., Tipple, B.J., Beasley, M.M., Prince-Buitenhuys, J.R., MacInnes, H., MacKinnon, A.T., Latham, K.E., 2018. Applications of stable isotope forensics for geolocating unidentified human remains from past conflict situations and large-scale humanitarian efforts. In: Latham, K.E., Bartelink, E.J., Finnegan, M. (Eds.), New Perspectives in Forensic Human Skeletal Identification. Academic Press, London, pp. 175–184.

Bataille, C.P., Bowen, G.J., 2012. Mapping $^{87}Sr/^{86}Sr$ variations in bedrock and water for large scale provenance studies. Chem. Geol. 304, 39–52.

Berg, G.E., Kenyhercz, M.W., 2017. Introducing human mandible identification [(hu) MANid]: a free, web-based GUI to classify human mandibles. J. Forensic Sci. 62 (6), 1592–1598.

Black, S., Aggrawal, A., Payne-James, J., 2010. Age Estimation in the Living: The practitioner's Guide. John Wiley & Sons, New York.

Bowen, G.J., Ehleringer, J.R., Chesson, L.A., Stange, E., Cerling, T.E., 2007. Stable isotope ratios of tap water in the contiguous United States. Water Resour. Res. 43 (3).

Brooks, S., Suchey, J.M., 1990. Skeletal age determination based on the os pubis: a comparison of the Acsádi-Nemeskéri and Suchey-Brooks methods. Hum. Evol. 5 (3), 227–238.

Buchholz, B.A., Alkass, K., Druid, H., Spalding, K.L., 2018. Bomb pulse radiocarbon dating of skeletal tissues. In: Latham, K.E., Bartelink, E.J., Finnegan, M. (Eds.), New Perspectives in Forensic Human Skeletal Identification. Academic Press, London, pp. 185–196.

Carson, E.A., 2006. Maximum likelihood estimation of human craniometric heritabilities. Am. J. Phys. Anthropol. 131, 169–180.

Cattaneo, C., DiMartino, S., Scali, S., Craig, O.E., Grandi, M., Sokol, R., 1999. Determining the human origin of fragments of burnt bone: a comparative study of histological, immunological, and DNA techniques. Forensic Sci. Int. 102 (2–3), 181–191.

Cerling, T.E., Barnette, J.E., Bowen, G.J., Chesson, L.A., Ehleringer, J.R., Remien, C.H., Shea, P., Tipple, B.J., West, J.B., 2016. Forensic stable isotope biogeochemistry. Annu. Rev. Earth Planet. Sci. 44, 175–206.

Chesson, L.A., Tipple, B.J., Youmans, L.V., O'Brien, M.A., Harmon, M.M., 2018a. Forensic identification of human skeletal remains using isotopes: a brief history of applications from archaeological dig sites to modern crime scenes. In: Latham, K.E., Bartelink, E.J., Finnegan, M. (Eds.), New Perspectives in Forensic Human Skeletal Identification. Academic Press, London, pp. 157–173.

Chesson, L.A., Tipple, B.J., Ehleringer, J.R., Park, T., Bartelink, E.J., 2018b. Forensic applications of isotope landscapes ("isoscapes"): a tool for predicting region-of-origin in forensic anthropology cases. In: Boyd, C.C., Boyd, D.C. (Eds.), Forensic Anthropology: Theoretical Framework and Scientific Basis. Wiley & Sons, Hoboken, pp. 127–148.

Christensen, A.M., Hatch, G.M., 2016. Quantification of radiologic identification (RADid) and the development of a population frequency data repository. J. Forensic Radiol. Imaging 1 (7), 14–16.

Christensen, A.M., Hatch, G.M., 2018. Advances in the use of frontal sinuses for human identification. In: Latham, K.E., Bartelink, E.J., Finnegan, M. (Eds.), New Perspectives in Forensic Human Skeletal Identification. Academic Press, London, pp. 227–240.

Clement, J.G., 2017. Odontology. In: Houck, M.M. (Ed.), Forensic Anthropology. Advanced Forensic Science Series, Elsevier, Amsterdam, pp. 287–295.

Crowder, C.M., Andronowski, J.M., Dominguez, V.M., 2018. Bone histology as an integrated tool in the process of human identification. In: Latham, K.E., Bartelink, E.J., Finnegan, M. (Eds.), New Perspectives in Forensic Human Skeletal Identification. Academic Press, London, pp. 201–213.

Cunningham, C., Scheuer, L., Black, S., 2016. Developmental Juvenile Osteology. Academic Press, Amsterdam.

Derrick, S.M., Raxter, M.H., Hipp, J.A., Goel, P., Chan, E.F., Love, J.C., Wiersema, J.M., Akella, N.S., 2015. Development of a computer-assisted forensic radiographic identification method using the lateral cervical and lumbar spine. J. Forensic Sci. 60 (1), 5–12.

Derrick, S.M., Hipp, J.A., Goel, P., 2018. The computer-assisted decedent identification method of computer-assisted radiographic identification. In: Latham, K.E., Bartelink, E.J., Finnegan, M. (Eds.), New Perspectives in Forensic Human Skeletal Identification. Academic Press, London, pp. 265–276.

Dirkmaat, D.C., Cabo, L.L., Ousley, S.D., Symes, S.A., 2008. New perspectives in forensic anthropology. Am. J. Phys. Anthropol. 137 (47), 33–52.

Edson, S.M., Root, K.A., Kahline, I.L., Dunn, C.A., Trotter, B.E., O'Rourke, J.A., 2018. Flexibility in testing skeletonized remains for DNA analysis can lead to increased success: suggestions and case studies. In: Latham, K.E., Bartelink, E.J., Finnegan, M. (Eds.), New Perspectives in Forensic Human Skeletal Identification. Academic Press, London, pp. 141–156.

Franklin, D., Flavel, A., Noble, J., Swift, L., Karkhanis, S., 2015. Forensic age estimation in living individuals: methodological considerations in the context of medico-legal practice. Res. Rep. Forensic Med. Sci. 5, 53–66.

Galloway, A., Stini, W.A., Fox, S.C., Stein, P., 1990. Stature loss among an older United States population and its relation to bone mineral status. Am. J. Phys. Anthropol. 83 (4), 467–476.

Getz, S.M., 2017. Improved Skeletal Age-at-Death Estimation and its Impact on Archaeological Analyses (Doctoral dissertation). Department of Anthropology, Pennsylvania State University.

Gibbon, V., Paximadis, M., Štrkalj, G., Ruff, P., Penny, C., 2009. Novel methods of molecular sex identification from skeletal tissue using the amelogenin gene. Forensic Sci. Int. Genet. 3 (2), 74–79.

Giles, E., 1991. Corrections for age in estimating older adults' stature from long bones. J. Forensic Sci. 36 (3), 898–901.

Hartnett, K.M., 2010a. Analysis of age-at-death estimation using data from a new, modern autopsy sample—Part I: pubic bone. J. Forensic Sci. 55 (5), 1145–1151.

Hartnett, K.M., 2010b. Analysis of age-at-death estimation using data from a new, modern autopsy sample—Part II: sternal end of the fourth rib. J. Forensic Sci. 55 (5), 1152–1156.

Hartnett-McCann, K., Fulginiti, L.C., Seidel, A.C., 2018. Adult age-at-death estimation in unknown decedents: new perspectives on an old problem. In: Latham, K.E., Bartelink, E.J., Finnegan, M. (Eds.), New Perspectives in Forensic Human Skeletal Identification. Academic Press, London, pp. 65–85.

Harvati, K., Weaver, T.D., 2006. Human cranial anatomy and the differential preservation of population history and climate signatures. Anat. Rec. A Discov. Mol. Cell. Evol. Bio. 288 (12), 1225–1233.

Hefner, J.T., 2009. Cranial nonmetric variation and estimating ancestry. J. Forensic Sci. 54, 985–995.

Hefner, J.T., Ousley, S.D., 2014. Statistical classification methods for estimating ancestry using morphoscopic traits. J. Forensic Sci. 59, 883–890.

Hülsemann, F., Lehn, C., Schneiders, S., Jackson, G., Hill, S., Rossmann, A., Scheid, N., Dunn, P.J., Flenker, U., Schänzer, W., 2015. Global spatial distributions of nitrogen and carbon stable isotope ratios of modern human hair. Rapid Commun. Mass Spectrom. 29 (22), 2111–2121.

İşcan, M.Y., Loth, S.R., Wright, R.K., 1984. Age estimation from the rib by phase analysis: white males. J. Forensic Sci. 29 (4), 1094–1104.

İşcan, M.Y., Loth, S.R., Wright, R.K., 1985. Age estimation from the rib by phase analysis: white females. J. Forensic Sci. 30 (3), 853–863.

Jantz, R.L., Ousley, S.D., 2005. FORDISC 3: Computerized Forensic Discriminant Functions (Version 3)., p. 292.

Keller, A.T., Regan, L.A., Lundstrom, C.C., Bower, N.W., 2016. Evaluation of the efficacy of spatiotemporal Pb isoscapes for provenancing of human remains. Forensic Sci. Int. 261, 83–92.

Kenyhercz, M.W., Berg, G.E., 2018. Evaluating mixture discriminant analysis to classify human mandibles with (hu) MANid, a free, R-based GUI. In: Latham, K.E., Bartelink, E.J., Finnegan, M. (Eds.), New Perspectives in Forensic Human Skeletal Identification. Academic Press, London, pp. 35–43.

Kenyhercz, M.W., Klales, A.R., Stull, K.E., McCormick, K.A., Cole, S.J., 2017. Worldwide population variation in pelvic sexual dimorphism: a validation and recalibration of the Klales et al. method. Forensic Sci. Int. 277 (259), e1.

Klales, A.R., Ousley, S.D., Vollner, J.M., 2012. A revised method of sexing the human innominate using Phenice's nonmetric traits and statistical methods. Am. J. Phys. Anthropol. 149 (1), 104–114.

Krogman, W.M., 1939. A guide to the identification of human skeletal material. FBI Law Enforc. Bull. 8, 3–31.

Krogman, W.M., 1962. The Human Skeleton in Forensic Medicine. Charles C. Thomas, Springfield, MA.

Langley, N.R., Dudzik, B., Cloutier, A., 2018. A decision tree for nonmetric sex assessment from the skull. J. Forensic Sci. 63 (1), 31–37.

Lewis, C.J., Garvin, H.M., 2016. Reliability of the Walker cranial nonmetric method and implications for sex estimation. J. Forensic Sci. 61 (3), 743–751.

Lewontin, R.C., 1972. The apportionment of human diversity. Evol. Biol. 6, 381–386.

Liebenberg, L., Krüger, G.C., L'Abbé, E.N., Stull, K.E., 2019. Postcraniometric sex and ancestry estimation in South Africa: a validation study. Int. J. Legal Med. 133 (1), 289–296.

Little, M.A., Kennedy, K.A., 2010. Introduction to the history of American physical anthropology. In: Little, M.A., Kennedy, K.A. (Eds.), Histories of American Physical Anthropology in the Twentieth Century. Lexington, Lanham, MD, pp. 1–24.

Lovejoy, C.O., Meindl, R.S., Pryzbeck, T.R., Mensforth, R.P., 1985. Chronological metamorphosis of the auricular surface of the ilium: a new method for the determination of adult skeletal age at death. Am. J. Phys. Anthropol. 68 (1), 15–28.

Meeusen, R.A., Christensen, A.M., Hefner, J.T., 2015. The use of femoral neck axis length to estimate sex and ancestry. J. Forensic Sci. 60 (5), 1300–1304.

Meier-Augenstein, W., 2018. Stable Isotope Forensics: Methods and Forensic Applications of Stable Isotope Analysis. Wiley & Sons, Hoboken, NJ.

Milligan, C.F., Finlayson, J.E., Cheverko, C.M., Zarenko, K.M., 2018. Advances in the use of craniofacial superimposition for human identification. In: Latham, K.E., Bartelink, E.J., Finnegan, M. (Eds.), New Perspectives in Forensic Human Skeletal Identification. Academic Press, London, pp. 241–250.

Milner, G.R., Boldsen, J.L., 2012. Transition analysis: a validation study with known-age modern American skeletons. Am. J. Phys. Anthropol. 148 (1), 98–110.

Murray, E.A., Anderson, B.E., Clark, S.C., Hanzlick, R.L., 2018. The history and use of the national missing and unidentified persons system (NamUs) in the identification of unknown persons. In: Latham, K.E., Bartelink, E.J., Finnegan, M. (Eds.), New Perspectives in Forensic Human Skeletal Identification. Academic Press, London, pp. 115–126.

Naji, S., Colard, T., Blondiaux, J., Bertrand, B., d'Incau, E., Bocquet-Appel, J.P., 2016. Cementochronology, to cut or not to cut? Int. J. Paleopathol. 15, 113–119.

National Research Council, 2009. Strengthening Forensic Science in the United States: A Path Forward. The National Academies Press, Washington, DC.

Osborne, D.L., Simmons, T.L., Nawrocki, S.P., 2004. Reconsidering the auricular surface as an indicator of age at death. J. Forensic Sci. 49 (5), 905–911.

Osborn-Gustavson, A.E., McMahon, T., Josserand, M., Spamer, B.J., 2018. The utilization of databases for the identification of human remains. In: Latham, K.E., Bartelink, E.J., Finnegan, M. (Eds.), New Perspectives in Forensic Human Skeletal Identification. Academic Press, London, pp. 129–139.

Ousley, S.D., 2012. Estimating stature. In: Dirkmaat, D.D. (Ed.), A Companion to Forensic Anthropology. Wiley-Blackwell, Oxford, pp. 330–334.

Parker, G.J., Yip, J.M., Eerkens, J.W., Salemi, M., Durbin-Johnson, B., Kiesow, C., et al., 2019. Sex estimation using sexually dimorphic amelogenin protein fragments in human enamel. J. Archaeol. Sci. 101, 169–180.

Phenice, T.W., 1969. A newly developed visual method of sexing the os pubis. Am. J. Phys. Anthropol. 30 (2), 297–301.

Pilloud, M.A., Hefner, J.T., 2016. Biological Distance Analysis: Forensic and Bioarchaeological Perspectives. Academic Press, San Diego, CA.

Pokines, J.T., Appel, N., Pollock, C., Eck, C.J., Maki, A.G., Joseph, A.S., et al., 2017. Anatomical taphonomy at the source: alterations to a sample of 84 teaching skulls at a medical school. J. Forensic Identif. 67 (4), 600–632.

Raxter, M.H., Ruff, C.B., 2018. Full skeleton stature estimation. In: Latham, K.E., Bartelink, E.J., Finnegan, M. (Eds.), New Perspectives in Forensic Human Skeletal Identification. Academic Press, London, pp. 105–113.

Schaefer, M., Geske, N., Cunningham, C., 2018. A decade of development in juvenile aging. In: Latham, K.E., Bartelink, E.J., Finnegan, M. (Eds.), New Perspectives in Forensic Human Skeletal Identification. Academic Press, London, pp. 45–60.

Scott, G.R., Pilloud, M.A., Navega, D., d'Oliveira, J., Cunha, E., Irish, J.D., 2018. rASUDAS: a new web-based application for estimating ancestry from tooth morphology. Forensic Anthropol. 1 (1), 18–31.

Slice, D.E., Algee-Hewitt, B.F., 2015. Modeling bone surface morphology: a fully quantitative method for age-at-death estimation using the pubic symphysis. J. Forensic Sci. 60 (4), 835–843.

Smith, H.F., 2011. The role of genetic drift in shaping modern human cranial evolution: a test using microevolutionary modeling. Int. J. Evol. Biol. 2011, 145262.

Spradley, M.K., 2014. Metric ancestry estimation from the postcranial skeleton. In: Berg, G.E., Ta'ala, S.C. (Eds.), Biological Affinity in Forensic Identification of Human Skeletal Remains: Beyond Black and White. CRC Press, Boca Raton, FL, pp. 83–94.

Spradley, M.K., Jantz, R.L., 2011. Sex estimation in forensic anthropology: skull versus postcranial elements. J. Forensic Sci. 56 (2), 289–296.

Stephan, C.N., Winburn, A.P., Christensen, A.F., Tyrrell, A.J., 2011. Skeletal identification by radiographic comparison: blind tests of a morphoscopic method using antemortem chest radiographs. J. Forensic Sci. 56 (2), 320–332.

Stephan, C.N., D'Alonzo, S.S., Wilson, E.K., Guyomarc'h, P., Berg, G.E., Byrd, J.E., 2018. Skeletal identification by radiographic comparison of the cervicothoracic region on chest radiographs. In: Latham, K.E., Bartelink, E.J., Finnegan, M. (Eds.), New Perspectives in Forensic Human Skeletal Identification. Academic Press, London, pp. 277–292.

Stewart, T.D., 1979. Essentials of Forensic Anthropology, Especially as Developed in the United States. Charles C. Thomas, Springfield, MA.

Stoyanova, D., Algee-Hewitt, B.F., Slice, D.E., 2015. An enhanced computational method for age-at-death estimation based on the pubic symphysis using 3D laser scans and thin plate splines. Am. J. Phys. Anthropol. 158 (3), 431–440.

Streetman, E., Fenton, T.W., 2018. Comparative medical radiography: practice and validation. In: Latham, K.E., Bartelink, E.J., Finnegan, M. (Eds.), New Perspectives in Forensic Human Skeletal Identification. Academic Press, London, pp. 251–264.

Stull, K.E., L'Abbé, E.N., Ousley, S.D., 2017. Subadult sex estimation from diaphyseal dimensions. Am. J. Phys. Anthropol. 163 (1), 64–74.

Ubelaker, D.H., 2018a. Species determination from fragmentary evidence. In: Latham, K.E., Bartelink, E.J., Finnegan, M. (Eds.), New Perspectives in Forensic Human Skeletal Identification. Academic Press, London, pp. 197–200.

Ubelaker, D.H., 2018b. Estimation of immature age from the dentition. In: Latham, K.E., Bartelink, E.J., Finnegan, M. (Eds.), New Perspectives in Forensic Human Skeletal Identification. Academic Press, London, pp. 61–64.

Viner, M., 2018. Overview of advances in forensic radiological methods of identification. In: Latham, K.E., Bartelink, E.J., Finnegan, M. (Eds.), New Perspectives in Forensic Human Skeletal Identification. Academic Press, London, pp. 217–226.

Walker, P.L., 2008. Sexing skulls using discriminant function analysis of visually assessed traits. Am. J. Phys. Anthropol. 136 (1), 39–50.

Wilson, R.J., Herrmann, N.P., Jantz, L.M., 2010. Evaluation of stature estimation from the database for forensic anthropology. J. Forensic Sci. 55 (3), 684–689.

Wittwer-Backofen, U., 2012. Age estimation using tooth cementum annulation. In: Bell, L.S. (Ed.), Forensic Microscopy for Skeletal Tissues. Humana Press, Totowa, NJ, pp. 129–143.

Yucha, J.M., Pokines, J.T., Bartelink, E.J., 2017. A comparative taphonomic analysis of 24 trophy skulls from modern forensic cases. J. Forensic Sci. 62 (5), 1266–1278.

Further reading

American Board of Forensic Anthropology, 2019. http://theabfa.org.

Nardoto, G.B., Silva, S., Kendall, C., Ehleringer, J.R., Chesson, L.A., Ferraz, E.S., Moreira, M.Z., Ometto, J.P., Martinelli, L.A., 2006. Geographical patterns of human diet derived from stable-isotope analysis of fingernails. Am. J. Phys. Anthropol. 131 (1), 137–146.

National Missing and Unidentified Persons System (NamUs). http://www.namus.gov.

CHAPTER

22

Generation of a personal chemical profile from skeletonized human remains

Suni M. Edson PhD

Assistant Technical Leader, Past Accounting Section, Armed Forces DNA Identification Laboratory (AFDIL), Armed Forces Medical Examiner System, Dover, DE, United States

Introduction

Skeletonized remains are persistently problematic in the identification of persons involved in mass fatality events. With the loss of tissue integrity or events in which there is a high degree of fragmentation, human remains can often be both commingled and skeletonized, further complicating identification efforts. Human identification (HID) is typically comprised of a combination of skills and disciplines, including fingerprinting, pathology, anthropology, archeology, family descriptions, toxicology, and DNA (to name just a few). With the loss of tissue and an increase in postmortem interval (PMI), those modalities that rely on soft tissue or scene investigation are lost, and investigators are left with a decreased spectrum of techniques to apply to the case.

DNA testing of dried skeletal specimens presents a particular challenge to researchers in that the skeletal matrix is both dense and comprised of minerals that may inhibit downstream DNA analysis. While DNA testing strategies have been optimized and are frequently successful in providing either autosomal short tandem repeat (auSTR), Y-chromosomal STR (Y-STR), or mitochondrial DNA (mtDNA) profiles, deconvolution of a highly commingled event requires testing of many (if not all) of the skeletal materials present. This is time-consuming and costly, often to the point of being prohibitive. Additional testing modalities for the separation of human remains that are inexpensive, fast, and effective can be useful.

This chapter describes the potential use of mass spectrometry to aid in identification efforts. With the selection of an optimal solvent for the suspension of any present materials, gas chromatography coupled with mass spectrometry (GC/MS) has shown itself to be useful for detection of materials present in skeletal remains, which allows for the creation of a personal

https://doi.org/10.1016/B978-0-12-815766-4.00024-8

chemical profile (Edson, 2017; Edson and Roberts, 2019). The chemical profile itself will not provide an identification; however, when paired with a DNA profile, it becomes unnecessary to DNA test every skeletal sample. Rather, GC/MS can be performed and the chemical profiles used to sort remains into discrete individuals (Edson and McMahon, 2019).

Protocol theory and development

The following section provides detail on the theory of the use of mass spectrometry for analysis of human remains. Readers should be able to use information provided in this section to develop their own protocol for use or could follow the protocol presented in Section "Protocol."

Mass spectrometry

Mass spectrometry (MS; also referred to as mass spec) is widely used in a variety of scientific applications. It is commonly used in forensics to evaluate trace materials in fires and explosions (e.g., Maurer et al., 2010; Dhabbah et al., 2014), in toxicology (e.g., Skender et al., 2002; Strano-Ross et al., 2010), and to determine ink composition (e.g., Yao et al., 2009; Koenig et al., 2015), to name just a few. MS is also widely used outside of the forensic science field for food safety evaluation (e.g., Gilbert-López et al., 2010; Jaffrès et al., 2011), pharmaceutical research (e.g., Meyer et al., 2013; Van den Broek et al., 2015), and health studies (e.g., Farré et al., 2007; Manning et al., 2015). Given the overall sensitivity of MS to detect and characterize intact proteins in biological fluids (Huang et al., 2006), it is somewhat peculiar that it has not previously been adapted for use in human identification.

Mass spectrometry is a fairly simple process that relies on vaporization of a substrate into a carrier gas. Once vaporized, the sample is injected into a column contained within a vacuum and subjected to an electrical field. The electrical field operates on ions of the analyte in much the same way as DNA separation is performed via electrophoresis, i.e., positively charged ions are lighter and will be detected first, while negatively charged ions are heavier and will come off the column last. Only charged particles (ions) are detected; therefore, other materials in the sample (e.g., bacteria) cannot be "seen" (or recorded) by the instrument.

As ions of the analyte move through the column and are separated, they pass a detection window and the pattern of ions inherent in the materials are captured as peaks. Fig. 1 shows a fairly simple pattern of peaks, often referred to as a *mass spectrograph* (or a "trace"). Each peak corresponds to a different compound within the sample and is determined by the mass-to-charge (m/z) ratio. Individual peaks can be further broken down into a spectrum pattern, from which the identity of the compound can be determined.

There are two main forms of mass spec: GC/MS (gas chromatography/mass spectrometry) and LC/MS (liquid chromatography/mass spectrometry). These two formats differ in how materials to be tested are volatilized and require different equipment. The same sample can be tested on both platforms with differing results. When developing a mass spectrometry protocol for regular use, the individual laboratory will need to decide which platform will provide the greatest amount of data.

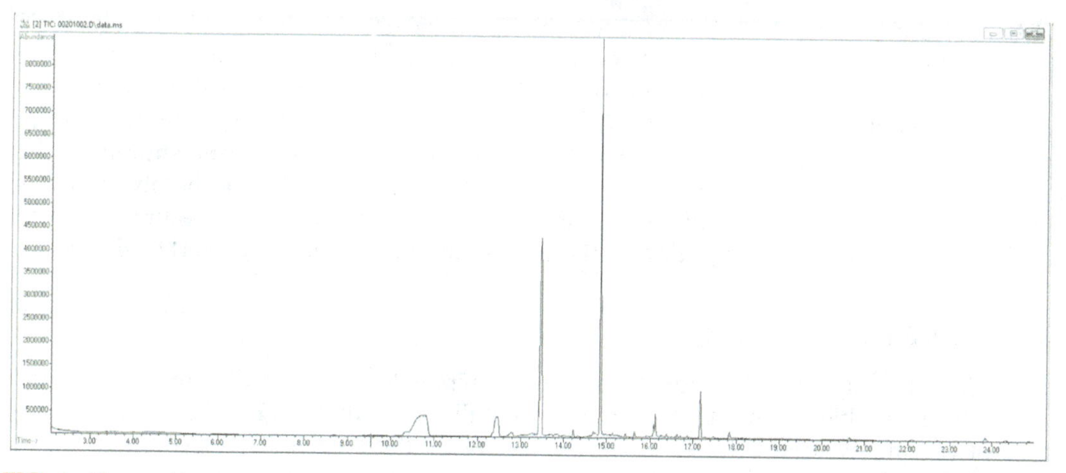

FIG. 1 The resulting gas chromatography/mass spectrometry (GC/MS) trace of a powdered tooth sample treated with dichloromethane.

Certain analytes may not be detected using GC/MS or LC/MS, and therefore, it may be optimal to use both types of equipment. Preliminary work done on a GC/MS (Agilent 7890B, Agilent, Santa Clara, California USA) versus an LC/MS–MS (Thermo Fisher Q-Exactive Plus, Thermo Fisher, Walther, Massachusetts USA) showed varying levels of detection of materials within the same sample (data not shown). Running the machinery in positive or negative mode may also change the types of materials detected. For example, some explosives can be detected in both positive and negative mode (Ifa et al., 2009), but others are not.

The analytes to be detected may not respond well to the differing forms of ionization and the preparation required. This is not necessarily a problem for targeted analysis, such as that which is used in pharmacology, toxicology, or even food safety. When the product to be detected is known, the settings on the equipment can be easily set based on the composition of the target. However, for human identification, it is recommended to use a broad-based screening approach in which all materials present in the remains are detected.

Preparation of materials

Given the complex matrix of osseous materials, the preparation and concentration of compounds within skeletal elements need to be considered by the laboratory. During initial development of the protocol presented in Section "Protocol," it was anticipated that preparation would be somewhat difficult. Trace analysis of accelerants and fuels generally requires a specialized setup involving the ability to concentrate the materials to be detected in an absorptive substrate, which is itself then exposed to the ionization source, rather than the source material. Rather than seek out a specialized technique for detection of materials, a standard chemical protocol for extracting and concentrating the materials within samples can be used, as this is the simplest approach.

During preparation of dried skeletal remains for DNA extraction, many laboratories remove the exterior of the bone in an attempt to reduce contamination by exogenous materials

(Edson and Christensen, 2012; Edson, 2019a, b). The osseous "powder" can be removed via sanding with a Dremel® tool (Dremel, Racine, Washington USA) or other abrasive substrate (e.g., sandpaper). This bone or tooth powder is typically discarded as medical (biohazardous) waste; however, it could be retained for GC/MS testing, thus allowing for additional testing without further destruction of the "parent" sample. Either a fraction or the entirety of the powder may be exposed to selected solvents. The powder itself will not dissolve into the solvent; rather, the solvent will dissolve chemicals or biological materials present that can be further concentrated prior to injection on the chosen instrumentation (e.g., Pert et al., 2006)

Solvent selection

Selection of a solvent is dependent upon the materials to be detected. Some compounds may not be detectable because they cannot be solubilized in what might otherwise be a preferred or commonly used solvent. Therefore, use of multiple solvents is recommended in order to capture as much information as possible about an unknown sample.

Methanol is a commonly used solvent for mass spectrometry; however, cholesterol cannot be readily ionized using this solvent as well as any material retaining water (Ifa et al., 2009). This reaction was observed during initial work with U.S.S. *Oklahoma* skeletal samples, which became irreversibly cloudy upon addition of methanol and water to the powdered materials in a "dilute and shoot" protocol recommended by a Thermo Fisher demo chemist. A "dilute and shoot" protocol involves addition of water directly to the osseous material, and then subsequently diluting it with methanol for injection on the instrument (R. Doyle, personal communication). Even with selection/use of multiple solvents, some materials (e.g., polycyclic aromatics) may not be detectible in any solvent-based ionization method (Domin et al., 1997). Selection of a single solvent is not recommended unless only a specific set of compounds are to be examined.

This concept is most simply illustrated by the samples shown in Fig. 2 and Fig. 3. Both figures show results from the same two samples; however, in Fig. 2, the skeletal materials were exposed to dichloromethane as the solvent of choice. The GC/MS traces are remarkably similar and, indeed, there is a great deal of overlap in the chemical composition. Fig. 3 shows the same two samples exposed to acetonitrile. While the profiles seem somewhat similar, they are distinctly different in chemical composition. Table 1 and Table 2 compare the compounds present in each sample and recovered by the different samples. Some of the materials may be inherent to the sample (e.g., lipids); however, others may originate from the environment surrounding the burial. Once DNA testing was completed on the samples, it was shown that the two skeletal elements were recovered from two separate individuals, with two separate loss locations.

Ambient ionization

Using an ionization method that is not dependent on the solvent being used is a possible alternative to achieve solubilization and concentration of target materials. These methods expose the sample itself to the ionization source and do not involve any lengthy sample preparation. Small molecules, such as those that may be found in skeletal remains, have been detected by such methods as DART (Direct Analysis in Real Time) (Cody et al., 2005; Pierce et al., 2007);

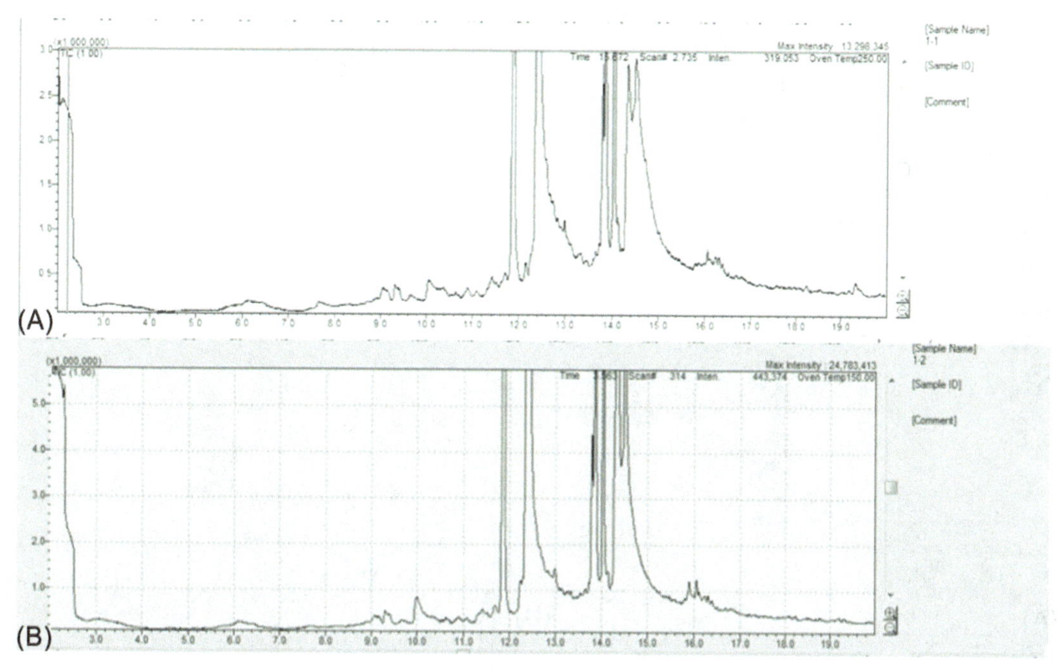

FIG. 2 The GC/MS traces of two different skeletal elements exposed to the solvent dichloromethane. The skeletal materials were returned to anthropologists investigating a Vietnam War loss location in Southeast Asia. Samples are designated as "a" and "b" to differentiate between the two individual samples. (Note that this designation is carried forward through Fig. 3, and in Tables 1 and 2.)

DESI (Desorption Electrospray Ionization) (Takáts et al., 2004); DAPCI (Desorption Atmospheric Pressure Chemical Ionization) (Takáts et al., 2005); MALDESI (Matrix-Assisted Laser Desorption Electrospray Ionization) (Sampson et al., 2006); and LAESI (Laser Ablation with Electrospray Ionization) (Nemes and Vertes, 2007). These ionization methods can be dependent on the type of sample to be tested. DESI has been used successfully on dehydrated samples (Huang et al., 2006) and for individualization of overlapping fingerprints (Ifa et al., 2008). LAESI has been successfully used for detection of large molecules (e.g., peptides) and has been used successfully on both animal and plant tissues; however, it requires a water-rich target.

DART seems to be most promising of these methods for targeted testing of powdered skeletal matrices. Moreno and McCord (2016) successfully used DART to determine the presence of inhibitors in DNA extracted from blood. The researchers successfully used a DART ion source coupled with a JEOL AccuTOF (JEOL, Peabody, Massachusetts USA) set in negative mode to determine the amount of indigo, phenol, bile salts, tannic acid, and EDTA remaining in blood samples spiked with known quantities of the inhibitors. DART has also been in nontargeted studies with beer (Cajka et al., 2011) and olive oil (Vaclavik et al., 2009), in a manner similar to what was done in the protocol described in Section "Protocol." However, DART has been shown to break down bonds in some metabolites, preventing detection of anything other than the "parent" material (Yu et al., 2009). Detection of only a parent compound may lead to faulty assignment of the source of the material.

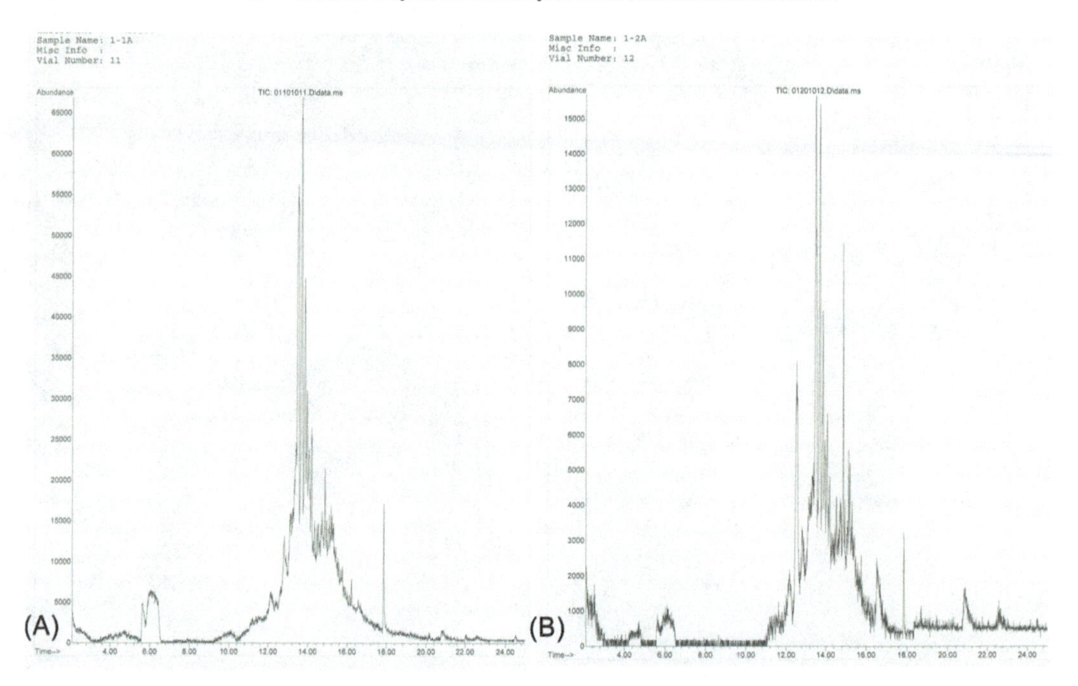

FIG. 3 The GC/MS traces of the same two skeletal elements presented in Fig. 2, but exposed to acetonitrile (instead of dichloromethane).

An ambient ionization source would seem to be a simpler testing strategy rather than a solvent-based technique. Ambient ionization could save approximately 24h and would eliminate handling of dangerous chemicals. However, ambient ionization has many drawbacks, some of which are noted in previous paragraphs. While efficient, most ambient ionization methods are also considered destructive in that a small amount of material is removed from the tested substrate. It is also possible that osseous material may prove to be too complex of a matrix for any ambient technique to appropriately ionize. Ambient ionization was not explored in the development of the protocol described in Section "Protocol" because the instrumentation was not available at the time of testing.

Protocol

After consideration of the different options and the available instrumentation, a protocol was developed to treat osseous material recovered from the exterior of skeletal elements using two solvents (acetonitrile and dichloromethane) sequentially. The recovered compounds were further concentrated and injected onto a GC/MS instrument after suspension in methanol. Development of the finalized protocol, including modifications to the GC/MS program and experiments with other solvents, is described in Edson and Roberts (2019). Included in this section are instructions on how to use the protocol and analyze the data generated.

TABLE 1 The compounds present in two sets of remains, recovered by treatment with dichloromethane. The GC/MS traces of both samples are shown in Fig. 2. Detected materials are presented in alphabetical order. Letter designations for the samples (a and b) refer to the associated traces in Fig. 2.

Sample a (from Fig. 2)			Sample b (from Fig. 2)		
Best Match	**Formula**	**Score**	**Best Match**	**Formula**	**Score**
1-Decanol, 2-hexyl-	$C_{16}H_{34}O$	82	17-Pentatriacontene	$C_{35}H_{70}$	80
1,2-Benzenedicarboxylic acid, bis(8-methylnonyl) ester	$C_{28}H_{46}O_4$	74	2-Methylhexacosane	$C_{27}H_{56}$	82
2-Isopropyl-5-methyl-1-heptanol	$C_{11}H_{24}O$	83	2-Methyltetracosane	$C_{25}H_{52}$	80
2-methyltetracosane	$C_{25}H_{52}$	82	9-Octadecenoic acid (Z)-, methyl ester	$C_{19}H_{36}O_2$	90
5-Fluoro-2-trifluoromethylbenzoic acid, 2-pentadecyl ester	$C_{23}H_{34}F_4O_2$	67	Bis(tridecyl)phthalate	$C_{34}H_{58}O_4$	72
8-Octadecenoic acid, methyl ester, (*E*)-	$C_{19}H_{36}O_2$	92	Cyclopentadecanone, 2-hydroxy-	$C_{15}H_{28}O_2$	80
9-Octadecenoic acid (Z)-, methyl ester	$C_{19}H_{36}O_2$	88	Heneicosanoic acid, methyl ester	$C_{22}H_{44}O_2$	80
Acetic acid, trichloro-, methyl ester	$C_3H_3Cl_3O_2$	88	Hexadecanoic acid, methyl ester	$C_{17}H_{34}O_2$	96
Bis(tridecyl) phthalate	$C_{34}H_{58}O_4$	70	l-(+)-Ascorbic acid, 2,6-dihexadecanoate	$C_{38}H_{68}O_8$	88
Docosanoic acid, 1,2,3-propanetriyl ester	$C_{69}H_{134}O_6$	75	Methyl stearate	$C_{19}H_{38}O_2$	94
E,E,Z-1,3,12-Nonadecatriene-5,14-diol	$C_{19}H_{34}O_2$	78	n-Hexadecanoic acid	$C_{16}H_{30}O_2$	91

Continued

TABLE 1 The compounds present in two sets of remains, recovered by treatment with dichloromethane. The GC/MS traces of both samples are shown in Fig. 2. Detected materials are presented in alphabetical order. Letter designations for the samples (a and b) refer to the associated traces in Fig. 2—cont'd

Sample a (from Fig. 2)			Sample b (from Fig. 2)		
Best Match	**Formula**	**Score**	**Best Match**	**Formula**	**Score**
Eicosane	$C_{20}H_{42}$	84	Octadecanoic acid	$C_{18}H_{36}O_2$	85
Hexadecanoic acid, methyl ester	$C_{17}H_{34}O_2$	97	Oleic acid	$C_{18}H_{34}O_2$	94
l-(+)-Ascorbic acid, 2,6-dihexadecanoate	$C_{38}H_{68}O_8$	89	Pentadecanoic acid	$C_{15}H_{30}O_2$	79
n-Hexadecanoic acid	$C_{16}H_{32}O_2$	92	Phenol, 3,5-bis(1,1-dimethylethyl)-	$C_{14}H_{22}O$	65
Octadecanoic acid	$C_{18}H_{36}O_2$	86	Tetradecanoic acid	$C_{14}H_{28}O_2$	89
Oleic acid	$C_{18}H_{34}O_2$	92	Tetrapentacontane, 1,54-dibromo-	$C_{54}H_{108}Br_2$	79
Oxiraneundecanoic acid, 3-pentyl-, methyl ester, trans-	$C_{19}H_{36}O_2$	78	trans-2-Hexadecenoic acid	$C_{16}H_{30}O_2$	79
Palmitic acid	$C_{16}H_{32}O_2$	78	Triacontanoic acid, methyl ester	$C_{31}H_{62}O_2$	84
Tetradecanoic acid	$C_{14}H_{28}O_2$	83	Trichloromethane	$CHCl_3$	78
Tetradecanoic acid, 2-phenyl-1,3dioxan-5-yl ester	$C_{24}H_{38}O_4$	66			
Tetrapentacontane, 1,54-dibromo-	$C_{54}H_{108}Br_2$	79			
Triacontanoic acid, methyl ester	$C_{31}H_{62}O_2$	86			
Trichloromethane	$CHCl_3$	80			

TABLE 2 The compounds present in two sets of remains, recovered by treatment with acetonitrile. The GC/MS traces of both samples are shown in Fig. 3. Detected materials are presented in alphabetical order. Letter designations for the samples (a and b) refer to the associated traces in Fig. 3.

Sample a (from Fig. 3)			Sample b (from Fig. 3)		
Best match	Formula	Score	Best match	Formula	Score
(1,2,3,3a,4,6a-Hexahydropentalen-2-yl)-dimethyl-amine	$C_{10}H_{17}N$	38	1-Decanol, 2-hexyl-	$C_{16}H_{34}O$	50
1,2-Benzenedicarboxylic acid, mono(2-ethylhexyl)ester	$C_{16}H2_{2}O_{4}$	90	1,1-Dodecanediol, diacetate	$C_{16}H_{30}O_{4}$	42
18-Norabietane	$C_{19}H_{34}$	96	1,7-Dimethyl-4-(1-methylethyl)cyclodecane	$C_{15}H_{30}$	35
2-Piperidinone, N-[4-bromo-n-butyl]-	$C_{19}H_{16}BrNO$	64	17-Pentatriacontene	$C_{35}H_{70}$	53
Anilazine	$C_{9}H_{5}C_{13}N_{4}$	25	Bicyclo[4.3.0]none, 2,2,6,7-tetramethyl-7-hydroxy-	$C_{13}H_{24}O$	50
Benzothiazole, 2-methyl-	$C_{8}H_{7}NS$	35	Caffeine	$C_{8}H_{10}N_{4}O_{2}$	80
Cyclohexene, 4-(4-ethylcyclohexyl)-1-pentyl-	$C_{19}H_{34}$	93	Heptadecanoic acid, 16-methyl, methyl ester	$C_{19}H_{38}O_{2}$	58
Diethyl phthalate	$C_{12}H_{14}O_{4}$	83	Octadecane, 1-(ethenyloxy)-	$C_{20}H_{40}O$	53
Naphthalene, 6-ethyl-1,2,3,4-tetrahydro-	$C_{19}H_{28}$	35	Oxalic acid, isobutyl tetradecyl ester	$C_{20}H_{38}O_{4}$	35
Octadecanoic acid, methyl ester	$C_{19}H_{38}O_{2}$	93	Pentadecanoic acid, 14-methyl-,methyl ester	$C_{17}H_{34}O_{2}$	62
Phthalic acid, butyl tricedyl ester	$C_{25}H_{40}O_{4}$	86	Pentatriacontane	$C_{35}H_{72}$	50
Phthalic acid, monoamide, N-ethyl-N-(3-methylphenyl)-, ethyl ester	$C_{19}H_{21}NO_{3}$	43	Phthalic acid, nonyl tridec-2-yn-1-yl ester	$C_{30}H_{46}O_{4}$	72
Spiro[4,5]decan-7-one, 1,8-dimethyl-8,9-epoxy-4-isopropyl	$C_{15}H_{24}O_{2}$	27	trans-2,3-Epoxydecane	$C_{10}H_{20}O$	50

The protocol described herein was developed using regular casework analysis of remains recovered by the Defense POW/MIA Accounting Agency (DPAA) and sent for testing at the Armed Forces Medical Examiner System – Armed Forces DNA Identification Laboratory (AFMES-AFDIL, often referred to simply as AFDIL). Osseous materials and teeth from unidentified individuals were recovered from worldwide burial locations. The postmortem interval (PMI) of the materials ranged from approximately 45 years to 75 + years.

Preparation of skeletal materials

All steps in this section should be performed according to standard forensic or ancient DNA protocols, both of which require decontamination of the instruments used as well as handling the samples within a sterile environment. While it is next to impossible to eliminate external chemical contamination from modern sources, the possibility can certainly be reduced within the laboratory.

To acquire the skeletal materials necessary for treatment, a sample of the exterior of the bone is required. Preferentially, this should be done after any remaining biological materials (i.e., desiccated or decomposed tissue, soil) have been removed from the immediate surface of the skeletal element, but prior to any further cleaning. In many forensics laboratories, the exposed surface of the bone is generally removed through abrasion, typically with a Dremel® sanding tool (Dremel, Racine, WI) (Edson and McMahon, 2019; Edson, 2019a) or other abrasive tool (e.g., sandpaper). While other techniques are certainly viable, abrasion generates a fine powder that can be collected for GC/MS testing. By collecting the osseous powder that is typically generated through normal cleaning of the sample and then discarded as medical waste, GC/MS analysis of skeletal materials is essentially nondestructive.

For ease of collection, the sample can be sanded directly over a large (140 × 140 × 22 mm) weigh boat (Thomas Scientific, Swedesboro, New Jersey USA), which collects the powder as it falls or onto the plastic-lined side of Benchkote (Whatman, GE Healthcare Life Sciences, Pittsburgh, Pennsylvania USA). Once sanding or abrasion is complete, the skeletal sample may continue through the traditional DNA testing pathway. The osseous powder can then be funneled into a sterile 15-mL polypropylene conical tube (Falcon, Corning, New York USA). A smaller conical tube (e.g., 1.7-mL polypropylene) may also be used; however, it can be challenging to accurately pour fine osseous powder into the narrow opening of such a small tube. Static can be a confounding issue.

Polypropylene should not be used for long-term storage of powdered material. By-products of the plastic may absorb into the osseous powder and provide erroneous signals in GC/MS analysis. Glass vials are preferred (and recommended) for storage of powder. However, the expense of glass vials can be cost-prohibitive. In addition, typically glass is not disposable and reusing glass vials may encourage transfer of materials from one sample to another, which can potentially result in generation of false (and/or inaccurate) signals. The less expensive alternative (i.e., plastic) is acceptable, as long as the analyst remains aware of possible plastic by-product "carryover" into the chemical analysis. These compounds or by-products, when detected, should be noted on reports as possible modern contamination.

Preparation of tooth materials

In addition to bone, teeth may also be used for GC/MS analysis. To maximize the usage of a tooth in a manner similar to that of osseous elements, the dental powder generated from drilling into the crown may be collected (Edson, 2019b). Alternatively, if a rasp is used to grind away the roots (Corrêa et al., 2017), or if the entire tooth is pulverized (Miloš et al., 2007), a portion may be collected prior to DNA extraction. Powder generated from teeth may be stored in the same fashion as that of the skeletal elements (with the same caveats and recommendations).

Solvent treatment

Typically, only a fraction of the osseous material recovered from sanding is required for testing. A portion as small as 0.1 g can be aliquoted into a 1.7-mL tube for testing. Lower quantities (less powder) may be used; however, this may result in a decrease in data recovery. The remainder of the powder may be stored for retesting, or discarded.

As discussed above, the choice of solvent is critical to the success of this protocol. More stringent solvents, such as acetonitrile or dichloromethane, were chosen for their ability to dissolve a wide variety of compounds. Water could also be used; however, many compounds of interest are insoluble in water, and therefore, would not be detected or visualized by the GC/MS system. If multiple solvents are used, the same fraction of osseous powder may be tested without loss of information. Solvents should progress from least stringent to most stringent. For example, if using the aforementioned recommended solvents, acetonitrile treatment should be performed first, followed by dichloromethane. Additionally, it is not necessary to dry the sample in between treatments.

Starting with acetonitrile (HPLC grade, Sigma-Aldrich, St. Louis, Missouri USA), in this protocol 1.0 mL of solvent is added to the fraction of osseous material to be tested. The sample should then be vigorously agitated and allowed to incubate at room temperature for at least 1 h. Ideally, the sample should be placed on a nutator (or another similar shaker system) and undergo agitation during the incubation period; however, this is not an absolute necessity. At completion of incubation, the sample should be centrifuged for 2 min at 13,000 rpm in order to pellet the osseous material.

After pelleting, the liquid should be poured into a clean 10-mL glass beaker or watch glass and then allowed to volatilize in a chemical fume hood (Fig. 4). Once the solvent has completely evaporated, the remaining material should be suspended in 500 μL of methanol (≥ 99.9%, HPLC grade, Sigma-Aldrich, St. Louis, Missouri, USA) and placed in 1.7-mL polypropylene tubes for storage (– 20°C) until loading on the GC/MS instrument. Next, the same process is repeated with the dichloromethane (HPLC grade; Pharmco AAPER, Brookfield, Connecticut USA) solvent.

Prior to loading on the instrument, the samples should again be centrifuged at 13,000 rpm. Pelleting of any flocculants or floating materials is critical at this stage. Introduction of solid materials into the injection port of the GC/MS instrument may cause damage to the instrument and/or destruction of the column. After pelleting, the liquid fraction is removed and added to 9-mm glass vials with crimp caps (Thermo Fisher, Walther, MA, USA).

FIG. 4 The drying fraction of dichloromethane recovered from treated osseous materials removed from skeletal elements exposed to high volumes of petrochemicals (which is responsible for the dark coloration of the materials in the beaker).

It is critical to have well-cleaned glassware for this step. During the use of this protocol (as described in Edson, 2017; Edson and McMahon, 2019), a set of new beakers should be dedicated for the testing. Glassware should be thoroughly scrubbed using a detergent such as Alconox or Tergazyme, followed by thorough rinsing with deionized water and ethanol. Regular tap water should be avoided as it may introduce minerals to the glassware and thus into the downstream testing. Glass vials used for instrument loading should not be reused.

GC/MS instrument loading

Samples are loaded onto an Agilent 7890A/5875C GC/MS System with a 20-m column (Agilent, Santa Clara, California USA). The starting oven temperature should be 150°C and should remain at that temperature for a 20-min hold before ramping at 20°C per minute to 250°C for an additional 30-min hold. The injection is splitless, meaning the entire fraction is injected into the run, which provides the best possible detection of low-level compounds. A full scan of the injection is performed with no subtraction of known elements. Other similar instruments may be used; however, performance checks should be done to verify the protocol operates accordingly and analogously.

Analysis

The analyst must be aware that, when performing a full-spectrum chemical analysis of a sample, one will be observing *any* material (biological or chemical) that will solubilize in the solvent of choice and be detectable with the chosen instrumentation. Different molecules will be detected using GC/MS vs. LC/MS, although it is unnecessary to run on both instruments.

A full scan may generate a trace that has numerous peaks (Fig. 5) or just a few peaks (Fig. 1). Each peak, or cluster of peaks, represents a chemical compound or biological material with a specific mass. The analytical software MassHunter coupled with ChemStation (Agilent, Santa Clara, California USA) compares the molecular weight of the components within each peak to a spectral library of known standards. The standards chosen are at the discretion of the laboratory, but a typical library of standards is generated by and available from the National Institute of Standards and Technology (NIST); and these standards are updated on a regular basis. More specialized sets of standards, such as those for illicit drugs (e.g., Cayman Spectral Library, Cayman Chemicals, Ann Arbor, Michigan USA), explosives, or pesticides, also exist. These libraries may be used in concert or individually, depending on the goals of the laboratory, although the NIST library is fairly comprehensive.

If the sample is fairly clean, the software used will automatically determine the identity of the molecule and provide the analyst with a printout. Unfortunately, most samples generated from skeletal materials are "noisy" simply by virtue of the volume of materials being detected. Fig. 5 is a good example of a "noisy" sample. The sample presented in this figure was recovered from the *USS Oklahoma*, in which all remains were exposed to an anaerobic environment of salt water and petrochemicals immediately postmortem (and until recovery).

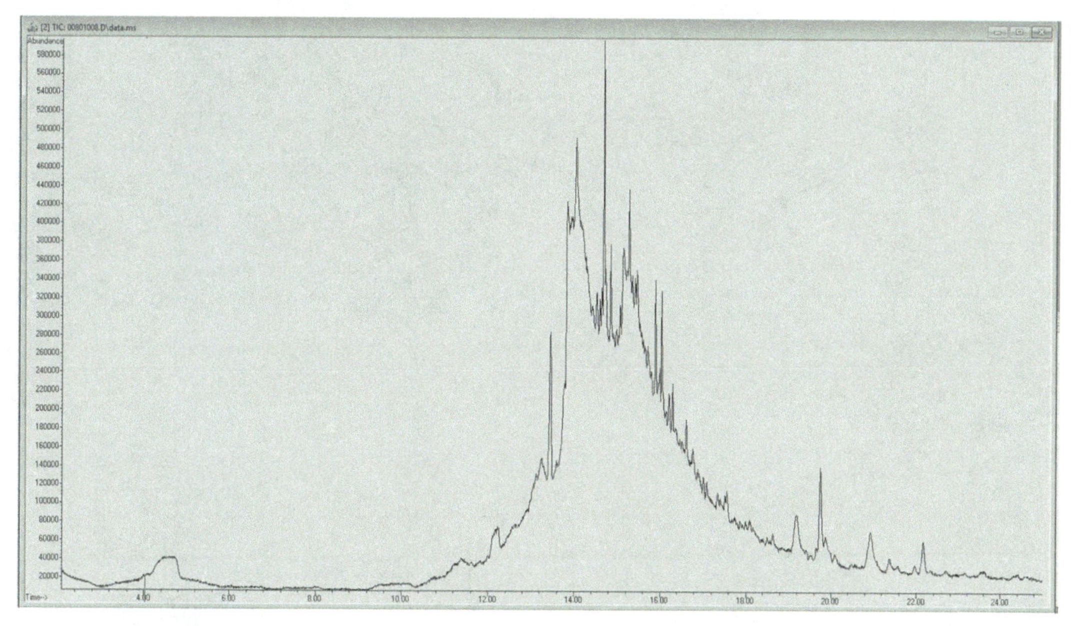

FIG. 5 The resulting GC/MS trace of an osseous sample treated with dichloromethane. The sample was recovered from the *USS Oklahoma* and was exposed to large amounts of petrochemicals immediately postmortem.

During GC/MS analysis, it was found that there is an intricate mixture of petrochemicals and lipids in the central peaks, which obscured many of the other compounds that might be found in the remains. Obfuscation of compounds by the overwhelming presence of others is not unheard of in GC/MS analysis (Lennard et al., 1995) and should be considered during reporting.

Should the instrument software not automatically identify the peaks present, the data should be manually examined with the aid of the software (McLafferty and Tureček, 1993). The analyst should evaluate the observed spectra for each peak and compare it to a published fragmentation library (e.g., to NIST2011). For samples with numerous peaks, this may be a time-consuming task. The following two figures use Fig. 1 as an example, to demonstrate how to break down and perform this type of analysis in ChemStation.

By selecting the indicated peak (Fig. 6), the analyst is provided with the spectra of that peak. The spectrum pattern determines the identity of the compound, and typically the largest peak corresponds to the relative molecular mass of the compound. In this example, the molecular weight should be close to 270 g/mol.

Selecting any point within the spectra will present the analyst with a list of choices for the possible identity of the compound (Fig. 7). The list is generated from known spectra of compounds contained within the library of choice. For a full-spectrum scan where the compounds are unknown, a large nonspecific spectral library (e.g., NIST 2011) is recommended over a compound-specific spectral library (e.g., SWGDRUG). However, nonspecific and specific libraries may be used in concert.

For each peak, the software assigns a score as to the probability that the material detected is the compound chosen. For example, during analysis of the peak in Fig. 6, the analyst is presented with a choice of 20–30 different compounds that are most likely to be the identity of the

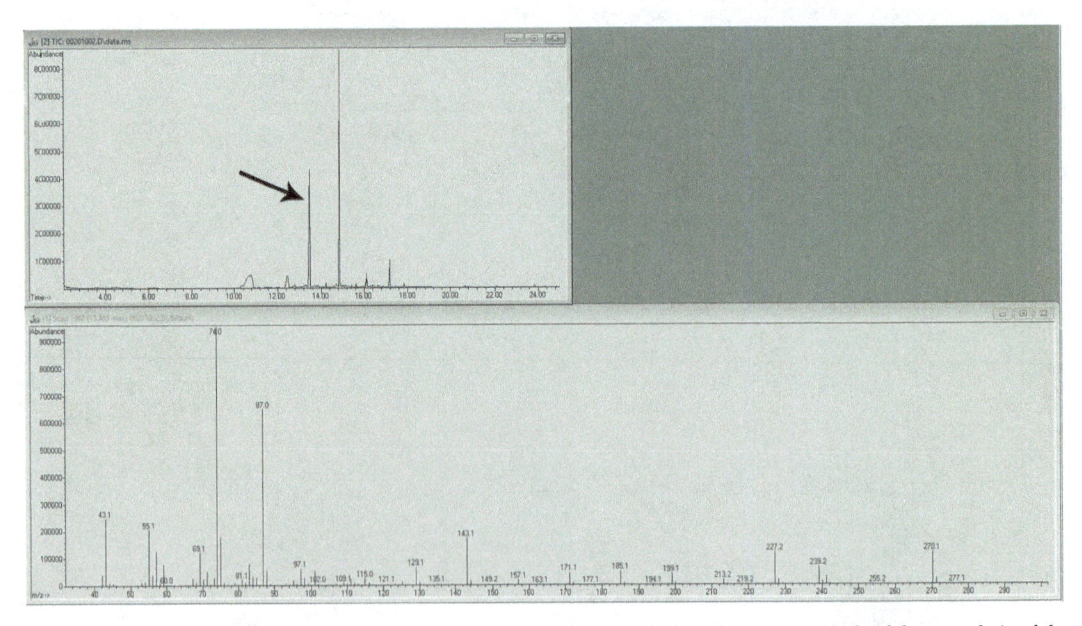

FIG. 6 A breakdown of a single peak in Fig. 1. The spectral pattern below the spectrograph of the sample is of the peak marked with the arrow.

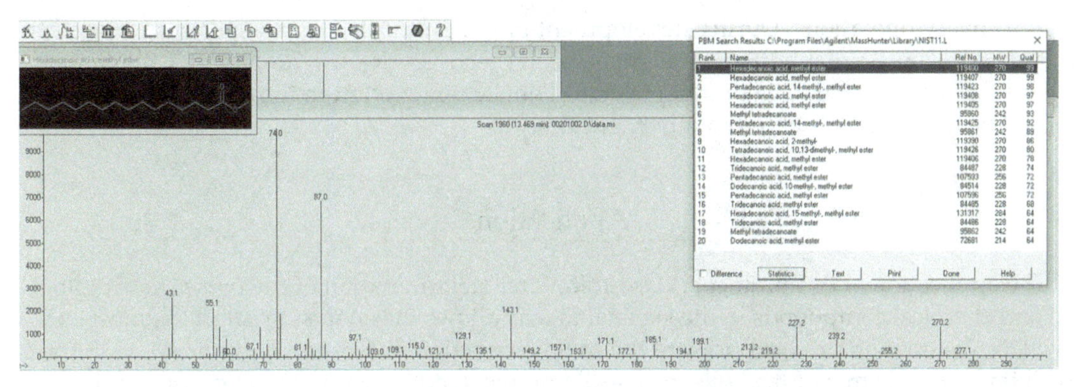

FIG. 7 The software determined list of compounds for the possible identity of the selected peak. The molecular structure in the upper left-hand corner is that of the first compound in the list of possible matches.

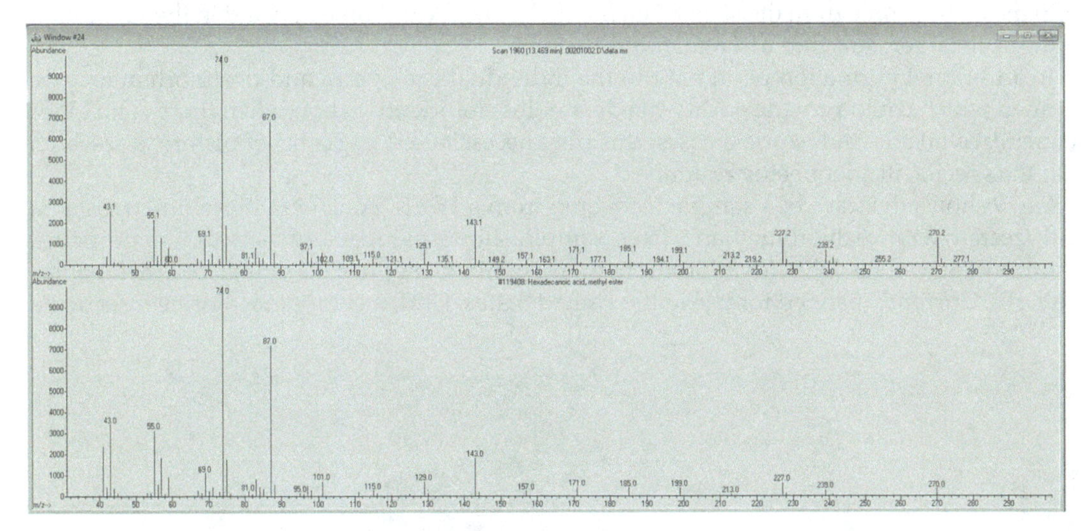

FIG. 8 The spectral profile of the sample *(top)* as compared to the most likely matching compound of Hexadecanoic acid, methyl ester.

compound detected (Fig. 7). Samples are ranked from best to worst, with scores between 100 and 1 (with 100 being the best ranking). Each compound may be evaluated for authenticity, and the most probable compound is then chosen (Fig. 8).

The compound with the highest score may not be the one chosen. There is some dependence on the comparison of the spectra between that of the sample and that of the library. Low scoring compounds may be chosen, but need to be evaluated as to the probability of that particular compound being present. For example, if an intact molecule of a compound (i.e., a "parent" molecule) is detected, but no corresponding metabolites are detected, the analyst must consider whether the compound is authentic to the skeletal materials being analyzed or perhaps due to modern contamination.

In fact, modern contaminants have been shown to be present in samples, not only from incompletely cleaned glassware (which may introduce soap or other residues), but also from field recovery operations. During development of the protocol, a broad-spectrum sunscreen was detected, presumably from a member of the recovery team who handled the remains without gloves. Modern contamination should not be removed from the chemical profile, but should be noted and observed in other samples.

Application

The application of this protocol is two-fold. First, determination of the presence of biological and chemical compounds in the skeletal matrix allows the analyst to select the most effective DNA extraction protocol prior to the initial extraction. For example, if the GC/MS trace exhibits a high degree of fats, oils, or petroleum products, a DNA extraction protocol using an organic purification method should be chosen as the oils will more efficiently removed. Secondly, a personal chemical profile of the individual may be developed. When coupled with the DNA profile from the skeletal element, the chemical profile may aid in the separation of the skeletal remains into discrete individuals.

In addition, information pertaining to the individual pre-, peri-, and postmortem may be gleaned which could provide additional data points for identification when there is no DNA reference available. In historical cases, this may necessitate doing a bit of historical research into the specific incident being examined.

Fig. 9 shows the trace of a sample recovered from a World War II era burial in Yugoslavia and treated with dichloromethane. This sample shows evidence of fats and some petrochemicals, as well as DDT (dichlorodiphenyltrichloroethane) and its associated metabolites (Table 3). Currently banned for use in the United States, DDT is an organochlorine insecticide

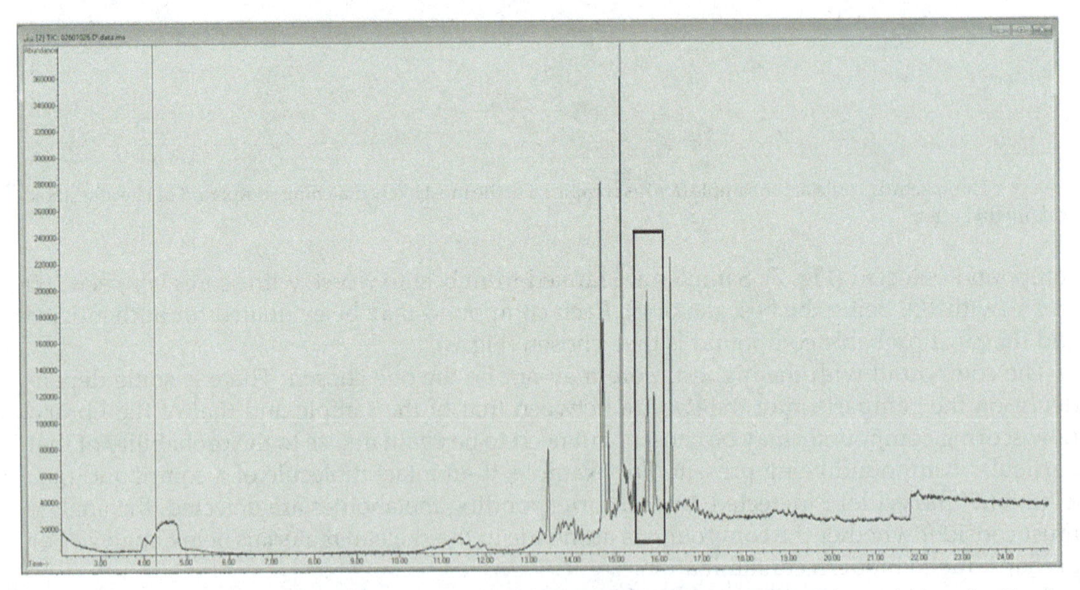

FIG. 9 Trace of a sample recovered from a World War II era gravesite in Yugoslavia. The peaks inscribed by the box are DDT (dichlorodiphenyltrichloroethane) and its associated metabolites.

TABLE 3 The chemical profile of the sample presented in Fig. 9. Only the compounds detected with dichloromethane are listed. Compounds are listed in alphabetical order. DDT and associated metabolites are indicated by the bold outlines.

Best match	Formula	Score
1-Chloroeicosane	$C_{20}H_{41}Cl$	66
1-Decanol, 2-hexyl-	$C_{16}H_{34}O$	72
1-Docosanethiol	$C_{22}H_{46}S$	27
1-Propanol, 2,3-bis[(3,7,11,15-tetramethylhexadecyl)oxy]-	$C_{43}H_{88}O_3$	38
1,1-Dichloro-2,2-bis(p-chlorophenyl)ethane	$C_{14}H_{10}Cl_4$	93
1,3-Doxolane, 4-ethyl-5-octyl-2,2-bis(trifluoromethyl)-, trans-	$C_{15}H_{24}F_6O_2$	38
2-Hexyl-1-octanol	$C_{14}H_{30}O$	47
7-Hexadecenal, (*Z*)-	$C_{16}H_{30}O$	52
9-Octadecenoic acid, methyl ester, (*E*)-	$C_{19}H_{36}O_2$	99
Benz(b)-1,4-oxazepine-4(5H)-thione, 2,3-dihydro-2,8-dimethyl-	$C_{11}H_{13}NOS$	38
Cyclohexan, 1,2,4-trimethyl-	C_9H_{18}	35
Cyclohexane, 1-ethyl-2-propyl-	$C_{11}H_{22}$	38
Cyclopentane, (2-methylbutyl)	$C_{10}H_{20}$	43
DDMU	$C_{14}H_9C_{13}$	90
Decane, 1,1′-oxybis-	$C_{20}H_{42}O$	38
Dodecane, 1-chloro-	$C_{12}H_{25}Cl$	23
Eicosane	$C_{20}H_{42}$	86
Ethyl acetate	$C_4H_8O_2$	59
Fumaric acid, 2-decyl tridecyl ester	$C_{15}H_{24}F_6O_2$	35
Heptadecyl heptafluorobutyrate	$C_{21}H_{35}F_7O_2$	50
Heptasiloxane, 1,1,3,3,5,5,7,7,9,9,11,11,13,13-tetradecamethyl-	$C_{14}H_{44}O_6Si_7$	72
Methyl stearate	$C_{19}H_{38}O_2$	99
n-Nonadecanol-1	$C_{19}H_{40}O$	43
o,p′-DDT	$C_{14}H_9Cl_5$	96
Octadecane, 3-methyl-5-(2-ethylbutyl)-	$C_{26}H_{54}$	38
Octadecanoic acid, 10-oxo-, methyl ester	$C_{19}H_{36}O_3$	64
Octasiloxane, 1,1,3,3,5,5,7,7,9,9,11,11,13,13,15,15-hexadecamethyl-	$C_{16}H_{50}O_7Si_8$	86
Oxalic acid, 6-ethyloct-3-yl heptyl ester	$C_{19}H_{36}O_4$	43
p,p′-DDE	$C_{14}H_8Cl_4$	99
p,p′-DDT	$C_{14}H_9Cl_5$	94
Pentadecanoic acid, 14-methyl-, methyl ester	$C_{17}H_{34}O_2$	97
Phenol, 2,5-bis(1,1-dimethylethyl)-	$C_{14}H_{22}O$	91
Squalane	$C_{30}H_{62}$	35

and was commonly used to treat body lice and malaria during World War II. It was often sprayed upon soldiers and civilian populations (World Health Organization (WHO), 1979; Frideman, 1992). DDT is easily absorbed through human skin and builds up within the fatty tissues of the body (Klaassen, 1996), so it stands to reason that it also would be detectable in the osseous materials.

Fig. 10 shows a fairly complicated trace of a skeletal sample recovered from the *USS Oklahoma*. While the bulk of the peaks are the aforementioned amalgam of lipids and

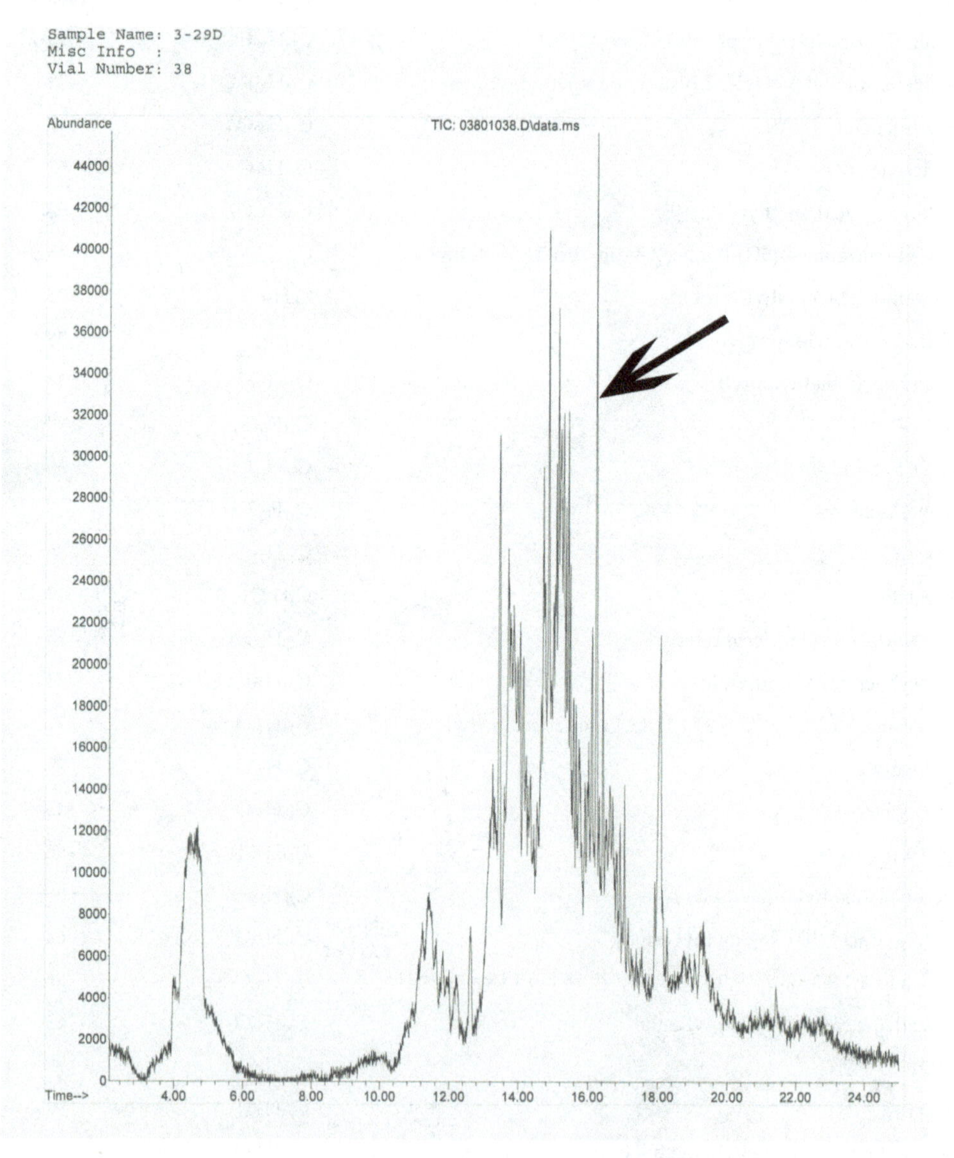

FIG. 10 Trace of a sample recovered from the USS *Oklahoma* treated with dichloromethane. The indicated peak is pine resin acid. The full chemical profile is not presented here (as it is quite extensive).

petrochemicals, the peaks indicated towards the end of the trace are pine resin acid. Pine resin, or wood tar, has been commonly used in maritime practices for the waterproofing and treatment of wooden decking and roping (Allen Jr., 2013). It is likely to have been present on the *USS Oklahoma* (L. Freas, personal communication), which had a fairly large carpentry shop onboard. While this is the most probable source of the pine tar observed on the trace, pine resin has also been widely used throughout history for the treatment of skin ailments (Barnes and Greive, 2017).

Summary

The protocol detailed above was originally developed to aid in the identification of potential PCR inhibitors present in osseous materials. The ability to develop a personal chemical profile from the same information is an additional benefit to this simple, cost-effective technique. Once the chemical profile is tied to a DNA profile, remains from a mass fatality incident may be sorted by the chemical profile, reducing the cost(s) of such efforts and increasing the speed of returning loved ones to their family.

As noted in Section "Application," GC/MS analysis of historical remains necessitates a bit of historical research to determine the potential source of compounds present. While knowing the source of the remains may introduce cognitive bias into the analysis, it will also reduce the chances of a false assignment of a modern compound to the chemical profile.

GC/MS analysis of skeletonized human remains holds much promise for human identification efforts. Some statistical analysis on the data set used to generate this protocol is ongoing, in order to generate a robust and reproducible model for the forensics community.

Disclaimer

The opinions or assertions presented are the private views of the author and should not be construed as official or as reflecting the views of the Department of Defense; the Defense Health Agency; the Armed Forces Medical Examiner System; the Defense POW/MIA Accounting Agency; SNA, International; or Flinders University.

References

Allen Jr., L.V., 2013. Basics of compounding with tars. Int. J. Pharm. Compd. 17, 400–410.

Barnes, T.M., Greive, K.A., 2017. Topical pine tar: history, properties, and use as a treatment for common skin conditions. Aust. J. Dermatol. 58, 80–85.

Cajka, T., Riddellova, R., Tomaniova, M., Hajslova, J., 2011. Ambient mass spectrometry employing a DART ion source for metabalomic fingerprinting/profiling: a powerful tool for beer origin recognition. Metabolomics 7 (4), 500–508.

Cody, R.B., Laramée, J.A., Durst, H.D., 2005. Versatile new ion source for the analysis of materials in open air under ambient conditions. Anal. Chem. 77, 2297–2302.

Corrêa, H.S.D., Pedro, F.L.M., Volpato, L.E.R., Pereira, T.M., Filho, G.S., Borges, A.E., 2017. Forensic DNA typing from teeth using demineralized root tips. Forensic Sci. Int. 280, 164–168.

Dhabbah, A.M., Al-Jaber, S.S., Al-Ghamdi, A.H., Aqel, A., 2014. Determination of gasoline residues on carpets by SPME-GC-MS. Arab. J. Sci. Eng. 39, 6749–6756.

Domin, M.A., Moreea, R., Lazaro, M.J., Herod, A.A., Kandiyoti, R., 1997. Choice of extraction voltage and matrix in the matrix-assisted laser desorption/ionization time-of-flight mass spectrometry of coal tar pitch—pyridine insolubles. Rapid Commun. Mass Spectrom. 11, 638–645.

Edson, S.M., 2017. DNA typing from skeletal remains: a study of inhibitors using mass spectrometry. Forensic Sci. Int. Genet. Suppl. Ser. 6, e337–e339.

Edson, S.M., 2019a. Extraction of DNA from skeletonized post-cranial remains: a discussion of protocols and testing modalities. J. Forensic Sci. 64 (5), 1312–1323.

Edson, S.M., 2019b. Getting ahead: extraction of DNA from skeletonized cranial material and teeth. J. Forensic Sci. 64 (6), 1646–1657.

Edson, S.M., Christensen, A.F., 2012. Field contamination of skeletonized human remains with exogenous DNA. J. Forensic Sci. 58 (1), 206–209.

Edson, S.M., McMahon, T.P., 2019. Testing of skeletonized human remains using GC/MS – development of a personal environmental profile. Aust. J. Forensic Sci. 51, S115–S118.

Edson, S.M., Roberts, M., 2019. Determination of materials present in skeletonized human remains and the associated DNA: development of a GC/MS protocol. Forensic Sci. Int. 1, 170–184.

Farré, M., Petrovic, M., Barceló, D., 2007. Recently developed GC/MS and LC/MS methods for determining NSAIDs in water samples. Anal. Bioanal. Chem. 387 (4), 1203–1214.

Frideman, H.B., 1992. DDT (dichlorodiphenyltrichloroethane): a chemist's tale. J. Chem. Educ. 69 (5), 362–365.

Gilbert-López, B., Robles-Molina, J., García-Reyes, J.F., Molina-Díaz, A., 2010. Rapid determination of BTEXS in olives and olive oil by headspace-gas chromatography/mass spectrometry (HS-GS-MS). Talanta 83, 391–399.

Huang, M.Z., Hsu, H.J., Lee, J.Y., Jeng, J., Shiea, J., 2006. Direct protein detection from biological media through electrospray-assisted laser desorption ionization/mass spectrometry. J. Proteome Res. 5 (5), 1107–1116.

Ifa, D.R., Manicke, N.E., Dill, A.L., Cooks, R.G., 2008. Latent fingerprint chemical imaging by mass spectrometry. Science 321 (5890), 802–805.

Ifa, D.R., Jackson, A.U., Paglia, G., Cooks, R.G., 2009. Forensic applications of ambient ionization mass spectrometry. Anal. Bioanal. Chem. 394 (8), 1995–2008.

Jaffrès, E., Lalanne, V., Macé, S., Coret, J., Cardinal, M., Sérot, T., Dousset, X., Joffraud, J.J., 2011. Sensory characteristics of spoilage and volatile compounds associated with bacteria isolated from cooked and peeled tropical shrimps using SMPE-GC-MS analysis. Int. J. Food Microbiol. 147, 195–202.

Klaassen, C.D. (Ed.), 1996. Casarett and Doull's Toxicology: The Basic Science of Poisons, fifth ed. McGraw-Hill, Health Professions Division, New York.

Koenig, A., Magnolon, S., Weyermann, C., 2015. A comparative study of ballpoint ink ageing parameters using GC/MS. Forensic Sci. Int. 252, 93–106.

Lennard, C.J., Rochaix, V.T., Margot, P., Huber, K., 1995. A GC-MS database of target compound chromatograms for the identification of arson accelerants. Sci. Justice 35 (1), 19–30.

Manning, T., Mikula, R., Wylie, G., Phillips, D., Jarvis, J., Zhang, F., 2015. Structural measurements and cell line studies of the copper-PEG-rifampicin complex against *Mycobacterium tuberculosis*. Bioorg. Med. Chem. Lett. 25, 451–458.

Maurer, M.K., Bukowski, M.R., Menacher, M.D., Zatorsky, A.R., 2010. Inquiry-based arson investigation for general chemistry using GC-MS. J. Chem. Educ. 87, 311–313.

McLafferty, F.W., Tureček, F., 1993. Interpretation of Mass Spectra, fourth ed. University Science Books, Sausalito, CA, USA.

Meyer, M.R., Bach, M., Welter, J., Bovens, M., Turcant, A., Maurer, H.H., 2013. Ketamine-derived designer drug methoxetamine: metabolism including isoenzyme kinetics and toxicological detectability using GC-MS and LC-(HR-) MSn. Anal. Bioanal. Chem. 405 (19), 6307–6321.

Miloš, A., Selmanović, A., Smajlović, L., Huel, R.L., Katzmarzyk, C., Rizvić, A., Parsons, T.J., 2007. Success rates of nuclear short tandem repeat typing from different skeletal elements. Croat. Med. J. 48 (4), 486–493.

Moreno, L.I., McCord, B.R., 2016. The use of direct analysis in real time (DART) to assess the levels of inhibitors co-extracted with DNA and the associated impact in quantification and amplification. Electrophoresis 37 (21), 2807–2816.

Nemes, P., Vertes, A., 2007. Laser ablation electrospray ionization for atmospheric pressure, in vivo, and imaging mass spectrometry. Anal. Chem. 79 (21), 8098–8106.

Pert, A.D., Baron, M.G., Birkett, J.W., 2006. Review of analytical techniques for arson residues. J. Forensic Sci. 51, 1033–1049.

Pierce, C.Y., Barr, J.R., Cody, R.B., Massung, R.F., Woolfitt, A.R., Maura, H., Thompson, H.A., Fernandez, F.M., 2007. Ambient generation of fatty acid methyl ester ions from bacterial whole cells by direct analysis in real time (DART) mass spectrometry. Chem. Commun. 8, 807–809.

Sampson, J.S., Hawkridge, A.M., Muddiman, D.C., 2006. Generation and detection of multiply-charged peptides and proteins by matrix-assisted laser desorption electrospray inonization (MALDESI) Fourier transform ion cyclotron resonance mass spectrometry. J. Am. Soc. Mass Spectrom. 17 (12), 1712–1716.

Skender, L., Karačić, V., Brčić, I., Bagarić, A., 2002. Quantitative determination of amphetamines, cocaine, and opiates in human hair by gas chromatography/mass spectrometry. Forensic Sci. Int. 125, 120–126.

Strano-Ross, S., Bermejo, A.M., de la Torre, X., Botré, F., 2010. Fast GC-MS method for the simultaneous screening of THC-COOH, cocaine, opiates and analogues including buprenorphine, and fentanyl, and their metabolites in urine. Anal. Bioanal. Chem. 399, 1623–1630.

Takáts, Z., Wiseman, J.M., Gologan, B., Cooks, R.G., 2004. Mass spectrometry sampling under ambient conditions with desorption electrospray ionization. Science 306 (5695), 471–473.

Takáts, Z., Cotte-Rodriguez, I., Talaty, N., Chen, H.W., Cooks, R.G., 2005. Direct, trace level detection of explosives on ambient surfaces by desorption electrospray ionization mass spectrometry. Chem. Commun. 15, 1950–1952.

Vaclavik, L., Cajka, T., Hrbek, V., Hajslova, J., 2009. Ambient mass spectrometry employing direct analysis in real time (DART) ion source for olive oil quality and authenticity assessment. Anal. Chim. Acta 645 (1–2), 56–63.

Van den Broek, I., Blokland, M., Nessen, M.A., Sterk, S., 2015. Current trends in mass spectrometry of peptides and proteins: application to veterinary and sports-doping control. Mass Spectrom. Rev. 34, 571–594.

World Health Organization (WHO), 1979. DDT and its derivatives. In: Environmental Health Criteria. vol. 9. Geneva, Switzerland.

Yao, Y.T., Song, J., Yu, J., Wang, X.F., Hou, F., Zhang, A.L., Liu, Y., Han, J., Xie, M.X., 2009. Differentiation and dating of red ink entries of seals on documents by HPLC and GC/MS. Sep. Sci. 32, 2919–2927.

Yu, S., Crawford, E., Tice, J., Musselman, B., Wu, J.T., 2009. Bionanalysis without sample clean-up or chromatography: the evaluation and initial implementation of direct analysis in real time ionization mass spectrometry for the quantification of drugs in biological matrices. Anal. Chem. 81 (1), 193–202.

CHAPTER

23

Forensic odontology: Historical perspectives and current applications for identification of human remains

Thomas J. David DDS, D-ABFO[a,b,c] *and*
James M. Lewis DMD, D-ABFO[b,c,d]

[a]Georgia Bureau of Investigation, Division of Forensic Sciences, Decatur, GA, United States [b]Department of General Dentistry, Division of Forensics, University of Tennessee Graduate School of Medicine, Knoxville, TN, United States [c]Institute for Human Identification, LMU College of Dental Medicine, Knoxville, TN, United States [d]Alabama Department of Forensic Sciences, Madison, AL, United States

Introduction

The National Missing and Unidentified Persons System (NamUs, 2018) estimates that in the United States, "4,400 unidentified bodies are recovered each year, with approximately 1,000 of those bodies remaining unidentified after one year" (https://namus.gov/). Providing the anthropological and dental biological profile of unidentified human remains to law enforcement investigators assists in narrowing the search parameters, thereby improving the odds of successful identification.

In the United States today, the Medical Examiner or Coroner in each jurisdiction has the legal responsibility of certifying the identification of human remains. Forensic odontologists provide their opinion and judgment regarding an individual's identity based upon their education, training, and expertise, in accordance with guidelines and best practices outlined by the American Board of Forensic Odontology (2018c). Forensic odontology is the application of dental science to legal and/or criminal investigations.

History of forensic dental identification of human remains

Forensic dental identification of human remains has been used for at least two thousand years. The first known dental identification occurred in the year AD 49 by Agrippina, the

Forensic Genetic Approaches for Identification of Human Skeletal Remains
https://doi.org/10.1016/B978-0-12-815766-4.00025-X

jealous fourth wife of the Roman Emperor Claudius (as recorded by the Roman historian Tacitus). Claudius became emperor in AD 41, following assassination of the notorious Emperor Caligula. In AD 48, Messalina—the third wife of Claudius—and her lover Silius attempted a *coup d'état*, which ultimately resulted in their own execution. Due to the attempted *coup d'état*, Claudius realized that he was in a position of political weakness and he became concerned that an heir to the throne did not exist; therefore, he surmised that he needed to remarry in an effort to solidify power. Advisors to Claudius presented three candidates as possible brides: (1) Caligula's third wife, Lollia Paulina, who had immense wealth; (2) Aelia Paetina, Claudius' divorced second wife; or (3) Agrippina, who was not only Caligula's sister, but Claudius' own niece. The latter was particularly favored because she had an adolescent son, Nero. Ultimately, Claudius perceived Agrippina as the best political alliance of the three candidates. He wed Agrippina and formally adopted her son Nero, who then became the missing heir to the throne. Although Agrippina was the Emperor's wife, she remained jealous of Lollia Paulina, accused her of sorcery, and tried to persuade Claudius through consultations with astrologers. As a result of these accusations, the Roman Senate confiscated Lollia Paulina's property and the majority of her vast fortune, and subsequently expelled her from Italy. This punishment, however, was not harsh enough for Agrippina. She later enlisted a Roman officer to locate the exiled woman and force her to commit suicide; the officer was ordered by Agrippina to return with the head of Lollia Paulina as evidence of her death (Tacitus, 2004). It is written that Agrippina did not recognize the woman's head when presented to her, until she opened the mouth and inspected the dentition (Dio, 2018).

The first text on forensic dentistry, *L'Art Dentaire en Medicine Legale*, was published in 1898 after forensic dental identifications were performed on 126 aristocrats who perished in a fire at a charity event in Paris in 1897 (Amoedo, 1898). The building hosting the charity event was largely comprised of wood, had a fabric ceiling with a gas-filled balloon suspended from the center, and was primarily decorated with canvas, cardboard, lace, ribbons and paper-mâché. The cause of the fire was determined to be ignition/explosion of a lamp in the building's new projection system (a lamp which utilized ether and oxygen in its operation). Due to the flammability of these accelerants and the building's construction, the fire spread rapidly, tragically trapping the majority of those in attendance inside (Lipton et al., 2013).

Scientific rationale for human dentition as a means of identification

The purpose of human dentition is to assist in mastication of food and in formation of speech. Humans are *diphyodonts*, meaning they develop two separate sets of teeth: (1) primary (deciduous) dentition, and (2) permanent dentition. Primary, or deciduous, dentition contains a total of twenty teeth, consisting of two central incisors, two lateral incisors, two cuspids (canines), and four molars in each of the two dental arches—i.e., the maxillary (upper) arch and the mandibular (lower) arch. By approximately age twelve, all primary deciduous teeth are replaced by the permanent teeth. Within each arch, the permanent dentition (32 teeth total) consists of four incisors (two central incisors, two lateral incisors), two cuspids, four bicuspids, and six molars (Fig. 1). Each tooth type has a specific masticatory function. The incisors (at the front of the mouth) are characterized by a thin chewing edge, and therefore, are useful in cutting or incising food. Immediately posterior to the incisors are the cuspids (also referred

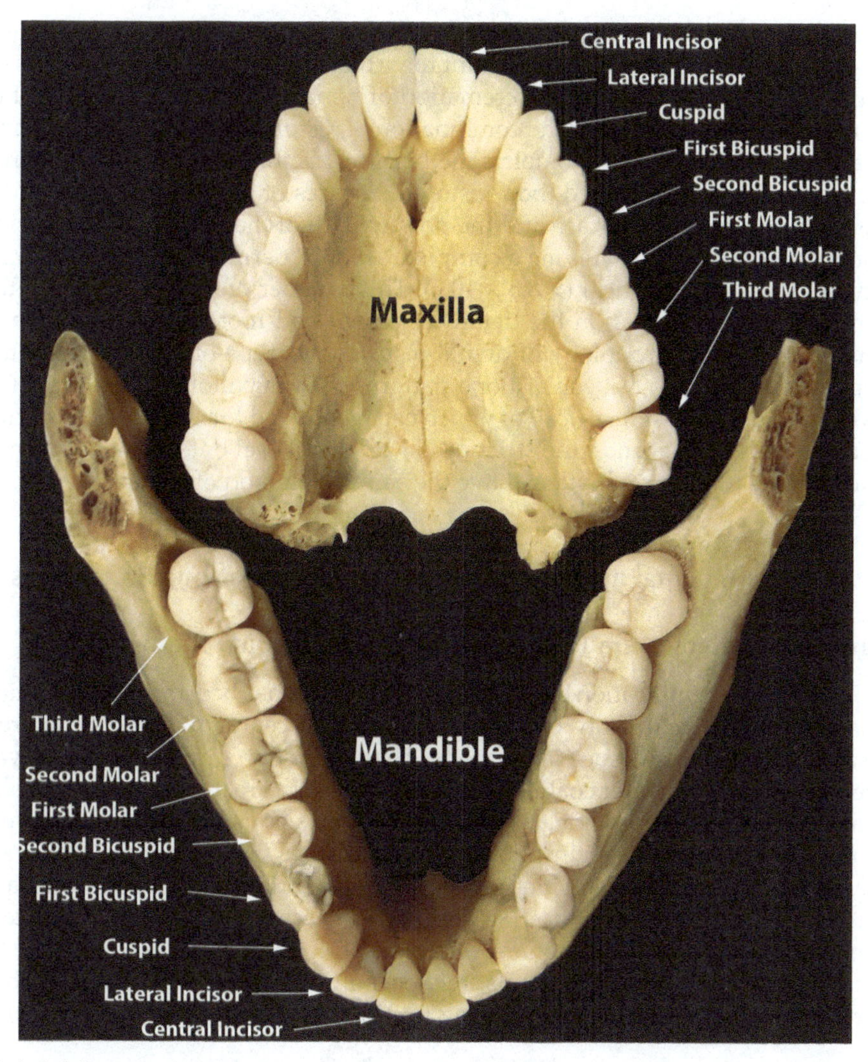

FIG. 1 Labeled diagram of adult human (permanent) dentition.

to as canines), long pointed teeth commonly associated with dogs and predatory animals. The purpose of cuspids is to tear. Bicuspids are transitional in shape between the cuspids and molars. Bicuspids have two cusps (as their name implies), while molars are larger and flatter, usually possessing four cusps. The function of both bicuspids and molars is to crush or grind food.

Loomis et al. (2018) stated that "the rationale for human identification is based on the hypothesis that every person has a unique phenotype, and thus a decedent's biometric data and physical changes that occur during their lifetime can be compared with antemortem biometric/medical/dental records of the presumed decedent or living amnesiac to establish a 'positive identification'." Importantly, the dentition of humans and animals is quite durable. The outer surface of a tooth's clinical crown is composed of enamel, which is approximately 96% mineral content and is by far the hardest and strongest component within the human body.

Dental restorative materials (e.g., fillings, crowns, veneers, implants, etc.) are all quite durable as well. These materials are resistant to virtually all environmental insults—can survive house fires and sometimes even cremation of remains. It is the durability of dentition and dental restorations, as well as the combination of restorative procedures, that make dental identification the preferred biometric method for identification of human remains. The possible combinations of restored/non-restored surfaces of a single tooth (considering the pulp chamber as a restored surface and the possibility that the tooth may be missing) result in 242 possibilities. In cases involving a complete set of 32 permanent teeth, the total number of combinations is astronomical ($> 8.5 \times 10^{39}$). This number of possible combinations does not even take into consideration the details of crown and root morphology, relative tooth position and rotation, morphology of dental restorations, type of restorative material, oral pathology and anomalies, trabecular bone patterns, and maxillofacial anatomy (including the maxillary and frontal sinuses), all of which can be utilized for comparison (Berman et al., 2013). Caution must be taken in the strict use of this statistical data, as not all dental restorative and missing tooth combinations have the same likelihood of occurrence.

Common dental approaches for identification of human remains

Comparison of antemortem (AM) and postmortem (PM) radiographs

Although dental identification began in ancient times, it has become more commonly used since the advent of dental X-ray technology in the 20th century. Dental structures are the hardest and most durable tissues in the body. They resist decomposition and elevated temperatures due to the hard outer coating of enamel, and thus, are the last tissue to be destroyed (Krishnan et al., 2015). Dental evidence persists long after soft tissue decomposition eliminates friction ridge patterns on fingertips. For this reason, aside from fingerprint analysis, dental identification is the most common method of forensic identification of human remains. If fingerprints are not available, identification efforts will proceed with dental evidence analysis, which relies on pattern recognition most often related to differences in restorative treatment patterns on teeth. For many individuals, reference images may be available in the form of antemortem (AM) radiographs taken during routine dental visits (Fig. 2). Postmortem (PM)

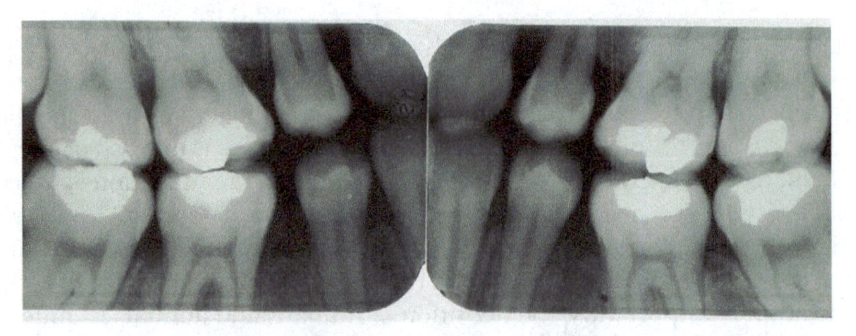

FIG. 2 Example of antemortem (AM) radiographs that can be used for comparison to X-rays of the dentition of decomposed or skeletonized remains.

X-rays are performed on decomposed or skeletonized remains either at a medical examiner's office, or in the field using portable handheld X-ray devices (Fig. 3).

Restorative treatment patterns recorded in postmortem (PM) radiographs then can be compared to antemortem radiographs to assist in the identification of a decedent (Fig. 4). However, in addition to restorative treatments, it should be noted that other patterns or characteristics for comparison (e.g., tooth morphology, sinus configuration, bone trabeculation) may be more subtle, but can be informative for identifications, especially when these features

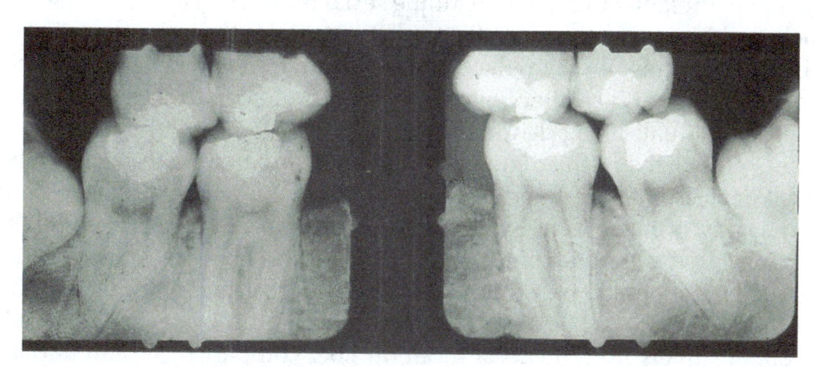

FIG. 3 Example of postmortem (PM) radiographs taken for comparison to antemortem (AM) dental X-rays to assist in identification of decomposed or skeletonized human remains.

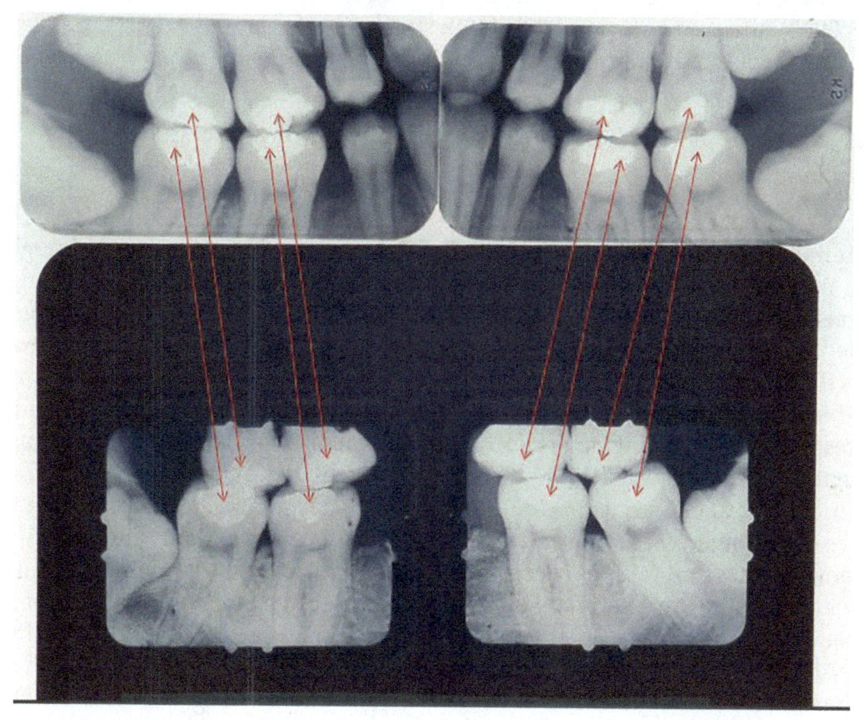

FIG. 4 Comparison of points of concordance in antemortem and postmortem radiographs.

are uncommon or unique in appearance. Ultimately, dental identification is a preferred initial approach because it is inexpensive, less time-consuming, and less labor-intensive compared to molecular-based methods (i.e., DNA typing).

Dental implant batch numbers and unique serial numbers on implanted medical devices

Generally, dental implants do not have unique serial numbers. However, some have batch numbers which can be of use in the identification of specific devices, their origin, etc. (Berketa et al., 2010). If manufacturers were to place unique serial numbers instead of batch numbers on dental implants, a new and easily utilized approach could be developed for the identification of deceased individuals using dental records.

Although not yet realized in the dental community, comparison of serial numbers on implanted medical devices is a rapid and definitive means of identification (Blessing and Lin, 2018). The practicality of this method was tested by retrospectively reviewing 608 consecutive forensic autopsies at a regional medical examiner office. Fifty-six of these cases required alternative methods of identification due to decomposition, gunshot, blunt trauma, or charring. Eight of these 56 cases (14.3%) had an implanted medical device. Five of these eight cases (63%) were positively identified via serial number comparisons. If an implanted medical device is present and medical records are available, identification by means of medical device serial number should be employed. Most implanted medical devices contain unique serial numbers as required by a Federal law passed in the United States in 2007 (Modern Healthcare, 2018); this law was passed to track surgically implanted medical devices in case something goes wrong with them and a recall must be performed.

Dental age assessment (DAA): Forensic estimation of a decedent's chronological age

Forensic assessment of age is a scientific process that estimates an individual's chronological age through the evaluation of skeletal/dental development and maturation. Although human growth and maturation is unique to each individual, dental techniques for estimating age are currently considered the best in assessing chronologic age, particularly during the age interval when the dentition is undergoing morphologic development, i.e., birth through approximately age 20 years (Lewis and Senn, 2015). An advantage in utilizing dental age assessment methodology is that the primary factors influencing dental growth and development are the individual's sex and ancestry. Environmental factors such as nutrition, disease, habits, addictions, and domicile affect dental development only in rare and unusually extreme circumstances.

The American Board of Forensic Odontology (ABFO) has established standards and guidelines for the dental age assessment and reporting (ABFO Diplomate's Reference Manual, 2018, Section IV). These guidelines recommend that a forensic dentist report an individual's age assessment as a statistical estimate of age and possible age interval, at a level of 95% confidence. In addition, the ABFO (American Board of Forensic Odontology, 2016, 2018a, b) has created three (3) supplemental age assessment charts to assist the forensic dentist in the selection of the best methodology to estimate dental age. As shown in Fig. 5, the *ABFO Dental Age Assessment Procedures* chart outlines technique selection based upon specific individual case

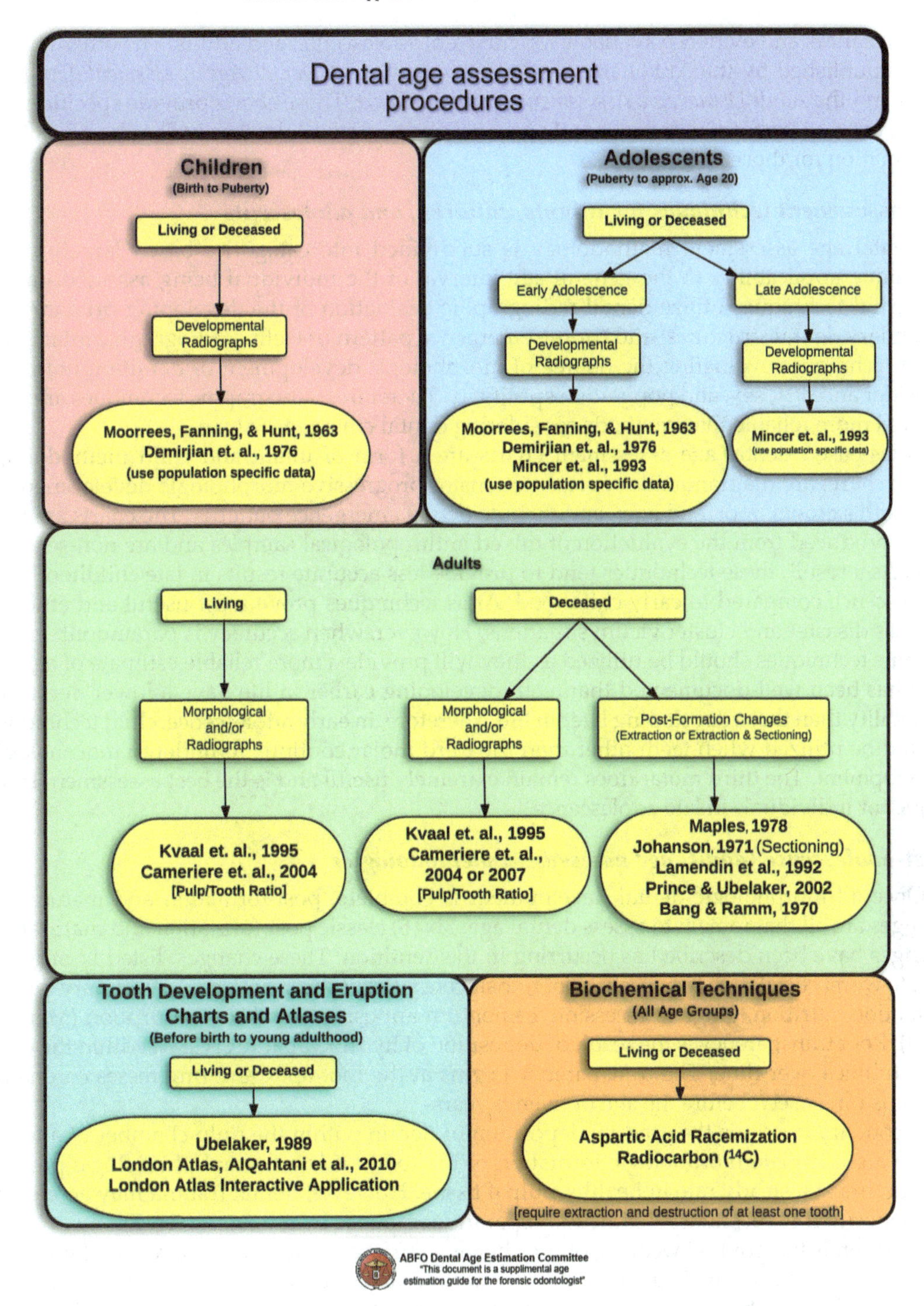

FIG. 5 The American Board of Forensic Odontology (ABFO) Dental Age Assessment Procedures Chart (ABFO, 2016).

circumstances and evidence available for children, adolescents, and adults. The other two (2) charts published by the ABFO are the *Child and Adolescent Dental Age Assessment Technique Chart* and the *Adult Dental Age Assessment Technique Chart*. These charts provide specific criteria required to perform various dental age assessment methodology as well as best practices information for those techniques.

Age assessment techniques for infants, children, and adolescents

Dental age assessment methodology is subdivided into categories based largely upon technique applicability to the general age interval of the individual being assessed. Infant and child techniques utilize dental radiographic evaluation of the developing primary and secondary dental structures, and/or the emergence pattern of teeth. Radiographic infant and child techniques evaluating the degree of morphologic development of dentition normally consider ancestry, sex, and population specificity. Therefore, radiographic techniques are considered more reliable than methods considering dental emergence alone.

Atlas-based dental age assessment charts are a form of infant and child methodology. These charts are diagrammatic representations of progressive morphologic development of the tooth's crown, root, and apex and include dental emergence patterns. The charts are typically produced from the evaluation of mixed anthropological samples and are non-sex specific. As a result, these techniques tend to provide less accurate results in late childhood and adolescence compared to early childhood. Atlas techniques prove most useful and efficient in mass disaster and cluster victim situations. However, when accuracy is paramount, radiographic techniques should be utilized as they will provide a more reliable estimate of age.

It has been well-documented that teeth developing earlier in life have a lower degree of variability than those developing later in life. Therefore, in early adolescence, child techniques should be utilized when teeth other than the third molar continue to undergo morphologic development. The third molar does remain extremely useful and is the best assessment technique for individuals in late adolescence.

Post-adolescence (adult) age assessment methodologies

Once all morphologic dental development is complete, post-formation and maturation changes are all that remain to assess dental age. Six (6) classic post-formation and maturation changes have been described as occurring in the dentition. These changes, listed in order of best to worst correlation to age, are: root translucency/transparency (Fig. 6), secondary dentin formation, attrition, gingival recession, cementum apposition, and root resorption (Maples, 1978). Root transparency is the result of deposition of hydroxyapatite crystals within the dentinal tubules over time. The transparency begins at the root apex and progresses coronally, and it is rarely seen before the age of twenty years.

Secondary dentin is the gradual deposition of dentin within the pulp chamber of a tooth following complete morphologic formation of the tooth. It is a normal physiologic process that occurs at a steady rate in healthy pulpal tissue. However, dental trauma may accelerate the deposition of secondary dentin as a reparative response by the tooth (Fig. 7).

Attrition is the gradual wearing of the chewing surface of the tooth as a result of mastication. However, the normal attrition rate of dentition can be accelerated as a result of bruxism, erosion, cultural habits, diet, or even dental treatment. As a result, today using attrition as an age estimation variable is mostly reserved for archaeological investigations.

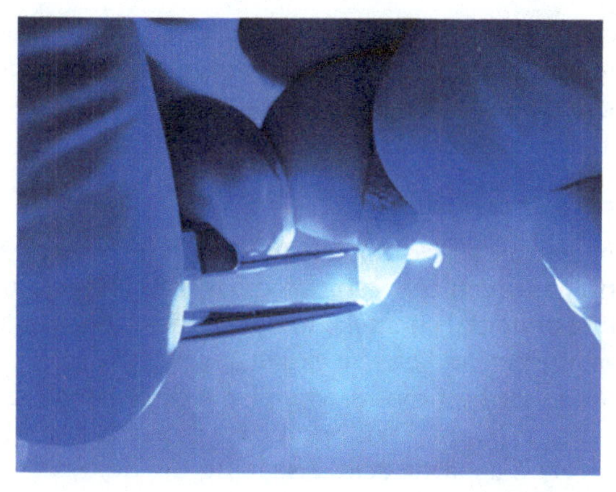

FIG. 6 Example of root translucency in a human tooth, a maturation change used in dental age assessment of unidentified human remains

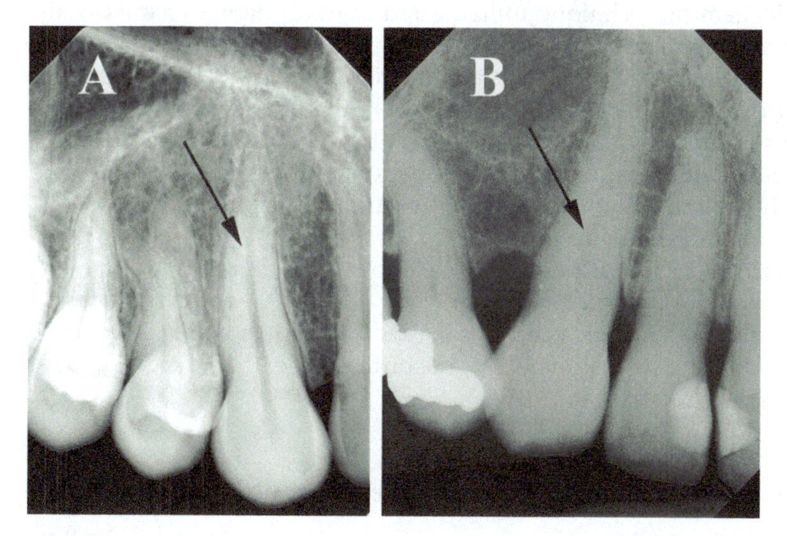

FIG. 7 Radiographs illustrating: (A) a tooth from a young individual with slight secondary dentin formation; and (B) a tooth from an older individual with narrowing of the pulpal space due to secondary dentin formation.

Gingival recession refers to gingival attachment loss and shrinkage of the gingiva and alveolar bone due to an inflammatory response caused by bacterial infection. Periodontal attachment loss can be difficult to accurately measure postmortem. Cementum is a poorly mineralized collagen layer on the outer tooth root surface that attaches the tooth to the alveolar bone. It is gradually deposited on the root surface throughout life. Root resorption refers to the progressive loss of the apical end of the root due to the action of osteoclasts. This is the same physiologic process that occurs in the exfoliation of the primary dentition. Traumatic dental injuries, prolonged localized inflammation, and orthodontic therapy can greatly affect the rate of root resorption. The effect of ancestry and sex is minimal on the post-formation progression of root translucency and secondary dentin apposition. However, dental restorations, endodontic and orthodontic treatments, trauma, decay, pathology, hyper-occlusion, and hypo-occlusion may affect dental maturation and post-formation criteria (with the exception of root translucency) (Johanson, 1971).

Biochemical age assessment techniques

There are two biochemical age assessment techniques which may be applied to individuals in any age group and offer relatively low estimated age intervals: (1) amino acid racemization and (2) radiocarbon dating of dental enamel. However, these tests require complex laboratory procedures, are time-consuming, and can be expensive.

Amino acid racemization is the process whereby a pure isomeric solution converts to a heterogeneous stereoisomer mixture. Tooth dentin is essentially metabolically inactive and at birth is rich in exclusively L-aspartic acid. Assaying the degree of aspartic acid racemization in dentin (i.e., conversion of L-aspartic acid to D-aspartic acid) at the time of extraction or death has provided an assessment of age to an error rate as low as ± 3 years (Ohtani and Yamamoto, 1991, 2010).

^{14}C is a biochemical technique utilizing radiocarbon dental enamel analysis to date the birth of humans within an error rate of ± 1.6 years (Spalding et al., 2005). This test is possible because tooth enamel is rich in carbon and metabolically inert; therefore, its carbon structure is equivalent to the atmospheric carbon levels at the time of crown formation of the tooth. However, this technique is limited by the first above-ground nuclear testing which occurred in 1955. The last tooth completing its clinical crown formation is the third molar and occurs approximately around age twelve. Thus, the earliest practical use of the technique and historic baseline for the technique is 1943.

When aspartic acid racemization and ^{14}C tooth enamel techniques are utilized in combination with other biochemical techniques (e.g., ^{13}C and ^{18}O) and DNA analysis of the human dentition, creation of a genetic profile of the individual is possible. This profile may include information indicating sex, estimated year of birth and death, DNA genotype, and geographic domicile of origin (Alkass et al., 2013).

Additional distinctive morphological features-of-interest in dental remains

Some dental characteristics are distinct and typify ethnic or cultural traits. Others are indicative of systemic disease, trauma, infection, or environmental factors. Nevertheless, the presence of these factors can sometimes help narrow the pool of potential victims when attempting to identify individuals.

cusp of Carabelli

The cusp of Carabelli, or Carabelli's tubercle, is an extra cusp on the mesiolingual cusp of a maxillary molar (Fig. 8). It may be present on either permanent or primary teeth and varies in its physical appearance, from a definite cusp to an indented pit or fissure (Neville et al., 1995a). It is commonly seen in individuals of European ancestry and rarely in those of Asian descent.

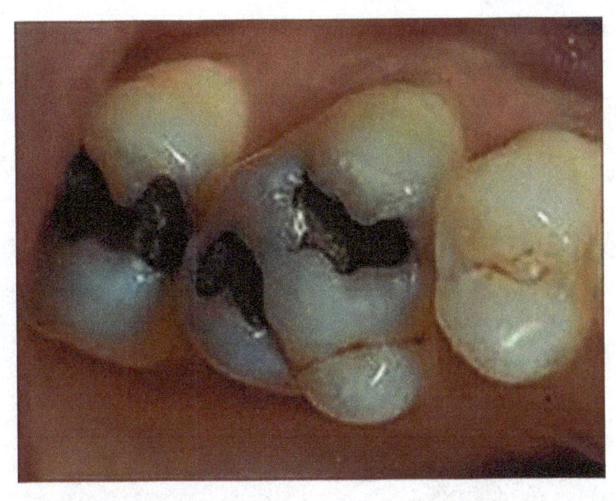

FIG. 8 Example of a cusp of Carabelli, an extra cusp on the mesiopalatal line of a maxillary molar tooth. The cusp of Carabelli is common in individuals of European ancestry, but rarely observed in individuals of Asian descent.

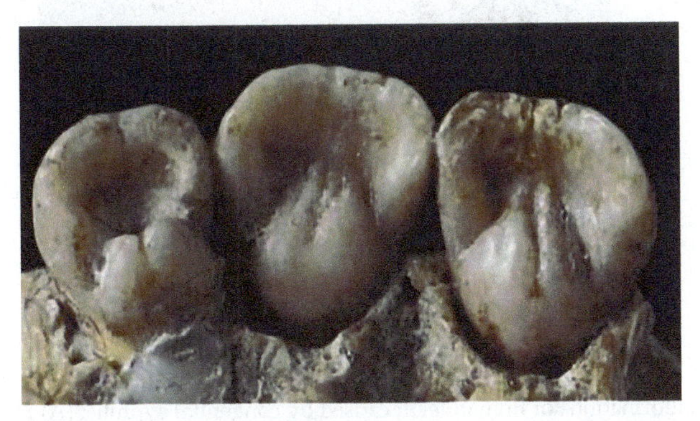

FIG. 9 Shovel-shaped incisors, a distinctive feature in individuals with Asian ancestry. This characteristic is observed in almost all Native Americans and native Alaskans.

Shovel-shaped incisors

Shovel-shaped incisors are a distinctive feature in which prominent lateral margins create a hollowed lingual surface in the incisors, similar to the scoop of a shovel (Fig. 9). This characteristic occurs primarily in individuals with Asian ancestry, with almost 100% occurrence in Native Americans and Alaskans. Shovel-shaped incisors appear in only ~ 15% of the European population (Neville et al., 1995a).

Hutchinson's incisors and Mulberry molars

Congenital syphilis causes distinctive malformations in both anterior and posterior teeth (referred to as Hutchinson's incisors and Mulberry molars, respectively). These tooth deformities are one of three key features (known as "Hutchinson's triad") observed in individuals with congenital syphilis (the other two pathologies being associated with the eyes and ears). Hutchinson's incisors display distinctive tapering of the incisal third of the tooth (Fig. 10A).

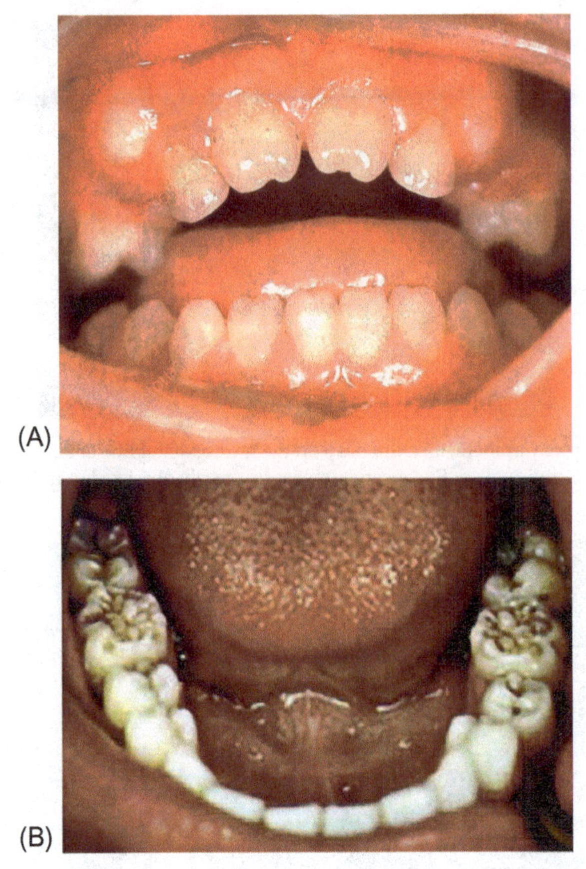

FIG. 10 Distinctive malformations of human teeth caused by congenital syphilis: (A) Hutchinson's incisors, and (B) Mulberry molars.

Mulberry molars taper toward the occlusal surface with a smaller occlusal table for chewing. The occlusal anatomy is also abnormal with random globular projections that resemble a mulberry (Fig. 10B) (Neville et al., 1995b).

Dental fluorosis

Ingestion of fluoride above 1 part-per-million (ppm) often results in enamel defects that stain human teeth. This staining is caused by hypomaturation of enamel that produces an increase in porosity. Staining occurs during the maturation phase of development and produces a brownish staining on the affected teeth. Most individuals that display this type of staining have been exposed to high levels of naturally occurring fluoride. High levels of naturally occurring fluoride are found in a specific region of the United States referred to as the "fluorosis belt." Part of this fluorosis belt includes the state of Colorado, and therefore, one of the first names given to this phenomenon was "Colorado brown stain" (Fig. 11) (Neville et al., 1995a).

Turner's hypoplasia

Enamel defects can be observed in permanent teeth as a result of inflammatory disease, infection, or trauma of the underlying primary tooth. The appearance of the defect, called Turner's hypoplasia, varies depending on the timing and severity of the injury process (Fig. 12). This phenomenon occurs most often in permanent bicuspids because they overlie primary molars, which are more frequently subjected to dental infection (Neville et al., 1995a).

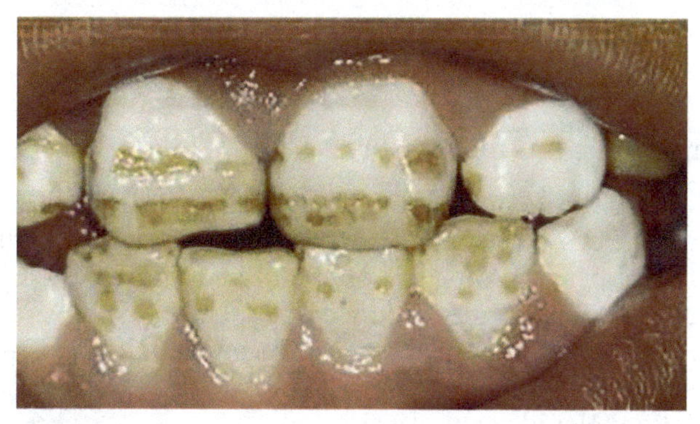

FIG. 11 Fluorosis staining of human teeth (often referred to as "Colorado brown stain"), caused by high levels of naturally occurring fluoride in drinking water in the "fluorosis belt" of the United States.

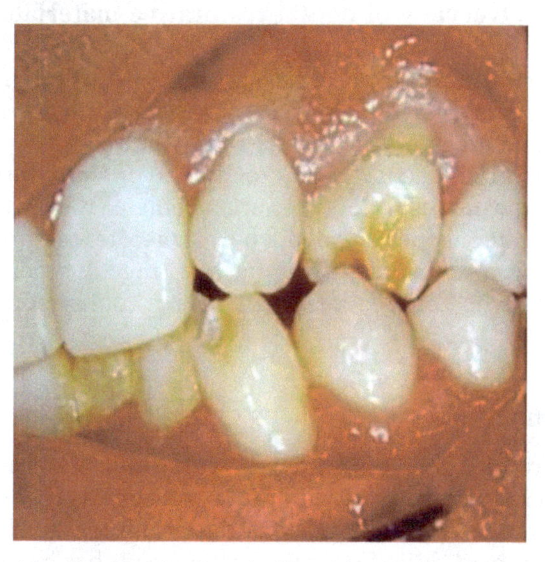

FIG. 12 Turner's hypoplasia, a defect in the enamel of human teeth (usually in permanent bicuspids), can be caused by inflammation, infection, or trauma of an underlying primary tooth.

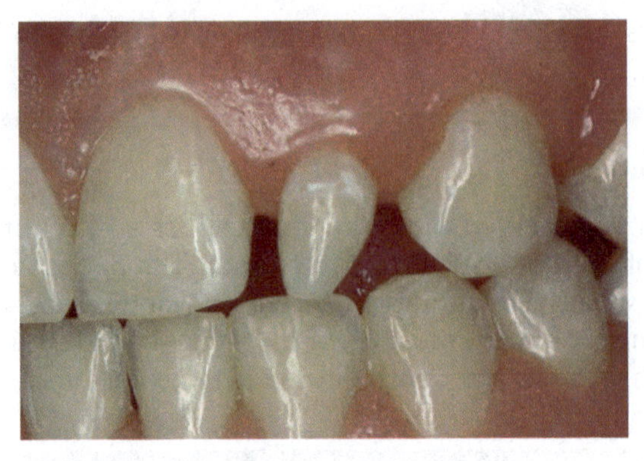

FIG. 13 Example of a peg lateral incisor (a form of isolated microdontia).

Isolated microdontia (peg lateral incisor)

Microdontia is a distinctive condition in which teeth appear smaller than normal. In isolated microdontia, underdevelopment of one type of tooth occurs, resulting in teeth that are comparatively small in appearance and are often pointed (like a cone). The maxillary lateral incisor is the most frequently undersized tooth and appears as a peg-shaped crown overlying a normal sized root. The diameter of the tooth is reduced, and the proximal surfaces converge toward the incisal edge of the tooth (Fig. 13). The prevalence of this feature varies from 0.8% to 8.4% of the population (Neville et al., 1995a).

Overview of dental restorative materials

Like many other health professions, dentistry has seen dramatic improvements in the care that is offered to dental patients today. Part of this improvement includes the types of restorative materials that are available to help maintain one's teeth. While gold and silver have been used in dentistry for centuries, more recent (contemporary) restorative materials include porcelain, composite resins, and ceramic. Some of the early materials have fallen into disuse as newer materials became available. Therefore, the presence of outdated materials can sometimes be of use in narrowing the age range of older dental remains that are found, i.e., in assessing whether recovered remains are contemporary or archaeological in origin.

Gold foil restorations

Direct gold, or gold foil, is the oldest dental restorative material and lasts longer than any other dental material. It is packaged in thin sheets or a vial of small pellets (Dentistry Today, 2006). It is placed in a tooth by means of annealing, which is a laborious process and requires placement by a skilled operator. Although it has been in use for centuries, it is rarely used today due to cost, as well as the time and skill required for proper placement (Fig. 14).

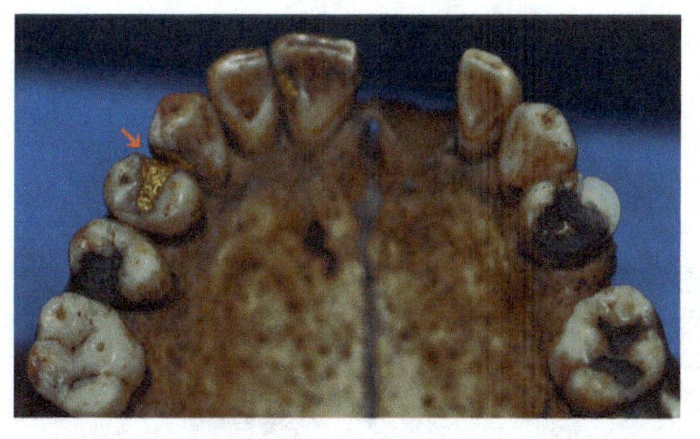

FIG. 14 Photograph illustrating a gold foil restoration (c.1850–1860).

Gold bridgework

Although gold bridgework has been used for a long time to replace missing teeth, it is usually placed after preparation of the teeth to receive it. However, prior to the 1890s when dental engines were electrified, it was a tedious process and was often not done. In those cases, the margins of the crowns were poorly adapted to the teeth, which left open margins that invited decay and ultimately resulted in tooth loss. Many examples of this type of dentistry are observed in the late 19th and early 20th centuries (Fig. 15) (Swank, 2018).

Vulcanite denture teeth

Prior to 1850, dentures were made out of swaged gold and ivory. These materials were very expensive and only available to the wealthy. However, in 1843, Charles Goodyear discovered how to make flexible rubber, which he named vulcanite. In 1851, his brother patented an improved process to make hard rubber. Vulcanite quickly replaced previously used materials due to its affordability (British Dental Association, 2018). Vulcanite continued to be used as denture base material until the 1960s, when it was replaced by acrylic (Fig. 16).

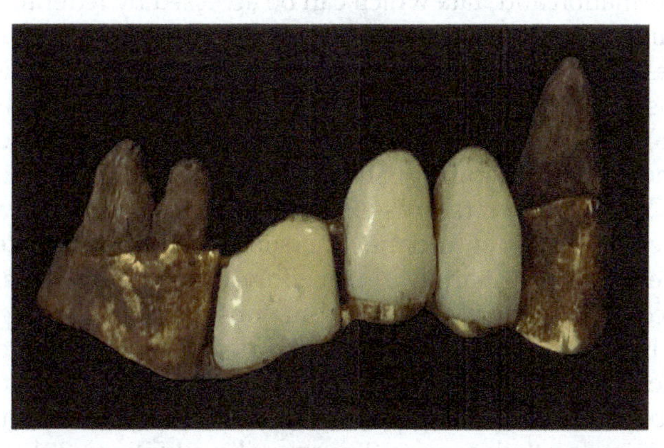

FIG. 15 Photograph illustrating swaged gold bridgework (circa late 19th/early 20th century).

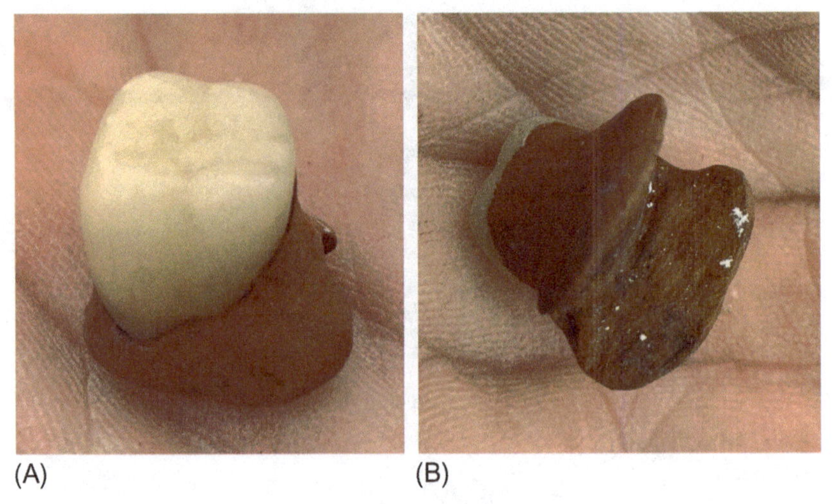

FIG. 16 Photographs of vulcanite dentures: (A) denture tooth with a vulcanite base (occlusal view), and (B) denture tooth with a vulcanite base (gingival view).

Identification networks for the missing and unidentified

There are a number of national and international software programs utilizing dental records as a resource to assist families, law enforcement, coroner and medical examiner offices, and governmental agencies in the identification of missing and unidentified individuals. Two of these programs are discussed below.

National Crime Information Center (NCIC)

The National Crime Information Center (NCIC) was established in 1967 as a part of the United States' Federal Bureau of Investigation (FBI). It is essentially an electronic clearinghouse of crime information and data which can be accessed by federal, state, and local agencies in both the United States and Canada. Among the data maintained is a "Missing Persons File," which houses records on individuals who have been reported missing to law enforcement where there is a reasonable concern for their safety (National Crime Information Center (NCIC)—FBI, 2018). However, this does not mean that all who have been reported missing are entered into the system. Criminal justice agencies have the sole authority to enter dental records into NCIC, and those records are entered in with a classification code specific to NCIC. Unfortunately, most law enforcement personnel lack sufficient dental knowledge and training to accurately code the dentition. In an effort to correct this issue, the Criminal Justice Information Services (CJIS) division of the FBI sponsors dental coding workshops to train dental professionals in the collection, analysis, coding, and submission of dental information to NCIC. Once a dental profile is entered, reports comparing dental profiles can be generated and agencies can be alerted regarding possible matches. Trained dental professionals assist these agencies in reviewing their reports (Kavanaugh and Filippi, 2013).

National Missing and Unidentified Persons System (NamUs)

The National Missing and Unidentified Persons System (NamUs) was established in 2009 and is a free online database system for unidentified remains and missing persons records (https://namus.gov). NamUs is managed by the University of North Texas Health Science Center in Fort Worth, Texas, and is funded with oversite by the National Institute of Justice (NIJ). NamUs permits families, victim advocates, law enforcement, coroners, and medical examiners to share and view information in an effort to establish victim identification. NamUs has established a network of trained forensic odontologists who provide assistance in securing dental records and proper dental coding, as well as scanning and uploading dental radiographs, treatment records, and charting into the system. These dental records are utilized to provide rapid and detailed comparison between unidentified decedents and missing persons. In addition to compiling a database of dental profiles for the missing and unidentified, NamUs also provides services for DNA, fingerprint, and anthropological skeletal examinations.

OdontoSearch

Ideally, forensic odontologists utilize antemortem dental and medical radiographs in comparison to postmortem radiographs to exclusively establish identification, or exclusion, of an individual. However, there are times in which antemortem radiographs are not available and comparisons must be made solely based upon the written dental/medical record and charting. Unfortunately, written dental records and charts cannot establish the identity of an individual to the exclusion of all others. OdontoSearch is a computer application providing an objective method of accessing the frequency of occurrence for an observed dental treatment. It compares the dental profile (i.e., pattern of missing, filled, and unrestored teeth) of twenty-eight teeth (excluding the third molars) to a representative U.S. sample, and then provides the frequency of occurrence for the dental profile pattern observed (Tabor and Schrader, 2010). Although OdontoSearch may assist in understanding the general frequency of a given dental profile, the creators caution that they "do not take responsibility for statistical inferences drawn by users of the program. All are encouraged to understand the strengths and limitations ... prior to its use" (OdontotoSearch, 2018). The latest version of this computer application available at the time of the writing of this chapter is Odontosearch 3.2; it is a free online application that can be found at www.odontosearch.com.

References

ABFO Diplomate's Reference Manual, 2018. ABFO Standards and Guidelines for Dental Age Assessment, Section IV. http://abfo.org/resources/abfo-manual/.

Alkass, K., et al., 2013. Analysis of radiocarbon, stable isotopes and DNA in teeth to facilitate identification of unknown decedents. PLoS One 8 (7), 1–12.

American Board of Forensic Odontology, 2016. Dental Age Assessment Procedures Chart. http://abfo.org/wp-content/uploads/2016/06/Dental-Age-Assessment-Procedures-Chart-2016.pdf.

American Board of Forensic Odontology, 2018a. Adult Dental Age Assessment (DAA) Technique Chart. http://abfo.org/wp-content/uploads/2016/06/Adult-DAA-Supplemental-Technique-Chart-2018.pdf.

American Board of Forensic Odontology, 2018b. Child and Adolescent Dental Age Assessment (DAA) Technique Chart. http://abfo.org/wp-content/uploads/2016/06/ChildandAdolescent-DAA-SupplementalTechnique-Chart-2018.pdf.

American Board of Forensic Odontology, 2018c. Body Identification Information & Guidelines. http://abfo.org/wp-content/uploads/2012/08/ABFO-Body-ID-Information-Guidelines-Feb-2017.pdf.

Amoedo, O., 1898. L'Art Dentaire du Medicine Legale. Masson & Cir. Editeurs, Paris.

Berketa, J., James, H., Marino, V., 2010. Survival of batch numbers within dental implants following incineration as an aid to identification. J. Forensic Odontostomatol. 28 (1), 1–4.

Berman, G.M., Bush, M.A., Bush, P.J., Freeman, A.J., Loomis, P.W., Miller, R.G., 2013. Dental identification. In: Senn, D.R., Weems, R.A. (Eds.), Manual of Forensic Odontology, fifth ed. Taylor and Francis Group, Boca Raton, FL, pp. 75–127.

Blessing, M.M., Lin, P.T., 2018. Identification of bodies by unique serial numbers on implanted medical devices. J. Forensic Sci. 63 (3).

British Dental Association, 2018. https://bda.org/museum/collections/teeth-and-dentures/vulcanite-dentures. (Accessed 6 July 2019).

Dentistry Today, 2006. Gold Foil and Its Use in Modern Dentistry. www.dentistrytoday.com/restorative-134/1838-sp-1547062389. (Accessed 6 July 2019).

Dio, C., 2018. Roman History. translated by H.B. Foster; Book 60, Vol. 4. http://www.gutenberg.org. (Accessed 6 October 2018).

Johanson, G., 1971. Age determinations from human teeth: a critical evaluation with special consideration of changes after fourteen years of age. Odontol. Revy 22, 1–126.

Kavanaugh, S.A., Filippi, J.E., 2013. Missing and unidentified persons. In: Senn, D.R., Weems, R.A. (Eds.), Manual of Forensic Odontology, fifth ed. Taylor and Francis Group, Boca Raton, FL, pp. 193–210.

Krishnan, K., Kanchan, T., Garg, A., 2015. Dental evidence in forensic identification—an overview, methodology, and present status. Open Dent. J. 9, 250–256.

Lewis, J.M., Senn, D.R., 2015. Forensic dental age estimation: an overview. J. Calif. Dent. Assoc. 43 (6), 315–319.

Lipton, B.E., Murmann, D.C., Pavlik, E.J., 2013. History of forensic odontology. In: Senn, D.R., Weems, R.A. (Eds.), Manual of Forensic Odontology, fifth ed. Taylor and Francis Group, Boca Raton, FL, pp. 1–39.

Loomis, P.W., Reid, J.S., Tabor, M.P., Weems, R.A., 2018. Dental identification & radiographic pitfalls. In: David, T.J., Lewis, J.M. (Eds.), Forensic Odontology: Principles and Practice. Academic Press, London, pp. 25–46.

Maples, W.R., 1978. An improved technique using dental histology for estimation of adult age. J. Forensic Sci. 23 (4), 764–770.

Modern Healthcare, 2018. Medical Devices Are Finally Marked With Unique IDs. https://www.modernhealthcare.com/article/20160917/MAGAZINE/309179999. (Accessed 18 November 2018).

NamUs @ UNT Center for Human Identification—Forensic Services Unit, 2018. https://www.untfsu.com/NamUs/Overview.html. (Accessed 15 November 2018).

National Crime Information Center (NCIC)—FBI, 2018. https://www.fbi.gov/services/cjis/ncic. (Accessed 15 November 2018).

Neville, B.W., Damm, D.D., Allen, C.M., Bouquot, J.E., 1995a. Abnormalities of the teeth. In: Neville, B.W., Damm, D.D., Allen, C.M., Bouquot, J.E. (Eds.), Oral and Maxillofacial Pathology, first ed. W.B. Saunders Company, Philadelphia, PA, pp. 44–95.

Neville, B.W., Damm, D.D., Allen, C.M., Bouquot, J.E., 1995b. Bacterial infections. In: Neville, B.W., Damm, D.D., Allen, C.M., Bouquot, J.E. (Eds.), Oral and Maxillofacial Pathology, first ed. W.B. Saunders Company, Philadelphia, PA, pp. 142–162.

OdontotoSearch, 2018. Odontosearch 3.2 Overview. http://www.odontosearch.com/en/3.2/overview.html. (Accessed 14 November 2018).

Ohtani, S., Yamamoto, T., 1991. Age estimation using the racemization of amino acid in human dentin. J. Forensic Sci. 36, 792–800.

Ohtani, S., Yamamoto, T., 2010. Age estimation by amino acid racemization in human teeth. J. Forensic Sci. 55 (6), 1630–1633.

Spalding, K.L., Buchholz, B.A., Bergman, L.E., Druid, H., Frisen, J., 2005. Age written in teeth by nuclear tests. Nature 437, 333.

Swank, S., 2018. Curator, National Museum of Dentistry. Personal Communication.

Tabor, M.P., Schrader, B.A., 2010. Forensic dental identification. In: Senn, D.R., Stimpson, P.G. (Eds.), Forensic Dentistry, second ed. Taylor and Frances Group, Boca Raton, FL, pp. 163–186.

Tacitus, C., 2004. The Annals/Tacitus. Translated, with introduction and notes by A.J. Woodman. Book 12, Hackett Publishing Co, Indianapolis, IN.

Further reading

Crime Scene Forensics, LLC, 2018. http://www.crimesceneforensics.com/History_of_Fingerprints.html. (Accessed 11 November 2018).

DNA Forensics, 2020. DNA Fingerprinting. http://www.dnaforensics.com/DNAFingerprinting.aspx. (Accessed 11 November 2018).

History, 2009. November 8, 1895: German Scientist Discovers X-Rays. https://www.history.com/this-day-in-history/german-scientist-discovers-x-rays. (Accessed 11 November 2018).

Kaushal, N., Kaushal, P., 2011. Human identification and fingerprints: a review. J. Biom. Biostat. 2 (4).

Quora, 2018. What Does "Bench Science" Mean? https://www.quora.com/What-does-bench-science-mean. (Accessed 17 November 2018).

Roewer, L., 2013. DNA fingerprinting in forensics: past, present, future. Investig. Genet. 4, 22.

Techtheday, 2018. How Differently Do Humans and Computers Visually Identify Objects? http://techtheday.com/how-differently-do-humans-and-computers-visually-identify-objects/. (Accessed 11 November 2018).

CHAPTER

24

Forensic facial reconstruction of skeletonized and highly decomposed human remains

Stephen Missal CFA

Maricopa County Medical Examiner Office, Forensic Science Center, Phoenix, AZ, United States; Art Institute of Phoenix, Phoenix, AZ, United States

Introduction

Forensic facial reconstruction is a discipline that has developed over the past 100 + years with the primary objective of taking cranial remains and reconstructing a three- (and now two-) dimensional, identifiable sculpture or drawing of the deceased subject. The history of its development is complicated, but included, among the main personalities involved in its evolution, Russian anthropologist and archaeologist Mikhail Gerasimov, American anthropologist Wilton Krogman, American anthropologist Clyde Snow (in collaboration with artist Betty Pat Gatliff), Richard Neave (British reconstruction artist on staff at the University of Manchester), forensic artist Karen T. Taylor (Texas Department of Public Safety), and British anthropologist Caroline Wilkinson. There have been many other worthy individuals who have also contributed substantially to this growing discipline. Forensic facial reconstruction is complex and still in development, and this chapter will direct potential artists to several books and online resources for further detailed information and instruction (Appendix).

Historically, the main objectives for forensic facial reconstruction are to identify current unidentified cranial remains and to reconstruct historic and prehistoric cranial remains as a sculpted or drawn image, in the most accurate manner possible. Corollary disciplines include composite sketching, which involves a forensic artist (often accompanied by law enforcement personnel) conducting a cognitive interview with one or more witnesses of an unidentified living subject, with the goal of constructing an identifiable sketch of this subject. Similar to this category is the technique of drawing a recognizable image of a subject from still and video closed circuit television (CCTV) and other related video technology, for the purpose of assisting law enforcement with general missing persons investigations.

Forensic Genetic Approaches for Identification of Human Skeletal Remains
https://doi.org/10.1016/B978-0-12-815766-4.00026-1

Other related techniques include age progression, where images of a known subject are drawn or digitally enhanced to accommodate the passage of various numbers of years of the subject and any resultant changes to the face. The final discipline is reconstruction of a deceased subject from a single cadaver image or set of cadaver images. This latter technique has its own unique set of anatomical, forensic, technical, and informational requirements.

For the purposes of this book, it must first be established that extensive practice in drawing and/or sculpting the human head and face is a necessary prerequisite for any forensic facial reconstruction artist. There is no substitute for this acquired skill set. This chapter is not an alternative for such experience. In addition, training in specific forensic sculpting and drawing techniques is highly recommended for the same reason. This chapter will, however, point out some issues and ideas about working on a case (taken from years of actual working experience), with the intention of augmenting any training the artist will procure on their own.

Standard reconstruction methods

Each of the three commonly used three-dimensional methods used to reconstruct a human likeness from cranial remains has slightly differing approaches. Gerasimov developed a means using only anatomical information (muscles, glands, cartilage) which were assembled on a skull with clay in layers, until a finished sculpture of the head and face, covered by a thin layer of clay representing skin tissue, emerged (Gupta et al., 2015). Either clay or taxidermy glass or plastic eyes (with best color estimation) are inserted into protected eye sockets. This is a slow and meticulous technique, requiring sound anatomical knowledge, and it has been recently less in vogue, although used still for fossilized and historical skulls. Taxidermy glass or plastic eyes (with best color prediction, or hazel as a generic color) are inserted into protected eye sockets.

The second method, actually conceived before Gerasimov's, was constructed by Krogman, and its main feature is the use of tissue depth markers, derived from extant charts of tissue depths of various ethnic and age groups (data which are continuously being updated), glued to appropriate locales on a skull (Gupta et al., 2015; Krogman, 1962). Clay is then applied, moving around the markers and using the precise depth of each marker as guidance, until (with foreknowledge of some subsurface anatomy) a facial likeness is produced. Again, clay, glass, or plastic eyes are inserted into the protected eye sockets. Although less anatomically rigorous than the final Manchester method, Krogman's approach is faster under experienced hands and is suitable for reconstructions. Betty Pat Gatliff essentially took this method and refined it into the one we are most familiar with today (Taylor, 2001).

The final three-dimensional technique, called the Manchester/British method, came from Neave and is the last of the physical three-dimensional methods to be developed (Gupta et al., 2015). In this technique, after assembling the skull with the cranium and mandible correctly connected [i.e., all on an adjustable stand with the skull in Frankfort Horizontal Plane (FHP)], with the lower edge of the external auditory meatus lined up with the bottom edge of the orbit, then the tissue depth markers are applied to the skull. The artist then builds, from clay, the muscles used for mastication and facial expression, placed correctly around and through the markers. As with the previously described techniques, clay (or more likely glass or plastic taxidermy) eyes are then inserted into the eye sockets.

The use of computerized three-dimensional reconstructions is in full development and mimics many of the physical techniques above. It is worthwhile in the hands of someone experienced in digital sculpting and who also possesses (or has access to) sound anthropological knowledge. This type of reconstruction can produce images in rotation and has the advantage of being able to be colorized and lighted as the artist desires. The drawbacks, however, are also obvious. Lack of access to digital means, as well as lack of specialized training, combines to make this approach less accessible for many forensic laboratories.

A two-dimensional version of these methods was created by Karen T. Taylor, as a means of dealing with logistic and time issues associated with three-dimensional reconstructions (Taylor, 2001). In Taylor's version, the skull is mounted on the stand in Frankfort Horizontal Plane (FHP), with tissue depth markers then applied (Fig. 1).

Then, using a camera aiming at midpoint and placed parallel to the frontal plane of the skull in FHP, a series of photographs (using a metric/inch guide for scale) are taken in *norma frontalis* and right profile. Other working photographs can be added as the artist needs. The measuring guide is lined up with the frontal plane for that view and with the sagittal plane for the profile (right) view. This author has not found that left profile is in any way less useful for a reconstruction, and therefore, it may be substituted or added for reconstructions. Three-quarter views may also be taken in order to help visualize protrusions and correct foreshortening on various facial bony elements. After photographs are printed at one hundred percent

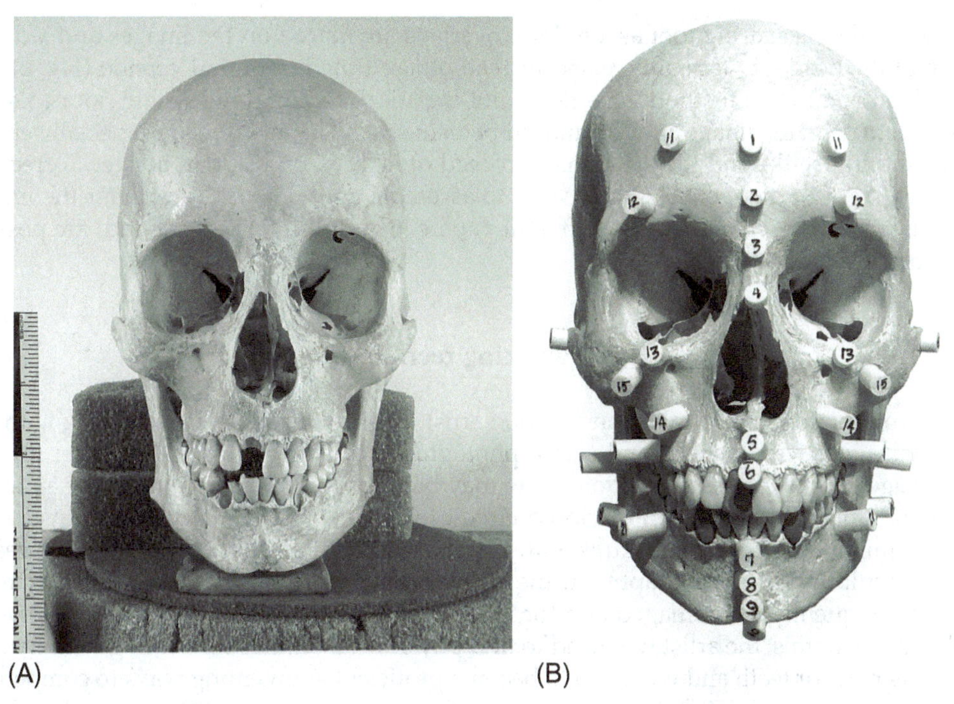

FIG. 1 Forensic photographic images of a human skull: (A) mounted in Frankfort Horizontal Plane (FHP); (B) with tissue depth markers applied. *Images courtesy of Maricopa County Medical Examiner's Office, Forensic Laboratory, Maricopa County, Arizona.*

FIG. 2 Pencil sketch derived from overlays of a forensic facial reconstruction of the human skull depicted in Fig. 1. *Sketch by Stephen Missal CFA; Image courtesy of Maricopa County Medical Examiner's Office, Forensic Laboratory, Maricopa County, Arizona.*

scale (using the measuring tool as a guide), overlays are placed on the images and a drawn version of the head and face are created instead of a sculpted or digital version (Fig. 2). The great advantage to this method is its flexibility logistically. Digital means are not necessary, other than to print the image at one hundred percent scale. The markers are especially easy to understand in profile, and the artist may proceed quickly to produce an image. Corrections require only eraser and pencil, and an easy to assemble light box to place under the images makes the process even more utile. Different papers (e.g., vellum, tracing, bond) are possible to use for the drawn images.

Basic working techniques

As a working forensic facial reconstruction artist, one will most likely be dealing with cranial remains, cadaver imagery, and extant photographs of living subjects (the latter for purposes of age progression). In addition, there may be a request to try to reconstruct an image from a sub-optimal photograph or closed circuit television (CCTV) frame. A forensic artist may be familiar with images of skulls with tissue depth markers and complex armatures for potential sculpting or photography. In the real working world of forensic art, the remains presented frequently are damaged and/or include missing structures of the skull, including teeth. Because of this, the artist will need to find physical substitutes for dealing with damage or missing bone or teeth and will need to become proficient at inventing ways to compensate for this circumstance. It may also be discovered that the remains are still encrusted with dirt or debris and may even have hair or some soft tissue still attached to the skull. It is possible that the forensic artist may encounter semi-mummified remains that have substantial tissue remnants on the skull. Additionally, there may be substantial odor present, emanating from the skull due to decomposition and absorption of gases. This may require the remains to be placed in a controlled (potentially outdoor) environment that is protected from human and animal traffic. An artist's first responsibility is to protect potential evidence and to also provide protection from contagion by wearing latex (or similar) gloves during the reconstruction. The artist will also need to obtain permission to clean the remains and must do so without jeopardizing the investigation of the case. Cleaning the skull can actually increase odor if the skull is from a recent death and might possibly remove valuable trace or DNA evidence.

When a forensic artist is working with a skull, he/she will need a clear working space, a stand for the cranial remains, and a soft boundary (e.g., a towel, pads, etc.) around the stand in case the skull accidentally dislodges or is bumped during transport or setup. Included later in this chapter are instructions for building a simple stand (armature) and extension that can be constructed with minimal tools and materials and will serve most forensic artists' needs under normal circumstances (Fig. 3). In scenarios in which funding is limited or nonexistent, and when specialized tools are not available, the stand/armature presented here is a relatively easy, economical, and efficient solution.

FIG. 3 Simplified armature (stand) for supporting cranial remains for forensic reconstruction and photography. *Image and design by Stephen Missal CFA.*

Other items necessary for forensic facial reconstruction are shims (constructed from cardboard, tile backboard, or similar material) to support and augment spacing of the skull from the stand. This is especially useful for asymmetric skulls, varying sizes of skulls in general, and damaged skulls. Additional tools needed include an X-acto knife (or equivalent), cutting board, eraser tubes or toothpicks, Duco cement or model cement, acetone (for cleaning), a metric measuring ruler, a pen (for marking), and probably a grid drawn on a sheet of Plexiglas, seated in a simple wood base (for lining up of elements and determining both the horizontal and vertical axis). It is also possible to use a small carpenter's level or metal right angle to assist with this task. Lastly, an artist's cutting board, or a heavy piece of cardboard or illustration board, is useful for cutting markers without marring underlying surfaces (Fig. 4).

Finally, the artist will need at least basic tissue depth marker charts, especially those that are common to specific ethnic and population group(s) of interest. These charts are available online, as well as in many of the more well-known forensic art textbooks. Artists should ensure that the working surface is not cluttered, and that the skull is firmly situated on the stand/armature with protective barriers around it before beginning reconstructive work or photography. It may be helpful to temporarily glue the shims to one another and to the base for stability, if shims have been deemed necessary for the current case. Additionally, the top shim(s) can be temporarily glued to the skull for the same reason. However, it is important to make sure that the glue used can be easily dissolved using available solvents (like acetone) and a soft brush or cotton cloth. Cotton balls for final clean-up are also a good idea.

If the goal is creation of a sculptural facial reconstruction, the artist will also need to have available some kind of non-drying clay, such as Plastalina (non-sulfur if possible, for toxicity reasons). To work with non-drying clay, a simple set of plastic ceramic tools will suffice,

FIG. 4 Forensic artist's cutting board. *Image courtesy of Stephen Missal CFA.*

although dental tools, nut extraction tools, and other similar devices can be incorporated as supplemental sculpting aids.

Sometimes law enforcement will require the sculpture to remain intact, in which case the artist will have to remove the skull and sculpture from the stand/armature and safely store it on another armature. In all other circumstances, the protocol typically involves removing the clay from the skull (or cast) after photographing and documenting the reconstruction.

Sculpting is inherently intrusive to the skull. It is quite easy to damage a skull while preparing it and adding clay. The eye sockets and nasal opening should be filled with cotton and then over-taped with artist's tape as a preventative measure against possible damage (and to provide a substrate for placement of the first layers of clay in those areas as the reconstruction commences). The artist will have to assess whether the skull is robust enough to accept a clay layer safely. If it is necessary to have had to build or replace missing parts of the skull (including teeth), one may find a clay overlay difficult to add without damaging the sculpted clay substitutes for teeth or bone. It can be useful to shield clay substitutions of missing bone with paper or plastic wrap to prevent them from sticking inadvertently to the reconstruction clay added later.

Drawn reconstructions only require several drawing pencils (HB to 4B), a pencil sharpener, a kneaded or rubber eraser, a good light source, a drawing stand with back lighting (which can be easily constructed if none are available), and vellum, tracing paper, or rag marker paper (or other similar substrates) that can be placed over photograph(s) of the skull when the forensic reconstruction is being drawn. A paper sheet with a grid placed underneath the drawing can be useful in helping place features and in keeping essential symmetry correct. The latter also is helpful in avoiding skewing the drawing on diagonals, which is a common occurrence for right-handed or left-handed artists.

Ideally, the forensic artist will have an anthropologist or anatomically trained individual or team to predict or assess the approximate age, gender, and ethnicity for the particular cranial remains in question. Other pertinent information, such as clothing size, hair color/length/texture, and so forth, would also be of benefit in a reconstruction. It is recommended that forensic artists keep a standard informational form for detailing the case number, measurements, and observations of skull properties, including a detailed dental chart. This should also have date(s) pertinent to the project and any other information or notes that would be valuable for legal and identification matters in the future. Additionally, there will also be a legal form called a chain of evidence (or chain of custody), granting the artist legal rights to hold and use the cranial remains until the reconstruction is finished and documented. The local legal representative and the artist will have to sign and date this form twice, both for initial acceptance of the remains and then later upon return of the remains when the project is concluded. Both the artist and the legal agency should keep separate copies of the chain of evidence (refer to examples in Appendix)

Building a simple cranial stand/armature

When doing forensic facial reconstructions from cranial remains, there are two essential ingredients to success. The first is a combination of artistic skill, basic anatomical knowledge, and experience. The second combines useful adjunct materials and a skull stand/armature.

As stated previously, success is highly dependent on the artist's careful handling of cranial remains so as not to break or mar any of the bony elements. A cranium should always be handled and moved with two hands, fingers should never be placed in the eye sockets (the walls of which are very thin and fragile), and suitable soft boundaries (such as rolled towels or padded long soft containers) should be strategically placed to protect the skull in case it rolls or is inadvertently moved. If necessary, the artist may hold the cranium by inserting a thumb in the foramen magnum and using the other hand as a brace on the back (occipital) portion of the skull. Handling should never be rushed, and it should never be assumed that the surface available for working on is adequate without first cross-checking it and preparing the materials and stand that are going to be used in the reconstruction. If working with more than one skull at a time, the forensic artist may require a separate surface for holding the other remains, unless there are additional skull stands/armatures available.

Although there are several varieties of sculptural artist armatures available to purchase, the forensic artist may wish to build his/her own, which can become technically challenging when trying to find parts and assemble them into an adjustable, sculptural stand. Because many sculptures are done in standard art classes using traditional armatures, this mode has been a default for many reconstruction artists (both in practice and in teaching). However, because an actual skull (whether complete or partial) is to be mounted on this apparatus, issues regarding the integrity of the skull become apparent when the reconstruction process begins. It is not the same as working on either an inert substrate or simply a clay core, since now the internal element is the skull itself. These matters can be compensated for to a certain degree, but issues such as clamping the skull to a metal armature stand via the foramen magnum invite damage from compression and the insertion and release of the clamping mechanism. Furthermore, a skull may move or slide unless the clamp is quite firm, which may be harsh and/or destructive to delicate bony tissue from decomposed or degraded remains. Some elements of a makeshift stand/armature that have been mentioned in the standard literature are difficult to locate or are simply not available anymore; hence, substitutes must be found to replace them. This is possible, but there are other ways to deal with this issue.

As an inexpensive and viable option, the author has created an alternative stand that utilizes simple materials, combined with shims of various depths and an extension support (for sculptures) if desired. The stand described herein uses readily available materials and is simple to assemble and easy to adjust. Further, it is stable, lightweight, sturdy, and does not pose risk of harming the skull (as some clamped versions of armatures are prone to doing). Several of these stands can be built in a relatively short period of time, and therefore, can address the need for equipment to support multiple reconstructions from multiple different cases simultaneously. Larger or smaller versions can be created as well, including the use of different PVC pipe diameters for child and adult skulls, which can have much different foramen magnum diameters. A simple blueprint for the parts described herein is shown in Fig. 5.

If using a material like tile backing board (which is illustrated for this book), only a box cutter and a metal metric straight edge are necessary for cutting apart the pieces before assembly. The design included in this chapter uses tile backing board, which has a 5 millimeter (mm) depth. This measurement should be considered and matched if alternative materials (e.g., stacking or laminating ragboard) are also used in the design. Other easily cut materials, including robust cardboard or illustration board, will also suffice for the external structural role. Once all external pieces have been cut, and after double-checking their measurements,

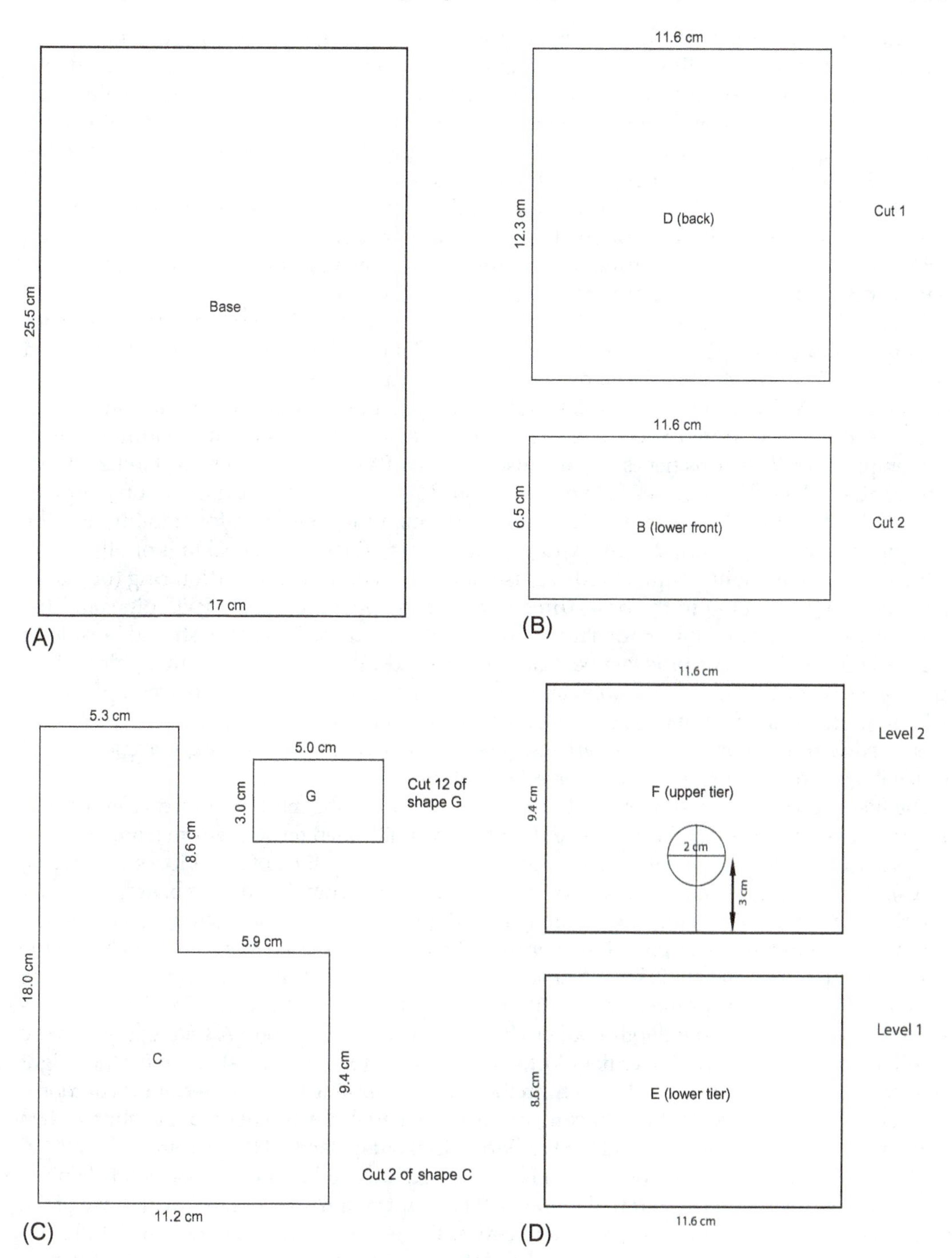

FIG. 5 Blueprints for constructing a simple armature (stand) to support cranial remains for facial reconstructions and forensic photography. *Design and image courtesy of Stephen Missal CFA.*

the stand may be assembled. First, starting from the base, place a foam or wood block that measures 10mm less than the width and length of the primary block shape of the stand and 5mm less than the vertical measurement for the back (template D). Allow a basically even border around the block and glue it down. The base will keep the assembly from tilting and will provide space for the metric measuring tool triangle. Once the core is firmly glued to the base, and after the glue has dried, the artist may move on with the assembly. Constructing the core first allows the other pieces to be adhered to the core without having to use devices to keep them in place. If the core piece of foam or wood is slightly too large, the extra amount can be adjusted by sanding, cutting, or trimming before actual gluing of the various faces commences (thus avoiding gapping at the joints).

To assemble the armature (refer to Fig. 5), the artist can begin gluing each side and surface on as the stand progresses upwards. Sides C1 and C2 should be first, and then, the top and lower steps (F and E) can be placed on top of the two tiers. Next, the front pieces (A and B) can be glued on. The top tier piece, F, should have a 21-mm hole drilled or cut into it with the center being 3cm from the edge of the front face and 5.8cm from either width terminus (see template for F). This is necessary for insertion of a PVC pipe in order to stabilize it in a vertical mode. The PVC pipe needs to extend about 13.3cm above the surface of top tier piece F. The hole should be drilled more than halfway through the core piece for stability, and the PVC pipe can then be inserted. Any extra gapping can be filled with spackling or other malleable materials that will dry to a hard consistency. Alternatively, an insertion ring (of plastic or metal) can be adhered to the base, directly below the opening for the PVC pipe, and the pipe can then be inserted into it for stabilization. In this case, the PVC pipe should be at least 28.2cm in full length. The pipe can be glued or screwed into immobility at this point (obviously one side or the back of the total assembly will have to be left open to accomplish this task). If no core is used, as in this scenario, then care must be taken to assure that all pieces are assembled with right angles at all vertices. Blocks of wood or carpentry right angles would suffice in guiding the pieces into the correct orientation.

The final pieces for the stand are the two support blocks for the mastoid processes and a contact group of three blocks for the mandible. These can be built as single pieces from foam or wood, or alternately, assembled from three 5-mm pieces per block (for a total of four blocks and twelve pieces). On tier F, the top level, take two of the stacks of three 5-mm tile backing board, or similar material (or wooden blocks that are 15mm tall), and glue them 15mm in from the outer margin and 5mm in from the face of piece B (lower frontal piece). This should leave a gap between the two blocks of approximately 2.5cm. A space is provided in the latter in the front to allow for the tissue depth marker for the menton (#10 tissue depth marker) location. (Note: the reader is advised to reference a tissue depth marker chart to verify this location). A final optional piece between the back of the two lower tier blocks can be added for increased stability for the height adjustment shims. The shims can be cut from the same tile backing board, or rectangles of robust card board, illustration board, Bristol board, or similar materials can be cut for use as shims when lifting mandible or mastoids to fulfill the Frankfort Horizontal Plane (FHP) placement.

The Frankfort Horizontal Plane (FHP) is a standard adopted in the late 1800s for displaying, studying, and comparing cranial remains. It is an approximation of the "level" base of the cranium in relation to the *x*-axis, *y*-axis, and *z*-axis. The remains must be in FHP to fulfill standard international protocol so that any other artists or agencies may view images of the remains and be confident that they are not distorted. The shims may also be temporarily glued together and to the base using commercially available gums, glue, or plastics (often used to adhere images to walls without damaging the walls) so that they will not move laterally once weight is placed upon them. Shims may also have other means of connecting them via small dowels or other devices created by the artist. A container for holding the shims would be useful. It is recommended that thicker and thinner shims be made in order to fine-tune the height of the skull in all directions depending upon the need of the forensic artist. It is even possible to have a recess built into the stand itself for storing these shims.

To complete the stand, take the final two 15-mm tall blocks and glue them in the front corners, aligning with the side and front, to support each mastoid process. The inside margin of each block should be approximately 2.5 cm from the center. Once the entire stand is constructed, check for any gross misalignments or level problems and, if found, disassemble that part and modify until the unit is vertically and horizontally square for practical purposes. The stand should be light and easy to move, but quite stable even with a heavy set of cranial remains on it. Small measurement errors are not crucial; the skull merely needs a stable platform that can have height adjustments in order for the remains to be photographed and/or sculpted upon.

A small metric and American measurement ruler can be glued to a right angle brace composed of the same material used to make the stand itself. This is useful for providing a mobile, vertical measuring tool that can be moved for the artist's convenience in photographing the cranial remains.

All items needed for mobile forensic reconstruction can be placed in a robust container and moved from studio or lab to other site locations and back easily and would weigh very little. To summarize, the artist would need the stand, shims, model glue and solvent, sticky pads or gums, artist's pencils and erasers, measuring ruler, a right angle device or carpenter's level to verify horizontality, tube erasers or toothpicks for tissue depth markers, an X-acto knife or box cutter and blades, a cutting board or surface, a pad of vellum, Bristol board or high-quality drawing paper, oil-based clay (optional), a small portable light box, a camera and a tripod or support for the camera, and the artist's forms for detailing in writing measurements and observations as the project unfolds. In addition, dental charts (adult and child) and tissue depth marker charts should accompany the artist. This entire package can be moved using a duffel bag or sturdy long box with handle suitable to hold the greatest width materials (stand). It would be practical to divide it into compartments that are padded for extra protection.

An alternative, simpler skull stand also is possible. In this version, the vertical PVC tube is extended using a dowel insert, and a simple "L" bracket is attached along with a foam form for fitting into the palate of the cranium for stability. This stand can be used if the skull must be entirely free-standing in images or for sculpting. Chicken wire or foil can fill the gap below the skull if the neck is to be included in the sculpture. The author has successfully used the simpler stand for this purpose, but each artist must choose what is most appropriate for the specific case at hand.

Two-dimensional (2D) reconstruction

If there are limitations inherent in the artist's circumstances (e.g., lack of materials and/or space, time constraints), the two-dimensional methodology for reconstruction pioneered by Taylor is recommended (Taylor, 2001). The technique is clearly outlined in her book and has been adopted by a large number of forensic artists with great success. It is recommended that the book be purchased (perhaps using funds from law enforcement) and that the techniques outlined within it be utilized for successful outcomes. The main point of the two-dimensional technique is to avoid the issues inherent in the application of clay to a cranium and to expedite the time necessary for accomplishing the reconstruction. To reiterate an earlier statement, forensic artists must use sound, tested forensic texts, and guides in order to accomplish a valid reconstruction. It is a task filled with nuance and interpretation based upon the latest research, and as such, requires at least a basic knowledge of these techniques and data sets. Onsite, tactile training with a competent forensic reconstruction artist is highly recommended; if this is not possible, one should seek online or digitally interactive ways to get the training.

The author recommends trying both the two-dimensional and three-dimensional (clay) technique to become familiar with the advantages and disadvantages of each approach. No matter the circumstances, it must be remembered that in photographing the skull (with or without tissue depth markers) a scale must be placed either at the frontal plane (with *norma frontalis* imagery) or at the sagittal plane for profile views. These images may then be transferred to a computer and printed at one hundred percent scale for use in constructing the final image. When printed at 100% life scale, measurements of various features can be made and transferred without fear that they have been distorted due to estimation. When doing two-dimensional drawings, the artist may use translucent vellum, heavy tracing paper, or even light drawing paper; however, this is dependent upon use of a sufficiently bright backlight placed underneath any drawing box or table, with a transparent pane of Plexiglas or glass attached to a simple wooden frame for supporting the photographs and overlying paper. All can be safely taped using low adhesion tape. The case number should be strategically placed on every image so that future scans, revisions, and analyses can easily be associated with the case without confusion.

The early part of the drawing should be light, involving controlling the large shapes and placing features properly (as, for example, putting the correct sized eyes in the orbits). As the drawing progresses, the artist may flip the overlay up and down in order to more clearly see the underlying cranial image for clarity sake. Smudges and preparatory lines can be erased as needed, and the final result can be scanned for a digital copy and back up. Again, each of these images should include the case number (at either the top or bottom of the image) and should have a folder or plastic slip-sleeve to place into for both protection and storage. The chain of evidence and any written notes about the case should also be stored along with each image.

Reconstructions are done in rigid sequence, and the artist should maintain this protocol. By doing so, they will ensure that no steps were forgotten or glossed over that might later invalidate the final product. This will also reassure law enforcement that any work done has followed general international protocol and that resultant images are acceptable for legal purposes. All mishaps and revisions should be documented if possible. The forensic anthropologist or lab technician should be consulted if any questions arise or if there are issues with the skull (e.g., post-acquisition damage). One common issue is the problem of loose or completely unattached teeth. These should be documented and photographed (if possible), and then reattached according to normal human dentition charts. If there is any question as to the reinsertion of the loose teeth, legal and/or professional dental authorities should be consulted.

Because cranial remains can also be photographed and drawn, the utilization of clay or similar products can be left as an unused option, should the lab and anthropologist feel comfortable with this alternative methodology. The main considerations are time, material cost, accuracy of outcome, and logistic ease. When photographing the skull, the camera's limitations will dictate distance from the skull while taking photographs. In general, the camera should be at least a couple of feet (roughly 1½m) from the skull, avoiding the worst of fish-eye lens or barrel lens distortion. It is recommended that optional back-up photographs be taken at slightly shorter or longer distances, with grids or measuring devices suitably placed alongside the skull, in order to ascertain which distance was optimal for the subsequent drawn or sculpted reconstruction.

Frontal, profile, and three-quarter views are useful for reconstructions. Although not commonly referred to in standard texts, three-quarter view photographs (both without and with tissue depth markers) can be useful to the artist in understanding the full nature of each skull form (i.e., nasal, orbital, zygomatic, etc.). With this view, issues involving foreshortening or overlapping forms can be resolved and are useful for others to see if questions about the reconstruction are raised. Because sculpted reconstructions will inevitably begin to hide the landmarks underneath as they progress, it is very practical to have the particular case skull photographs available for reference should any questions arise. If the reconstruction takes more than one day, especially in the setup and photographing stage, a suitable covering and storage arrangement should be made to protect the cranial remains until work resumes.

Cadaver image reconstruction

Because much of the workload of forensic artists is not strictly from cranial remains, but instead from postmortem imagery, there are an entire set of guidelines and issues that evolve from this part of the forensic artist's needs in order to produce a viable and useful reconstruction. In this arena, the variety and difficulty with utility of images expands exponentially. Many photographs are taken in situ (either at the crime scene or discovery location), and sometimes additionally at the morgue, medical examiner's building, or in a forensic laboratory itself. Photographs may be taken by law enforcement or individuals untrained in optimal imagery, which can introduce a host of challenges for the forensic artist when collected and provided for a case reconstruction.

Among the variances from ideal are the following: fish-eye or barrel lens distortion; rotation of the deceased in x-axis, y-axis, and/or z-axis (which produces difficult to interpret foreshortening); obscuring materials between camera and subject; poor lighting or flash lighting; bloating from outgassing and decomposition; insect activity; predator activity; injuries sustained ante- or postmortem; fungal or other living entities extruding from or obscuring parts of the remains; clothing or hair obscuring the image; detached parts of remains (e.g., eyes or teeth); inadvertent shadows; blood; color distortion; burned tissue; and mummified tissue. The possibilities are endless, and the artist will encounter most of them eventually.

Burn victim images (if the damage is extensive) should be avoided unless there are no other alternatives. It is literally impossible to accurately predict what the living tissue would have looked like when extensively burned tissue is the only available visual resource. If permission can be obtained to extract the skull from the remains, the artist should opt for this choice. Often images are from cold cases, and there may be no extant remains to modify or examine. In that case, the only possible solution involves educated approximation and analysis of head and face shape (if determinable), any protruding or obvious bone structure, orbit size and placement, and teeth size and placement (if visible). These are poor substitutes for standard evaluation, but can be used as a last resort if circumstances require it.

Drowning victims present a puzzle all their own. The facial features of drowning victims become severely distorted as the body decomposes, and reconstruction should not be attempted until the gas has dissipated and the features have at least nominally reduced in scale to their normal stature. If this is not possible, features should be chosen as defining landmarks with which to approximate the living appearance of the victim. For example, the scale and shape of the eye orbits might be possible to determine (along with a general idea of the scale of the eye) and, combined with the length and breadth of the nose, can be used to begin a first initial drawn image. The distance between the nose and upper lip, and between the mouth opening and the base of the chin, should be added, as well as the height to the apex of the forehead. If present, hair (which often sloughs off during decomposition) can help determine identity as well. Issues that can affect the accuracy and quality of images are multiple. First, decomposition may be so advanced that the remains would be better de-fleshed to access the skull for reconstruction purposes. Second, photographic images of the victim must be done correctly, or at least as closely to protocol (i.e., FHP) as possible. This avoids unnecessary guesswork on the part of the forensic artist, in terms of possible foreshortening or rotational image distortion. Additionally, flash photography can reduce detail and make certain features appear either raised or recessed more than they actually are; hence, it must be used with caution and without overwhelming the facial features.

A successful forensic reconstruction is indeed possible from a drowning victim. Fig. 6 depicts a Hispanic female drowning victim. In this case, there was just enough information available in the original image (even with bloating) to "reverse engineer" the final image.

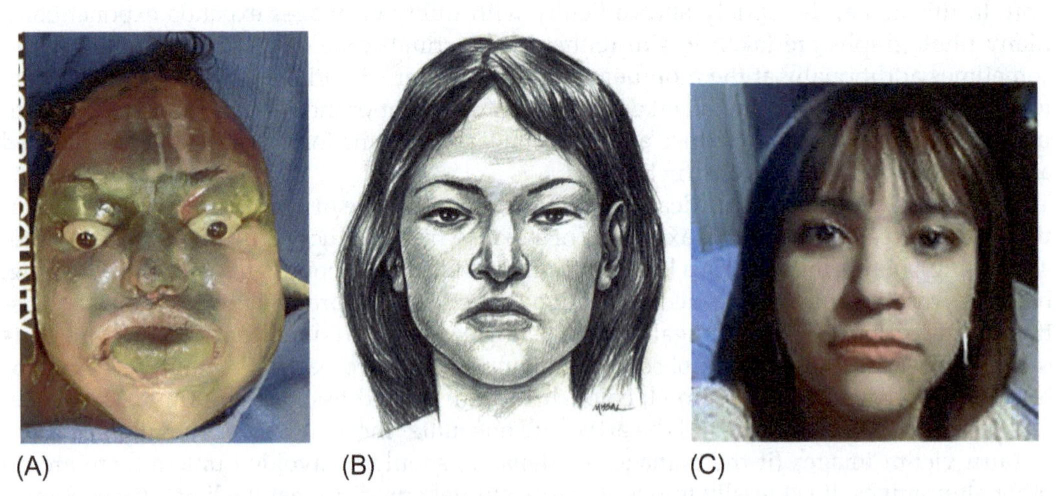

FIG. 6 Forensic reconstruction of a Hispanic female drowning victim: (A) postmortem cadaver image; (B) reverse-engineered pencil sketch (by Stephen Missal CFA); (C) actual photograph in life. *Images courtesy of Maricopa County Medical Examiner's Office, Forensic Laboratory, Maricopa County, Arizona.*

The level of decomposition was not too advanced, and despite inflation of the features, the features were distinguishable enough for a drawing to be completed and an identification to be made.

There are, however, instances where the best the reconstruction artist can do is provide a general sense of the head proportions, the locations of the facial features, their approximate scale, and the shape and size of the nose. Ears may have to be extrapolated, as they often are distorted or partially missing by the time the artist begins analysis. The most common mistake is to assume the face is larger than it actually is, especially when there is still some bloating from outgassing. It might be safer to err on the side of slightly too small rather than too inflated.

Victims in cadaver imagery can display a wide variety of tissue and substructure damage. There may be a point of no return where the artist decides, based upon available visual cues, that a reconstruction is not possible. If the information in the image or images seems sufficient, the artist must then begin assembling the most important information and do preliminary sketches to determine if the interpretation of a particular visual datum is correct.

As addressed previously, quality of photographic images is crucial. The forensic artist must determine as best they can whether the camera was in alignment correctly with the remains, or whether there are one or both of the following problems. First, if the camera was too close to the head when taking the photographs, the area closest to the camera will be enlarged and distorted in a "fish-eye" lens manner, protruding too far into the near "*z*" axis and larger than it should be. In most cases, this determination will have to be an estimate, based upon imagery and cases that do not exhibit this problem. Usually the issue will be overt enough to warrant modification of the forensic drawing. In a sense, the artist will (as in other cadaver images) "reverse engineer" the image, shrinking the frontal area until it seems to fit parallel images in terms of proportion. The change must be greatest (in reverse) where the image was most enlarged, and less where it was not so great. The entire head may be subject to this exercise, and it will test the drawing skills of the artist to the utmost. The second issue is complex and involves rotation of the head in any of the three cardinal axes (*xyz*), i.e., horizontally, vertically, and finally to and from the viewer spatially. It is entirely possible that in the image in question the head exhibits all three rotations. There now exist two intertwining challenges: (1) correction of the foreshortening in each of the rotated directions and (2) understanding how the rotation might affect such things as the size and depth of any damaged areas. Lividity and rigor mortis also are important considerations. Lividity, the pooling of blood to the lowest point of gravity, produces a faux bruising or discoloration that can be mistaken for injury. Rigor mortis, the gradual (impermanent) stiffening of muscles in the body as the muscles lose adenosine triphosphate (ATP) during initial chemical changes after death, can distort facial features before the condition eases. Finally, natural sagging and laxness of muscle tissue after death can drag masses towards the gravitational source, again changing the manner in which the living tissue would have existed and been observed in life. Any rotation in each of the three axes can distort any or all features and tissue, and so the artist must be especially alert to cues that would indicate this.

To change a rotation back to normal FHP orientation, a forensic artist must understand how basic foreshortening works in perspective. In true perspective, we must remember that the habit of seeing and interpreting the world in a flat manner, with true right angles, is in fact not entirely true. Because we see the world with the kind of eye and lens that nature provided, what appears to the naïve eye as true squares and rectangles are in fact already very slightly distorted in a mild "fish-eye" lens manner. This will allow the pairs of parallel lines to move away from the center of the rectilinear figure (i.e., the x and y axes), converging towards their natural vanishing points. Because we look at things and convert data into a simple picture, distortion of the curve of each side of the rectilinear figure is not in our awareness. One outcome of this is to note that any regular, bisymmetrical shape will have a very slight reduction in scale as we focus on the receding areas that are in the 45 degree zone (corners of a square or rectangle encasing the image), and that both the vertical and horizontal axes will be diminished by the movement towards vanishing points. This is a subtle and not widely known fact, even among trained artists. It clarifies, for example, why looking at a skull in person seems to especially diminish the y-axis in length, despite appearances from photographs (which will also exhibit this phenomenon to a certain extent). That is, the vertical measurement will seem slightly shortened, due to our emphasis on this direction in viewing faces. When doing diagrams of skulls in profile and matching them up with anterior views, and then subsequently comparing these to actual photographs and live examination, the artist may be frustrated by the change between what "should" be a vertical matching of dimensions and what is actually observed. Because we are drawing what humans see when they look at other humans in real life, distortion or diminishing of these axes is normal and should be taken into account. The effect is more and more apparent as the viewer gets closer to the subject, which is why photography should be done at a distance where this is minimized.

Now, in order to correct rotation, the artist should estimate the foreshortening (that is, shrinkage in the direction of the axis) for each direction of rotation, remembering that the parts farther from the eye will grow in scale and in the direction of the axis, and that the parts nearer to the eye will slightly shrink in a similar manner. Since there is a virtual impossibility of determining the conditions of the production of the cadaver images (e.g., type of camera, lens type, distance from subject), the artist will have to rely again on experience and comparison to similar subjects to best estimate each direction of rotation. One of the obvious changes will be to the length of the nose if the y-axis was involved in rotation. As the nose is a key feature in reconstructions and identification, it must be understood that the length will be dependent upon the corrected position. Two factors can assist with this: (1) the tops of the ears should align approximately with the top of the orbits, and (2) the eyes themselves should be beside the nasal bone (encased in the orbits). If the ears and eyes do not seem to line up properly as described, then it is a signal for rotation (however slight). Once this area is defined, at least vertically, then it will become easier to analyze other parts of the face and head as a result. The ears at their base will be slightly below the bottom of the nose in many subjects, and this can be added to the analysis.

It should be clear now that, using an overlay, lines can be drawn showing the centerline where bisymmetry is defined, and lines through the orbital tops and the base of the nose can be drawn to determine whether there is rotation. If the bisymmetry is disturbed in any direction, it is a cue that rotation has occurred. Again, the same corrections apply, i.e., enlarge the farthest receding parts and diminish the nearer, until bisymmetry and alignment of features appear to be in Frankfort Horizontal Plane (FHP).

After this process, which can be tedious, the artist is ready to begin dealing with tissue distortion, lighting effects (e.g., flaring or flash photography loss of detail, shadows obscuring visual data), injury modification of facial features, and so on. Merely tracing a photograph of a deceased person and opening their eyes is in no way a viable reconstruction of what that person actually looked like in real life.

The first and foremost piece of data is head shape and the closest approximate ratio between height and width. Here, width will be usually at its maximum at the parietal bones, but some individuals may have the zygomatica or even the gonial region of the face claiming this distinction. A forensic artist must be careful not to be a slave to facial formulas, but must concentrate on the actual data in front of them. If there are any cues from the surroundings, or even a metric ruler held next to the head, then a relatively accurate scale may be determined. The artist should be cognizant that the human eye is approximately 24–25 mm in diameter and, if all else fail, analysis can begin at this region. Once the orbital region scale is determined based upon approximate eye size, this information can be applied to the rest of the face. Remember that the eye is 1–2 mm out from orbital center, and about 1 mm or more up from center, when making a diagram of this region. The eye itself is not exactly circular, so estimations are just that, approximate.

The next priority is to locate the base of the malar region, the cheekbones, which should approximately line up with the base of the nose. This also will help with correction of rotational errors. After this is completed, the sides of the nostrils should be determined in order to get an estimate of the nasal width; one can look below and above to see where these features line up with other parts of the face. Because the skull is not available for study, the vertical depth of the upper jaw must be determined by correcting any rotation, and then by comparing the space to living subjects. The majority of humans have an upper jaw (i.e., area between the upper lip and septum) that is shorter vertically than the space between the base of the lower lip and the base of the mandible (fleshed). In some people, this is reversed, but should be fairly obvious upon examination. At this point, the shape of the chin can be recorded, noting the degree of flare (if any) of the jaw line from the chin to the gonial region of the jaw. This is part of the primal shape of the face and is crucial to identification.

The philtrum, the depression between the septum and the upper lip, next is added to the mix, with the artist noting width and depth (if the lighting allows). The vertical dimension of the upper and lower lip should likewise be noted and sketched, with the obvious caveat of being wary of swelling or injury. Some deceased individuals will exhibit asymmetric lips (left to right) due to injury. As best as possible, the artist should try to determine whether the asymmetry is natural or due to impact.

The mouth itself is a challenge all its own. The artist should remember that lips are actually everted mucus membranes and will exhibit wrinkling as well as color and tone changes from the skin surrounding them. In addition, a sort of light "halo" effect of the transition zone between the lips and the surrounding skin should be observed and, if obvious, drawn. The lower lip is often slightly shorter in the x-axis than the upper lip, but this is not universal. In cases of prognathic upper and lower jaws, the lips and mouth will be distorted in various ways depending upon the degree of protrusion. A very rough approximation is that the corners of the mouth line up with the pupils of the eyes; however, this has been modified since the original formula was derived, and it is advised that the artist use direct observational skill to determine actuality in the case being worked upon. Some mouth termini are actually lined up with the nostril lateral borders, which gives the appearance of a very narrow and child-like mouth. Again, observation is key.

The eyes may be closed, distorted (e.g., as in the case of drowning victims), or even minus the coloration of the iris. The surrounding tissue and overall orbit size should be carefully inspected to help determine whether the upper lid should be separate and visible, or whether it should be partly hidden. In the case of epicanthic folds, this may be less obvious, but there are still wide variations in the vertical overt dimension of the upper eyelid. Once this is done, the iris may be inserted, with awareness that the upper portion is usually more overlapped by the upper lid than is the lower half of the iris. The eye should be opened in what would be a standard width vertically according to ethnicity, as best as can be estimated. The lower lid will, in most cases, slightly overlap the iris. A few individuals have eyes in which one or both of the lids naturally do not overlap the iris, giving the individual a rather surprised or startled look. This is fairly rare and should not be applied to most reconstructions. With the overall scale of the eye sketched, the iris can be added as a slightly less-than-half diameter scale element within the eye. However, there will be shadows and highlights, even in the sclera (white) of the eye. The eyebrows, if present, should be relatively easy to reproduce, although the artist should ensure that they are positioned with a corrected FHP view.

Hair is a subject entirely too complex to cover in a simple introduction. The best advice is to initially determine the hairline and its relationship to the forehead and ears, and then treat the whole mass in terms of light and dark prior to adding linear texture in moderation, to simulate the various kinds of hair the artist might encounter. The shape of the head and nature of the hair are among the first, large cues to identification and should not be underestimated. Large shapes and sub-shapes, with highlights and mid-tones, should be laid in first, and then more nuanced modifications can be added as the drawing resolves.

The human ear is relatively simple, yet many artists find it difficult to draw. This is entirely due to unfamiliarity. From the front, information is even more reduced, and the artist is urged to copy and draw many ears to get comfortable with the variation that exists in shapes and tones of the human ear. It should be noted that many hairstyles overlap the ears to a certain extent, sometimes completely obstructing visualization of the ear. If available, photographs can provide clues as to whether this is present in the case being studied. The same practice with clay will produce increasingly confident results in the understanding and construction of the ear. There are many visuals of each facial feature available online or in texts, and the artist is urged to continue solidifying their understanding and technical mastery over the construction of all of these.

Some examples of reconstructions (now often called facial approximations) follow. Accompanying each reconstruction is a brief commentary regarding specific challenges associated with the case and success indicators.

As previously introduced at the beginning of this section on forensic facial reconstruction from cadaver images, the first example is a Hispanic female drowning victim (Fig. 6). The author received the image and had a very tight deadline for delivery of a reconstruction. The reason for this is that immediate circulation of an image may connect with current knowledge of someone who is missing or generally unaccounted for. The turnaround time was approximately one hour. Usually the artist needs a more extensive period of time in which to do the analysis, but with this case the result was sufficient to trigger an actual identification. The young woman had fallen or jumped into a canal in the Phoenix metropolitan area and did not survive. Note both the swelling and discoloration of the face (Fig. 6A). In addition, the eyes were protruding out of the sockets, the tongue was swollen and forced out of the mouth, and the general configuration of the shape of the face was difficult to ascertain. However, using guidelines presented earlier in this chapter, a general sense of the width of the face before bloating was estimated, and a fairly clear sense of the height could be established as well. Swelling on the chin was not severe enough to totally obscure the basic shape. The lips were assumed to be stretched in the x-axis due to swelling, and after virtually removing the tongue, seemed to display average vertical height, with the upper lip probably being moderately thin. The portion of the eyebrows still extant in the cadaver image provided data as to their direction, density, and slope; after analysis, these assessments were added to the drawing (Fig. 6B). Note that the artist overestimated the closure of the eyelids, but the general likeness was good. Hair was difficult to predict, so a generic dark Latina hairstyle was applied to the face. The gonion region was somewhat over everted, but not enough to deter recognition. The decedent's nose had little damage or swelling, and therefore, was more easily transferable to the reconstruction. Fig. 6C depicts a photograph of the drowning victim when she was still alive.

A second example of a cadaver image shows a more decomposed face, with eyes missing and the mouth open and distorted (Fig. 7). However, the general aspect ratio (height versus width), scale, and shape of the head were relatively clear. The age of the girl (a Hispanic) was estimated to be approximately fifteen years old. She had dark braided hair, a small chin, prominent cheekbones, fairly large eye orbitals in relation to the overall face, and fairly visible eyebrows in the cadaver image (Fig. 7A). In addition, the hairline was intact. Comparing the drawn image (Fig. 7B) to the photograph, the scale and placement of the features coincides quite well, with the distended nostrils in the cadaver image reduced to normal size, and the general shape of the jawline, malar region, and eye placement all fairly accurate. Although in the life image (i.e., the photograph) the glasses and hair dominate (Fig. 7C), it is clear that the basic analysis of the features and head shape was successful.

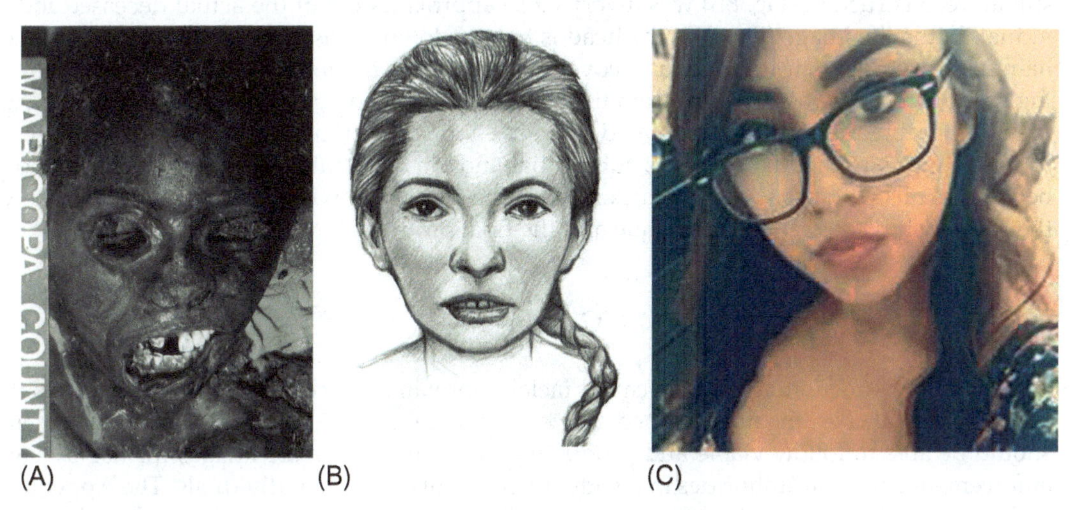

FIG. 7 Forensic reconstruction of decomposed human (Hispanic female) remains: (A) postmortem cadaver image; (B) reverse-engineered pencil sketch (by Stephen Missal CFA); (C) actual photograph in life. *Images courtesy of Maricopa County Medical Examiner's Office, Forensic Laboratory, Maricopa County, Arizona.*

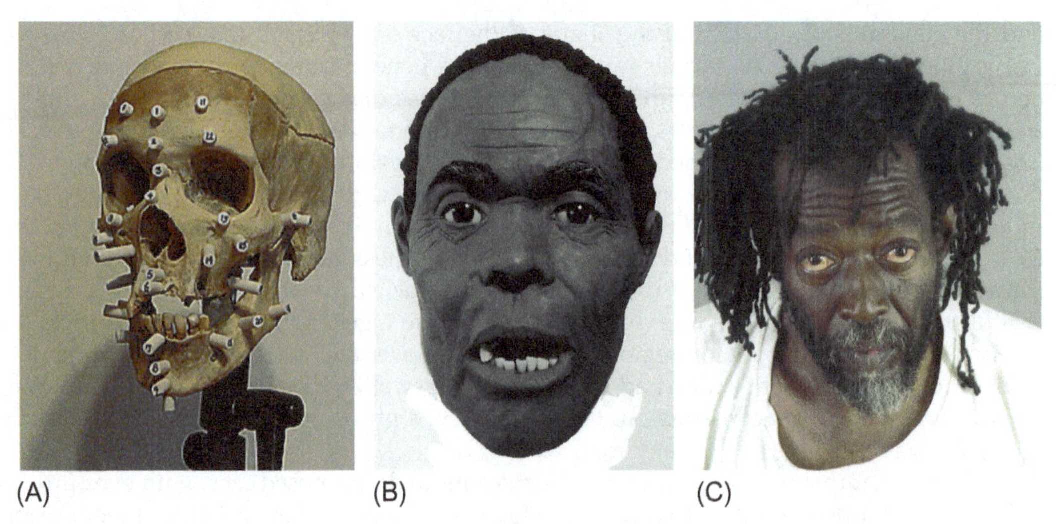

FIG. 8 Forensic reconstruction from the cranial remains of an African American male: (A) skull (in norma frontalis) with tissue depth markers applied; (B) sculpted reconstruction (by Stephen Missal CFA); (C) actual photograph in life. *Images courtesy of Maricopa County Medical Examiner's Office, Forensic Laboratory, Maricopa County, Arizona.*

The final example is a reconstruction completed from an actual set of cranial remains (Fig. 8A). In this case, the artist chose to do a sculpted version, following guidelines provided in Taylor's book (and during face-to-face training from the same source) (Taylor, 2001). The teeth were unusual; hence, the choice of showing the mouth slightly open in the sculpture. After application of tissue depth markers, the orbits and nasal openings were protected, and artificial taxidermy glass eyes were inserted into the orbits. Clay was applied, and the resultant reconstruction (Fig. 8B) was a very close approximation of the actual deceased individual. In the life image (Fig. 8C), the head is turned downwards, the mouth closed, and the hairstyle and facial hair is longer and covering portions of the forehead and chin. A furrowing of the forehead was apparent from the quizzical expression the subject displayed during the photographic session. When rotated upwards and with removal of facial hair and excess forehead wrinkles, the likeness was substantial and was generally a success. The transition between skulls with tissue depth markers to final reconstruction is worth looking at, to show the relationship between the technique and the final product.

Conclusion

Facial reconstruction, now often called facial approximation, is an evolving and fascinating field. As data become more refined and widely disseminated, forensic artists everywhere should be able to follow viable and productive procedures to create images for use by law enforcement and local authorities in the identification of deceased individuals. The Appendix at the end of this chapter lists various resources that are available for learning and application of forensic facial reconstruction information and techniques. The reader is urged to access as many of these as possible and to carefully read and apply the information supplied in the texts. Ultimately, results in forensic facial reconstruction improve with practice, and over time, the nascent reconstruction artist will begin to develop the skills and analytical techniques necessary for successful outcomes.

Appendix

George, R.M., 2007. *Facial Geometry.* Charles C. Thomas Publisher, Springfield, Illinois.
Gibson, L., 2008. *Forensic Art Essentials.* Academic Press (Elsevier), Burlington, Massachusetts.
Gordon, L., 1977. *How to Draw the Human Head.* Penguin Press, New York, New York.
Hayes, S., 2016. *Faces in the museum: revising the methods of facial reconstructions*. Museum Management and Curatorship 31(3), 218–244. Available at: https://ro.uow.edu.au/smhpapers/3508/.
Nafte, M., 2000. *Flesh and Bone.* Carolina Academic Press, Durham, North Carolina.
Neave, R., Prag, J., 1997. *Making Faces.* British Museum Press, London, England.
Sampson, H.W., Montgomery, J.L., Henryson, G.L., 1991. *Atlas of the Human Skull.* Texas A&M University Press, College Station, Texas.
Verze, L., 2009. *History of facial reconstruction*. Acta. Biomed. 80(1), 5–12.
Wilkinson, C., 2008. *Forensic Facial Reconstruction.* Cambridge University Press, Cambridge, England.

References

Gupta, S., Gupta, V., Vij, H., Vij, R., Tyagi, N., 2015. Forensic facial reconstruction: the final frontier. J. Clin. Diagn. Res. 9 (9), ZE26–ZE28.
Krogman, W.M., 1962. The Human Skeleton in Forensic Medicine. Charles C. Thomas, Publisher, Springfield, Illinois.
Taylor, K., 2001. Forensic Art and Illustration. CRC Press, Boca Raton, Florida.

Index

Note: Page numbers followed by *f* indicate figures and *t* indicate tables.

E

F

G

N

T